For c

For cor

PRINCIPLES OF COMMUNICATION SYSTEMS

McGraw-Hill Series in Electrical Engineering

Consulting Editor
Stephen W. Director, Carnegie-Mellon University

Circuits and Systems
Communications and Signal Processing
Control Theory
Electronics and Electronic Circuits
Power and Energy
Electromagnetics
Computer Engineering
Introductory
Radar and Antennas
VLSI

PRINCIPLES OF COMMUNICATION SYSTEMS

Second Edition

Herbert Taub

Donald L. Schilling

Professors of Electrical Engineering
The City College of New York

McGraw-Hill Book Company

New York St. Louis San Francisco Auckland Bogotá Hamburg
Johannesburg London Madrid Mexico Montreal New Delhi
Panama Paris São Paulo Singapore Sydney Tokyo Toronto

PRINCIPLES OF COMMUNICATION SYSTEMS
INTERNATIONAL EDITION

2nd printing 1987.

This book was set in Times Roman.
The editor was Sanjeev Rao.
The cover was designed by John Hite.

Library of Congress Cataloging in Publication Data

Taub, Herbert, 1918—
Principles of communication systems.

(McGraw-Hill series in electrical engineering. Communications and signal processing)
Includes bibliographies.
1. Telecommunication systems. I. Schilling, Donald L.
II. Title. III. Series
TK5101.T28 1986 621.38 85-11638
ISBN 0–07–062955–2 (text)
ISBN 0–07–062956–0 (solutions manual)

When ordering this title use ISBN 0–07–100313–4

Printed in Singapore by
Singapore National Printers Ltd

To
ESTHER and ANNETTE

CONTENTS

PREFACE

The first edition of this text was intended to serve as a one-semester text for a senior-level or first-year graduate course in communications. This second edition can be used as a two-semester course in which the Instructor has great latitude in the selection of topics. As in the previous edition, consistent with the level of presentation, every effort has been made to ensure that the material included represents the present state of the art and current expectations of the direction of future developments. Accordingly, although analog communication systems are accorded a full and complete treatment, the emphasis is placed on digital systems, and chapters are provided in telephone switching, computer communications and spread spectrum.

This book is an outgrowth of courses, both undergraduate and graduate, given at The City College of New York, and also to engineering personnel at the many Government Agencies, Services and Corporations to whom we have presented this material. Thus, for the undergraduate student and beginning graduate students the book provides the excitement of the communications field as well as the background required for advanced study in communication systems. The practicing engineer will find it of service to update his knowledge in the field.

Considerable thought and effort were devoted to the pedagogy of presentation and to the clarity of presentation. Great care has been exercised in the development of approximately 600 homework problems. These problems serve to elucidate the text, in some cases to extend the discussion, and are an integral and important part of the book.

The introductory chapters cover the mathematical background required for the remainder of the text. It is assumed that the reader has had some previous exposure to the elementary notions of spectral analysis including such topics as Fourier series and the Fourier transform. Spectral analysis is a mathematical tool of such wide applicability that we have judged it rather unlikely that the senior engineering student, employing the text, will not have had some experience with the subject. Hence Chap. 1, which deals with spectral analysis, is intended for the most part as a review and to refresh the students' knowledge. Certain more advanced topics, such as the concepts of power spectral density, the correlation between waveforms and representation of signals by orthogonal functions are dealt with more fully.

The importance of probabilistic concepts in the analysis of communication systems cannot be overemphasized. A discussion of communications that does not take into account the presence of noise omits a basic and essential feature. On the other hand, it is all too easy to devote so much time to an introduction to probabilistic concepts that, within the constraint of a single semester, hardly any time remains to cover the subject of communications itself. We believe that we have steered a reasonable and effective middle course. All of the background in random variables and processes required in this text is included in a single chapter (Chap. 2).

The discussion of communication systems begins in Chap. 3, where amplitude-modulation systems are covered, and Chap. 4, which presents angle-modulation systems. Chapter 5 discusses the sampling theorem, the concept of quantization and pulse-code-modulation systems as well as other analog to digital conversion techniques such as adaptive delta-modulation, encoders and linear predictive coding. This chapter also considers $T1$, $T2$, etc. hierarchy, multiplexing, signaling and synchronization. Chapter 6 discusses transmission of quantized and coded messages by means of M-ary phase-shift keying, quadrature AM, M-ary frequency-shift keying as well as some of the newer modulation techniques such as MSK and Tamed FM. In these four chapters (3 through 6), communication systems are discussed without reference to the manner in which noise affects their performance. We concentrate instead on comparing spectrum and relative complexity. We have found that such an initial presentation, excluding considerations of noise, is pedagogically very effective.

A mathematical representation of noise, using the concepts introduced in Chap. 2, is presented in Chap. 7. In Chaps. 8 and 9 the discussion of AM and FM is extended to the analysis of the influence of noise on system performance.

Chapter 10 discusses the subject of threshold and threshold extension in frequency modulation. The phase-locked loop, carrier synchronization, and bit synchronization, extremely important components in any communication system, are here discussed.

Chapters 11 and 12 discuss the performance of digital communication systems in the presence of noise. The matched-filter concept is studied, and the notion of probability of error is introduced as a criterion for system comparison. In Chap. 11, using the Union Bound, the student is shown how to calculate the error rate of the systems described in Chap. 6 and is shown how to compare the performance of such systems. Chapter 12 is concerned with the SNR of PCM and DM systems.

The underlying unity of the theory of communication systems is presented in Chap. 13 which includes a discussion of information theory and coding for error correction and detection. Although the presentation is introductory, we believe it is more extensive and complete than comparable discussions in other introductory texts. Block codes including BCN and Reed–Solomon codes are considered as well as convolutional coding and sequential and Viterbi algorithm decoding. In addition, we include here a discussion of trellis decoded modulation, a new modulation technique.

Chapter 14 serves to provide an overall view of a communication system. Sources of noise are discussed and concepts such as noise figure, noise temperature, and path loss are introduced.

Chapters 15, 16 and 17 are new to this text and are not usually found in senior/first year graduate level texts. Chapter 15, Telephone Switching, is an introduction to telephone switching, covering Strowger, Crossbar and Digital Switching. Chapter 16, Computer Communications, provides insight to the reasons that makes this area so fascinating. Topics, such as, Design considerations, ARPANET, Packet Radio (ALOHA and CSMA), as well as Protocols are introduced. Finally Chap. 17 looks at Spread Spectrum applications.

We are pleased to acknowledge our indebtedness to Dr. Adam Lender who reviewed the entire manuscript. His most gracious encouragement and very valued constructive criticism are most sincerely appreciated. In addition, we would like to thank Mr. David Monela for his criticism of the text and for his preparation of the Solutions Manual. Particular thanks is also due to Professors S. Davilorici, L. Milstein, R. L. Pickholtz and T. Saadarvi for reading and criticizing selected portions of this text. Their contributions were invaluable. We express our appreciation to Mrs. Joy Rubin of the Electrical Engineering Department at The City College, Miss Victoria Benzinqu and Mrs Annalisle Speckeritch for their most skillful service in the preparation of the manuscript. We also thank all of our colleagues and students for their assistance.

Herbert Taub
Donald L. Schilling

PRINCIPLES OF COMMUNICATION SYSTEMS

CHAPTER

ONE

SPECTRAL ANALYSIS

INTRODUCTION

Suppose that two people, separated by a considerable distance, wish to communicate with one another. If there is a pair of conducting wires extending from one location to another, and if each place is equipped with a microphone and earpiece, the communication problem may be solved. The microphone, at one end of the wire communications channel, impresses an electric signal voltage on the line, which voltage is then received at the other end. The received signal, however, will have associated with it an erratic, random, unpredictable voltage waveform which is described by the term *noise.* The origin of this noise will be discussed more fully in Chaps. 7 and 14. Here we need but to note that at the atomic level the universe is in a constant state of agitation, and that this agitation is the source of a very great deal of this noise. Because of the length of the wire link, the received message signal voltage will be greatly attenuated in comparison with its level at the transmitting end of the link. As a result, the message signal voltage may not be very large in comparison with the noise voltage, and the message will be perceived with difficulty or possibly not at all. An amplifier at the receiving end will not solve the problem, since the amplifier will amplify signal and noise alike. As a matter of fact, as we shall see, the amplifier itself may well be a source of additional noise.

A principal concern of communication theory and a matter which we discuss extensively in this book is precisely the study of methods to suppress, as far as possible, the effect of noise. We shall see that, for this purpose, it may be better not to transmit directly the original signal (the microphone output in our example). Instead, the original signal is used to generate a different signal waveform, which new signal waveform is then impressed on the line. This processing of the original signal to generate the transmitted signal is called *encoding* or *modulation*. At the receiving end an inverse process called *decoding* or *demodulation* is required to recover the original signal.

It may well be that there is a considerable expense in providing the wire communication link. We are, therefore, naturally led to inquire whether we may use the link more effectively by arranging for the simultaneous transmission over the link of more than just a single waveform. It turns out that such multiple transmission is indeed possible and may be accomplished in a number of ways. Such simultaneous multiple transmission is called *multiplexing* and is again a principal area of concern of communication theory and of this book. It is to be noted that when wire communications links are employed, then, at least in principle, separate links may be used for individual messages. When, however, the communications medium is free space, as in radio communication from antenna to antenna, multiplexing is essential.

In summary then, *communication theory* addresses itself to the following questions: Given a communication channel, how do we arrange to transmit as many simultaneous signals as possible, and how do we devise to suppress the effect of noise to the maximum extent possible? In this book, after a few mathematical preliminaries, we shall address ourselves precisely to these questions, first to the matter of multiplexing, and thereafter to the discussion of noise in communications systems.

A branch of mathematics which is of inestimable value in the study of communications systems is *spectral analysis*. Spectral analysis concerns itself with the description of waveforms in the *frequency domain* and with the correspondence between the frequency-domain description and the time-domain description. It is assumed that the reader has some familiarity with spectral analysis. The presentation in this chapter is intended as a review, and will serve further to allow a compilation of results which we shall have occasion to use throughout the remainder of this text.

1.1 FOURIER SERIES[1]

A periodic function of time $v(t)$ having a fundamental period T_0 can be represented as an infinite sum of sinusoidal waveforms. This summation, called a *Fourier series*, may be written in several forms. One such form is the following:

$$v(t) = A_0 + \sum_{n=1}^{\infty} A_n \cos \frac{2\pi nt}{T_0} + \sum_{n=1}^{\infty} B_n \sin \frac{2\pi nt}{T_0} \tag{1.1-1}$$

The constant A_0 is the average value of $v(t)$ given by

$$A_0 = \frac{1}{T_0} \int_{-T_0/2}^{T_0/2} v(t)\, dt \tag{1.1-2}$$

while the coefficients A_n and B_n are given by

$$A_n = \frac{2}{T_0} \int_{-T_0/2}^{T_0/2} v(t) \cos \frac{2\pi nt}{T_0} \, dt \tag{1.1-3}$$

and

$$B_n = \frac{2}{T_0} \int_{-T_0/2}^{T_0/2} v(t) \sin \frac{2\pi nt}{T_0} \, dt \tag{1.1-4}$$

An alternative form for the Fourier series is

$$v(t) = C_0 + \sum_{n=1}^{\infty} C_n \cos \left(\frac{2\pi nt}{T_0} - \phi_n \right) \tag{1.1-5}$$

where C_0, C_n, and ϕ_n are related to A_0, A_n, and B_n by the equations

$$C_0 = A_0 \tag{1.1-6a}$$

$$C_n = \sqrt{A_n^2 + B_n^2} \tag{1.1-6b}$$

and

$$\phi_n = \tan^{-1} \frac{B_n}{A_n} \tag{1.1-6c}$$

The Fourier series of a periodic function is thus seen to consist of a summation of harmonics of a fundamental frequency $f_0 = 1/T_0$. The coefficients C_n are called *spectral amplitudes*; that is, C_n is the amplitude of the *spectral component* $C_n \cos (2\pi n f_0 t - \phi_n)$ at frequency nf_0. A typical *amplitude spectrum* of a periodic waveform is shown in Fig. 1.1-1*a*. Here, at each harmonic frequency, a vertical line has been drawn having a length equal to the spectral amplitude associated with each harmonic frequency. Of course, such an amplitude spectrum, lacking the phase information, does not specify the waveform $v(t)$.

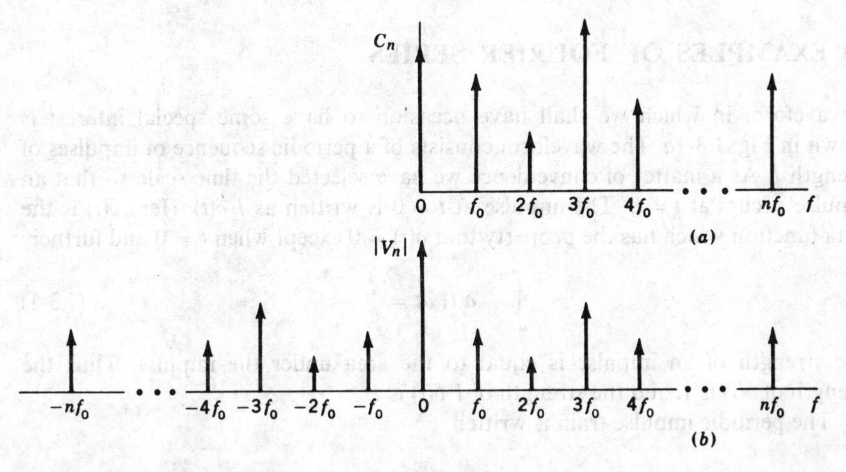

Figure 1.1-1 (*a*) A one-sided plot of spectral amplitude of a periodic waveform. (*b*) The corresponding two-sided plot.

1.2 EXPONENTIAL FORM OF THE FOURIER SERIES

The exponential form of the Fourier series finds extensive application in communication theory. This form is given by

$$v(t) = \sum_{n=-\infty}^{\infty} V_n e^{j2\pi nt/T_0} \tag{1.2-1}$$

where V_n is given by

$$V_n = \frac{1}{T_0} \int_{-T_0/2}^{T_0/2} v(t) e^{-j2\pi nt/T_0}\, dt \tag{1.2-2}$$

The coefficients V_n have the property that V_n and V_{-n} are complex conjugates of one another, that is, $V_n = V^*_{-n}$. These coefficients are related to the C_n's in Eq. (1.1-5) by

$$V_0 = C_0 \tag{1.2-3a}$$

$$V_n = \frac{C_n}{2} e^{-j\phi_n} \tag{1.2-3b}$$

The V_n's are the *spectral amplitudes* of the *spectral components* $V_n e^{j2\pi n f_0 t}$. The amplitude spectrum of the V_n's shown in Fig. 1.1-1*b* corresponds to the amplitude spectrum of the C_n's shown in Fig. 1.1-1*a*. Observe that while $V_0 = C_0$, otherwise each spectral line in 1.1-1*a* at frequency f is replaced by the 2 spectral lines in 1.1-1*b*, each of half amplitude, one at frequency f and one at frequency $-f$. The amplitude spectrum in 1.1-1*a* is called a *single-sided* spectrum, while the spectrum in 1.1-1*b* is called a *two-sided* spectrum. We shall find it more convenient to use the two-sided amplitude spectrum and shall consistently do so from this point on.

1.3 EXAMPLES OF FOURIER SERIES

A waveform in which we shall have occasion to have some special interest is shown in Fig. 1.3-1*a*. The waveform consists of a periodic sequence of impulses of strength I. As a matter of convenience we have selected the time scale so that an impulse occurs at $t = 0$. The impulse at $t = 0$ is written as $I\,\delta(t)$. Here $\delta(t)$ is the delta function which has the property that $\delta(t) = 0$ except when $t = 0$ and further

$$\int_{-\infty}^{\infty} \delta(t)\, dt = 1 \tag{1.3-1}$$

The strength of an impulse is equal to the area under the impulse. Thus the strength of $\delta(t)$ is 1, and the strength of $I\,\delta(t)$ is I.

The periodic impulse train is written

$$v(t) = I \sum_{k=-\infty}^{\infty} \delta(t - kT_0) \tag{1.3-2}$$

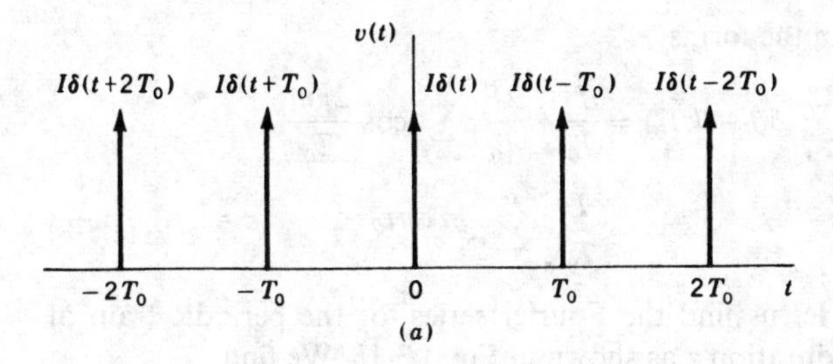

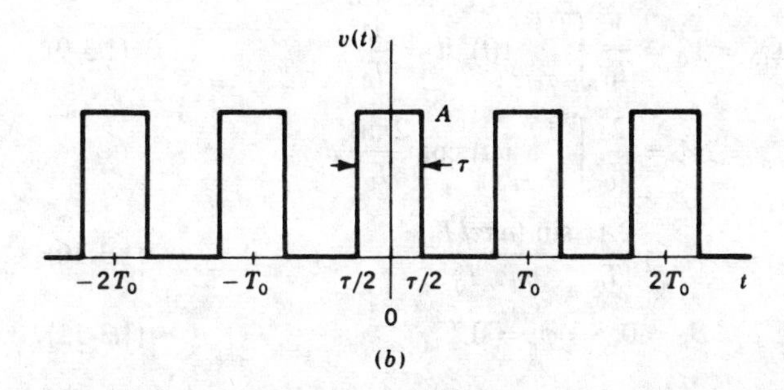

Figure 1.3-1 Examples of periodic functions. (*a*) A periodic train of impulses. (*b*) A periodic train of pulses of duration τ.

We then find, using Eqs. (1.1-2) and (1.1-3),

$$A_0 = \frac{I}{T_0} \int_{-T_0/2}^{T_0/2} \delta(t)\, dt = \frac{I}{T_0} \tag{1.3-3}$$

$$A_n = \frac{2I}{T_0} \int_{-T_0/2}^{T_0/2} \delta(t) \cos \frac{2\pi nt}{T_0}\, dt = \frac{2I}{T_0} \tag{1.3-4}$$

and, using Eq. (1.1-4),

$$B_n = \frac{2I}{T_0} \int_{-T_0/2}^{T_0/2} \delta(t) \sin \frac{2\pi nt}{T_0}\, dt = 0 \tag{1.3-5}$$

Further we have, using Eq. (1.1-6),

$$C_0 = \frac{I}{T_0} \qquad C_n = \frac{2I}{T_0} \qquad \phi_n = 0 \tag{1.3-6}$$

and from Eq. (1.2-3)

$$V_0 = V_n = \frac{I}{T_0} \tag{1.3-7}$$

Hence $v(t)$ may be written in the forms

$$v(t) = I \sum_{k=-\infty}^{\infty} \delta(t - kT_0) = \frac{I}{T_0} + \frac{2I}{T_0} \sum_{n=1}^{\infty} \cos \frac{2\pi nt}{T_0}$$
$$= \frac{I}{T_0} \sum_{n=-\infty}^{\infty} e^{j2\pi nt/T_0} \tag{1.3-8}$$

As a second example, let us find the Fourier series for the periodic train of pulses of amplitude A and duration τ as shown in Fig. 1.3-1*b*. We find

$$A_0 = C_0 = V_0 = \frac{1}{T_0} \int_{-T_0/2}^{T_0/2} v(t)\, dt = \frac{A\tau}{T_0} \tag{1.3-9}$$

$$A_n = C_n = 2V_n = \frac{2}{T_0} \int_{-T_0/2}^{T_0/2} v(t) \cos \frac{2\pi nt}{T_0}\, dt$$
$$= \frac{2A\tau}{T_0} \frac{\sin (n\pi\tau/T_0)}{n\pi\tau/T_0} \tag{1.3-10}$$

and

$$B_n = 0 \qquad \phi_n = 0 \tag{1.3-11}$$

Thus

$$v(t) = \frac{A\tau}{T_0} + \frac{2A\tau}{T_0} \sum_{n=1}^{\infty} \frac{\sin (n\pi\tau/T_0)}{n\pi\tau/T_0} \cos \frac{2\pi nt}{T_0} \tag{1.3-12a}$$

$$= \frac{A\tau}{T_0} \sum_{n=-\infty}^{\infty} \frac{\sin (n\pi\tau/T_0)}{n\pi\tau/T_0} e^{j2\pi nt/T_0} \tag{1.3-12b}$$

Suppose that in the waveform of Fig. 1.3-1*b* we reduce τ while adjusting A so that $A\tau$ is a constant, say $A\tau = I$. We would expect that in the limit, as $\tau \to 0$, the Fourier series for the pulse train in Eq. (1.3-12) should reduce to the series for the impulse train in Eq. (1.3-8). It is readily verified that such is indeed the case since as $\tau \to 0$

$$\frac{\sin (n\pi\tau/T_0)}{n\pi\tau/T_0} \to 1 \tag{1.3-13}$$

1.4 THE SAMPLING FUNCTION

A function frequently encountered in spectral analysis is the sampling function $Sa(x)$ defined by

$$Sa(x) \equiv \frac{\sin x}{x} \tag{1.4-1}$$

[A closely related function is sinc x defined by sinc $x = (\sin \pi x)/\pi x$.] The function $Sa(x)$ is plotted in Fig. 1.4-1. It is symmetrical about $x = 0$, and at $x = 0$ has

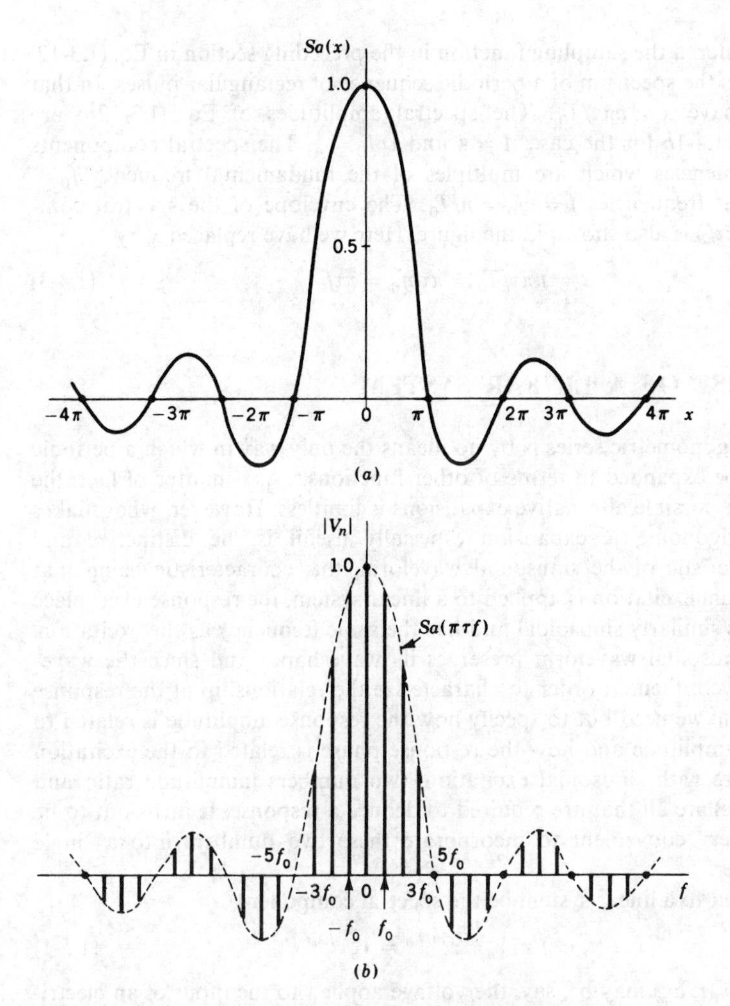

Figure 1.4-1 (*a*) The function $Sa(x)$. (*b*) The spectral amplitudes V_n of the two-sided Fourier representation of the pulse train of Fig. 1.3-1*b* for $A = 4$ and $\tau/T_0 = \frac{1}{4}$.

the value $Sa(0) = 1$. It oscillates with an amplitude that decreases with increasing x. The function passes through zero at equally spaced intervals at values of $x = \pm n\pi$, where n is an integer other than zero. Aside from the peak at $x = 0$, the maxima and minima occur *approximately* midway between the zeros, i.e., at $x = \pm(n + \frac{1}{2})\pi$, where $|\sin x| = 1$. This approximation is poorest for the minima closest to $x = 0$ but improves as x becomes larger. Correspondingly, the approximate value of $Sa(x)$ at these extremal points is

$$Sa[\pm(n + \tfrac{1}{2})\pi] = \frac{2(-1)^n}{(2n + 1)\pi} \tag{1.4-2}$$

We encountered the sampling function in the preceding section in Eq. (1.3-12) which expresses the spectrum of a periodic sequence of rectangular pulses. In that equation we have $x = n\pi\tau/T_0$. The spectral amplitudes of Eq. (1.3-12*b*) are plotted in Fig. 1.4-1*b* for the case $A = 4$ and $\tau/T_0 = \frac{1}{4}$. The spectral components appear at frequencies which are multiples of the fundamental frequency $f_0 = 1/T_0$, that is, at frequencies $f = nf_0 = n/T_0$. The envelope of the spectral components of $Sa(\pi\tau f)$ is also shown in the figure. Here we have replaced x by

$$x = n\pi\tau/T_0 = \pi\tau n f_0 = \pi\tau f \tag{1.4-3}$$

1.5 RESPONSE OF A LINEAR SYSTEM

The Fourier trigonometric series is by no means the only way in which a periodic function may be expanded in terms of other functions.[2] As a matter of fact, the number of such possible alternative expansions is limitless. However, what makes the Fourier trigonometric expansion especially useful is the distinctive and unique characteristic of the sinusoidal waveform, this characteristic being that when a sinusoidal excitation is applied to a linear system, the response everyplace in the system is similarly sinusoidal and has the same frequency as the excitation. That is, the sinusoidal waveform preserves its waveshape. And since the waveshape is preserved, then, in order to characterize the relationship of the response to the excitation, we need but to specify how the response amplitude is related to the excitation amplitude and how the response phase is related to the excitation phase. Therefore with sinusoidal excitation, two numbers (amplitude ratio and phase difference) are all that are required to deduce a response. It turns out to be possible and very convenient to incorporate these two numbers into a single complex number.

Let the input to a linear system be the spectral component

$$v_i(t, \omega_n) = V_n e^{j2\pi nt/T_0} = V_n e^{j\omega_n t} \tag{1.5-1}$$

The waveform $v_i(t, \omega_n)$ may be, say, the voltage applied to the input of an electrical filter as in Fig. 1.5-1. Then the filter output $v_o(t, \omega_n)$ is related to the input by a complex transfer function

$$H(\omega_n) = |H(\omega_n)| e^{-j\theta(\omega_n)} \tag{1.5-2}$$

that is, the output is

$$\begin{aligned} v_o(t, \omega_n) = H(\omega_n) v_i(t, \omega_n) &= |H(\omega_n)| e^{-j\theta(\omega_n)} V_n e^{j\omega_n t} \\ &= |H(\omega_n)| V_n e^{j[\omega_n t - \theta(\omega_n)]} \end{aligned} \tag{1.5-3}$$

Actually, the spectral component in Eq. (1.5-1) is not a physical voltage. Rather, the physical input voltage $v_{ip}(t)$ which gives rise to this spectral component is the sum of this spectral component and its complex conjugate, that is,

$$v_{ip}(t, \omega_n) = V_n e^{j\omega_n t} + V_{-n} e^{-j\omega_n t} = V_n e^{j\omega_n t} + V_n^* e^{-j\omega_n t} = 2Re(V_n e^{j\omega_n t}) \tag{1.5-4}$$

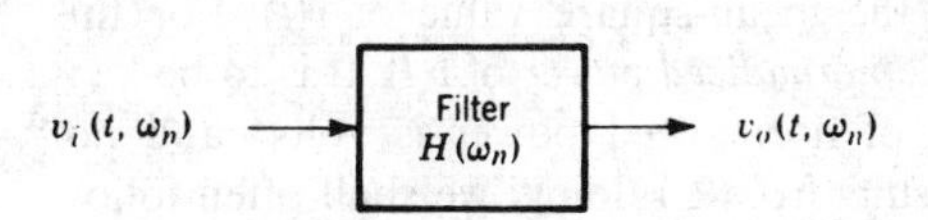

Figure 1.5-1 A sinusoidal waveform $v_i(t, \omega_n)$ of angular frequency ω_n is applied at the input to a network (filter) whose transfer characteristic at ω_n is $H(\omega_n)$. The output $v_o(t, \omega_n)$ differs from the input only in amplitude and phase.

The corresponding *physical* output voltage is $v_{op}(t, \omega_n)$ given by

$$v_{op}(t, \omega_n) = H(\omega_n)V_n e^{j\omega_n t} + H(-\omega_n)V_n^* e^{-j\omega_n t} \tag{1.5-5}$$

Since $v_{op}(t, \omega_n)$ must be real, the two terms in Eq. (1.5-5) must be complex conjugates, and hence we must have that $H(\omega_n) = H^*(-\omega_n)$. Therefore, since $H(\omega_n) = |H(\omega_n)| e^{-j\theta(\omega_n)}$, we must have that

$$|H(\omega_n)| = |H(-\omega_n)| \tag{1.5-6}$$

and

$$\theta(\omega_n) = -\theta(-\omega_n) \tag{1.5-7}$$

that is, $|H(\omega_n)|$ must be an even function and $\theta(\omega_n)$ an odd function of ω_n.

If, then, an excitation is expressed as a Fourier series in exponential form as in Eq. (1.2-1), the response is

$$v_o(t) = \sum_{n=-\infty}^{\infty} H(\omega_n)V_n e^{j2\pi nt/T_0} \tag{1.5-8}$$

If the form of Eq. (1.1-5) is used, the response is

$$v_o(t) = H(0)C_0 + \sum_{n=1}^{\infty} |H(\omega_n)| C_n \cos\left[\frac{2\pi nt}{T_0} - \phi_n - \theta(\omega_n)\right] \tag{1.5-9}$$

Given a periodic waveform, the coefficients in the Fourier series may be evaluated. Thereafter, if the transfer function $H(\omega_n)$ of a system is known, the response may be written out formally as in, say, Eq. (1.5-8) or Eq. (1.5-9). Actually these equations are generally of small value from a *computational* point of view. For, except in rather special and infrequent cases, we should be hard-pressed indeed to recognize the waveform of a response which is expressed as the sum of an infinite (or even a large) number of sinusoidal terms. On the other hand, the *concept* that a response may be written in the form of a linear superposition of responses to individual spectral components, as, say, in Eq. (1.5-8), is of inestimable value.

1.6 NORMALIZED POWER

In the analysis of communication systems, we shall often find that, given a waveform $v(t)$, we shall be interested in the quantity $\overline{v^2(t)}$ where the bar indicates the time-average value. In the case of periodic waveforms, the time averaging is done over one cycle. If, in our mind's eye, we were to imagine that the waveform $v(t)$ appear across a 1-ohm resistor, then the power dissipated in that resistor would be $\overline{v^2(t)}$ volts^2/1 ohm = W watts, where the number W would be numerically equal to the numerical value of $\overline{v^2(t)}$, the mean-square value of $v(t)$. For this reason $\overline{v^2(t)}$ is generally referred to as the *normalized power* of $v(t)$. It is to be kept in mind, however, that the dimension of normalized power is volts^2 and not watts. When, however, no confusion results from so doing, we shall often follow the generally accepted practice of dropping the word "normalized" and refer instead simply to "power." We shall often have occasion also to calculate *ratios* of normalized powers. In such cases, even if the dimension "watts" is applied to normalized power, no harm will have been done, for the dimensional error in the numerator and the denominator of the ratio will cancel out.

Suppose that in some system we encounter at one point or another normalized powers S_1 and S_2. If the ratio of these powers is of interest, we need but to evaluate, say, S_2/S_1. It frequently turns out to be more convenient not to specify this ratio directly but instead to specify the quantity K defined by

$$K \equiv 10 \log \frac{S_2}{S_1} \tag{1.6-1}$$

Like the ratio S_2/S_1, the quantity K is dimensionless. However, in order that one may know whether, in specifying a ratio we are stating the number S_2/S_1 or the number K, the term *decibel* (abbreviated dB) is attached to the number K. Thus, for example, suppose $S_2/S_1 = 100$, then $\log S_2/S_1 = 2$ and $K = 20$ dB. The advantages of the use of the decibel are twofold. First, a very large power ratio may be expressed in decibels by a much smaller and therefore often more convenient number. Second, if power ratios are to be multiplied, such multiplication may be accomplished by the simpler arithmetic operation of addition if the ratios are first expressed in decibels. Suppose that S_2 and S_1 are, respectively, the normalized power associated with sinusoidal signals of amplitudes V_2 and V_1. Then $S_2 = V_2^2/2$, $S_1 = V_1^2/2$, and

$$K = 10 \log \frac{V_2^2/2}{V_1^2/2} = 20 \log \frac{V_2}{V_1} \tag{1.6-2}$$

The use of the decibel was introduced in the early days of communications systems in connection with the transmission of signals over telephone lines. (The "bel" in decibel comes from the name Alexander Graham Bell.) In those days the decibel was used for the purpose of specifying ratios of *real* powers, not normalized powers. Because of this early history, occasionally some confusion occurs in the meaning of a ratio expressed in decibels. To point out the source of this confusion and, we hope, thereby to avoid it, let us consider the situation represented in Fig. 1.6-1. Here a waveform $v_i(t) = V_i \cos \omega t$ is applied to the input of a linear amplifier of input impedance R_i. An output signal $v_o(t) = V_o \cos (\omega t + \theta)$ then appears across the load resistor R_o. A real power $P_i = V_i^2/2R_i$ is supplied to the input, and the real power delivered to the load is $P_o = V_o^2/2R_o$. The real power gain P_o/P_i of the amplifier expressed in decibels is

$$K_{\text{real}} = 10 \log \frac{V_o^2/2R_o}{V_i^2/2R_i} \tag{1.6-3}$$

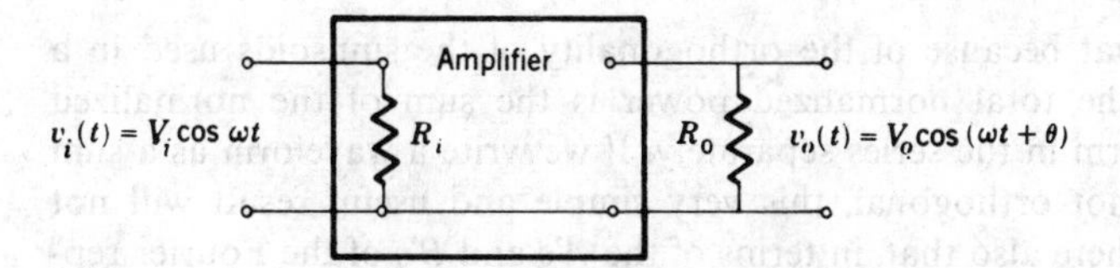

Figure 1.6-1 An amplifier of input impedance R_i with load R_o.

If it should happen that $R_i = R_o$, then K_{real} may be written as in Eq. (1.6-2).

$$K_{\text{real}} = 20 \log \frac{V_o}{V_i} \tag{1.6-4}$$

But if $R_i \neq R_o$, then Eq. (1.6-4) does not apply. On the other hand, if we calculate the normalized power gain, then we have

$$K_{\text{norm}} = 10 \log \frac{V_o^2/2}{V_i^2/2} = 20 \log \frac{V_o}{V_i} \tag{1.6-5}$$

So far as the normalized power gain is concerned, the impedances R_i and R_o in Fig. 1.6-1 are *absolutely irrelevant.* If it should happen that $R_i = R_o$, then $K_{\text{real}} = K_{\text{norm}}$, but otherwise they would be different.

1.7 NORMALIZED POWER IN A FOURIER EXPANSION

Let us consider two typical terms of the Fourier expansion of Eq. (1.1-5). If we take, say, the fundamental and first harmonic, we have

$$v'(t) = C_1 \cos\left(\frac{2\pi t}{T_0} - \phi_1\right) + C_2 \cos\left(\frac{4\pi t}{T_0} - \phi_2\right) \tag{1.7-1}$$

To calculate the normalized power S' of $v'(t)$, we must square $v'(t)$ and evaluate

$$S' = \frac{1}{T_0} \int_{-T_0/2}^{T_0/2} [v'(t)]^2 \, dt \tag{1.7-2}$$

When we square $v'(t)$, we get the square of the first term, the square of the second term, and then the cross-product term. However, the two cosine functions in Eq. (1.7-1) are *orthogonal.* That is, when their product is integrated over a complete period, the result is zero. Hence in evaluating the normalized power, we find no term corresponding to this cross product. We find actually that S' is given by

$$S' = \frac{C_1^2}{2} + \frac{C_2^2}{2} \tag{1.7-3}$$

By extension it is apparent that the normalized power associated with the entire Fourier series is

$$S = C_0^2 + \sum_{n=1}^{\infty} \frac{C_n^2}{2} \tag{1.7-4}$$

Hence we observe that because of the orthogonality of the sinusoids used in a Fourier expansion, the total normalized power is the sum of the normalized power due to each term in the series separately. If we write a waveform as a sum of terms which are not orthogonal, this very simple and useful result will not apply. We may note here also that, in terms of the A's and B's of the Fourier representation of Eq. (1.1-1), the normalized power is

$$S = A_0^2 + \sum_{n=1}^{\infty} \frac{A_n^2}{2} + \sum_{n=1}^{\infty} \frac{B_n^2}{2} \tag{1.7-5}$$

It is to be observed that power and normalized power are to be associated with a *real* waveform and not with a *complex* waveform. Thus suppose we have a term $A_n \cos(2\pi nt/T_0)$ in a Fourier series. Then the normalized power contributed by this term is $A_n^2/2$ quite independently of all other terms. And this normalized power comes from averaging, over time, the product of the term $A_n \cos(2\pi nt/T)$ by *itself*. On the other hand, in the complex Fourier representation of Eq. (1.2-1) we have terms of the form $V_n e^{j2\pi nt/T_0}$. The average value of the square of such a term is zero. We find as a matter of fact that the contributions to normalized power come from product terms

$$V_n e^{j2\pi nt/T_0} V_{-n} e^{-j2\pi nt/T_0} = V_n V_{-n} = V_n V_n^* \tag{1.7-6}$$

The total normalized power is

$$S = \sum_{n=-\infty}^{n=+\infty} V_n V_n^* \tag{1.7-7}$$

Thus, in the complex representation, the power associated with a particular *real* frequency $n/T_0 = nf_0$ (f_0 is the fundamental frequency) is associated neither with the spectral component at nf_0 nor with the component at $-nf_0$, but rather with the *combination* of spectral components, one in the positive-frequency range and one in the negative-frequency range. This power is

$$V_n V_n^* + V_{-n} V_{-n}^* = 2V_n V_n^* \tag{1.7-8}$$

It is nonetheless a procedure of great convenience to associate one-half of the power in this combination of spectral components (that is, $V_n V_n^*$) with the frequency nf_0 and the other half with the frequency $-nf_0$. Such a procedure will always be valid provided that we are careful to use the procedure to calculate only the *total* power associated with frequencies nf_0 and $-nf_0$. Thus we may say that the power associated with the spectral component at nf_0 is $V_n V_n^*$ and the power associated with the spectral component at $-nf_0$ is similarly $V_n V_n^*$ ($= V_{-n} V_{-n}^*$). If we use these associations only to arrive at the result that the total power is $2V_n V_n^*$, we shall make no error.

In correspondence with the one-sided and two-sided spectral amplitude pattern of Fig. 1.1-1 we may construct one-sided and two-sided spectral (normalized) power diagrams. A two-sided power spectral diagram is shown in Fig. 1.7-1. The vertical axis is labeled S_n, the power associated with each spectral

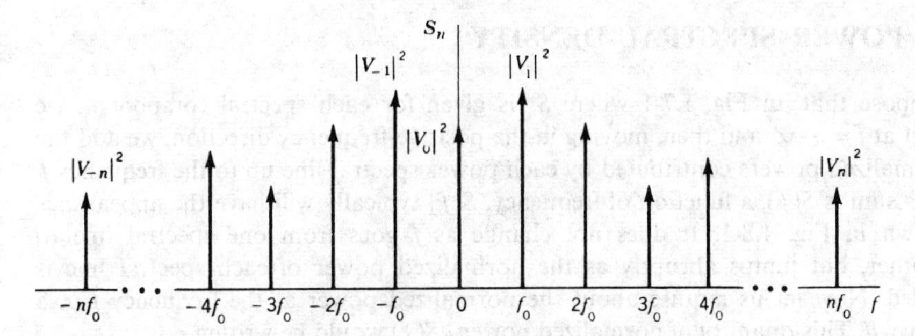

Figure 1.7-1 A two-sided power spectrum.

component. The height of each vertical line is $|V_n|^2$. Because of its greater convenience and because it lends a measure of systemization to very many calculations, we shall use the two-sided amplitude and power spectral pattern exclusively throughout this text.

We shall similarly use a two-sided representation to specify the transmission characteristics of filters. Thus, suppose we have a low-pass filter which transmits without attenuation all spectral components up to a frequency f_M and transmits nothing at a higher frequency. Then the magnitude of the transfer function will be given as in Fig. 1.7-2. The transfer characteristic of a bandpass filter will be given as in Fig. 1.7-3.

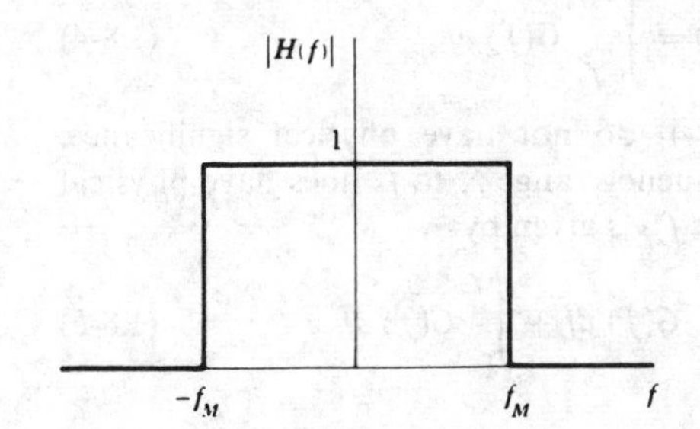

Figure 1.7-2 The transfer characteristic of an idealized low-pass filter.

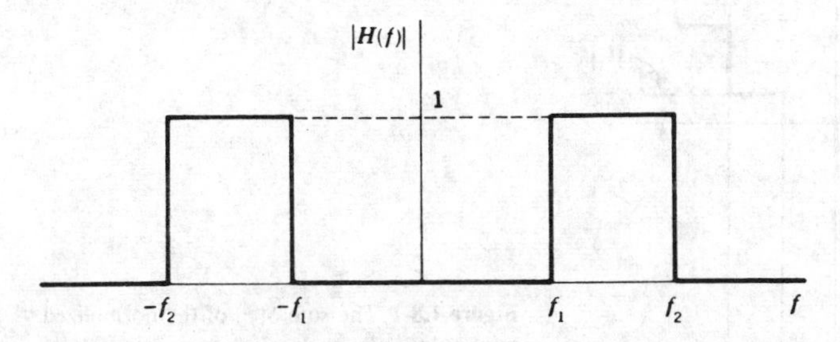

Figure 1.7-3 The transfer characteristic of an idealized bandpass filter with passband from f_1 to f_2.

1.8 POWER SPECTRAL DENSITY

Suppose that, in Fig. 1.7-1 where S_n is given for each spectral component, we start at $f = -\infty$ and then, moving in the positive-frequency direction, we add the normalized powers contributed by each power spectral line up to the frequency f. This sum is $S(f)$, a function of frequency. $S(f)$ typically will have the appearance shown in Fig. 1.8-1. It does not change as f goes from one spectral line to another, but jumps abruptly as the normalized power of each spectral line is added. Now let us inquire about the normalized power at the frequency f in a range df. This quantity of normalized power $dS(f)$ would be written

$$dS(f) = \frac{dS(f)}{df}\, df \tag{1.8-1}$$

The quantity $dS(f)/df$ is called the (normalized) *power spectral density* $G(f)$; thus

$$G(f) \equiv \frac{dS(f)}{df} \tag{1.8-2}$$

The power in the range df at f is $G(f)\, df$. The power in the positive-frequency range f_1 to f_2 is

$$S(f_1 \leq f \leq f_2) = \int_{f_1}^{f_2} G(f)\, df \tag{1.8-3}$$

The power in the negative-frequency range $-f_2$ to $-f_1$ is

$$S(-f_2 \leq f \leq -f_1) = \int_{-f_2}^{-f_1} G(f)\, df \tag{1.8-4}$$

The quantities in Eqs. (1.8-3) and (1.8-4) do not have physical significance. However, the total power in the *real* frequency range f_1 to f_2 does have physical significance, and this power $S(f_1 \leq |f| \leq f_2)$ is given by

$$S(f_1 \leq |f| \leq f_2) = \int_{-f_2}^{-f_1} G(f)\, df + \int_{f_1}^{f_2} G(f)\, df \tag{1.8-5}$$

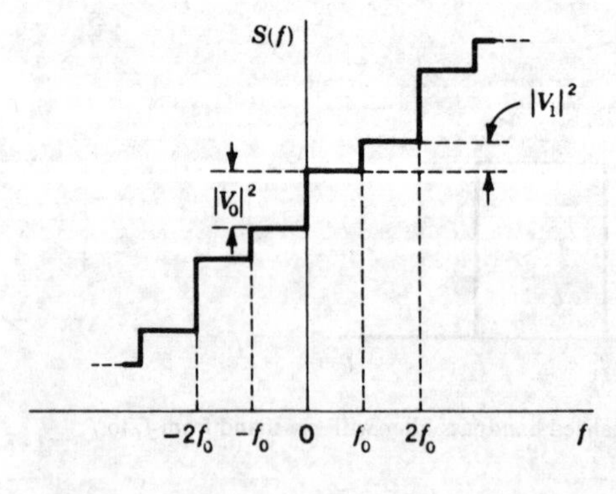

Figure 1.8-1 The sum $S(f)$ of the normalized power in all spectral components from $f = -\infty$ to f.

To find the power spectral density, we must differentiate $S(f)$ in Fig. 1.8-1. Between harmonic frequencies we would have $G(f) = 0$. At a harmonic frequency, $G(f)$ would yield an impulse of strength equal to the size of the jump in $S(f)$. Thus we would find

$$G(f) = \sum_{n=-\infty}^{\infty} |V_n|^2 \, \delta(f - nf_0) \tag{1.8-6}$$

If, in plotting $G(f)$, we were to represent an impulse by a vertical arrow of height proportional to the impulse strength, then a plot of $G(f)$ versus f as given by Eq. (1.8-6) would have exactly the same appearance as the plot of S_n shown in Fig. 1.7-1.

1.9 EFFECT OF TRANSFER FUNCTION ON POWER SPECTRAL DENSITY

Let the input signal $v_i(t)$ to a filter have a power spectral density $G_i(f)$. If V_{in} are the spectral amplitudes of this input signal, then, using Eq. (1.8-6)

$$G_i(f) = \sum_{n=-\infty}^{\infty} |V_{in}|^2 \, \delta(f - nf_0) \tag{1.9-1}$$

where, from Eq. (1.2-2),

$$V_{in} = \frac{1}{T_0} \int_{-T_0/2}^{T_0/2} v_i(t) e^{-j2\pi nt/T_0} \, dt \tag{1.9-2}$$

Let the output signal of the filter be $v_o(t)$. If V_{on} are the spectral amplitudes of this output signal, then the corresponding power spectral density is

$$G_0(f) = \sum_{n=-\infty}^{\infty} |V_{on}|^2 \, \delta(f - nf_0) \tag{1.9-3}$$

where

$$V_{on} = \frac{1}{T_0} \int_{-T_0/2}^{T_0/2} v_o(t) e^{-j2\pi nt/T_0} \, dt \tag{1.9-4}$$

As discussed in Sec. 1.5, if the transfer function of the filter is $H(f)$, then the output coefficient V_{on} is related to the input coefficient by

$$V_{on} = V_{in} H(f = nf_0) \tag{1.9-5}$$

Hence,

$$|V_{on}|^2 = |V_{in}|^2 \, |H(f = nf_0)|^2 \tag{1.9-6}$$

Substituting Eq. (1.9-6) into Eq. (1.9-3) and comparing the result with Eq. (1.9-1) yields the important result

$$G_0(f) = G_i(f) \, |H(f)|^2 \tag{1.9-7}$$

Equation (1.9-7) is of the greatest importance since it relates the power spectral density, and hence the power, at one point in a system to the power spectral density at another point in the system. Equation (1.9-7) was derived for the special case of periodic signals; however, it applies to nonperiodic signals and signals represented by random processes (Sec. 2.23) as well.

As a special application of interest of the result given in Eq. (1.9-7), assume that an input signal $v_i(t)$ with power spectral density $G_i(f)$ is passed through a differentiator. The differentiator output $v_o(t)$ is related to the input by

$$v_o(t) = \tau \frac{d}{dt} v_i(t) \tag{1.9-8}$$

where τ is a constant. The operation indicated in Eq. (1.9-8) multiplies each spectral component of $v_i(t)$ by $j2\pi f\tau = j\omega\tau$. Hence $H(f) = j\omega\tau$, and $|H(f)|^2 = \omega^2\tau^2$. Thus from Eq. (1.9-7) the spectral density of the output is

$$G_0(f) = \omega^2\tau^2 G_i(f) \tag{1.9-9}$$

1.10 THE FOURIER TRANSFORM

A periodic waveform may be expressed, as we have seen, as a sum of spectral components. These components have finite amplitudes and are separated by finite frequency intervals $f_0 = 1/T_0$. The normalized power of the waveform is finite, as is also the normalized energy of the signal in an interval T_0. Now suppose we increase without limit the period T_0 of the waveform. Thus, say, in Fig. 1.3-1*b* the pulse centered around $t = 0$ remains in place, but all other pulses move outward away from $t = 0$ as $T_0 \to \infty$. Then eventually we would be left with a single-pulse nonperiodic waveform.

As $T_0 \to \infty$, the spacing between spectral components becomes infinitesimal. The frequency of the spectral components, which in the Fourier series was a discontinuous variable with a one-to-one correspondence with the integers, becomes instead a continuous variable. The normalized energy of the nonperiodic waveform remains finite, but, since the waveform is not repeated, its normalized power becomes infinitesimal. The spectral amplitudes similarly become infinitesimal. The Fourier series for the periodic waveform

$$v(t) = \sum_{n=-\infty}^{\infty} V_n e^{j2\pi n f_0 t} \tag{1.10-1}$$

becomes (see Prob. 1.10-1)

$$v(t) = \int_{-\infty}^{\infty} V(f) e^{j2\pi f t}\, df \tag{1.10-2}$$

The finite spectral amplitudes V_n are analogous to the infinitesimal spectral

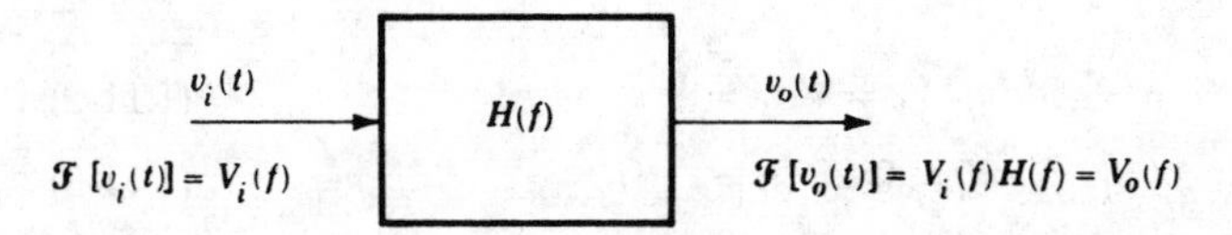

Figure 1.10-1 A waveform $v_i(t)$ of transform $V_i(f)$ is transmitted through a network of transfer function $H(f)$. The output waveform $v_o(t)$ has a transform $V_o(f) = V_i(f)H(f)$.

amplitudes $V(f)\,df$. The quantity $V(f)$ is called the *amplitude spectral density* or more generally the *Fourier transform* of $v(t)$. The Fourier transform is given by

$$V(f) = \int_{-\infty}^{\infty} v(t)e^{-j2\pi ft}\,dt \tag{1.10-3}$$

in correspondence with V_n, which is given by

$$V_n = \frac{1}{T_0}\int_{-T_0/2}^{T_0/2} v(t)e^{-j2\pi nf_0t}\,dt \tag{1.10-4}$$

Again, in correspondence with Eq. (1.5-8), let $H(f)$ be the transfer function of a network. If the input signal is $v_i(t)$, then the output signal will be $v_o(t)$ given by

$$v_o(t) = \int_{-\infty}^{\infty} H(f)V(f)e^{j2\pi ft}\,df \tag{1.10-5}$$

Comparing Eq. (1.10-5) with Eq. (1.10-2), we see that the Fourier transform $V_o(f) \equiv \mathcal{F}[v_o(t)]$ is related to the transform $V_i(f)$ of $v_i(t)$ by

$$\mathcal{F}[v_o(t)] = H(f)\mathcal{F}[v_i(t)] \tag{1.10-6}$$

or

$$V_o(f) = H(f)V_i(f) \tag{1.10-7}$$

as indicated in Fig. 1.10-1.

1.11 EXAMPLES OF FOURIER TRANSFORMS

In this section we shall evaluate the transforms of a number of functions both for the sake of review and also because we shall have occasion to refer to the results.

Example 1.11-1 If $v(t) = \cos \omega_0 t$, find $V(f)$.

SOLUTION The function $v(t) = \cos \omega_0 t$ is periodic, and therefore has a Fourier series representation as well as a Fourier transform.

The exponential Fourier series representation of $v(t)$ is

$$v(t) = \tfrac{1}{2}e^{+j\omega_0 t} + \tfrac{1}{2}e^{-j\omega_0 t} \qquad \omega_0 = \frac{2\pi}{T_0} \tag{1.11-1}$$

Thus

$$V_1 = V_{-1} = \tfrac{1}{2} \tag{1.11-2}$$

and

$$V_n = 0 \qquad n \neq \pm 1 \tag{1.11-3}$$

The Fourier transform $V(f)$ is found using Eq. (1.10-3):

$$V(f) = \int_{-\infty}^{\infty} \cos \omega_0 t e^{-j2\pi f t}\, dt = \frac{1}{2}\int_{-\infty}^{\infty} e^{-j2\pi(f - f_0)t}\, dt + \frac{1}{2}\int_{-\infty}^{\infty} e^{-j2\pi(f + f_0)t}\, dt \tag{1.11-4a}$$

$$= \tfrac{1}{2}\delta(f - f_0) + \tfrac{1}{2}\delta(f + f_0) \tag{1.11-4b}$$

[See Eq. (1.11-22) below.]

From Eqs. (1.11-1) and (1.11-3) we draw the following conclusion: The Fourier transform of a sinusoidal signal (or other periodic signal) consists of impulses located at each harmonic frequency of the signal, i.e., at $f_n = n/T_0 = nf_0$. The strength of each impulse is equal to the amplitude of the Fourier coefficient of the exponential series.

Example 1.11-2 A signal $m(t)$ is multiplied by a sinusoidal waveform of frequency f_c. The product signal is

$$v(t) = m(t) \cos 2\pi f_c t \tag{1.11-5}$$

If the Fourier transform of $m(t)$ is $M(f)$, that is,

$$M(f) = \int_{-\infty}^{\infty} m(t) e^{-j2\pi f t}\, dt \tag{1.11-6}$$

find the Fourier transform of $v(t)$.

SOLUTION Since

$$m(t) \cos 2\pi f_c t = \tfrac{1}{2} m(t) e^{j2\pi f_c t} + \tfrac{1}{2} m(t) e^{-j2\pi f_c t} \tag{1.11-7}$$

then the Fourier transform $V(f)$ is given by

$$V(f) = \frac{1}{2}\int_{-\infty}^{\infty} m(t) e^{-j2\pi(f + f_c)t}\, dt + \frac{1}{2}\int_{-\infty}^{\infty} m(t) e^{-j2\pi(f - f_c)t}\, dt \tag{1.11-8}$$

Comparing Eq. (1.11-8) with Eq. (1.11-6), we have the result that

$$V(f) = \tfrac{1}{2} M(f + f_c) + \tfrac{1}{2} M(f - f_c) \tag{1.11-9}$$

The relationship of the transform $M(f)$ of $m(t)$ to the transform $V(f)$ of $m(t) \cos 2\pi f_c t$ is illustrated in Fig. 1.11-1*a*. In Fig. 1.11-1*b* we see the spectral pattern of $M(f)$ replaced by two patterns of the same form. One is shifted to the right and one to the left, each by amount f_c. Further, the amplitudes of each of these two spectral patterns is one-half the amplitude of the spectral pattern $M(f)$.

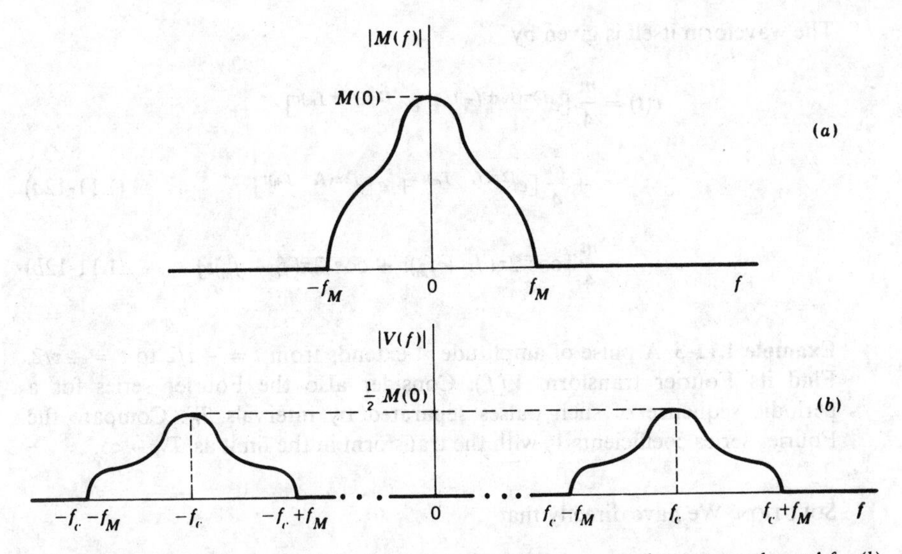

Figure 1.11-1 (*a*) The amplitude spectrum of a waveform with no spectral component beyond f_M. (*b*) The amplitude spectrum of the waveform in (*a*) multiplied by $\cos 2\pi f_c t$.

A case of special interest arises when the waveform $m(t)$ is itself sinusoidal. Thus assume

$$m(t) = m \cos 2\pi f_m t \tag{1.11-10}$$

where m is a constant. We then find that $V(f)$ is given by

$$V(f) = \frac{m}{4}\,\delta(f + f_c + f_m) + \frac{m}{4}\,\delta(f + f_c - f_m) + \frac{m}{4}\,\delta(f - f_c + f_m) + \frac{m}{4}\,\delta(f - f_c - f_m) \tag{1.11-11}$$

This spectral pattern is shown in Fig. 1.11-2. Observe that the pattern has four spectral lines corresponding to two real frequencies $f_c + f_m$ and $f_c - f_m$.

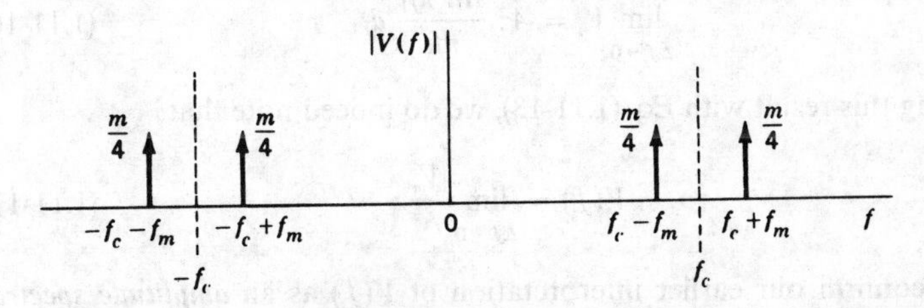

Figure 1.11-2 The two-sided amplitude spectrum of the product waveform $v(t) = m \cos 2\pi f_c t$.

The waveform itself is given by

$$v(t) = \frac{m}{4} [e^{j2\pi(f_c + f_m)t} + e^{-j2\pi(f_c + f_m)t}]$$

$$+ \frac{m}{4} [e^{j2\pi(f_c - f_m)t} + e^{-j2\pi(f_c - f_m)t}] \tag{1.11-12a}$$

$$= \frac{m}{4} [\cos 2\pi(f_c + f_m)t + \cos 2\pi(f_c - f_m)t] \tag{1.11-12b}$$

Example 1.11-3 A pulse of amplitude A extends from $t = -\tau/2$ to $t = +\tau/2$. Find its Fourier transform $V(f)$. Consider also the Fourier series for a periodic sequence of such pulses separated by intervals T_0. Compare the Fourier series coefficients V_n with the transform in the limit as $T_0 \to \infty$.

SOLUTION We have directly that

$$V(f) = \int_{-\tau/2}^{\tau/2} Ae^{-j2\pi ft}\, dt = A\tau \frac{\sin \pi f\tau}{\pi f\tau} \tag{1.11-13}$$

The Fourier series coefficients of the periodic pulse train are given by Eq. (1.3-12*b*) as

$$V_n = \frac{A\tau}{T_0} \frac{\sin (n\pi\tau/T_0)}{n\pi\tau/T_0} \tag{1.11-14}$$

The fundamental frequency in the Fourier series is $f_0 = 1/T_0$. We shall set $f_0 \equiv \Delta f$ in order to emphasize that $f_0 \equiv \Delta f$ is the frequency interval between spectral lines in the Fourier series. Hence, since $1/T_0 = \Delta f$, we may rewrite Eq. (1.11-14) as

$$V_n = A\tau \frac{\sin (\pi n\, \Delta f\tau)}{\pi n\, \Delta f\tau} \Delta f \tag{1.11-15}$$

In the limit, as $T_0 \to \infty$, $\Delta f \to 0$. We then may replace Δf by df and replace $n\, \Delta f$ by a continuous variable f. Equation (1.11-15) then becomes

$$\lim_{\Delta f \to 0} V_n = A\tau \frac{\sin \pi ft}{\pi ft} df \tag{1.11-16}$$

Comparing this result with Eq. (1.11-13), we do indeed note that

$$V(f) = \lim_{\Delta f \to 0} \frac{V_n}{\Delta f} \tag{1.11-17}$$

Thus we confirm our earlier interpretation of $V(f)$ as an *amplitude spectral density*.

Example 1.11-4

(*a*) Find the Fourier transform of $\delta(t)$, an impulse of unit strength.

(*b*) Given a network whose transfer function is $H(f)$. An impulse $\delta(t)$ is applied at the input. Show that the response $v_o(t) \equiv h(t)$ at the output is the inverse transform of $H(f)$, that is, show that $h(t) = \mathscr{F}^{-1}[H(f)]$.

SOLUTION

(*a*) The impulse $\delta(t) = 0$ except at $t = 0$ and, further, has the property that

$$\int_{-\infty}^{\infty} \delta(t)\, dt = 1 \tag{1.11-18}$$

Hence

$$V(f) = \int_{-\infty}^{\infty} \delta(t) e^{-j2\pi f t}\, dt = 1 \tag{1.11-19}$$

Thus the spectral components of $\delta(t)$ extend with uniform amplitude and phase over the entire frequency domain.

(*b*) Using the result given in Eq. (1.10-7), we find that the transform of the output $v_o(t) \equiv h(t)$ is $V_o(f)$ given by

$$V_o(f) = 1 \times H(f) \tag{1.11-20}$$

since the transform of $\delta(t)$, $\mathscr{F}[\delta(t)] = 1$. Hence the inverse transform of $V_o(f)$, which is the function $h(t)$, is also the inverse transform of $H(f)$. Specifically, for an impulse input, the output is

$$h(t) = \int_{-\infty}^{\infty} H(f) e^{j2\pi f t}\, df \tag{1.11-21}$$

We may use the result given in Eq. (1.11-21) to arrive at a useful representation of $\delta(t)$ itself. If $H(f) = 1$, then the response $h(t)$ to an inpulse $\delta(t)$ is the impulse itself. Hence, setting $H(f) = 1$ in Eq. (1.11-21), we find

$$\delta(t) = \int_{-\infty}^{\infty} e^{j2\pi f t}\, df = \int_{-\infty}^{\infty} e^{-j2\pi f t}\, df \tag{1.11-22}$$

1.12 CONVOLUTION

Suppose that $v_1(t)$ has the Fourier transform $V_1(f)$ and $v_2(t)$ has the transform $V_2(f)$. What then is the waveform $v(t)$ whose transform is the product $V_1(f)V_2(f)$? This question arises frequently in spectral analysis and is answered by the *convolution theorem*, which says that

$$v(t) = \int_{-\infty}^{\infty} v_1(\tau) v_2(t - \tau)\, d\tau \tag{1.12-1}$$

or equivalently

$$v(t) = \int_{-\infty}^{\infty} v_2(\tau) v_1(t - \tau)\, d\tau \tag{1.12-2}$$

The integrals in Eq. (1.12-1) or (1.12-2) are called *convolution integrals*, and the process of evaluating $v(t)$ through these integrals is referred to as *taking the convolution* of the functions $v_1(t)$ and $v_2(t)$.

To prove the theorem, we begin by writing

$$v(t) = \mathscr{F}^{-1}[V_1(f)V_2(f)] \tag{1.12-3a}$$

$$= \frac{1}{2\pi}\int_{-\infty}^{\infty} V_1(f)V_2(f)e^{j\omega t}\,d\omega \tag{1.12-3b}$$

By definition we have

$$V_1(f) = \int_{-\infty}^{\infty} v_1(\tau)e^{-j\omega\tau}\,d\tau \tag{1.12-4}$$

Substituting $V_1(f)$ as given by Eq. (1.12-4) into the integrand of Eq. (1.12-3*b*), we have

$$v(t) = \frac{1}{2\pi}\int_{-\infty}^{\infty}\int_{-\infty}^{\infty} v_1(\tau)e^{-j\omega\tau}\,d\tau\,V_2(f)e^{j\omega t}\,d\omega \tag{1.12-5}$$

Interchanging the order of integration, we find

$$v(t) = \int_{-\infty}^{\infty} v_1(\tau)\left[\frac{1}{2\pi}\int_{-\infty}^{\infty} V_2(f)e^{j\omega(t-\tau)}\,d\omega\right]d\tau \tag{1.12-6}$$

We recognize that the expression in brackets in Eq. (1.12-6) is $v_2(t-\tau)$, so that finally

$$v(t) = \int_{-\infty}^{\infty} v_1(\tau)v_2(t-\tau)\,d\tau \tag{1.12-7}$$

Examples of the use of the convolution integral are given in Probs. 1.12-2 and 1.12-3. These problems will also serve to recall to the reader the relevance of the term *convolution*.

A special case of the convolution theorem and one of very great utility is arrived at through the following considerations. Suppose a waveform $v_i(t)$ whose transform is $V_i(f)$ is applied to a linear network with transfer function $H(f)$. The transform of the output waveform is $V_i(f)H(f)$. What then is the waveform of $v_o(t)$?

In Eq. (1.12-7) we identify $v_1(\tau)$ with $v_i(\tau)$ and $v_2(t)$ with the inverse transform of $H(f)$. But we have seen [in Eq. (1.11-21)] that the inverse transform of $H(f)$ is $h(t)$, the impulse response of the network. Hence Eq. (1.12-7) becomes

$$v_o(t) = \int_{-\infty}^{\infty} v_i(\tau)h(t-\tau)\,d\tau \tag{1.12-8}$$

in which the output $v_o(t)$ is expressed in terms of the input $v_i(t)$ and the inpulse response of the network.

1.13 PARSEVAL'S THEOREM

We saw that for periodic waveforms we may express the normalized power as a summation of powers due to individual spectral components. We also found for periodic signals that it was appropriate to introduce the concept of *power spectral density*. Let us keep in mind that we may make the transition from the periodic to the nonperiodic waveform by allowing the period of the periodic waveform to approach infinity. A nonperiodic waveform, so generated, has a finite *normalized energy*, while the normalized power approaches zero. We may therefore expect that for a nonperiodic waveform the energy may be written as a continuous summation (integral) of energies due to individual spectral components in a continuous distribution. Similarly we should expect that with such nonperiodic waveforms it should be possible to introduce an *energy spectral density*.

The normalized energy of a periodic waveform $v(t)$ in a period T_0 is

$$E = \int_{-T_0/2}^{T_0/2} [v(t)]^2 \, dt \tag{1.13-1}$$

From Eq. (1.7-7) we may write E as

$$E = T_0 S = T_0 \sum_{n=-\infty}^{n=+\infty} V_n V_n^* \tag{1.13-2}$$

Again, as in the illustrative Example 1.11-3, we let $\Delta f \equiv 1/T_0 = f_0$, where f_0 is the fundamental frequency, that is, Δf is the spacing between harmonics. Then we have

$$V_n V_n^* = \frac{V_n}{\Delta f} \frac{V_n^*}{\Delta f} (\Delta f)^2 \tag{1.13-3}$$

and Eq. (1.13-2) may be written

$$E = \sum_{n=-\infty}^{\infty} \frac{V_n}{\Delta f} \frac{V_n^*}{\Delta f} \Delta f \tag{1.13-4}$$

In the limit, as $\Delta f \to 0$, we replace Δf by df, we replace $V_n/\Delta f$ by the transform $V(f)$ [see Eq. (1.11-17)], and the summation by an integral. Equation (1.13-4) then becomes

$$E = \int_{-\infty}^{\infty} V(f)V^*(f) \, df = \int_{-\infty}^{\infty} |V(f)|^2 \, df = \int_{-\infty}^{\infty} [v(t)]^2 \, dt \tag{1.13-5}$$

This equation expresses *Parseval's theorem*. Parseval's theorem is the extension to the nonperiodic case of Eq. (1.7-7) which applies for periodic waveforms. Both results simply express the fact that the power (periodic case) or the energy (nonperiodic case) may be written as the superposition of power or energy due to individual spectral components separately. The validity of these results depends on the fact that the spectral components are orthogonal. In the periodic case the spectral components are orthogonal over the interval T_0. In the nonperiodic case this interval of orthogonality extends over the entire time axis, i.e., from $-\infty$ to $+\infty$.

From Eq. (1.13-5), we find that the *energy density* is $G_E(f)$ given by

$$G_E(f) \equiv \frac{dE}{df} = |V(f)|^2 \tag{1.13-6}$$

in correspondence with Eq. (1.8-6), which gives the power spectral density for a periodic waveform.

Parseval's theorem may be derived more formally by a direct application of the convolution theorem and by other methods (see Prob. 1.13-1).

1.14 POWER AND ENERGY TRANSFER THROUGH A NETWORK

Suppose that a periodic signal $v_i(t)$ is applied to a network of transfer function $H(f)$ which yields an output $v_o(t)$. If $v_i(t)$ is written

$$v_i(t) = \sum_{n=-\infty}^{\infty} V_n e^{j2\pi n f_0 t} \tag{1.14-1}$$

then

$$v_o(t) = \sum_{n=-\infty}^{\infty} H(nf_0) V_n e^{j2\pi n f_0 t} \tag{1.14-2}$$

The power of $v_i(t)$ is

$$S_i = \sum_{n=-\infty}^{\infty} |V_n|^2 \tag{1.14-3}$$

while the power of $v_o(t)$ is

$$S_0 = \sum_{n=-\infty}^{\infty} |H(nf_0)|^2 |V_n|^2 \tag{1.14-4}$$

Thus, it is seen from Eq. (1.14-4) that, depending on $H(nf_0)$, the power associated with particular spectral components may be increased or decreased. Suppose for example that $H(nf_0) = h$ (a constant) for values of n between $n = i$ and $n = j$, and that $H(nf_0) = 0$ otherwise. Then the power associated with spectral components outside the range $i \leq |n| \leq j$ will be *lost*, and the network output waveform will be

$$v_o(t) = \sum_{n=-j}^{-i} hV_n e^{j2\pi n f_0 t} + \sum_{n=i}^{j} hV_n e^{j2\pi n f_0 t} \tag{1.14-5}$$

while the power output will be

$$S_0 = \sum_{n=-j}^{-i} h^2 |V_n|^2 + \sum_{n=i}^{j} h^2 |V_n|^2 = 2 \sum_{n=i}^{j} h^2 |V_n|^2 \tag{1.14-6}$$

Alternatively, as may be verified, S_0 may be written in terms of the output-power spectral density as

$$S_0 = \int_{-jf_0}^{-if_0} G(f)\, df + \int_{if_0}^{jf_0} G(f)\, df = 2 \int_{if_0}^{jf_0} G(f)\, df \qquad (1.14\text{-}7)$$

Similarly, suppose the signal is nonperiodic and is transmitted through a network for which $H(f) = 1$ between the real frequencies f_1 and f_2 and $H(f) = 0$ otherwise. Then in correspondence with Eq. (1.14-7) the output energy is

$$E_0 = \int_{-f_2}^{-f_1} G_E(f)\, df + \int_{f_1}^{f_2} G_E(f)\, df = 2 \int_{f_1}^{f_2} G_E(f)\, df \qquad (1.14\text{-}8)$$

where G_E is the energy spectral density.

1.15 BANDLIMITING OF WAVEFORMS

Rather generally, waveforms, periodic or nonperiodic, have spectral components which extend, at least in principle, to infinite frequency. Periodic waveforms may or may not have a dc component. Nonperiodic waveforms usually have spectral components extending to zero or near-zero frequency. Therefore, again, at least in principle, if an arbitrary waveform is to pass through a network without changing its shape, the transfer function of the network must not discriminate among spectral components. All spectral amplitudes must be increased or decreased by the same amount, and each spectral component must be delayed equally. A network which introduces no distortion, therefore, has the transfer function

$$H(f) = h_0\, e^{-j\omega T_D} \qquad (1.15\text{-}1)$$

in which h_0 is a constant and T_D is the delay time introduced by the network. An example of a network which produces no distortion is a length of transmission line of uniform cross section, having no losses and properly terminated.

A signal may have spectral components which extend in the upper-frequency direction only up to a maximum frequency f_M. Such a signal is described as being *bandlimited* (at the high-frequency end) to f_M. If such a signal is passed through a network (filter) for which $H(f) = 1$ for $f \leq f_M$, the signal will be transmitted without distortion. Suppose, however, the signal is not precisely bandlimited to f_M but nonetheless, a very large part of its power or energy lies in spectral components below f_M. For example, suppose 99 percent of the power or energy lies below f_M. Then we may reasonably expect that the filter will introduce no serious distortion. Similarly, a signal may be bandlimited at the low-frequency end to a frequency f_L. And, again, if the signal is not precisely so bandlimited, but if a negligible fraction of the signal energy or power lies below f_L, a filter with low-frequency cutoff at f_L will produce negligible distortion.

We shall frequently have occasion to deal with waveforms which are not of themselves bandlimited but are passed through filters, introducing bandlimiting and hence producing distortion. The effect of such bandlimiting may be seen in a general sort of way by considering the response of the networks shown in Fig. 1.15-1 to a step function. We select the step function because it represents a combination of the fastest possible rate of change of voltage (rise time equal to zero) and the slowest posssible (zero) rate of change of voltage after the abrupt rise. We select the circuit of Fig. 1.15-1*a* because it produces bandlimiting (albeit not abrupt) at the high-frequency end, while the circuit in Fig. 1.15-1*b* produces bandlimiting at the low-frequency end.

The low-pass *RC* circuit of Fig. 1.15-1*a* has a transfer function

$$H(f) = \frac{1}{1 + jf/f_2} \tag{1.15-2}$$

where $f_2 = 1/(2\pi RC)$. The magnitude of $H(f)$ is plotted in Fig. 1.15-1*c*. At the frequency $f = f_2$, $|H(f)|$ has fallen to the value $1/\sqrt{2} = 0.707$, corresponding to a reduction of 3 dB from its value at $f = 0$. The frequency f_2 is called the 3-dB fre-

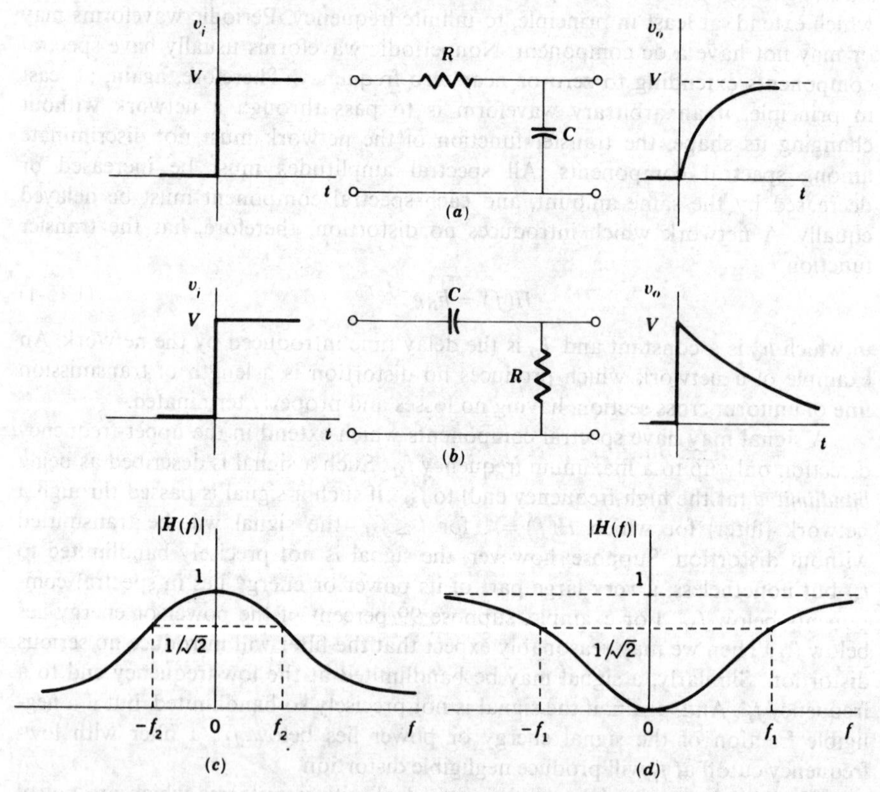

Figure 1.15-1 (*a*) A low-pass *RC* circuit. (*b*) A high-pass *RC* circuit. (*c*) $|H(f)|$ for the circuit in (*a*). (*d*) $|H(f)|$ for the circuit in (*b*).

quency of the network and is sometimes referred to as the *passband* of the network. For a step input of amplitude V_0 the output is

$$v_o(t) = V_0(1 - e^{-t/RC}) = V_0(1 - e^{-2\pi f_2 t}) \tag{1.15-3}$$

The essential distortion which has been introduced by the frequency discrimination of the network is that the output rises gradually rather than abruptly as does the input. We thus associate with the output waveform a *rise time*. This rise time may be defined in a variety of ways. One definition commonly employed is that the rise time is the time required for the output $v_o(t)$ to change from $0.1V_0$ to $0.9V_0$. It may be verified from Eq. (1.15-3) that this rise time t_r is related to f_2 by

$$t_r f_2 = 0.35 \tag{1.15-4}$$

This rise time is a measure of the promptness with which the output responds to a voltage change at the input. A long rise time indicates a sluggish response. We note then from Eq. (1.15-4) that the more the upper-frequency range of the transfer function is restricted, the longer the rise time becomes.

The principle enunciated in Eq. (1.15-4) is of general validity. Even for filter circuits of much greater complexity and sharper cutoff than the simple *RC* circuit of Fig. 1.15-1*a*, the product $t_r f_2$ remains approximately constant. By way of example, suppose we postulate an ideal (unrealizable) filter. Such a filter has an arbitrarily sharp cutoff. That is, $H(f) = 1$ for $0 \leq f \leq f_M$, and $H(f) = 0$ for $f > f_M$. Then in this case it turns out that $t_r f_2 = 0.44$. On the other hand, Eq. (1.15-4) will often be encountered in texts and in literature with a rather different constant. For example, the result for the ideal filter is often given as $t_r f_2 = 1$. These differences result from different definitions of the rise time.

If the input to the *RC* low-pass filter is a pulse, then there will be a rise time associated with the leading and trailing edges of the pulse. Let the pulse duration be τ. Then as a rule of thumb it is generally considered that to preserve the waveshape of the pulse with reasonable fidelity, it is necessary that the bandwidth f_2 be at least large enough to satisfy the condition $f_2 \tau = 1$. For combining this condition with Eq. (1.15-4), we find $t_r = 0.35\tau$. In this case, with the rise time about one-third the pulse duration, the output pulse waveform has the form shown in Fig. 1.15-2*a*.

The high-pass *RC* circuit of Fig. 1.15-1*b* has a transfer function

$$H(f) = \frac{1}{1 - jf_1/f} \tag{1.15-5}$$

where

$$f_1 = \frac{1}{2\pi RC} \tag{1.15-6}$$

The magnitude of $H(f)$ is plotted in Fig. 1.15-1*d*. The frequency f_1 is the 3-dB frequency of the network and is generally described as the low-frequency cutoff of the network. For a step input of amplitude V_0, the output is

$$v_o(t) = V_0 e^{-t/RC} = V_0 e^{-2\pi f_1 t} \tag{1.15-7}$$

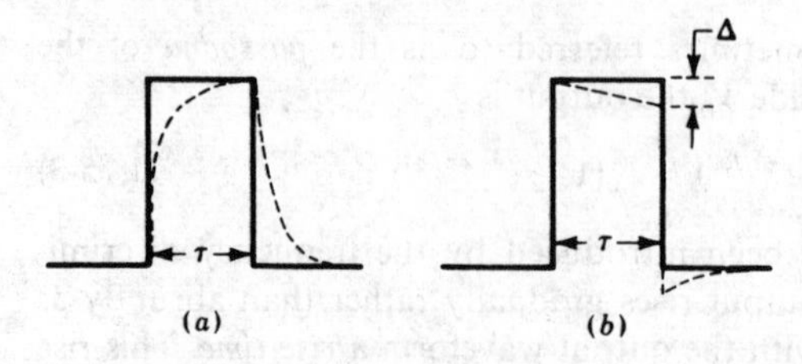

Figure 1.15-2 (*a*) A rectangular pulse (solid) and the response (dashed) of a low-pass *RC* circuit. The rise time $t_r \approx 0.35\tau$. (*b*) The response of a high-pass *RC* circuit for $f_1\tau = 0.02$.

The essential distortion introduced by the frequency discrimination of this network is that the output does not sustain a constant voltage level when the input is constant. Instead, the output begins immediately to decay toward zero, which is its asymptotic limit. From Eq. (1.15-7) we have

$$\frac{1}{v_o}\frac{dv_o}{dt} = -\frac{1}{RC} = -2\pi f_1 \tag{1.15-8}$$

Thus the low-frequency cutoff f_1 alone determines the percentage drop in voltage per unit time. Again the importance of Eq. (1.15-8) is that, at least as a reasonable approximation, it applies to high-pass networks quite generally, even when the network is very much more complicated than the simple *RC* circuit.

If the input to the *RC* high-pass circuit is a pulse of duration τ, then the output has the waveshape shown in Fig. 1.15-2*b*. The output exhibits a tilt and an undershoot. As a rule of thumb, we may assume that the pulse is reasonably faithfully reproduced if the tilt Δ is no more than $0.1V_0$. Correspondingly, this condition requires that f_1 be no higher than given by the condition

$$f_1\tau \approx 0.02 \tag{1.15-9}$$

1.16 CORRELATION BETWEEN WAVEFORMS

The correlation between waveforms is a measure of the similarity or relatedness between the waveforms. Suppose that we have waveforms $v_1(t)$ and $v_2(t)$, not necessarily periodic nor confined to a finite time interval. Then the correlation between them, or more precisely the *average cross correlation* between $v_1(t)$ and $v_2(t)$, is $R_{12}(\tau)$ defined as

$$R_{12}(\tau) \equiv \lim_{T\to\infty} \frac{1}{T}\int_{-T/2}^{T/2} v_1(t)v_2(t+\tau)\,dt \tag{1.16-1}$$

If $v_1(t)$ and $v_2(t)$ are periodic with the same fundamental period T_0, then the average cross correlation is

$$R_{12}(\tau) = \frac{1}{T_0}\int_{-T_0/2}^{T_0/2} v_1(t)v_2(t+\tau)\,dt \tag{1.16-2}$$

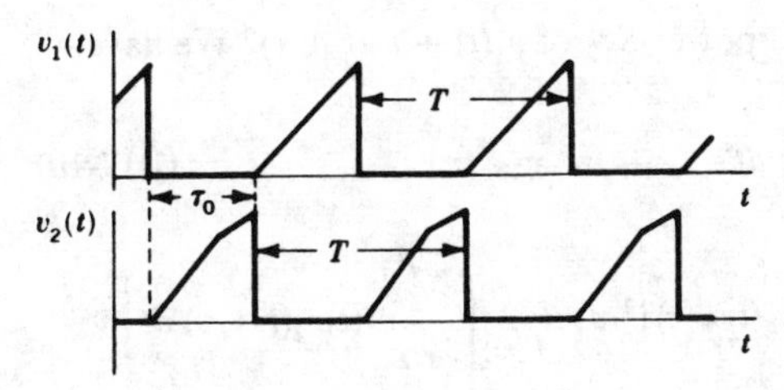

Figure 1.16-1 Two related waveforms. The timing is such that the product $v_1(t)v_2(t) = 0$.

If $v_1(t)$ and $v_2(t)$ are waveforms of finite energy (for example, nonperiodic pulse-type waveforms), then the cross correlation is defined as

$$R_{12}(\tau) = \int_{-\infty}^{\infty} v_1(t)v_2(t + \tau)\, dt \tag{1.16-3}$$

The need for introducing the parameter τ in the definition of cross correlation may be seen by the example illustrated in Fig. 1.16-1. Here the two waveforms, while different, are obviously related. They have the same period and nearly the same form. However, the integral of the product $v_1(t)v_2(t)$ is zero since at all times one or the other function is zero. The function $v_2(t + \tau)$ is the function $v_2(t)$ shifted to the left by amount τ. It is clear from the figure that, while $R_{12}(0) = 0$, $R_{12}(\tau)$ will increase as τ increases from zero, becoming a maximum when $\tau = \tau_0$. Thus τ is a "searching" or "scanning" parameter which may be adjusted to a proper time shift to reveal, to the maximum extent possible, the relatedness or correlation between the functions. The term *coherence* is sometimes used as a synonym for correlation. Functions for which $R_{12}(\tau) = 0$ for all τ are described as being *uncorrelated* or *noncoherent*.

In scanning to see the extent of the correlation between functions, it is necessary to specify which function is being shifted. In general, $R_{12}(\tau)$ is not equal to $R_{21}(\tau)$. It is readily verified (Prob. 1.16-2) that

$$R_{21}(\tau) \equiv \lim_{T \to \infty} \frac{1}{T} \int_{-T/2}^{T/2} v_1(t + \tau)v_2(t)\, dt = R_{12}(-\tau) \tag{1.16-4}$$

with identical results for periodic waveforms or waveforms of finite energy.

1.17 POWER AND CROSS CORRELATION

Let $v_1(t)$ and $v_2(t)$ be waveforms which are not periodic nor confined to a finite time interval. Suppose that the normalized power of $v_1(t)$ is S_1 and the normalized power of $v_2(t)$ is S_2. What, then, is the normalized power of $v_1(t) + v_2(t)$? Or, more generally, what is the normalized power S_{12} of $v_1(t) + v_2(t + \tau)$? We have

$$S_{12} = \lim_{T \to \infty} \frac{1}{T} \int_{-T/2}^{T/2} [v_1(t) + v_2(t + \tau)]^2 \, dt \tag{1.17-1a}$$

$$= \lim_{T \to \infty} \frac{1}{T} \left\{ \int_{-T/2}^{T/2} v_1^2(t) \, dt + \int_{-T/2}^{T/2} [v_2(t + \tau)]^2 \, dt + 2 \int_{-T/2}^{T/2} v_1(t) v_2(t + \tau) \, dt \right\} \tag{1.17-1b}$$

$$= S_1 + S_2 + 2R_{12}(\tau) \tag{1.17-1c}$$

In writing Eq. (1.17-1*c*), we have taken account of the fact that the normalized power of $v_2(t + \tau)$ is the same as the normalized power of $v_2(t)$. For, since the integration in Eq. (1.17-1*b*) extends eventually over the entire time axis, a time shift in v_2 will clearly not affect the value of the integral.

From Eq. (1.17-1*c*) we have the important result that if two waveforms are uncorrelated, that is, $R_{12}(\tau) = 0$ for all τ, then no matter how these waveforms are time-shifted with respect to one another, the normalized power due to the superposition of the waveforms is the sum of the powers due to the waveforms individually. Similarly if a waveform is the sum of any number of mutually uncorrelated waveforms, the normalized power is the sum of the individual powers. It is readily verified that the same result applies for periodic waveforms. For finite energy waveforms, the result applies to the normalized energy.

Suppose that two waveforms $v_1'(t)$ and $v_2'(t)$ are uncorrelated. If dc components V_1 and V_2 are added to the waveforms, then the waveforms $v_1(t) = v_1'(t) + V_1$ and $v_2(t) = v_2'(t) + V_2$ will be correlated with correlation $R_{12}(\tau) = V_1 V_2$. In most applications where the correlations between waveforms is of concern, there is rarely any interest in the dc component. It is customary, then, to continue to refer to waveforms as being uncorrelated if the only source of the correlation is the dc components.

1.18 AUTOCORRELATION

The correlation of a function with itself is called the *autocorrelation*. Thus with $v_1(t) = v_2(t)$, $R_{12}(\tau)$ becomes $R(\tau)$ given, in the general case, by

$$R(\tau) = \lim_{T \to \infty} \frac{1}{T} \int_{-T/2}^{T/2} v(t) v(t + \tau) \, dt \tag{1.18-1}$$

A number of the properties of $R(\tau)$ are listed in the following:

(*a*)
$$R(0) = \lim_{T \to \infty} \frac{1}{T} \int_{-T/2}^{T/2} [v(t)]^2 \, dt = S \tag{1.18-2}$$

That is, the autocorrelation for $\tau = 0$ is the average power S of the waveform.

(*b*) $$R(0) \geq R(\tau) \tag{1.18-3}$$

This result is rather intuitively obvious since we would surely expect that similarity between $v(t)$ and $v(t + \tau)$ be a maximum when $\tau = 0$. The student is guided through a more formal proof in Prob. 1.18-1.

(*c*) $$R(\tau) = R(-\tau) \tag{1.18-4}$$

Thus the autocorrelation function is an even function of τ. To prove Eq. (1.18-4), assume that the axis $t = 0$ is moved in the negative t direction by amount τ. Then the integrand in Eq. (1.18-1) would become $v(t - \tau)v(t)$, and $R(\tau)$ would become $R(-\tau)$. Since, however, the integration eventually extends from $-\infty$ to ∞, such a shift in time axis can have no effect on the value of the integral. Thus $R(\tau) = R(-\tau)$.

The three characteristics given in Eqs. (1.18-2) to (1.18-4) are features not only of $R(\tau)$ defined by Eq. (1.18-1) but also for $R(\tau)$ as defined by Eqs. (1.16-2) and (1.16-3) for the periodic case and the non-periodic case of finite energy. In the latter case, of course, $R(0) = E$, the energy rather than the power.

1.19 AUTOCORRELATION OF A PERIODIC WAVEFORM

When the function is periodic, we may write

$$v(t) = \sum_{n=-\infty}^{\infty} V_n e^{j2\pi nt/T_0} \tag{1.19-1}$$

and, using the correlation integral in the form of Eq. (1.16-2), we have

$$R(\tau) = \frac{1}{T_0} \int_{-T_0/2}^{T_0/2} \left(\sum_{m=-\infty}^{\infty} V_m e^{j2\pi mt/T_0} \right) \left(\sum_{n=-\infty}^{\infty} V_n e^{j2\pi n(t+\tau)/T_0} \right) dt \tag{1.19-2}$$

The order of integration and summation may be interchanged in Eq. (1.19-2). If we do so, we shall be left with a double summation over m and n of terms $I_{m,n}$ given by

$$I_{m,n} = \frac{1}{T_0} e^{j2\pi n\tau/T_0} \int_{-T_0/2}^{T_0/2} V_m V_n e^{j2\pi(m+n)t/T_0} \, dt \tag{1.19-3a}$$

$$= V_m V_n e^{j2\pi n\tau/T_0} \frac{[\sin \pi(m+n)]}{\pi(m+n)} \tag{1.19-3b}$$

Since m and n are integers, we see from Eq. (1.19-3*b*) that $I_{m,n} = 0$ except when $m = -n$ or $m + n = 0$. To evaluate $I_{m,n}$ in this latter case, we return to Eq. (1.19-3*a*) and find

$$I_{m,n} = I_{-n,n} = V_n V_{-n} e^{j2\pi n\tau/T_0} \tag{1.19-4}$$

Finally, then,

$$R(\tau) = \sum_{n=-\infty}^{\infty} V_n V_{-n} e^{j2\pi n\tau/T_0} = \sum_{n=-\infty}^{\infty} |V_n|^2 e^{j2\pi n\tau/T_0} \tag{1.19-5a}$$

$$= |V_0|^2 + 2 \sum_{n=1}^{\infty} |V_n|^2 \cos 2\pi n \frac{\tau}{T_0} \tag{1.19-5b}$$

We note from Eq. (1.19-5*b*) that $R(\tau) = R(-\tau)$ as anticipated, and we note as well that for this case of a periodic waveform the correlation $R(\tau)$ is also periodic with the same fundamental period T_0.

We shall now relate the correlation function $R(\tau)$ of a periodic waveform to its power spectral density. For this purpose we compute the Fourier transform of $R(\tau)$. We find, using $R(\tau)$ as in Eq. (1.19-5*a*), that

$$\mathscr{F}[R(\tau)] = \int_{-\infty}^{\infty} \left(\sum_{n=-\infty}^{\infty} |V_n|^2 e^{j2\pi n\tau/T_0} \right) e^{-j2\pi f\tau}\, d\tau \tag{1.19-6}$$

Interchanging the order of integration and summation yields

$$\mathscr{F}[R(\tau)] = \sum_{n=-\infty}^{\infty} |V_n|^2 \int_{-\infty}^{\infty} e^{-j2\pi(f-n/T_0)\tau}\, d\tau \tag{1.19-7}$$

Using Eq. (1.11-22), we may write Eq. (1.19-7) as

$$\mathscr{F}[R(\tau)] = \sum_{n=-\infty}^{\infty} |V_n|^2\, \delta\!\left(f - \frac{n}{T_0}\right) \tag{1.19-8}$$

Comparing Eq. (1.19-8) with Eq. (1.8-6), we have the interesting result that for a periodic waveform

$$G(f) = \mathscr{F}[R(\tau)] \tag{1.19-9}$$

and, of course, conversely

$$R(\tau) = \mathscr{F}^{-1}[G(f)] \tag{1.19-10}$$

Expressed in words, we have the following: *The power spectral density and the correlation function of a periodic waveform are a Fourier transform pair.*

1.20 AUTOCORRELATION OF NONPERIODIC WAVEFORM OF FINITE ENERGY

For pulse-type waveforms of finite energy there is a relationship between the correlation function of Eq. (1.18-1) and the energy spectral density which corresponds to the relationship given in Eq. (1.19-9) for the periodic waveform. This relationship is that the correlation function $R(\tau)$ and the *energy* spectral density are a Fourier transform pair. This result is established as follows.

We use the convolution theorem. We combine Eqs. (1.12-1) and (1.12-3) for the case where the waveforms $v_1(t)$ and $v_2(t)$ are the same waveforms, that is, $v_1(t) = v_2(t) = v(t)$, and get

$$\mathscr{F}^{-1}[V(f)V(f)] = \int_{-\infty}^{\infty} v(\tau)v(t - \tau)\, d\tau \tag{1.20-1}$$

Since $V(-f) = V^*(f) = \mathscr{F}[v(-t)]$, Eq. (1.20-1) may be written

$$\mathscr{F}^{-1}[V(f)V^*(f)] = \mathscr{F}^{-1}[|V(f)|^2] = \int_{-\infty}^{\infty} v(\tau)v(\tau - t)\, d\tau \tag{1.20-2}$$

The integral in Eq. (1.20-2) is a function of t, and hence this equation expresses $\mathscr{F}^{-1}[V(f)V^*(f)]$ as a function of t. If we want to express $\mathscr{F}^{-1}[V(f)V^*(f)]$ as a function of τ without changing the form of the function, we need but to interchange t and τ. We then have

$$\mathscr{F}^{-1}[V(f)V^*(f)] = \int_{-\infty}^{\infty} v(t)v(t - \tau)\, dt \tag{1.20-3}$$

The integral in Eq. (1.20-3) is precisely $R(\tau)$, and thus

$$\mathscr{F}[R(\tau)] = V(f)V^*(f) = |V(f)|^2 \tag{1.20-4}$$

which verifies that $R(\tau)$ and the energy spectral density $|V(f)|^2$ are Fourier transform pairs.

1.21 AUTOCORRELATION OF OTHER WAVEFORMS

In the preceding sections we discussed the relationship between the autocorrelation function and power or energy spectral density of *deterministic* waveforms. We use the term "deterministic" to indicate that at least, in principle, it is possible to write a function which specifies the value of the function at all times. For such deterministic waveforms, the availability of an autocorrelation function is of no particularly great value. The autocorrelation function does not include within itself complete information about the function. Thus we note that the autocorrelation function is related only to the amplitudes and not to the phases of the spectral components of the waveform. The waveform cannot be reconstructed from a knowledge of the autocorrelation functions. Any characteristic of a deterministic waveform which may be calculated with the aid of the autocorrelation function may be calculated by direct means at least as conveniently.

On the other hand, in the study of communication systems we encounter waveforms which are not deterministic but are instead random and unpredictable in nature. Such waveforms are discussed in Chap. 2. There we shall find that for such random waveforms no explicit function of time can be written. The waveforms must be described in statistical and probabilistic terms. It is in connection with such waveforms that the concepts of correlation and autocorrelation find their true usefulness. Specifically, it turns out that even for such random waveforms, the autocorrelation function and power spectral density are a Fourier transform pair. The proof[3] that such is the case is formidable and will not be undertaken here.

1.22 EXPANSIONS IN ORTHOGONAL FUNCTIONS

Let us consider a set of functions $g_1(x), g_2(x), \ldots, g_n(x) \cdots$, defined over the interval $x_1 \leq x < x_2$ and which are related to one another in the very special way that any two different ones of the set satisfy the condition

$$\int_{x_1}^{x_2} g_i(x) g_j(x)\, dx = 0 \tag{1.22-1}$$

That is, when we multiply two different functions and then integrate over the interval from x_1 to x_2 the result is zero. A set of functions which has this property is described as being *orthogonal over the interval from* x_1 *to* x_2. The term "orthogonal" is employed here in correspondence to a similar situation which is encountered in dealing with vectors. The *scalar* product of two vectors $\mathbf{V}_i$ and $\mathbf{V}$ (also referred to as the *dot* product or as the *inner* product) is a scalar quantity V_{ij} defined as

$$V_{ij} = |\mathbf{V}_i||\mathbf{V}_j| \cos(\mathbf{V}_i, \mathbf{V}_j) = V_{ji} \tag{1.22-2}$$

In Eq. (1.22-2) $|\mathbf{V}_i|$ and $|\mathbf{V}_j|$ are the magnitudes of the respective vectors and $\cos(\mathbf{V}_i, \mathbf{V}_j)$ is the cosine of the angle between the vectors. If it should turn out that $V_{ij} = 0$ then (ignoring the trivial cases in which $\mathbf{V}_1 = 0$ or $\mathbf{V}_j = 0$) $\cos(\mathbf{V}_i, \mathbf{V}_j)$ must be zero and correspondingly it means that the vectors $\mathbf{V}_i$ and $\mathbf{V}_j$ are perpendicular (i.e., orthogonal) to one another. Thus vectors whose scalar product is zero are physically orthogonal to one another and, in correspondence, functions whose integrated product, as in Eq. (1.22-1) is zero are also orthogonal to one another.

Now consider that we have some arbitrary function $f(x)$ and that we are interested in $f(x)$ only in the range from x_1 to x_2, i.e., in the interval over which the set of functions $g(x)$ are orthogonal. Suppose further that we undertake to write $f(x)$ as a linear sum of the functions $g_n(x)$. That is, we write

$$f(x) = C_1 g_1(x) + C_2 g_2(x) + \cdots + C_n g_n(x) + \cdots \tag{1.22-3}$$

in which the C's are numerical coefficients. Assuming that such an expansion is indeed possible, the orthogonality of the g's makes it very easy to compute the coefficients C_n. Thus to evaluate C_n we multiply both sides of Eq. (1.22-3) by $g_n(x)$ and integrate over the interval of orthogonality. We have

$$\int_{x_1}^{x_2} f(x) g_n(x)\, dx = C_1 \int_{x_1}^{x_2} g_1(x) g_n(x)\, dx$$
$$+ C_2 \int_{x_1}^{x_2} g_2(x) g_n(x)\, dx + \cdots + C_n \int_{x_1}^{x_2} g_n^2(x)\, dx + \cdots \tag{1.22-4}$$

Because of the orthogonality, *all of the terms on the right-hand side of Eq. (1.22-4) become zero with a single exception* and we are left with

$$\int_{x1}^{x2} f(x)g_n(x)\,dx = C_n \int_{x1}^{x2} g_n^2(x)\,dx \tag{1.22-5}$$

so that the coefficient that we are evaluating becomes

$$C_n = \frac{\int_{x1}^{x2} f(x)g_n(x)\,dx}{\int_{x1}^{x2} g_n^2(x)\,dx} \tag{1.22-6}$$

The mechanism by which we use the orthogonality of the functions to "drain" away all the terms except the term that involves the coefficient we are evaluating is often called the "orthogonality sieve."

Next suppose that each $g_n(x)$ is selected so that the denominator of the right-hand member of Eq. (1.22-6) (which is a numerical constant) has the value

$$\int_{x1}^{x2} g_n^2(x)\,dx = 1 \tag{1.22-7}$$

In this case

$$C_n = \int_{x1}^{x2} f(x)g_n(x)\,dx \tag{1.22-8}$$

When the orthogonal functions $g_n(x)$ are selected as in (1.22-7) they are described as being *normalized.* The use of normalized functions has the merit that the C_n's can then be calculated from Eq. (1.22-8) and thereby avoids the need to evaluate $\int_{x1}^{x2} g_n^2(x)\,dx$ in each case as called for in Eq. (1.22-6). A set of functions which is both orthogonal and normalized is called an *orthonormal* set.

1.23 COMPLETENESS OF AN ORTHOGONAL SET: THE FOURIER SERIES

Suppose on the one hand we expand a function $f(x)$ in terms of orthogonal functions as

$$f(x) = C_1 s_1(x) + C_2 s_2(x) + C_3 s_3(x) + \cdots \tag{1.23-1}$$

and on the other hand, we expand it as

$$f(x) = C_1 s_1(x) + C_3 s_3(x) + \cdots \tag{1.23-2}$$

That is, in the second case, we have deliberately omitted one term. A moment's review of the procedure, described in the previous section, for evaluating coefficients makes it apparent that all the coefficients C_1, C_3, etc., that appear in both expansions will turn out to be *the same.* Hence if one expansion is correct the other is in error. We might be suspicious of the expansion of Eq. (1.23-2) on the grounds that a term is missing. But then how do we know that a term is not missing in the expansion of Eq. (1.23-1)? The point is that simply having a set of orthogonal functions and having a procedure for evaluating coefficients does not guarantee that the series so developed can represent an arbitrary function. Such can well be the case even when the orthogonal set consists of an infinite number of independent functions. When, however, the orthogonal set does indeed include all the functions necessary to allow an error free expansion of an arbitrary function then the set is said to be *complete*.

A most important orthogonal set which is complete is the set of sinusoidal functions (both sines and cosines) which generate the *Fourier series*. In this case, because of the periodicity of the functions, it is not necessary to specify the *end points* of the interval over which the expansion is to be valid but only to specify the *length* of the interval. (It may, however, be useful to specify the interval end points for the sake of computational convenience in connection with evaluating the coefficients.) Specifically, if the variable of interest is time t and the length of time interval is T, then the Fourier expansion of a function $v(t)$ is

$$v(t) = \sum_{n=0}^{\infty} A_n \cos \frac{2\pi nt}{T} + \sum_{n=0}^{\infty} B_n \sin \frac{2\pi nt}{T} \tag{1.23-3}$$

We take account of the fact that $\cos 0 = 1$ and $\sin 0 = 0$, and applying the normalized procedure of Sec. 1.22 we can express the expansion of $v(t)$ in terms of orthogonal functions as

$$v(t) = \frac{A_0}{\sqrt{T}} + \sum_{n=1}^{\infty} A_n \sqrt{2/T} \cos \frac{2\pi nt}{T} + \sum_{n=1}^{\infty} B_n \sqrt{2/T} \sin \frac{2\pi nt}{T} \tag{1.23-4}$$

The orthonormal functions are

$$1/\sqrt{T}; \quad \sqrt{2/T} \cos 2\pi nt/T, n \neq 0; \quad \text{and} \quad \sqrt{2/T} \sin 2\pi nt/T.$$

Any two such different functions when multiplied and integrated over T yields zero, and any function squared and integrated over T yields unity.

In the general case an expansion is valid only over the finite interval T of orthogonality. In the present case we observe from Eq. (1.23-4) that the expansion is periodic with period T. If it should happen that $v(t)$ is also periodic with period T then the expansion is valid for all t. Thus a periodic function with period T can be expanded into a Fourier series as in Eq. (1.23-4) in which the coefficients are given by

$$A_0 = \frac{1}{\sqrt{T}} \int_T v(t)\, dt \tag{1.23-5a}$$

$$A_n = \sqrt{\frac{2}{T}} \int_T v(t) \cos \frac{2\pi nt}{T}\, dt \qquad n \neq 0 \tag{1.23-5b}$$

$$B_n = \sqrt{\frac{2}{T}} \int_T v(t) \cos \frac{2\pi nt}{T}\, dt \tag{1.23-5c}$$

1.24 THE GRAM–SCHMITT PROCEDURE

Orthogonal sets of functions that are complete for the purpose of representing an arbitrary function inevitably have an infinite number of components. It turns out however that it is useful, on occasion, to be able to construct an orthogonal set which is complete for the purpose of allowing valid expansions of only a finite number of functions. In such cases the orthogonal set itself has only a finite number of functions and the Gram–Schmitt procedure which we now describe allows us to construct this orthogonal (or orthonormal) set.

Let us call the time functions which are to be expanded: $s_1(t), s_2(t), \ldots, s_N(t)$, and the orthonormal functions $u_1(t), u_2(t), \ldots, u_N(t)$. Then we require, given the functions $s(t)$, to find the functions $u(t)$ and to evaluate the coefficients s_{ij} in the expansions:

$$s_1(t) = s_{11}u_1(t) + s_{12}u_2(t) + \cdots + s_{1N}u_N(t) \qquad (1.24\text{-}1a)$$

$$s_2(t) = s_{21}u_1(t) + s_{22}u_2(t) + \cdots + s_{2N}u_N(t) \qquad (1.24\text{-}1b)$$

$$\vdots \qquad\qquad \vdots \qquad\qquad \vdots \qquad\qquad \vdots$$

$$s_N(t) = s_{N1}u_1(t) + s_{N2}u_2(t) + \cdots + s_{NN}u_N(t) \qquad (1.24\text{-}1n)$$

or, using a shorter notation

$$s_i(t) = \sum_{j=1}^{N} s_{ij}u_j(t) \qquad i = 1, 2, \ldots, N \qquad (1.24\text{-}2)$$

The orthogonality of the functions $u(t)$ over the interval T is expressed by

$$\int_T u_j(t)u_k(t)\,dt = \begin{cases} 1 & \text{if } j = k \\ 0 & \text{if } j \neq k \end{cases} \qquad (1.24\text{-}3)$$

The Gram–Schmitt method proceeds as follows:

Step 1. In Eq. (1.24-1a) we set to zero all the coefficients except s_{11}. We then have

$$s_1(t) = s_{11}u_1(t) \qquad (1.24\text{-}4a)$$

Since $u_1(t)$ is to be a normalized function we find that

$$s_{11} = \left[\int_T s_1^2(t)\,dt\right]^{1/2} \qquad (1.24\text{-}4b)$$

and $u_1(t) \doteq s_1(t)/s_{11}$ is now determined.

Step 2. In Eq. (1.24-1b) we set to zero all coefficients except the first two, s_{21} and s_{22}. We then have

$$s_2(t) = s_{21}u_1(t) + s_{22}u_2(t) \qquad (1.24\text{-}5)$$

Multiplying both sides of Eq. (1.24-5) by $u_1(t)$, integrating over the interval T, and taking account of the orthonormality of $u_1(t)$ and $u_2(t)$ we find that

$$s_{21} = \int_T s_2(t)u_1(t)\,dt \tag{1.24-6}$$

Now that s_{21} is known we can evaluate s_{22}. We rewrite Eq. (1.24-5) as

$$s_2(t) - s_{21}u_1(t) = s_{22}\,u_2(t) \tag{1.24-7a}$$

Squaring and integrating we have

$$\int_T [s_2(t) - s_{21}u_1(t)]^2\,dt = s_{22}^2 \int_T u_2^2(t)\,dt = s_{22}^2 \tag{1.24-7b}$$

so that

$$s_{22} = \left\{\int_T [s_2(t) - s_{21}u_1(t)]^2\,dt\right\}^{1/2} \tag{1.24-8}$$

Finally, since now both s_{21} and s_{22} are known, we can again use Eq. (1.24-5) to find $u_2(t)$ with the result:

$$u_2(t) = \frac{1}{s_{22}}[s_2(t) - s_{21}u_1(t)]$$

$$= \frac{1}{s_{22}}\left[s_2(t) - \frac{s_{21}s_1(t)}{s_{11}}\right] \tag{1.24-9}$$

Step 3. Continuing the pattern as above we write $s_3(t)$ in Eq. (1.24-1c) as

$$s_3(t) = s_{31}u_1(t) + s_{32}\,u_2(t) + s_{33}\,u_3(t) \tag{1.24-10}$$

that is, we set s_{34}, etc., to zero. We then find, as can be verified, that

$$s_{31} = \int_T s_3(t)u_1(t)\,dt \tag{1.24-11a}$$

$$s_{32} = \int_T s_3(t)u_2(t)\,dt \tag{1.24-11b}$$

and

$$s_{33} = \left\{\int_T [s_3(t) - s_{31}u_1(t) - s_{32}\,u_2(t)]^2\,dt\right\}^{1/2} \tag{1.24-11c}$$

and finally

$$u_3(t) = \frac{s_3(t) - s_{31}u_1(t) - s_{32}\,u_2(t)}{s_{33}} \tag{1.24-12}$$

Step 4. We continue the procedure until we have used all N equations and shall finally have N orthonormal functions $u_1(t)$, $u_2(t)$, ..., $u_N(t)$. We shall also have evaluated the coefficients s_{ij} needed to express the functions $s_1(t)$, $s_2(t)$, ..., $s_N(t)$ in terms of the $u(t)$'s.

It is assumed in this discussion that the N functions $s(t)$ are linearly independent, that is that no one of the $s(t)$'s can be expressed as a linear sum of the other $s(t)$'s. Suppose, on the other hand, that such is not the case. Assume, for example, that $s_3(t) = C_1 s_1(t) + C_2 s_2(t)$, C_1 and C_2 being constants. Then since $s_1(t)$ and $s_2(t)$ can be expressed in terms of $u_1(t)$ and $u_2(t)$ so also would $s_3(t)$ be expressible in terms of $u_1(t)$ and $u_2(t)$. Thus in Eq. (1.24-10) we would find that $s_{33} = 0$ and our procedure would not generate a new function $u_3(t)$. In general, if there are N functions $s_i(t)$ but only M of them are linearly independent, then the procedure will generate $M(\leq N)$ orthonormal functions in terms of which we shall be able to express all N functions $s(t)$.

Given N functions $s(t)$ we are, of course, at liberty to number them in any manner we please. Correspondingly, $u_1(t)$ will be determined by which of the functions we please to call $s_1(t)$, $u_2(t)$ will then be determined by this selection of $s_1(t)$ and the subsequent selection of $s_2(t)$, etc. In short, the set of orthonormal functions generated by our procedure is not *unique*. There are, in general, as many sets as there are ways to assign the numbers 1 through N to N functions.

Example 1.24-1 Two functions $s_1(t)$ and $s_2(t)$ are given in Fig. 1.24-1*a*. Here the interval of interest extends from $t = 0$ to $t = T$. Use the Gram–Schmitt procedure to express these functions in terms of orthonormal components.

SOLUTION We find from Eq. (1.24-4*b*) that

$$s_{11} = \left[\int_0^T s_1^2(t)\,dt\right]^{1/2} = \left(\int_0^T 2^2\,dt\right)^{1/2} = 2\sqrt{T}$$

$$u_1(t) = \frac{s_1(t)}{s_{11}} = \frac{s_1(t)}{2\sqrt{T}} \tag{1.24-13}$$

From Eq. (1.24-6) we have

$$s_{21} = \int_0^T s_2(t)u_1(t)\,dt = \int_0^{T/2} 4\left(\frac{2}{\sqrt{4T}}\right)dt = 2\sqrt{T}$$

From Eq. (1.24-8)

$$s_{22} = \left\{\int_0^T [s_2(t) - s_{21}u_1(t)]^2\,dt\right\}^{1/2}$$

The function $[s_2(t) - s_{21}u_1(t)]$ and $[s_2(t) - s_{21}u_1(t)]^2$ and drawn in Fig. 1.24-1*b*. We have, accordingly,

$$s_{22} = \left(\int_0^T 4\,dt\right)^{1/2} = 2\sqrt{T}$$

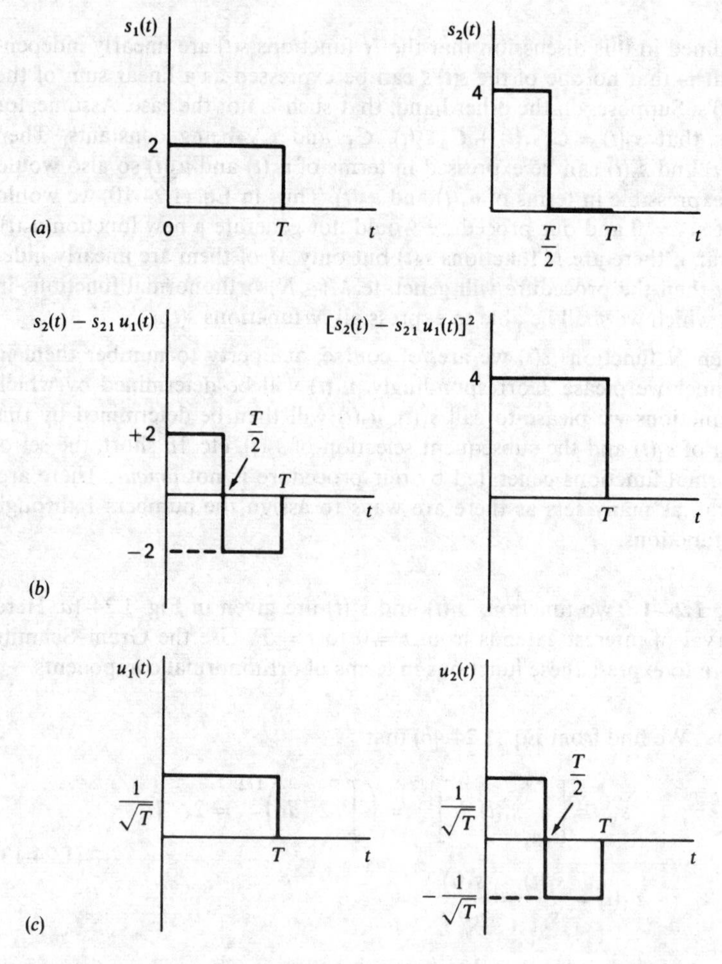

Figure 1.24-1 Decomposition of two signals using the Gram-Schmitt technique. (*a*) The original signals. (*b*) Evaluating s_{22}. (*c*) Orthonormal functions $u_1(t)$ and $u_2(t)$.

and finally from Eq. (1.24-9) we find $u_2(t)$ to be

$$u_2(t) = \frac{1}{2\sqrt{T}}\left[s_2(t) - \frac{2\sqrt{T}\,s_1(t)}{2\sqrt{T}}\right] = \frac{1}{2\sqrt{T}}\,[s_2(t) - s_1(t)]$$

The orthonormal functions $u_1(t)$ and $u_2(t)$ are shown in Fig. 1.24-1*c*. Finally, we have

$$s_2(t) = 2\sqrt{T}\,u_1(t) + 2\sqrt{T}\,u_2(t) \tag{1.24-14}$$

An alternative set of orthonormal functions can be generated by interchanging $s_1(t)$ and $s_2(t)$ (see Prob. 1.24-1).

1.25 CORRESPONDENCE BETWEEN SIGNALS AND VECTORS

In this section we shall draw an analogy and show a correspondence between the specification of a vector in terms of orthonormal vector components and of a signal in terms of orthonormal functions.

In Fig. 1.25-1*a* a vector **A** is represented in an XYZ coordinate system. We have defined three unit vectors i, j, and k in the X, Y, and Z directions, respectively. These unit vectors have the property that the scalar products of a unit vector with itself are

$$i \cdot i = j \cdot j = k \cdot k = 1 \tag{1.25-1}$$

We note a correspondence between the characteristic noted in Eq. (1.24-3) and the characteristic of the orthonormal function discussed here that

$$\int_T u_i(t) \cdot u_i(t)\, dt = 1 \tag{1.25-2}$$

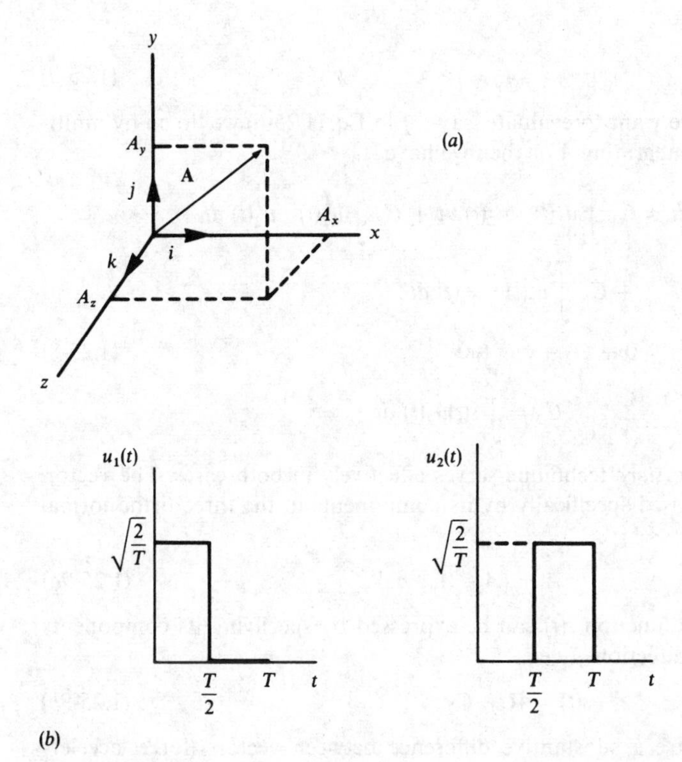

Figure 1.25-1 (*a*) Expressing a vector **A** in a three dimensional cartesian coordinate system. (*b*) Unit vectors u_1 and u_2.

Further, the unit vectors have the property that

$$i \cdot j = j \cdot k = k \cdot i = 0 \tag{1.25-3}$$

in correspondence with the property of the orthonormal functions

$$\int_T u_i(t) \cdot u_j(t)\, dt = 0 \qquad (i \neq j) \tag{1.25-4}$$

An arbitrary vector **A** can be written in terms of its components as

$$\mathbf{A} = A_x i + A_y j + A_z k \tag{1.25-5}$$

While an arbitrary function $s(t)$ can be written as

$$s(t) = C_1 u_1(t) + C_2 u_2(t) + \cdots + C_N u_N(t) \tag{1.25-6}$$

If we want to evaluate a vector component, say A_y, we can do so by multiplying Eq. (1.25-5) by j. For then we have

$$\begin{aligned} j \cdot \mathbf{A} &= A_n j \cdot i + A_y j \cdot j + A_z j \cdot k \\ &= 0 + A_y + 0 \end{aligned}$$

so that

$$A_y = j \cdot \mathbf{A} \tag{1.25-7}$$

correspondingly if we want to evaluate, say C_2 in Eq. (1.25-6) we do so by multiplying by $u_2(t)$ and integrating. For then we have

$$\begin{aligned} \int_T s(t)u_2(t)\, dt &= C_1 \int_T u_1(t) \cdot u_2(t)\, dt + C_2 \int_T u_2(t) \cdot u_2(t)\, dt + \cdots \\ &\quad + C_N \int u_n(t) \cdot u_2(t)\, dt \\ &= 0 + C_2 + \cdots + 0 \end{aligned} \tag{1.25-8}$$

so that

$$C_2 = \int_T s(t)u_2(t)\, dt$$

Thus we see that the *sieve* technique serves effectively in both cases. The vector, finally, can be expressed specifically by its components in the three orthonormal directions, i.e.,

$$A = \{A_x, A_y, A_z\} \tag{1.25-9a}$$

Correspondingly the function $s(t)$ can be expressed by specifying its components in N orthonormal "directions," i.e.,

$$s(t) = \{C_1, C_2, \ldots, C_N\} \tag{1.25-9b}$$

There is of course a substantive difference between vectors (force, acceleration, etc.) and functions. In the case of vectors the coordinate system, in which we specify our vectors, is always a system in which the axes extend in three orthogonal directions in a physical space. With functions, the coordinates extend in N "directions" in some "function space" and not in a physical space. Still, continuing our comparisons we note that a vector **A** has a magnitude $|A|$ which is calculated

$$|\mathbf{A}| = [\mathbf{A} \cdot \mathbf{A}]^{1/2} = [A_x^2 + A_y^2 + A_z^2]^{1/2} \tag{1.25-10a}$$

Correspondingly we can take the "magnitude" of $s(t)$, $|s(t)|$ to be defined by

$$|s(t)| = \left[\int_T s^2(t)\, dt\right]^{1/2} \tag{1.25-10b}$$

Note that the magnitude squared of $s(t)$ is the *signal energy*. From Eq. (1.25-6), taking into account the orthonormality of the functions $u_1(t), u_2(t), \ldots, u_N(t)$ it is readily verified (Prob. 1.25-1) that

$$|s(t)|^2 = [C_1^2 + C_2^2 + \cdots + C_N^2] \tag{1.25-10c}$$

In summary: To represent any vector we first establish and define a coordinate system by introducing three orthogonal unit vectors i, j, and k. The vector is expressed as a linear combination of these unit vectors as in Eq. (1.25-5). The vector can then be represented by the coefficients A_x, A_y, and A_z which are the *components* of the vector in the specified coordinate system. In a second coordinate system, oriented differently from the first, with unit vectors i', j', and k', the components will be different from what they were in the first coordinate system. Of course, however, intrinsic properties of the vector such as magnitude will be invariant to the selection of the coordinate system.

In function space, to represent N signals $s_i(t)$, we first establish and define a coordinate system by introducing N orthonormal functions, $u_1(t), u_2(t), \ldots, u_N(t)$. The signals are then expressed as a linear combination of these orthonormal functions as in Eq. (1.25-6). Thus, the signal is represented by the coefficients C_1, $C_2, \ldots, C_N$ which are the *components* of the function in the specified coordinate system. In a second coordinate system, "oriented" differently from the first, with "unit functions" $u'_1(t), u'_2(t), \ldots, u'_N(t)$ the components will be different from what they were in the first coordinate system. Of course, intrinsic properties of the functions $s_i(t)$, such as the energy of each signal, are invariant to the choice of coordinate system.

In the Gram–Schmitt procedure we start by orienting the function space coordinate system in such manner that the first function $s_1(t)$ [see Eq. (1.24-4a)] points in the direction of one of the coordinate axes. Hence the first function $s_1(t)$ has only a single component s_{11}, all other components being zero. Next we rotate the coordinate system about the $u_1(t)$ axis until the second function lies in a "plane" defined by the two axes $u_1(t)$ and $u_2(t)$. Hence $s_2(t)$ has only two components s_{21} and s_{22}. Proceeding in this manner we find that each successive function $s_i(t)$ has one additional component provided that it is independent of those which have preceded it in the procedure. We have already noted that we are free to number the functions $s_i(t)$ as we please. Each new numbering order yields a different function space coordinate system. These different coordinate systems are "rotated" with respect to one another and correspondingly have different components $C_1, C_2, \ldots, C_N$ for the same function $s(t)$.

The application of the Gram–Schmitt procedure yields a function-space coordinate system which has a special orientation with respect to the functions $s(t)$. However, we can, if we please, easily generate a limitless number of function-space coordinate systems which have no special orientation with respect to the functions $s(t)$ so that, in the general case, every $s(t)$ has as many non-zero components as there are coordinates. The procedure for generating such coordinate systems with no special orientation (having again an exact correspondence with the procedure applicable to vectors in real space) will be illustrated by the following example involving just two functions. The generalization to an arbitrary number of functions will be readily apparent.

Example 1.25-1 In Example 1.24-1 we started with the functions $s_1(t)$ and $s_2(t)$ of Fig. 1.24-1*a* and found the expansions for $s_1(t)$ and $s_2(t)$ to be as given in Eqs. (1.24-12) and (1.24-13), that is,

$$s_1(t) = 2\sqrt{T}\, u_1(t)$$

$$s_2(t) = 2\sqrt{T}\, u_1(t) + 2\sqrt{T}\, u_2(t)$$

where $u_1(t)$ and $u_2(t)$ are given in Fig. 1.24-1*c*. That is, $s_1(t)$ and $s_2(t)$ expressed in terms of components are

$$s_1(t) = \{2\sqrt{T}, 0\} \tag{1.25-11a}$$

$$s_2(t) = \{2\sqrt{T}, 2\sqrt{T}\} \tag{1.25-11b}$$

Find more general orthonormal functions $u'_1(t)$ and $u'_2(t)$ which generate coordinate systems which have no special orientation with respect to the function $s_1(t)$ and $s_2(t)$.

SOLUTION We take $u'_1(t)$ to be a linear combination of the orthonormal functions $u_1(t)$ and $u_2(t)$, say α parts of $u_2(t)$ to one part of $u_1(t)$. We multiply this linear combination by a number N_1 which we shall later adjust to arrange that $u'_1(t)$ be normalized. Thus, altogether we write

$$u'_1(t) = N_1[u_1(t) + \alpha u_2(t)] \tag{1.25-12a}$$

Correspondingly we set

$$u'_2(t) = N_2[u_1(t) + \beta u_2(t)] \tag{1.25-12b}$$

It is essential that α and β not be the same for otherwise $u'_1(t)$ and $u'_2(t)$ would be the same functions except for a multiplicative constant. Since $u_1(t)$ and $u_2(t)$ are orthonormal and we require that $u'_1(t)$ and $u'_2(t)$ also be orthonormal, we find that (see Prob. 1.25-3)

$$1 + \alpha\beta = 0 \tag{1.25-13}$$

Equation (1.25-13) constitutes the only constraint on α and β. Within this constraint we are at liberty to select α and β as we please, and thereby are able to generate a limitless number of different function-space coordinate systems. Having selected α and β we can then adjust N_1 and N_2 to normalize $u_1'(t)$ and $u_2'(t)$.

As a specific example, let us assume $\alpha = 1$ so that $\beta = -1$. In this case we readily find, using $u_1(t)$ and $u_2(t)$ as given in Fig. 1.25-2 that $N_1 = N_2 = 1/\sqrt{2}$ and $u_1'(t)$ and $u_2'(t)$ appear as shown. We find further that $s_1(t)$ and $s_2(t)$ are expressed in terms of these new orthonormal functions as

$$s_1(t) = \sqrt{2T}\, u_1'(t) + \sqrt{2T}\, u_2'(t) \tag{1.25-14a}$$

$$s_2(t) = 2\sqrt{2T}\, u_1'(t) \tag{1.25-14b}$$

In terms of components in this coordinate system in which the unit functions are $u_1'(t)$ and $u_2'(t)$ we find

$$s_1(t) = \{\sqrt{2T}, \sqrt{2T}\} \tag{1.25-15a}$$

$$s_2(t) = \{2\sqrt{2T}, 0\} \tag{1.25-15b}$$

In conclusion, we have drawn in Fig. 1.25-2 a coordinate system in which the unit functions are $u_1(t)$ and $u_2(t)$ and, in this coordinate system, using the

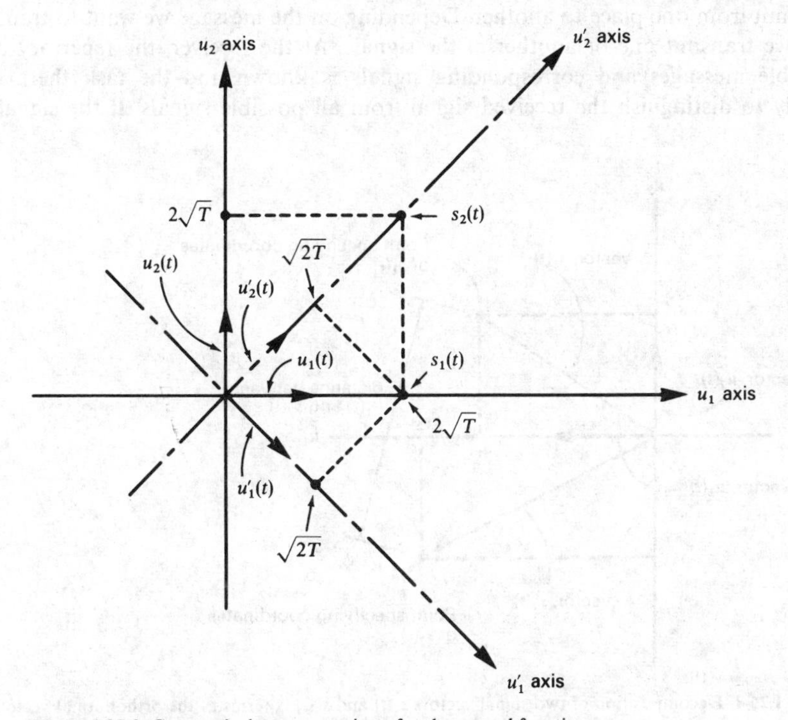

Figure 1.25-2 Geometrical representation of orthonormal functions.

components given in Eq. (1.25-11) we have located the functions $s_1(t)$ and $s_2(t)$. We have also drawn the coordinate system in which the unit functions are $u_1'(t)$ and $u_2'(t)$ in which the components of the functions $s_1(t)$ and $s_2(t)$ are given in Eqs. (1.25-15).

1.26 DISTINGUISHABILITY OF SIGNALS

In the general case, then (for two independent functions $s_1(t)$ and $s_2(t)$), the expansion of the function $s_i(t)$ into an orthonormal series yields the geometric representation of Fig. 1.26-1. A coordinate system is defined by the unit orthonormal vectors $u_1(t)$ and $u_2(t)$. The functions $s_1(t)$ and $s_2(t)$ are then represented by their components, that is the coordinates (s_{11}, s_{12}) and (s_{21}, s_{22}) or by the function vectors themselves drawn from the origin of the system to the coordinate point. Thus s_{11} is the component of $s_1(t)$ in the direction of $u_1(t)$, etc. A change in the selection of the unit vectors would generate a rotation of the coordinate system. Under such a rotation, however, the magnitude of each vector $s_1(t)$ and $s_2(t)$, and the angle between them would remain invariant just as in the case of real vectors in physical space.

As we shall describe more fully in Chap. 6 the set of N waveforms $s_1(t)$, $s_2(t)$, ..., $s_N(t)$ can be used to represent N different messages which we may want to transmit from one place to another. Depending on the message we want to transmit, we transmit one or another of the signals. At the receiver, the repertory of possible messages and corresponding signals is known and the task there is simply to distinguish the received signal from all possible signals. If the signals

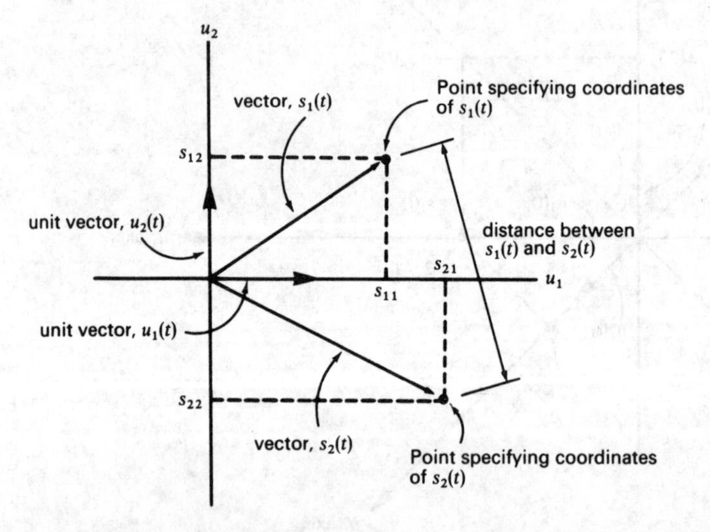

Figure 1.26-1 Decomposition of two signal vectors $s_1(t)$ and $s_2(t)$ is terms of the orthonormal vectors $u_1(t)$ and $u_2(t)$.

were received as transmitted, there would be no problem at all. More generally however, in transmission, the signal will become mixed with noise. In Chap. 2 we show that because of the characteristics of the noise, no matter how little is added to the signal, the noise will generate a finite probability that an error will be made at the receiver in judging which signal has been transmitted. It seems intuitively reasonable that to minimize the probability of error, it would be advantageous to make the individual signals $s_i(t)$ as "distinguishable" from one another as possible. The question is now how to measure distinguishability. Referring to Fig. 1.26-1 it again seems reasonable to suggest that distinguishability is measured by the "distance" between the two coordinate points representing $s_1(t)$ and $s_2(t)$, that is by the magnitude of the difference between the two signal vectors.

$$|s_1(t) - s_2(t)| = [(s_{11} - s_{21})^2 + (s_{12} - s_{22})^2]^{1/2} \tag{1.26-1}$$

In Chap. 11 we shall see that this surmise is indeed valid. Applying the principle to the present two signal case, it then appears that we should make the magnitude of the two signals as large as possible and select signals where the angle between them is 180°.

Example 1.26-1 Consider the four signals $s_1(t)$, $s_2(t)$, $s_3(t)$, and $s_4(t)$ given by the relation:

$$s_i(t) = \sqrt{2P_s} \cos \left[\omega_0 t + (2i - 1)\frac{\pi}{4} \right] \tag{1.26-2}$$

$i = 1, 2, 3, 4$ for $0 \leq t \leq T$. Assume $2\omega_0 T = n\pi$.

(*a*) Find a set of orthonormal coordinates.

(*b*) Plot the four signals using the orthonormal coordinates.

Solution $s_i(t)$ can be expanded and written as

$$s_i(t) = \sqrt{2P_s}\left[\cos (2i-1)\frac{\pi}{4}\right] \cos \omega_0 t - \sqrt{2P_s}\left[\sin (2i-1)\frac{\pi}{4}\right] \sin \omega_0 t$$

We shall choose

$$u_1(t) = \sqrt{\frac{2}{T}} \cos \omega_0 t \tag{1.26-3}$$

and

$$u_2(t) = \sqrt{\frac{2}{T}} \sin \omega_0 t \tag{1.26-4}$$

Then $s_i(t)$ can be rewritten as

$$s_i(t) = \sqrt{P_s T}\left[\cos (2i-1)\frac{\pi}{4}\right] u_1(t) - \sqrt{P_s T}\left[\sin (2i-1)\frac{\pi}{4}\right] u_2(t) \tag{1.26-5}$$

The results for $i = i$, 2, 3, and 4 are shown in Fig. 1.26-2. Note that the distance d between, say, $s_1(t)$ and $s_2(t)$ is $d = \sqrt{2P_s T}$.

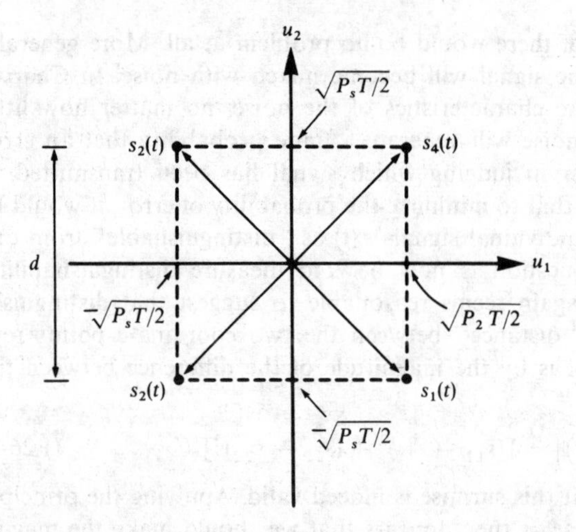

Figure 1.26-2 Signal vectors representing the four signals of Eq. (1.26-2).

It is important to note that the magnitude of the signal vectors shown in Fig. 1.26-2 is equal to the square root of the signal energy. Thus, in Eq. (1.26-2) each signal $s_i(t)$ has a power P_s and a duration T. Thus the signal energy is $P_s T$, which is equal to the magnitude squared of the signal vectors as shown in Fig. 1.26-2. This result simplifies the drawing of the signal vectors in signal space.

For example, consider the signal

$$s_i(t) = \sqrt{2P_s} \cos \left[\omega_0 t + (2i - 1) \frac{\pi}{8} \right] \qquad \begin{cases} i = 1, 2, \ldots, 8 \\ 0 \leq t \leq T \end{cases} \tag{1.26-6}$$

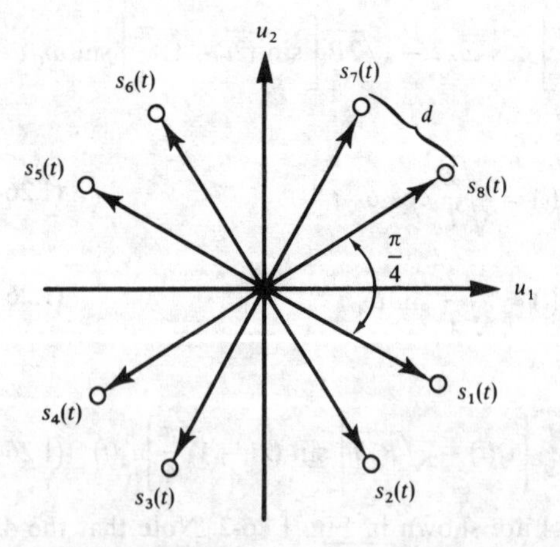

Figure 1.26-3 Signal vectors representing the four signals of Eq. (1.26-6).

Then each signal has the power P_s and duration T. Thus the magnitude of each signal vector is $\sqrt{P_s T}$. Furthermore, each vector is rotated $\pi/4$ radians from the previous vector as shown in Fig. 1.26-3. The reader should verify (Prob. 1.26-1) that Fig. 1.26-3 is correct by finding the two orthonormal components $u_1(t)$ and $u_2(t)$ as in Eqs. (1.26-3) and (1.26-4), expanding Eq. (1.26-6) in a form similar to Eq. (1.26-5) and then plotting the result.

REFERENCES

1. Javid, M., and E. Brenner: "Analysis of Electric Circuits," 2d ed., McGraw-Hill Book Company, New York, 1967.
   Papoulis, A.: "The Fourier Integral and Applications," McGraw-Hill Book Company, New York, 1963.
2. Churchil, R. V.: "Fourier Series," McGraw-Hill Book Company, New York, 1941.
3. Papoulis, A.: " Probability, Random Variables, and Stochastic Processes," McGraw-Hill Book Company, New York, 1965.

PROBLEMS

1.1-1. Verify the relationship of C_n and ϕ_n to A_n and B_n as given by Eq. (1.1-6).

1.1-2. Calculate A_n, B_n, C_n, and ϕ_n for a waveform $v(t)$ which is a symmetrical square wave and which makes peak excursions to $+\frac{3}{4}$ volt and $-\frac{1}{4}$ volt, and has a period $T = 1$ sec. A positive going transition occurs at $t = 0$.

1.2-1. Verify the relationship of the complex number V_n to C_n and ϕ_n as given by Eq. (1.2-3).

1.2-2. The function

$$p(t) = \begin{cases} e^{-t} & 0 \le t \le 1 \\ 0 & \text{elsewhere} \end{cases}$$

is repeated every $T = 1$ sec. Thus, with $u(t)$ the unit step function

$$v(t) = \sum_{n=-\infty}^{\infty} p(t-n)u(t-n)$$

Find V_n and the exponential Fourier series for $v(t)$.

1.3-1. The impulse function can be defined as:

$$\delta(t) = \lim_{a \to \infty} \frac{a}{2} e^{-a|t|}$$

Discuss.

1.3.2. In Eq. (1.3-8) we see that

$$v(t) = I \sum_{k=-\infty}^{\infty} \delta(t - kT_0) = \frac{I}{T_0} + \frac{2I}{T_0} \sum_{n=1}^{\infty} \cos \frac{2\pi nt}{T_0}$$

Set $I = 1$, $T_0 = 1$. Show by plotting

$$v(t) = 1 + 2 \sum_{n=1}^{3} \cos 2\pi nt$$

that the Fourier series does approximate the train of impulses.

1.4-1. $Sa(x) \equiv (\sin x)/x$. Determine the maxima and minima of $Sa(x)$ and compare your result with the approximate maxima and minima obtained by letting $x = (2n + 1)\pi/2$, $n = 1, 2, \ldots$.

1.4-2. A train of rectangular pulses, making excursions from zero to 1 volt, have a duration of 2 μs and are separated by intervals of 10 μs.

(*a*) Assume that the center of one pulse is located at $t = 0$. Write the exponential Fourier series for this pulse train and plot the spectral amplitude as a function of frequency. Include at least 10 spectral components on each side of $f = 0$, and draw also the envelope of these spectral amplitudes.

(*b*) Assume that the left edge of a pulse rather than the center is located at $t = 0$. Correct the Fourier expansion accordingly. Does this change affect the plot of spectral amplitudes? Why not?

1.5-1. (*a*) A periodic waveform $v_i(t)$ is applied to the input of a network. The output $v_o(t)$ of the network is $v_o(t) = \tau[dv_i(t)/dt]$, where τ is a constant. What is the transfer function $H(\omega)$ of this network?

(*b*) A periodic waveform $v_i(t)$ is applied to the *RC* network shown, whose time constant is $\tau \equiv RC$. Assume that the highest frequency spectral component of $v_i(t)$ is of a frequency $f \ll 1/\tau$. Show that, under these circumstances, the output $v_o(t)$ is approximately $v_o(t) \simeq \tau[dv_i(t)/dt]$.

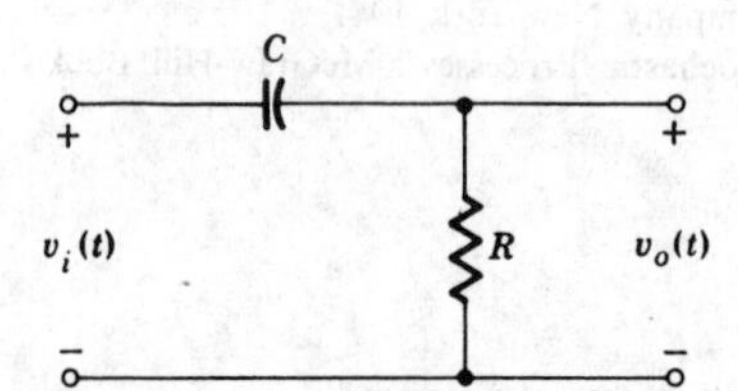

Figure P1.5-1

1.5.2. A voltage represented by an impulse train of strength I and period T is filtered by a low-pass *RC* filter having a 3-dB frequency f_c.

(*a*) Find the Fourier series of the output voltage across the capacitor.

(*b*) If the third harmonic of the output is to be attenuated by 1000, find $f_c T$.

1.6-1. Measurements on a voltage amplifier indicate a gain of 20 dB.

(*a*) If the input voltage is 1 volt, calculate the output voltage.

(*b*) If the input power is 1 mw, calculate the output power.

1.6-2. A voltage gain of 0.1 is produced by an attenuator.

(*a*) What is the gain in decibels?

(*b*) What is the power gain (not in decibels)?

1.7-1. A periodic triangular waveform $v(t)$ is defined by

$$v(t) = \frac{2t}{T} \quad \text{for} \quad -\frac{T}{2} < t < \frac{T}{2}$$

and

$$v(t \pm T) = v(t)$$

and has the Fourier expansion

$$v(t) = \frac{2}{\pi} \sum_{n=1}^{\infty} \frac{(-1)^{n+1}}{n} \sin 2\pi n \frac{t}{T}$$

Calculate the fraction of the normalized power of this waveform which is contained in its first three harmonics.

1.7-2. The complex spectral amplitudes of a periodic waveform are given by

$$V_n = \frac{1}{|n|} e^{-j \arctan (n/2)} \qquad n = \pm 1, \pm 2, \ldots$$

Find the ratio of the normalized power in the second harmonic to the normalized power in the first harmonic.

1.8-1. Find $G(f)$ for the following voltages:

(*a*) An impulse train of strength I and period T.

(*b*) A pulse train of amplitude A, duration $\tau = I/A$, and period T.

1.8-2. Plot $G(f)$ for a voltage source represented by an impulse train of strength I and period nT for $n = 1, 2, 10$, infinity. Comment on this limiting result.

1.9-1. $G_i(f)$ is the power spectral density of a *square-wave* voltage of peak-to-peak amplitude 1 and period 1. The square wave is filtered by a low-pass RC filter with a 3-dB frequency 1. The output is taken across the capacitor.

(*a*) Calculate $G_i(f)$.

(*b*) Find $G_0(f)$.

1.9-2. (*a*) A symmetrical square wave of zero mean value, peak-to-peak voltage 1 volt, and period 1 sec is applied to an ideal low-pass filter. The filter has a transfer function $|H(f)| = \frac{1}{2}$ in the frequency range $-3.5 \leq f \leq 3.5$ Hz, and $H(f) = 0$ elsewhere. Plot the power spectral density of the filter output.

(*b*) What is the normalized power of the input square wave? What is the normalized power of the filter output?

1.10-1. In Eqs. (1.2-1) and (1.2-2) write $f_0 \equiv 1/T_0$. Replace f_0 by Δf, i.e., Δf is the frequency interval between harmonics. Replace $n\,\Delta f$ by f, i.e., as $\Delta f \to 0$, f becomes a continuous variable ranging from 0 to ∞ as n ranges from 0 to ∞. Show that in the limit as $\Delta f \to 0$, so that Δf may be replaced by the differential df, Eq. (1.2-1) becomes

$$v(t) = \int_{-\infty}^{\infty} V(f) e^{j2\pi f t}\, df$$

in which $V(f)$ is

$$V(f) = \lim_{\substack{f_0 = \Delta f \to 0 \\ nf_0 \to f}} \int_{-1/2f_0}^{1/2f_0} v(t) e^{-j2\pi n f_0 t}\, dt = \int_{-\infty}^{\infty} v(t) e^{-j2\pi f t}\, dt$$

1.10-2. Find the Fourier transform of $\sin \omega_0 t$. Compare with the transform of $\cos \omega_0 t$. Plot and compare the power spectral densities of $\cos \omega_0 t$ and $\sin \omega_0 t$.

1.10-3. The waveform $v(t)$ has the Fourier transform $V(f)$. Show that the waveform delayed by time t_d, i.e., $v(t - t_d)$ has the transform $V(f)e^{-j\omega t_d}$.

1.10-4. (*a*) The waveform $v(t)$ has the Fourier transform $V(f)$. Show that the time derivative $(d/dt)v(t)$ has the transform $(j2\pi f)V(f)$.

(*b*) Show that the transform of the integral of $v(t)$ is given by

$$\mathscr{F}\left[\int_{-\infty}^{t} v(\lambda)\, d\lambda\right] = \frac{V(f)}{j2\pi f}$$

1.12-1. Derive the convolution formula in the frequency domain. That is, let $V_1(f) = \mathscr{F}[v_1(t)]$ and $V_2(f) = \mathscr{F}[v_2(t)]$. Show that if $V(f) = \mathscr{F}[v_1(t)v_2(t)]$, then

$$V(f) = \frac{1}{2\pi} \int_{-\infty}^{\infty} V_1(\lambda) V_2(f - \lambda)\, d\lambda$$

or

$$V(f) = \frac{1}{2\pi} \int_{-\infty}^{\infty} V_2(\lambda) V_1(f - \lambda)\, d\lambda$$

1.12-2. (*a*) A waveform $v(t)$ has a Fourier transform which extends over the range from $-f_M$ to $+f_M$. Show that the waveform $v^2(t)$ has a Fourier transform which extends over the range from $-2f_M$ to $+2f_M$. (*Hint:* Use the result of Prob. 1.12-1.)

(*b*) A waveform $v(t)$ has a Fourier transform $V(f) = 1$ in the range $-f_M$ to $+f_M$ and $V(f) = 0$ elsewhere. Make a plot of the transform of $v^2(t)$.

1.12-3. A filter has an impulse response $h(t)$ as shown. The input to the network is a pulse of unit amplitude extending from $t = 0$ to $t = 2$. By graphical means, determine the output of the filter.

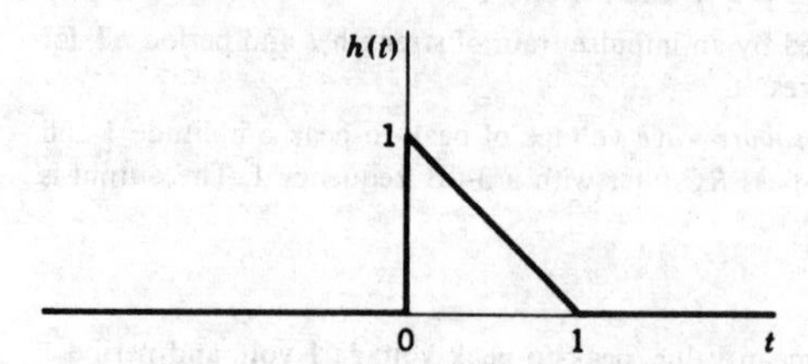

Figure P1.12-3

1.13-1. The energy of a nonperiodic waveform $v(t)$ is

$$E = \int_{-\infty}^{\infty} v^2(t)\, dt$$

(a) Show that this can be written as

$$E = \int_{-\infty}^{\infty} dt v(t) \int_{-\infty}^{\infty} V(f) e^{j2\pi ft}\, df$$

(b) Show that by interchanging the order of integration we have

$$E = \int_{-\infty}^{\infty} V(f)V^*(f)\, df = \int_{-\infty}^{\infty} |V(f)|^2\, df$$

which proves Eq. (1.13-5). This is an alternate proof of Parseval's theorem.

1.13-2. If $V(f) = AT \sin 2\pi fT/2\pi fT$, find the energy E contained in $v(t)$.

1.13-3. A waveform $m(t)$ has a Fourier transform $M(f)$ whose magnitude is as shown.

(a) Find the normalized energy content of the waveform.

(b) Calculate the frequency f_1 such that one-half of the normalized energy is in the frequency range $-f_1$ to f_1.

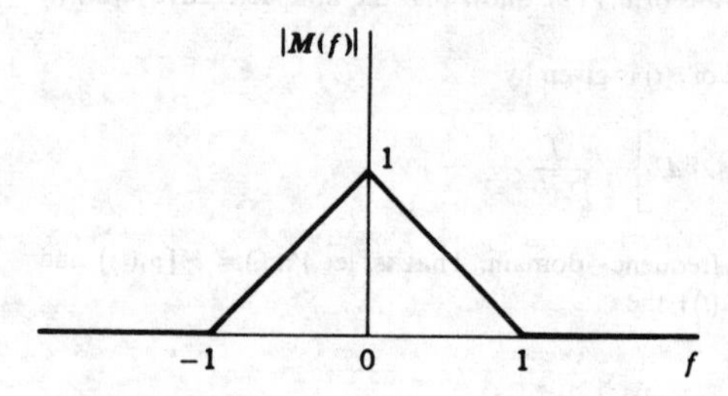

Figure P1.13-3

1.14-1. The signal $v(t) = \cos \omega_0 t + 2 \sin 3\omega_0 t + 0.5 \sin 4\omega_0 t$ is filtered by an RC low-pass filter with a 3-dB frequency $f_c = 2f_0$.

(a) Find $G_i(f)$.

(b) Find $G_0(f)$.

(c) Find S_0.

1.14-2. The waveform $v(t) = e^{-t/\tau}u(t)$ is passed through a high-pass RC circuit having a time constant equal to τ.

(a) Find the energy spectral density at the output of the circuit.

(b) Show that the total output energy is one-half the input energy.

1.15-1. (*a*) An impulse of strength I is applied to a low-pass RC circuit of 3-dB frequency f_2. Calculate the output waveform.

(*b*) A pulse of amplitude A and duration τ is applied to the low-pass RC circuit. Show that, if $\tau \ll 1/f_2$, the response at the output is approximately the response that would result from the application to the circuit of an impulse of strength $I' = A\tau$. Generalize this result by considering any voltage waveform of area I' and duration τ.

1.15-2. A pulse extending from 0 to A volts and having a duration τ is applied to a high-pass RC circuit. Show that the area under the response waveform is zero.

1.16-1. Find the cross correlation of the functions sin ωt and cos ωt.

1.16-2. Prove that $R_{21}(\tau) = R_{12}(-\tau)$.

1.17-1. Find the cross-correlation function $R_{12}(\tau)$ of the two periodic waveforms shown.

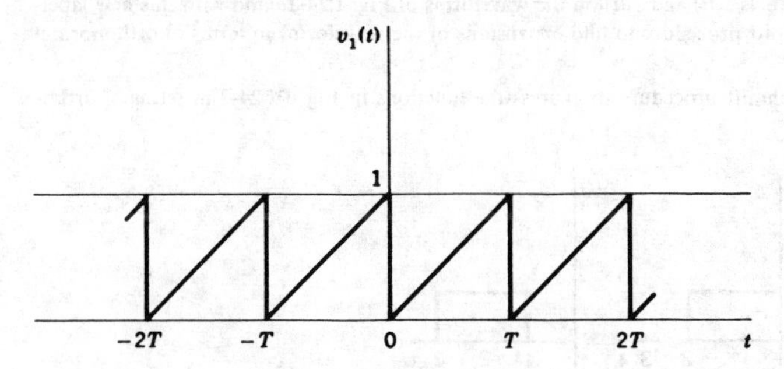

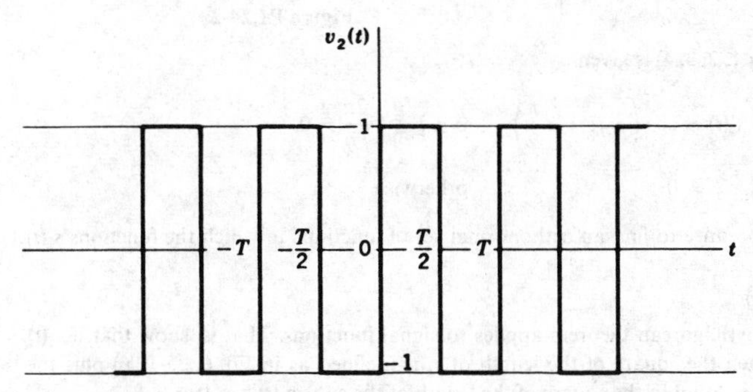

Figure P1.17-1

1.18-1.

$$R(\tau) = \lim_{T \to \infty} \frac{1}{T} \int_{-T/2}^{T/2} v(t)v(t + \tau)\, dt$$

Prove that $R(0) \geq R(\tau)$. *Hint:* Consider

$$I = \lim_{T \to \infty} \frac{1}{T} \int_{-T/2}^{T/2} [v(t) - v(t + \tau)]^2\, dt$$

Is $I \geq 0$? Expand and integrate term by term. Show that

$$I = 2[R(0) - R(\tau)] \geq 0$$

1.18-2. Determine an expression for the correlation function of a square wave having the values 1 or 0 and a period T.

1.19-1. (*a*) Find the power spectral density of a square-wave voltage by Fourier transforming the correlation function. Use the results of Prob. 1.18-2.

(*b*) Compare the answer to (*a*) with the spectral density obtained from the Fourier series (Prob. 1.9-1*a*) itself.

1.19-2. If $v(t) = \sin \omega_0 t$,

(*a*) Find $R(\tau)$.

(*b*) If $G(f) = \mathscr{F}[R(\tau)]$, find $G(f)$ directly and compare.

1.20-1. A waveform consists of a single pulse of amplitude A extending from $t = -\tau/2$ to $t = \tau/2$.

(*a*) Find the autocorrelation function $R(\tau)$ of this waveform.

(*b*) Calculate the energy spectral density of this pulse by evaluating $G_E(f) = \mathscr{F}[R(\tau)]$.

(*c*) Calculate $G_E(f)$ directly by Parseval's theorem and compare.

1.24-1. Interchange the labels $s_1(t)$ and $s_2(t)$ on the waveforms of Fig. 1.24-1*a* and with this new labeling apply the Gram–Schmitt procedure to find expansions of the waveforms in terms of orthonormal functions.

1.24-2. Use the Gram–Schmitt procedure to express the functions in Fig. P1.24-2 in terms of orthonormal components.

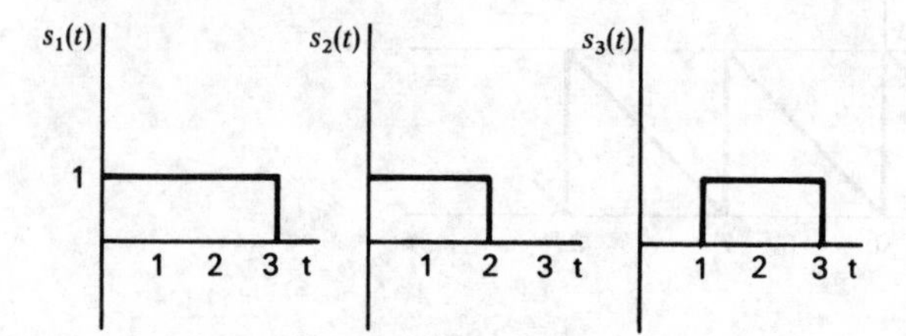

Figure P1.24-2

1.24-3. A set of signals ($k = 1, 2, 3, 4$) is given by

$$s_k(t) = \cos\left(\omega_0 t + k\frac{\pi}{2}\right) \qquad 0 \le t \le k\frac{2\pi}{\omega_0} = 0,$$

$$= 0 \qquad \text{otherwise}$$

Use the Gram–Schmitt procedure to find an orthonormal set of functions in which the functions $s_k(t)$ can be expanded.

1.25-1. Verify Eq. (1.25-10*c*).

1.25-2. (*a*) Show that the pythagorean theorem applies to signal functions. That is, show that if $s_1(t)$ and $s_2(t)$ are orthogonal then the square of the length of $s_1(t)$ (defined as in Eq. (1.25-10*b*)) plus the square of the length of $s_2(t)$ is equal to the square of the length of the sum $s_1(t) + s_2(t)$.

(*b*) Let $s_1(t)$ and $s_2(t)$ be the signals shown in Fig. P1.25-2. Draw the signal $s_1(t) + s_2(t)$. Show that $s_1(t)$ and $s_2(t)$ are orthogonal. Show that $|s_1(t) + s_2(t)|^2 = |s_1(t)|^2 + |s_2(t)|^2$.

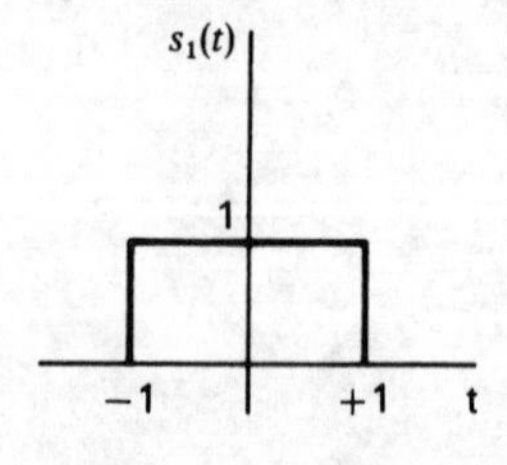

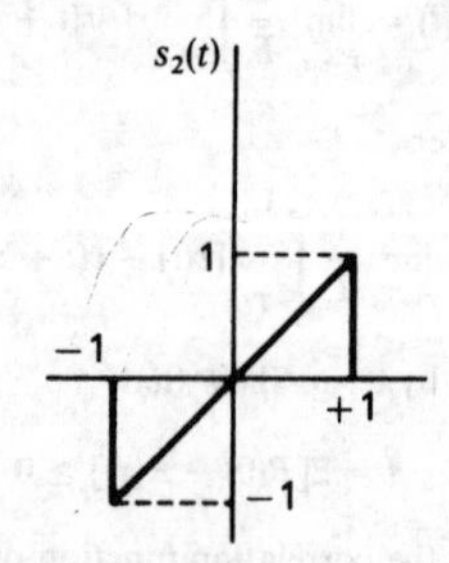

Figure P1.25-2

1.25-3. Verify Eq. (1.25-13).

1.25-4. (*a*) Refer to Fig. 1.25-2 and Eq. (1.25-13). Verify that the parameter α in Eq. (1.25-13) is the tangent of the angle between the axes of the u_1, u_2 coordinate system and the $u'_1 u'_2$ coordinate system.

(*b*) Expand the functions $s_1(t)$ and $s_2(t)$ of Fig. 1.24-1 in terms of orthonormal functions $u'_1(t)$ and $u'_2(t)$ which are the axes of a coordinate system which is rotated 60° counterclockwise from the coordinate system whose axes are $u_1(t)$ and $u_2(t)$ of the same figure.

1.26-1. Apply the Gram–Schmitt procedure to the eight signals of Eq. (1.26-6). Verify that two orthonormal components suffice to allow a representation of any of the signals. Show that in a coordinate system of these two orthonormal signals, the geometric representation of the eight signals is as shown in Fig. 1.26-3.

1.26-2. A set of signals is $s_{a1}(t) = \sqrt{2P_a} \sin \omega_0 t$, $s_{a2} = -\sqrt{2P_a} \sin \omega_0 t$. A second set of signals is $s_{b2} = \sqrt{2P_b} \sin \omega_0 t$, $s_{b2} = \sqrt{2P_b} \sin (\omega_0 t + \pi/2)$. If the two sets of signals are to have the same distinguishability, what is the ratio P_b/P_a?

1.26-3. Given two signals

$$s_1(t) = f(t) \qquad 0 \le t \le T; \qquad s_2(t) = -f(t) \qquad 0 \le t \le T$$

Show that independently of the form of $f(t)$ the distinguishability of the signals is given by $\sqrt{2E}$ where E is the normalized energy of the waveform $f(t)$.

1.26-4. (*a*) Use the Gram–Schmitt procedure to express the functions in Fig. P1.26-4 in terms of orthonormal components. In applying the procedure, involve the functions in the order $s_1(t)$, $s_2(t)$, $s_3(t)$, and $s_4(t)$. Plot the functions as points in a coordinate system in which the coordinate axes are measured in units of the orthonormal functions.

(*b*) Repeat part (*a*) except that the functions are to be involved in the order $s_1(t)$, $s_4(t)$, $s_3(t)$, and $s_2(t)$. Plot the functions.

(*c*) Show that the procedures of parts (*a*) and (*b*) yield the same distances between function points.

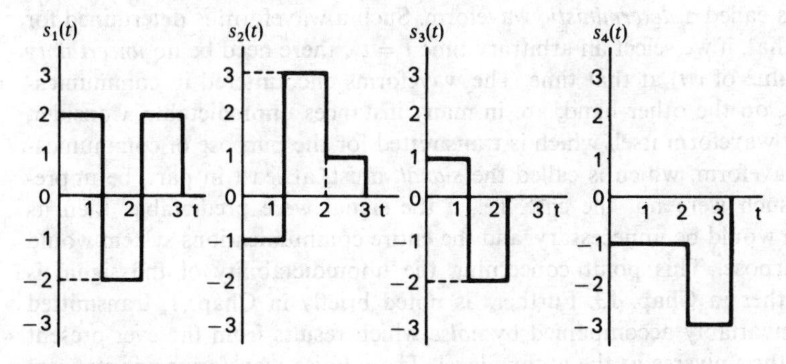

Figure P1.26-4

CHAPTER

TWO

RANDOM VARIABLES AND PROCESSES

A waveform which can be expressed, at least in principle, as an explicit function of time $v(t)$ is called a *deterministic* waveform. Such a waveform is determined for all times in that, if we select an arbitrary time $t = t_1$, there need be no *uncertainty* about the value of $v(t)$ at that time. The waveforms encountered in communication systems, on the other hand, are in many instances unpredictable. Consider, say, the very waveform itself which is transmitted for the purpose of communication. This waveform, which is called the *signal*, must, at least in part, be unpredictable. If such were not the case, i.e., if the signal were predictable, then its transmission would be unnecessary, and the entire communications system would serve no purpose. This point concerning the unpredictability of the signal is explored further in Chap. 13. Further, as noted briefly in Chap. 1, transmitted signals are invariably accompanied by *noise* which results from the ever-present agitation of the universe at the atomic level. These noise waveforms are also not predictable. Unpredictable waveforms such as a signal voltage $s(t)$ or a noise voltage $n(t)$ are examples of *random processes*. [Note that in writing symbols like $s(t)$ and $n(t)$ we do not imply that we can write explicit functions for these time functions.]

While random processes are not predictable, neither are they completely unpredictable. It is generally possible to predict the future performance of a random process with a certain *probability* of being correct. Accordingly, in this chapter we shall present some elemental ideas of probability theory and apply them to the description of random processes. We shall rather generally limit our discussion to the development of only those aspects of the subject which we shall have occasion to employ in this text.

2.1 PROBABILITY[1]

The concept of probability occurs naturally when we contemplate the possible outcomes of an experiment whose outcome is not always the same. Suppose that one of the possible outcomes is called A and that when the experiment is repeated N times the outcome A occurs N_A times. The relative frequency of occurrence of A is N_A/N, and this ratio N_A/N is not predictable unless N is very large. For example, let the experiment consist of the tossing of a die and let the outcome A correspond to the appearance of, say, the number 3 on the die. Then in 6 tosses the number 3 may not appear at all, or it may appear 6 times, or any number of times in between. Thus with $N = 6$, N_A/N may be 0 or 1/6, etc., up to $N_A/N = 1$. On the other hand, we know from experience that when an experiment, whose outcomes are determined by chance, is repeated *very many times*, the relative frequency of a particular outcome approaches a fixed limit. Thus, if we were to toss a die very many times we would expect that N_A/N would turn out to be very close to 1/6. This limiting value of the relative frequency of occurrence is called the probability of outcome A, written $P(A)$, so that

$$P(A) = \lim_{N \to \infty} \frac{N_A}{N} \tag{2.1-1}$$

In many cases the experiments needed to determine the probability of an event are done more in thought than in practice. Suppose that we have 10 balls in a container, the balls being identical in every respect except that 8 are white and 2 are black. Let us ask about the probability that, in a single draw, we shall select a black ball. If we draw blindly, so that the color has no influence on the outcome, we would surely judge that the probability of drawing the black ball is 2/10. We arrive at this conclusion on the basis that we have postulated that there is absolutely nothing which favors one ball over another. There are 10 possible outcomes of the experiment, that is, any of the 10 balls may be drawn; of these 10 outcomes, 2 are favorable to our interest. The only reasonable outcome we can imagine is that, in very many drawings, 2 out of 10 will be black. Any other outcome would immediately suggest that either the black or the white balls had been favored. These considerations lead to an alternative definition of the probability of occurrence of an event A, that is

$$P(A) = \frac{\text{number of possible favorable outcomes}}{\text{total number of possible equally likely outcomes}} \tag{2.1-2}$$

It is apparent from either definition, Eq. (2.1-1) or (2.1-2), that the probability of occurrence of an event P is a positive number and that $0 \leq P \leq 1$. If an event is not possible, then $P = 0$, while if an event is certain, $P = 1$.

2.2 MUTUALLY EXCLUSIVE EVENTS

Two possible outcomes of an experiment are defined as being *mutually exclusive* if the occurrence of one outcome precludes the occurrence of the other. In this case, if the events are A_1 and A_2 with probabilities $P(A_1)$ and $P(A_2)$, then the probability of occurrence of *either* A_1 or A_2 is written $P(A_1 \text{ or } A_2)$ and given by

$$P(A_1 \text{ or } A_2) = P(A_1) + P(A_2) \tag{2.2-1}$$

This result follows directly from Eq. (2.1-1). For suppose that in a very large number N of repetitions of the experiment, outcome A_1 had occurred N_1 times and outcome A_2 had occurred N_2 times. Then A_1 or A_2 will have occurred $N_1 + N_2$ times and

$$P(A_1 \text{ or } A_2) = \frac{N_1 + N_2}{N} = \frac{N_1}{N} + \frac{N_2}{N} = P(A_1) + P(A_2) \tag{2.2-2}$$

Equation (2.2-2) may be extended, of course, to more than two mutually exclusive outcomes. Thus

$$P(A_1 \text{ or } A_2 \text{ or } \cdots \text{ or } A_L) = \sum_{j=1}^{L} P(A_j) \tag{2.2-3}$$

and if it should happen that there are only L possible events, then, of course,

$$\sum_{j=1}^{L} P(A_j) = 1 \tag{2.2-4}$$

As an example of the calculation of the probability of mutually exclusive events, we ask, in connection with the tossing of a die, about the probability that either a 1 or a 2 will appear. Since the probability of either event is 1/6 and since, one having occurred the other cannot take place, the probability of one or the other is $1/6 + 1/6 = 1/3$.

2.3 JOINT PROBABILITY OF RELATED AND INDEPENDENT EVENTS

Suppose that we contemplate two experiments A and B with outcomes A_1, A_2, ... and $B_1, B_2, \ldots$. The probability of the joint occurrence of, say, A_j and B_k is written $P(A_j \text{ and } B_k)$ or more simply $P(A_j, B_k)$.

It may be that the probability of event B_k depends on whether A_j does indeed occur. For example, imagine 4 balls in a box, 2 white and 2 black. The probability of drawing a white ball is 1/2. If, having drawn a ball, we replace it and draw again, then we are repeating the same experiment and the probability of drawing a white ball is again 1/2. Suppose, however, that the first ball drawn is not replaced. Then if we draw a second ball, we shall be performing a new experiment. If the first ball drawn was white, the probability of a white draw in the second experiment is 1/3. If the first ball drawn was black, the probability of a white draw in the second experiment is 2/3. Here, then, we have a situation in which the outcome of the second experiment is *conditional* on the outcome of the first experiment. The probability of the outcome B_k, given that A_j is known to have occurred, is called the *conditional probability* and written $P(B_k \mid A_j)$.

Suppose that we perform N times (N a very large number) the experiment of determining which pairs of outcomes of experiment A and B occur jointly. Let N_j be the number of times A_j occurs with or without B_k, N_k the number of times B_k occurs with or without A_j, and N_{jk} the number of times of joint occurrence. Then

$$P(B_k \mid A_j) = \frac{N_{jk}}{N_j} = \frac{N_{jk}/N}{N_j/N} = \frac{P(A_j, B_k)}{P(A_j)} \tag{2.3-1}$$

Similarly we have, since $N_{jk} = N_{kj}$,

$$P(A_j \mid B_k) = \frac{N_{kj}}{N_k} = \frac{N_{kj}/N}{N_k/N} = \frac{P(A_j, B_k)}{P(B_k)} \tag{2.3-2}$$

From Eqs. (2.3-1) and (2.3-2) we have

$$P(A_j, B_k) = P(A_j)P(B_k \mid A_j) = P(B_k)P(A_j \mid B_k) \tag{2.3-3}$$

so that

$$P(A_j \mid B_k) = \frac{P(A_j)}{P(B_k)} P(B_k \mid A_j) \tag{2.3-4}$$

which result is known as *Bayes' theorem.*

It is apparent that Eq. (2.3-4) applies equally well to the case of a *single* experiment whose outcome is characterized by two events. This result follows from a simple restatement of the considerations leading up to Bayes' theorem. We need only view the successive performance of experiment A followed by experiment B as a single joint experiment. If we consider the example of the 2 white and 2 black balls in a box, it is obvious that the probability of picking a black ball (second experiment) after having picked a white ball (first experiment) is the same as the probability that, in picking 2 balls (joint experiment), we find the first white and the second black.

2.4 STATISTICAL INDEPENDENCE

Suppose, as before, that A_j and B_k are the possible outcomes of two successive experiments or the joint outcome of a single experiment. And suppose that it turns out that the probability of the occurrence of outcome B_k simply does not depend at all on which outcome A_j accompanies it. Then we say that the outcomes A_j and B_k are *independent*. In this case of complete independence

$$P(B_k \mid A_j) = P(B_k) \tag{2.4-1}$$

and Eq. (2.3-3) yields

$$P(A_j, B_k) = P(A_j)P(B_k) \tag{2.4-2}$$

Expressed in words: when outcomes are independent, the probability of a joint occurrence of particular outcomes is the product of the probabilities of the individual independent outcomes. This result may be extended to any arbitrary number of outcomes. Thus

$$P(A_j, B_k, C_l, \ldots) = P(A_j)P(B_k)P(C_l) \cdots \tag{2.4-3}$$

2.5 RANDOM VARIABLES

The term random variable is used to signify a *rule* by which a real number is assigned to each possible outcome of an experiment. Let the possible outcomes of an experiment be identified by the symbols λ_i. These symbols λ_i need not be numbers. For example, the experiment might consist of running a horse race where the outcome of interest is the name of the winner. The symbols λ_i might then be the names of the horses. Let us now arbitrarily establish some rule by which we assign real numbers $X(\lambda_i)$ to each possible outcome. Then the *rule* or *functional relationship* represented by the symbol $X(\;)$ is called a *random variable.* When used in this sense, the term random variable is a misnomer, since $X(\;)$ is not a variable at all and is in no way random, being a perfectly definite rule.

By a rather easy extension of meaning, the term random variable is also used to refer to a *variable* which may assume any of the *numbers* $X(\lambda_i)$. When used in this sense, the random variable may be represented by the symbol $X(\lambda_i)$.

In many cases, the identifying symbols λ_i may turn out to be numbers. To illustrate this point, consider that we perform the experiment of measuring a random voltage V between a set of terminals and find a number of possible outcomes v_1, v_2, or v_3. Then for identifying symbols for the outcomes we would rather naturally be led to use the numbers themselves; that is, we would use $\lambda_1 = v_1$, $\lambda_2 = v_2$, and $\lambda_3 = v_3$. Further, we might very naturally use a rule in which the number assigned to an outcome is the very same number used for identifications. We would then have $X(\lambda_i) = X(v_i) = v_i$. With these considerations in mind, we might well refer to the random voltage itself as a random variable and represent it by the symbol V.

A random variable may be *discrete* or *continuous.* If in any finite interval $X(\lambda)$ assumes only a finite number of distinct values, then the random variable is *discrete.* An example of an experiment which yields such a discrete random variable is the tossing of a die. If, however, $X(\lambda)$ can assume any value within an interval, the random variable is continuous. Thus suppose we fire a bullet at a target. Because of wind currents and other unpredictable influences the bullet may miss its mark, and the magnitude of the miss will be a *continuous random variable.*

2.6 CUMULATIVE DISTRIBUTION FUNCTION

The *cumulative distribution function* associated with a random variable is defined as the probability that the outcome of an experiment will be one of the outcomes for which $X(\lambda) \leq x$, where x is any given number. This probability will depend on the number x and also on $X(\ )$, that is, on the rule by which we assign numbers to outcomes. Thus we use the symbol $F_{X(\ )}(x)$ to represent the cumulative distribution function, and we have the definition.

$$F_{X(\ )}(x) \equiv P[X(\lambda) \leq x] \tag{2.6-1}$$

Observe that in Eq. (2.6-1) we used the symbol $X(\ )$ to represent a rule and the symbol $X(\lambda)$ to represent a variable which ranges over the numbers assigned to possible outcomes. If a different rule $Y(\ )$ were used to assign numbers, $F_{Y(\ )}(x)$ would differ from $F_{X(\ )}(x)$ for the same x. It is for this reason that the subscript $X(\ )$ is included in the notation $F_{X(\ )}(x)$. Generally, for simplicity of notation, Eq. (2.6-1) is written more simply in the form

$$F_X(x) \equiv P(X < x) \tag{2.6-2}$$

Even further, when there is only a single random variable under discussion and no ambiguity will result, we shall drop the subscript X and use the symbol $F(x)$ to represent the cumulative distribution function.

The cumulative distribution function has the properties

$$0 \leq F(x) \leq 1 \tag{2.6-3a}$$

$$F(-\infty) = 0 \qquad F(\infty) = 1 \tag{2.6-3b}$$

$$F(x_1) \leq F(x_2) \qquad \text{if } x_1 < x_2 \tag{2.6-3c}$$

The property in Eq. (2.6-3*a*) follows from the fact that $F(x)$ is a probability. The property in Eq. (2.6-3*b*) follows from the fact that $F(-\infty)$ includes no possible events, while $F(\infty)$ includes all possible events. Finally Eq. (2.6-3*c*) holds since for $x_1 < x_2$, $F(x_2)$ includes as many or more of the possible outcomes as does $F(x_1)$.

The features of the cumulative distribution function, as well as the concept of a random variable, will be clarified in the course of the following illustrative discussion: We consider the experiment which consists in the rolling of 2 dice. There are 36 possible outcomes since each die may show the numbers 1 through 6. These 36 outcomes may be represented by 36 symbols λ_{ij}, where, say, i is the number that shows on the first die, and j the number that shows on the second die. Suppose however that our interest in the outcome extends only to knowing the sum of the numbers appearing on the dice. Then we may use as our random variable the function $N(\lambda_{ij})$ defined by $N(\lambda_{ij}) = i + j$. Observe that in so doing we are assigning the same number to $N(\lambda_{ij})$ and to $N(\lambda_{ji})$. Let

$$n \equiv N(\lambda_{ij}) = i + j$$

Then each integral value of n from $n = 2$ to $n = 12$ corresponds to outcomes which we care to distinguish from one another. The probabilities $P(n)$ are readily calculated. We find $P(1) = 0$, $P(2) = P(12) = 1/36$, $P(3) = P(11) = 2/36$, $P(4) = P(10) = 3/36$, $P(5) = P(9) = 4/36$, $P(6) = P(8) = 5/36$, and $P(7) = 6/36$.

Let us calculate, as an example, the cumulative distribution function $F(n)$ for, say, $n = 3$. We have from the definition of Eq. (2.6-2)

$$F(3) = P[N(\lambda_{ij}) \leq 3] = P(n \leq 3) \tag{2.6-4}$$

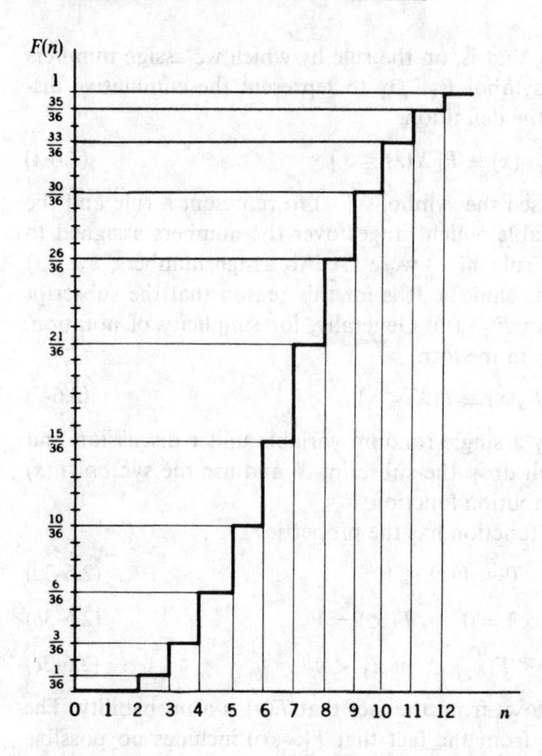

Figure 2.6-1 Cumulative distribution function associated with rolling two dice.

since $N(\lambda_{ij}) \equiv n$. We must now add up the probabilities of all outcomes corresponding to $n \leq 3$. We find that

$$F(3) = P(1) + P(2) + P(3) = 0 + 1/36 + 2/36 = 3/36 \tag{2.6-5}$$

In a similar way $F(n)$ for other values of n may be determined. $F(n)$ is plotted as a function of n in Fig. 2.6-1. Note that $F(n)$ satisfies all the conditions given in Eq. (2.6-3*a* to *c*) above. Observe also that there is a significance to $F(n)$ for nonintegral values of n. Note also that if the function $N(\lambda_{ij})$ had been selected in a different manner, say, $N'(\lambda_{ij})$, $F_{N'}(n)$ would be different from $F_N(n)$.

2.7 PROBABILITY DENSITY FUNCTION

The *probability density function* (pdf) $f_X(x)$ is defined in terms of the cumulative distribution function $F_X(x)$ as

$$f_X(x) = \frac{d}{dx} F_X(x) \tag{2.7-1}$$

[Having made the point in Eq. (2.7-1) that $f_X(x)$, like $F_X(x)$, requires a subscript in principle, we shall again omit it where no confusion will result.] Thus $f(x)$ is simply the derivative of the cumulative distribution function $F(x)$. The pdf has the following properties:

(*a*)
$$f(x) \geq 0 \qquad \text{for all } x \tag{2.7-2}$$

This results from the fact that $F(x)$ increases monotonically, for as x increases, more outcomes are included in the probability of occurrence represented by $F(x)$.

(*b*)
$$\int_{-\infty}^{\infty} f(x)\, dx = 1 \tag{2.7-3}$$

This result is to be seen from the fact that

$$\int_{-\infty}^{\infty} f(x)\, dx = F(\infty) - F(-\infty) = 1 - 0 = 1 \tag{2.7-4}$$

(*c*)
$$F(x) = \int_{-\infty}^{x} f(x)\, dx \tag{2.7-5}$$

This result follows directly from Eq. (2.7-1).

The pdf corresponding to the cumulative distribution function of Fig. 2.6-1 is shown in Fig. 2.7-1. Since the derivative of a step of amplitude A is an impulse of strength $I = A$, we have $f(2) = 1/36\ \delta(n-2)$, $f(3) = 2/36\ \delta(n-3)$, etc. In Fig. 2.7-1 the height of the arrow is proportional to the strength of the impulse.

As we have already noted, a random variable may be discrete or continuous. The random variable associated, in the discussion above, with the tossing of 2 dice is a discrete random variable. The corresponding cumulative distribution function increases by steps as in Fig. 2.6-1, and the probability density displays impulses as in Fig. 2.7-1. With a continuous random variable, on the other hand, the cumulative distribution function and the probability density will generally be *smooth* as shown by the typical plots of Fig. 2.7-2.

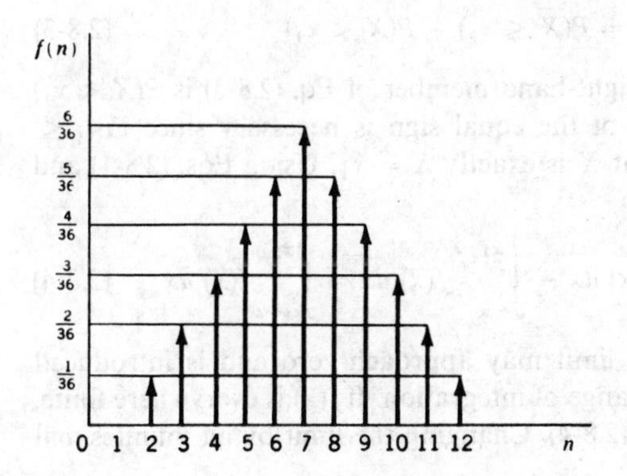

Figure 2.7-1 The probability density function corresponding to the cumulative distribution function of Fig. 2.6-1. The heights of the arrows represent the strengths of impulses.

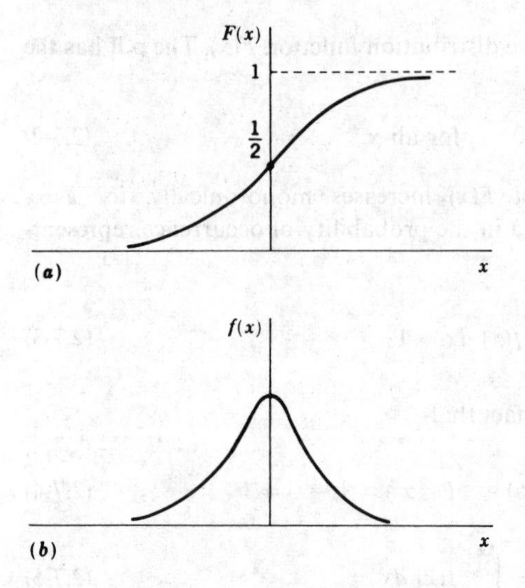

Figure 2.7-2 (*a*) A continuous cumulative distribution function and (*b*) the corresponding probability density function.

2.8 RELATION BETWEEN PROBABILITY AND PROBABILITY DENSITY

From Eqs. (2.6-1) and (2.7-5) we find that the probability of the outcome X being less than or equal to x_1 is

$$P(X \leq x_1) = F(x_1) = \int_{-\infty}^{x_1} f(x)\, dx \tag{2.8-1}$$

Similarly, the probability that the outcome X is less than or equal to x_2 is

$$P(X \leq x_2) = F(x_2) = \int_{-\infty}^{x_2} f(x)\, dx \tag{2.8-2}$$

The probability that the outcome lies in the range $x_1 \leq X \leq x_2$ is

$$P(x_1 \leq X \leq x_2) = P(X \leq x_2) - P(X < x_1) \tag{2.8-3}$$

Note that the second term of the right-hand member of Eq. (2.8-3) is $P(X < x_1)$ and not $P(X \leq x_1)$. The exclusion of the equal sign is necessary since $P(x_1 \leq X \leq x_2)$ includes the possibility that X is exactly $X = x_1$. Using Eqs. (2.8-1) and (2.8-2) with Eq. (2.8-3), we have

$$P(x_1 \leq X \leq x_2) = \int_{-\infty}^{x_2} f(x)\, dx - \int_{-\infty}^{x_1 - \varepsilon} f(x)\, dx = \int_{x_1 - \varepsilon}^{x_2} f(x)\, dx \tag{2.8-4}$$

where ε is a number which in the limit may approach zero and is introduced simply to exclude $x = x_1$ from the range of integration. If $f(x)$ is everywhere finite, then ε need not be included in Eq. (2.8-4). Changing the limit by an infinitesimal amount will change the integral only infinitesimally. If, however, the probability density contains impulses as in Fig. 2.7-1, and if there is an impulse at $x = x_1$, then the presence of the ε reminds us to include this impulse when evaluating Eq. (2.8-4). For the case where $f(x)$ has no impulses, we have the result which expresses the most important property of the probability density, namely,

$$P(x_1 \le X \le x_2) = \int_{x_1}^{x_2} f(x)\,dx \tag{2.8-5}$$

or

$$P(x \le X \le x + dx) = f(x)\,dx \tag{2.8-6}$$

Example 2.8-1 Consider the probability density $f(x) = ae^{-b|x|}$, where X is a random variable whose allowable values range from $x = -\infty$ to $x = +\infty$. Find (*a*) the cumulative distribution function $F(x)$, (*b*) the relationship between a and b, and (*c*) the probability that the outcome X lies between 1 and 2.

SOLUTION (*a*) The cumulative distribution function is

$$F(x) = P(X \le x) = \int_{-\infty}^{x} f(x)\,dx = \int_{-\infty}^{x} ae^{-b|x|}\,dx \tag{2.8-7}$$

$$= \begin{cases} \dfrac{a}{b} e^{bx} & x \le 0 \\ \dfrac{a}{b}(2 - e^{-bx}) & x \ge 0 \end{cases} \tag{2.8-8}$$

(*b*) In order that $f(x)$ be a probability density, it is necessary that

$$\int_{-\infty}^{\infty} f(x)\,dx = \int_{-\infty}^{\infty} ae^{-b|x|}\,dx = \frac{2a}{b} = 1 \tag{2.8-9}$$

so that $a/b = \frac{1}{2}$.

(*c*) The probability that X lies in the range between 1 and 2 is

$$P(1 \le X \le 2) = \frac{b}{2}\int_{1}^{2} e^{-b|x|}\,dx = \tfrac{1}{2}(e^{-b} - e^{-2b}) \tag{2.8-10}$$

2.9 JOINT CUMULATIVE DISTRIBUTION AND PROBABILITY DENSITY

It may be necessary to identify the outcome of an experiment by two (or more) random variables. These random variables may or may not be independent of one another. The concepts of a cumulative distribution and a probability density are readily extended to such cases.

For the case of a single random variable, Eq. (2.8-6) expresses, in terms of the probability density, the probability that the random variable X lies in the range $x \leq X \leq x + dx$. Correspondingly, for two random variables X and Y, the probability that $x \leq X \leq x + dx$ while at the same time $y \leq Y \leq y + dy$ is written

$$P(x \leq X \leq x + dx, y \leq Y \leq y + dy) = f_{XY}(x, y)\, dx\, dy \tag{2.9-1}$$

A comparison of a probability density $f_X(x)$, involving a single random variable, and a density $f_{XY}(x, y)$ involving two variables, is indicated in Fig. 2.9-1. Part (*a*) shows a portion of a *curve* which is a plot of $f_X(x)$ as a function of x. Part (*b*) shows a portion of a *surface* which is a plot of $f_{XY}(x, y)$ as a function of the two variables x and y. The *area* indicated in (*a*) is the probability that $x \leq X \leq x + dx$, while the *volume* indicated in (*b*) is the probability that $x \leq X \leq x + dx$ and that $y \leq Y \leq y + dy$.

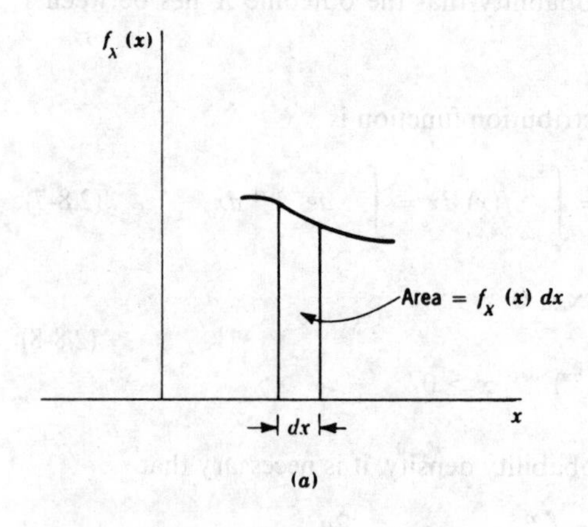

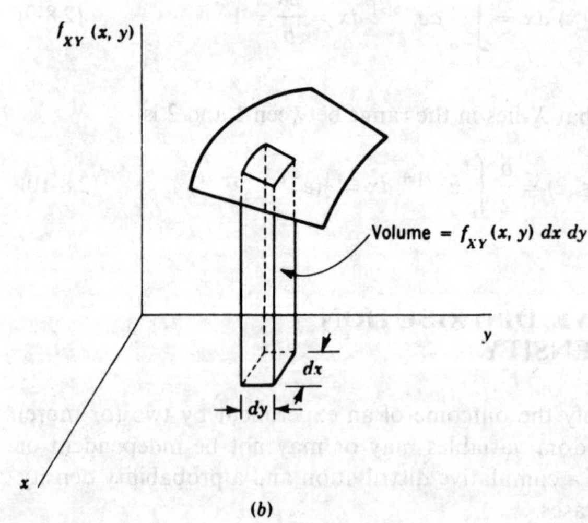

Figure 2.9-1 A comparison between a probability density function involving a simple random variable in (*a*) and a density function involving two random variables as in (*b*).

Extending Eq. (2.9-1) to a finite interval, we have, in correspondence with Eq. (2.8-5),

$$P(x_1 \le X \le x_2, y_1 \le Y \le y_2) = \int_{y_1}^{y_2} \int_{x_1}^{x_2} f_{XY}(x, y)\, dx\, dy \tag{2.9-2}$$

The cumulative distribution function is

$$F_{XY}(x, y) = P(X \le x, Y \le y) = \int_{-\infty}^{y} \int_{-\infty}^{x} f_{XY}(x, y)\, dx\, dy \tag{2.9-3}$$

If we should be concerned only with the cumulative probability up to, say, some value of x quite independently of y, we would write

$$F_X(x) = P(X \le x, -\infty \le Y \le \infty) = \int_{-\infty}^{\infty} \int_{-\infty}^{x} f_{XY}(x, y)\, dx\, dy \tag{2.9-4}$$

The probability density corresponding to $F_X(x)$ in Eq. (2.9-4) is

$$f_X(x) = \frac{d}{dx} F_X(x) = \int_{-\infty}^{\infty} f_{XY}(x, y)\, dy \tag{2.9-5}$$

If the random variables X and Y are *independent*, then by an easy extension of the considerations of Sec. 2.4, the probability given in Eq. (2.9-1) may be written as a product in which each factor involves only one random variable. That is, we may write

$$P(x \le X \le x + dx, y \le Y \le y + dy) = [f_X(x)\, dx][f_Y(y)\, dy] \tag{2.9-6}$$

The function $f_X(x)$, which is given by Eq. (2.9-5), depends only on x and need bear no simple relationship to $f_{XY}(x, y)$ given in Eq. (2.9-1). A similar comment applies, of course, to $f_Y(y)$. We then have, from Eqs. (2.9-1) and (2.9-6), that, if X and Y are independent,

$$f_{XY}(x, y) = f_X(x) f_Y(y) \tag{2.9-7}$$

Further, we have, when X and Y are independent,

$$P(x_1 \le X \le x_2, y_1 \le Y \le y_2) = \left[\int_{x_1}^{x_2} f_X(x)\, dx\right]\left[\int_{y_1}^{y_2} f_Y(y)\, dy\right] \tag{2.9-8}$$

Henceforth, where no confusion will be caused thereby, we shall drop the subscripts from the probability density functions even when more than one such function is involved. Thus we shall write $f_X(x) = f(x)$, $f_Y(y) = f(y)$, and $f_{XY}(x, y) = f(x, y)$. Thus the symbol $f(\ \ )$ represents not a particular probability density but rather a probability density function in general. And the particular random variable or variables referred to are to be determined by the argument of the function.

Example 2.9-1 The joint probability density of the random variables X and Y is

$$f(x, y) = \tfrac{1}{4}e^{-|x|-|y|} \qquad -\infty < x < \infty, \; -\infty < y < \infty$$

(*a*) Are X and Y statistically independent random variables?
(*b*) Calculate the probability that $X \leq 1$ and $Y \leq 0$.

SOLUTION (*a*) Since $f(x, y)$ can be written as

$$f(x, y) = \tfrac{1}{2}e^{-|x|}\tfrac{1}{2}e^{-|y|} = f(x)f(y)$$

X and Y are statistically independent.

(*b*)
$$P(X \leq 1, Y \leq 0) = \int_{-\infty}^{1} dx \int_{-\infty}^{0} dy f(x, y)$$
$$= \int_{-\infty}^{1} \tfrac{1}{2}e^{-|x|}\, dx \int_{-\infty}^{0} \tfrac{1}{2}e^{-|y|}\, dy$$
$$= \left(\frac{2 - e^{-1}}{2}\right)\frac{1}{2} = \tfrac{1}{4}(2 - e^{-1})$$

2.10 A COMMUNICATIONS EXAMPLE

As we noted in the introductory paragraphs of Chap. 1, one of our principal interests is the transmission of signals from place to place in a manner which suppresses the effects of noise to the maximum extent possible. In the simplest case the medium over which signals are transmitted is simply a length of a pair of wires, the signal being applied directly at one end and being received at the other. In the more general case, before application to the transmitting medium, the signal may be elaborately processed and modified (a process called *modulation*) in a manner calculated to suppress noise. Correspondingly at the receiving end it will be necessary to undo the modulation (by a process called *demodulation*) to recover the original signal. As the signal, modulated or not, traverses the transmission medium (wires, or free space as in radio, etc.) noise will be added to the signal. Generally, as well, the process of demodulation will also add noise. Hence as finally received and perceived, the signal will be "corrupted" or "contaminated" by noise from a number of sources. Everything that intervenes between the very original signal at the transmitting end and the final recovered signal at the receiving end, including the modulation equipment, the demodulation equipment, the transmission medium and all the noise sources that corrupt the signal is referred to as the *channel*.

Often the transmitted signal is a more or less continuous waveform as in the case where speech or the music output of a microphone is applied directly to the transmitting end of a channel. In other cases we may want to use a channel to transmit *digital* information, that is, a sequence in time of bits-logical 0's and 1's. Thus in successive intervals we want to transmit one of two possible *messages*, the message m_0 that the bit 0 is intended, or the message m_1 that the bit 1 is intended. The two possible messages might be represented at the transmitting end by two distinct waveforms, each limited in time duration to the interval allocated to a bit. At the receiving end we might devise a system whereby the message m_0, when received, generates some voltage, say r_0, which may be as simple as a dc voltage, while m_1 when received generates a voltage r_1.

In the absence of noise, the message m_0 generates r_0 and m_1 generates r_1, each with complete certainty. However, because of noise, errors may occur. Even though the message sent is m_0 the received indication might be r_1, and correspondingly the message m_1 may generate r_0. We shall assume, for generality, that the probability of an error is dependent on which message was sent and we introduce the following conditional probabilities called the *transition* probabilities:

$P(r_0 \mid m_0)$ = probability that r_0 is received given that m_0 is sent,
$P(r_1 \mid m_0)$ = probability that r_1 is received given that m_0 is sent,
$P(r_0 \mid m_1)$ = probability that r_0 is received given that m_1 is sent,
$P(r_1 \mid m_1)$ = probability that r_1 is received given that m_1 is sent.

We also allow for the general case that the messages m_1 and m_0 do not occur with equal frequency and we introduce the probabilities $P(m_1)$ and $P(m_0)$ which are the probabilities that m_1 and m_0 are the messages respectively intended in an arbitrary message interval. These probabilities $P(m_1)$ and $P(m_0)$ are called the *a priori* probabilities. With these probability definitions the two-message system we have described can be represented as in Fig. 2.10-1.

From an observed response, r_1 or r_0, we cannot determine with certainty which of the messages was sent. Let us then develop an algorithm which will serve to allow an opinion about the message intended with *maximum probability that our opinion is correct*. Consider then that say r_0 is received. We have then to compare the conditional probabilities, called the *a posteriori* probabilities:

$P(m_0 \mid r_0)$ = probability that m_0 is the message given that r_0 is received;
$P(m_1 \mid r_0)$ = probability that m_1 is the message given that r_0 is received.

Clearly if $P(m_0 \mid r_0) > P(m_1 \mid r_0)$ then we should decide that m_0 is intended and if

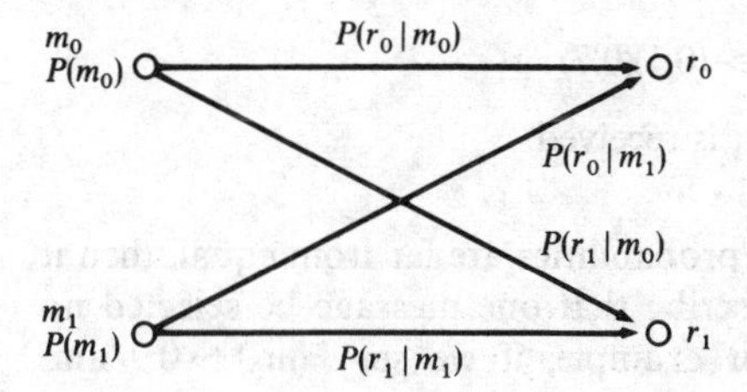

Figure 2.10-1 A representation of a two-message communication system.

the inequality is reversed we should decide for m_1. Altogether then our algorithm should be:

If r_0 is received:

$$\text{choose } m_0 \text{ if } P(m_0 \mid r_0) > P(m_1 \mid r_0) \tag{2.10-1a}$$

$$\text{choose } m_1 \text{ if } P(m_1 \mid r_0) > P(m_0 \mid r_0) \tag{2.10-1b}$$

When the inequality is reversed, the choice is reversed and if the inequality is replaced by an equality we are at liberty to make either choice.

If r_1 is received:

$$\text{choose } m_0 \text{ if } P(m_0 \mid r_1) > P(m_1 \mid r_1) \tag{2.10-2a}$$

$$\text{choose } m_1 \text{ if } P(m_1 \mid r_i) > P(m_0 \mid r_1) \tag{2.10-2b}$$

A receiver which operates in accordance with this algorithm is said to "maximize the *a posteriori* probability" (m.a.p.) of a correct decision and is called an *optimum* receiver. The algorithm can be expressed in terms of the a priori and the transition probabilities. For example, starting with Eq. (2.10-1*a*) and multiplying both sides by $P(r_0)$ we have the result that, if r_0 is received, m_0 should be chosen if

$$P(m_0 \mid r_0)P(r_0) > P(m_1 \mid r_0)P(r_0) \tag{2.10-3}$$

From the general result given by Eq. (2.3-3) we have that Eq. (2.10-3) can be rewritten

$$P(r_0 \mid m_0)P(m_0) > P(r_0 \mid m_1)P(m_1) \tag{2.10-4}$$

Correspondingly if r_1 is received we choose m_1 only if

$$P(r_1 \mid m_1)P(m_1) > P(r_1 \mid m_0)P(m_0) \tag{2.10-5}$$

Example 2.10-1 Apply the optimum-receiver algorithm to the case $P(m_0) = 0.7$, $P(m_1) = 0.3$, $P(r_0 \mid m_0) = 0.9$, $P(r_1 \mid m_0) = 0.1$, $P(r_0 \mid m_1) = 0.4$, $P(r_1 \mid m_1) = 0.6$.

SOLUTION We find that Eq. (2.10-4) is valid since

$$(0.9)(0.7) > (0.4)(0.3)$$

Hence we would select m_0 whenever r_0 is received. Also we find that Eq. (2.10-5) is valid since

$$(0.6)(0.3) > (0.1)(0.7)$$

Hence we should select m_1 whenever r_1 is received.

It is interesting to note that if the a priori probabilities are far from equal, then it may turn out that the algorithm will prescribe that one message be selected no matter what the received indication. For example, if we set $P(m_0) = 0.9$ and

$P(m_1) = 0.1$ the optimum receiver will *always* select m_0 regardless of whether r_0 or r_1 is received.

Having established an algorithm for the optimum receiver, it is next of interest to know how effective the algorithm will be in yielding the correct determination. We accordingly want to calculate the probability $P(c)$ that a *correct* determination will result. The probability of an error is $1 - P(c) \equiv P(\varepsilon)$.

Example 2.10-2 Referring to Example 2.10-1 calculate $P(c)$ and $P(\varepsilon)$.

SOLUTION The probability that the transmitted signal is correctly read at the receiver is equal to the probability that m_0 was sent when r_0 was read plus the probability that m_1 was sent when r_1 was read. Hence

$$P(c) = P(r_0 \mid m_0)P(m_0) + P(r_1 \mid m_1)P(m_1) \tag{2.10-6}$$

In the present case we find

$$P(c) = (0.9)(0.7) + (0.6)(0.3) = 0.81$$

and

$$P(\varepsilon) = 0.19$$

It can happen that even in a case in which only two messages are contemplated, that because of noise, the receiver may generate more than just two responses. Such a situation is considered in the next example.

Example 2.10-3 In the system represented in Fig. 2.10-2, m_0 would generate r_0, and m_1 would generate r_1 with certainty if there were no noise and r_2 would never occur. For the situation depicted in Fig. 2.10-2, (*a*) find the optimum receiver and (*b*) calculate the probability of error.

SOLUTION (*a*) We find that

$$P(r_0 \mid m_0)P(m_0) > P(r_0 \mid m_1)P(m_1)$$

since

$$(0.6)(0.6) > (0)(0.4)$$

Hence we select m_0 whenever r_0 is received. We also find that

$$P(r_1 \mid m_1)P(m_1) > P(r_1 \mid m_0)P(m_0)$$

since

$$(0.7)(0.4) > (0.2)(0.6)$$

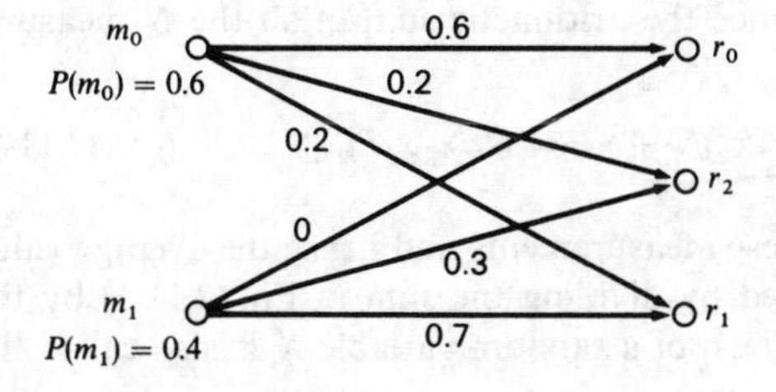

Figure 2.10-2 Communication system for Example 2.10-3. Transition probabilities $P(r \mid m)$ and a priori probabilities $P(m)$ are given.

Hence we select m_1 whenever r_1 is received. To decide how to assign r_2 we compare $P(r_2 | m_0)P(m_0)$ with $P(r_2 | m_1)P(m_1)$. We find

$$P(r_2 | m_0)P(m_0) = P(r_2 | m_1)P(m_1)$$

since
$$(0.2)(0.6) = (0.3)(0.4)$$

We can accordingly make either assignment and we arbitrarily associate r_2 with m_0.

(*b*) The probability of being correct is

$$\begin{aligned} P(c) &= P(r_0 | m_0)P(m_0) + P(r_1 | m_1)P(m_1) + P(r_2 | m_0)P(m_0) \\ &= (0.6)(0.6) + (0.7)(0.4) + (0.2)(0.6) \\ &= 0.76 \end{aligned} \tag{2.10-7}$$

and $P(\varepsilon) = 0.24$.

Alternatively we can calculate the probability of error by adding the probabilities that incorrect decisions are made. We then have, by simply interchanging m_1 and m_0 in Eq. (2.10-7)

$$\begin{aligned} P(\varepsilon) &= (0)(0.4) + (0.2)(0.6) + (0.3)(0.4) \\ &= 0.24 \end{aligned}$$

In the general case there can be K messages $m_1, \ldots, m_K$ and J received responses $r_1, \ldots, r_J$. The optimum receiver is, of course, still designed in accordance with the rule:

If r_j is received, choose m_k if

$$P(m_k | r_j) > P(m_i | r_j) \qquad \text{for all } i \neq k \tag{2.10-8}$$

2.11 AVERAGE VALUE OF A RANDOM VARIABLE

Consider now that we have the values and their associated probabilities of a discrete random variable. The possible numerical values of the random variable X are $x_1, x_2, x_3, \ldots$, with probabilities of occurrence $P(x_1), P(x_2), P(x_3) \ldots$. As the number of measurements N of X becomes very large, we would expect that we would find the outcome $X = x_1$ would occur $NP(x_1)$ times, the outcome $X = x_2$ would occur $NP(x_2)$ times, etc. Hence the arithmetic sum of all the N measurements would be

$$x_1 P(x_1)N + x_2 P(x_2)N + \cdots = N \sum_i x_i P(x_i) \tag{2.11-1}$$

The *mean* or *average value* of all these measurements and hence the average value of the random variable is calculated by dividing the sum in Eq. (2.11-1) by the number of measurements N. The mean of a random variable X is also called the *expectation* of X and is represented either by the notation $\bar{X}$ or by $E(X)$. We shall use these notations interchangeably. Thus, using m to represent the value of the average or expectation of X, we have, from Eq. (2.11-1),

$$\bar{X} \equiv E(X) = m = \sum_i x_i P(x_i) \tag{2.11-2}$$

To calculate the average for a continuous random variable, let us divide the range of the variable into small intervals Δx. Then from Eq. (2.8-6) the probability that X lies in the range between x_i and $x_i + \Delta x$ is $P(x_i \leq X \leq x_i + \Delta x) \equiv P(x_i)$ is given approximately by

$$P(x_i) = f(x_i)\,\Delta x \tag{2.11-3}$$

Substituting Eq. (2.11-3) into Eq. (2.11-2), we have

$$m = \sum_i x_i f(x_i)\,\Delta x \tag{2.11-4}$$

In the limit, as $\Delta x \to 0$ and is replaced by dx, the summation in Eq. (2.11-4) becomes an integral, and

$$m = \int_{-\infty}^{\infty} x f(x)\,dx \tag{2.11-5}$$

In general, the average value, or expectation, of a function $g(X)$ of the random variable X is

$$\overline{g(X)} = E[g(X)] = \int_{-\infty}^{\infty} g(x) f(x)\,dx \tag{2.11-6}$$

If the function $g(X)$ is X raised to a power, that is, $g(X) = X^n$, the average value $E(X^n)$ is referred to as the *nth moment* of the random variable. For this reason, the average value $\bar{X}$ is also called the *first moment* of X.

If a random variable Z is a function of two random variables X and Y, say $Z = w(X, Y)$, then by an extension of the above discussion it may be shown that

$$\bar{Z} = \int_{-\infty}^{\infty}\int_{-\infty}^{\infty} w(x, y) f_{XY}(x, y)\,dx\,dy \tag{2.11-7}$$

In particular, if $Z = XY$, then

$$\bar{Z} = \int_{-\infty}^{\infty} xy f_{XY}(x, y)\,dx\,dy \tag{2.11-8}$$

and if X and Y are independent random variables, then from Eq. (2.9-7)

$$\bar{Z} = \int_{-\infty}^{\infty} xy f_X(x) f_Y(y)\,dx\,dy \tag{2.11-9a}$$

$$= \int_{-\infty}^{\infty} x f_X(x)\,dx \int_{-\infty}^{\infty} y f_Y(y)\,dy = \bar{X}\bar{Y} = m_x m_y \tag{2.11-9b}$$

2.12 VARIANCE OF A RANDOM VARIABLE

In Fig. 2.12-1 are shown two probability density functions $f(x)$ and $f'(x)$ for two random variables X and X'. As a matter of simplicity we have drawn them of the same general form and have drawn them symmetrically about a common average value m. But these features are not essential to the ensuing discussion. Rather, the important point is that $f(x)$ is *narrower* than is $f'(x)$. Suppose, then, that experimental determinations were made of X and X' yielding numerical outcomes x and x'. We would surely find that, on the average, x would be closer to m than x' would be to m'. Thus in comparing X and X', we find that the outcomes of X have a higher probability of occurring in a smaller range. In other words, if a number of determinations were made of X and X', we would expect to find that the outcomes of X would *cluster* more closely around m than would be the case for X'.

It is convenient to have a number which serves as a measure of the "width" of a probability density function. We might suggest as a candidate for such a number the average value of $(X - m)$ that is, $\overline{X - m}$. However, $\overline{X - m} = 0$, since positive and negative contributions from the portions of $f(x)$ above and below x cancel. A second possibility is $\overline{|X - m|}$, since taking the absolute value of $X - m$ would avoid the cancellation. However, a more useful measure is the square root of the average value of $(X - m)^2$, that is, of the second moment of $X - m$. This second moment is represented by the symbol σ^2 and is called the *variance* of the random variable. Thus

$$\sigma^2 \equiv E[(X - m)^2] = \int_{-\infty}^{\infty} (x - m)^2 f(x)\, dx \tag{2.12-1}$$

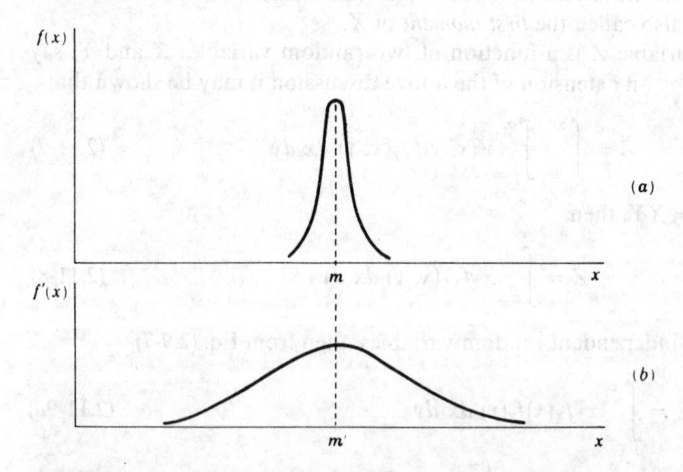

Figure 2.12-1 Two probability density functions corresponding to random variables with different variances.

Writing $(x - m)^2 = x^2 - 2mx + m^2$ in the integral of Eq. (2.12-1) and integrating term by term, we find

$$\sigma^2 = E(X^2) - 2m^2 + m^2 \tag{2.12-2a}$$

$$= E(X^2) - m^2 \tag{2.12-2b}$$

The quantity σ itself is called the standard deviation and is the *root mean square* (rms) value of $(X - m)$. If the average value $m = 0$, then

$$\sigma^2 = E(X^2) \tag{2.12-3}$$

2.13 TCHEBYCHEFF'S INEQUALITY

We have noted that the variance of a random variable measures the "width" of the probability density function. Tchebycheff's rule expresses this same quality of the variance in an alternative manner. In Fig. 2.13-1 we have drawn a density function which, for simplicity, we have assumed to be symmetrical about a mean value of zero; these features however are not essential to the discussion. The probability that a sample of the random variable X should have a value x such that $|x| \geq \varepsilon$ is equal to the area of the shaded regions. We should, of course, expect that the probability $P(|x| \geq \varepsilon)$ will decrease as ε increases. Further, since the variance measures the width of the density function about the mean we further expect that $P(|x| \geq \varepsilon)$ should decrease as σ^2 decreases. Both of these expectations are to be seen in Tchebycheff's rule which, however, yields not an exact value for $P(|x| \geq \varepsilon)$ in terms of σ^2 and ε but rather an upper bound (or limit). Specifically Tchebycheff's rule is an inequality given by

$$P(|x| \geq \varepsilon) \leq \frac{\sigma^2}{\varepsilon^2} \tag{2.13-1}$$

To prove this inequality we start with Eq. (2.12-1) which defines σ^2. Assuming $m = \bar{x} = 0$ we have

$$\sigma^2 = \int_{-\infty}^{-\infty} x^2 f(x)\, dx \tag{2.13-2a}$$

$$= \int_{-\infty}^{-\varepsilon} x^2 f(x)\, dx + \int_{-\varepsilon}^{+\varepsilon} x^2 f(x)\, dx + \int_{+\varepsilon}^{-\infty} x^2 f(x)\, dx \tag{2.13-2b}$$

Since $x^2 \geq 0$ and $f(x) \geq 0$ for all x we have that

$$\int_{-\varepsilon}^{\varepsilon} x^2 f(x)\, dx \geq 0 \tag{2.13-3}$$

so that Eq. (2.13-2b) can be written

$$\sigma^2 \geq \int_{-\infty}^{-\varepsilon} x^2 f(x)\, dx + \int_{+\varepsilon}^{\infty} x^2 f(x)\, dx \tag{2.13-4}$$

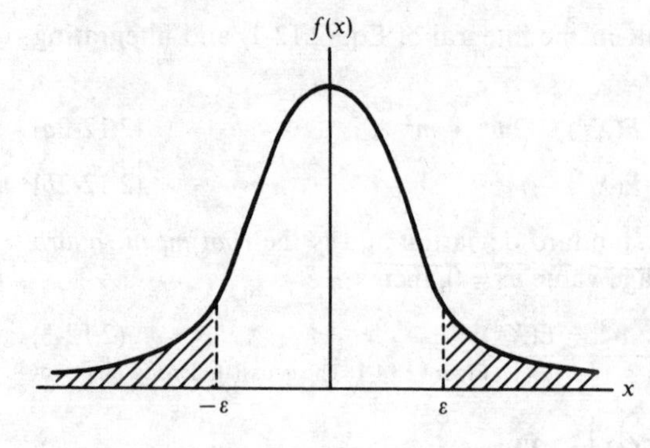

Figure 2.13-1 A probability density function.

In the ranges $-\infty \leq x \leq -\varepsilon$ and $\varepsilon \leq x \leq \infty$

$$x^2 \geq \varepsilon^2 \tag{2.13-5}$$

Therefore if we replace x^2 by ε^2 in Eq. (2.13-4) the sum of the two integrals in the right-hand member cannot be increased. Hence, with this substitution, the inequality persists and we have

$$\sigma^2 \geq \varepsilon^2\left[\int_{-\infty}^{-\varepsilon} f(x)\,dx + \int_{+\varepsilon}^{+\infty} f(x)\,dx\right] \tag{2.13-6}$$

But

$$P(x \leq -\varepsilon) = \int_{-\infty}^{-\varepsilon} f(x)\,dx \tag{2.13-7}$$

and

$$P(x \geq +\varepsilon) = \int_{\varepsilon}^{\infty} f(x)\,dx \tag{2.13-8}$$

Thus Eq. (2.13-6) can be written as

$$\frac{\sigma^2}{\varepsilon^2} \geq P(x \leq -\varepsilon) + P(x \geq \varepsilon) = P(|x| \geq \varepsilon) \tag{2.13-9}$$

and the rule is proved. In Sec. 2.21 we shall apply the inequality in a communications example.

2.14 THE GAUSSIAN PROBABILITY DENSITY

The *gaussian* (also called *normal*) probability density function is of the greatest importance because many naturally occurring experiments are characterized by random variables with a gaussian density. It is of special relevance to us because the random variables of concern to us will be described by, almost exclusively, the gaussian density function. A further importance is attached to the gaussian

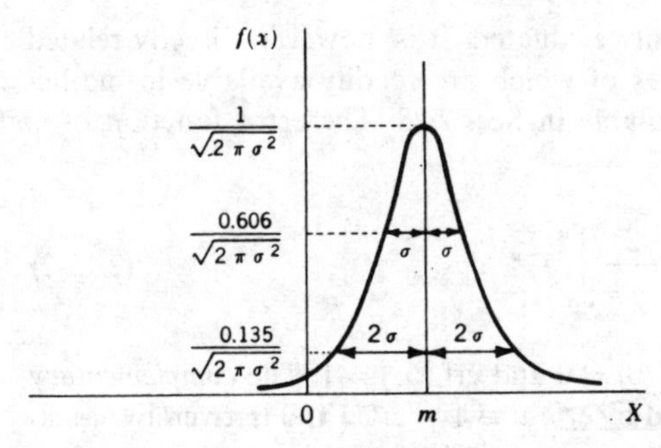

Figure 2.14-1 The gaussian density function.

density because it is involved in a remarkable theorem called the *central-limit theorem* which we discuss in Sec. 2.20.

The gaussian probability density function is defined as

$$f(x) = \frac{1}{\sqrt{2\pi\sigma^2}} e^{-(x-m)^2/2\sigma^2} \tag{2.14-1}$$

and is plotted in Fig. 2.14-1. In using the symbols m and σ^2 in Eq. (2.14-1), we have taken cognizance of the fact that m and σ^2 are indeed the average value and variance associated with $f(x)$. Thus we find that

$$\bar{X} = \int_{-\infty}^{\infty} \frac{xe^{-(x-m)^2/2\sigma^2}}{\sqrt{2\pi\sigma^2}} dx = m \tag{2.14-2}$$

and

$$E[(X-m)^2] = \int_{-\infty}^{\infty} \frac{(x-m)^2 e^{-(x-m)^2/2\sigma^2}}{\sqrt{2\pi\sigma^2}} dx = \sigma^2 \tag{2.14-3}$$

It may also be verified that

$$\int_{-\infty}^{\infty} f(x)\, dx = 1 \tag{2.14-4}$$

as is required for a probability density function.

As is indicated in Fig. 2.14-1, when $x - m = \pm\sigma$, that is, at values of x separated from m by the standard deviation, $f(x)$ has fallen to 0.606 of its peak value. When $x - m = \pm 2\sigma$, $f(x)$ falls to 0.135 of the peak value, and at $x - m = 3\sigma$ (not shown in the figure), $f(x)$ has fallen to 0.01 of the peak value.

2.15 THE ERROR FUNCTION

The cumulative distribution corresponding to the gaussian probability density, for $m = 0$, is

$$P(X \le x) = F(x) = \int_{-\infty}^{x} \frac{e^{-x^2/2\sigma^2}}{\sqrt{2\pi\sigma^2}} dx \tag{2.15-1}$$

The integral in Eq. (2.15-1) is not easily evaluated. It is, however, directly related to the *error function*, tabulated values of which are readily available in mathematical tables.[2] A short table is available in Sec. 2.28. The error function of u, written erf u, is defined as

$$\operatorname{erf} u \equiv \frac{2}{\sqrt{\pi}} \int_0^u e^{-u^2}\, du \tag{2.15-2}$$

The error function has the values erf (0) = 0 and erf (∞) = 1. The *complementary error function*, written erfc u, is defined by erfc $u \equiv 1 - \operatorname{erf} u$ and is given by

$$\operatorname{erfc} u \equiv 1 - \operatorname{erf} u = \frac{2}{\sqrt{\pi}} \int_u^\infty e^{-u^2}\, du \tag{2.15-3}$$

The cumulative distribution $F(x)$ of Eq. (2.15-1) may be expressed in terms of the error function and the complementary error function. For $x \geq 0$ we write

$$F(x) = \int_{-\infty}^{x} \frac{e^{-x^2/2\sigma^2}}{\sqrt{2\pi\sigma^2}}\, dx = \int_{-\infty}^{\infty} \frac{e^{-x^2/2\sigma^2}}{\sqrt{2\pi\sigma^2}}\, dx - \int_{x}^{\infty} \frac{e^{-x^2/2\sigma^2}}{\sqrt{2\pi\sigma^2}}\, dx \tag{2.15-4}$$

The first term of the right-hand member of Eq. (2.15-4) is the integral from $-\infty$ to $+\infty$ of the probability density $f(x)$ and hence has the value 1. If we let $u \equiv x/\sqrt{2}\,\sigma$, Eq. (2.15-4) becomes

$$F(x) = 1 - \frac{1}{2}\left(\frac{2}{\sqrt{\pi}} \int_{x/\sqrt{2}\,\sigma}^{\infty} e^{-u^2}\, du\right) = 1 - \tfrac{1}{2} \operatorname{erfc}\left(\frac{x}{\sqrt{2}\,\sigma}\right) \tag{2.15-5}$$

For $x \leq 0$, since tabulated values of erfc $(x/\sqrt{2}\,\sigma) = \operatorname{erfc} u$ are readily available only for positive u, we proceed as follows:

$$F(x) = F(-|x|) = \int_{-\infty}^{-|x|} \frac{e^{-x^2/2\sigma^2}}{\sqrt{2\pi\sigma^2}}\, dx = \frac{1}{\sqrt{\pi}} \int_{-\infty}^{-|x|/\sqrt{2}\,\sigma} e^{-u^2}\, du \tag{2.15-6}$$

Letting $\xi = -u$ yields

$$F(x) = \frac{1}{2}\left(\frac{2}{\sqrt{\pi}} \int_{|x|/\sqrt{2}\,\sigma}^{\infty} e^{-\xi^2}\, d\xi\right) = \tfrac{1}{2} \operatorname{erfc}\left(\frac{|x|}{\sqrt{2}\,\sigma}\right) \tag{2.15-7}$$

The error function and complementary error function may be used for many additional useful calculations in connection with the gaussian density. A matter of interest, for example, is the probability that a measurement will yield an outcome that falls within a certain range about the average value of the random variable. Since the "width" of the probability density depends on the standard deviation σ, we ask for the probability $P(m - k\sigma \leq X \leq m + k\sigma)$, that is, the probability that the random variable X is not further from $x = m$ than $k\sigma$, where k is a constant number. It may be shown (Prob. 2.15-3) that

$$P_{\pm k\sigma} \equiv P(m - k\sigma \leq X \leq m + k\sigma) = \operatorname{erf}\left(\frac{k}{\sqrt{2}}\right) \tag{2.15-8}$$

For future reference some values of $P_{\pm k\sigma}$ are tabulated here.

k	$P_{\pm k\sigma}$	k	$P_{\pm k\sigma}$
0.5	0.383	2.5	0.988
1.0	0.683	3.0	0.997
1.5	0.866	3.5	0.9995
2.0	0.955	4.0	0.99994

Observe how small is the likelihood that a measured value will fall outside the range $\pm 3\sigma$.

As a further example of a probability calculation, consider the following situation: Across a set of terminals there appears a voltage V which is either $V = v_0$ volts or $V = 0$ volt. One of the two possible constant voltages is transmitted over wires from a distance point to alert us as to which of two possible situations prevails there. We are to determine the situation by performing the experiment of making an instantaneous measurement of the voltage across the terminals. A difficulty arises because superimposed on the constant 0 or v_0 volts is noise. We assume that the noise has a gaussian probability density and zero average value. Because of the noise our measurement will, in general, yield neither the result 0 volt nor the result v_0 volts. What procedure are we to use to make a decision? What is the probability that our decision will be in error?

When the transmitted voltage is 0 volt, the terminal voltage is $V = N$, where N is the gaussian random variable representing the noise. Hence V is also a gaussian random variable with probability density as indicated in Fig. 2.15-1. When the transmitted voltage is v_0, then $V = v_0 + N$. In this latter case, V is a gaussian random variable with mean v_0 but with the same variance as in the former case (Prob. 2.14-3). The probability density of $V = v_0 + N$ is also shown in Fig. 2.15-1. On the basis of the symmetry in Fig. 2.15-1 it is apparent that the decision level should be $V = v_0/2$. That is, when we measure V to be $V < v_0/2$, we should decide that the transmitted voltage is 0 volt, and when $V > v_0/2$, we should decide that v_0 volts was transmitted.

When v_0 volts is transmitted, the probability of making an error is the probability $P(V = v_0 + N < v_0/2)$, for, if $v_0 + N < v_0/2$, the v_0-volt transmission will be mistaken for a 0-volt transmission. This *probability of error* is

$$P_{\text{error}} = P\left(V = v_0 + N < \frac{v_0}{2}\right) = P\left(N < -\frac{v_0}{2}\right) = \int_{-\infty}^{-v_0/2} \frac{e^{-n^2/2\sigma^2}}{\sqrt{2\pi\sigma^2}}\, dn \tag{2.15-9}$$

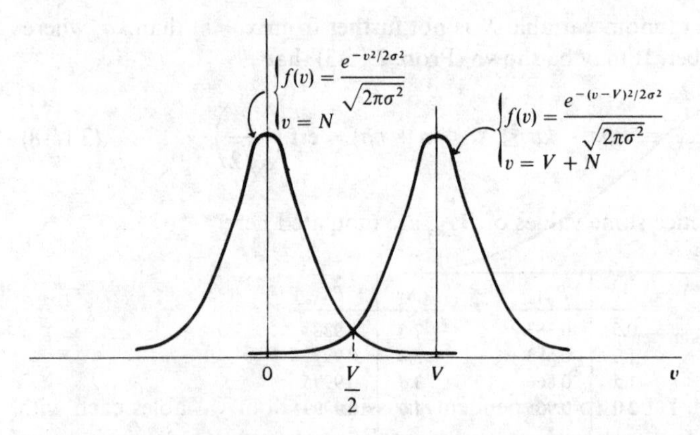

Figure 2.15-1 Probability density function of a gaussian random variable N and of the sum $v_0 + N$.

Letting $u \equiv -n/\sqrt{2}\sigma$, we find

$$P_{\text{error}} = \frac{1}{\sqrt{\pi}} \int_{v_0/(2\sqrt{2}\,\sigma)}^{\infty} e^{-u^2}\, du = \tfrac{1}{2} \operatorname{erfc} \frac{v_0}{2\sqrt{2}\,\sigma} \tag{2.15-10}$$

Similarly 0 volt will be mistaken for v_0 volts if the noise is *positive* and greater in magnitude than $v_0/2$. Since the gaussian density is symmetrical, the probability that either transmission will be mistaken for the other is the same and is given by Eq. (2.15-10). Finally, then, P_{error} in Eq. (2.15-10) gives the probability of an error *without reference* to which voltage is transmitted.

We may note that, because of the manner in which the error probabilities are related to the error function, this function is appropriately named.

2.16 THE RAYLEIGH PROBABILITY DENSITY

We consider now the Rayleigh probability density function. The Rayleigh density is of interest to us principally because of a special relationship which it holds to the gaussian density. For reasons which will be apparent shortly, we use the symbol R to represent the random variable and r to represent the value assumed by the variable. The Rayleigh density is defined by

$$f(r) = \begin{cases} \dfrac{r}{\alpha^2} e^{-r^2/2\alpha^2} & 0 \le r \le \infty \\ 0 & r < 0 \end{cases} \tag{2.16-1}$$

Note particularly that $f(r)$ is nonzero only for positive values of r. A plot of $f(r)$ as a function of r is shown in Fig. 2.16-1. It attains a maximum value $1/(\alpha\sqrt{e})$ at $r = \alpha$. It may be verified that the mean value $\bar{R} = \sqrt{\pi/2}\,\alpha$, the mean-square value $\overline{R^2} = 2\alpha^2$, and the variance $\sigma_r^2 = (2 - \pi/2)\alpha^2$.

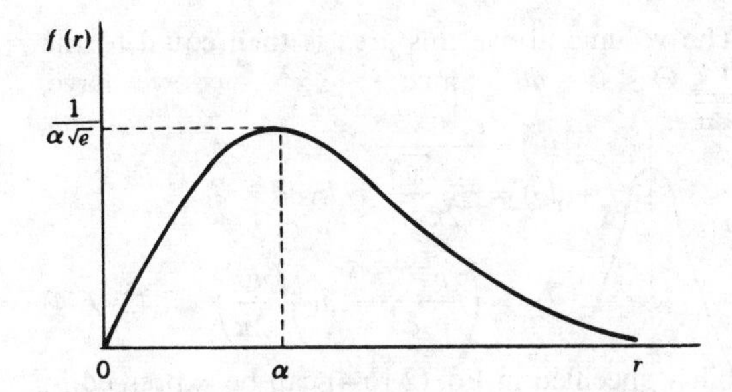

Figure 2.16-1 The Rayleigh density.

Now let X and Y be two independent gaussian random variables each with average value zero and each with variance σ^2. The joint density function is, using Eq. (2.9-7),

$$f(x, y) = f(x)f(y) = \frac{e^{-x^2/2\sigma^2}}{\sqrt{2\pi\sigma^2}} \frac{e^{-y^2/2\sigma^2}}{\sqrt{2\pi\sigma^2}} = \frac{e^{-(x^2+y^2)/2\sigma^2}}{2\pi\sigma^2} \tag{2.16-2}$$

If we should now make a plot of $f(x, y)$ as a function of x and y, we should find a bell-shaped surface above the xy plane. The probability

$$P(x \leq X \leq x + dx, y \leq Y \leq y + dy) = f(x, y)\, dx\, dy \tag{2.16-3}$$

has the significance indicated in Fig. 2.9-1b. That is, to find the probability specified in Eq. (2.16-3), we mark off an area $dx\, dy$ on the xy plane, and the probability specified is equal to the volume above the surface area $dx\, dy$ and below the surface $f(x, y)$.

Suppose, however, that to locate a point on the xy plane we use not the x and y coordinates but rather the coordinates r and θ, where r is the length of the radius vector to the point from the coordinate system origin and θ is the angle measured from the x axis as shown in Fig. 2.16-2. A differential area in such an r,

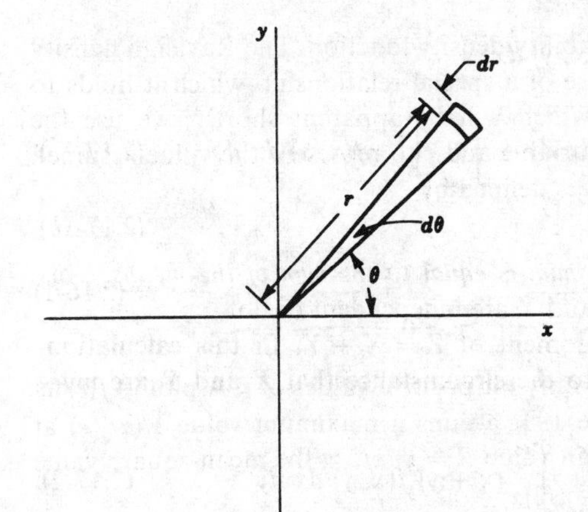

Figure 2.16-2 Cartesian and polar representation of an area.

θ coordinate system is $r\,dr\,d\theta$. The volume above this area is then equal to the probability $P(r \le R \le r + dr,\ \theta \le \Theta \le \theta + d\theta)$. Since $r^2 = x^2 + y^2$, we have, from Eqs. (2.16-2) and (2.16-3), that

$$P(r \le R \le r + dr, \theta \le \Theta \le \theta + d\theta) = \frac{e^{-r^2/2\sigma^2}}{2\pi\sigma^2} r\,dr\,d\theta$$

$$= \left(\frac{re^{-r^2/2\sigma^2}}{\sigma^2} dr\right)\left(\frac{d\theta}{2\pi}\right) \tag{2.16-4}$$

Hence it appears that the probability specified in Eq. (2.16-4) can be expressed in terms of the probability density functions associated with two *independent* random variables R and Θ. The probability densities are

$$f_R(r) = \begin{cases} \dfrac{re^{-r^2/2\sigma^2}}{\sigma^2} & r \ge 0 \\ 0 & r < 0 \end{cases} \tag{2.16-5}$$

and
$$f_\Theta(\theta) = \frac{1}{2\pi} \qquad -\pi \le \theta \le \pi \tag{2.16-6}$$

The independence is apparent from the fact, as appears in Eq. (2.16-4), that the joint density $f_{R,\Theta}(r, \theta)$ appears as the product $f_R(r)f_\Theta(\theta)$.

We note that $f_R(r)$ is precisely the Rayleigh density with $\sigma^2 = \alpha^2$, while $f_\Theta(\theta)$ is a uniform density independent of the angle θ. We observe in Fig. 2.16-1 that the most probable value of r is $r = \alpha = \sigma$.

2.17 MEAN AND VARIANCE OF THE SUM OF RANDOM VARIABLES

Let X and Y be two random variables with means m_x and m_y. Let $Z = X + Y$ and have the mean m_z. Then $m_z = m_x + m_y$. This result follows directly from the definition of the mean. We have, using Eqs. (2.11-7) and (2.11-5), that

$$m_z = \int_{-\infty}^{\infty}\int_{-\infty}^{\infty} (x + y)f(x, y)\,dx$$

$$= \int_{-\infty}^{\infty}\int_{-\infty}^{\infty} xf(x, y)\,dx\,dy + \int_{-\infty}^{\infty}\int_{-\infty}^{\infty} yf(x, y)\,dx\,dy \tag{2.17-1a}$$

$$= m_x + m_y \tag{2.17-1b}$$

Expressed in words, *the mean of the sum is equal to the sum of the means.* This result holds whether the variables X and Y are independent or not.

We calculate next the second moment of $Z = X + Y$. In this calculation, however, we shall restrict ourselves to the circumstance that X and Y are *independent*. We have

$$\overline{Z^2} = \overline{(X + Y)^2} = \int_{-\infty}^{\infty}\int_{-\infty}^{\infty} (x + y)^2 f(x, y)\,dx\,dy \tag{2.17-2}$$

Because of the *independence* of X and $Y, f(x, y) = f(x)f(y)$ so that

$$\overline{Z^2} = \int_{-\infty}^{\infty} x^2 f(x)\,dx \int_{-\infty}^{\infty} f(y)\,dy + \int_{-\infty}^{\infty} y^2 f(y)\,dy \int_{-\infty}^{\infty} f(x)\,dx$$
$$+ 2\int_{-\infty}^{\infty} xf(x)\,dx \int_{-\infty}^{\infty} yf(y)\,dy \qquad (2.17\text{-}3)$$

However

$$\int_{-\infty}^{\infty} f(x)\,dx = \int_{-\infty}^{\infty} f(y)\,dy = 1 \qquad (2.17\text{-}4)$$

so that

$$\overline{Z^2} = \overline{X^2} + \overline{Y^2} + 2\bar{X}\bar{Y} \qquad (2.17\text{-}5)$$

If either $\bar{X}$ or $\bar{Y}$ or both are zero, then $\overline{Z^2} = \overline{X^2} + \overline{Y^2}$. In this case we also have from Eq. (2.12-3) that

$$\sigma_z^2 = \sigma_x^2 + \sigma_y^2 \qquad (2.17\text{-}6)$$

This result can be obtained directly by calculating $\sigma_z^2 = \overline{(Z - m_z)^2}$.

2.18 PROBABILITY DENSITY OF $Z = X + Y$

We now calculate the probability density $f(z)$ of $Z = X + Y$ in terms of the joint density $f(x, y)$. Assume an arbitrary value of Z and call it z. Then the region $Y \leq z - X$ is shown as the shaded region in Fig. 2.18-1. Hence the probability

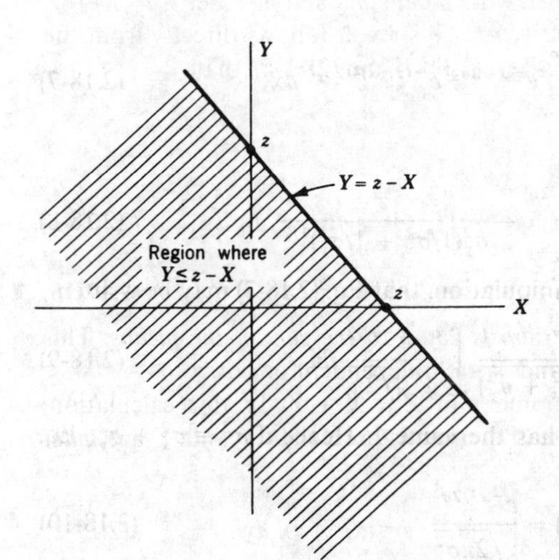

Figure 2.18-1 Related to the calculation of the probability density of a sum of random variables.

that $Z \leq z$ is the same as the probability that $Y \leq z - X$ independently of the value of X, that is, for $-\infty \leq X \leq +\infty$. This probability is

$$F(z) = P(Z \leq z) = P(X \leq \infty, Y \leq z - X) \tag{2.18-1}$$

Using Eq. (2.9-3), we have

$$F(z) = \int_{-\infty}^{\infty} dx \int_{-\infty}^{z-x} f(x, y)\, dy \tag{2.18-2}$$

The probability density of Z is found by differentiating $F(z)$ with respect to z. We then have

$$f(z) = \frac{dF(z)}{dz} = \int_{-\infty}^{\infty} f(x, z - x)\, dx \tag{2.18-3}$$

which expresses $f_Z(z)$ in terms of $f_{XY}(x, y = z - x)$.

When X and Y are *independent*, $f(x, y) = f(x)f(y)$, and Eq. (2.18-3) may be written

$$f(z) = \int_{-\infty}^{\infty} f(x)f(z - x)\, dx \tag{2.18-4}$$

Comparing Eq. (2.18-4) with Eq. (1.12-1), we recognize that $f(z)$ is the *convolution* of $f(x)$ and $f(y)$.

As a most important consequence of the result given in Eq. (2.18-4), let us consider the case where $f(x)$ and $f(y)$ are both gaussian densities.

$$f(x) = \frac{e^{-x^2/2\sigma_x^2}}{\sqrt{2\pi\sigma_x^2}} \tag{2.18-5}$$

$$f(y) = \frac{e^{-y^2/2\sigma_y^2}}{\sqrt{2\pi\sigma_y^2}} \tag{2.18-6}$$

Then $f(z)$ is given by

$$f(z) = \frac{1}{2\pi\sigma_x\sigma_y} \int_{-\infty}^{\infty} e^{-x^2/2\sigma_x^2} e^{-(z-x)^2/2\sigma_y^2}\, dx \tag{2.18-7}$$

If we define a variable u by

$$u \equiv \left(\frac{1}{\sigma_x^2} + \frac{1}{\sigma_y^2}\right)^{1/2} x - \frac{z}{\sigma_y^2(1/\sigma_x^2 + 1/\sigma_y^2)^{1/2}} \tag{2.18-8}$$

then we find, after some algebraic manipulation, that Eq. (2.18-7) may be written

$$f(z) = \frac{e^{-z^2/2(\sigma_x^2 + \sigma_y^2)}}{\sqrt{2\pi(\sigma_x^2 + \sigma_y^2)}} \int_{-\infty}^{\infty} \frac{e^{-u^2/2}}{\sqrt{2\pi}}\, du \tag{2.18-9}$$

The definite integral in Eq. (2.18-9) has the value 1. Hence if $\sigma^2 \equiv \sigma_x^2 + \sigma_y^2$, Eq. (2.18-9) becomes

$$f(z) = \frac{e^{-z^2/2\sigma^2}}{\sqrt{2\pi\sigma^2}} \tag{2.18-10}$$

which is a gaussian density function. Thus we have the extremely interesting and important result: Given two independent gaussian random variables, the sum of these variables is itself a gaussian random variable. We note, of course, that the variance σ^2 of the sum variable is the sum $\sigma_x^2 + \sigma_y^2$ of the individual variables. This result, $\sigma^2 = \sigma_x^2 + \sigma_y^2$, applies for independent random variables generally, not just gaussian variables. Further, in the discussion above we assumed, for simplicity, that the variables X and Y had zero mean. However, this assumption does not make the result above any less general. For if we had assumed average values m_x and m_y for X and Y, then we would have found that Z had an average value $m_z = m_x + m_y$ and Eq. (2.18-10) would become

$$f(z) = \frac{e^{-(z-m_z)^2/2\sigma^2}}{\sqrt{2\pi\sigma^2}} \tag{2.18-11}$$

This result, concerning the sum of independent gaussian random variables, may be extended somewhat. If X_1 is a gaussian random variable, then, as may be easily verified, if c_1 is a constant, $c_1 X_1$ is also a gaussian variable. It follows, by extension of the result above for two independent gaussian random variables, that a linear combination of independent gaussian variables is also gaussian. Explicitly, if $X_1, X_2, X_3, \ldots$ are independent gaussian variables and $c_1, c_2, c_3, \ldots$ are constants, then

$$X = c_1 X_1 + c_2 X_2 + c_3 X_3 \cdots \tag{2.18-12}$$

is also a gaussian random variable. Our proof does not extend to the case of *dependent* gaussian variables. However, it is interesting to note that, even in the case where the gaussian variables are dependent, a linear combination of such variables is still gaussian. In general, the probability density function of a sum of variables of like probability density does not preserve the form of the probability density of the individual variables. Indeed, as shown in Sec. 2.20, X tends to become gaussian if a large number of random variables are summed, regardless of their probability density.

2.19 CORRELATION BETWEEN RANDOM VARIABLES

The *covariance* μ of two random variables X and Y is defined as

$$\mu \equiv E\{(X - m_x)(Y - m_y)\} \tag{2.19-1}$$

If X and Y are independent random variables, we find, using Eq. (2.9-7) and (2.11-7), that

$$\begin{aligned} \mu &= E\{(X - m_x)(Y - m_y)\} \\ &= \int_{-\infty}^{\infty} (x - m_x) f(x)\, dx \int_{-\infty}^{\infty} (y - m_y) f(y)\, dy \\ &= (m_x - m_x)(m_y - m_y) = 0 \end{aligned} \tag{2.19-2}$$

This result is rather to have been expected on intuitive grounds. For, as assumed, when an experiment is performed to determine a joint outcome specified by the set of numbers x and y, the outcome y is in no way conditional on the outcome x. Hence we would find that a particular outcome x_1 would occur jointly sometimes with a value y which was positive with respect to its average m_y, and sometimes with a value y negative with respect to m_y. In the course of many experiments with the outcome x_1, the sum of the numbers $x_1(y - m_y)$ would add up to zero.

On the other hand, suppose X and Y were dependent. Suppose, for example, that outcome y was conditioned on the outcome x in such manner that there was an enhanced probability that $(y - m_y)$ was of the same sign as $(x - m_x)$. In such a case we would anticipate that the expected value $E\{(X - m_x)(Y - m_y)\} > 0$. Similarly, if there were an enhanced probability that $(x - m_x)$ and $(y - m_y)$ were of opposite sign, we would expect $E\{(X - m_x)(Y - m_y)\} < 0$.

As an extreme case, let us assume the maximum possible dependency between X and Y. Let us assume that $X = Y$ or $X = -Y$. In these cases we would find (with $m_x = m_y = 0$)

$$E\{XY\} = E\{X^2\} = E\{Y^2\} = \sigma_x^2 = \sigma_y^2 = \sigma_x \sigma_y \tag{2.19-3}$$

or

$$E\{XY\} = E\{-X^2\} = E\{-Y^2\} = -\sigma_x^2 = -\sigma_y^2 = -\sigma_z \sigma_y \tag{2.19-4}$$

Consider then the quantity ρ defined by

$$\rho \equiv \frac{\mu}{\sigma_x \sigma_y} = \frac{E\{XY\}}{\sigma_x \sigma_y} \tag{2.19-5}$$

This number ρ is called the *correlation coefficient* between the variables X and Y and serves as a measure of the extent to which X and Y are dependent. From Eqs. (2.19-3) to (2.19-5) we have that ρ falls in the range

$$-1 \leq \rho \leq +1 \tag{2.19-6}$$

When X and Y are independent, $\rho = 0$. When $X = Y$, $\rho = 1$; when $X = -Y$, $\rho = -1$. If X and Y are neither identical nor independent, then ρ will have a magnitude between 0 and 1. When $\rho = 0$, the random variables X and Y are said to be *uncorrelated*.

When random variables are independent, they are uncorrelated. However, the fact that they are uncorrelated does not ensure that they are independent. A simple illustrative example will establish this point.

Example 2.19-1 Let Z be a random variable with probability density $f(z) = \frac{1}{2}$ in the range $-1 \leq z \leq 1$. Let the random variable $X = Z$ and the random variable $Y = Z^2$. Obviously X and Y are not independent since $X^2 = Y$. Show, nonetheless, that X and Y are uncorrelated.

SOLUTION We have

$$E\{Z\} = \int_{-1}^{1} \tfrac{1}{2}\, dz = 0 \tag{2.19-7}$$

Since $X = Z$, $E\{X\} = E\{Z\} = 0$. Since $Y = Z^2$, $E\{Y\} = E\{Z^2\}$, so that

$$E\{Y\} = \int_{-1}^{1} \tfrac{1}{2} z^2\, dz = \tfrac{1}{3} \tag{2.19-8}$$

The covariance μ is

$$\mu = E\{(X - m_x)(Y - m_y)\} = E\{(X)(Y - \tfrac{1}{3})\}$$

$$= E\{XY - \tfrac{1}{3}X\} \tag{2.19-9a}$$

$$= E\left\{Z^3 - \frac{Z}{3}\right\} = \int_{-1}^{1} \frac{1}{2}\left(z^3 - \frac{z}{3}\right) dz = 0 \tag{2.19-9b}$$

Hence $\rho = 0$, and the variables are uncorrelated.

As we have seen, it is easy enough to invent random variables which are uncorrelated but are nevertheless dependent. It is even possible to do so when the random variables are gaussian. On the other hand, the random variables of interest to us will be variables which occur in the description of such natural physical processes as noise. As already noted, for the most part, such variables are gaussian. It does indeed turn out that, with such gaussian random variables, an absence of correlation does imply independence. We shall therefore generally assume that two gaussian random variables, for which the covariance $\mu = 0$, are independent variables.

2.20 THE CENTRAL-LIMIT THEOREM[3]

The result of the previous section concerning the probability density of the sum of gaussian random variables is a special case of the *central-limit theorem*. The so-called central-limit theorem is actually a group of related theorems which are collectively grouped under a single name. We shall not even undertake to state these theorems precisely, let alone undertake to prove them. For our purposes it will be adequate to note that the central-limit theorem indicates that the probability density of a sum of N independent random variables tends to approach a gaussian density as the number N increases. The mean and variance of this gaussian density are respectively the sum of the means and the sum of the variances of the N independent random variables. The theorem applies even when (with a few special exceptions) the individual random variables are not gaussian. In addition, the theorem applies in certain special cases even when the individual random variables are not independent.

As an illustration of the central-limit theorem consider the case where the individual random variable has a uniform (constant) probability density as shown in Fig. 2.20-1*a*. Note here that the area under the density has been adjusted to unity. Then, as indicated by Eq. (2.18-4), if there were two terms in the sum, the density of the sum would be determined as the convolution of the density in Fig. 2.20-1*a* with itself. The result of such a convolution is indicated in Fig. 2.20-1*b*. Similarly, the density of a sum of the three random variables is the convolution of the density in Fig. 2.20-1*b* with the density in Fig. 2.20-1*a*. The result (Prob. 2.20-1) is shown in Fig. 2.20-1*c*. Note that even for this sum of only three terms the result suggests a gaussian density. In the limit, as more and more terms are added, the density would indeed become gaussian.

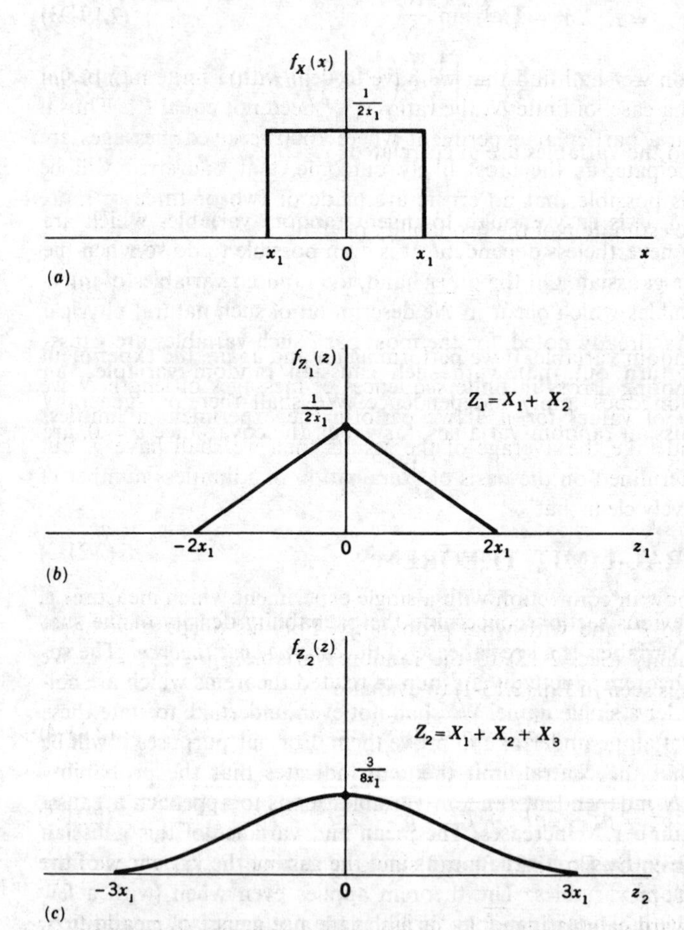

Figure 2.20-1 (*a*) A random variable X has a uniform probability density. (*b*) The probability density of the random variable $X_1 + X_3$. (*c*) The density of the random variable $X_1 + X_2 + X_3$.

2.21 ERROR PROBABILITY AS MEASURED BY FINITE SAMPLES

We noted in Sec. 2-10 how a communications channel is used to transmit, in successive intervals, a sequence of messages. We noted there, further, that because of noise, some errors will be made at the receiver as messages will be incorrectly read and will be confused with one another. If we were to undertake to establish the probability of error P_e by experiment we would read a very large number N of messages and take note of the number of instances N_e in which errors were made. In principle it would be required that N be infinite since by definition (see Eq. 2.1-1) the error probability is

$$P_e \equiv \lim_{N \to \infty} \frac{N_e}{N} \tag{2.21-1}$$

In a real situation we shall find that we have to deal with a finite number of messages N. In such a case, of finite N, the ratio N_e/N need not equal P_e. Thus if $P_e = 1/1000$, then, in a particular experiment where 1000 received messages are read we would anticipate, as the most likely outcome, that one error will be made. However, it is possible that no errors are made or two or three or more errors are made. The estimate p of the probability of error

$$p \equiv \frac{N_e}{N} \tag{2.21-2}$$

for a finite N is a random variable. If we perform again and again, the experiment of measuring p by noting errors in finite sequences of messages of length N we shall have an array of values for p. If we perform the experiment a limitless number of times and take the average of the results then we shall have $\bar{p}$. But since $\bar{p}$ like P_e is determined on the basis of examination of a limitless number of messages it is intuitively clear that

$$\bar{p} = P_e \tag{2.21-3}$$

It is useful to know in connection with a single experiment which measures p, how close p comes to P_e and with what probability. For this purpose we apply Tchebycheff's inequality (Sec. 2-13) to the random variable $|p - P_e| \equiv x$. We need accordingly, as is seen in Eq. (2.13-1) to evaluate

$$\sigma_x^2 = \overline{(p - P_e)^2} \tag{2.21-4}$$

We have

$$\sigma_x^2 = \overline{p^2 - 2pP_e + P_e^2} = \overline{p^2} - 2\bar{p}P_e + P_e^2 \tag{2.21-5}$$

The quantity $\bar{p}$ is given by Eq. (2.21-3) and since we assume P_e known we have now only to calculate $\overline{p^2}$.

With a view toward calculating $\overline{p^2}$ let us associate a sequence of messages by a corresponding sequence of numerical digits d_i; the digit $d_i = 0$ if the corresponding message is received correctly and $d_i = 1$ if an error is made. The usefulness of this association is that in a sequence of length N, the sum of the digits d_i divided by N is precisely p, i.e.,

$$p = \frac{1}{N}\sum_{i=1}^{N} d_i \tag{2.21-6}$$

we then have

$$p^2 = \frac{1}{N^2}\left(\sum_{i=1}^{N} d_i\right)\left(\sum_{j=1}^{N} d_j\right) \tag{2.21-7a}$$

$$= \frac{1}{N^2}\sum_{i=1}^{N} d_i^2 + \frac{1}{N^2}\sum_{i=1}^{N}\sum_{j=1}^{N} d_i d_j \qquad i \neq j \tag{2.21-7b}$$

The first member on the right-hand side of Eq. (2.21-7*b*) is $1/N^2$ times the sum which occurs due to the multiplication of each digit by itself, while the second term is $1/N^2$ times the sum which accrues when each digit is multiplied by every *other* digit. We now have

$$\overline{p^2} = \frac{1}{N^2}\sum_{i=1}^{N} \overline{d_i^2} + \frac{1}{N^2}\sum_{i=1}^{N}\sum_{j=1}^{N} \overline{d_i d_j} \qquad i \neq j \tag{2.21-8}$$

We calculate $\overline{d_i^2}$ by writing

$$\overline{d_i^2} = (1)^2 P(d_i = 1) + (0)^2 P(d_i = 0) = P(d_i = 1) = P_e \tag{2.21-9}$$

We further consider that the events of errors being made in two messages are entirely independent of one another. In this case, using the result given in Eq. (2.11-9) we find

$$\begin{aligned}\overline{d_i d_j} = (\bar{d}_i)(\bar{d}_j) &= [1P(d_i = 1) + 0P(d_i = 0)] \\ &\quad \times [1P(d_j = 1) + 0P(d_j = 0)] \\ &= P_e^2\end{aligned} \tag{2.21.10}$$

Substituting the results of Eqs. (2.21-9) and (2.21-10) into Eq. (2.21-8) and taking account of the number of terms in the summations we find that

$$\overline{p^2} = \frac{1}{N^2}(N)P_e + \frac{1}{N^2}(N)(N-1)P_e^2 \tag{2.21-11}$$

Substituting the results of Eq. (2.21-11) and (2.21-3) into Eq. (2.21-5) we find, after simplification that

$$\sigma_x^2 = \frac{P_e - P_e^2}{N} \tag{2.21-12}$$

Rather inevitably, any communications system that has any practical value has $P_e \ll 1$ so that Eq. (2.21-12) may be written

$$\sigma_x^2 \simeq \frac{P_e}{N} \tag{2.21-13}$$

and Tchebycheff's inequality becomes

$$P(|p - P_e| > \varepsilon) \leq \frac{P_e}{N\varepsilon^2} \tag{2.21-14}$$

As an example consider a system in which $P_e = 10^{-4}$. Let us perform an experiment to find the number of errors in 4×10^5 messages. (In this case if p were equal to P_e we would find 40 errors.) Let us ask: what is the probability that the actual measured value of p does not differ from P_e by more than 50 percent, that is $\varepsilon = 50$ percent of $10^{-4} = 0.5 \times 10^{-4}$. In this case we find

$$P(|p - 10^{-4}| \geq 0.5 \times 10^{-4}) \leq \frac{10^{-4}}{4 \times 10^5 \times 0.25 \times 10^{-8}} = 10\% \tag{2.21-15}$$

Rather generally, values of $\varepsilon = P_e/2$ and $P(|p - P_e| \geq \varepsilon) \leq 10\%$ are acceptable. In this case, as is readily verified, the number of messages which must be included in a sequence is

$$N \cong 40/P_e \tag{2.21-16}$$

In conclusion, the Tchebycheff inequality provides an upper bound on the probability of p differing from P_e by more than an amount ε. However, tighter bounds, such as the Chernov bound,[4] do exist. Using this bound, however, is often quite complicated, since the result depends on parameters other than σ^2. For this reason, the Tchebycheff inequality is very popular among communications engineers.

2.22 SIGNAL DETERMINATION WITH NOISE DESCRIBED BY A DISTRIBUTION FUNCTION

In Sec. 2.10 we considered a communication channel in which one or another of a discrete number of messages is presented at the input sending end. Some fixed time interval T is allocated to the transmission of a single message, the messages being distinguished from one another through the fact that different waveforms are used to represent the different messages. The observed response generated at the receiving end by virtue of the processing performed by the channel is quite generally a fixed voltage which persists nominally for the time T allocated to a message. Because of noise, the observed response does not establish with certainty which message was transmitted. We have, however, determined a procedure for estimating which message was transmitted, the procedure assuring minimum likelihood of error. For a response r_j we formed the quantities $P(m_k)P(r_j | m_k)$ for all values of k, that is, for all possible messages, and we estimated that the message sent was the one for which $P(m_k)P(r_j | m_k)$ was a maximum.

Now let us consider a situation in which, in the *absence* of noise, each message m_k causes the generation of a signal s_k and a unique response r_k.

However, in this case, when noise is superimposed on s_k, the response is a continuous random variable which we call R (with sample values r).

When R is a continuous random variable, the probability is zero that the generated response is precisely some fixed number r. We ask instead about the probability that the response is in the range r to $r + dr$. Having observed a response in the range r to $r + dr$ we shall decide which is the transmitted message by comparing values of

$$P(m_k)P(r \leq R \leq r + dr \mid m_k) = P(m_k)f_R(r \mid m_k)\, dr \tag{2.22-1}$$

Suppose that there were an interval during which no signal was transmitted. In this case, at the receiver the total receiver response R would be due only to the noise; that is, $R = N$. This noise N can be characterized by a probability density function $f_N(n)$. Very frequently the noise has a gaussian density function as illustrated in Fig. 2.22-1a. The shaded area at $N = n$ represents the probability that N is in the range n to $n + dn$.

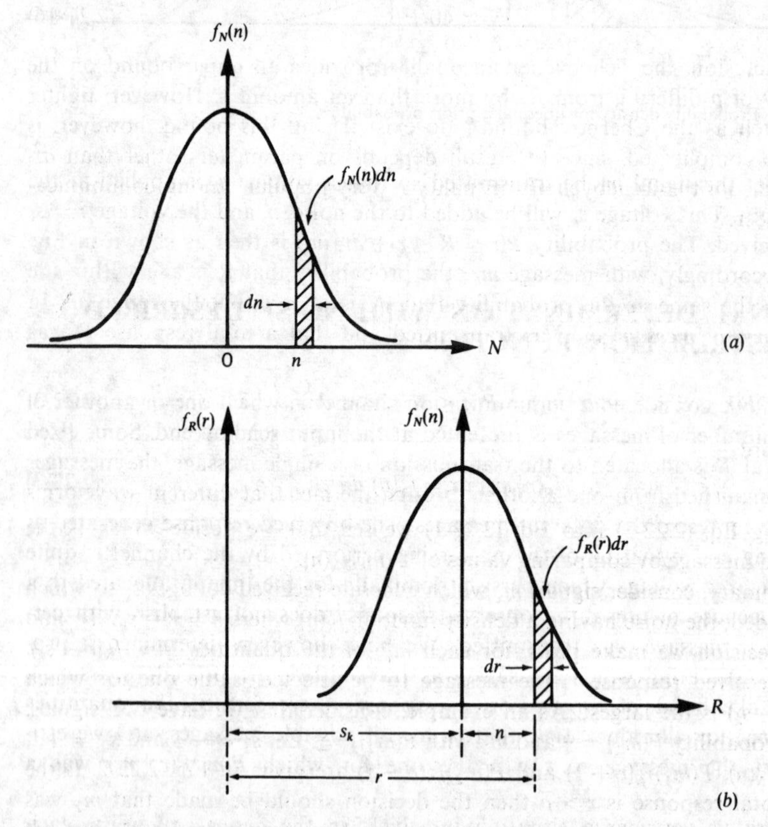

Figure 2.22-1 (a) Probability density of noise, N. (b) Probability density of the received voltage $R = s_k + N$.

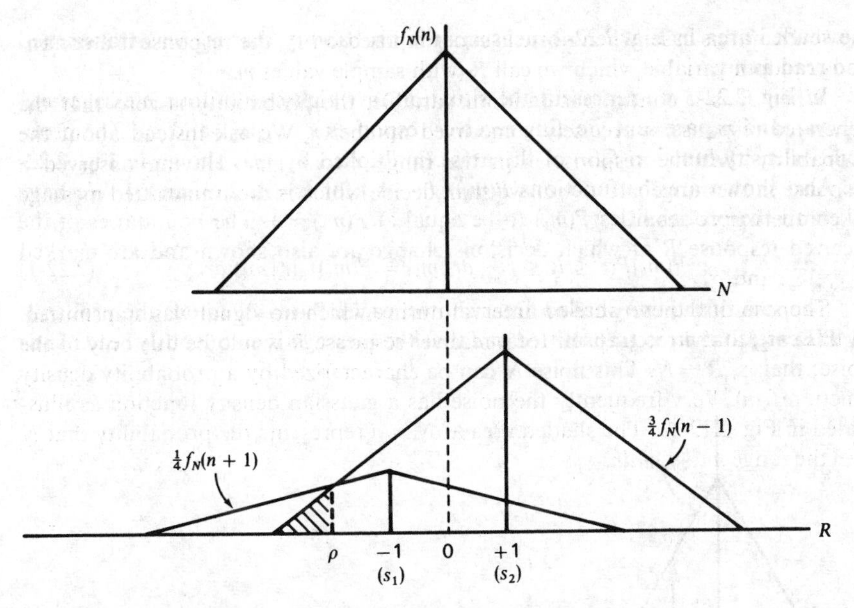

Figure 2.22-2 Probability density for unequal a priori probabilities.

Next, let the signal m_k be transmitted so that a voltage s_k is applied at the channel input. This voltage s_k will be added to the noise, n, and the voltage $r = s_k + n$ is received. The probability $P(r \leq R \leq r + dr \,|\, m_k)$ is then as shown in Fig. 2.22-1*b*. Accordingly, with message m_k, the probability that R is at r within the range dr is the same as the probability that, N is at n within the range dn. In short, *given that message* m_k was transmitted and that a total response R was received,

$$P(r \leq R \leq r + dr \,|\, m_k) = P(n \leq N \leq n + dn) \qquad n + s_k = r \qquad (2.22\text{-}2)$$

and therefore

$$f_R(r \,|\, m_k)\, dr = f_N(n)\, dn \qquad (2.22\text{-}3)$$

Substituting Eq. (2.22-3) into Eq. (2.22-1) we see that we decide which is the transmitted message by comparing values of $P(m_k)f_N(n)$.

In summary, consider signals m_k which generate received voltages s_k to which noise is added, the noise having a density function $f_N(n)$ where $n = r - s_k$. To find the best decision we make plots, for each m_k, of the quantities $P(m_k)f_N(r - s_k)$. For any received response r, the message to be selected is the one for which $P(m_k)f_N(r - s_k)$ is the largest. As an example, consider that we have two signals, m_1 with probability $P(m_1) = \frac{1}{4}$ and m_2 with $P(m_2) = \frac{3}{4}$. Let $s_1 = -1$ and $s_2 = +1$. Plots for $f_N(n)$, $P(m_1)f_N(r + 1)$ and $P(m_2)f_N(r - 1)$ are given in Fig. 2.22-2. If the received total response is $r > \rho$ then the decision should be made that m_2 was sent and if $r < \rho$ that m_1 was sent. Note that if m_2 was indeed sent, but the response is $r < \rho$, then we would mistakenly judge that m_1 had been sent. Thus the shaded area in Fig. 2.22-2 measures the probability that m_2 was transmitted and read as m_1.

In Fig. 2.22-3 we represent the situation in which one of four messages m_1, m_2, m_3, and m_4 are sent yielding receiver responses s_1, s_2, s_3, and s_4. The probability density function $f_N(n)$ of the noise (multiplied by $\frac{1}{4}$) is shown in Fig. 2.22-3*a*. Also shown are the functions $P(m_k)f_N(r - s_k)$ for $k = 1, 2, 3$, and 4. We have taken all the probabilities $P(m_k)$ to be equal at $P(m_k) = \frac{1}{4}$. The boundaries of the received response R at which decisions change are also shown and are marked ρ_{12}, ρ_{23}, and ρ_{34}.

The sum of the two shaded areas shown, each of area Δ, equals the probability $P(E, m_2)$ that m_2 is transmitted *and* that the message is erroneously read. Thus

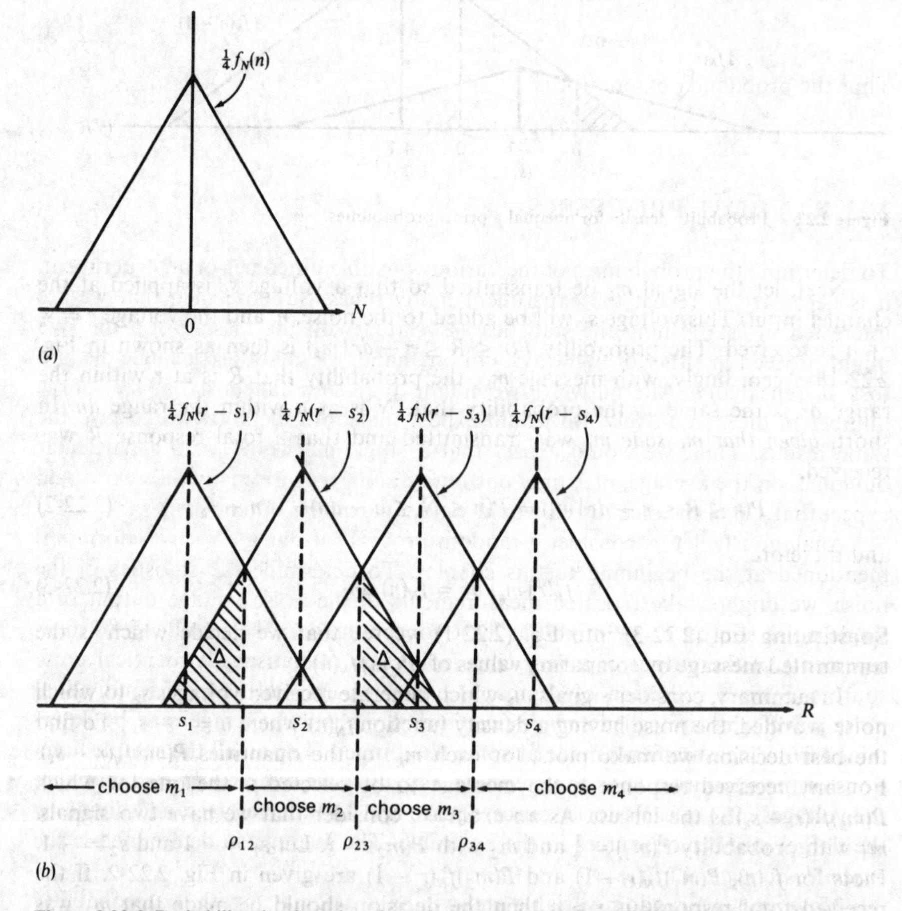

Figure 2.22-3 Probability density when four signals are transmitted. (*a*) Probability density of $f_2(r)$ when noise alone is received. (*b*) Probability density when one of four signals is transmitted and received in noise.

$P(E, m_2) = 2\Delta$. The remaining area under $\frac{1}{4}f_N(r - s_2)$ is the probability that m_2 is transmitted and is correctly read. This probability is $P(C, m_2) = \frac{1}{4} - 2\Delta$. We find of course that $P(E, m_3) = P(E, m_2)$. But it is most interesting to note that when we evaluate $P(E, m_1)$ or $P(E, m_4)$ there is only a single area Δ involved. Hence $P(E, m_1) = P(E, m_4) = \frac{1}{2}P(E, m_2) = \frac{1}{2}P(E, m_3)$. Thus, we arrive at the most interesting result that given four equally likely messages, when we make boundary decisions which assure minimum average likelihood of error, the probability of making an error is not the same for all messages. Of course, this result applies for any number of equally likely messages greater than two.

The probability that a message is read correctly is

$$\begin{aligned} P(C) &= P(C, m_1) + P(C, m_2) + P(C, m_3) + P(C, m_4) \\ &= (\tfrac{1}{4} - \Delta) + (\tfrac{1}{4} - 2\Delta) + (\tfrac{1}{4} - 2\Delta) + (\tfrac{1}{4} - \Delta) \\ &= 1 - 6\Delta \end{aligned} \tag{2.22-4}$$

Thus the probability of an error is

$$P(E) = 1 - P(C) = 6\Delta \tag{2.22-5}$$

2.23 RANDOM PROCESSES

To determine the probabilities of the various possible outcomes of an experiment, it is necessary to repeat the experiment many times. Suppose then that we are interested in establishing the statistics associated with the tossing of a die. We might proceed in either of two ways. On one hand, we might use a single die and toss it repeatedly. Alternatively, we might toss simultaneously a very large number of dice. Intuitively, we would expect that both methods would give the same results. Thus, we would expect that a single die would yield a particular outcome, on the average, of 1 time out of 6. Similarly, with many dice we would expect that 1/6 of the dice tossed would yield a particular outcome.

Analogously, let us consider a random process such as a noise waveform $n(t)$ mentioned at the beginning of this chapter. To determine the statistics of the noise, we might make repeated measurements of the noise voltage output of a single noise source, or we might, at least conceptually, make simultaneous measurements of the output of a very large collection of statistically identical noise sources. Such a collection of sources is called an *ensemble*, and the individual noise waveforms are called *sample functions*. A statistical average may be determined from measurements made at some fixed time $t = t_1$ on all the sample functions of the ensemble. Thus to determine, say, $\overline{n^2(t)}$, we would, at $t = t_1$, measure the voltages $n(t_1)$ of each noise source, square and add the voltages, and divide by the (large) number of sources in the ensemble. The average so determined is the *ensemble average* of $n^2(t_1)$.

Now $n(t_1)$ is a random variable and will have associated with it a probability density function. The ensemble averages will be identical with the statistical averages computed earlier in Secs. 2.11 and 2.12 and may be represented by the same symbols. Thus the statistical or ensemble average of $n^2(t_1)$ may be written $E[n^2(t_1)] = \overline{n^2(t_1)}$. The averages determined by measurements on a single sample function at successive times will yield a *time average*, which we represent as $\langle n^2(t) \rangle$.

In general, ensemble averages and time averages are not the same. Suppose, for example, that the statistical characteristics of the sample functions in the ensemble were changing with time. Such a variation could not be reflected in measurements made at a fixed time, and the ensemble averages would be different at different times. When the statistical characteristics of the sample functions do not change with time, the random process is described as being *stationary*. However, even the property of being stationary does not ensure that ensemble and time averages are the same. For it may happen that while each sample function is stationary the individual sample functions may differ statistically from one another. In this case, the time average will depend on the particular sample function which is used to form the average. When the nature of a random process is such that ensemble and time averages are identical, the process is referred to as *ergodic*. An ergodic process is stationary, but, of course, a stationary process is not necessarily ergodic.

Throughout this text we shall assume that the random processes with which we shall have occasion to deal are ergodic. Hence the ensemble average $E\{n(t)\}$ is the same as the time average $\langle n(t) \rangle$, the ensemble average $E\{n^2(t)\}$ is the same as the time average $\langle n^2(t) \rangle$, etc.

Example 2.23-1 Consider the random process

$$V(t) = \cos(\omega_0 t + \Theta) \tag{2.23-1}$$

where Θ is a random variable with a probability density

$$f(\theta) = \frac{1}{2\pi} \qquad -\pi \le \theta \le \pi$$

$$= 0 \qquad \text{elsewhere} \tag{2.23-2}$$

(*a*) Show that the first and second moments of $V(t)$ are independent of time.
(*b*) If the random variable Θ in Eq. (2.23-1) is replaced by a fixed angle θ_0, will the ensemble mean of $V(t)$ be time independent?

SOLUTION (*a*) Choose a fixed time $t = t_1$. Then

$$E\{V(t_1)\} = \int_{-\pi}^{\pi} \frac{1}{2\pi} \cos(\omega_0 t_1 + \theta)\, d\theta = 0 \tag{2.23-3}$$

and

$$E\{V^2(t_1)\} = \int_{-\pi}^{\pi} \frac{1}{2\pi} \cos^2(\omega_0 t_1 + \theta)\, d\theta = \tfrac{1}{2} \tag{2.23-4}$$

We note that these moments are independent of t_1 and hence independent of time.

In a similar manner it can be established that all of the moments and all other statistical characteristics of $V(t)$ are independent of time. *Hence $V(t)$ is a stationary process.*

(*b*) Since θ is known, $V(t)$ is deterministic. For example, if $\theta = 30°$

$$E\{V(t) = V(t) = \cos(\omega_0 t + 30°) \neq \text{constant} \tag{2.23-5}$$

Thus, $V(t)$ is not stationary.

Example 2.23-2 A voltage $V(t)$, which is a gaussian ergodic random process with a mean of zero and a variance of 4 volt^2, is measured by a dc meter, a true rms meter, and a meter which first squares $V(t)$ and then reads its dc component.

Find the output of each meter.

SOLUTION (*a*) The dc meter reads

$$\langle V(t)\rangle = E\{V(t)\}$$

since $V(t)$ is ergodic. Since $E\{V(t)\} = 0$, the dc meter reads zero.

(*b*) The true rms meter reads

$$\sqrt{\langle V^2(t)\rangle} = \sqrt{E\{V^2(t)\}}$$

since $V(t)$ is ergodic. Since $V(t)$ has a zero mean, the true rms meter reads $\sigma = 2$ volts.

(*c*) The square and average meter (a full-wave rectifier meter) yields a deflection proportional to

$$\langle v^2(t)\rangle = E\{V^2(t)\} = \sigma^2 = 4$$

2.24 AUTOCORRELATION

A random process $n(t)$, being neither periodic nor of finite energy has an autocorrelation function defined by Eq. (1.18-1). Thus

$$R(\tau) = \lim_{T\to\infty} \frac{1}{T}\int_{-T/2}^{T/2} n(t)n(t+\tau)\, dt \tag{2.24-1}$$

In connection with deterministic waveforms we were able to give a physical significance to the concept of a power spectral density $G(f)$ and to show that $G(f)$ and $R(\tau)$ constitute a Fourier transform pair. As an extension of that result we shall *define* the power spectral density of a random process in the same way. Thus for a random process we take $G(f)$ to be

$$G(f) = \mathscr{F}[R(\tau)] = \int_{-\infty}^{\infty} R(\tau)e^{-j\omega\tau}\, d\tau \tag{2.24-2}$$

It is of interest to inquire whether $G(f)$ defined in Eq. (2.24-2) for a random process has a physical significance which corresponds to the physical significance of $G(f)$ for deterministic waveforms.

For this purpose consider a *deterministic* waveform $v(t)$ which extends from $-\infty$ to ∞. Let us select a section of this waveform which extends from $-T/2$ to $T/2$. This waveform $v_T(t) = v(t)$ in this range, and otherwise $v_T(t) = 0$. The waveform $v_T(t)$ has a Fourier transform $V_T(f)$. We recall that $|V_T(f)|^2$ is the energy spectral density; that is, $|V_T(f)|^2 \, df$ is the normalized energy in the spectral range df. Hence, over the interval T the normalized power density is $|V_T(f)|^2/T$. As $T \to \infty$, $v_T(t) \to v(t)$, and we then have the result that the physical significance of the power spectral density $G(f)$, at least for a deterministic waveform, is that

$$G(f) = \lim_{T \to \infty} \frac{1}{T} |V_T(f)|^2 \tag{2.24-3}$$

Correspondingly, we state, without proof, that when $G(f)$ is defined for a random process, as in Eq. (2.24-2), as the transform of $R(\tau)$, then $G(f)$ has the significance that

$$G(f) = \lim_{T \to \infty} E\left\{\frac{1}{T} |N_T(f)|^2\right\} \tag{2.24-4}$$

where $E\{\ \}$ represents the ensemble average or expectation and $N_T(f)$ represents the Fourier transform of a truncated section of a sample function of the random process $n(t)$.

The autocorrelation function $R(\tau)$ is, as indicated in Eq. (2.24-1), a *time average* of the product $n(t)$ and $n(t + \tau)$. Since we have assumed an ergodic process, we are at liberty to perform the averaging over any sample function of the ensemble, since every sample function will yield the same result. However, again because the noise process is ergodic, we may replace the time average by an ensemble average and write, instead of Eq. (2.24-1),

$$R(\tau) = E\{n(t)n(t + \tau)\} \tag{2.24-5}$$

The averaging indicated in Eq. (2.24-5) has the following significance: At some *fixed* time t, $n(t)$ is a random variable, the possible values for which are the values $n(t)$ assumed at time t by the individual sample functions of the ensemble. Similarly, at the *fixed* time $t + \tau$, $n(t + \tau)$ is also a random variable. It then appears that $R(\tau)$ as expressed in Eq. (2.24-5) is the covariance between these two random variables.

Suppose then that we should find that for some τ, $R(\tau) = 0$. Then the random variables $n(t)$ and $n(t + \tau)$ are uncorrelated, and for the gaussian process of interest to us, $n(t)$ and $n(t + \tau)$ are independent. Hence, if we should select some sample function, a knowledge of the value of $n(t)$ at time t would be of no assistance in improving our ability to predict the value attained by that same sample function at time $t + \tau$.

The physical fact about the noise, which is of principal concern in connection with communications systems, is that such noise has a power spectral density

$G(f)$ which is uniform over all frequencies. Such noise is referred to as "white" noise in analogy with the consideration that white light is a combination of all colors, that is, colors of all frequencies. Actually as is pointed out in Sec. 14.5, there is an upper-frequency limit beyond which the spectral density falls off sharply. However, this upper-frequency limit is so high that we may ignore it for our purposes.

Now, since the autocorrelation $R(\tau)$ and the power spectral density $G(f)$ are a Fourier transform pair, they have the properties of such pairs. Thus when $G(f)$ extends over a wide frequency range, $R(\tau)$ is restricted to a narrow range of τ. In the limit, if $G(f) = I$ (a constant) for all frequencies from $-\infty \leq f \leq +\infty$, then $R(\tau)$ becomes $R(\tau) = I\ \delta(\tau)$, where $\delta(\tau)$ is the delta function with $\delta(\tau) = 0$ except for $\tau = 0$. Since, then, for white noise, $R(\tau) = 0$ except for $\tau = 0$, Eq. (2.24-5) says that $n(t)$ and $n(t + \tau)$ are uncorrelated and hence independent, no matter how small τ.

2.25 POWER SPECTRAL DENSITY OF A SEQUENCE OF RANDOM PULSES

We shall occasionally need to have information about the power spectral density of a sequence of random pulses such as is indicated in Fig. 2.25-1. The pulses are of the same form but have random amplitudes and statistically independent random times of occurrence. The waveform (the random process) is stationary so that the statistical features of the waveforms are time invariant. Correspondingly, there is an invariant *average* time of separation T_s between pulses. We further assume that there is no overlap between pulses.

If the Fourier transform of a single sample pulse $P_1(t)$ is $P_1(f)$ then Parseval's theorem [Eq. (1.13-5)] states that the normalized energy of the pulse is

$$E_1 = \int_{-\infty}^{\infty} P_1(f)P_1^*(f)\ df = \int_{-\infty}^{\infty} |P_1(f)|^2\ df \tag{2.25-1}$$

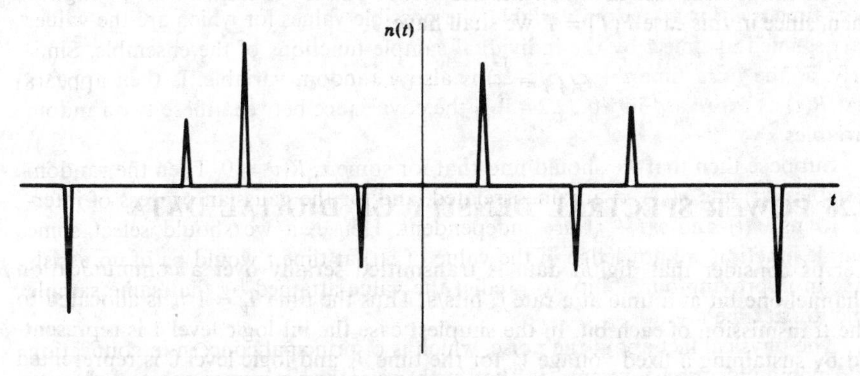

Figure 2.25-1 Pulses of random amplitude and time of occurrence.

The energy in the range df at a frequency f is

$$dE_1 = |P_1(f)|^2 \, df \tag{2.25-2}$$

Now consider a sequence of n successive pulses. Since we assume that the pulses do not overlap, the energy in the range df at the frequency f due to the n pulses is:

$$dE = dE_1 + dE_2 + \cdots + dE_n = \{|P_1(f)|^2 + |P_2(f)|^2 + \cdots + |P_n(f)|^2\} \, df \tag{2.25-3}$$

The average value $\overline{|P(f)|^2}$ of the sequence of n pulses is, by definition

$$\overline{|P(f)|^2} \equiv \frac{1}{n}\{|P_1(f)|^2 + |P_2(f)|^2 + \cdots + |P_n(f)|^2\} \tag{2.25-4}$$

so that dE in Eq. (2.25-3) can be written

$$dE = n\overline{|P(f)|^2} \, df \tag{2.25-5}$$

The average time of separation between pulses is T_s so that n pulses will occur in a time nT_s. The differential energy in the band df contained in the time interval nT_s is, from Eq. (2.25-5)

$$\frac{dE}{nT_s} = \frac{1}{nT_s} n\overline{|P(f)|^2} \, df = \frac{1}{T_s}\overline{|P(f)|^2} \, df \tag{2.25-6}$$

The power spectral density in the frequency range df is $G(f) = (dE/nT_s)/df$. Hence, from Eq. (2.25-6), $G(f)$ is:

$$G(f) = \frac{1}{T_s}\overline{|P(f)|^2} \tag{2.25-7}$$

Hence, whenever we make an observation or measurement of the pulse waveform which extends over a duration long enough so that the average observed pulse shape, such as their amplitudes, widths, and spacings are representative of the waveform generally, we shall find that Eq. (2.25-7) applies.

In the special case in which the individual pulses are impulses of strength I, then, since in this case $P(f) = I$, we shall have:

$$G(f) = \frac{I^2}{T_s} \qquad -\infty < f < \infty \tag{2.25-8}$$

2.26 POWER SPECTRAL DENSITY OF DIGITAL DATA

Let us consider that digital data is transmitted serially over a communication channel one bit at a time at a rate f_b bits/s. Thus the time $T_b = 1/f_b$ is allocated to the transmission of each bit. In the simplest case the bit logic level 1 is represented by sustaining a fixed voltage V_b for the time T_b and logic level 0 is represented by the fixed voltage $-V_b$ for the time T_b.

Digital transmission systems are usually *synchronous*, that is, they operate in conjunction with a *clock* waveform. A clock is a regular periodic waveform, whose period is equal to the bit time T_b. The clock serves many purposes in a digital system but, so far as our present concerns are involved, we need but note that the clock serves both to establish the duration T_b and also to mark the beginning and end of the interval allocated to each bit. In a transmission system there must be a clock waveform at both the sending end and at the receiving end and these clocks must be synchronized with one another.

In Fig. 2.26-1 we have indicated a clock waveform which marks off the bit intervals and we have also indicated two formats by which the bits (logic 1 and 0) are distinguished from one another. In (*a*) and (*b*) we show the scheme already referred to. Here logic 1 is represented by the voltage V_b held for the duration of a clock cycle and logic 0 by the voltage $-V_b$ held for an equal time. This scheme is characterized as a *non-return-to-zero* (NRZ) representation for a reason that will be apparent shortly. A second scheme identified by the name *biphase* is shown in (*c*) and (*d*). Here the digital bits are represented by waveforms which are at V_b for half the cycle and at $-V_b$ for the other half.

In Fig. 2.26-2 we again show the clock waveform and show also the waveform generated in the transmission of the sequence ...10111001... using the NRZ and the biphase formats. Observe that in the NRZ format, when a sequence of like bits is encountered, the signal sustains a fixed level $+V_b$ or $-V_b$, the

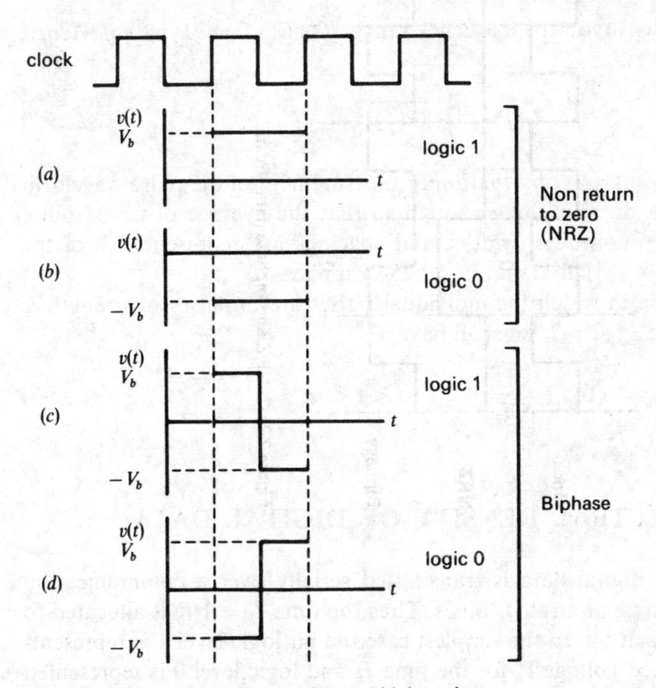

Figure 2.26-1 Clock waveform, NRZ and biphase data.

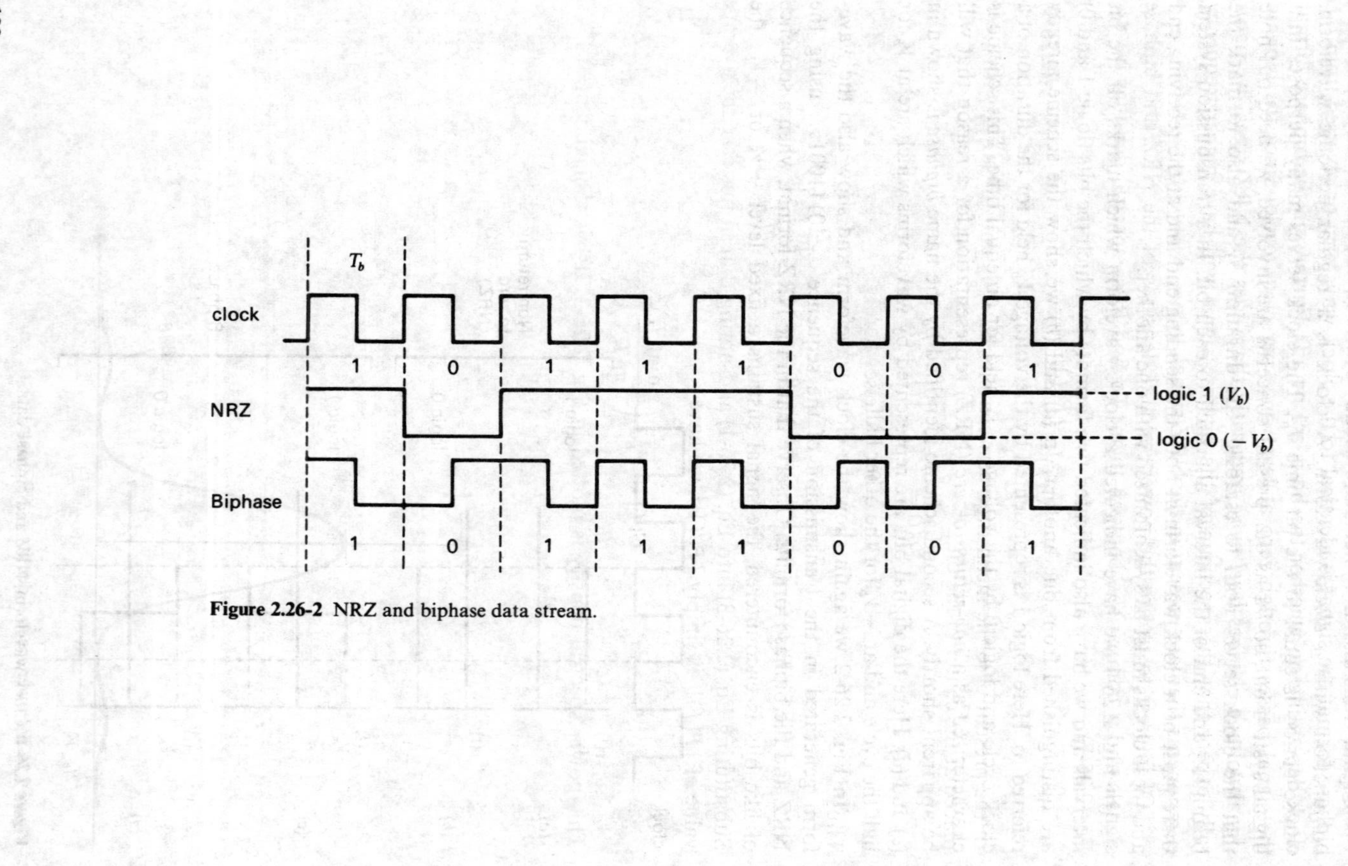

Figure 2.26-2 NRZ and biphase data stream.

voltage never returning to the zero voltage level, hence the name NRZ. In the biphase format the signal is identical to the clock waveform or reversed from the clock depending on whether the bit is a 1 or a 0. Further, in the biphase format the signal passes through zero at least once per clock cycle. Since the biphase signal voltage *changes level more frequently* than does the NRZ signal we naturally expect that the spectrum of the biphase signal will have higher frequency components than are present in the NRZ signal.

Of interest to us are the power spectral densities of the NRZ and biphase signals and a comparison between them. To find these spectral densities we use Eq. (2.25-7):

$$G(f) = \frac{\overline{|P(f)|^2}}{T_s} \tag{2.26-1}$$

In the present case the mean time between bits, $T_s = T_b$ and for NRZ, with $t = 0$ placed at the center of the bit,

$$\begin{aligned} p(t) &= \pm V_b \qquad |t| \le T_b/2 \\ &= 0 \qquad \text{elsewhere} \end{aligned} \tag{2.26-2}$$

Thus

$$P(f) = \pm V_b \int_{-T_b/2}^{T_b/2} e^{-j2\pi f t}\, dt = \pm V_b T_b \frac{\sin \pi f T_b}{\pi f T_b} \tag{2.26-3}$$

Substituting Eq. (2.26-3) into Eq. (2.26-1) and noting that in this case the average value of $|P(f)|^2$ is simply $|P(f)|^2$, we have

$$G(f) = V_b^2 T_b \left(\frac{\sin \pi f T_b}{\pi f T_b} \right)^2 \tag{2.26-4}$$

This power spectral density is plotted in Fig. 2.26-3. It has a maximum at $f = 0$

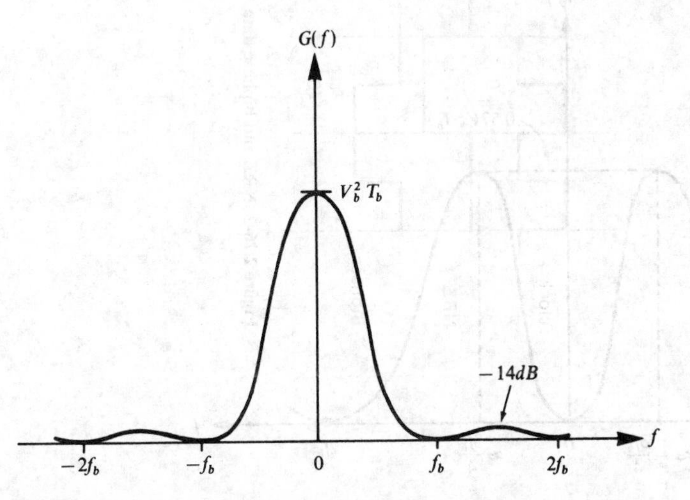

Figure 2.26-3 Power spectral density of NRZ data.

and is $G(f) = 0$ at all multiples of $f_b(= 1/T_b)$. The peak between f_b and $2f_b$ occurs at $f = 1.5f_b$ and is 14 dB lower than the peak at $f = 0$. It is left as a student exercise (by consulting a table of integrals) to verify that the total power in an NRZ signal is reduced by only 10 percent if the signal is passed through an ideal lower-pass filter with cutoff at $f = f_b$. Thus 90 percent of the power is in the main lobe centered around $f = 0$. It is to be noted however that the NRZ signal has no dc component. In the frequency range from $f = 0$ to $f = \pm\Delta f$ the power is $2G(f)\,\Delta f$ and in the limit as $\Delta f \to 0$ this power becomes zero. (If there were a dc component $G(f)$ would be impulsive at $f = 0$.) This result is, of course, to be anticipated since we would expect the NRZ signal to spend, on the average, equal time at the two levels.

In the biphase case we have

$$P(f) = \int_{-T_b/2}^{0} \pm V_b e^{-j2\pi f t}\,dt - \int_{0}^{T_b/2} \pm V_b e^{-j2\pi f t}\,dt \tag{2.26-5a}$$

$$= j(\pm V_b)T_b \frac{\sin^2 \pi f T_b/2}{\pi f T_b/2} \tag{2.26-5b}$$

so that

$$\overline{|P(f)|^2} = V_b^2 T_b^2 \left[\frac{\sin^2 \pi f T_b/2}{\pi f T_b/2}\right]^2 \tag{2.26-6}$$

and the power spectral density is

$$G(f) = V_b^2 T_b \left[\frac{\sin^2 \pi f T_b/2}{\pi f T_b/2}\right]^2 \tag{2.26-7}$$

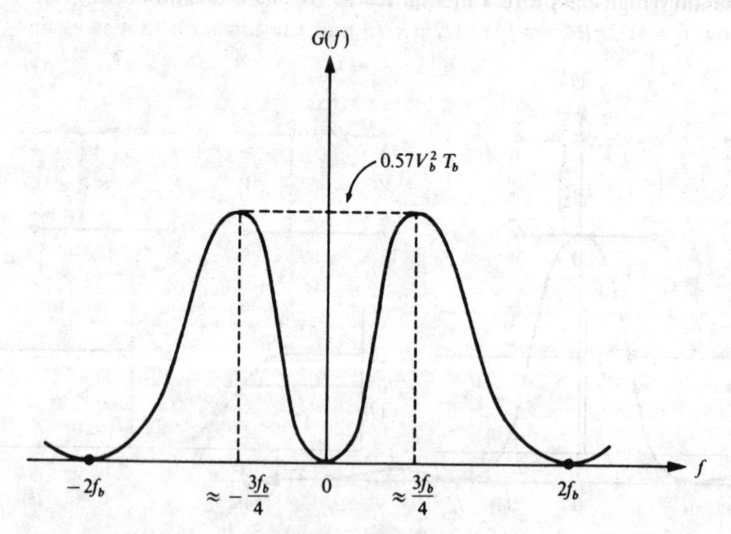

Figure 2.26-4 Power spectral density of biphase data.

$G(f)$, as given in Eq. (2.26-7) is plotted in Fig. 2.26-4. Note that here the principal lobes extend from $f = 0$ to $f = 2f_b$ and have peaks at approximately $\pm 3f_b/4$. At $f = 0$, $G(f) = 0$. It can be verified that if biphase data is transmitted through an ideal low-pass filter, with cutoff at frequency $2f_b$, 95 percent of the power will be passed. However, if the filter cuts off at f_b then only approximately 70 percent of the power is passed.

2.27 EFFECT OF RUDIMENTARY FILTERS ON DIGITAL DATA

In the previous section we noted that NRZ data has a power spectral density whose principal lobe has a peak at $f = 0$ while for biphase data the principal lobe has a peak at $f \simeq 3f_b/4$. We therefore anticipate that, for effective transmission of NRZ data, greater demands are made on the ability of the channel to transmit lower frequencies than for biphase data. To explore the matter in an easy manner we consider now the effect on digital transmission of the most rudimentary high-pass and low-pass filters.

The rudimentary low-pass filters are shown in Fig. 2.27-1*a* and *b*. They are characterized in the frequency domain, as shown in Fig. 2.27-1*c* by a 3-dB frequency f_1 at which the transmission ratio V_o/V_i falls to $0.707 = (1/\sqrt{2})$. In the *R-C* circuit $f_1 = 1/2\pi RC$ and in the *R-L* circuit $f_1 = 1/2\pi(R/L)$. In the time domain, as indicated in Fig. 2.27-1*d* the filters are characterized by a time constant τ which is $\tau = RC$ or $\tau = L/R$. In either case $f_1\tau = 1/2\pi$. The response of the filters to an applied step voltage is, as shown, an exponential decay to zero voltage with time constant τ.

The rudimentary high-pass filters are shown in Fig. 2.27-1*e* and *f*. The 3-dB frequency is now $f_2 = 1/2\pi RC$ or $f_2 = 1/2\pi(R/L)$ and the time constant is again

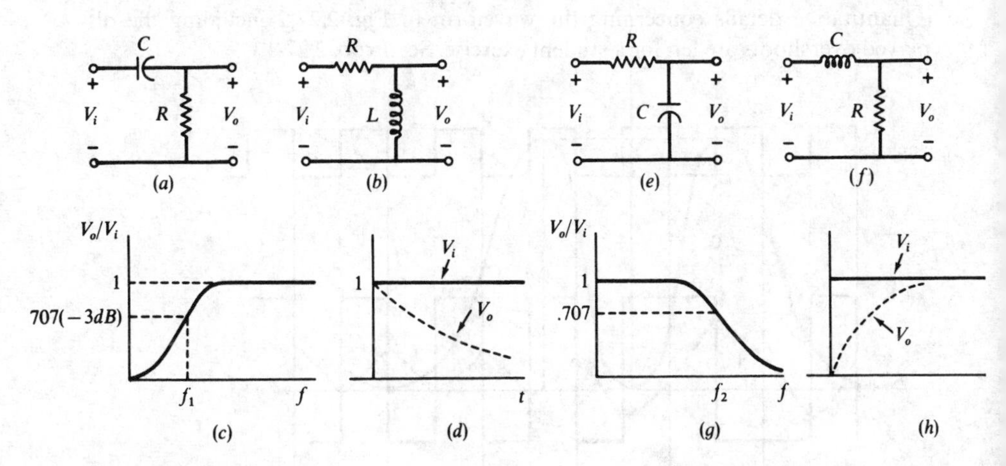

Figure 2.27-1 Response of filters.

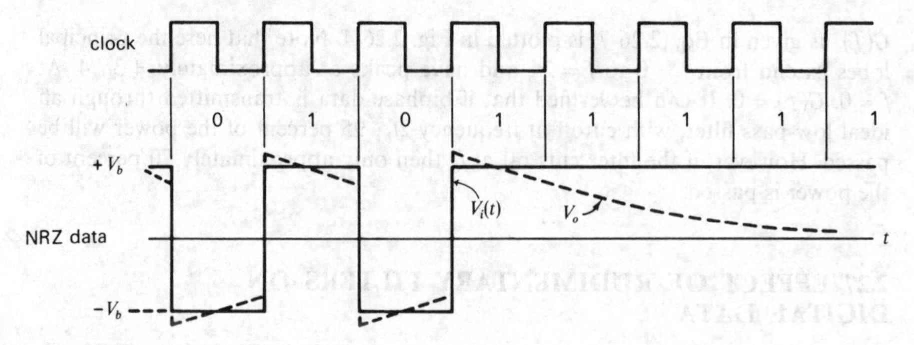

Figure 2.27-2 Tilt of NRZ signal produced by high-pass filter.

$\tau = RC$ or $\tau = L/R$. The response to a step voltage is an exponential rise to the voltage of the step with a time constant τ.

In Fig. 2.27-2, against the timing reference of a clock waveform, we have drawn an NRZ data stream which, at the transmitter, consists of alternating 1's and 0's and then a sustained sequence of 1's. We consider that the channel can be approximated by the high-pass filter shown in Fig. 2.27-1*a*. The dashed waveform is the received signal response. Note that when the transmitted signal V_i sustains a fixed level $+V_b$ or $-V_b$ the received signal V_o decays toward zero. So long as the fixed level is sustained for only one clock interval and assuming, as we have, that τ and T_b are comparable, there is no serious deterioration of the signal. What starts out as a flat top develops a tilt but there will clearly be no difficulty in distinguishing a bit of logic level 1 from a bit of logic level 0. On the other hand, after a long sequence of 1's (or 0's) the received signal may approach so close to zero that, particularly in the presence of noise, a correct reading of the bit is not certain. This problem would not develop in biphase transmission where the transmitted waveform never sustains a fixed level for more than one clock cycle. (Quantitative details concerning the waveform in Fig. 2.27-2 including the displayed overshoots are left for a student exercise. See Prob. 2.27-1.)

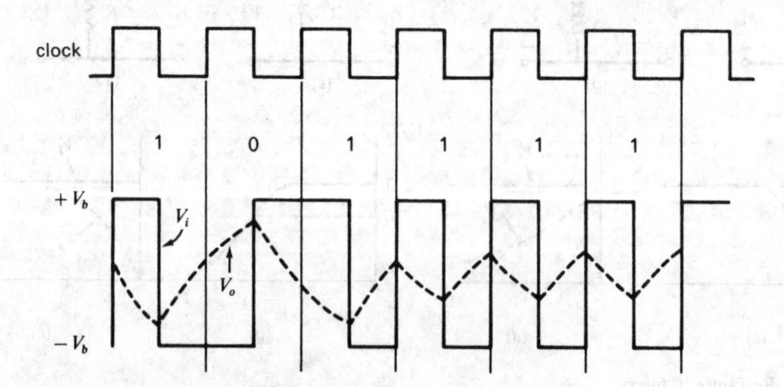

Figure 2.27-3 Effect of NRZ passed through a low-pass filter.

In Fig. 2.27-3 we show a biphase data stream consisting of alternating 1's and 0's and then a sustained sequence of 1's. Let us here consider that the channel can be approximated by the low-pass filter shown in Fig. 2.27-1*e*. Again we see that the waveform is most adversely affected during a sequence of similar bits. Between transitions of the original waveform V_i the received waveform V_o heads asymptotically toward $+V_b$ or $-V_b$. When 1's and 0's alternate this asymptotic rise or fall can persist for a full clock period T_b. When 1's or 0's are sustained available time is reduced to one-half period.

Altogether, then, it appears that if a channel is able to transmit *direct current* and is limited at high frequencies, NRZ transmission may be advantageous. If, however, *dc* transmission is not possible, biphase transmission will be needed and we shall also have to take into account that a higher frequency cutoff will have to be provided. We may also note, incidentally, that the low-pass filter of Fig. 2.27-1*b* serves as an equivalent circuit for devices that involve magnetic circuits such as transformers, tape recorders, etc.

2.28 THE COMPLEMENTARY ERROR FUNCTION

$$\text{erfc}(x) \equiv \frac{2}{\sqrt{\pi}} \int_x^\infty e^{-u^2}\, du$$

x†	erfc (x)	x	erfc (x)
0.0	1.000	2.2	1.86×10^{-3}
0.2	0.777	2.4	6.9×10^{-4}
0.4	0.572	2.6	2.4×10^{-4}
0.6	0.396	2.8	7.5×10^{-5}
0.8	0.258	3.0	2.2×10^{-5}
1.0	0.157	3.3	3.06×10^{-6}
1.2	8.97×10^{-2}	3.7	1.67×10^{-7}
1.4	4.77×10^{-2}	4.0	1.54×10^{-8}
1.6	2.37×10^{-2}	5.0	1.54×10^{-12}
1.8	1.09×10^{-2}		
2.0	4.38×10^{-3}		

† For large values of x, erfc $(x) \simeq e^{-x^2}/(x\sqrt{\pi})$

REFERENCES

1. Mood, A., and F. Graybill: "Introduction to the Theory of Statistics," McGraw-Hill Book Company, New York, 1963.
2. Peirce, B. O.: "A Short Table of Integrals," Ginn and Company, Boston, 1956.
3. Papoulis, A.: "Probability, Random Variables, and Stochastic Processes," McGraw-Hill Book Company, 1965.
4. Wozencraft, J., and I. Jacobs: "Communications Engineering," John Wiley and Sons, New York, 1966.

PROBLEMS

2.1-1. Six dice are thrown simultaneously. What is the probability that at least 1 die shows a 3?

2.1-2. A card is drawn from a deck of 52 cards.

(*a*) What is the probability that a 2 is drawn?
(*b*) What is the probability that a 2 of clubs is drawn?
(*c*) What is the probability that a spade is drawn?

2.2-1. A card is picked from each of four 52-card decks of cards.

(*a*) What is the probability of selecting at least one 6 of spades?
(*b*) What is the probability of selecting at least 1 card larger than an 8?

2.3-1. A card is drawn from a 52-card deck, and without replacing the first card a second card is drawn. The first and second cards are not replaced and a third card is drawn.

(*a*) If the first card is a heart, what is the probability of the second card being a heart?
(*b*) If the first and second cards are hearts, what is the probability that the third card is the king of clubs?

2.3-2. Two factories produce identical clocks. The production of the first factory consists of 10,000 clocks of which 100 are defective. The second factory produces 20,000 clocks of which 300 are defective. What is the probability that a particular defective clock was produced in the first factory?

2.3-3. One box contains two black balls. A second box contains one black and one white ball. We are told that a ball was withdrawn from one of the boxes and that it turned out to be black. What is the probability that this withdrawal was made from the box that held the two black balls?

2.4-1. Two dice are tossed.

(*a*) Find the probability of a 3 and a 4 appearing.
(*b*) Find the probability of a 7 being rolled.

2.4-2. A card is drawn from a deck of 52 cards, then replaced, and a second card drawn.

(*a*) What is the probability of drawing the same card twice?
(*b*) What is the probability of first drawing a 3 of hearts and then drawing a 4 of spades?

2.6-1. A die is tossed and the number appearing is n_i. Let N be the random variable identifying the outcome of the toss defined by the specifications $N = n_i$ when n_i appears. Make a plot of the probability $P(N \leq n_i)$ as a function of n_i.

2.6-2. A coin is tossed four times. Let H be the random variable which identifies the number of heads which occur in these four tosses. It is defined by $H = h$, where h is the number of heads which appear. Make a plot of the probability $P(H \leq h)$ as a function of h.

2.6-3. A coin is tossed until a head appears. Let T be the random variable which identifies the number of tosses t required for the appearance of this first head. Make a plot of the probability $P(T \leq t)$ as a function of t up to $t = 5$.

2.7-1. An important probability density function is the Rayleigh density

$$f(x) = \begin{cases} xe^{-x^2/2} & x \geq 0 \\ 0 & x < 0 \end{cases}$$

(*a*) Prove that $f(x)$ satisfies Eqs. (2.7-2) and (2.7-3).
(*b*) Find the distribution function $F(x)$.

2.8-1. Refer to Fig. 2.6-1.

(*a*) Find $P(2 < n \leq 11)$.
(*b*) Find $P(2 \leq n < 11)$.
(*c*) Find $P(2 \leq n \leq 11)$.
(*d*) Find $F(9)$.

2.8-2. Refer to the Rayleigh density function given in Prob. 2.7-1. Find the probability $P(x_1 < x \leq x_2)$, where $x_2 - x_1 = 1$, so that $P(x_1 < x \leq x_2)$ is a maximum. *Hint:* Find $P(x_1 < x \leq x_2)$; replace x_2 by $1 + x_1$, and maximize P with respect to x_1.

2.8-3. Refer to the Rayleigh density function given in Prob. 2.7-1. Find
(*a*) $P(0.5 < x \le 2)$.
(*b*) $P(0.5 \le x < 2)$.

2.9-1. The joint probability density of the random variables X and Y is $f(x, y) = ke^{-(x+y)}$ in the range $0 \le x \le \infty, 0 \le y \le \infty$, and $f(x, y) = 0$ otherwise.
(*a*) Find the value of the constant k.
(*b*) Find the probability density $f(x)$, the probability density of X independently of Y.
(*c*) Find the probability $P(0 \le X \le 2; 2 \le Y \le 3)$.
(*d*) Are the random variables dependent or independent?

2.9-2. X is a random variable having a gaussian density. $E(X) = 0$, $\sigma_x^2 = 1$. V is a random variable having the values 1 or -1, each with probability $\frac{1}{2}$.
(*a*) Find the joint density $f_{X,V}(x, v)$.
(*b*) Show that $f_V(v) = \int_{-\infty}^{\infty} f_{XV}(x, v)\, dx$.

2.9-3. The joint probability density of the random variables X and Y is $f(x, y) = xe^{-x(y+1)}$ in the range $0 \le x \le \infty, 0 \le y \le \infty$, and $f(x, y) = 0$ otherwise.
(*a*) Find $f(x)$ and $f(y)$, the probability density of X independently of Y and Y independently of X.
(*b*) Are the random variables dependent or independent?

2.10-1. (*a*) In a communication channel as represented in Fig. 2.10-1 $P(r_0 | m_0) = 0.9$, $P(r_1 | m_0) = 0.1$, $P(r_0 | m_1) = 0.4$, and $P(r_1 | m_1) = 0.6$. On a single set of coordinate axes make plots of $P(m_0 | r_0)$, $P(m_1 | r_0)$, $P(m_0 | r_1)$ and $P(m_1 | r_1)$ as a function of $P(m_0)$. Mark the range of m_0 for which the algorithm of Eq. (2.10-1) prescribes that we choose m_0 if r_0 is received and m_1 if r_1 is received, the range for which we choose m_0 no matter what is received and the range for which we choose m_1 no matter what is received.
(*b*) Calculate and plot the probability of error as a function of $P(m_0)$.

2.10-2. (*a*) For the channel and message probabilities given in Fig. P2.10-2 determine the best decisions about the transmitted message for each possible received response.
(*b*) With decisions made as in part (*a*) calculate the probability of error.
(*c*) Suppose the decision-making apparatus at the receiver were inoperative so that at the receiver nothing could be determined except that a message had been received. What would be the best strategy for determining what message had been transmitted and what would be the corresponding error probability?

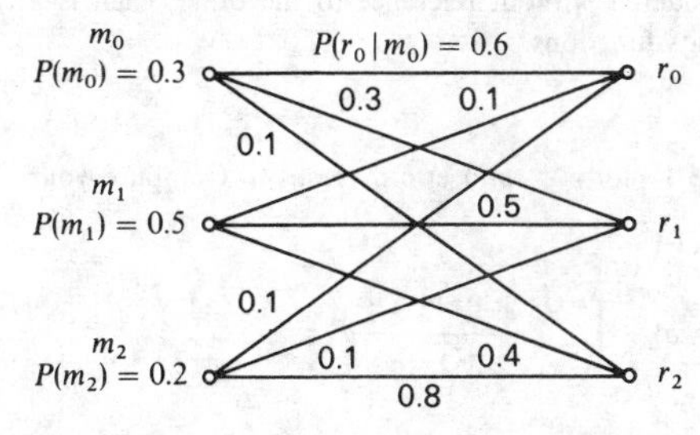

Figure P2.10-2

2.11-1. If $f_X(x) = \dfrac{1}{\sqrt{2\pi}} e^{-x^2/2}$ for all x, show that
(*a*) $E(X^{2n}) = 1 \cdot 3 \cdot 5 \cdots (n-1), n = 1, 2, \ldots$.
(*b*) $E(X^{2n-1}) = 0, n = 1, 2, \ldots$.

2.11-2. Compare the most probable [$f(x)$ is a maximum] and the average value of X when

(*a*) $f_{X_1}(x) = \dfrac{1}{\sqrt{2\pi}} e^{-(x-m)^2/2}$ for all x

(*b*) $f_{X_2}(x) = \begin{cases} xe^{-x^2/2} & \text{for } x \ge 0 \\ 0 & \text{elsewhere} \end{cases}$

2.12-1. Calculate the variance of the random variables having densities:

(a) The gaussian density $f_{X_1}(x) = \dfrac{1}{\sqrt{2\pi}} e^{-(x-m)^2/2}$, all x.

(b) The Rayleigh density $f_{X_2}(x) = xe^{-x^2/2}$, $x \geq 0$.

(c) The uniform density $f_{X_3}(x) = 1/a$, $-a/2 \leq x \leq a/2$.

2.12-2. Consider the Cauchy density function

$$f(x) = \frac{K}{1 + x^2} \qquad -\infty \leq x \leq \infty$$

(a) Find K so that $f(x)$ is a density function.

(b) Find $E(X)$.

(c) Find the variance of X. Comment on the significance of this result.

2.12-3. The random variable X has a variance σ^2 and a mean m. The random variable Y is related to X by $Y = aX + b$, where a and b are constants. Find the mean and variance of Y.

2.13-1. A probability density function $f(x)$ is uniform over the range from $-L$ to $+L$.

(a) Calculate and plot the probability $P[|x| \leq \varepsilon]$ as a function of ε.

(b) Calculate and plot the quantity σ^2/ε^2 and verify that Tchebycheff's rule is valid in this case.

2.14-1. Refer to the gaussian density given in Eq. (2.14-1).

(a) Show that $E((X - m)^{2n-1}) = 0$.

(b) Show that $E((X - m)^{2n}) = 1 \cdot 3 \cdot 5 \cdots (n - 1)\sigma^2$.

2.14-2. Given a gaussian probability function $f(x)$ of mean value zero and variance σ^2.

(a) As a function of σ, plot the probability $P[|x| \geq \varepsilon]$.

(b) Calculate and plot the quantity σ^2/ε^2 and verify that Tchebycheff's rule is valid in this case.

2.14-3. A random variable $V = b + X$, where X is a gaussian distributed random variable with mean 0 and variance σ^2, and b is a constant. Show that V is a gaussian distributed random variable with mean b and variance σ^2.

2.14-4. The joint density function of two dependent variables X and Y is

$$f(x, y) = \frac{1}{\pi\sqrt{3}} e^{-2(x^2 - xy + y^2)/3}$$

(a) Show that, when X and Y are each considered without reference to the other, each is a gaussian variable, i.e., $f(x)$ and $f(y)$ are gaussian density functions.

(b) Find σ_x^2 and σ_y^2.

2.15-1. Obtain values for and plot erf u versus u.

2.15-2. On the same set of axes used in Prob. 2.15-1 plot e^{-u^2} and erfc u versus u. Compare your results.

2.15-3. The probability

$$P_{\pm k\sigma} \equiv P(m - k\sigma \leq X \leq m + k\sigma) = \int_{m-k\sigma}^{m+k\sigma} \frac{e^{-(x-m)^2/2\sigma^2}}{\sqrt{2\pi}\,\sigma} dx$$

(a) Change variables by letting $u = x - m/\sqrt{2}\,\sigma$.

(b) Show that $P_{\pm k\sigma} = \text{erf}\,(k/\sqrt{2})$.

2.16-1. Show that a random variable with a Rayleigh density as in Eq. (2.16-1) has a mean value $R = \sqrt{(\pi/2)}\,\alpha$, a mean square value $R^2 = 2\alpha^2$, and a variance $\sigma^2 = (2 - \pi/2)\alpha^2$.

2.16-2. (a) A voltage V is a function of time t and is given by

$$V(t) = X \cos \omega t + Y \sin \omega t$$

in which ω is a constant angular frequency and X and Y are independent gaussian variables each with zero mean and variance σ^2. Show that $V(t)$ may be written

$$V(t) = R \cos (\omega t + \Theta)$$

in which R is a random variable with a Rayleigh probability density and Θ is a random variable with uniform density.

(*b*) If $\sigma^2 = 1$, what is the probability that $R \geq 1$?

2.17-1. Derive Eq. (2.17-6) directly from the definition $\sigma_z^2 = E\{(Z - m_z)^2\}$.

2.17-2. $Z = X_1 + X_2 + \cdots + X_N$, $E(X_i) = m$.

(*a*) Find $E(Z)$.

(*b*) If $E(X_i X_j) = \begin{cases} 1 & j = i \\ \rho & j = i \pm 1 \\ 0 & \text{otherwise} \end{cases}$

find (1) $E(Z^2)$ and (2) σ_z^2.

2.18-1. The independent random variables X and Y are added to form Z. If

$$f_X(x) = xe^{-x^2/2} \qquad 0 \leq x \leq \infty \qquad \text{and} \qquad f_Y(y) = \tfrac{1}{2}e^{-|y|} \qquad |y| < \infty$$

find $f_Z(z)$.

2.18-2. The independent random variables X and Y have the probability densities

$$f(x) = e^{-x} \qquad 0 \leq x \leq \infty$$

$$f(y) = 2e^{-2y} \qquad 0 \leq y \leq \infty$$

Find and plot the probability density of the variable $Z = X + Y$.

2.18-3. The random variable X has a probability density uniform in the range $0 \leq x \leq 1$ and zero elsewhere. The independent variable Y has a density uniform in the range $0 \leq x \leq 2$ and zero elsewhere. Find and plot the density of $Z = X + Y$.

2.18-4. The N independent gaussian random variables $X_1, \ldots, X_N$ are added to form Z. If the mean of X_i is 1 and its variance is 1, find $f_Z(z)$.

2.19-1. Two gaussian random variables X and Y, each with mean zero and variance σ^2, between which there is a correlation coefficient ρ, have a joint probability density given by

$$f(x, y) = \frac{1}{2\pi\sigma^2\sqrt{1 - \rho^2}} \exp - \left[\frac{x^2 - 2\rho xy + y^2}{2\sigma^2(1 - \rho^2)}\right]$$

(*a*) Verify that the symbol ρ in the expression for $f(x, y)$ is indeed the correlation coefficient. That is, evaluate $E\{XY\}/\sigma^2$ and show that the result is ρ as required by Eq. (2.19-5).

(*b*) Show that the case $\rho = 0$ corresponds to the circumstance where X and Y are independent.

2.19-2. The random variables X and Y are related to the random variable Θ by $X = \sin \Theta$ and $Y = \cos \Theta$. The variable Θ has a uniform probability density in the range from 0 to 2π. Show that X and Y are not independent but that, nonetheless, $E(XY) = 0$ so that they are uncorrelated.

2.19-3. The random variables $X_1, X_2, X_3, \ldots$ are dependent but uncorrelated. $Z = X_1 + X_2 + X_3 + \cdots$. Show that $\sigma_Z^2 = \sigma_1^2 + \sigma_2^2 + \sigma_3^2 + \cdots$.

2.20-1. The random variables X_1 X_2, X_3 are independent and each has a uniform probability density in the range $0 \leq x \leq 1$. Find and plot the probability density of $X_1 + X_2$ and of $X_1 + X_2 + X_3$.

2.21-1. Verify Eq. (2.21-16).

2.21-2. In a communication system used to transmit a sequence of messages, it is known that the error probability is 4×10^{-5}. A sample survey of N messages is made. It is required that, with a probability not to exceed 0.05, the error rate in the sample is not to be larger than 5×10^{-5}. How many messages must be included in the sample?

2.22-1. A communication channel transmits, in random order, two messages m_1 and m_2. The message m_1 occurs three times more frequently than m_2. Message m_1 generates a receiver response $r_1 = -1V$ while m_2 generates $r_2 = +1V$. The channel is corrupted by noise n with a uniform probability density which extends from $n = -1.5V$ to $n = +1.5V$.

(*a*) Find the probability that m_1 is mistaken for m_2 and the probability that m_2 is mistaken for m_1.

(*b*) What is the probability that the receiver will determine a message correctly?

2.22-2. A communication channel transmits, in random order, two messages m_1 and m_2 with equal likelihood. Message m_1 generates response $r_1 = -1V$ at the receiver and message m_2 generates $r_2 = +1V$. The channel is corrupted with gaussian noise with variance $\sigma^2 = 1$ volt^2. Find the probability that the receiver will determine a message correctly.

2.22-3. A communication channel transmits, in random order, two messages m_1 and m_2. The message m_1 occurs three times more frequently than m_2. Message m_1 generates a receiver response $r_1 = +1V$ while m_2 generates $r_2 = -1V$. The channel is corrupted by noise n whose probability density has the triangular form shown in Fig. 2.22-2*a* with $f_N(n) = 0$ at $f_N(n) = -2V$ and at $f_N(n) = +2V$.

(*a*) What are the ranges of r for which the decision is to be made that m_1 was transmitted and for which the decision is to be made that m_2 was transmitted?

(*b*) What is the probability that the message will be read correctly?

2.23-1. The function of time $Z(t) = X_1 \cos \omega_0 t - X_2 \sin \omega_0 t$ is a random process. If X_1 and X_2 are independent gaussian random variables each with zero mean and variance σ^2, find

(*a*) $E(Z)$, $E(Z^2)$, σ_z^2, and

(*b*) $f_Z(z)$.

2.23-2. $Z(t) = M(t) \cos (\omega_0 t + \Theta)$. $M(t)$ is a random process, with $E(M(t)) = 0$ and $E(M^2(t)) = M_0^2$.

(*a*) If $\Theta = 0$, find $E(Z^2)$. Is $Z(t)$ stationary?

(*b*) If Θ is an independent random variable such that $f_\Theta(\theta) = 1/2\pi$, $-\pi \leq \theta \leq \pi$, show that $E(Z^2(t)) = E(M^2(t))E(\cos^2 (\omega_0 t + \Theta)) = M_0^2/2$. Is $Z(t)$ now stationary?

2.24-1. Refer to Prob. 2.23-1. Find $R_z(\tau)$.

2.24.2. A random process $n(t)$ has a power spectral density $G(f) = \eta/2$ for $-\infty \leq f \leq \infty$. The random process is passed through a low-pass filter which has a transfer function $H(f) = 2$ for $-f_M \leq f \leq f_M$ and $H(f) = 0$ otherwise. Find the power spectral density of the waveform at the output of the filter.

2.24-3. White noise $n(t)$ with $G(f) = \eta/2$ is passed through a low-pass *RC* network with a 3-dB frequency f_c.

(*a*) Find the autocorrelation $R(\tau)$ of the output noise of the network.

(*b*) Sketch $\rho(\tau) = R(\tau)/R(0)$.

(*c*) Find $\omega_c \tau$ such that $\rho(\tau) \leq 0.1$.

2.25-1. Consider a train of rectangular pulses. The kth pulse has a width τ and a height A_k. A_k is a random variable which can have the values 1, 2, 3, ... 10 with equal probability. Assuming statistical independence between amplitudes and assuming that the average separation between pulses is T_s, find the power spectral density $G_n(f)$ of the pulse train.

2.25-2. A pulse train consists of rectangular pulses having an amplitude of 2 volts and widths which are either 1 μs or 2 μs with equal probability. The mean time between pulses is 5 μs. Find the power spectral density $G_n(f)$ of the pulse train.

2.26-1. Consider the power spectral density of an NRZ waveform as given by Eq. (2.26-4) and as shown in Fig. 2.26-3. By consulting a table of integrals, show that the power of the NRZ waveform is reduced by only 10 percent if the waveform is passed through an ideal low-pass filter with cutoff at $f = f_b$.

2.26-2. Consider the power spectral density of a biphase waveform as given by Eq. (2.26-7) and as shown in Fig. 2.26-4. By consulting a table of integrals, show that, if the waveform is passed through an ideal low-pass filter with cutoff at $2f_b$, 95 percent of the power will be passed. Show also that, if the filter cutoff is set at f_b, then only 70 percent of the power is transmitted.

2.27-1. An NRZ waveform consists of alternating 0's and 1's. The bit interval is 1 μs and the waveform makes excursions between $+1V$ and $-1V$. The waveform is transmitted through an *RC* high-pass filter of time constant 1 μs. Draw the output waveform and calculate numerical values of all details of the waveform.

2.27-2. An NRZ waveform consists of alternating 0's and 1's. The bit interval is 1 μs and the waveform makes excursions between $+1V$ and $-1V$. The waveform is transmitted through an *RC* low-pass filter of time constant 1 μs. Draw the output waveform and calculate numerical values of all details of the waveform.

CHAPTER

THREE

AMPLITUDE-MODULATION SYSTEMS

One of the basic problems of communication engineering is the design and analysis of systems which allow many individual messages to be transmitted simultaneously over a single communication channel. A method by which such multiple transmission, called *multiplexing*, may be achieved consists in translating each message to a different position in the *frequency* spectrum. Such multiplexing is called *frequency multiplexing*. The individual message can eventually be separated by filtering. Frequency multiplexing involves the use of an auxiliary waveform, usually sinusoidal, called a *carrier*. The operations performed on the signal to achieve frequency multiplexing result in the generation of a waveform which may be described as the carrier modified in that its amplitude, frequency, or phase, individually or in combination, varies with time. Such a modified carrier is called a *modulated* carrier. In some cases the modulation is related simply to the message; in other cases the relationship is quite complicated. In this chapter, we discuss the generation and characteristics of amplitude-modulated carrier waveforms.[1]

3.1 FREQUENCY TRANSLATION

It is often advantageous and convenient, in processing a signal in a communications system, to translate the signal from one region in the frequency domain to another region. Suppose that a signal is bandlimited, or nearly so, to the frequency range extending from a frequency f_1 to a frequency f_2. The process of frequency translation is one in which the original signal is replaced with a new signal whose spectral range extends from f'_1 to f'_2 and which *new* signal bears, in recoverable form, the same *information* as was borne by the original signal. We discuss now a number of useful purposes which may be served by frequency translation.

Frequency Multiplexing

Suppose that we have several different signals, all of which encompass the same spectral range. Let it be required that all these signals be transmitted along a single communications channel in such a manner that, at the receiving end, the signals be separately recoverable and distinguishable from each other. The single channel may be a single pair of wires or the free space that separates one radio antenna from another. Such multiple transmissions, i.e., multiplexing, may be achieved by translating each one of the original signals to a different frequency range. Suppose, say, that one signal is translated to the frequency range f'_1 to f'_2, the second to the range f''_1 to f''_2, and so on. If these new frequency ranges do not overlap, then the signal may be separated at the receiving end by appropriate bandpass filters, and the outputs of the filters processed to recover the original signals.

Practicability of Antennas

When free space is the communications channel, antennas radiate and receive the signal. It turns out that antennas operate effectively only when their dimensions are of the order of magnitude of the wavelength of the signal being transmitted. A signal of frequency 1 kHz (an audio tone) corresponds to a wavelength of 300,000 m, an entirely impractical length. The required length may be reduced to the point of practicability by translating the audio tone to a higher frequency.

Narrowbanding

Returning to the matter of the antenna, just discussed, suppose that we wanted to transmit an audio signal directly from the antenna, and that the inordinate length of the antenna were no problem. We would still be left with a problem of another type. Let us assume that the audio range extends from, say, 50 to 10^4 Hz. The ratio of the highest audio frequency to the lowest is 200. Therefore, an antenna suitable for use at one end of the range would be entirely too short or too long for the other end. Suppose, however, that the audio spectrum were translated so that it occupied the range, say, from $(10^6 + 50)$ to $(10^6 + 10^4)$ Hz. Then the ratio of highest to lowest frequency would be only 1.01. Thus the processes of frequency translation may be used to change a "wideband" signal into a "narrowband" signal which may well be more conveniently processed. The terms "wideband" and "narrowband" are being used here to refer not to an absolute range of frequencies but rather to the fractional change in frequency from one band edge to the other.

Common Processing

It may happen that we may have to process, in turn, a number of signals similar in general character but occupying different spectral ranges. It will then be necessary, as we go from signal to signal, to adjust the frequency range of our processing apparatus to correspond to the frequency range of the signal to be processed. If the processing apparatus is rather elaborate, it may well be wiser to leave the processing apparatus to operate in some fixed frequency range and instead to translate the frequency range of each signal in turn to correspond to this fixed frequency.

3.2 A METHOD OF FREQUENCY TRANSLATION

A signal may be translated to a new spectral range by *multiplying* the signal with an auxiliary sinusoidal signal. To illustrate the process, let us consider initially that the signal is sinusoidal in waveform and given by

$$v_m(t) = A_m \cos \omega_m t = A_m \cos 2\pi f_m t \tag{3.2-1a}$$

$$= \frac{A_m}{2}(e^{j\omega_m t} + e^{-j\omega_m t}) = \frac{A_m}{2}(e^{j2\pi f_m t} + e^{-j2\pi f_m t}) \tag{3.2-1b}$$

in which A_m is the constant amplitude and $f_m = \omega_m/2\pi$ is the frequency. The two-sided spectral amplitude pattern of this signal is shown in Fig. 3.2-1*a*. The pattern consists of two lines, each of amplitude $A_m/2$, located at $f = f_m$ and at $f = -f_m$. Consider next the result of the multiplication of $v_m(t)$ with an auxiliary sinusoidal signal

$$v_c(t) = A_c \cos \omega_c t = A_c \cos 2\pi f_c t \tag{3.2-2a}$$

$$= \frac{A_c}{2}(e^{j\omega_c t} + e^{-j\omega_c t}) = \frac{A_c}{2}(e^{j2\pi f_c t} + e^{-j2\pi f_c t}) \tag{3.2-2b}$$

in which A_c is the constant amplitude and f_c is the frequency. Using the trigonometric identity $\cos\alpha \cos\beta = \frac{1}{2}\cos(\alpha+\beta) + \frac{1}{2}\cos(\alpha-\beta)$, we have for the product $v_m(t)v_c(t)$

$$v_m(t)v_c(t) = \frac{A_m A_c}{2}[\cos(\omega_c + \omega_m)t + \cos(\omega_c - \omega_m)t] \tag{3.2-3a}$$

$$= \frac{A_m A_c}{4}(e^{j(\omega_c+\omega_m)t} + e^{-j(\omega_c+\omega_m)t} + e^{j(\omega_c-\omega_m)t} + e^{-j(\omega_c-\omega_m)t}) \tag{3.2-3b}$$

The new spectral amplitude pattern is shown in Fig. 3.2-1*b*. Observe that the two original spectral lines have been *translated*, both in the positive-frequency direction by amount f_c and also in the negative-frequency direction by the same amount. There are now four spectral components resulting in two sinusoidal waveforms, one of frequency $f_c + f_m$ and the other of frequency $f_c - f_m$. Note that

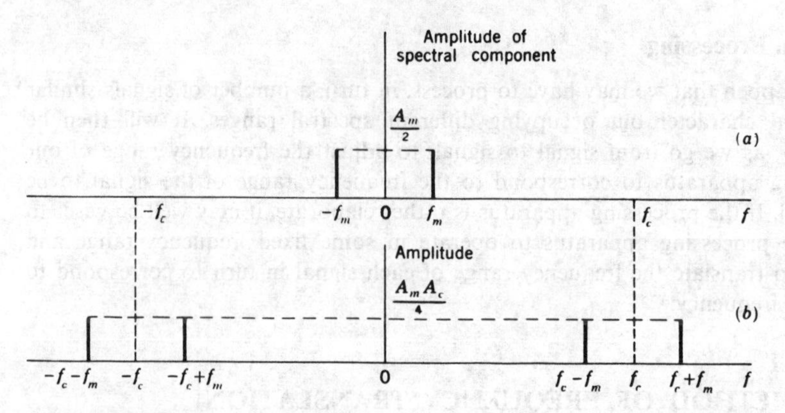

Figure 3.2-1 (*a*) Spectral pattern of the waveform $A_m \cos \omega_m t$. (*b*) Spectral pattern of the product waveform $A_m A_c \cos \omega_m t \cos \omega_c t$.

while the product signal has four spectral components each of amplitude $A_m A_c/4$, there are only two frequencies, and the amplitude of each sinusoidal component is $A_m A_c/2$.

A generalization of Fig. 3.2-1 is shown in Fig. 3.2-2. Here a signal is chosen which consists of a superposition of four sinusoidal signals, the highest in frequency having the frequency f_M. Before translation by multiplication, the two-sided spectral pattern displays eight components centered around zero frequency. After multiplication, we find this spectral pattern translated both in the positive- and the negative-frequency directions. The 16 spectral components in this two-sided spectral pattern give rise to eight sinusoidal waveforms. While the original signal extends in range up to a frequency f_M, the signal which results from multiplication has sinusoidal components covering a range $2f_M$, from $f_c - f_M$ to $f_c + f_M$.

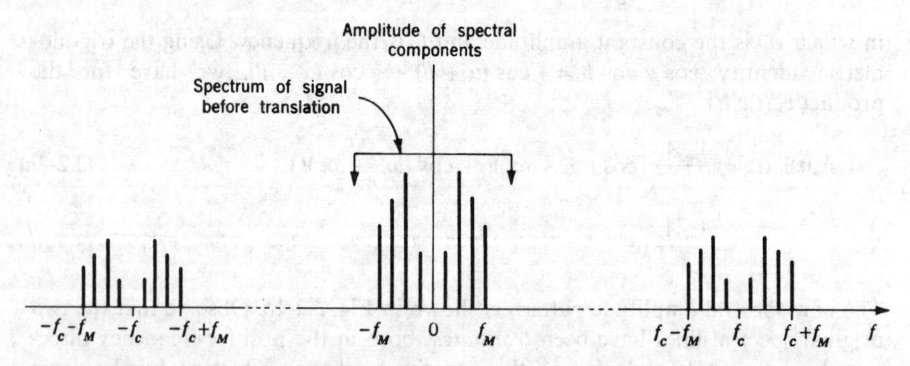

Figure 3.2-2. An original signal consisting of four sinusoids of differing frequencies is translated through multiplication and becomes a signal containing eight frequencies symmetrically arranged about f_c.

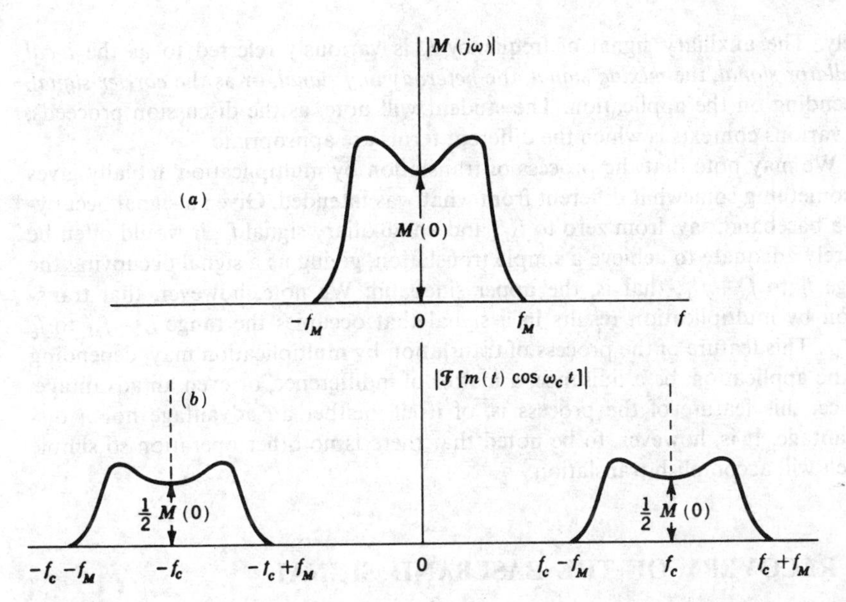

Figure 3.2-3 (*a*) The spectral density $|M(j\omega)|$ of a nonperiodic signal $m(t)$. (*b*) The spectral density of $m(t) \cos 2\pi f_c t$.

Finally, we consider in Fig. 3.2-3 the situation in which the signal to be translated may not be represented as a superposition of a number of sinusoidal components at sharply defined frequencies. Such would be the case if the signal were of finite energy and nonperiodic. In this case the signal is represented in the frequency domain in terms of its Fourier transform, that is, in terms of its spectral density. Thus let the signal $m(t)$ be bandlimited to the frequency range 0 to f_M. Its Fourier transform is $M(j\omega) = \mathscr{F}[m(t)]$. The magnitude $|M(j\omega)|$ is shown in Fig. 3.2-3*a*. The transform $M(j\omega)$ is symmetrical about $f = 0$ since we assume that $m(t)$ is a real signal. The spectral density of the signal which results when $m(t)$ is multiplied by $\cos \omega_c t$ is shown in Fig. 3.2-3*b*. This spectral pattern is deduced as an extension of the results shown in Figs. 3.2-1 and 3.2-2. Alternatively, we may easily verify (Prob. 3.2-2) that if $M(j\omega) = \mathscr{F}[m(t)]$, then

$$\mathscr{F}[m(t) \cos \omega_c t] = \tfrac{1}{2}[M(j\omega + j\omega_c) + M(j\omega - j\omega_c)] \tag{3.2-4}$$

The spectral range occupied by the original signal is called the *baseband frequency range* or simply the *baseband*. On this basis, the original signal itself is referred to as the baseband signal. The operation of multiplying a signal with an auxiliary sinusoidal signal is called *mixing* or *heterodyning*. In the translated signal, the part of the signal which consists of spectral components *above* the auxiliary signal, in the range f_c to $f_c + f_M$, is called the *upper-sideband* signal. The part of the signal which consists of spectral components below the auxiliary signal, in the range $f_c - f_M$ to f_c, is called the *lower-sideband* signal. The two sideband signals are also referred to as the *sum* and the *difference* frequencies, respectively. The auxiliary signal of frequency f_c is variously referred to as the *local oscillator signal*, the *mixing signal*, the *heterodyning signal*, or as the *carrier signal*, depending on the application. The student will note, as the discussion proceeds, the various contexts is which the different terms are appropriate.

We may note that the process of translation by multiplication actually gives us something somewhat different from what was intended. Given a signal occupying a baseband, say, from zero to f_M, and an auxiliary signal f_c, it would often be entirely adequate to achieve a simple translation, giving us a signal occupying the range f_c to $f_c + f_M$, that is, the upper sideband. We note, however, that translation by multiplication results in a signal that occupies the range $f_c - f_M$ to $f_c + f_M$. This feature of the process of translation by multiplication may, depending on the application, be a nuisance, a matter of indifference, or even an advantage. Hence, this feature of the process is, of itself, neither an advantage nor a disadvantage. It is, however, to be noted that there is no other operation so simple which will accomplish translation.

3.3 RECOVERY OF THE BASEBAND SIGNAL

Suppose a signal $m(t)$ has been translated out of its baseband through multiplication with $\cos \omega_c t$. How is the signal to be recovered? The recovery may be achieved by a reverse translation, which is accomplished simply by multiplying the translated signal with $\cos \omega_c t$. That such is the case may be seen by drawing spectral plots as in Fig. 3.2-2 or 3.2-3 and noting that the difference-frequency signal obtained by multiplying $m(t) \cos \omega_c t$ by $\cos \omega_c t$ is a signal whose spectral range is back at baseband. Alternatively, we may simply note that

$$[m(t) \cos \omega_c t] \cos \omega_c t = m(t) \cos^2 \omega_c t = m(t)(\tfrac{1}{2} + \tfrac{1}{2} \cos 2\omega_c t) \qquad (3.3\text{-}1a)$$

$$= \frac{m(t)}{2} + \frac{m(t)}{2} \cos 2\omega_c t \qquad (3.3\text{-}1b)$$

Thus, the baseband signal $m(t)$ reappears. We note, of course, that in addition to the recovered baseband signal there is a signal whose spectral range extends from $2f_c - f_M$ to $2f_c + f_M$. As a matter of practice, this latter signal need cause no difficulty. For most commonly $f_c \gg f_M$, and consequently the spectral range of this double-frequency signal and the baseband signal are widely separated. Therefore the double-frequency signal is easily removed by a low-pass filter.

This method of signal recovery, for all its simplicity, is beset by an important inconvenience when applied in a physical communication system. Suppose that the auxiliary signal used for recovery differs in phase from the auxiliary signal used in the initial translation. If this phase angle is θ, then, as may be verified (Prob. 3.3-1), the recovered baseband waveform will be proportional to $m(t) \cos \theta$. Therefore, unless it is possible to maintain $\theta = 0$, the signal strength at recovery will suffer. If it should happen that $\theta = \pi/2$, the signal will be lost entirely. Or consider, for example, that θ drifts back and forth with time. Then in this case the signal strength will wax and wane, in addition, possibly, to disappearing entirely from time to time.

Alternatively, suppose that the recovery auxiliary signal is not precisely at frequency f_c but is instead at $f_c + \Delta f$. In this case we may verify (Prob. 3.3-2) that the recovered baseband signal will be proportional to $m(t) \cos 2\pi \Delta f t$, resulting in a signal which will wax and wane or even be entirely unacceptable if Δf is comparable to, or larger than, the frequencies present in the baseband signal. This latter contingency is a distinct possibility in many an instance, since usually $f_c \gg f_M$ so that a small percentage change in f_c will cause a Δf which may be comparable or larger than f_M. In telephone or radio systems, an offset $\Delta f \leq 30$ Hz is deemed acceptable.

We note, therefore, that signal recovery using a second multiplication requires that there be available at the recovery point a signal which is precisely *synchronous* with the corresponding auxiliary signal at the point of the first multiplication. In such a *synchronous* or *coherent* system a *fixed* initial phase discrepancy is of no consequence since a simple phase shifter will correct the matter. Similarly it is not essential that the recovery auxiliary signal be sinusoidal (see Prob. 3.3-3). What is essential is that, in any time interval, the number of cycles executed by the two auxiliary-signal sources be the same. Of course, in a physical system, where some signal distortion is tolerable, some lack of synchronism may be allowed.

When the use of a common auxiliary signal is not feasible, it is necessary to resort to rather complicated means to provide a synchronous auxiliary signal at the location of the receiver. One commonly employed scheme is indicated in Fig. 3.3-1. To illustrate the operation of the synchronizer, we assume that the baseband signal is a sinusoidal $\cos \omega_m t$. The received signal is $s_i(t) = A \cos \omega_m t \cos \omega_c t$, with A a constant amplitude. This signal $s_i(t)$ does not have a spectral component at the angular frequency ω_c. The output of the squaring circuit is

$$s_i^2(t) = A^2 \cos^2 \omega_m t \cos^2 \omega_c t \tag{3.3-2a}$$

$$= A^2(\tfrac{1}{2} + \tfrac{1}{2} \cos 2\omega_m t)(\tfrac{1}{2} + \tfrac{1}{2} \cos 2\omega_c t) \tag{3.3-2b}$$

$$= \frac{A^2}{4} [1 + \tfrac{1}{2} \cos 2(\omega_c + \omega_m)t + \tfrac{1}{2} \cos 2(\omega_c - \omega_m)t$$

$$+ \cos 2\omega_m t + \cos 2\omega_c t] \tag{3.3-2c}$$

The filter selects the spectral component $(A^2/4) \cos 2\omega_c t$, which is then applied to a circuit which divides the frequency by a factor of 2. (See Prob. 3.3-4.) This fre-

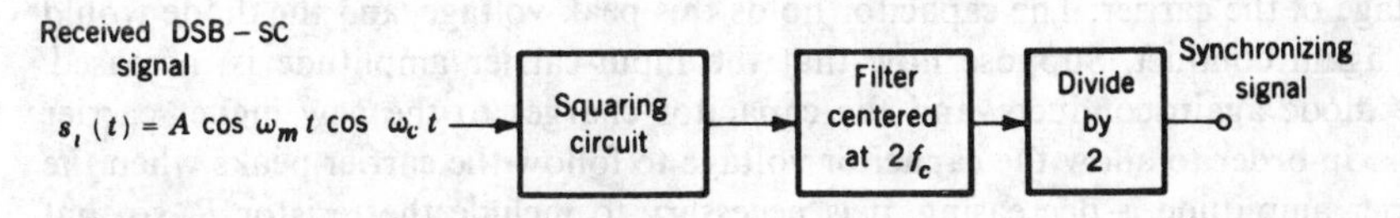

Figure 3.3-1 A simple squaring synchronizer.

quency division may be accomplished by using, for example, a bistable multivibrator. The output of the divider is used to demodulate (multiply) the incoming signal and thereby recover the baseband signal $\cos \omega_m t$.

We turn our attention now to a modification of the method of frequency translation, which has the great merit of allowing recovery of the baseband signal by an extremely simple means. This technique is called *amplitude modulation.*

3.4 AMPLITUDE MODULATION

A frequency-translated signal from which the baseband signal is easily recoverable is generated by adding, to the product of baseband and carrier, the carrier signal itself. Such a signal is shown in Fig. 3.4-1. Figure 3.4-1*a* shows the carrier signal with amplitude A_c, in Fig. 3.4-1*b* we see the baseband signal. The translated signal (Fig. 3.4-1*c*) is given by

$$v(t) = A_c[1 + m(t)] \cos \omega_c t \tag{3.4-1}$$

We observe, from Eq. (3.4-1) as well as from Fig. 3.4-1*c*, that the resultant waveform is one in which the carrier $A_c \cos \omega_c t$ is *modulated in amplitude*. The process of generating such a waveform is called *amplitude modulation*, and a communication system which employs such a method of frequency translation is called an *amplitude-modulation* system, or *AM* for short. The designation "carrier" for the auxiliary signal $A_c \cos \omega_c t$ seems especially appropriate in the present connection since this signal now "carries" the baseband signal as its envelope. The term "carrier" probably originated, however, in the early days of radio when this relatively high-frequency signal was viewed as the *messenger* which actually "carried" the baseband signal from one antenna to another.

The very great merit of the amplitude-modulated carrier signal is the ease with which the baseband signal can be recovered. The recovery of the baseband signal, a process which is referred to as *demodulation* or *detection*, is accomplished with the simple circuit of Fig. 3.4-2*a*, which consists of a diode D and the resistor-capacitor RC combination. We now discuss the operation of this circuit briefly and qualitatively. For simplicity, we assume that the amplitude-modulated carrier which is applied at the input terminals is supplied by a voltage source of zero internal impedance. We assume further that the diode is ideal, i.e., of zero or infinite resistance, depending on whether the diode current is positive or the diode voltage negative.

Let us initially assume that the input is of fixed amplitude and that the resistor R is not present. In this case, the capacitor charges to the peak positive voltage of the carrier. The capacitor holds this peak voltage, and the diode would not again conduct. Suppose now that the input-carrier amplitude is increased. The diode again conducts, and the capacitor charges to the new higher carrier peak. In order to allow the capacitor voltage to follow the carrier peaks when the carrier amplitude is decreasing, it is necessary to include the resistor R, so that the capacitor may discharge. In this case the capacitor voltage v_c has the form

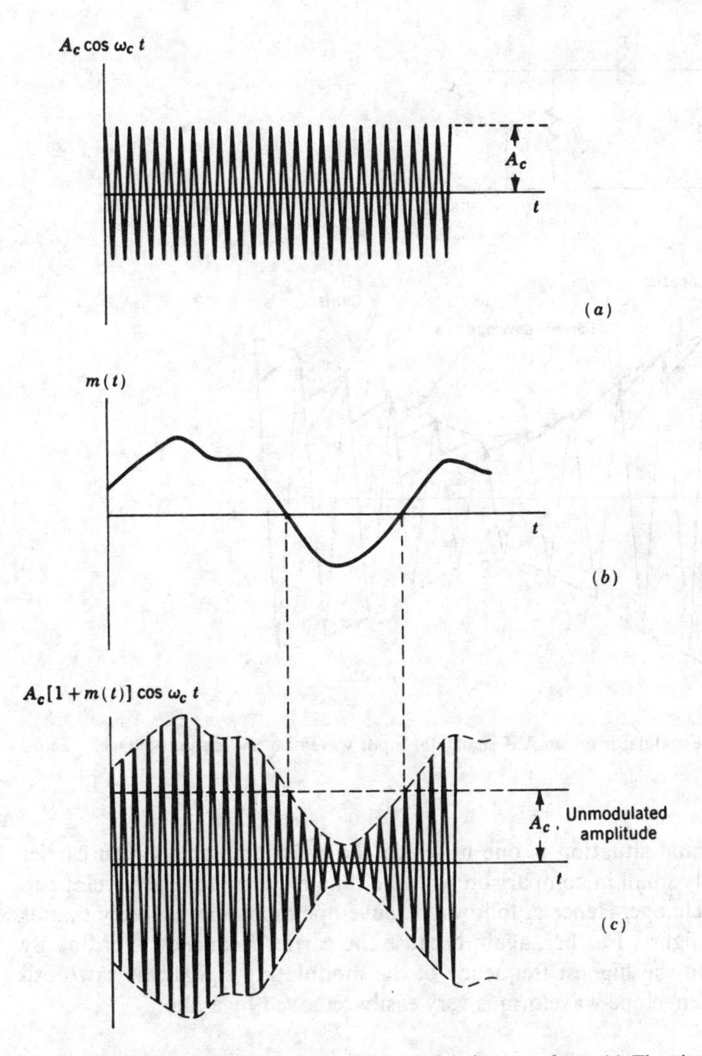

Figure 3.4-1 (*a*) A sinusoidal carrier. (*b*) A modulating waveform. (*c*) The sinusoidal carrier in (*a*) modulated by the waveform in (*b*).

shown in Fig. 3.4-2*b*. The capacitor charges to the peak of each carrier cycle and decays slightly between cycles. The time constant RC is selected so that the change in v_c between cycles is at least equal to the decrease in carrier amplitude between cycles. This constraint on the time constant RC is explored in Probs. 3.4-1 and 3.4-2.

It is seen that the voltage v_c follows the carrier envelope except that v_c also has superimposed on it a sawtooth waveform of the carrier frequency. In Fig. 3.4-2*b* the discrepancy between v_c and the envelope is greatly exaggerated. In

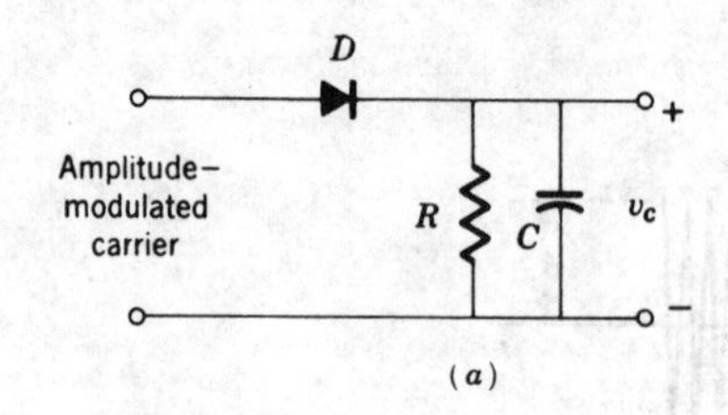

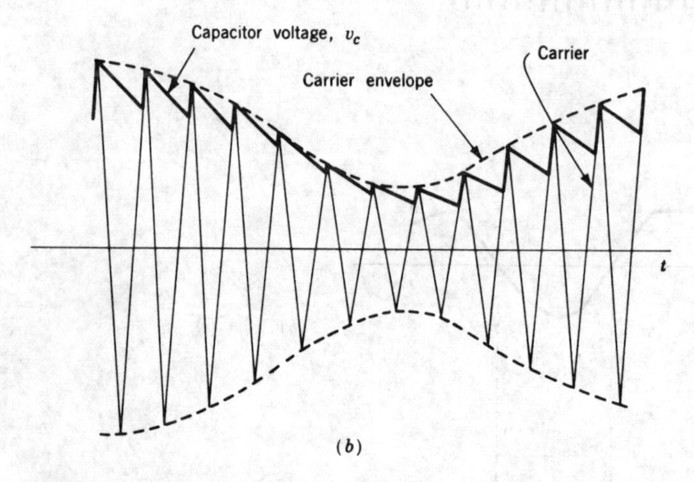

Figure 3.4-2 (*a*) A demodulator for an AM signal. (*b*) Input waveform and output voltage v_c across capacitor.

practice, the normal situation is one in which the time interval between carrier cycles is extremely small in comparison with the time required for the envelope to make a sizeable change. Hence v_c follows the envelope much more closely than is suggested in the figure. Further, again because the carrier frequency is ordinarily much higher than the highest frequency of the modulating signal, the sawtooth distortion of the envelope waveform is very easily removed by a filter.

3.5 MAXIMUM ALLOWABLE MODULATION

If we are to avail ourselves of the convenience of demodulation by the use of the simple diode circuit of Fig. 3.4-2*a*, we must limit the extent of the modulation of the carrier. That such is the case may be seen from Fig. 3.5-1. In Fig. 3.5-1*a* is shown a carrier modulated by a sinusoidal signal. It is apparent that the envelope of the carrier has the waveshape of the modulating signal. The modulating signal is sinusoidal; hence $m(t) = m \cos \omega_m t$, where m is a constant. Equation (3.4-1) becomes

$$v(t) = A_c(1 + m \cos \omega_m t) \cos \omega_c t \tag{3.5-1}$$

In Fig. 3.5-1*b* we have shown the situation which results when, in Eq. (3.5-1), we adjust $m > 1$. Observe now that the diode demodulator which yields as an output the positive envelope (a negative envelope if the diode is reversed) will not reproduce the sinusoidal modulating waveform. In this latter case, where $m > 1$, we may recover the modulating waveform but not with the diode modulator. Recovery would require the use of a coherent demodulation scheme such as was employed in connection with the signal furnished by a multiplier.

It is therefore necessary to restrict the excursion of the modulating signal in the direction of decreasing carrier amplitude to the point where the carrier amplitude is just reduced to zero. No such similar restriction applies when the modulation is increasing the carrier amplitude. With sinusoidal modulation, as in Eq. (3.5-1), we require that $|m| \leq 1$. More generally in Eq. (3.4-1) we require that the maximum negative excursion of $m(t)$ be -1.

The extent to which a carrier has been amplitude-modulated is expressed in terms of a *percentage modulation*. Let A_c, A_c(max), and A_c(min), respectively, be

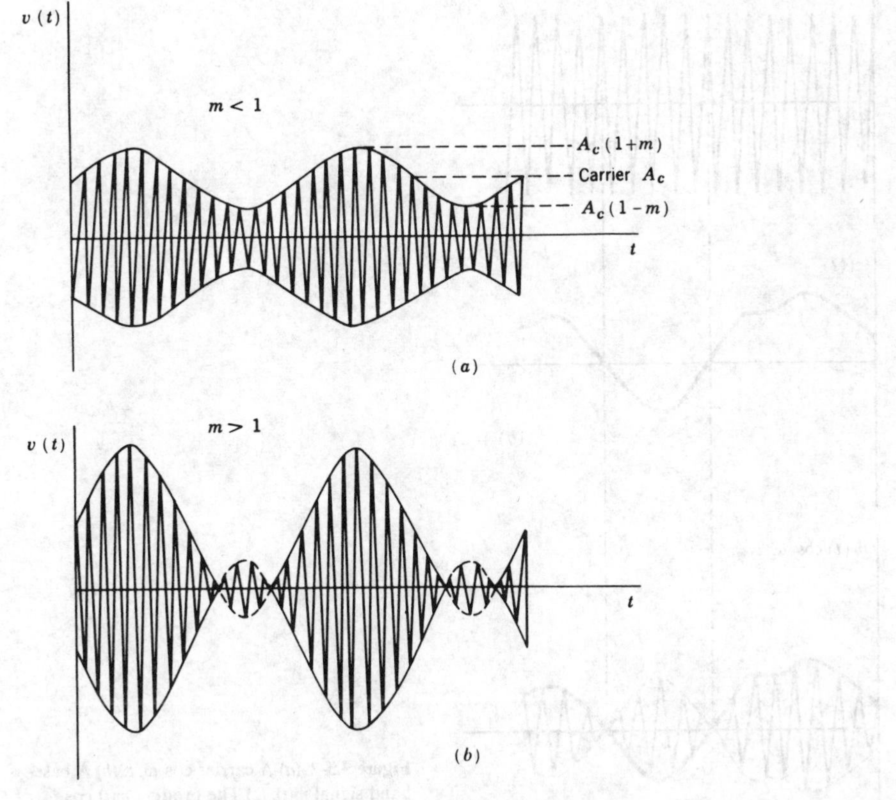

Figure 3.5-1 (*a*) A sinusoidally modulated carrier ($m < 1$). (*b*) A carrier overmodulated ($m > 1$) by a sinusoidal modulating waveform.

the unmodulated carrier amplitude and the maximum and minimum carrier levels. Then if the modulation is symmetrical, the percentage modulation is defined as P, given by

$$\frac{P}{100\%} = \frac{A_c(\max) - A_c}{A_c} = \frac{A_c - A_c(\min)}{A_c} = \frac{A_c(\max) - A_c(\min)}{2A_c} \tag{3.5-2}$$

In the case of sinusoidal modulation, given by Eq. (3.5-1) and shown in Fig. 3.5-1*a*, $P = m \times 100$ percent.

Having observed that the signal $m(t)$ may be recovered from the waveform $A_c[1 + m(t)] \cos \omega_c t$ by the simple circuit of Fig. 3.4-2*a*, it is of interest to note that a similar easy recovery of $m(t)$ is not possible from the waveform $m(t) \cos \omega_c t$. That such is the case is to be seen from Fig. 3.5-2. Figure 3.5-2*a* shows the carrier signal. The modulation or baseband signal $m(t)$ is shown in Fig.

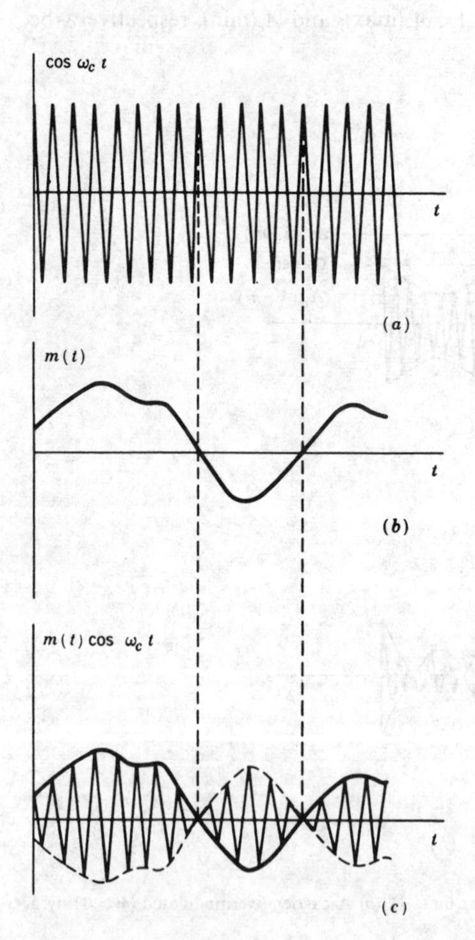

Figure 3.5-2 (*a*) A carrier $\cos \omega_c t$. (*b*) A baseband signal $m(t)$. (*c*) The product $m(t) \cos \omega_c t$ and its envelope.

3.5-2*b*, and the product $m(t) \cos \omega_c t$ is shown in Fig. 3.5-2*c*. We note that the envelope in Fig. 3.5-2*c* has the waveform not of $m(t)$ but rather of $|m(t)|$, the *absolute value* of $m(t)$. Observe the reversal of phase of the carrier in Fig. 3.5-2*c* whenever $m(t)$ passes through zero.

3.6 THE SQUARE-LAW DEMODULATOR

An alternative method of recovering the baseband signal which has been superimposed as an amplitude modulation on a carrier is to pass the AM signal through a nonlinear device. Such demodulation is illustrated in Fig. 3.6-1. We assume here for simplicity that the device has a square-law relationship between input signal x (current or voltage) and output signal y (current or voltage). Thus $y = kx^2$, with k a constant. Because of the nonlinearity of the transfer characteristic of the device, the output response is different for positive and for negative excursions of the carrier away from the quiescent operating point O of the device.

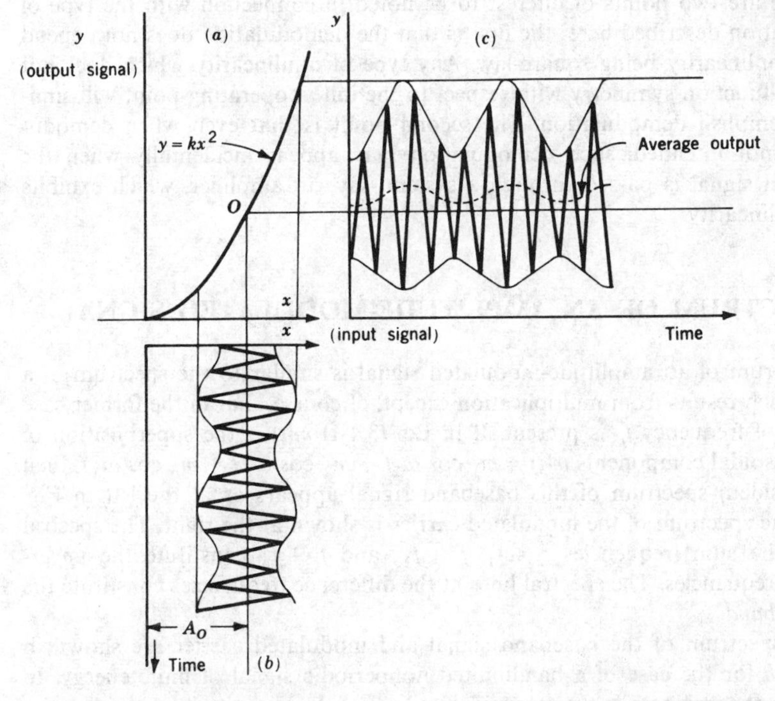

Figure 3.6-1 Illustrating the operation of a square-law demodulator. The output is the value of y averaged over many carrier cycles.

As a result, and as is shown in Fig. 3.6-1*c*, the output, when averaged over a time which encompasses many carrier cycles but only a very small part of the modulation cycle, has the waveshape of the envelope.

The applied signal is

$$x = A_o + A_c[1 + m(t)] \cos \omega_c t \tag{3.6-1}$$

Thus the output of the squaring circuit is

$$y = k\{A_o + A_c[1 + m(t)] \cos \omega_c t\}^2 \tag{3.6-2}$$

Squaring, and dropping dc terms as well as terms whose spectral components are located near ω_c and $2\omega_c$, we find that the output signal $s_o(t)$, that is, the signal output of a low-pass filter located after the squaring circuit, is

$$s_o(t) = kA_c^2[m(t) + \tfrac{1}{2}m^2(t)] \tag{3.6-3}$$

Observe that the modulation $m(t)$ is indeed recovered but that $m^2(t)$ appears as well. Thus the total recovered signal is a distorted version of the original modulation. The distortion is small, however, if $\frac{1}{2}m^2(t) \ll |m(t)|$ or if $|m(t)| \ll 2$.

There are two points of interest to be noted in connection with the type of demodulation described here; the first is that the demodulation does not depend on the nonlinearity being square-law. Any type of nonlinearity which does not have odd-function symmetry with respect to the initial operating point will similarly accomplish demodulation. The second point is that even when demodulation is not intended, such demodulation may appear incidentally when the modulated signal is passed through a system, say, an amplifier, which exhibits some nonlinearity.

3.7 SPECTRUM OF AN AMPLITUDE-MODULATED SIGNAL

The spectrum of an amplitude-modulated signal is similar to the spectrum of a signal which results from multiplication except, of course, that in the former case a carrier of frequency f_c is present. If in Eq. (3.4-1) $m(t)$ is the superposition of three sinusoidal components $m(t) = m_1 \cos \omega_1 t + m_2 \cos \omega_2 t + m_3 \cos \omega_3 t$, then the (one-sided) spectrum of this baseband signal appears as at the left in Fig. 3.7-1*a*. The spectrum of the modulated carrier is shown at the right. The spectral lines at the sum frequencies $f_c + f_1$, $f_c + f_2$, and $f_c + f_3$ constitute the *upper-sideband* frequencies. The spectral lines at the difference frequencies constitute the *lower sideband*.

The spectrum of the baseband signal and modulated carrier are shown in Fig. 3.7-1*b* for the case of a bandlimited nonperiodic signal of finite energy. In this figure the ordinate is the spectral density, i.e., the magnitude of the Fourier transform rather than the spectral amplitude, and consequently the carrier is represented by an impulse.

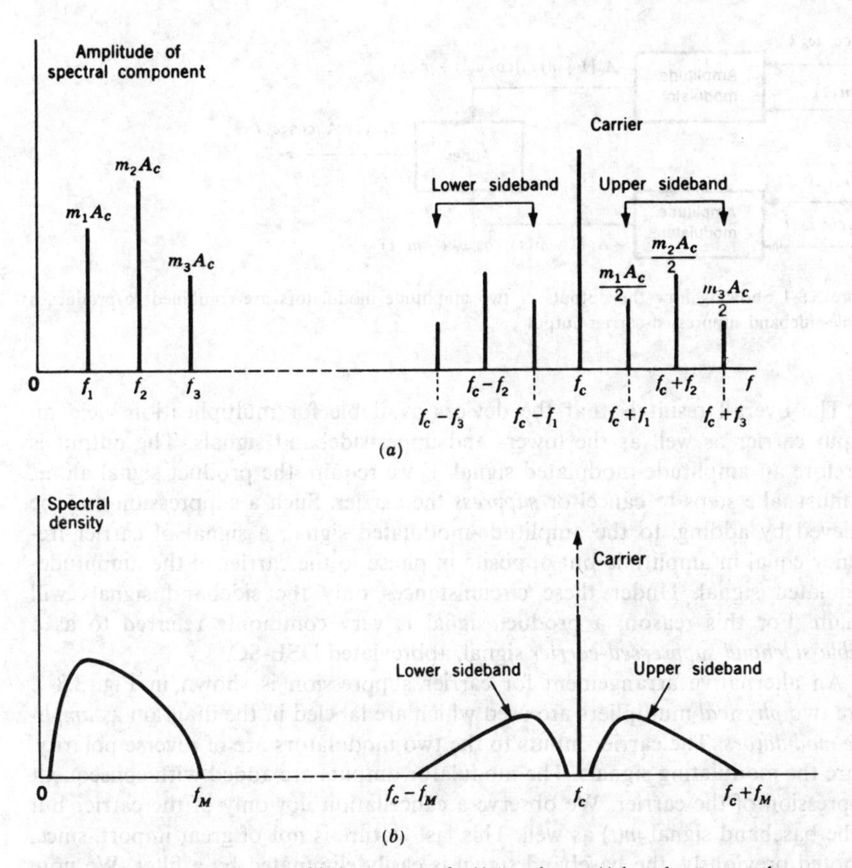

Figure 3.7-1 (*a*) At left the one-sided spectrum of $m(t)A_c$, where $m(t)$ has three spectral components. At right the spectrum of $A_c[1 + m(t)] \cos 2\pi f_c t$. (*b*) Same as in (*a*) except $m(t)$ is a nonperiodic signal and the vertical axis is spectral density rather than spectral amplitude.

3.8 MODULATORS AND BALANCED MODULATORS

We have described a "multiplier" as a device that yields as an output a signal which is the product of two input signals. Actually no simple physical device now exists which yields the product alone. On the contrary, all such devices yield, at a minimum, not only the product but the input signals themselves. Suppose, then, that such a device has as inputs a carrier $\cos \omega_c t$ and a modulating baseband signal $m(t)$. The device output will then contain the product $m(t) \cos \omega_c t$ and also the signals $m(t)$ and $\cos \omega_c t$. Ordinarily, the baseband signal will be bandlimited to a frequency range very much smaller than $f_c = \omega_c/2\pi$. Suppose, for example, that the baseband signal extends from zero frequency to 1000 Hz, while $f_c = 1$ MHz. In this case, the carrier and its sidebands extend from 999,000 to 1,001,000 Hz, and the baseband signal is easily removed by a filter.

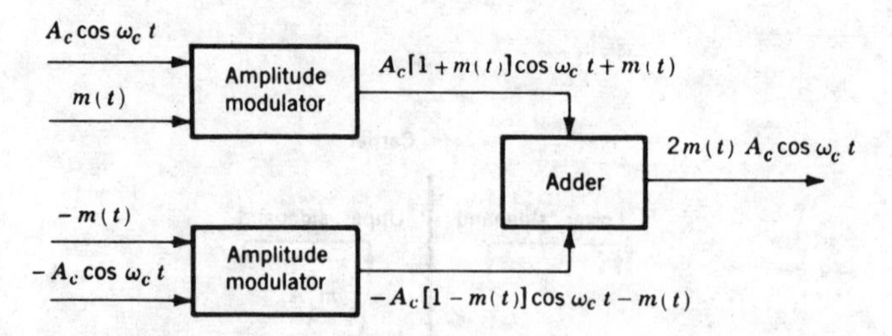

Figure 3.8-1 Showing how the outputs of two amplitude modulators are combined to produce a double-sideband suppressed-carrier output.

The overall result is that the devices available for multiplication yield an output carrier as well as the lower- and upper-sideband signals. The output is therefore an amplitude-modulated signal. If we require the product signal alone, we must take steps to cancel or *suppress* the carrier. Such a suppression may be achieved by adding, to the amplitude-modulated signal, a signal of carrier frequency equal in amplitude but opposite in phase to the carrier of the amplitude-modulated signal. Under these circumstances only the sideband signals will remain. For this reason, a product signal is very commonly referred to as a *double-sideband suppressed-carrier* signal, abbreviated DSB-SC.

An alternative arrangement for carrier suppression is shown in Fig. 3.8-1. Here two *physical* multipliers are used which are labeled in the diagram as *amplitude modulators.* The carrier inputs to the two modulators are of reverse polarity, as are the modulating signals. The modulator outputs are added with consequent suppression of the carrier. We observe a cancellation not only of the carrier but of the baseband signal $m(t)$ as well. This last feature is not of great import, since, as noted previously, the baseband signal is easily eliminated by a filter. We note that the product terms of the two modulators reinforce. The arrangement of Fig. 3.8-1 is called a *balanced modulator*.

3.9 SINGLE-SIDEBAND MODULATION

We have seen that the baseband signal may be recovered from a double-sideband suppressed-carrier signal by multiplying a second time, with the same carrier. It can also be shown that the baseband signal can be recovered in a similar manner even if only one sideband is available. For suppose a spectral component of the baseband signal is multiplied by a carrier $\cos \omega_c t$, giving rise to an upper sideband at $\omega_c + \omega$ and a lower sideband at $\omega_c - \omega$. Now let us assume that we have filtered out one sideband and are left with, say, only the upper sideband at $\omega_c + \omega$. If now this sideband signal is again multiplied by $\cos \omega_c t$, we shall generate a signal at $2\omega_c + \omega$ and the original baseband spectral component at ω. If we had used the lower sideband at $\omega_c - \omega$, the second multiplication would have yielded a signal at $2\omega_c - \omega$ and again restored the baseband spectral component.

Since it is possible to recover the baseband signal from a single sideband, there is an obvious advantage in doing so, since spectral space is used more economically. In principle, two single-sideband (abbreviated SSB) communications systems can now occupy the spectral range previously occupied by a single amplitude-modulation system or a double-sideband suppressed-carrier system.

The baseband signal may *not* be recovered from a single-sideband signal by the use of a diode modulator. That such is the case is easily seen by considering, for example, that the modulating signal is a sinusoid of frequency f. In this case the single-sideband signal will consist also of a single sinusoid of frequency, say, $f_c + f$, and there is no amplitude variation at all at the baseband frequency to which the diode modulator can respond.

Baseband recovery is achieved at the receiving end of the single-sideband communications channel by heterodyning the received signal with a local carrier signal which is synchronous (coherent) with the carrier used at the transmitting end to generate the sideband. As in the double-sideband case it is necessary, in principle, that the synchronism be exact and, in practice, that synchronism be maintained to a high order of precision. The effect of a lack of synchronism is different in a double-sideband system and in a single-sideband system. Suppose that the carrier received is $\cos \omega_c t$ and that the local carrier is $\cos (\omega_c t + \theta)$. Then with DSB-SC, as noted in Sec. 3.3, the spectral component $\cos \omega t$ will, upon demodulation, reappear as $\cos \omega t \cos \theta$. In SSB, on the other hand, the spectral component, $\cos \omega t$ will reappear (Prob. 3.9-2) in the form $\cos (\omega t - \theta)$. Thus, in one case a phase offset in carriers affects the amplitude of the recovered signal and, for $\phi = \pi/2$, may result in a total loss of the signal. In the other case the offset produces a phase change but not an amplitude change.

Alternatively, let the local oscillator carrier have an angular frequency offset $\Delta\omega$ and so be of the form $\cos (\omega_c + \Delta\omega)t$. Then as already noted, in DSB-SC, the recovered signal has the form $\cos \omega t \cos \Delta\omega t$. In SSB, however, the recovered signal will have the form $\cos (\omega + \Delta\omega)t$. Thus, in one case the recovered spectral component $\cos \omega t$ reappears with a "warble," that is, an amplitude fluctuation at the rate $\Delta\omega$. In the other case the amplitude remains fixed, but the frequency of the recovered signal is in error by amount $\Delta\omega$.

A phase offset between the received carrier and the local oscillator will cause distortion in the recovered baseband signal. In such a case each spectral component in the baseband signal will, upon recovery, have undergone the *same* phase shift. Fortunately, when SSB is used to transmit voice or music, such *phase distortion* does not appear to be of major consequence, because the human ear seems to be insensitive to the phase distortion.

A frequency offset between carriers in amount of Δf will cause each recovered spectral component of the baseband signal to be in error by the same amount Δf. Now, if it had turned out that the frequency error were proportional to the frequency of the spectral component itself, then the recovered signal would sound like the original signal except that it would be at a higher or lower pitch. Such, however, is not the case, since the frequency error is fixed. Thus frequencies in the original signal which were harmonically related will no longer be so related after recovery. The overall result is that a frequency offset between carriers adversely affects the intelligibility of spoken communication and is not well tolerated in connection with music. As a matter of experience, it turns out that an error Δf of less than 30 Hz is acceptable to the ear.

The need to keep the frequency offset Δf between carriers small normally imposes severe restrictions on the frequency stabilities of the carrier signal generators at both ends of the communications system. For suppose that we require to keep Δf to 10 Hz or less, and that our system uses a carrier frequency of 10 MHz. Then the sum of the frequency drift in the two carrier generators may not exceed 1 part in 10^6. The required equality in carrier frequency may be maintained through the use of quartz crystal oscillators using crystals cut for the same frequency at transmitter and receiver. The receiver must use as many crystals (or equivalent signals derived from crystals) as there are channels in the communications system.

It is also possible to tune an SSB receiver manually and thereby reduce the frequency offset. To do this, the operator manually adjusts the frequency of the receiver carrier generator until the received signal sounds "normal." Experienced operators are able to tune correctly to within 10 or 20 Hz. However, because of oscillator drift, such tuning must be readjusted periodically.

When the carrier frequency is very high, even quartz crystal oscillators may be hard pressed to maintain adequate stability. In such cases it is necessary to transmit the carrier itself along with the sideband signal. At the receiver the carrier may be separated by filtering and used to synchronize a local carrier generator. When used for such synchronization, the carrier is referred to as a "pilot carrier" and may be transmitted at a substantially reduced power level.

It is interesting to note that the squaring circuit used to recover the frequency and phase information of the DSB-SC system cannot be used here. In any event, it is clear that a principal complication in the way of more widespread use of single sideband is the need for supplying an accurate carrier frequency at the receiver.

3.10 METHODS OF GENERATING AN SSB SIGNAL

Filter Method

A straightforward method of generating an SSB signal is illustrated in Fig. 3.10-1. Here the baseband signal and a carrier are applied to a balanced modulator. The output of the balanced modulator bears both the upper- and lower-sideband signals. One or the other of these signals is then selected by a filter. The filter is a bandpass filter whose passband encompasses the frequency range of the sideband selected. The filter must have a cutoff sharp enough to separate the selected sideband from the other sideband. The frequency separation of the sidebands is twice the frequency of the lowest frequency spectral components of the baseband signal. Human speech contains spectral components as low as about 70 Hz.

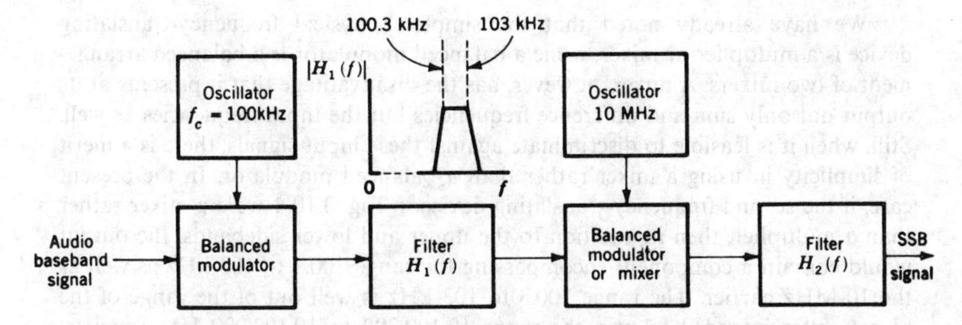

Figure 3.10-1 Block diagram of the filter method of generating a single-sideband signal.

However, to alleviate the sideband filter selectivity requirements in an SSB system, it is common to limit the lower spectral limit of speech to about 300 Hz. It is found that such restriction does not materially affect the intelligibility of speech. Similarly, it is found that no serious distortion results if the upper limit of the speech spectrum is cut off at about 3000 Hz. Such restriction is advantageous for the purpose of conserving bandwidth. Altogether, then, a typical sideband filter has a passband which, measured from f_c, extends from about 300 to 3000 Hz and in which range its response is quite flat. Outside this passband the response falls off sharply, being down about 40 dB at 4000 Hz and rejecting the unwanted sideband also be at least 40 dB. The filter may also serve, further, to suppress the carrier itself. Of course, in principle, no carrier should appear at the output of a balanced modulator. In practice, however, the modulator may not balance exactly, and the precision of its balance may be subject to some variation with time. Therefore, even if a pilot carrier is to be transmitted, it is well to suppress it at the output of the modulator and to add it to the signal at a later point in a controllable manner.

Now consider that we desire to generate an SSB signal with a carrier of, say, 10 MHz. Then we require a passband filter with a selectivity that provides 40 dB of attenuation within 600 Hz at a frequency of 10 MHz, a percentage frequency change of 0.006 percent. Filters with such sharp selectivity are very elaborate and difficult to construct. For this reason, it is customary to perform the translation of the baseband signal to the final carrier frequency in several stages. Two such stages of translation are shown in Fig. 3.10-1. Here we have selected the first carrier to be of frequency 100 kHz. The upper sideband, say, of the output of the balanced modulator ranges from 100.3 to 103 kHz. The filter following the balanced modulator which selects this upper sideband need now exhibit a selectivity of only a hundredth of the selectivity (40 dB in 0.6 percent frequency change) required in the case of a 10-MHz carrier. Now let the filter output be applied to a second balanced modulator, supplied this time with a 10-MHz carrier. Let us again select the upper sideband. Then the second filter must provide 40 dB of attenuation in a frequency range of 200.6 kHz, which is nominally 2 percent of the carrier frequency.

We have already noted that the simplest physical frequency-translating device is a multiplier or mixer, while a balanced modulator is a balanced arrangement of two mixers. A mixer, however, has the disadvantage that it presents at its output not only sum and difference frequencies but the input frequencies as well. Still, when it is feasible to discriminate against these input signals, there is a merit of simplicity in using a mixer rather than a balanced modulator. In the present case, if the second frequency-translating device in Fig. 3.10-1 were a mixer rather than a multiplier, then in addition to the upper and lower sidebands, the output would contain a component encompassing the range 100.3 to 103 kHz as well as the 10-MHz carrier. The range 100.3 to 103 kHz is well out of the range of the second filter intended to pass the range 10,100,300 to 10,103,000 Hz. And it is realistic to design a filter which will suppress the 10-MHz carrier, since the carrier frequency is separated from the lower edge of the upper sideband (10,100,300) by nominally a 1-percent frequency change.

Altogether, then, we note in summary that when a single-sideband signal is to be generated which has a carrier in the megahertz or tens-of-megahertz range, the frequency translation is to be done in more than one stage—frequently two but not uncommonly three. If the baseband signal has spectral components in the range of hundreds of hertz or lower (as in an audio signal), the first stage invariably employs a balanced modulator, while succeeding stages may use mixers.

Phasing Method

An alternative scheme for generating a single-sideband signal is shown in Fig. 3.10-2. Here two balanced modulators are employed. The carrier signals of angular frequency ω_c which are applied to the modulators differ in phase by 90°.

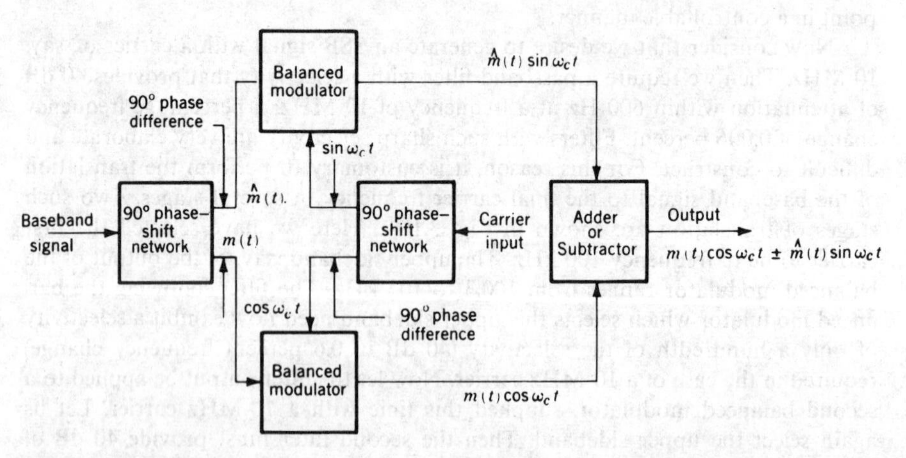

Figure 3.10-2 A method of generating a single-sideband signal using balanced modulators and phase shifters.

Similarly the baseband signal, before application to the modulators, is passed through a 90° phase-shifting network so that there is a 90° phase shift between any spectral component of the baseband signal applied to one modulator and the like-frequency component applied to the other modulator.

To see most simply how the arrangement of Fig. 3.10-2 operates, let us assume that the baseband signal is sinusoidal and appears at the input to one modulator as $\cos \omega_m t$ and hence as $\sin \omega_m t$ at the other. Also, let the carrier be $\cos \omega_c t$ at one modulator and $\sin \omega_c t$ at the other. Then the outputs of the balanced modulators (multipliers) are

$$\cos \omega_m t \cos \omega_c t = \tfrac{1}{2}[\cos (\omega_c - \omega_m)t + \cos (\omega_c + \omega_m)t] \qquad (3.10\text{-}1)$$

$$\sin \omega_m t \sin \omega_c t = \tfrac{1}{2}[\cos (\omega_c - \omega_m)t - \cos (\omega_c + \omega_m)t] \qquad (3.10\text{-}2)$$

If these waveforms are added, the lower sideband results; if subtracted, the upper sideband appears at the output. In general, if the modulation $m(t)$ is given by

$$m(t) = \sum_{i=1}^{m} A_i \cos (\omega_i t + \theta_i) \qquad (3.10\text{-}3)$$

then, using Fig. 3.10-2, we see that the output of the SSB modulator is in general

$$m(t) \cos \omega_c t \pm \hat{m}(t) \sin \omega_c t \qquad (3.10\text{-}4)$$

where

$$\hat{m}(t) \equiv \sum_{i=1}^{m} A_i \sin (\omega_i t + \theta_i) \qquad (3.10\text{-}5)$$

The single-sideband generating system of Fig. 3.10-2 generally enjoys less popularity than does the filter method. The reason for this lack of favor is that the present phasing method requires, for satisfactory operation, that a number of constraints be rather precisely met if the carrier and one sideband are adequately to be suppressed. It is required that each modulator be rather carefully balanced to suppress the carrier. It requires also that the baseband signal phase-shifting network provide to the modulators signals in which equal frequency spectral components are of exactly equal amplitude and differ in phase by precisely 90°. Such a network is difficult to construct for a baseband signal which extends over many octaves. It is also required that each modulator display equal sensitivity to the baseband signal. Finally, the carrier phase-shift network must provide exactly 90° of phase shift. If any of these constraints is not satisfied, the suppression of the rejected sideband and of the carrier will suffer. The effect on carrier and sideband suppression due to a failure precisely to meet these constraints is explored in Probs. 3.10-3 and 3.10-4. Of course, in any physical system a certain level of carrier and rejected sideband is tolerable. Still, there seems to be a general inclination to achieve a single sideband by the use of passive filters rather than by a method which requires many exactly maintained balances in passive and active circuits. There is an alternative single-sideband generating scheme[2] which avoids the need for a wideband phase-shifting network but which uses four balanced modulators.

3.11 VESTIGIAL-SIDEBAND MODULATION

As a preliminary to a discussion of vestigial-sideband modulation, let us consider the situation when a single-sideband signal is accompanied by its carrier. Suppose a carrier of angular frequency ω_c is amplitude modulated by a sinusoid of angular frequency ω_m to the extent where the resultant signal displays a percentage modulation m. Then the waveform is

$$f_1(t) = A(1 + m \cos \omega_m t) \cos \omega_c t \tag{3.11-1}$$

$$= A \cos \omega_c t + \frac{mA}{2} [\cos (\omega_c + \omega_m)t + \cos (\omega_c - \omega_m)t \tag{3.11-2}$$

If one of the sidebands is removed, leaving, however, the carrier, we have

$$f_2(t) = A \cos \omega_c t + \frac{mA}{2} \cos (\omega_c + \omega_m)t \tag{3.11-3}$$

To calculate the response of a diode demodulator to $f_2(t)$ we need to have the form of the envelope of $f_2(t)$. We have

$$f_2(t) = A \cos \omega_c t + \frac{mA}{2} \cos \omega_c t \cos \omega_m t - \frac{mA}{2} \sin \omega_c t \sin \omega_m t$$

$$= A\left(1 + \frac{m}{2} \cos \omega_m t\right) \cos \omega_c t - \frac{mA}{2} \sin \omega_m t \sin \omega_c t \tag{3.11-4}$$

The amplitude $A(t)$ of $f_2(t)$ is

$$A(t) = \sqrt{A^2\left(1 + \frac{m}{2} \cos \omega_m t\right)^2 + \left(\frac{mA}{2} \sin \omega_m t\right)^2}$$

$$= \sqrt{A^2\left(1 + \frac{m^2}{4}\right) + A^2 m \cos \omega_m t} \tag{3.11-5}$$

and for $m \ll 1$,

$$A(t) \cong A\left(1 + \frac{m}{2} \cos \omega_m t\right) \tag{3.11-6}$$

We note that if m is small, the diode demodulator does demodulate a signal which is lacking one sideband. Comparing the amplitude $A(t)$ given in Eq. (3.11-6) with the factor in parentheses in Eq. (3.11-1), we observe that the baseband signal output with one sideband suppressed is half as large as it would be if both sidebands had been present, a result to have been anticipated.

The diode-demodulator method for SSB application is of interest since it allows recovery of the baseband signal with a receiving system intended for double-sideband amplitude-modulation signals. Many amplitude-modulation "communications" type receivers are equipped with an adjustable oscillator which can be adjusted in frequency to serve as the local carrier and then added to a single-sideband signal. Hence such AM receivers may demodulate SSB signals, with, however, some distortion.

This technique should be compared with the use of the synchronous demodulator. Although the synchronous demodulator yields no distortion when the carrier phase is perfectly adjusted, the diode demodulator introduces some distortion (see Prob. 3.11-1). However, for synchronous demodulation we need, at the receiver, information about both the frequency and phase of the carrier. With a diode demodulator we need to know only the frequency.

Turning now to vestigial-sideband modulation, we shall take account of the fact that the principal application of this modulation method is to be found in commercial television broadcasting. Therefore, by way of illustration and for the sake of motivation, we refer specifically to that application.

The television picture signal, i.e., the video signal, in accordance with practice that prevails in the United States, occupies a bandwidth of nominally 4.5 MHz. A carrier, amplitude-modulated with such a signal, would give rise to a signal extending over 9 MHz. Since this bandwidth is about 9 times the frequency range which encompasses *all* standard AM radio broadcasting stations, some means of conserving bandwidth is surely needed. Single sideband is not feasible because of the complexity it must introduce into each of the millions of receivers. A workable compromise between the spectrum-conserving characteristic of single-sideband modulation and the demodulation simplicity of double-sideband amplitude modulation is found in the vestigial-sideband system which is standard in television broadcasting.

In the vestigial-sideband system, an amplitude-modulated signal, carrier plus double sideband, is passed through a filter before transmission to the receiving end. The response of the filter is plotted in Fig. 3.11-1*a* as a function of the deviation Δf of the frequency from the carrier frequency. (The sound which accompanies the video signal is transmitted by frequency modulation, discussed in the next chapter, on a carrier located 4.5 MHz above the picture carrier. A frequency range of 100 kHz is allowed on each side of the sound carrier for the sound sidebands.) The upper sideband of the picture carrier is transmitted without attenuation up to 4 MHz. Thereafter the sideband is attenuated so that it does not interfere with the lower sideband of the sound carrier. The lower sideband of the picture carrier is transmitted without attenuation over the range 0.75 MHz and is entirely attenuated at 1.25 MHz. Thus the picture signal is transmitted double sideband over the range 0 to 0.75 MHz, single sideband over the range 1.25 MHz and above, while in the intermediate range, 0.75 to 1.25 MHz, the transition is made from one to the other. Altogether, however, the entire transmission is confined to a range of about 6 MHz, a saving of *one-third* of the bandwidth that would be required for full double-sideband transmission.

We noted above that when only a single sideband is present, the output of a diode demodulator is half the output yielded when both sidebands are present. Therefore, with a vestigial-sideband signal, the relative demodulator output, plotted against frequency for a fixed percentage modulation, has the form shown in Fig. 3.11-1*b*. This lack of uniformity is corrected at the receiver by passing the

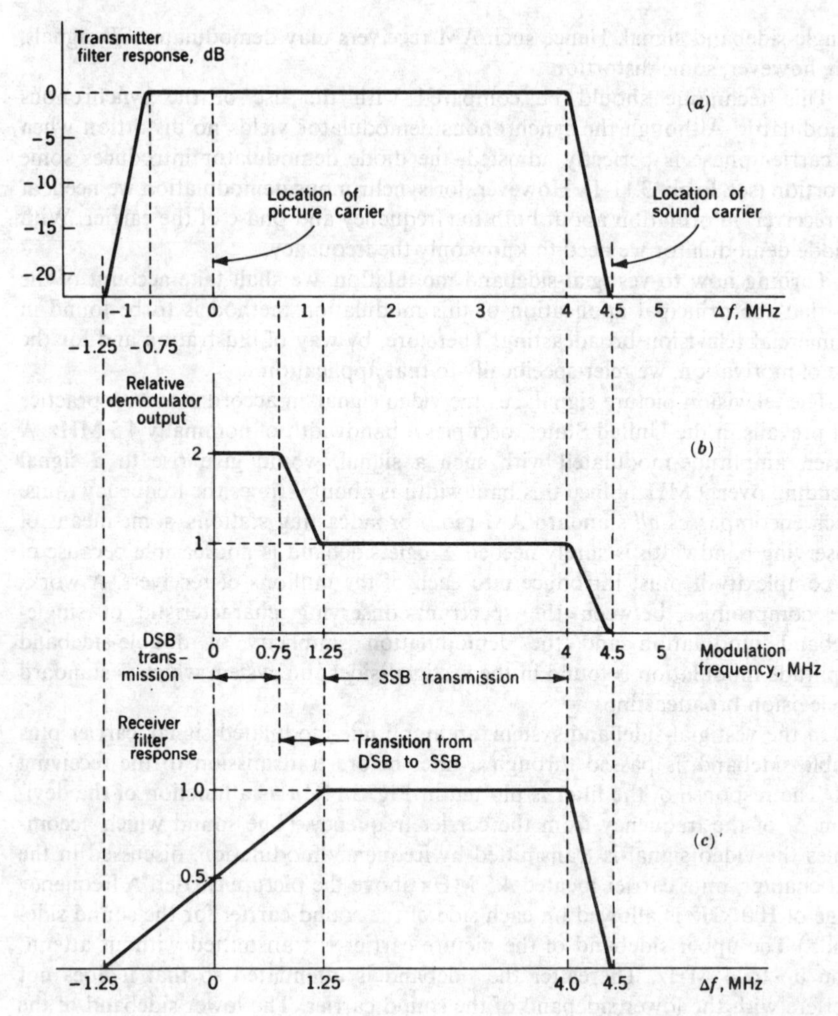

Figure 3.11-1 Vestigial-sideband transmission. (*a*) Transmitter filter response. (*b*) Relative diode demodulator output. (*c*) Receiver filter response.

received signal through a filter before demodulation. The relative response of this filter is shown in Fig. 3.11-1*c*. Over the range 1.25 MHz on either side of the picture carrier the response varies linearly as shown. As a result, for modulating frequencies up to 1.25 MHz, the *sum* of the amplitudes of the two sidebands, and hence of the demodulator output, is the same as is yielded by the single sideband above 1.25 MHz. This result is easily verified. For example, the received signal frequency component at $\Delta f = 0$ is attenuated by a factor of 2. Referring to Fig. 3.11-1*b*, we see that this reduces the modulation to 1 when $\Delta f = 0$. As a second illustration, let $\Delta f = \pm 0.75$. The sideband amplitude due to $\Delta f = -0.75$ is 0.2, while the sideband amplitude due to $\Delta f = +0.75$ is 0.8. The sum is again unity, as expected.

It is, of course, to be anticipated that the vestigial-sideband system will introduce some distortion into the demodulated signal, especially at high-percentage modulation. The experimental fact is, however, that in the transmission of picture information the distortion can be kept within tolerable levels. We may even wonder why the cutoff frequency of the filter which removes the lower sideband is not adjusted to be even closer to the carrier frequency. thereby conserving additional bandwidth. It is found rather generally, in real filters, that spectral components which lie within the passband but close to the cutoff frequency suffer distortion-producing phase shifts, even when the amplitude response of the filter is maintained uniform. The nature of the picture signal is such that its waveforms suffer substantial distortion from relatively small phase shifts of its low-frequency components. Thus the decision to leave a *vestige* of the nominally suppressed sideband is dictated by an engineering compromise between bandwidth economy and faithfulness of reproduction of the picture.

3.12 COMPATIBLE SINGLE SIDEBAND

We are now rather naturally led to inquire whether it is possible to generate an amplitude-modulated signal whose bandwidth is f_M Hz, for a baseband signal whose highest spectral component is f_M Hz, and from which the baseband signal may be faithfully recovered by the simple diode demodulator, even when no carrier signal is transmitted. Such a single-sideband signal would be *compatible* for reception by an AM radio receiver and is referred to as a *compatible single-sideband* signal, abbreviated CSSB. It has been shown[3] that such a signal exists. However, the involved signal processing required for the production of such a CSSB signal makes the system presently impractical for commercial use.

3.13 MULTIPLEXING

We have explored the principles of operation of a number of amplitude-modulation *modems* (systems of modulation and demodulation). The manner in which such systems are used for multiplexing (i.e., transmitting many baseband signals over a common communications channel) is shown in Fig. 3.13-1. The individual baseband signals $m_1(t)$, $m_2(t)$, ..., $m_\lambda(t)$, each bandlimited to f_M, are applied to individual modulators, each modulator being supplied as well with carrier waveforms of frequency $f_1, f_2, \ldots, f_\lambda$. The individual modulator-output signals extend over a limited range in the neighborhood of the individual carrier frequencies. Most importantly the carrier frequencies are selected to the spectral ranges of the modulator-output signals do not overlap. This separation in frequency is precisely the feature that allows the eventual recovery of the individual signals, and for this reason this multiplexing system is referred to as *frequency*

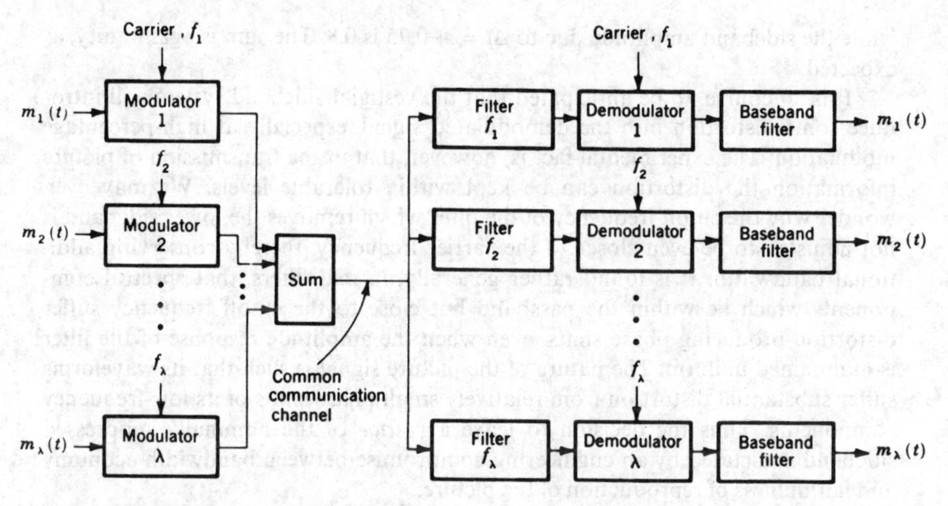

Figure 3.13-1 Multiplexing many baseband signals over a single communications channel.

multiplexing. As a matter of fact, to facilitate this separation of the individual signals, the carrier frequencies are selected to leave a comfortable margin (guard band) between the limit of one frequency range and the beginning of the next. The combined output of all the modulators, i.e., the *composite* signal, is applied to a common communications channel. In the case of radio transmission, the channel is free space, and coupling to the channel is made by means of an antenna. In other cases wires are used.

At the receiving end the composite signal is applied to each of a group of bandpass filters whose passbands are in the neighborhood $f_1, f_2, \ldots, f_\lambda$. The filter f_1 is a bandpass filter which passes only the spectral range of the output of modulator 1 and similarly for the other bandpass filters. The signals have thus been separated. They are then applied to individual demodulators which extract the baseband signals from the carrier. The carrier inputs to the demodulators are required only for synchronous demodulation and are not used otherwise.

The final operation indicated in Fig. 3.13-1 consists in passing the demodulator output through a baseband filter. The baseband filter is a low-pass filter with cutoff at the frequency f_M to which the baseband signal is limited. This baseband filter will pass, without modification, the baseband signal output of the modulator and in this sense serves no function in the system as described up to the present point. We shall, however, see in Chaps. 8 and 9 that such baseband filters are essential to suppress the noise which invariably accompanies the signal.

REFERENCES

1. Bell Telephone Laboratories: "Transmission Systems for Communications," Western Electric Company, Tech. Pub., Winston-Salem, N.C., 1964.

2. Norgaard, D. E.: A Third Method of Generation and Detection of Single-sideband Signals, *Proc. IRE*, December, 1956.
3. Voelcker, H.: Demodulation of Single-sideband Signals Via Envelope Detection, *IEEE Trans. on Communication Technology*, pp. 22–30, February, 1966.

PROBLEMS

3.2-1. A signal $v_m(t)$ is bandlimited to the frequency range 0 to f_M. It is frequency-translated by multiplying it by the signal $v_c(t) = \cos 2\pi f_c t$. Find f_c so that the bandwidth of the translated signal is 1 percent of the frequency f_c.

3.2-2. The Fourier transform of $m(t)$ is $\mathscr{F}[m(t)] \equiv M(f)$. Show that

$$\mathscr{F}[m(t) \cos 2\pi f_c t] = \tfrac{1}{2}[M(f + f_c) + M(f - f_c)]$$

3.2-3. The signals

$$v_1(t) = 2 \cos \omega_1 t + \cos 2\omega_1 t$$

and

$$v_2(t) = \cos \omega_2 t + 2 \cos 2\omega_2 t$$

are multiplied. Plot the resultant amplitude-frequency characteristic, assuming that $\omega_2 > 2\omega_1$, but is not a harmonic of ω_1. Repeat for $\omega_2 = 2\omega_1$.

3.3-1. The baseband signal $m(t)$ in the frequency-translated signal $v(t) = m(t) \cos 2\pi f_c t$ is recovered by multiplying $v(t)$ by the waveform $\cos(2\pi f_c t + \theta)$.

(*a*) The product waveform is transmitted through a low-pass filter which rejects the double-frequency signal. What is the output signal of the filter?

(*b*) What is the maximum allowable value for the phase θ if the recovered signal is to be 90 percent of the maximum possible value?

(*c*) If the baseband signal $m(t)$ is bandlimited to 10 kHz, what is the minimum value of f_c for which it is possible to recover $m(t)$ by filtering the product waveform $v(t) \cos(2\pi f_c t + \theta)$?

3.3-2. The baseband signal $m(t)$ in the frequency-translated signal $v(t) = m(t) \cos 2\pi f_c t$ is recovered by multiplying $v(t)$ by the waveform $\cos 2\pi(f_c + \Delta f)t$. The product waveform is transmitted through a low-pass filter which rejects the double-frequency signal. Find the output signal of the filter.

3.3-3. (*a*) The baseband signal $m(t)$ in the frequency-translated signal $v(t) = m(t) \cos 2\pi f_c t$ is to be recovered. There is available a waveform $p(t)$ which is periodic with period $1/f_c$. Show that $m(t)$ may be recovered by appropriately filtering the product waveform $p(t)v(t)$.

(*b*) Show that $m(t)$ may be recovered as well if the periodic waveform has a period n/f_c, where n is an integer. Assume $m(t)$ bandlimited to the frequency range from 0 to 5 kHz and let $f_c = 1$ MHz. Find the largest n which will allow $m(t)$ to be recovered. Are all periodic waveforms acceptable?

3.3-4. The signal $m(t)$ in the DSB-SC signal $v(t) = m(t) \cos(\omega_c t + \theta)$ is to be reconstructed by multiplying $v(t)$ by a signal derived from $v^2(t)$.

(*a*) Show that $v^2(t)$ has a component at the frequency $2f_c$. Find its amplitude.

(*b*) If $m(t)$ is bandlimited to f_M and has a probability density

$$f(m) = \frac{1}{\sqrt{2\pi}} e^{-m^2/2} \qquad -\infty \le m \le \infty$$

find the expected value of the amplitude of the component of $v^2(t)$ at $2f_c$.

3.4-1. The envelope detector shown in Fig. 3.4-2*a* is used to recover the signal $m(t)$ from the AM signal $v(t) = [1 + m(t)] \cos \omega_c t$, where $m(t)$ is a square wave taking on the values 0 and -0.5 volt and having a period $T \gg 1/f_c$. Sketch the *recovered* signal if $RC = T/20$ and $4T$.

3.4-2. (*a*) The waveform $v(t) = (1 + m \cos \omega_m t) \cos \omega_c t$, with m a constant ($m \leq 1$), is applied to the diode demodulator of Fig. 3.4-2*a*. Show that, if the demodulator output is to follow the envelope of $v(t)$, it is required that at any time t_0:

$$\frac{1}{RC} \geq \omega_m \left(\frac{m \sin \omega_m t_0}{1 + m \cos \omega_m t_0} \right)$$

(*b*) Using the result of part (*a*), show that if the demodulator is to follow the envelope at all times then m must be less than or equal to the value of m_0, determined from the equation

$$RC = \frac{1}{\omega_m} \frac{\sqrt{1 - m_0^2}}{m_0}$$

(*c*) Draw, qualitatively, the form of the demodulator output when the condition specified in part (*b*) is not satisfied.

3.5-1. The signal $v(t) = (1 + m \cos \omega_m t) \cos \omega_c t$ is detected using a diode envelope detector. Sketch the detector output when $m = 2$.

3.6-1. The signal $v(t) = [1 + 0.2 \cos (\omega_M/3)t] \cos \omega_c t$ is demodulated using a square-law demodulator having the characteristic $v_o = (v + 2)^2$. The output $v_o(t)$ is then filtered by an ideal low-pass filter having a cutoff frequency at f_M Hz. Sketch the amplitude-frequency characteristics of the output waveform in the frequency range $0 \leq f \leq f_M$.

3.6-2. Repeat Prob. 3.6-1 if the square-law demodulator is centered at the origin so that $v_o = v^2$.

3.6-3. The signal $v(t) = [1 + m(t)] \cos \omega_c t$ is square-law detected by a detector having the characteristic $v_o = v^2$. If the Fourier transform of $m(t)$ is a constant M_0 extending from $-f_M$ to $+f_M$, sketch the Fourier transform of $v_o(t)$ in the frequency range $-f_M < f < f_M$. *Hint:* Convolution in the frequency domain is needed to find the Fourier transform of $m^2(t)$. See Prob. 1.12-2.

3.6-4. The signal $v(t) = (1 + 0.1 \cos \omega_1 t + 0.1 \cos 2\omega_1 t) \cos \omega_c t$ is detected by a square-law detector, $v_o = 2v^2$. Plot the amplitude-frequency characteristic of $v_o(t)$.

3.9-1. (*a*) Show that the signal

$$v(t) = \sum_{i=1}^{N} [\cos \omega_c t \cos (\omega_i t + \theta_i) - \sin \omega_c t \sin (\omega_i t + \theta_i)]$$

is an SSB-SC signal ($\omega_c \gg \omega_N$). Is it the upper or lower sideband?

(*b*) Write an expression for the missing sideband.

(*c*) Obtain an expression for the total DSB-SC signal.

3.9-2. The SSB signal in Prob. 3.9-1 is multiplied by $\cos \omega_c t$ and then low-pass filtered to recover the modulation.

(*a*) Show that the modulation is completely recovered if the cutoff frequency of the low-pass filter f_0 is $f_M < f_0 < 2f_c$.

(*b*) If the multiplying signal were $\cos (\omega_c t + \theta)$, find the recovered signal.

(*c*) If the multiplying signal were $\cos (\omega_c + \Delta\omega)t$, find the recovered signal. Assume that $\Delta\omega \ll \omega_1$.

3.9-3. Show that the squaring circuit shown in Fig. 3.3-1 will not permit the generation of a local oscillator signal capable of demodulating an SSB-SC signal.

3.10-1. A baseband signal, bandlimited to the frequency range 300 to 3000 Hz, is to be superimposed on a carrier of frequency of 40 MHz as a single-sideband modulation using the filter method. Assume that bandpass filters are available which will provide 40 dB of attenuation in a frequency interval which is about 1 percent of the filter center frequency. Draw a block diagram of a suitable system. At each point in the system draw plots indicating the spectral range occupied by the signal present there.

3.10-2. The system shown in Fig. 3.10-2 is used to generate a single-sideband signal. However, an ideal 90° phase-shifting network which is independent of a frequency is unattainable. The 90° phase shift is approximated by a lattice network having the transfer function

$$H(f) = e^{-j \arctan(f/30)}$$

The input to this network is $m(t)$, given by Eq. (3.10-3). If $f_1 = 300$ Hz and $f_M = 3000$ Hz show that $H(f) \cong e^{-j\pi/2} e^{j30/f}$, for $f_1 \le f \le f_M$.

3.10-3. In the SSB generating system of Fig. 3.10-2, the carrier phase-shift network produces a phase shift which differs from 90° by a *small* angle α. Calculate the output waveform and point out the respects in which the output no longer meets the requirements for an SSB waveform. Assume that the input is a single spectral component $\cos \omega_m t$.

3.10-4. Repeat Prob. 3.10-3 except, assume instead, that the baseband phase-shift network produces a phase shift differing from 90° by a *small* angle α.

3.11-1. A received SSB signal in which the modulation is a single spectral component has a normalized power of 0.5 volt^2. A carrier is added to the signal, and the carrier plus signal are applied to a diode demodulator. The carrier amplitude is to be adjusted so that at the demodulator output 90 percent of the normalized power is in the recovered modulating waveform. Neglect dc components. Find the carrier amplitude required.

CHAPTER

FOUR

FREQUENCY-MODULATION SYSTEMS

In the amplitude-modulation systems described in Chap. 3, the modulator output consisted of a carrier which displayed variations in its amplitude. In the present chapter we discuss modulation systems in which the modulator output is of constant amplitude and in which the signal information is superimposed on the carrier through variations of the carrier frequency.

4.1 ANGLE MODULATION[1]

All the modulation schemes considered up to the present point have two principal features in common. In the first place, each spectral component of the baseband signal gives rise to one or two spectral components in the modulated signal. These components are separated from the carrier by a frequency difference equal to the frequency of the baseband component. Most importantly, the nature of the modulators is such that the spectral components which they produce depend only on the carrier frequency and the baseband frequencies. The amplitudes of the spectral components of the modulator output may depend on the amplitude of the input signals; however, the frequencies of the spectral components do not. In the second place, all the operations performed on the signal (addition, subtraction, and multiplication) are linear operations so that superposition applies. Thus, if a baseband signal $m_1(t)$ introduces one spectrum of components into the modulated signal and a second signal $m_2(t)$ introduces a second spectrum, the application of the sum $m_1(t) + m_2(t)$ will introduce a spectrum which is the sum of the spectra separately introduced. All these systems are referred to under the designation "amplitude or linear modulation." This terminology must be taken with some reservation, for we have noted that, at least in the special case of single sideband using modulation with a single sinusoid, there is no amplitude variation at all. And even more generally, when the amplitude of the modulated signal does vary, the carrier envelope need not have the waveform of the baseband signal.

We now turn our attention to a new type of modulation which is not characterized by the features referred to above. The spectral components in the modulated waveform depend on the amplitude as well as the frequency of the spectral components in the baseband signal. Furthermore, the modulation system is *not* linear and superposition does *not* apply. Such a system results when, in connection with a carrier of constant amplitude, the phase angle is made to respond in some way to a baseband signal. Such a signal has the form

$$v(t) = A \cos [\omega_c t + \phi(t)] \tag{4.1-1}$$

in which A and ω_c are constant but in which the phase angle $\phi(t)$ is a function of the baseband signal. Modulation of this type is called *angle modulation* for obvious reasons. It is also referred to as *phase modulation* since $\phi(t)$ is the phase angle of the argument of the cosine function. Still another designation is *frequency modulation* for reasons to be discussed in the next section.

4.2 PHASE AND FREQUENCY MODULATION

To review some elementary ideas in connection with sinusoidal waveforms, let us recall that the function $A \cos \omega_c t$ can be written as

$$A \cos \omega_c t = \text{real part } (Ae^{j\omega_c t}) \tag{4.2-1}$$

The function $Ae^{j\theta}$ is represented in the complex plane by a phasor of length A and an angle θ measured counterclockwise from the real axis. If $\theta = \omega_c t$, then the phasor rotates in the counterclockwise direction with an angular velocity ω_c. With respect to a coordinate system which also rotates in the counterclockwise direction with angular velocity ω_c, the phasor will be stationary. If in Eq. (4.1-1) ϕ is actually not time-dependent but is a constant, then $v(t)$ is to be represented precisely in the manner just described. But suppose $\phi = \phi(t)$ does change with time and makes positive and negative excursions. Then $v(t)$ would be represented by a phasor of amplitude A which runs ahead of and falls behind the phasor representing $A \cos \omega_c t$. We may, therefore, consider that the angle $\omega_c t + \phi(t)$, of $v(t)$, undergoes a *modulation* around the angle $\theta = \omega_c t$. The waveform of $v(t)$ is, therefore, a representation of a signal which is *modulated in phase*.

If the phasor of angle $\theta + \phi(t) = \omega_c t + \phi(t)$ alternately runs ahead of and falls behind the phasor $\theta = \omega_c t$, then the first phasor must alternately be rotating more, or less, rapidly than the second phasor. Therefore we may consider that the angular velocity of the phasor of $v(t)$ undergoes a modulation around the nominal angular velocity ω_c. The signal $v(t)$ is, therefore, an angular-velocity-modulated waveform. The angular velocity associated with the argument of a sinusoidal function is equal to the time rate of change of the argument (i.e., the angle) of the function. Thus we have that the instantaneous radial frequency $\omega = d(\theta + \phi)/dt$, and the corresponding frequency $f = \omega/2\pi$ is

$$f = \frac{1}{2\pi}\frac{d}{dt}[\omega_c t + \phi(t)] = \frac{\omega_c}{2\pi} + \frac{1}{2\pi}\frac{d}{dt}\phi(t) \tag{4.2-2}$$

The waveform $v(t)$ is, therefore, *modulated in frequency*.

In initial discussions of the sinusoidal waveform it is customary to consider such a waveform as having a fixed frequency and phase. In the present discussion we have generalized these concepts somewhat. To acknowledge this generalization, it is not uncommon to refer to the frequency f in Eq. (4.2-2) as the *instantaneous frequency* and $\phi(t)$ as the *instantaneous phase*. If the frequency variation about the nominal frequency ω_c is small, that is, if $d\phi(t)/dt \ll \omega_c$, then the resultant waveform will have an appearance which is readily recognizable as a "sine wave," albeit with a period which changes somewhat from cycle to cycle. Such a waveform is represented in Fig. 4.2-1. In this figure the modulating signal is a square wave. The frequency-modulated signal changes frequency whenever the modulation changes level.

Among the possibilities which suggest themselves for the design of a modulator are the following. We might arrange that the phase $\phi(t)$ in Eq. (4.1-1) be directly proportional to the modulating signal, or we might arrange a direct proportionality between the modulating signal and the derivative, $d\phi(t)/dt$. From Eq. (4.2-2), with $f_c = \omega_c/2\pi$

$$\frac{d\phi(t)}{dt} = 2\pi(f - f_c) \tag{4.2-3}$$

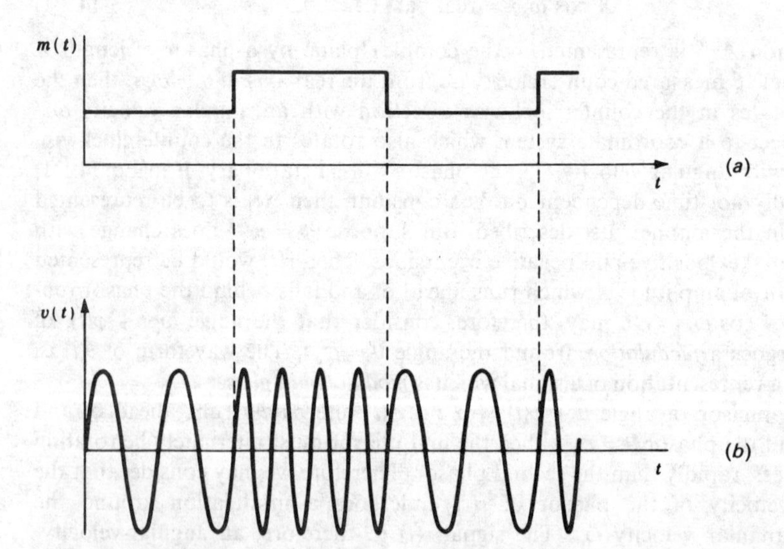

Figure 4.2-1 An angle-modulated waveform. (*a*) Modulating signal. (*b*) Frequently-modulated sinusoidal carrier signal.

where f is the instantaneous frequency. Hence in this latter case the proportionality is between modulating signal and the departure of the instantaneous frequency from the carrier frequency. Using standard terminology, we refer to the modulation of the first type as *phase* modulation, and the term *frequency modulation* refers only to the second type. On the basis of these definitions it is, of course, not possible to determine which type of modulation is involved simply from a visual examination of the waveform or from an analytical expression for the waveform. We would also have to be given the waveform of the modulating signal. This information is, however, provided in any practical communication system.

4.3 RELATIONSHIP BETWEEN PHASE AND FREQUENCY MODULATION

The relationship between phase and frequency modulation may be visualized further by a consideration of the diagrams of Fig. 4.3-1. In Fig. 4.3-1*a* the phase-modulator block represents a device which furnishes an output $v(t)$ which is a carrier, phase-modulated by the input signal $m_i(t)$. Thus

$$v(t) = A \cos [\omega_c t + k' m_i(t)] \tag{4.3-1}$$

k' being a constant. Let the waveform $m_i(t)$ be derived as the integral of the modulating signal $m(t)$ so that

$$m_i(t) = k'' \int_{-\infty}^{t} m(t)\, dt \tag{4.3-2}$$

in which k'' is also a constant. Then with $k = k'k''$ we have

$$v(t) = A \cos [\omega_c t + k \int_{-\infty}^{t} m(t)\, dt] \tag{4.3-3}$$

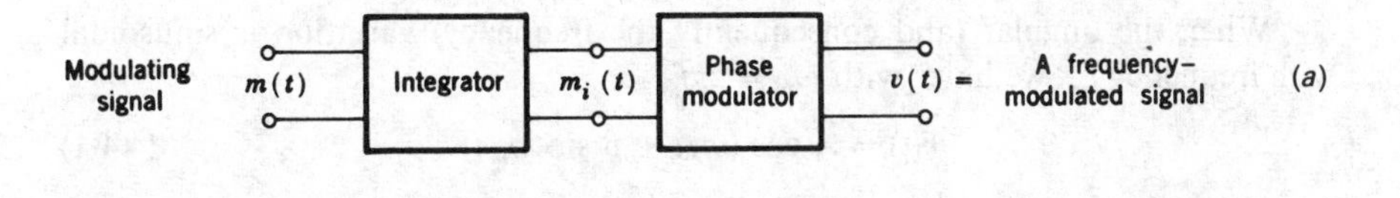

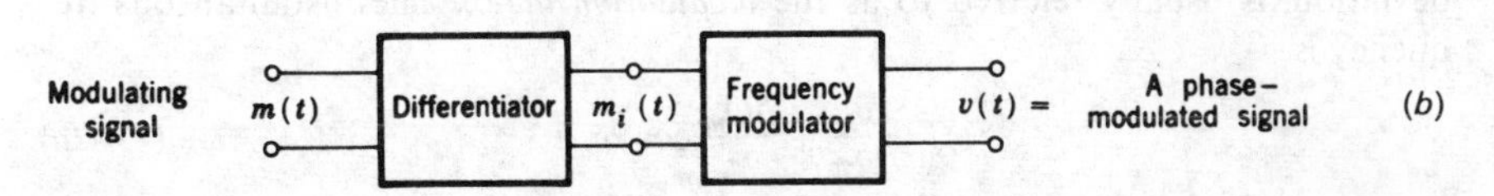

Figure 4.3-1 Illustrating the relationship between phase and frequency modulation.

The instantaneous angular frequency is

$$\omega = \frac{d}{dt}\left[\omega_c t + k \int_{-\infty}^{t} m(t)\, dt\right] = \omega_c + km(t) \tag{4.3-4}$$

The deviation of the instantaneous frequency from the carrier frequency $\omega_c/2\pi$ is

$$\nu \equiv f - f_c = \frac{k}{2\pi} m(t) \tag{4.3-5}$$

Since the deviation of the instantaneous frequency is directly proportional to the modulating signal, the combination of *integrator* and *phase modulator* of Fig. 4.3-1*a* constitutes a device for producing a *frequency-modulated* output. Similarly, the combination in Fig. 4.3-1*b* of the differentiator and frequency modulator generates a *phase-modulated output*, i.e., a signal whose phase departure from the carrier is proportional to the modulating signal.

In summary, we have referred generally to the waveform given by Eq. (4.1-1) as an *angle-modulated* waveform, an appropriate designation when we have no interest in, or information about, the modulating signal. When $\phi(t)$ is proportional to the modulating signal $m(t)$, we use the designation *phase modulation* or PM. When the time derivative of $\phi(t)$ is proportional to $m(t)$, we use the term *frequency modulation* or FM. In an FM waveform, the form of Eq. (4.3-3) is of special interest, since here the instantaneous frequency deviation is directly proportional to the signal $m(t)$ which appears explicitly in the expression. In general usage, however, we find that such precision of language is not common. Very frequently the terms angle modulation, phase modulation, and frequency modulation are used rather interchangeably and without reference to, or even interest in, the modulating signal.

4.4 PHASE AND FREQUENCY DEVIATION

In the waveform of Eq. (4.1-1) the maximum value attained by $\phi(t)$, that is, the maximum phase deviation of the total angle from the carrier angle $\omega_c t$, is called the *phase deviation*. Similarly the maximum departure of the instantaneous frequency from the carrier frequency is called the *frequency deviation*.

When the angular (and consequently the frequency) variation is sinusoidal with frequency f_m, we have, with $\omega_m = 2\pi f_m$

$$v(t) = A \cos(\omega_c t + \beta \sin \omega_m t) \tag{4.4-1}$$

where β is the peak amplitude of $\phi(t)$. In this case β which is the maximum phase deviation, is usually referred to as the *modulation index*. The instantaneous frequency is

$$f = \frac{\omega_c}{2\pi} + \frac{\beta\omega_m}{2\pi} \cos \omega_m t \tag{4.4-2a}$$

$$= f_c + \beta f_m \cos \omega_m t \tag{4.4-2b}$$

The maximum frequency deviation is defined as Δf and is given by

$$\Delta f = \beta f_m \tag{4.4-3}$$

Equation (4.4-1) can, therefore, be written

$$v(t) = A \cos\left(\omega_c t + \frac{\Delta f}{f_m} \sin \omega_m t\right) \tag{4.4-4}$$

While the instantaneous frequency f lies in the range $f_c \pm \Delta f$, it should not be concluded that all spectral components of such a signal lie in this range. We consider next the spectral pattern of such an angle-modulated waveform.

4.5 SPECTRUM OF AN FM SIGNAL: SINUSOIDAL MODULATION

In this section we shall look into the frequency spectrum of the signal

$$v(t) = \cos(\omega_c t + \beta \sin \omega_m t) \tag{4.5-1}$$

which is the signal of Eq. (4.4-1) with the amplitude arbitrarily set at unity as a matter of convenience. We have

$$\cos(\omega_c t + \beta \sin \omega_m t) = \cos \omega_c t \cos(\beta \sin \omega_m t) - \sin \omega_c t \sin(\beta \sin \omega_m t) \tag{4.5-2}$$

Consider now the expression $\cos(\beta \sin \omega_m t)$ which appears as a factor on the right-hand side of Eq. (4.5-2). It is an *even*, periodic function having an angular frequency ω_m. Therefore it is possible to expand this expression in a Fourier series in which $\omega_m/2\pi$ is the fundamental frequency. We shall not undertake the evaluation of the coefficients in the Fourier expansion of $\cos(\beta \sin \omega_m t)$ but shall instead simply write out the results. The coefficients are, of course, functions of β, and, since the function is *even*, the coefficients of the odd harmonics are zero. The result is

$$\cos(\beta \sin \omega_m t) = J_0(\beta) + 2J_2(\beta) \cos 2\omega_m t + 2J_4(\beta) \cos 4\omega_m t + \cdots + 2J_{2n}(\beta) \cos 2n\omega_m t + \cdots \tag{4.5-3}$$

while for $\sin(\beta \sin \omega_m t)$, which is an *odd* function, we find the expansion contains only odd harmonics and is given by

$$\sin(\beta \sin \omega_m t) = 2J_1(\beta) \sin \omega_m t + 2J_3(\beta) \sin 3\omega_m t + \cdots + 2J_{2n-1}(\beta) \sin(2n-1)\omega_m t + \cdots \tag{4.5-4}$$

The functions $J_n(\beta)$ occur often in the solution of engineering problems. They are known as Bessel functions of the first kind and of order n. The numerical values of $J_n(\beta)$ are tabulated in texts of mathematical tables.[2]

Putting the results given in Eqs. (4.5-3) and (4.5-4) back into Eq. (4.5-2) and using the identities

$$\cos A \cos B = \tfrac{1}{2} \cos (A - B) + \tfrac{1}{2} \cos (A + B) \tag{4.5-5}$$

$$\sin A \sin B = \tfrac{1}{2} \cos (A - B) - \tfrac{1}{2} \cos (A + B) \tag{4.5-6}$$

we find that $v(t)$ in Eq. (4.5-1) becomes

$$\begin{aligned} v(t) = J_0(\beta) \cos \omega_c t &- J_1(\beta)[\cos (\omega_c - \omega_m)t - \cos (\omega_c + \omega_m)t] \\ &+ J_2(\beta)[\cos (\omega_c - 2\omega_m)t + \cos (\omega_c + 2\omega_m)t] \\ &- J_3(\beta)[\cos (\omega_c - 3\omega_m)t - \cos (\omega_c + 3\omega_m)t] \\ &+ \cdots \end{aligned} \tag{4.5-7}$$

Observe that the spectrum is composed of a carrier with an amplitude $J_0(\beta)$ and a set of sidebands spaced symmetrically on either side of the carrier at frequency separations of ω_m, $2\omega_m$, $3\omega_m$, etc. In this respected the result is unlike that which prevails in the amplitude-modulation systems discussed earlier, since in AM a sinusoidal modulating signal gives rise to only one sideband or one pair of sidebands. A second difference, which is left for verification by the student (Prob. 4.5-1), is that the present modulation system is nonlinear, as anticipated from the discussion of Sec. 4.1.

4.6 SOME FEATURES OF THE BESSEL COEFFICIENTS

Several of the Bessel functions which determine the amplitudes of the spectral components in the Fourier expansion are plotted in Fig. 4.6-1. We note that, at $\beta = 0$, $J_0(0) = 1$, while all other J_n's are zero. Thus, as expected when there is no modulation, only the carrier, of normalized amplitude unity, is present, while all sidebands have zero amplitude. When β departs slightly from zero, $J_1(\beta)$ acquires a magnitude which is significant in comparison with unity, while all higher-order J's are negligible in comparison. That such is the case may be seen either from Fig. 4.6-1 or from the approximations[3] which apply when $\beta \ll 1$, that is,

$$J_0(\beta) \cong 1 - \left(\frac{\beta}{2}\right)^2 \tag{4.6-1}$$

$$J_n(\beta) \cong \frac{1}{n!}\left(\frac{\beta}{2}\right)^n \qquad n \neq 0 \tag{4.6-2}$$

Accordingly, for β very small, the FM signal is composed of a carrier and a single pair of sidebands with frequencies $\omega_c \pm \omega_m$. An FM signal which is so constituted, that is, a signal where β is small enough so that only a single sideband pair is of significant magnitude, is called a *narrowband* FM signal. We see further, in Fig. 4.6-1, as β becomes somewhat larger, that the amplitude J_1 of the first sideband pair increases and that also the amplitude J_2 of the second sideband pair

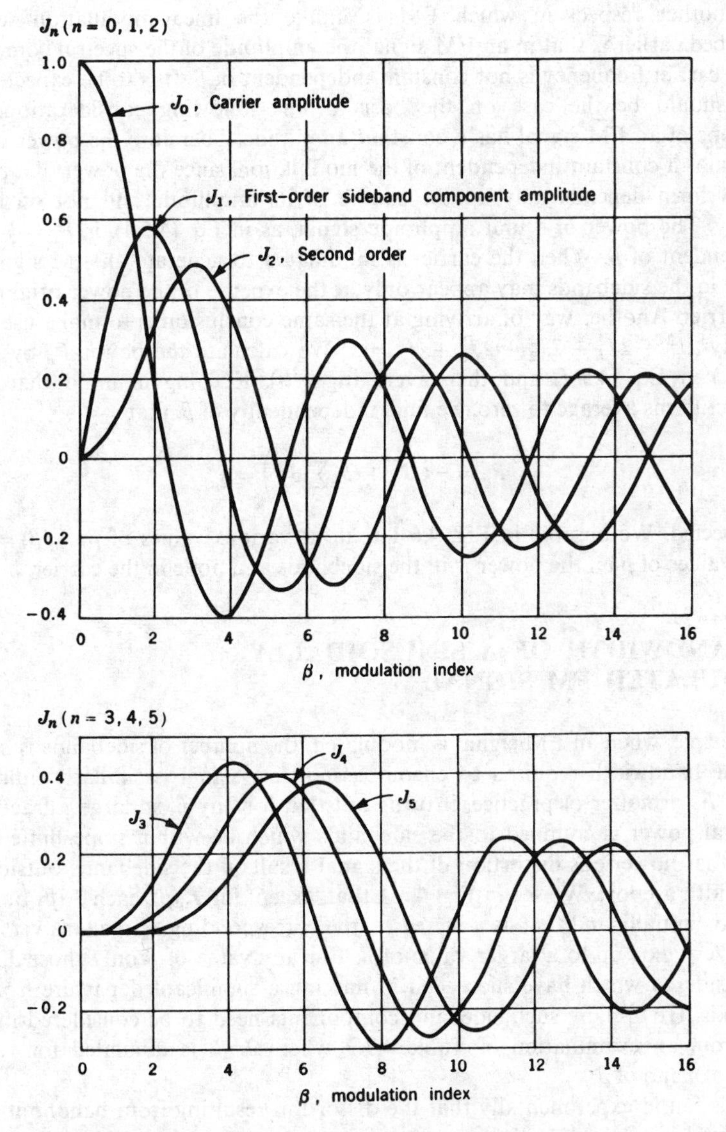

Figure 4.6-1 The Bessel functions $J_n(\beta)$ plotted as a function of β for $n = 0, 1, 2, \ldots, 5$.

becomes significant. Further, as β continues to increase, J_3, J_4, etc. begin to acquire significant magnitude, giving rise to sideband pairs at frequencies $\omega_c \pm 2\omega_m$, $\omega_c \pm 3\omega_m$, etc.

Another respect in which FM is unlike the linear-modulation schemes described earlier is that in an FM signal the amplitude of the spectral component at the carrier frequency is not constant independent of β. It is to be expected that such should be the case on the basis of the following considerations. The envelope of an FM signal has a constant amplitude. Therefore the power of such a signal is a constant independent of the modulation, since the power of a periodic waveform depends only on the square of its amplitude and not on its frequency. The power of a unit amplitude signal, as in Eq. (4.5-1), is $P_v = \frac{1}{2}$ and is independent of β. When the carrier is modulated to generate an FM signal, the power in the sidebands may appear only at the expense of the power originally in the carrier. Another way of arriving at the same conclusion is to make use of the identity[3] $J_0^2 + 2J_1^2 + 2J_2^2 + 2J_3^2 + \cdots = 1$. We calculate the power P_v by squaring $v(t)$ in Eq. (4.5-7) and then averaging $v^2(t)$. Keeping in mind that cross-product terms average to zero, we find, independently of β, that

$$P_v = \frac{1}{2}\left(J_0^2 + 2\sum_{n=1}^{\infty} J_n^2\right) = \frac{1}{2} \tag{4.6-3}$$

as expected. We observe in Fig. 4.6-1 that, at various values of β, $J_0(\beta) = 0$. At these values of β all the power is in the sidebands and none in the carrier.

4.7 BANDWIDTH OF A SINUSOIDALLY MODULATED FM SIGNAL

In principle, when an FM signal is modulated, the number of sidebands is infinite and the bandwidth required to encompass such a signal is similarly infinite in extent. As a matter of practice, it turns out that for any β, so large a fraction of the total power is confined to the sidebands which lie within some finite bandwidth that no serious distortion of the signal results if the sidebands outside this bandwidth are lost. We see in Fig. 4.6-1 that, except for $J_0(\beta)$, each $J_n(\beta)$ hugs the zero axis initially and that as n increases, the corresponding J_n remains very close to the zero axis up to a larger value of β. For any value of β only those J_n need be considered which have succeeded in making a significant departure from the zero axis. How many such sideband components need to be considered may be seen from an examination of Table 4.7-1 where $J_n(\beta)$ is tabulated for various values of n and of β.

It is found experimentally that the distortion resulting from bandlimiting an FM signal is tolerable as long as 98 percent or more of the power is passed by the bandlimiting filter. This definition of the bandwidth of a filter is, admittedly, somewhat vague, especially since the term "tolerable" means different things in different applications. However, using this definition for bandwidth, one can proceed with an initial tentative design of a system. When the system is built, the

Table 4.7-1 Values of the Bessel functions $J_n(\beta)$ for various orders n and integral values of β

n \ β	1	2	3	4	5	6	7	8	9	10
0	0.7652	0.2239	−0.2601	−0.3971	−0.1776	0.1506	0.3001	0.1717	−0.09033	−0.2459
1	0.4401	0.5767	0.3391	−0.06604	−0.3276	−0.2767	−0.004683	0.2346	0.2453	0.04347
2	0.1149	0.3528	0.4861	0.3641	0.04657	−0.2429	−0.3014	−0.1130	0.1448	0.2546
3	0.01956	0.1289	0.3091	0.4302	0.3648	0.1148	−0.1676	−0.2911	−0.1809	0.05838
4	0.002477	0.03400	0.1320	0.2811	0.3912	0.3576	0.1578	−0.1054	−0.2655	−0.2196
5		0.007040	0.04303	0.1321	0.2611	0.3621	0.3479	0.1858	−0.05504	−0.2341
6		0.001202	0.01139	0.04909	0.1310	0.2458	0.3392	0.3376	0.2043	−0.01446
7			0.002547	0.01518	0.05338	0.1296	0.2336	0.3206	0.3275	0.2167
8				0.004029	0.01841	0.05653	0.1280	0.2235	0.3051	0.3179
9					0.005520	0.02117	0.05892	0.1263	0.2149	0.2919
10					0.001468	0.006964	0.02354	0.06077	0.1247	0.2075
11						0.002048	0.008335	0.02560	0.06222	0.1231
12							0.002656	0.009624	0.02739	0.06337
13								0.003275	0.01083	0.02897
14								0.001019	0.003895	0.01196
15									0.001286	0.004508
16										0.001567

bandwidth may thereafter be readjusted, if necessary. In each column of Table 4.7-1, a line has been drawn after the entries which account for at least 98 percent of the power. To illustrate this point, consider $\beta = 1$. Then the power contained in the terms $n = 0$, 1, and 2 is

$$P = \tfrac{1}{2}J_0^2(1) + J_1^2(1) + J_2^2(1)$$

$$= 0.289 + 0.193 + 0.013 = 0.495 \tag{4.7-1}$$

The sum 0.495 is 99 percent of the power in the FM signal, which is $\frac{1}{2}$.

We note that the horizontal lines in Table 4.7-1, which indicate the value of n for 98 percent power transmission, always occur just after $n = \beta + 1$. Thus, for sinusoidal modulation the bandwidth required to transmit or receive the FM signal is

$$B = 2(\beta + 1)f_m \tag{4.7-2}$$

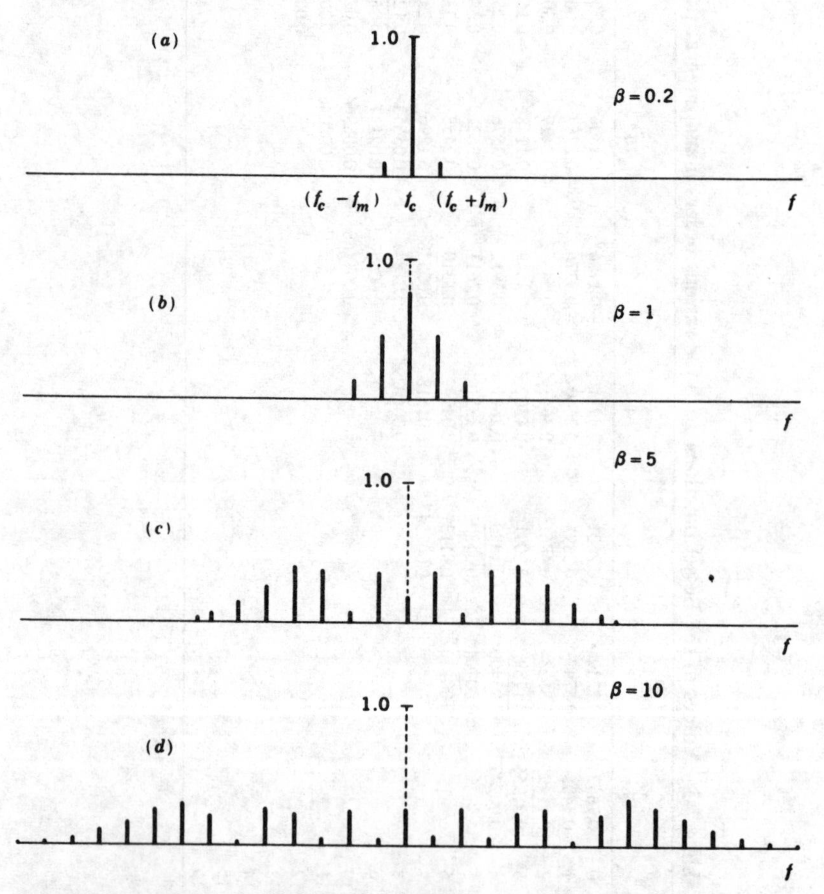

Figure 4.7-1 The spectra of sinusoidally modulated FM signals for various values of β.

By way of example, when $\beta = 5$, the sideband components furthest from the carrier which have adequate amplitude to require consideration are those which occur at frequencies $f_c \pm 6f_m$. From the table of Bessel functions published in Jahnke and Emde[2] it may be verified on a numerical basis that the rule given in Eq. (4.7-2) holds without exception up to $\beta = 29$, which is the largest value of β for which J_m is tabulated there.

Using Eq. (4.4-3), we may put Eq. (4.7-2) in a form which is more immediately significant. We find

$$B = 2(\Delta f + f_m) \tag{4.7-3}$$

Expressed in words, *the bandwidth is twice the sum of the maximum frequency deviation and the modulating frequency*. This rule for bandwidth is called *Carson's rule*.

We deduced Eqs. (4.7-2) and (4.7-3) as a generalization from Table 4.7-1, which begins with $\beta = 1$. We may note, however, that the bandwidth approximation applies quite well even when $\beta \ll 1$. For, in that case, we find that Eq. (4.7-2) gives $B = 2f_m$, which we know to be correct from our earlier discussion of narrowband FM.

The spectra of several FM signals with sinusoidal modulation are shown in Fig. 4.7-1 for various values of β. These spectra are constructed directly from the entries in Table 4.7-1 except that the signs of the terms have been ignored. The spectral lines have, in every case, been drawn upward even when the corresponding entry is negative. Hence, the lines represent the *magnitudes* only of the spectral components. Not all spectral components have been drawn. Those, far removed from the carrier, which are too small to be drawn conveniently to scale, have been omitted.

4.8 EFFECT OF THE MODULATION INDEX β ON BANDWIDTH

The modulation index β plays a role in FM which is not unlike the role played by the parameter m in connection with AM. In the AM case, and for sinusoidal modulation, we established that to avoid distortion we must observe $m = 1$ as an upper limit. It was also apparent that when it is feasible to do so, it is advantageous to adjust m to be close to unity, that is, 100 percent modulation; by so doing, we keep the magnitude of the recovered baseband signal at a maximum. On this same basis we expect the advantage to lie with keeping β as large as possible. For, again, the larger is β, the stronger will be the recovered signal. While in AM the constraint that $m \leq 1$ is imposed by the necessity to avoid distortion, there is no similar absolute constraint on β.

There is, however, a constraint which needs to be imposed on β for a different reason. From Eq. (4.7-2) for $\beta \gg 1$ we have $B \cong 2\beta f_m$. Therefore the maximum value we may allow for β is determined by the maximum allowable bandwidth and the modulation frequency. In comparing AM with FM, we may then note, in review, that in AM the recovered modulating signal may be made progressively larger subject to the onset of distortion in a manner which keeps the occupied bandwidth constant. In FM there is no similar limit on the modulation, but increasing the magnitude of the recovered signal is achieved at the expense of bandwidth. A more complete comparison is deferred to Chaps. 8 and 9, where we shall take account of the presence of noise and also of the relative power required for transmission.

4.9 SPECTRUM OF "CONSTANT BANDWIDTH" FM

Let us consider that we are dealing with a modulating signal voltage $v_m \cos 2\pi f_m t$ with v_m the peak voltage. In a phase-modulating system the phase angle $\phi(t)$ would be proportional to this modulating signal so that $\phi(t) = k' v_m \cos 2\pi f_m t$, with k' a constant. The phase deviation is $\beta = k' v_m$, and, for constant v_m, the bandwidth occupied increases linearly with modulating frequency since $B \cong 2\beta f_m = 2k' v_m f_m$. We may avoid this variability of bandwidth with modulating frequency by arranging that $\phi(t) = (k/2\pi f_m) v_m \sin 2\pi f_m t$ (k a constant). For, in this case

$$\beta = \frac{k v_m}{2\pi f_m} \tag{4.9-1}$$

and the bandwidth is $B \cong (2k/2\pi) v_m$, independently of f_m. In this latter case, however, the instantaneous frequency is $\omega = \omega_c + k v_m \cos 2\pi f_m t$. Since the instantaneous frequency is proportional to the modulating signal, the initially angle-modulated signal has become a frequency-modulated signal. Thus a signal intended to occupy a nominally constant bandwidth is a frequently-modulated rather than an angle-modulated signal.

In Fig. 4.9-1 we have drawn the spectrum for three values of β for the condition that βf_m is kept constant. The nominal bandwidth $B \cong 2\,\Delta f = 2\beta f_m$ is consequently constant. The amplitude of the unmodulated carrier at f_c is shown by a dashed line. Note that the extent to which the actual bandwidth extends beyond the nominal bandwidth is greatest for small β and large f_m and is least for large β and small f_m.

In commercial FM broadcasting, the Federal Communications Commission allows a frequency deviation $\Delta f = 75$ kHz. If we assume that the highest audio frequency to be transmitted is 15 kHz, then at this frequency $\beta = \Delta f / f_m = 75/15 = 5$. For all other modulation frequencies β is larger than 5. When $\beta = 5$, there are $\beta + 1 = 6$ significant sideband pairs so that at $f_m = 15$ kHz the bandwidth required is $B = 2 \times 6 \times 15 = 180$ kHz, which is to be compared with $2\,\Delta f = 150$ kHz. When $\beta = 20$, there are 21 significant sideband pairs, and $B = 2 \times 21 \times 15/4 = 157.5$ kHz. In the limiting case of very large β and correspondingly very small f_m, the actual bandwidth becomes equal to the nominal bandwidth $2\,\Delta f$.

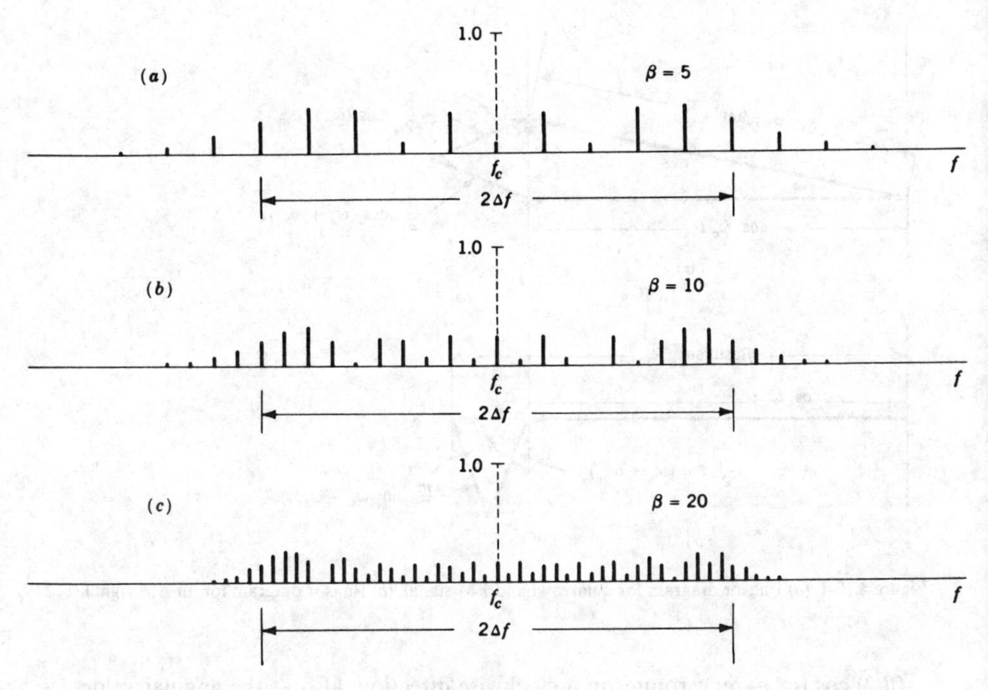

Figure 4.9-1 Spectra of sinusoidally modulated FM signals. The nominal bandwidth $B \approx 2\beta f_m = 2\ \Delta f$ is kept fixed.

4.10 PHASOR DIAGRAM FOR FM SIGNALS

With the aid of a phasor diagram we shall be able to arrive at a rather physically intuitive understanding of how so odd an assortment of sidebands as in Eq. (4.5-7) yields an FM signal of constant amplitude. The diagram will also make clear the difference between AM and narrowband FM (NBFM). In both of these cases there is only a single pair of sideband components.

Let us consider first the case of narrowband FM. From Eqs. (4.4-1), (4.6-1), and (4.6-2) we have for $\beta \ll 1$ that

$$v(t) = \cos(\omega_c t + \beta \sin \omega_m t) \tag{4.10-1a}$$

$$\cong \cos \omega_c t - \frac{\beta}{2} \cos(\omega_c - \omega_m)t + \frac{\beta}{2} \cos(\omega_c + \omega_m)t \tag{4.10-1b}$$

Refer to Fig. 4.10-1*a*. Assuming a coordinate system which rotates counterclockwise at an angular velocity ω_c, the phasor for the carrier-frequency term in Eq. (4.10-1) is fixed and oriented in the horizontal direction. In the same coordinate system, the phasor for the term $(\beta/2) \cos(\omega_c + \omega_m)t$ rotates in a counterclockwise direction at an angular velocity ω_m, while the phasor for the term

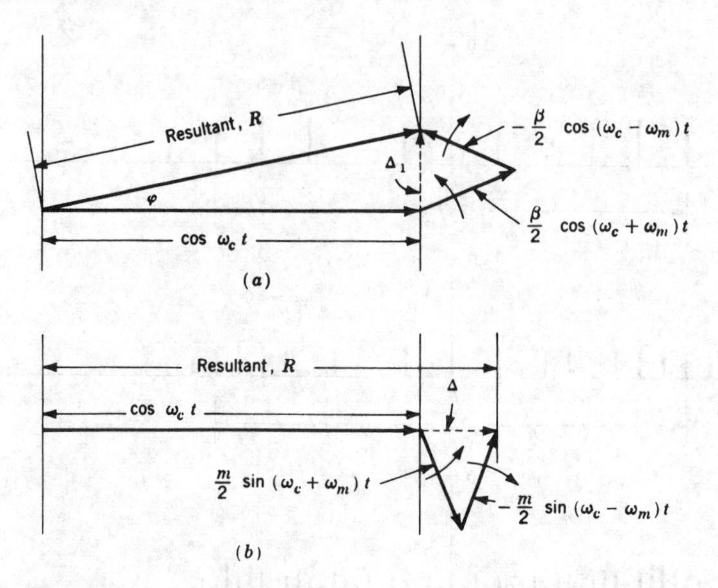

Figure 4.10-1 (*a*) Phasor diagram for a narrowband FM signal. (*b*) Phasor diagram for an AM signal.

$-(\beta/2)\cos(\omega_c - \omega_m)t$ rotates in a clockwise direction, also at the angular velocity ω_m. At the time $t = 0$, both phasors, which represent the sideband components, have maximum projections in the horizontal direction. At this time one is parallel to, and one is antiparallel to, the phasor representing the carrier, so that the two cancel. The situation depicted in Fig. 4.10-1*a* corresponds to a time shortly after $t = 0$. At this time, the rotation of the sideband phasors which are in opposite directions, as indicated by the curved arrows, have given rise to a sum phasor Δ_1. In the coordinate system in which the carrier phasor is stationary, the phasor Δ_1 always stands perpendicularly to the carrier phasor and has the magnitude

$$\Delta_1 = \beta \sin \omega_m t \tag{4.10-2}$$

The carrier, now slightly reduced in amplitude, and Δ_1 combine to give rise to a resultant R. The angular departure of R from the carrier phasor is ϕ. It is readily seen from Fig. 4.10-1*a* that since $\beta \ll 1$, the maximum value of $\phi \simeq \tan \phi = \beta$, as is to be expected. The small variation in the amplitude of the resultant which appears in Fig. 4.10-1*a* is only the result of the fact that we have neglected higher-order sidebands.

Now let us consider the phasor diagram for AM. The AM signal is

$$(1 + m \sin \omega_m t)\cos \omega_c t = \cos \omega_c t + \frac{m}{2}\sin(\omega_c + \omega_m)t - \frac{m}{2}\sin(\omega_c - \omega_m)t \tag{4.10-3}$$

and the individual terms are represented as phasors in Fig. 4.10-1*b*. Comparing Eqs. (4.10-1) and (4.10-3), we see that there is a 90° phase shift in the phases of the sidebands between the FM and AM cases. In Fig. 4.10-1*b* the sum Δ of the sideband phasors is given by

$$\Delta = m \sin \omega_m t \tag{4.10-4}$$

The important difference between the FM and AM cases is that in the former the sum Δ_1 is always perpendicular to the carrier phasor, while in the latter the sum Δ is always parallel to the carrier phasor. Hence in the AM case, the resultant R does not rotate with respect to the carrier phasor but instead varies in amplitude between $1 + m$ and $1 - m$.

Another way of looking at the difference between AM and NBFM is to note that in NBFM where $\beta \ll 1$

$$v(t) \approx \cos \omega_c t - \beta \sin \omega_m t \sin \omega_c t \tag{4.10-5}$$

while in AM

$$v(t) = \cos \omega_c t + m \sin \omega_m t \cos \omega_c t \tag{4.10-6}$$

Note that in NBFM the first term is $\cos \omega_c t$, while the second term involves $\sin \omega_c t$, a *quadrature* relationship. In AM both first and second terms involve $\cos \omega_c t$, an *in-phase* relationship.

To return now to the FM case and to Fig. 4.10-1*a*, the following point is worth noting. When the angle ϕ completes a full cycle, that is, Δ_1 varies from $+\beta$ to $-\beta$ and back again to $+\beta$, the magnitude of the resultant R will have executed *two* full cycles. For R is a maximum at $\Delta_1 = \beta$, a minimum at $\Delta_1 = 0$, a maximum again when $\Delta_1 = -\beta$, and so on. On this basis, it may well be expected that if an additional sideband pair is to be added to the first to make R more nearly constant, this new pair must give rise to a resultant Δ_2 which varies at the frequency $2\omega_m$. Thus, we are not surprised to find that as the phase deviation β increases, a sideband pair comes into existence at the frequencies $\omega_c \pm 2\omega_m$.

As long as we depend on the first-order sideband pair only, we see from Fig. 4.10-1*a* that ϕ cannot exceed 90°. A deviation of such magnitude is hardly adequate. For consider, as above, that $\Delta f = 75$ kHz and that $f_m = 50$ Hz. Then $\omega_m = 75{,}000/50 = 1500$ rad, and the resultant R must, in this case, spin completely about $1500/2\pi$ or about 240 times. Such wild whirling is made possible through the effect of the higher-order sidebands. As noted, the first-order sideband pair gives rise to a phasor $\Delta_1 = J_1(\beta) \sin \omega_m t$, which phasor is perpendicular to the carrier phasor. It may also be established by inspection of Eq. (4.5-7) that the second-order sideband pair gives rise to a phasor $\Delta_2 = J_2(\beta) \cos 2\omega_m t$ and that this phasor is *parallel* to the carrier phasor. Continuing, we easily establish that all odd-numbered sideband pairs give rise to phasors

$$\Delta_n = J_n(\beta) \sin n\omega_m t \qquad n \text{ odd} \tag{4.10-7}$$

which are perpendicular to the carrier phasor, while all even-numbered sideband pairs give rise to phasors

$$\Delta_n = J_n(\beta) \cos n\omega_m t \qquad n \text{ even} \tag{4.10-8}$$

which are parallel to the carrier phasor. Thus, phasors Δ_1, Δ_2, Δ_3, etc., alternately perpendicular and parallel to the carrier phasor, are added to carry the end point of the resultant phasor R completely around as many times as may be required, while maintaining R at constant magnitude. It is left as an exercise for the student to show by typical examples how the superposition of a carrier and sidebands may swing a constant-amplitude resultant around through an arbitrary angle (Prob. 4.10-2).

4.11 SPECTRUM OF NARROWBAND ANGLE MODULATION: ARBITRARY MODULATION

Previously we considered the spectrum, in NBFM, which is produced by sinusoidal modulation. We found that, just as in AM, such modulation gives rise to two sidebands at frequencies $\omega_c + \omega_m$ and $\omega_c - \omega_m$. We extend the result now to an arbitrary modulating waveform.

We may readily verify (Prob. 4.11-1) that superposition applies in narrowband angle modulation just as it does to AM. That is, if $\beta_1 \sin \omega_1 t + \beta_2 \sin \omega_2 t$ is substituted in Eq. (4.10-1a) in place of $\beta \sin \omega_m t$, the sidebands which result are the sum of the sidebands that would be yielded by either modulation alone. Hence even if a modulating signal of waveform $m(t)$, with a continuous distribution of spectral components, is used in either AM or narrowband angle modulation the forms of the sideband spectra will be the same in the two cases.

More formally, we have in AM, when the modulating waveform is $m(t)$, the signal is

$$v_{\text{AM}}(t) = A[1 + m(t)] \cos \omega_c t = A \cos \omega_c t + Am(t) \cos \omega_c t \tag{4.11-1}$$

Let us assume, for simplicity, that $m(t)$ is a finite energy waveform with a Fourier transform $M(j\omega)$. We use the theorem that if the Fourier transform $\mathscr{F}[m(t)] = M(j\omega)$, then $\mathscr{F}[m(t) \cos \omega_c t]$ is as given in Eq. (3.2-4). We then find that

$$\mathscr{F}[v_{\text{AM}}(t)] = \frac{A}{2}[\delta(\omega + \omega_c) + \delta(\omega - \omega_c)] + \frac{A}{2}[M(j\omega + j\omega_c) + M(j\omega - j\omega_c)] \tag{4.11-2}$$

The narrowband angle-modulation signal of Eq. (4.10-1), except of amplitude A and with phase modulation $m(t)$, may be written, for $|m(t)| \ll 1$,

$$v_{\text{PM}}(t) \cong A \cos \omega_c t - Am(t) \sin \omega_c t \tag{4.11-3}$$

so that

$$\mathscr{F}[v_{\text{PM}}(t)] = \frac{A}{2}[\delta(\omega+\omega_c)+\delta(\omega-\omega_c)] + \frac{A}{2}e^{-j\pi/2}[M(j\omega+j\omega_c)-M(j\omega-j\omega_c)] \tag{4.11-4}$$

Comparing Eq. (4.11-2) with Eq. (4.11-4), we observe that

$$|\mathscr{F}[v_{\text{AM}}]|^2 = |\mathscr{F}[v_{\text{PM}}]|^2 \tag{4.11-5}$$

Thus, if we were to make plots of the energy spectral densities of $v_{\text{AM}}(t)$ and of $v_{\text{PM}}(t)$, we would find them identical. Similarly, if $m(t)$ were a signal of finite power, we would find that plots of power spectral density would be the same.

4.12 SPECTRUM OF WIDEBAND FM (WBFM): ARBITRARY MODULATION[4]

In this section we engage in a heuristic discussion of the spectrum of a wideband FM signal. We shall not be able to deduce the spectrum with the precision that is possible in the NBFM case described in the previous section. As a matter of fact, we shall be able to do no more than to deduce a means of expressing approximately the power spectral density of a WBFM signal. But this result is important and useful.

Previously, to characterize an FM signal as being narrowband or wideband, we had used the parameter $\beta \equiv \Delta f/f_m$, where Δf is the frequency deviation and f_m the frequency of the sinusoidal modulating signal. The signal was then NBFM or WBFM depending on whether $\beta \ll 1$ or $\beta \gg 1$. Alternatively we distinguished one from the other on the basis of whether one or very many sidebands were produced by each spectral component of the modulating signal, and on the basis of whether or not superposition applies. We consider now still another alternative.

Let the symbol $\nu \equiv f - f_c$ represent the frequency difference between the instantaneous frequency f and the carrier frequency f_c; that is, $\nu(t) = (k/2\pi)m(t)$ [see Eq. (4.3-5)]. The period corresponding to ν is $T = 1/\nu$. As f varies, so also will ν and T. The frequency ν is the frequency with which the resultant phasor R in Fig. 4.10-1 rotates in the coordinate system in which the carrier phasor is fixed. In WBFM this resultant phasor rotates through many complete revolutions, and its speed of rotation does not change radically from revolution to revolution. Since, the resultant R is constant, then if we were to examine the plot as a function of time of the projection of R in, say, the horizontal direction, we would recognize it as a sinusoidal waveform because its frequency would be changing very slowly. No appreciable change in frequency would take place during the course of a cycle. Even a long succession of cycles would give the appearance of being of rather constant frequency. In NBFM, on the other hand, the phasor R simply oscillates about the position of the carrier phasor. Even though, in this case, we may still formally calculate a frequency ν, there is no corresponding time interval during which the phasor makes complete revolutions at approximately a constant rate ν.

Now let us consider that a carrier is wideband FM-modulated by a signal $m(t)$ such as, say, an audio signal. Let the modulation $m(t)$ be characterized by the probability density function $f(m)$. Then the fraction of the time that $m(t)$ spends in the range between m_1 and $m_1 + dm$ is the probability that $m(t)$ lies between m_1 and $m_1 + dm$, that is, $f(m_1)\,dm$. Corresponding to each value of $m(t)$, the value of the frequency deviation $\nu(t) = (k/2\pi)m(t)$. Hence, during the time $m(t)$ is in the range m_1 and $m_1 + dm$, ν is in the range ν_1 and $\nu_1 + d\nu$. As we have seen in WBFM, the frequency ν changes only relatively slowly. Thus the assignment of a frequency ν to a waveform during the interval when $m(t)$ has a value corresponding to ν has a physical as well as a purely mathematical significance. On this basis, it is reasonable to say that, of the total power in the FM waveform, the fraction of the power in the frequency range between ν_1 and $\nu_1 + d\nu$ is proportional to the time $m(t)$ spends in the range m_1 to $m_1 + dm$. With $G(\nu)$, the power spectral density of the FM waveform, we have the result that $G(\nu_1)\,d\nu$ is proportional to $f(m_1)\,dm$. Finally, since $d\nu$ is proportional to dm, we have the most important result that $G(\nu)$ is proportional to $f(m)$. Expressed in words, the *power spectral density $G(\nu)$ of a WBFM waveform is determined by, and has the same form as, the density function $f(m)$ of the modulating waveform.*

Example 4.12-1 In the WBFM signal

$$v(t) = A\cos\left[2\pi f_c t + k\int_{-\infty}^{t} m(\lambda)\,d\lambda\right] \tag{4.12-1}$$

$m(t)$ is an ergodic random process having a probability density

$$f(m) = \begin{cases} \dfrac{1}{M} & -\dfrac{M}{2} \le m \le \dfrac{M}{2} \\ 0 & \text{elsewhere} \end{cases} \tag{4.12-2}$$

Obtain an expression for $G(f)$, the power spectral density of $v(t)$.

SOLUTION Since $G(\nu)$ (with $\nu \equiv f - f_c$) is proportional to $f(m)$, we have

$$G(\nu) = \alpha f(m) \tag{4.12-3}$$

where α is a constant of proportionality. Since $\nu(t) = km(t)/2\pi$,

$$G(\nu) = \begin{cases} \dfrac{\alpha}{M} & -\dfrac{kM}{4\pi} \le \nu \le \dfrac{kM}{4\pi} \\ 0 & \text{elsewhere} \end{cases} \tag{4.12-4}$$

Replacing ν by $f - f_c$, and expressing the power spectral density for both positive and negative frequencies (i.e., a two-sided density), we have

$$G(f) = \begin{cases} \dfrac{\alpha}{2M} & f_c - \dfrac{kM}{4\pi} \le |f| \le f_c + \dfrac{kM}{4\pi} \\ 0 & \text{elsewhere} \end{cases} \tag{4.12-5}$$

To evaluate α, we note that the power of the FM waveform is $A^2/2$. Hence,

$$\int_{-\infty}^{\infty} G(f)\, df = \frac{A^2}{2} \tag{4.12-6}$$

From Eqs. (4.12-5) and (4.12-6) we find $\alpha = \pi A^2/k$, so that

$$G(f) = \begin{cases} \dfrac{\pi A^2}{2kM} & f_c - \dfrac{kM}{4\pi} \le |f| \le f_c + \dfrac{kM}{4\pi} \\ 0 & \text{elsewhere} \end{cases} \tag{4.12-7}$$

4.13 BANDWIDTH REQUIRED FOR A GAUSSIAN MODULATED WBFM SIGNAL

Earlier, we found that when a carrier was sinusoidally modulated, Carson's rule $B = 2(\Delta f + f_m)$ given in Eq. (4.7-3) specifies the bandwidth required to transmit enough of the power (98 percent) so that the modulation may be recovered without distortion. We now make a similar calculation for the case where the modulation has a gaussian distribution. The result is extremely important since many physically encountered signals, while not precisely gaussian, are reasonably approximated as gaussian.

Using the result stated at the end of Sec. 4-12, we note that if $m(t)$ is gaussian, so also is $G(f)$. Therefore the two-sided spectral density $G(f)$ has the form

$$G(f) = \frac{A^2}{4\sqrt{2\pi}\,\Delta f_{\text{rms}}} [e^{-(f-f_c)^2/2(\Delta f_{\text{rms}})^2} + e^{-(f+f_c)^2/2(\Delta f_{\text{rms}})^2}] \tag{4.13-1}$$

as shown in Fig. 4.13-1 where A is the amplitude of the FM waveform and Δf_{rms} is the variance of the gaussian power spectrum density. An FM waveform of

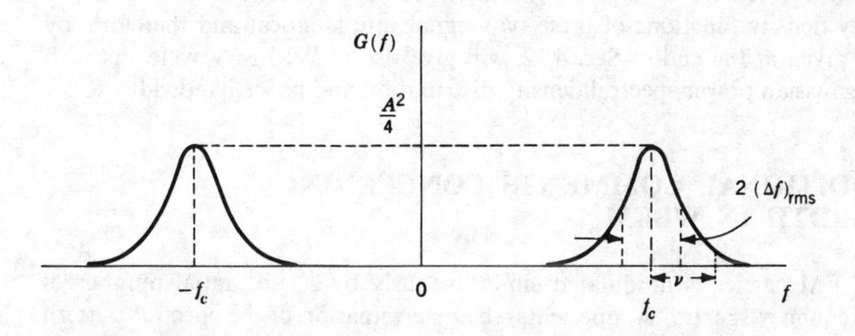

Figure 4.13-1 The power spectral density of a carrier f_c frequency-modulated by a baseband signal with a gaussian amplitude distribution. The variable ν measures the departure of frequency from the carrier frequency.

amplitude A has a power $A^2/2$, and the student may verify that, with $G(f)$ as given in Eq. (4.13-1),

$$\int_{-\infty}^{\infty} G(f)\, df = \frac{A^2}{2} \tag{4.13-2}$$

as required.

We now ask what must be the bandwidth B of a rectangular bandpass filter centered at f_c which will pass 98 percent of the power of the FM waveform. Recognizing that each of the terms in Eq. (4.13-1) makes equal contributions to the power and using the variable $v \equiv f \pm f_c$, we find that B is determined by the equation

$$\frac{1}{\sqrt{2\pi}\,\Delta f_{\text{rms}}} \int_{-B/2}^{B/2} e^{-v^2/2(\Delta f_{\text{rms}})^2}\, dv = 0.98 \tag{4.13-3}$$

Letting $x \equiv v/(\sqrt{2}\,\Delta f_{\text{rms}})$ yields

$$0.98 = \frac{2}{\sqrt{\pi}} \int_0^{B/(2\sqrt{2}\,\Delta f_{\text{rms}})} e^{-x^2}\, dx = \operatorname{erf} \frac{B}{2\sqrt{2}\,\Delta f_{\text{rms}}} \tag{4.13-4}$$

From a table of values of the error functions we find

$$B = 2\sqrt{2}(1.645)\,\Delta f_{\text{rms}}$$

$$= 4.6\,\Delta f_{\text{rms}} \tag{4.13-5}$$

To recapitulate, we find that a modulating signal with a gaussian amplitude distribution gives rise to an FM waveform with a gaussian power spectral density. If the variance of the spectral density is $(\Delta f_{\text{rms}})^2$, the bandwidth required to pass 98 percent of the power of the waveform is given by Eq. (4.13-5).

There is another useful result that may be deduced for the case of a gaussian modulating signal. Suppose we have two gaussian modulating signals $m_1(t)$ and $m_2(t)$ which are related in that $\overline{m_1^2(t)} = \overline{m_2^2(t)}$ but are otherwise arbitrary. The probability density functions of these two signals are identical and therefore, by the result given at the end of Sec. 4.12, will give rise to WBFM waveforms with the same gaussian power spectral density distribution and hence bandwidth B.

4.14 ADDITIONAL COMMENTS CONCERNING BANDWIDTH IN WBFM

When an FM carrier is modulated simultaneously by a substantial number of discrete frequency spectral components, the determination of the spectral pattern that results is a very formidable task. For this reason, we find in the literature at least two *rule-of-thumb* estimates of bandwidth which are widely used. Suppose that the individual modulating signals acting alone would produce frequency deviations $(\Delta f)_1$, $(\Delta f)_2$, etc. Then one, rather pessimistic, rule says $B = 2[(\Delta f)_1 + (\Delta f)_2 + \cdots]$, while a second, rather optimistic, rule says $B = 2$[rms value of (Δf)'s]. As is readily verified (Prob. 4.14-1), these two rules may give widely different results.

When the modulating signal has a continuous spectral density and gives rise to an FM signal with sidebands which have a continuous spectral density, a bandwidth definition often used is defined by

$$B \equiv 2\left[\frac{\int_{-\infty}^{\infty} v^2 G(v)\, dv}{\int_{-\infty}^{\infty} G(v)\, dv}\right]^{1/2} \tag{4.14-1}$$

This definition yields an *rms bandwidth* which is twice the *radius of gyration* of the area under the power spectral density plot. This definition often finds favor with mathematicians because of its computational ease. A comparison of B as given by Eq. (4.14-1) with the bandwidth given by other definitions is explored in Probs. 4.14-1 and 4.14-2.

4.15 FM GENERATION: PARAMETER-VARIATION METHOD

The generator which produces the carrier of an FM waveform is, in many instances, a tuned circuit oscillator. Such oscillator circuits furnish a sinusoidal waveform whose frequency is very largely determined by, and is very nearly equal to, the resonant frequency of an inductance-capacitance combination. Thus the frequency of oscillation is $f = (2\pi\sqrt{LC})^{-1}$, in which L is the inductance and C the capacitance. Such an LC combination, a parallel combination in this case, is shown in Fig. 4.15-1. The capacitor consists here of a fixed capacitor C_0, which is shunted by a voltage-variable capacitor C_v. A voltage-variable capacitor, commonly called a *varicap*, is one whose capacitance value depends on the dc biasing voltage maintained across its electrodes. Semiconductor diodes, when operated with a reverse bias, have characteristics suitable to permit their use as voltage-variable capacitors.

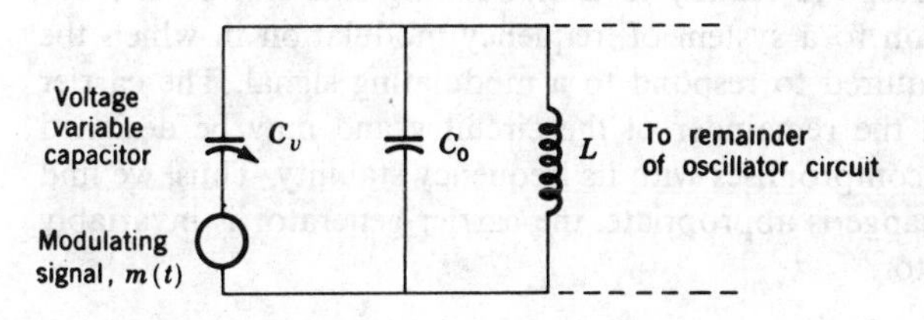

Figure 4.15-1 A voltage-variable capacitor is used to frequency-modulate an LC oscillator.

In the circuit of Fig. 4.15-1 the modulating signal varies the voltage across C_v. As a consequence, the capacitance of C_v changes and causes a corresponding change in the oscillator frequency. Ordinarily the modulating frequency is very small in comparison with the oscillator frequency. Therefore the fractional change in C_v may be very small during the course of many cycles of the oscillator signal. We may consequently expect that even with this variable capacitance, the instantaneous oscillator frequency will be given by $f = (2\pi\sqrt{LC})^{-1}$. Then we have the result that the system suggested in Fig. 4.15-1 will generate an oscillator output signal whose instantaneous frequency depends on the instantaneous value of the modulating signal. Any oscillator whose frequency is controlled by the modulating-signal voltage is called a *voltage-controlled oscillator*, or VCO.

Frequency modulation may be achieved by the variation of any element or parameter on which the frequency depends. If the frequency variation is to occur in response to a modulating signal $m(t)$, then a component must be available, capacitor, resistor, or inductor, whose value can be varied with an electrical signal. We have noted that reversed-biased junctions may serve as voltage-variable capacitors. Similarly PIN diodes and FETs (field-effect transistors) have found application as variable resistors. The inductance of a magnetic-cored inductor, called a saturable reactor, can be varied by changing the dc biasing current through the winding, and thereby changing the core permeability.

There are occasions when it is appropriate to use a signal source such as a multivibrator as the carrier generator. In this case, of course, the waveform generated will not be sinusoidal. There are situations (more frequently in laboratory equipment than in a communications channel) where a sinusoidal form is not required, or, if a sinusoidal waveform is needed, it is often possible to convert the waveform to such a sinusoidal waveform by filtering, or by the use of nonlinear shaping circuits. In such cases, where the prime timing generator is a multivibrator-type circuit, there is available an additional mechanism for frequency modulation. For, in such circuits, the frequency depends not only on the values of the passive components but also on the supply voltages which are used to bias the active devices. Thus frequency modulation may be achieved by using the modulating signal to control these biasing voltages.

A principal difficulty with the parameter-variation method of frequency modulation is the difficulty it entails when we require that the carrier frequency [the signal frequency when the modulation signal $m(t) = 0$] be maintained constant to a high order of precision over extended periods of time. There is a certain measure of inconsistency in requiring that a device have *long-time* frequency stability and yet be able to respond readily to a modulating signal. We turn our attention in the next section to a system of frequency modulation in which the carrier generator is *not* required to respond to a modulating signal. The carrier generator is isolated from the remainder of the circuitry and may be designed without the need to make compromises with its frequency stability. Thus, we find that, when the frequency range is appropriate, the carrier generator is invariably a crystal-controlled oscillator.

4.16 AN INDIRECT METHOD OF FREQUENCY MODULATION (ARMSTRONG SYSTEM)

A phase-modulated waveform in which the modulating waveform is $m(t)$ is written $\cos [\omega_c t + m(t)]$. If the modulation is narrowband $[|m(t)| \ll 1]$, then we may use the approximation

$$\cos [\omega_c t + m(t)] \cong \cos \omega_c t - m(t) \sin \omega_c t \tag{4.16-1}$$

The term $m(t) \sin \omega_c t$ is a DSB-SC waveform in which $m(t)$ is the modulating waveform and $\sin \omega_c t$ the carrier. We note that the carrier of the FM waveform, that is, $\cos \omega_c t$, and the carrier of the DSB-SC waveform are in quadrature. We may note in passing that if the two carriers are in phase, the result is an AM signal since

$$\cos \omega_c t + m(t) \cos \omega_c t = [1 + m(t)] \cos \omega_c t \tag{4.16-2}$$

A technique used in commercial FM systems to generate NBFM, which is based on our observation in connection with Eq. (4.16-1), is shown in Fig. 4.16-1. Here a balanced modulator is employed to generate the DSB-SC signal using $\sin \omega_c t$ as the carrier of the modulator. This carrier is then shifted in phase by 90° and, when added to the balanced modulator output, thereby forms an NBFM signal. However, the signal so generated will be phase-modulated rather than frequency-modulated. If we desire that the frequency rather than the phase be proportional to the modulation $m(t)$, then, as discussed in Sec. 4.3 and illustrated in Fig. 4.3-1, we need merely integrate the modulating signal before application to the modulator.

If the system of Fig. 4.16-1 is to yield an output signal whose phase deviation is directly proportional to the amplitude of the modulating signal, then the phase deviation must be kept small. That such is the case is readily to be seen in Fig. 4.10-1*a*. If we neglect the small second-order correction in the carrier amplitude and assume it to be of unit magnitude, we have $\tan \phi = \Delta_1$. Since, however, Δ_1 $(= \beta \sin \omega_m t)$ is proportional to the modulating signal, we actually require that

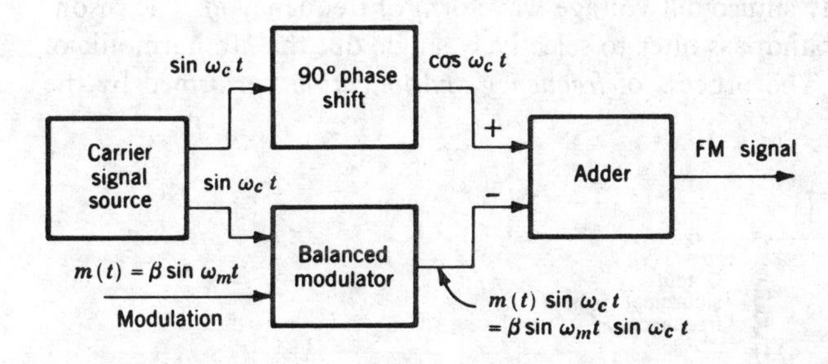

Figure 4.16-1 Illustrating the principle of the Armstrong system of generating a PM signal.

$\phi = \Delta_1$. In order that we may replace tan ϕ by ϕ, we require that at all times $\phi \ll 1$. In this case $\beta \ll 1$, and then $\phi = \beta \sin \omega_m t$.

The restriction that $\beta \ll 1$ imposes a similar constraint on the allowable frequency deviation $\Delta f (= \beta\omega_m/2\pi)$ when the system of Fig. 4.16-1 is adapted for use as a frequency-modulation system by the addition of an integrator. In the next section we discuss how the frequency deviation, and the phase deviation as well, of a narrowband signal may be increased by the process of frequency multiplication.

4.17 FREQUENCY MULTIPLICATION

A *frequency multiplier* is a combination of a nonlinear element and a bandpass filter. One such possible combination is shown in Fig. 4.17-1. We consider the operation, qualitatively, in order to see the relevance of the process to our present interest of increasing the frequency deviation of an FM signal.

Assume that the input signal to the transistor in the circuit of Fig. 4.17-1 is a periodic signal, possibly sinusoidal but not necessarily so. The amplitude of the input signal is large enough, and the biasing (not shown) is such that the transistor operates nonlinearly. Typically, the transistor operates in the *class C* mode. In this mode of operation the transistor is in the cutoff region for more than half of the period of the input signal. During intervals in the neighborhood of the peak positive excursions of the input signal, the transistor is driven into the active region, possibly even into saturation. Collector current flows, not continuously, but rather in spurts, forming pulses, one pulse for each cycle of the input driving signal. The collector-current waveform has the same fundamental period as has the driving signal but is rich in higher-frequency harmonics. The *LC* parallel resonant circuit is tuned to resonance at the *n*th harmonic of the frequency f of the input signal. The sharpness of the resonance is such that the impedance presented by the resonant circuit is very small at all harmonic frequencies except the *n*th. All components of collector current except the component at frequency nf pass through the resonant circuit without developing appreciable voltage. However, in response to this *n*th harmonic current component, there appears across the resonant circuit a very nearly sinusoidal voltage waveform of frequency nf. The resonant circuit serves as a bandpass filter to selectively single out the *n*th harmonic of the driving waveform. The process of *frequency multiplication* performed by the

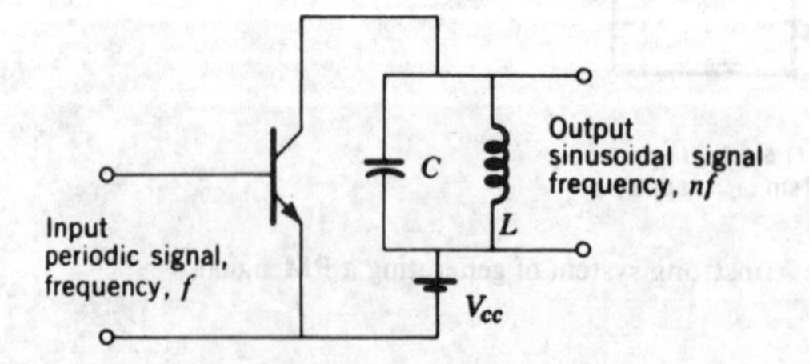

Figure 4.17-1 A frequency-multiplier circuit.

multiplier under consideration is one in which a periodic signal of frequency f serves to generate a second periodic signal of frequency nf, with n an integer.

In principle, we may multiply by an arbitrary integral number n by simply tuning the resonant circuit to nf. In practice, of course, most periodic waveforms encountered in engineering applications, such as the collector-current pulse waveform of the transistor of Fig. 4.17-1, are characterized by a progressive decrease of amplitude of harmonic with increasing harmonic number. As a result, we find that as the order of multiplication increases, the output signal becomes progressively smaller. Circuits such as in Fig. 4.17-1 are commonly used for multiplication by factors from about 2 to 5. Where higher orders of multiplication are required, multipliers may be cascaded. Cascaded multipliers of order $n_1, n_2, n_3, \ldots$ yield an overall multiplication of order $n_1 n_2 n_3 \cdots$.

4.18 FREQUENCY MULTIPLICATION APPLIED TO FM SIGNALS

Now let us consider the result of the application to a frequency multiplier of an FM signal. If we think of an FM signal as a "sinusoidal" signal in which the frequency changes from moment to moment, then at each instant we may expect that the output frequency will be n times the input frequency. Hence, if the input consists of a carrier of frequency f_c which ranges through a frequency deviation $\pm \Delta f$, the output will have a carrier frequency nf_c and will range through the deviation $\pm n\,\Delta f$. The multiplier multiplies both the carrier and the deviation frequency. Also, since the modulation index is proportional to the frequency deviation, for a fixed modulation frequency, the multiplier increases the modulation index by the same factor n.

By way of example, consider the case of commercial FM broadcasting in the United States. Here, the allowable frequency deviation is 75 kHz, so that for a modulation frequency $f_m = 50$ Hz, $\beta = \Delta f/f_m = 1500$. Even if we allow ϕ in Fig. 4.10-1*a* to attain a maximum value as large as $\phi = 0.5$, then the multiplication needed is $1500/0.5 = 3000$. On the other hand, at the high-frequency end of the baseband spectrum, say, $f_m = 15$ kHz, $\beta = 75/15 = 5$. Correspondingly, with a multiplication by a factor of 3000, the phase ϕ in Fig. 4.10-1*a* need attain only a maximum value $5/3000 = 1.7 \times 10^{-3}$. Thus, it is to be seen that the required multiplication is determined by the low-frequency limit of the baseband spectrum.

4.19 AN EXAMPLE OF AN ARMSTRONG FM SYSTEM

The block diagram of Fig. 4.19-1 represents an Armstrong FM system which supplies a signal whose carrier is at 96 MHz (which is near the center of the commercial FM broadcasting band). It allows direct *phase modulation* of the carrier,

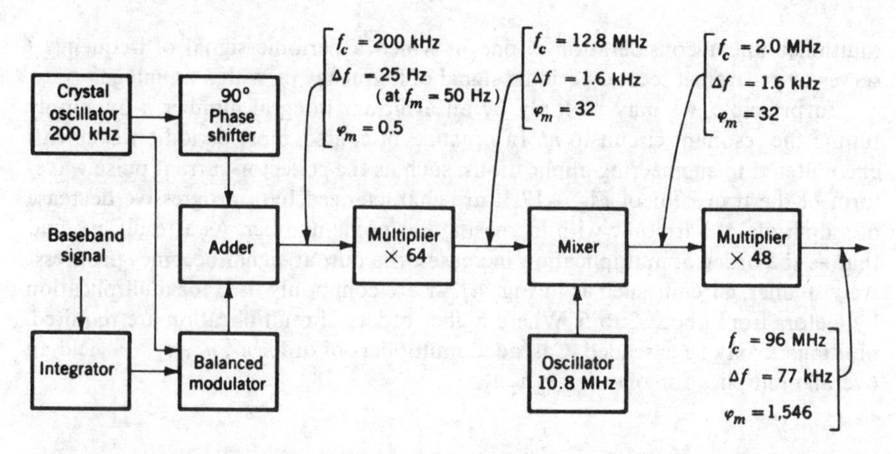

Figure 4.19-1 Block diagram of an Armstrong system of generating an FM signal using multipliers to increase the frequency deviation.

before multiplication, to the extent of $\phi_m = 0.5$. Thus, at $f_m = 50$ Hz, we have $\Delta f = 25$ Hz. Note that at higher modulating frequencies, ϕ_m is less than 0.5 rad. The carrier frequency before multiplication has been selected at 200 kHz, a frequency at which very stable crystal oscillators and balanced modulators are readily constructed.

As already noted, if we require that $\Delta f \approx 75$ kHz, then a multiplication by a factor of 3000 is required. In Fig. 4.19-1 the multiplication is actually $3072(= 64 \times 48)$. The values were selected so that the multiplication may be done by factors of 2 and 3, that is, $64 = 2^6$, $48 = 3 \times 2^4$. Direct multiplication would yield a signal of carrier frequency

$$200 \text{ kHz} \times 3072 = 614.4 \text{ MHz}$$

This signal might then be heterodyned with a signal of frequency, say, $614.4 - 96.0 = 518.4$ MHz. The difference signal output of such a mixer would be a signal of carrier frequency 96 MHz. Note particularly that a mixer, since it yields sum and difference frequencies, will translate the frequency spectrum of an FM signal but will have no effect on its frequency deviation. In the system of Fig. 4.19-1, in order to avoid the inconvenience of heterodyning at a frequency in the range of hundreds of megahertz, the frequency translation has been accomplished at a point in the chain of multipliers where the frequency is only in the neighborhood of approximately 10 MHz.

A feature, not indicated in Fig. 4.19-1 but which may be incorporated, is to derive the 10.8 MHz mixing signal not from a separate oscillator but rather through multipliers from the 0.2 MHz crystal oscillator. The multiplication required is $10.8/0.2 = 54 = 2 \times 3^3$. Such a derivation of the 10.8-MHz signal will suppress the effect of any drift in the frequency of this signal (see Prob. 4.19-1).

4.20 FM DEMODULATORS

With a view toward describing how we can recover the modulating signal from a frequency-modulated carrier we consider the situation represented in Fig. 4.20-1. Here a waveform of frequency f_0 and input amplitude A_i is applied to a frequency selective network which then yields an output of amplitude A_o. The ratio of amplitudes A_o/A_i is the absolute value of the transfer function of the network, that is, $|H(j\omega)|$. This output waveform is then applied to a diode AM demodulator (see Fig. 3.4-2). The diode demodulator generates an output which is equal to the peak value of the sinusoidal input so that the diode demodulator output is equal to A_o. Suppose now that the input waveform, instead of being of fixed frequency f_0, is actually a frequency modulated waveform. Then even for a fixed input amplitude A_i, the output amplitude A_o will not remain fixed but instead be modulated because of the frequency selectivity of the transmission network. Correspondingly the diode demodulator output will follow the variation of A_o.

In short, a fixed amplitude, frequency-modulated input will generate, at the output of the frequency selective network, a waveform which is not only frequency modulated but also amplitude modulated. The diode demodulator will ignore the frequency modulation but will respond to the amplitude modulation. (In general, the frequency-selective network will not only give rise to an amplitude change but will also generate a frequency-dependent phase change, as noted in Fig. 4.20-1. But such a phase change is simply additional angle modulation which the diode demodulator will ignore.)

What we require of an FM demodulator is that the instantaneous output signal A_o be proportional to the instantaneous frequency deviation of the received signal from the carrier frequency. If the carrier frequency is f_0 then we require a linear relationship between A_o and $(f - f_0)$ where f is the instantaneous frequency. Such a linear relationship is indicated in the plot of Fig. 4.20-1b. We

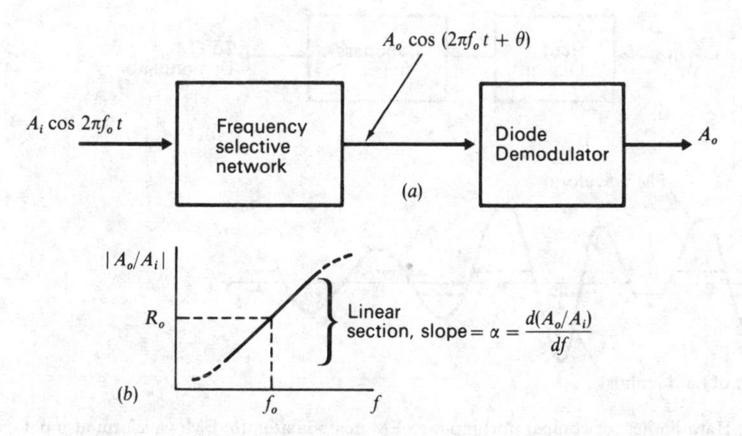

Figure 4.20-1 (a) FM demodulation. (b) Frequency selective network, typically an LC circuit.

require actually that the linearity extend only as far as is necessary to accommodate the maximum frequency deviation to which the carrier is subject.

As indicated in Fig. 4.20-1*b* let us consider that the frequency selective network has a linear transfer characteristic of slope α over an adequate range in the neighborhood of the carrier frequency f_0, and that, at $f = f_0$, $|A_o/A_i| = R_0 = |H(f = f_0)|$. Then we shall have, as required

$$A_o = R_0 A_i + \alpha A_i(f - f_0) \tag{4.20-1}$$

the term $R_0 A_i$ in Eq. (4.20-1) is a term of fixed value which displays no response to frequency deviation and the second term $\alpha A_i(f - f_0)$ provides the required response to the instantaneous frequency deviation of the input frequency-modulated signal.

We observe however from Eq. (4.20-1) that if the amplitude A_i of the input signal is not fixed then the demodulator output will respond to the input amplitude variations as well as to frequency deviations. Ordinarily in a frequency-modulated communication system the amplitude of the transmitted signal will not be modulated deliberately so that any such modulation which does appear will be due to noise. Hence it is of advantage, for the purpose of suppressing the noise, to reduce the dependence of A_o on A_i in Eq. (4.20-1). This is accomplished by passing the incoming signal through a hard limiter as shown in Fig. 4.20-2. The purpose of the hard limiter (or comparator) is to insure that variations of $A_i(t)$ are removed. Figure 4.20-2 shows the original FM waveform with amplitude variations at the output of the hard limiter. Since the waveform has been reduced to a frequency modulated "square wave" a bandpass filter is inserted to extract to first harmonic frequency f_0. The resulting FM waveform is now applied to the FM demodulator.

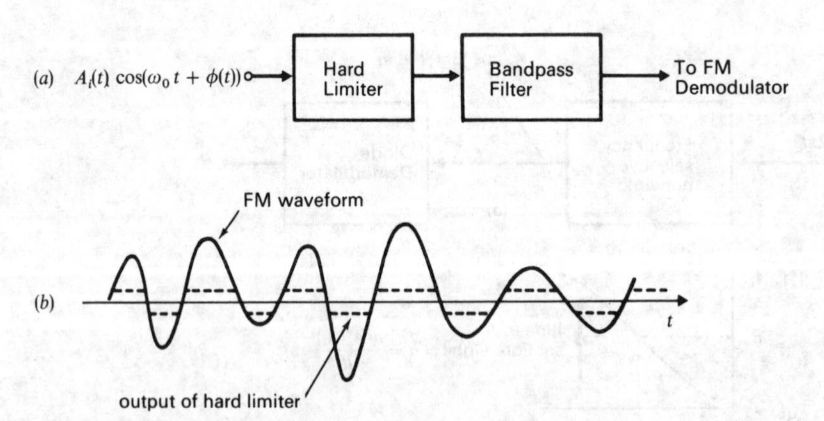

Figure 4.20-2 (*a*) Hard limiter (or comparator) input to FM demodulator. (*b*) FM waveform at input and output of hard limiter.

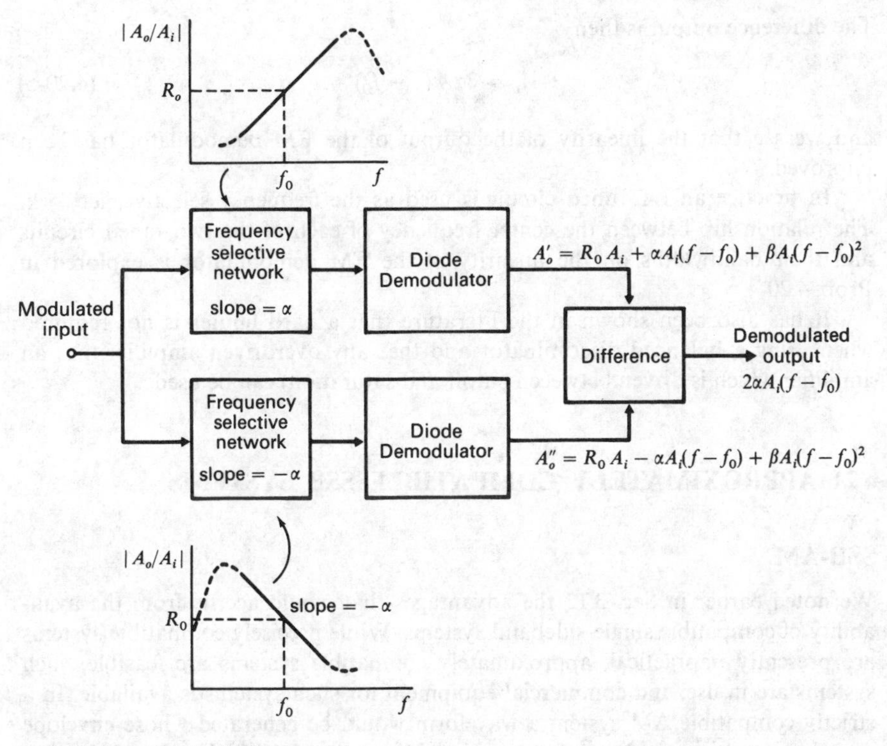

Figure 4.20-3 A balanced FM demodulator.

Unfortunately, one cannot construct a network having the precisely linear transfer characteristic shown in Eq. (4.20-1). Indeed, in practical networks the output amplitude appears as

$$A_o = R_0 A_i + \alpha A_i(f - f_0) + \beta A_i(f - f_0)^2 + \cdots \tag{4.20-2}$$

The balanced FM demodulator shown in Fig. 4.20-3 can be used to remove the constant term $R_0 A_i$ and all even harmonics, thereby reducing the distortion produced by the nonlinearity of the bandpass filters. Here two demodulators are employed, differing only in that in one case the frequency-selective network has a slope α and in the other the slope is $-\alpha$. The output provided by this balanced modulator is the difference between the two individual demodulators. These individual outputs are (assuming a second order nonlinearity):

$$A'_o = R_0 A_i + \alpha A_i(f - f_0) + \beta A_i(f - f_0)^2 \tag{4.20-3}$$

and

$$A''_o = R_0 A_i - \alpha A_i(f - f_0) + \beta A_i(f - f_0)^2 \tag{4.20-4}$$

The difference output is then

$$A_o = 2\alpha A_i(f - f_0) \tag{4.20-5}$$

and we see that the linearity of the output of the FM demodulator has been improved.

In practice an LC tuned circuit is used as the frequency selective network. The relationship between the center frequency of each of the two tuned circuits and their bandwidths on the linearity of the FM demodulator is explored in Prob. 4.20-4.

It has also been shown in the literature that a hard limiter is not required when using a balanced discriminator and that any overdriven amplifier (i.e., an amplifier which is driven betweeen cutoff and saturation) can be used.

4.21 APPROXIMATELY COMPATIBLE SSB SYSTEMS

SSB-AM

We noted earlier in Sec. 3.12 the advantages that would accrue from the availability of compatible single-sideband systems. While precisely compatible systems are presently impractical, approximately compatible systems are feasible, such systems are in use, and commercial equipment for such systems is available. In a strictly compatible AM system a waveform would be generated whose envelope exactly reproduces the baseband signal. Additionally if the highest baseband frequency component is f_M, the frequency range encompassed by the compatible signal would extend from f_c, the carrier frequency, to, say, $f_c + f_M$. An approximately compatible system is one in which there is some relaxation of these specifications concerning envelope shape or spectral range. On the basis of our discussion of FM-type waveforms we are now able briefly and qualitatively to discuss the principle of operation of one type of approximately compatible AM system.

We saw in Sec. 3.11 that if we suppressed one sideband component of an amplitude-modulated waveform, the envelope of the resultant waveform would still have the form of the modulating signal, provided the percentage modulation was kept small. Consider now the phasor diagram of the AM signal shown in Fig. 4.10-1*b*. From this phasor diagram it is apparent that when one of the sidebands is suppressed, the resultant is a waveform which is modulated in *both amplitude and phase*. The amplitude will vary between $1 + m/2$ and $1 - m/2$. The resultant R will no longer always be parallel to the carrier phasor. Instead it will rotate clockwise by an angle ϕ such that $\tan \phi = m/2$ and rotate counterclockwise by a similar angle.

Thus we find that a carrier of fixed frequency (or phase), when amplitude-modulated, gives rise to two sidebands. But an angle-modulated carrier, when amplitude-modulated, may give rise to a single sideband. Suppose then that we arrange to amplitude-modulate a carrier in such manner that its envelope faithfully reproduces the modulating signal. Is it then possible to also angle-modulate that carrier so that a single sideband results? It turns out that, to a good approximation, the answer is yes!

In the system described by Kahn[5] the modulating signal modulates not only the amplitude but the carrier phase as well. The relationship between phase and modulating signal is *nonlinear* and has been determined, at least in part experimentally, on the basis of system performance. An analysis of the system is very involved because of the two nonlinearities involved: the inherent nonlinearity of FM and the additional nonlinearity between phase and modulating signal. The system is not strictly single sideband in the sense that a modulating tone gives rise not to a single side tone but to a spectrum of side tones. However, all the side tones are on the *same side* of the carrier. The predominant side tone is separated from the carrier by the tone frequency, and the others are separated by multiples of the modulating-tone frequency.

The system is able to operate, in effect, as a single-sideband system because of the characteristics of speech or music for which it is intended, and because the side tones, other than the predominant one, fall off sharply in power content with increasing harmonic number. Most of the power in sound is in the lower-frequency ranges. High-power low-frequency spectral components in sound may give rise to numerous harmonic side tones, but because of the low frequency of the fundamental, the harmonics will still fall in the audio spectrum. On the other hand, high-frequency tones, which may give rise to harmonic side tones that may fall outside the allowed spectral range, are of very small energy. The overall result is that there may well be spectral components that fall outside the spectral range allowable in a spectrum-conserving single-sideband system. However, it has been shown experimentally that such components are small enough to cause no interference with the signal in an adjacent single-sideband channel.

SSB-FM

A compatible SSB-FM signal has frequency components extending either above or below the carrier frequency. In addition it can be demodulated using a standard limiter-discriminator. The compatible FM signal is constructed by adding amplitude modulation to the frequency-modulated signal. Since the limiter removes any and all amplitude modulation, the addition of the AM does not affect the recovery of the modulation by the discriminator. By properly adjusting the amplitude modulation, either the upper or lower sideband can be removed.

4.22 STEREOPHONIC FM BROADCASTING

In *monophonic* broadcasting of sound, a single audio baseband signal is transmitted from broadcasting studio to a home receiver. At the receiver, the audio signal is applied to a loudspeaker which then reproduces the original sound. The original source of the baseband waveform is a single microphone at the broadcasting studio. (If more than one microphone is used, as, for example, where a large orchestra is involved, the outputs of the individual microphones are combined to generate a single audio baseband signal.) In stereophonic broadcasting, two microphones (or two groups of microphones) are used. Two audio baseband signals are transmitted to the receiver where they are applied to two individual loudspeakers. At the broadcast studio, the microphones are located some distance apart from one another, and at the receiver the two loudspeakers are also physically separated. The advantage of such stereophonic broadcasting is that it yields at the receiver a more "natural" sound. The sound heard at the receiver is more nearly what the listener would hear if he were located at the broadcasting studio itself, where his two ears would receive somewhat different sounds.

The earliest commercial FM broadcasts were monophonic and conformed in transmission characteristics to standards established by the Federal Communications Commission. These standards required that the carrier frequencies of stations occupying adjacent frequency channels be separated by 200 kHz. To give reasonable assurance that there would be no interference between adjacent stations, the maximum allowable instantaneous-frequency deviation was limited by the FCC to 75 kHz. With sinusoidal modulation of the carrier frequency at the modulating frequency f_m and a frequency deviation Δf, the bandwidth requirement is $B = 2(\Delta f_m + f_m)$. If we assume a maximum audio frequency $f_m = 20$ kHz, and even if we imagine the extreme case that all the allowable baseband power is located at that frequency, then we find $B = 2(75 + 20) = 190$ kHz.

By the time commercial stereo broadcasting began to be contemplated, monophonic broadcasting was well established, and there were already many millions of monophonic FM receivers is use. Accordingly the FCC ruled that no proposed stereo scheme would be acceptable unless it were entirely *compatible* in the sense that a standard FM receiver, without modification, would be able to receive a monophonic version of a stereo transmission. Additionally the FCC required that the bandwidth occupied by a stereo transmission be no greater than the bandwidth already allocated for monophonic transmission. Many possible stereo systems were considered. We discuss now the system finally adopted in 1961 and presently in use.

Transmitted Signal

At the broadcast studio two microphones or microphone groups generate a left-hand audio signal $L(t)$ and a right-hand signal $R(t)$, as indicated in Fig. 4.22-1*a*. These signals are added and subtracted to generate $L(t) + R(t)$ and $L(t) - R(t)$. These sum and difference signals are each bandlimited to 15 kHz by filters not explicitly indicated in Fig. 4.22-1*a*. An oscillator makes available a sinusoidal waveform, referred to as a *pilot* carrier at a frequency $f_p = 19$ kHz. The pilot carrier is applied to a frequency doubler which generates a sinusoidal *subcarrier* at the frequency $f_{sc} = 2 \times f_p = 38$ kHz. The subcarrier and the difference signal are applied to the input of a balanced modulator, and the output of that modula-

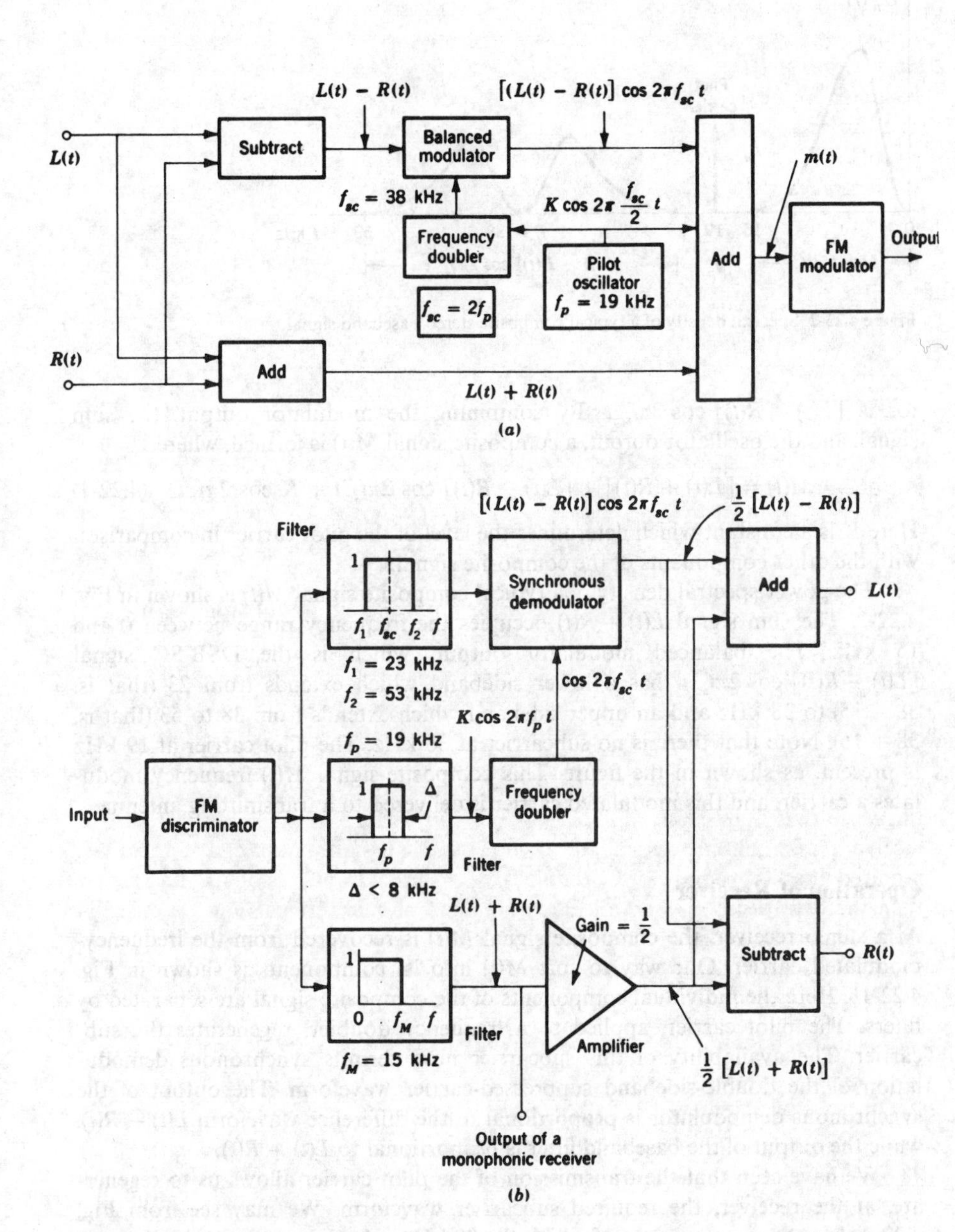

Figure 4.22-1 Stereophonic broadcasting system. (*a*) Transmitter. (*b*) Receiver.

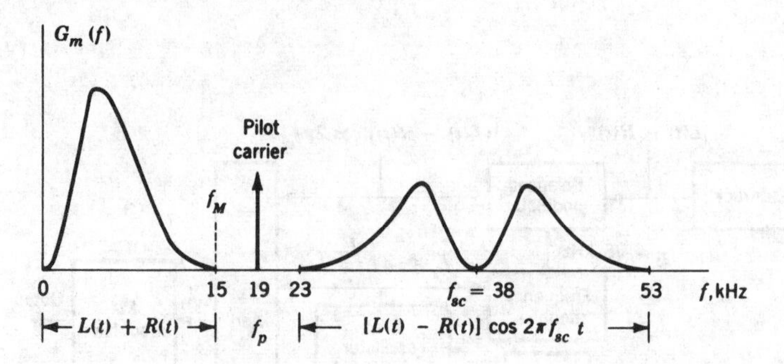

Figure 4.22-2 Spectral density of a typical composite stereo baseband signal.

tor is $[L(t) - R(t)] \cos 2\pi f_{sc} t$. By combining the modulator output, the sum signal, and the oscillator output, a composite signal $M(t)$ is formed, where

$$M(t) = [L(t) + R(t)] + [L(t) - R(t)] \cos 2\pi f_{sc} t + K \cos 2\pi f_p t \quad (4.22\text{-}1)$$

Here K is a constant which determines the level of the pilot carrier in comparison with the other components of the composite signal.

The power spectral density of a typical composite signal $M(t)$ is shown in Fig. 4.22-2. The sum signal $L(t) + R(t)$ occupies the frequency range between 0 and 15 kHz. The balanced modulator output, which is the DSB-SC signal $[L(t) - R(t)] \cos 2\pi f_{sc} t$, has a lower sideband which extends from 23 (that is, 38 − 15) to 38 kHz and an upper sideband which extends from 38 to 53 (that is, 38 + 15). Note that there is no subcarrier at 38 kHz. The pilot carrier at 19 kHz is present, as shown in the figure. This composite-signal $M(t)$ frequency modulates a carrier, and this modulated carrier is delivered to a transmitting antenna.

Operation of Receiver

At a stereo receiver, the composite signal $M(t)$ is recovered from the frequency-modulated carrier. One way to sort $M(t)$ into its components is shown in Fig. 4.22-1*b*. Here the individual components of the composite signal are separated by filters. The pilot carrier, applied to a frequency doubler, regenerates the subcarrier. The availability of this subcarrier now permits synchronous demodulation of the double-sideband suppressed-carrier waveform. The output of the synchronous demodulator is proportional to the difference waveform $L(t) - R(t)$, while the output of the baseband filter is proportional to $L(t) + R(t)$.

We have seen that the transmission of the pilot carrier allows us to regenerate, at the receiver, the required subcarrier waveform. We may see from Fig. 4.22-2 the reason on account of which the 38-kHz subcarrier was not itself transmitted directly. Such a subcarrier is not separated by any appreciable frequency interval from the spectral components of its accompanying sidebands. Hence, to extract such a subcarrier would require a very narrow and sharply tuned filter.

On the other hand, the pilot carrier occupies an isolated place in the spectrum, there being no other spectral components present over a range of 4 kHz on either side.

Now having available the sum signal $L(t) + R(t)$, and the difference signal $L(t) - R(t)$, then, as indicated in Fig. 4.22-1*b*, the individual signals $L(t)$ and $R(t)$ respectively, are recovered by adding and subtracting.

The system is entirely compatible with the requirements of a monophonic receiver. In such a receiver, the sum signal $L(t) + R(t)$ passes through the baseband filter while the pilot carrier and the suppressed-carrier signal do not. Hence these latter two signals contribute nothing to the output of a monophonic receiver and neither do they interfere with the receiver's operation.

Interleaving

A monophonic receiver makes use only of the $L + R$ signal in the stereo transmission. In order that the monophonic receiver output be as loud and disturbance-free as possible, it is necessary that the sum signal $V_s = L + R$, which modulates the FM carrier, be as large as possible. If the only component of the modulation waveform were the sum signal V_s, we would be at liberty to increase the peak excursion of V_s to the point where the corresponding peak instantaneous-frequency deviation of the carrier would be ± 75 kHz. However, in order to accommodate to the requirements of the stereo receiver, the composite modulating waveform must include as well the DSB-SC signal $V_d = (L - R) \cos 2\pi f_{sc} t$. We now require that the peak excursion of $V_s + V_d$ be no larger than the peak excursion previously allowed to V_s alone, i.e., the peak must correspond to a frequency deviation no larger than ± 75 kHz. These considerations suggest that, when V_d is added to V_s, the level of V_s needs to be reduced and that, as a consequence, monophonic reception of a stereo transmission would be inferior to monophonic reception of a monophonic transmission. We shall now show that, as a matter of practice, such is not the case. We shall show that with V_s itself adjusted to produce a peak allowable frequency deviation, V_d may be added without exceeding the allowable frequency deviation. This characteristic of the stereo system under consideration is known as *interleaving*.

We recognize at the outset that, although $L(t)$ and $R(t)$ are different, they will ordinarily not be greatly different. After all, the microphones which generate the two signals are intended to represent a person's two ears. Hence, if naturalness in sound is to be achieved, presumably the placement of the microphones at the studio will take this fact into account. Thus we may expect that the levels of the signal outputs of the two microphones will be comparable. The maximum excursion L_m of $L(t)$ and the maximum excursion R_m of $R(t)$ will be about the same. We must allow for the fact that from time to time both $L(t)$ and $R(t)$ will attain a maximum at about the same time. Therefore we set the maximum of the sum signal V_{sm} at

$$V_{sm} = L_m + R_m \approx 2L_m \approx 2R_m \tag{4.22-2}$$

Turning now to the composite signal $M(t)$ in Eq. (4.22-1), we note that $\cos 2\pi f_{sc} t$ oscillates rapidly between $+1$ and -1, and, ignoring temporarily the pilot carrier, we have that $M(t)$ oscillates rapidly between $M(t) = 2L(t)$ and $M(t) = 2R(t)$. The maximum attained by $M(t)$ is then $M_m = 2L_m$ or $M_m = 2R_m$. From Eq. (4.22-2) $M_m = V_{sm}$. Hence, in summary, we find that the addition of the difference signal V_d to the sum signal V_s does not increase the peak signal excursion.

Effect of the Pilot Carrier

Unlike the DSB-SC signal, the pilot carrier, when added to the other components of the composite modulating signal, does produce an increase in peak excursion. Hence the addition of the pilot carrier calls for a reduction in the sound signal modulation level. A low-level pilot carrier allows greater sound signal modulation, while a high-level pilot carrier eases the burden of extracting the pilot carrier at the receiver. As an engineering compromise, the FCC standards call for a pilot carrier of such level that the peak sound modulation amplitude has to be reduced to about 90 percent of what would be allowed in the absence of a carrier. This 10 percent reduction corresponds to a loss in signal level of less than 1 dB.

REFERENCES

1. Bell Telephone Laboratories: "Transmission Systems for Communications," Western Electric Company, Tech. Pub., Winston-Salem, N.C., 1964.
2. Jahnke, E., and F. Emde: "Tables of Functions," Dover Publications Inc., New York, 1945.
3. Pipes, L. A.: "Applied Mathematics for Engineers and Physicists," McGraw-Hill Book Company, New York, 1958.
4. Blachman, N.: Calculation of the Spectrum of an FM Signal Using Woodwards Theorem, *IEEE Trans. Communication Technology*, August, 1969.
5. Kahn, L. R.: Compatible Single Sideband. *Proc. IRE*, vol. 49, pp. 1503–1527, October, 1961.

PROBLEMS

4.2-1. Consider the signal $\cos [\omega_c t + \phi(t)]$ where $\phi(t)$ is a square wave taking on the values $\pm\pi/3$ every $2/f_c$ sec.

(*a*) Sketch $\cos [\omega_c t + \phi(t)]$.

(*b*) Plot the phase as a function of time.

(*c*) Plot the frequency as a function of time.

4.2-2. If the waveform $\cos (\omega_c t + k \sin \omega_m t)$ is a phase-modulated carrier, sketch the waveform of the modulating signal. Sketch the waveform of the modulating signal if the carrier is frequency-modulated.

4.3-1. What are the dimensions of the constants k', k'', and k that appear in Eqs. (4.3-1), (4.3-2), and (4.3-3)?

4.4-1. An FM signal is given by

$$v(t) = \cos \left[\omega_c t + \sum_{k=1}^{K} \beta_k \cos (k\omega_0 t + \theta_k) \right]$$

(*a*) If $\theta_k = 0$ and $K = 1, 2$, find the maximum frequency deviations.

(*b*) If each θ_k is an independent random variable, uniformly distributed between $-\pi$ and π, find the rms frequency deviation.

(*c*) Under the condition of (*b*) calculate the rms phase deviation.

4.4-2. If $v(t) = \cos \left[\omega_c t + k \int_{-\infty}^{t} m(\lambda)\, d\lambda \right]$, where $m(t)$ has a probability density

$$f(m) = \frac{1}{\sqrt{2\pi}} e^{-m^2/2}$$

calculate the rms frequency deviation.

4.4-3. A carrier which attains a peak voltage of 5 volts has a frequency of 100 MHz. This carrier is frequency-modulated by a sinusoidal waveform of frequency 2 kHz to such extent that the frequency deviation from the carrier frequency is 75 kHz. The modulated waveform passes through zero and is increasing at time $t = 0$. Write an expression for the modulated carrier waveform.

4.4-4. A carrier of frequency 10^6 Hz and amplitude 3 volts is frequency-modulated by a sinusoidal modulating waveform of frequency 500 Hz and of peak amplitude 1 volt. As a consequence, the frequency deviation is 1 kHz. The level of the modulating waveform is changed to 5 volts peak, and the modulating frequency is changed to 2 kHz. Write the expression for the new modulated waveform.

4.5-1. A carrier is angle-modulated by two sinusoidal modulating waveforms simultaneously so that

$$v(t) = A \cos (\omega_c t + \beta_1 \sin \omega_1 t + \beta_2 \sin \omega_2 t)$$

Show that this waveform has sidebands separated from the carrier not only at multiples of ω_1 and of ω_2 but also has sidebands as well at separations of multiples of $\omega_1 + \omega_2$ and of $\omega_1 - \omega_2$.

4.6-1. Bessel functions are said to be *almost periodic* with a period of almost 2π. Demonstrate this by recording the values of β, for $J_0(\beta)$ and $J_1(\beta)$, required to make these functions equal to zero.

4.6-2. The primary difference between the Bessel functions and the sine wave is that the envelope of the Bessel function decreases.

(*a*) Tabulate the magnitude of all peak values of $J_0(\beta)$, positive and negative peaks, as a function of β.

(*b*) Plot the magnitude of the peak values obtained in part (*a*) versus β and draw a smooth curve through the points.

(*c*) Show that the magnitude decreases as $\dfrac{1}{\sqrt{\beta}}$.

4.6-3. An FM carrier is sinusoidally modulated. For what values of β does all the power lie in the sidebands (i.e., no power in the carrier)?

4.7-1. A bandwidth rule sometimes used for space communication systems is $B = (2\beta + 1)f_M$. What fraction of the signal power is included in that frequency band. Consider $\beta = 1$ and 10.

4.7-2. A carrier is frequency-modulated by a sinusoidal modulating signal of frequency 2 kHz, resulting in a frequency deviation of 5 kHz. What is the bandwidth occupied by the modulated waveform? The amplitude of the modulating sinusoid is increased by a factor of 3 and its frequency lowered to 1 kHz. What is the new bandwidth?

4.7-3. Plot the spectrum of $\cos (2\pi \times 4t + 5 \sin 2\pi t)$. Note that the spectrum indicates the presence of a dc component. Plot the waveform as a function of time to indicate that the dc component is to have been expected.

4.10-1. $v(t) = \cos \omega_c t + 0.2 \cos \omega_m t \sin \omega_c t$.

(*a*) Show that $v(t)$ is a combination AM-FM signal.

(*b*) Sketch the phasor diagram at $t = 0$.

4.10-2. Consider the angle-modulated waveform $\cos(\omega_c t + 2 \sin \omega_m t)$, i.e., $\beta = 2$, so that the waveform may be approximated by a carrier and three pairs of sidebands. In a coordinate system in which the carrier phasor Δ_0 is at rest, determine the phasors Δ_1, Δ_2, and Δ_3, representing respectively the first, second, and third sideband pairs. Draw diagrams combining the four phasors for the cases $\omega_m t = 0, \pi/4, \pi/2, 3\pi/4$, and π. For each case calculate the magnitude of the resultant phasor.

4.10-3. (*a*) Show with a phasor diagram that $v(t)$ given by

$$v(t) = \cos(2\pi \times 10^6 t) + 0.02 \cos[2\pi \times (10^6 + 10^3)t]$$

represents a carrier which is modulated both in amplitude and frequency.

(*b*) Show that, on the basis of the relative magnitudes of the two terms in $v(t)$, the amplitude and the frequency variations both vary approximately sinusoidally with time with frequency 10^3 Hz.

(*c*) Express $v(t)$ in the form

$$v(t) \approx (1 + m \cos 2\pi \times 10^3 t) \cos(2\pi \times 10^6 t + \beta \sin 2\pi \times 10^3 t)$$

Find m and β. Write an expression for the instantaneous frequency as a function of time.

4.10-4. Consider the angle-modulated waveform $\cos(\omega_c t + 6 \sin \omega_m t)$, i.e., $\beta = 6$, so that the waveform may be approximated by a carrier and seven pairs of sidebands. In a coordinate system in which the carrier phasor Δ_0 is at rest, determine the phasor Δ_1, Δ_2, etc., representing the first, second, etc., sideband pairs at a time when $\sin \omega_m t = 1$. For this time draw a phasor diagram showing each phasor Δ_i and the resultant phasor.

4.11-1. Verify the comment made in Sec. 4.11 that superposition applies in NBFM. To do this consider $v(t) = \cos[\omega_c t + \phi(t)]$ where $\phi(t) = \beta_1 \sin \omega_1 t + \beta_2 \sin \omega_2 t$. Let β_1 and β_2 be sufficiently small so that $|\phi(t)| \ll \pi/2$. Show that $v(t) \simeq \cos \omega_c t - (\beta_1 \sin \omega_1 t + \beta_2 \sin \omega_2 t) \sin \omega_c t$.

4.12-1. The frequency of a laboratory oscillator is varied back and forth extremely slowly and at a uniform rate between the frequencies of 99 and 101 kHz. The amplitude of the oscillator output is constant at 2 volts. Make a plot of the two-sided power spectral density of the oscillator output waveform.

4.12-2. If the probability density of the amplitude of $m(t)$ is Rayleigh:

$$f(m) = \begin{cases} me^{-m^2/2} & m \geq 0 \\ 0 & \text{elsewhere} \end{cases}$$

Find the power spectral density $G_v(f)$ of the FM signal

$$v(t) = \cos\left[\omega_c t + k \int_{-\infty}^{t} m(\lambda)\, d\lambda\right]$$

4.12-3. Repeat Prob. 4.12-2 if the probability density of $m(t)$ is $f(m) = \frac{1}{2}e^{-|m|}$.

4.12-4. The frequency of a laboratory oscillator is varied back and forth extremely slowly and in such a manner that the instantaneous frequency of the oscillator varies sinusoidally with time between the limits of 99 and 101 kHz. The amplitude of the oscillator output is constant at 2 volts.

(*a*) Find a function of frequency $g(f)$ such that $g(f)\, df$ is the fraction of the time that the instantaneous frequency is in the range between f and $f + df$.

(*b*) Make a plot of the two-sided power spectral density of the oscillator output waveform.

4.13-1. Consider that the WBFM signal having the power spectral density of Eq. (4.13-1) is filtered by a gaussian filter having the bandpass characteristic

$$|H(f)|^2 = e^{-(f-f_c)^2/2B^2} + e^{-(f+f_c)^2/2B^2}$$

Assume $f_c \gg B$:

(*a*) Sketch $|H(f)|^2$ as a function of f.

(*b*) Calculate the 3-dB bandwidth of the filter in terms of B.

(*c*) Find B so that 98 percent of the signal power of the WBFM signal is passed.

4.13-2. The two independent modulating signals $m_1(t)$ and $m_2(t)$ are both gaussian and both of zero mean and variance 1 volt^2. The modulating signal $m_1(t)$ is connected to a source which can be frequency-modulated in such manner that, when $m_1(t) = 1$ volt (constant), the source frequency, initially 1 MHz, increases by 3 kHz. The modulating signal $m_2(t)$ is connected in such manner that, when $m_2(t) = 1$ volt (constant), the source frequency decreases by 4 kHz. The carrier amplitude is 2 volts. The two modulating signals are applied simultaneously. Write an expression for the power spectral density of the output of the frequency-modulated source.

4.14-1. Consider the FM signal

$$v(t) = \cos\left[\omega_c t + \sum_{k=1}^{K} \beta_k \cos(\omega_k t + \theta_k)\right]$$

Let $\beta_k \omega_k = 1$ for each k.

(*a*) Find B if $B \equiv 2[(\Delta f)_1 + (\Delta f)_2 + \cdots + (\Delta f)_K]$.

(*b*) Find B if $B \equiv 2[(\Delta f)_{1_{\text{rms}}} + (\Delta f)_{2_{\text{rms}}} + \cdots + (\Delta f)_{K_{\text{rms}}}]$.

4.14-2. If $G(f)$ is gaussian and is given by Eq. (4.13-1), find the rms bandwidth B. Compare your result with the value $B = 4.6\ \Delta f_{\text{rms}}$ given in Eq. (4.13-5).

4.15-1. In Fig. 4.15-1 the voltage-variable capacitor is a reversed-biased *pn* junction diode whose capacitance is related to the reverse-biasing voltage v by $C_v = (100/\sqrt{1 + 2v})$ pF. The capacitance $C_0 = 200$ pF and L is adjusted for resonance at 5 MHz when a fixed reverse voltage $v = 4$ volts is applied to the capacitor C_v. The modulating voltage is $m(t) = 4 + 0.045 \sin 2\pi \times 10^3 t$. If the oscillator amplitude is 1 volt, write an expression for the angle-modulated output waveform which appears across the tank circuit.

4.17-1. (*a*) In the multiplier circuit of Fig. 4.17-1 assume that the transistor acts as a current source and is so biased and so driven that the collector current consists of alternate half-cycles of a sinusoidal waveform with a peak value of 50 mA. The input frequency of the driving signal is 1 MHz, and the multiplication by a factor of 3 is to be accomplished. If $C = 200$ pF and the inductor $Q = 30$, find the inductance of the inductor and calculate the amplitude of the third harmonic voltage across the tank.

(*b*) If multiplication by 10 is to be accomplished, calculate the amplitude of the tank voltage. Assume that the resonant impedance of the tank remains the same as in part (*a*).

4.18-1. (*a*) Consider the narrowband waveform $v(t) = \cos(\omega_c t + \beta \sin \omega_m t)$, with $\beta \ll 1$ and with $\omega_m \ll \omega_c$. Show that $v(t)$, which has a frequency deviation $\Delta f = \beta f_m$, may be written approximately as

$$v(t) = \cos \omega_c t - \beta/2 \cos(\omega_c - \omega_m)t + \beta/2 \cos(\omega_c + \omega_m)t$$

and that this approximation is consistent with the general expansion for an angle-modulated waveform as given by Eq. (4.5-7). Use the approximations of Eqs. (4.6-1) and (4.6-2).

(*b*) Let $v(t)$ be applied as the input to a device whose output is $v^2(t)$ (i.e., the device is nonlinear and is to be used for frequency multiplication by a factor of 2). Square the approximate expression for $v(t)$ as given in part (*a*). Compare the spectrum of $v^2(t)$ so calculated with the exact spectrum for an angle-modulated waveform with frequency deviation $2\beta f_m$.

4.19-1. Assume that the 10.8-MHz signal in Fig. 4.19-1 is derived from the 200-kHz oscillator by multiplying by 54 and that the 200-kHz oscillator drifts by 0.1 Hz.

(*a*) Find the drift, in hertz, in the 10.8-MHz signal.

(*b*) Find the drift in the carrier of the resulting FM signal.

4.19-2. In an Armstrong modulator, as shown in Fig. 4.19-1, the crystal-oscillator frequency is 200 kHz. It is desired, in order to avoid distortion, to limit the maximum angular deviation to $\phi_m = 0.2$. The system is to accommodate modulation frequencies down to 40 Hz. At the output of the modulator the carrier frequency is to be 108 MHz and the frequency deviation 80 kHz. Select multiplier and mixer oscillator frequencies to accomplish this end.

4.20-1. The narrowband phase modulator of Fig. 4.16-1 is converted to a frequency modulator by preceding the balanced modulator with an integrator. The input signal is a sinusoid of angular frequency ω_m.

(*a*) Show that, unless the frequency deviation is kept small, the modulator output, when demodulated, will yield not only the input signal but also its odd harmonics.

(*b*) If the modulation frequency is 50 Hz, find the allowable frequency deviation if the normalized power associated with the third harmonic is to be no more than 1 percent of the fundamental power.

4.20-2. (*a*) Consider the FM demodulator of Fig. 4.20-1. Let the frequency selective network be an RC integrating network. The 3-dB frequency of the network is $f_2 (= 1/2\pi RC)$. If the carrier frequency of the FM waveform is f_c, how should f_2 be selected so that the demodulator has the greatest sensitivity (i.e., greatest change in output per change in input frequency)?

(*b*) With f_2 selected for maximum sensitivity and with $f_c = 1$ MHz, find the change in demodulator output for a 1-Hz change in input frequency.

4.20-3. A "zero-crossing" FM discriminator operates in the following manner. The modulated waveform

$$v(t) = A \cos \left[2\pi f_c t + k \int_{-\infty}^{t} m(\lambda)\, d\lambda \right]$$

is applied to an electronic circuit which generates a narrow pulse on each occasion when $v(t)$ passes through zero. The pulses are of fixed polarity, amplitude, and duration. This pulse train is applied to a low-pass filter, say an RC low-pass network of 3-dB frequency f_2. Assume that the bandwidth of the baseband waveform $m(t)$ is f_M. Discuss the operation of this discriminator. Show that if $f_c \gg f_2 \gg f_M$ the output of the low-pass network is indeed proportional to the instantaneous frequency of $v(t)$.

4.20-4. (*a*) A frequency selective network is shown in Fig. P4.20-4. Calculate the ratio $|V_o(f)/V_i(f)|$, i.e., the ratio of the amplitude of the output to the amplitude of the input, as a function of frequency. Verify that

$$\left| \frac{V_o(f)}{V_i(f)} \right| = \left\{ 1 + \frac{Q_0^2}{(f/f_0)^2} [(f/f_0)^2 - 1]^2 \right\}^{1/2}$$

in which $f_0 = 1/2\pi\sqrt{1/LC}$ is the resonant frequency and $Q_0 = R/2\pi f_0 L$ is the energy storage factor of the resonant network at f_0.

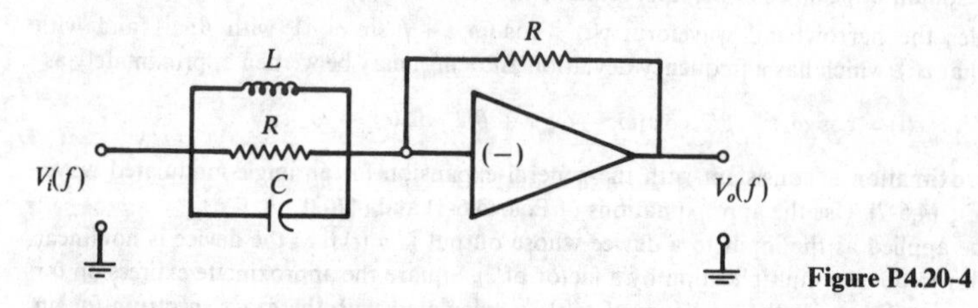

Figure P4.20-4

(*b*) Find the frequencies at which $|V_o/V_i|$ is higher by 3 dB than its value at the resonant frequency. Show that, for large Q_0, these frequencies are $f = f_0(1 \pm 1/Q_0)$. Calculate the slope of $|V_o/V_i|$ at the frequency $f_H = f_0(1 + 1/Q_0)$. A frequency modulated waveform with carrier frequency f_H is applied as $V_i(f)$. What is the ratio of the change in amplitude $\Delta V_o(f)$ to the change, Δf, of the instantaneous frequency of the input?

(*c*) In the discriminator of Fig. 4.20-1 let the two frequency selective networks be the network shown in Fig. P4.20-4. The resonant frequencies of the two networks are to be $f_H = f_c(1 + 1/Q_0)$ and $f_L = f_c(1 - 1/Q_0)$, f_c being the carrier frequency of an input frequency modulated waveform. Make a plot of the discriminator output as a function of the instantaneous frequency of the input. Use $Q_0 = 50$.

CHAPTER

FIVE

ANALOG-TO-DIGITAL CONVERSION

PULSE-MODULATION SYSTEMS

In Chaps. 3 and 4 we described systems with which we can transmit many signals simultaneously over a single communications channel. We found that at the transmitting end the individual signals were translated in frequency so that each occupied a separate and distinct frequency band. It was then possible at the receiving end to separate the individual signals by the use of filters.

In the present chapter we shall discuss a second method of multiplexing. This second method depends on the fact that a bandlimited signal, even if it is a continuously varying function of time, may be specified exactly by samples taken sufficiently frequently. Multiplexing of several signals is then achieved by interleaving the samples of the individual signals. This process is called time-division multiplexing. Since the sample is a pulse, the systems to be discussed are called pulse-amplitude modulation systems.

Pulse-amplitude modulation (PAM) systems are analog systems and share a common problem with the AM and FM modulation systems studied earlier. That is, each of these analog modulation systems are extremely sensitive to the noise present in the receiver. In this chapter we consider how to convert the analog signal into a digital signal prior to modulation. Such a conversion is called *source encoding*. We shall see in Chap. 12 that when a digital signal is modulated and transmitted, the received signal is far less sensitive to receiver noise than is analog modulation.

Noisy Communications Channels

We consider a basic problem associated with the transmission of a signal over a noisy communication channel. For the sake of being specific, suppose we require that a telephone conversation be transmitted from New York to Los Angeles. If the signal is transmitted by radio, then, when the signal arrives as its destination, it will be greatly attenuated and also combined with noise due to thermal noise present in all receivers (Chap. 14), and to all manner of random electrical disturbances which are added to the radio signal during its propagation across country. (We neglect as irrelevant, for the present discussion, whether such direct radio communication is reliable over such long channel distances.) As a result, the received signal may not be distinguishable against its background of noise. The situation is not fundamentally different if the signal is transmitted over wires. Any physical wire transmission path will both attenuate and distort a signal by an amount which increases with path length. Unless the wire path is completely and perfectly shielded, as in the case of a perfect coaxial cable, electrical noise and crosstalk disturbances from neighboring wire paths will also be picked up in amounts increasing with the path length. In this connection it is of interest to note that even coaxial cable does not provide complete freedom from crosstalk. External low-frequency magnetic fields will penetrate the outer conductor of the coaxial cable and thereby induce signals on the cable. In telephone cable, where coaxial cables are combined with parallel wire signal paths, it is common practice to wrap the coax in Permalloy for the sake of magnetic shielding. Even the use of fiber optic cables which are relatively immune to such interference, does not significantly alter the problem since receiver noise is often the noise source of largest power.

One attempt to resolve this problem is simply to raise the signal level at the transmitting end to so high a level that, in spite of the attenuation, the received signal substantially overrides the noise. (Signal distortion may be corrected separately by equalization.) Such a solution is hardly feasible on the grounds that the signal power and consequent voltage levels at the transmitter would be simply astronomical and beyond the range of amplifiers to generate, and cables to handle. For example, at 1 kHz, a telephone cable may be expected to produce an attenuation of the order of 1 dB per mile. For a 3000-mile run, even if we were satisfied with a received signal of 1 mV, the voltage at the transmitting end would have to be 10^{147} volts.

An amplifier at the receiver will not help the above situation, since at this point both signal and noise levels will be increased together. But suppose that a repeater (repeater is the term used for an amplifier in a communications channel) is located at the midpoint of the long communications path. This repeater will raise the signal level; in addition, it will raise the level of only the noise introduced in the first half of the communications path. Hence, such a midway repeater, as contrasted with an amplifier at the receiver, has the advantage of improving the received signal-to-noise ratio. This midway repeater will relieve the burden imposed on transmitter and cable due to higher power requirements when the repeater is not used.

The next step is, of course, to use additional repeaters, say initially at the one-quarter and three-quarter points, and thereafter at points in between. Each added repeater serves to lower the maximum power level encountered on the communications link, and each repeater improves the signal-to-noise ratio over what would result if the corresponding gain were introduced at the receiver.

In the limit we might, conceptually at least, use an infinite number of repeaters. We could even adjust the gain of each repeater to be infinitesimally greater than unity by just the amount to overcome the attenuation in the infinitesimal section between repeaters. In the end we would thereby have constructed a channel which had no attenuation. The signal at the receiving terminal of the channel would then be the unattenuated transmitted signal. We would then, in addition, have at the receiving end all the noise introduced at all points of the channel. This noise is also received without attenuation, no matter how far away from the receiving end the noise was introduced. If now, with this finite array of repeaters, the signal-to-noise ratio is not adequate, there is nothing to be done but to raise the signal level or to make the channel quieter.

The situation is actually somewhat more dismal than has just been intimated, since each repeater (transistor amplifier) introduces some noise on its own accord. Hence, as more repeaters are cascaded, each repeater must be designed to more exacting standards with respect to noise figure (see Sec. 14.10). The limitation of the system we have been describing for communicating over long channels is that once noise has been introduced any place along the channel, we are "stuck" with it.

If we now were to transmit a digital signal over the same channel we would find that significantly less signal power would be needed in order to obtain the same performance at the receiver. The reason for this is that the significant parameter is now not the signal-to-noise ratio but the probability of mistaking a digital signal for a different digital signal. In practice we find that signal-to-noise ratios of 40–60 dB are required for analog signals while 10–12 dB are required for digital signals. This reason and others, to be discussed subsequently, have resulted in a large commercial and military switch to digital communications.

5.1 THE SAMPLING THEOREM. LOW-PASS SIGNALS

We consider at the outset the fundamental principle of digital communications; the sampling theorem:

Let $m(t)$ be a signal which is bandlimited such that its highest frequency spectral component is f_M. Let the values of $m(t)$ be determined at regular intervals separated by times $T_s \leq 1/2f_M$, that is, the signal is periodically *sampled* every T_s seconds. Then these samples $m(nT_s)$, where n is an integer, uniquely determine the signal, and the signal may be reconstructed from these samples with no distortion.

The time T_s is called the *sampling time.* Note that the theorem requires that the *sampling rate* be rapid enough so that at least two samples are taken during the course of the period corresponding to the highest-frequency spectral com-

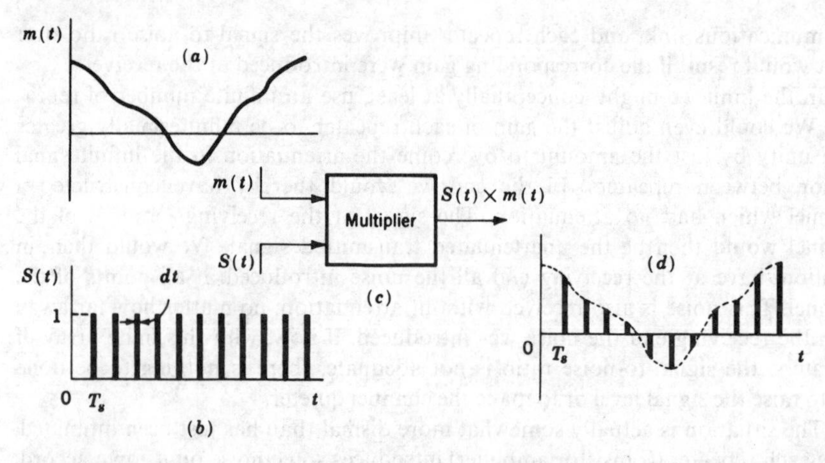

Figure 5.1-1 (*a*) A signal $m(t)$ which is to be sampled. (*b*) The sampling function $S(t)$ consists of a train of very narrow unit amplitude pulses. (*c*) The sampling operation is performed in a multiplier. (*d*) The samples of the signal $m(t)$.

ponent. We shall now prove the theorem by showing how the signal may be reconstructed from its samples.

The baseband signal $m(t)$ which is to be sampled is shown in Fig. 5.1-1*a*. A periodic train of pulses $S(t)$ of unit amplitude and of period T_s is shown in Fig. 5.1-1*b*. The pulses are arbitrarily narrow, having a width dt. The two signals $m(t)$ and $S(t)$ are applied to a multiplier as shown in Fig. 5.1-1*c*, which then yields as an output the product $S(t)m(t)$. This product is seen in Fig. 5.1-1*d* to be the signal $m(t)$ *sampled* at the occurrence of each pulse. That is, when a pulse occurs, the multiplier output has the same value as does $m(t)$, and at all other times the multiplier output is zero.

The signal $S(t)$ is periodic, with period T_s, and has the Fourier expansion [see Eq. (1.3-8) with $I = dt$ and $T_0 = T_s$]

$$S(t) = \frac{dt}{T_s} + \frac{2\,dt}{T_s}\left(\cos 2\pi \frac{t}{T_s} + \cos 2 \times 2\pi \frac{t}{T_s} + \cdots\right) \tag{5.1-1}$$

For the case $T_s = 1/2f_M$, the product $S(t)m(t)$ is

$$S(t)m(t) = \frac{dt}{T_s}\, m(t) + \frac{dt}{T_s}\,[2m(t) \cos 2\pi(2f_M)t + 2m(t) \cos 2\pi(4f_M)t + \cdots] \tag{5.1-2}$$

We now observe that the first term in the series is, aside from a constant factor, the signal $m(t)$ itself. Again, aside from a multiplying factor, the second term is the product of $m(t)$ and a sinusoid of frequency $2f_M$. This product then, as discussed in Sec. 3.2, gives rise to a double-sideband suppressed-carrier signal with carrier frequency $2f_M$. Similarly, succeeding terms yield DSB-SC signals with carrier frequencies $4f_M$, $6f_M$, etc.

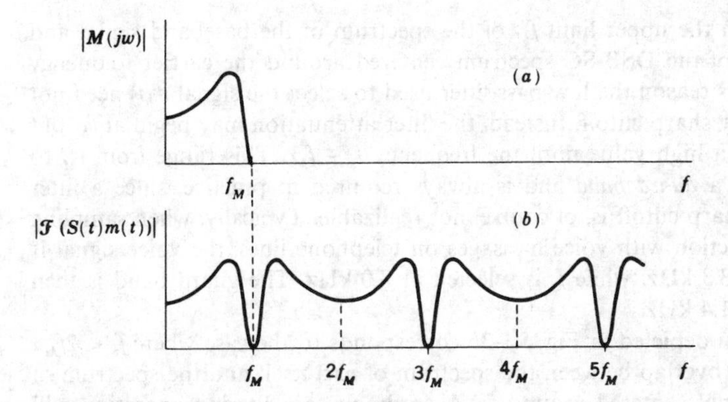

Figure 5.1-2 (*a*) The magnitude plot of the spectral density of a signal bandlimited to f_M. (*b*) Plot of amplitude of spectrum of sampled signal.

Let the signal $m(t)$ have a spectral density $M(j\omega) = \mathscr{F}[m(t)]$ which is as shown in Fig. 5.1-2*a*. Then $m(t)$ is bandlimited to the frequency range below f_M. The spectrum of the first term in Eq. (5.1-2) extends from 0 to f_M. The spectrum of the second term is symmetrical about the frequency $2f_M$ and extends from $2f_M - f_M = f_M$ to $2f_M + f_M = 3f_M$. Altogether the spectrum of the sampled signal has the appearance shown in Fig. 5.1-2*b*. Suppose then that the sampled signal is passed through an ideal low-pass filter with cutoff frequency at f_M. If the filter transmission were constant in the passband and if the cutoff were infinitely sharp at f_M, the filter would pass the signal $m(t)$ and nothing else.

The spectral pattern corresponding to Fig. 5.1-2b is shown in Fig. 5.1-3*a* for the case in which the sampling rate $f_s = 1/T_s$ is larger than $2f_M$. In this case there

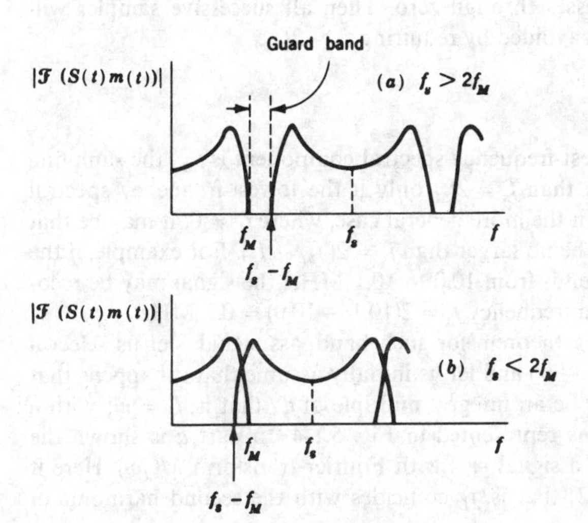

Figure 5.1-3 (*a*) A guard band appears when $f_s > 2f_M$. (*b*) Overlapping of spectra when $f_s < 2f_M$.

is a gap between the upper limit f_M of the spectrum of the baseband signal and the lower limit of the DSB-SC spectrum centered around the carrier frequency $f_s > 2f_M$. For this reason the low-pass filter used to select the signal $m(t)$ need not have an infinitely sharp cutoff. Instead, the filter attenuation may begin at f_M but need not attain a high value until the frequency $f_s - f_M$. This range from f_M to $f_s - f_M$ is called a *guard band* and is always required in practice, since a filter with infinitely sharp cutoff is, of course, not realizable. Typically, when sampling is used in connection with voice messages on telephone lines, the voice signal is limited to $f_M = 3.3$ kHz, while f_s is selected at 8.0 kHz. The guard band is then $8.0 - 2 \times 3.3 = 1.4$ kHz.

The situation depicted in Fig. 5.1-3*b* corresponds to the case where $f_s < 2f_M$. Here we find an overlap between the spectrum of $m(t)$ itself and the spectrum of the DSB-SC signal centered around f_s. Accordingly, no filtering operation will allow an exact recovery of $m(t)$.

We have just proved the *sampling theorem* since we have shown that, in principle, the sampled signal can be recovered exactly when $T_s \leq 1/2f_M$. It has also been shown why the minimum allowable sampling rate is $2f_M$. This minimum sampling rate is known as the *Nyquist rate.* An increase in sampling rate above the Nyquist rate increases the width of the guard band, thereby easing the problem of filtering. On the other hand, we shall see that an increase in rate extends the bandwidth required for transmitting the sampled signal. Accordingly an engineering compromise is called for.

An interesting special case is the sampling of a sinusoidal signal having the frequency f_M. Here, *all* the signal power is concentrated precisely at the cutoff frequency of the low-pass filter, and there is consequently some ambiguity about whether the signal frequency is inside or outside the filter passband. To remove this ambiguity, we require that $f_s > 2f_M$ rather than that $f_s \geq 2f_M$. To see that this condition is necessary, assume that $f_s = 2f_M$ but that an initial sample is taken at the moment the sinusoid passes through zero. Then all successive samples will also be zero. This situation is avoided by requiring $f_s > 2f_M$.

Bandpass Signals

For a signal $m(t)$ whose highest-frequency spectral component is f_M, the sampling frequency f_s must be no less than $f_s = 2f_M$ only if the lowest-frequency spectral component of $m(t)$ is $f_L = 0$. In the more general case, where $f_L \neq 0$, it may be that the sampling frequency need be no larger than $f_s = 2(f_M - f_L)$. For example, if the spectral range of a signal extends from 10.0 to 10.1 MHz, the signal may be recovered from samples taken at a frequency $f_s = 2(10.1 - 10.0) = 0.2$ MHz.

To establish the sampling theorem for such bandpass signals, let us select a sampling frequency $f_s = 2(f_M - f_L)$ and let us initially assume that it happens that the frequency f_L turns out to be an integral multiple of f_s, that is, $f_L = nf_s$ with n an integer. Such a situation is represented in Fig. 5.1-4. In part *a* is shown the two-sided spectral pattern of a signal $m(t)$ with Fourier transform $M(j\omega)$. Here it has been arranged that $n = 2$; that is, f_L coincides with the second harmonic of

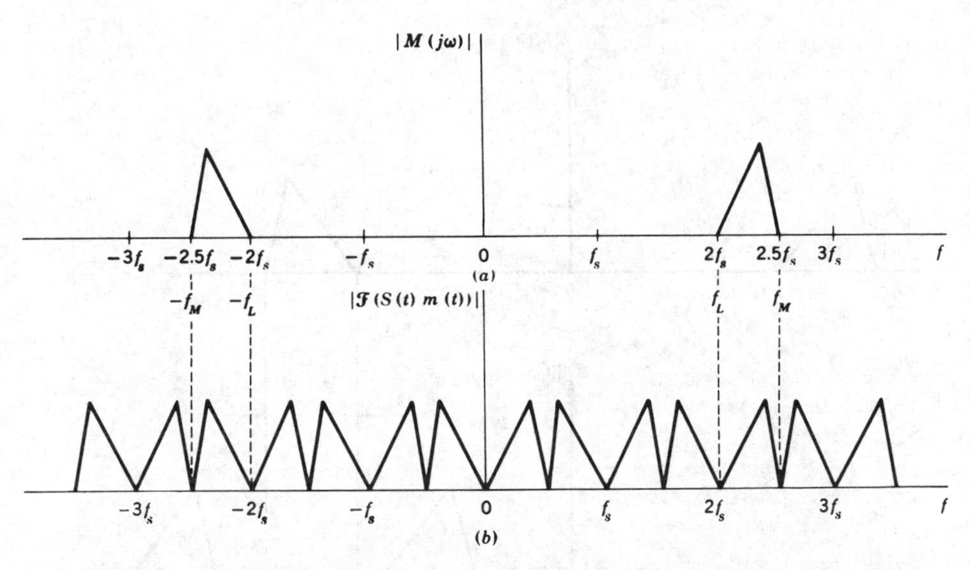

Figure 5.1-4 (*a*) The spectrum of a bandpass signal. (*b*) The spectrum of the sampled bandpass signal.

the sampling frequency, while the sampling frequency is exactly $f_s = 2(f_M - f_L)$. In part *b* is shown the spectral pattern of the sampled signal $S(t)m(t)$. The product of $m(t)$ and the dc term of $S(t)$ [Eq. (5.1-1)] duplicates in part *b* the form of the spectral pattern in part *a* and leaves it in the same frequency range from f_L to f_M. The product of $m(t)$ and the spectral component in $S(t)$ of frequency $f_s(= l/T_s)$ gives rise in part *b* to a spectral pattern derived from part *a* by shifting the pattern in part *a* to the right and also to the left by amount f_s. Similarly, the higher harmonics of f_s in $S(t)$ give rise to corresponding shifts, right and left, of the spectral pattern in part *a*. We now note that if the sampled signal $S(t)m(t)$ is passed through a bandpass filter with arbitrarily sharp cutoffs and with passband from f_L to f_M, the signal $m(t)$ will be recovered exactly.

In Fig. 5.1-4 the spectrum of $m(t)$ extends over the first half of the frequency interval between harmonics of the sampling frequency, that is, from $2.0f_s$ to $2.5f_s$. As a result, there is no spectrum overlap, and signal recovery is possible. It may also be seen from the figure that if the spectral range of $m(t)$ extended over the second half of the interval from $2.5f_s$ to $3.0f_s$, there would similarly be no overlap. Suppose, however, that the spectrum of $m(t)$ were confined neither to the first half nor to the second half of the interval between sampling-frequency harmonics. In such a case, there would be overlap between the spectrum patterns, and signal recovery would not be possible. Hence the minimum sampling frequency allowable is $f_s = 2(f_M - f_L)$ provided that either f_M or f_L is a harmonic of f_s.

If neither f_M nor f_L is a harmonic of f_s, a more general analysis is required. In Fig. 5.1-5*a* we have reproduced the spectral pattern of Fig. 5.1-4. The positive-frequency part and the negative-frequency part of the spectrum are called PS and NS respectively. Let us, for simplicity, consider separately PS and NS and the

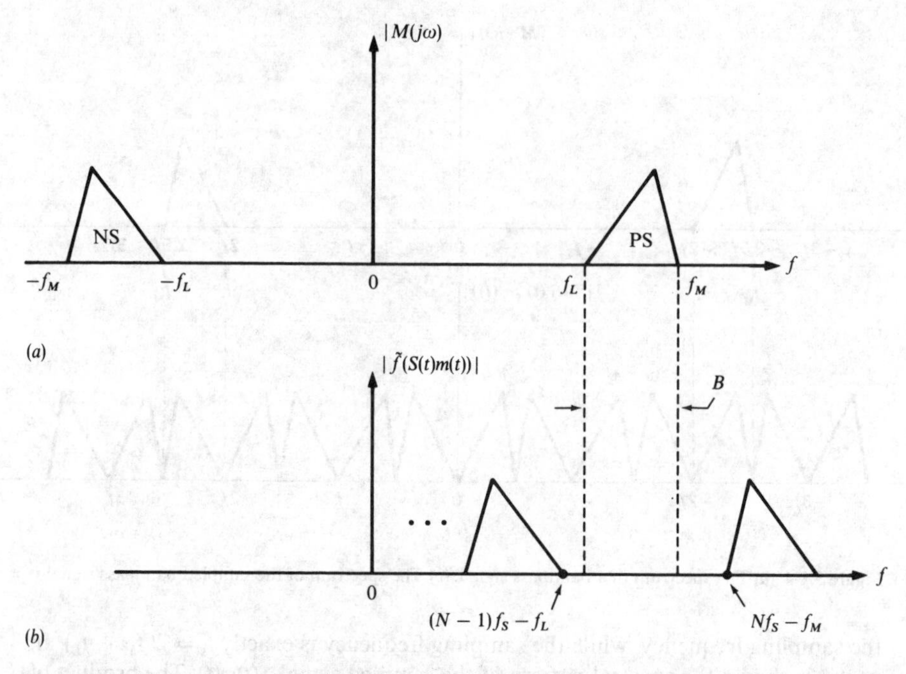

Figure 5.1-5 (*a*) Spectrum of the bandpass signal. (*b*) Spectrum of NS shifted by the $(N-1)$st and the Nth harmonic of the sampling waveform.

manner in which they are shifted due to the sampling and let us consider initially what constraints must be imposed so that we cause no overlap over, say, PS.

The product of $m(t)$ and the dc component of the sampling waveform leaves PS unmoved and it is this part of the spectrum which we propose to selectively draw out to reproduce the original signal. If we select the minimum value of f_s to be $f_s = 2(f_H - f_L) = 2B$ then the shifted PS patterns will not overlap PS. The NS will also generate a series of shifted patterns to the left and to the right. The left shiftings cannot cause an overlap of PS. However, the right shiftings of NS might cause an overlap and these right shiftings of NS are the only possible source of such overlap over PS.

Shown in Fig. 5.1-5*b* are the right shifted patterns of NS due to the (N-1)st and Nth harmonics of the sampling waveform. It is clear that to avoid overlap it is necessary that

$$(N-1)f_s - f_L \leq f_L \tag{5.1-3}$$

and

$$Nf_s - f_M \geq f_M \tag{5.1-4}$$

so that, with $B \equiv f_M - f_L$ we have

$$(N-1)f_s \leq 2(f_M - B) \tag{5.1-5}$$

and

$$Nf_s \geq 2f_M \tag{5.1-6}$$

If we let $k \equiv f_M/B$, Eqs. (5.1-5) and (5.1-6) become

$$f_s \leq 2B\left(\frac{k-1}{N-1}\right) \tag{5.1-7}$$

and

$$f_s \geq 2B\left(\frac{k}{N}\right) \tag{5.1-8}$$

in which $k \geq N$ since $f_s \geq 2B$. Equations (5.1-7) and (5.1-8) establish the constraint which must be observed to avoid an overlap on PS. It is clear from the symmetry of the initial spectrum and the symmetry of the shiftings required that this same constraint assures that there will be no overlap on NS.

Equations (5.1-7) and (5.1-8) have been plotted in Fig. 5.1-6 for several values of N. The shaded regions are the regions where the constraints are satisfied, while in the unshaded regions the constraints are not satisfied and overlap will occur. As an example of the use of these plots consider a case in which a baseband signal has a spectrum which extends from $f_L = 2.5$ kHz to $f_M = 3.5$ kHz. Here $B = 1$ kHz and $k = f_M/B = 3.5$. On the plot of Fig. 5.1-6 we have accordingly erected a dashed vertical line at $k = 3.5$. We observe that for this value of k, the

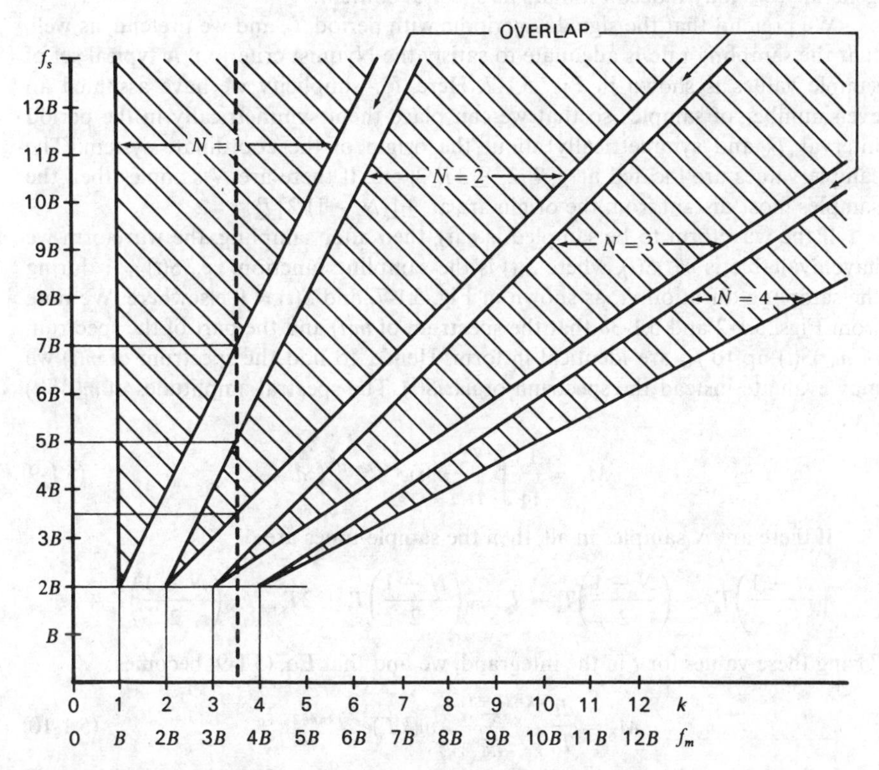

Figure 5.1-6 Showing the regions (shaded) in which both Eqs (5.1-7) and (5.1-8) are satisfied.

selection of a sampling frequency $f_s = 2B = 2$ kHz brings us to a point in an overlap region. As f_s is increased there is a small range of f_s, corresponding to $N = 3$, where there is no overlap. Further increase in f_s again takes us to an overlap region, while still further increase in f_s provides a nonoverlap range, corresponding to $N = 2$ (from $f_s = 3.5B$ to $f_s = 5B$). Increasing f_s further we again enter an overlap region while at $f_s = 7B$ we enter the nonoverlap region for $N = 1$. When $f_s \geq 7B$ we do not again enter an overlap region. (This is the region where $f_s \geq 2f_M$; that is, we assume we have a lowpass rather than a bandpass signal.)

The Discrete Fourier Transform

There are occasions when the only information we have available about a signal is a set of sample values, N in number, taken at regularly spaced intervals T_s over a period of time T_0. From this sampled data we should often like to be able to arrive at some reasonable approximation of the spectral content of the signal. If the sample period and the number of samples is adequate to give us some confidence that what has been observed of the signal is representative of the signal generally, we may indeed estimate its spectral content.

We pretend that the signal is periodic with period T_0 and we pretend, as well, that the sampling rate is adequate to satisfy the Nyquist criterion. A typical set of sample values is shown in Fig. 5.1-7. Here, for simplicity we have assumed an even number of samples so that we can place them symmetrically in the period interval T_0 and symmetrically about the origin of the coordinate system. The sample values are located at $\pm T_s/2$, $\pm 3T_s/2$, etc. If there are N samples then the samples most distant from the origin are at $\pm[(N-1)/2]T_s$.

If the waveform to be sampled is $m(t)$, then after sampling, the waveform we have available is $m(t)S(t)$, where $S(t)$ is the sampling function, i.e., $S(t) = 1$ during the sampling duration dt, as shown in Fig. 5.1-7, and $S(t) = 0$ elsewhere. We note from Figs. 5.1-2 and 5.1-3a that the spectrum of $m(t)$ and the part of the spectrum of $m(t)S(t)$ up to f_M are identical in form. Hence, to find the spectrum of $m(t)$ we may evaluate instead the spectrum of $m(t)S(t)$. The spectral amplitudes of $m(t)S(t)$ are:

$$M_n = \frac{1}{T_0} \int_{-T_0/2}^{T_0/2} m(t) e^{-j2\pi nt/T_0}\, dt \tag{5.1-9}$$

If there are N samples in all, then the sample times are:

$$-\left(\frac{N-1}{2}\right)T_s,\ -\left(\frac{N-1}{2}\right)T_s + T_s,\ -\left(\frac{N-1}{2}\right)T_s + 2T_s, \ldots, \left(\frac{N-1}{2}\right)T_s + \cdots$$

Using these values for t in the integrand, we find that Eq. (5.1-9) becomes:

$$M_n = \frac{dt}{T_0} \sum_{k=-(N-1)/2}^{k=(N-1)/2} m(kT_s) e^{-j2\pi nkT_s/T_0} \tag{5.1-10}$$

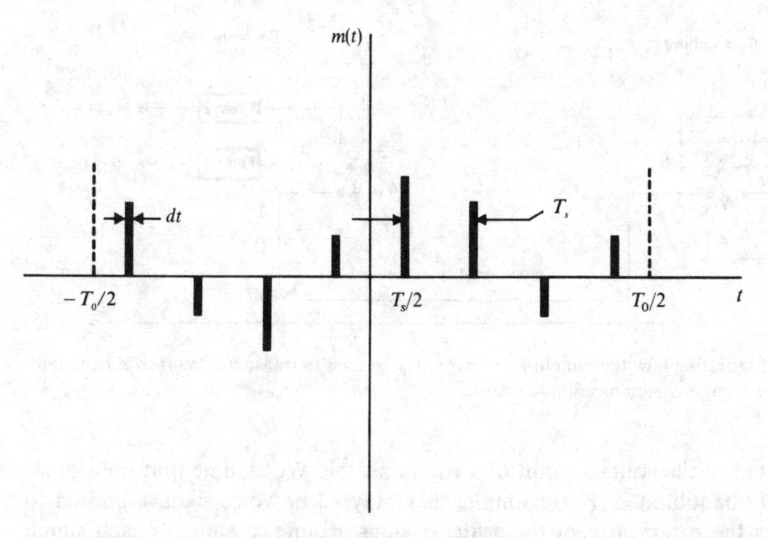

Figure 5.1-7 A possible set of sample values of a waveform $m(t)$, taken every T_s over a time interval T_0.

As a purely mathematical exercise we could use Eq. (5.1-10) to calculate spectral components v_n for any value of n. But consistently with our assumptions, the largest value of n allowable is determined by the Nyquist criterion. The highest frequency component should have a period which is $2T_s$ so that:

$$f_n(\text{max}) = \frac{1}{2T_s} \tag{5.1-11}$$

The fundamental period is T_0 so that the fundamental frequency is therefore $f_0 = 1/T_0$. Since $T_0 = NT_s$, we have:

$$f_n(\text{max}) = \frac{N}{2}\frac{1}{T_0} = \frac{N}{2} f_0 \tag{5.1-12}$$

Hence, the highest value of n for which Eq. (5.1-10) should be used is $n = N/2$.

5.2 PULSE-AMPLITUDE MODULATION

A technique by which we may take advantage of the sampling principle for the purpose of time-division multiplexing is illustrated in the idealized representation of Fig. 5.2-1. At the transmitting end on the left, a number of bandlimited signals

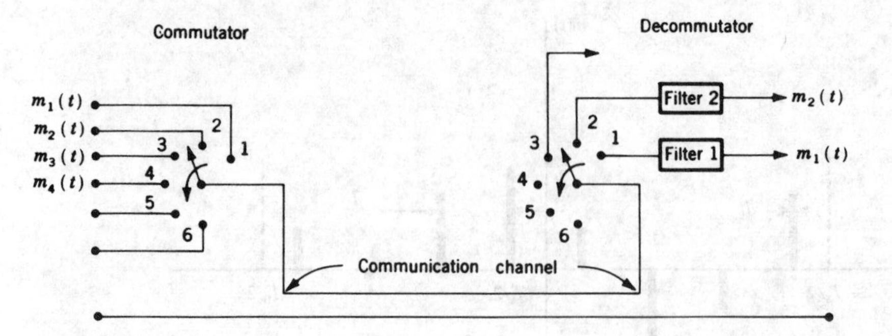

Figure 5.2-1 Illustrating how the sampling principle may be used to transmit a number of bandlimited signals over a single communications channel.

are connected to the contact point of a rotary switch. We assume that the signals are similarly bandlimited. For example, they may all be voice signals, limited to 3.3 kHz. As the rotary arm of the switch swings around, it samples each signal sequentially. The rotary switch at the receiving end is in synchronism with the switch at the sending end. The two switches make contact simultaneously at similarly numbered contacts. With each revolution of the switch, one sample is taken of each input signal and presented to the correspondingly numbered contact of the receiving-end switch. The train of samples at, say, terminal 1 in the receiver, pass through low-pass filter 1, and, at the filter output, the original signal $m_1(t)$ appears reconstructed. Of course, if f_M is the highest-frequency spectral component present in any of the input signals, the switches must make at least $2f_M$ revolutions per second.

When the signals to be multiplexed vary slowly with time, so that the sampling rate is correspondingly slow, mechanical switches, indicated in Fig. 5.2-1, may be employed. When the switching speed required is outside the range of

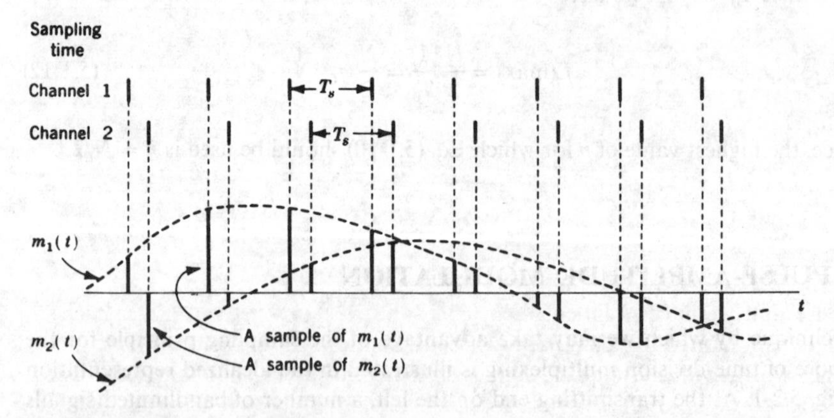

Figure 5.2-2 The interlacing of two baseband signals.

mechanical switches, electronic switching systems may be employed. In either event, the switching mechanism, corresponding to the switch at the left in Fig. 5.2-1, which samples the signals, is called the *commutator*. The switching mechanism which performs the function of the switch at the right in Fig. 5.2-1 is called the *decommutator*. The commutator samples and combines samples, while the decommutator separates samples belonging to individual signals so that these signals may be reconstructed.

The interlacing of the samples that allows multiplexing is shown in Fig. 5.2-2. Here, for simplicity, we have considered the case of the multiplexing of just two signals $m_1(t)$ and $m_2(t)$. The signal $m_1(t)$ is sampled regularly at intervals of T_s and at the times indicated in the figure. The sampling of $m_2(t)$ is similarly regular, but the samples are taken at a time different from the sampling time of $m_1(t)$. The input waveform to the filter numbered 1 in Fig. 5.2-1 is the train of samples of $m_1(t)$, and the input to the filter numbered 2 is the train of samples of $m_2(t)$. The timing in Fig. 5.2-2 has been deliberately drawn to suggest that there is room to multiplex more than two signals. We shall see shortly, in principle, how many signals may be multiplexed.

We observe that the train of pulses corresponding to the samples of each signal are *modulated in amplitude* in accordance with the signal itself. Accordingly, the scheme of sampling is called *pulse-amplitude modulation* and abbreviated PAM.

Multiplexing of several PAM signals is possible because the various signals are kept distinct and are separately recoverable by virtue of the fact that they are sampled at different times. Hence this system is an example of a *time-division multiplex* (TDM) system. Such systems are the counterparts in the time domain of the systems of Chap. 3. There, the signals were kept separable by virtue of their translation to different portions of the frequency domain, and those systems are called frequency-division multiplex (FDM) systems.

If the multiplexed signals are to be transmitted directly, say, over a pair of wires, no further signal processing need be undertaken. Suppose, however, we require to transmit the TDM-PAM signal from one antenna to another. It would then be necessary to amplitude-modulate or frequency-modulate a high-frequency carrier with the TDM-PAM signal; in such a case the overall system would be referred to, respectively, as PAM-AM or PAM-FM. Note that the same terminology is used whether a single signal or many signals (TDM) are transmitted.

5.3 CHANNEL BANDWIDTH FOR A PAM SIGNAL

Suppose that we have N independent baseband signals $m_1(t)$, $m_2(t)$, etc., each of which is bandlimited to f_M. What must be the bandwidth of the communications channel which will allow all N signals to be transmitted simultaneously using PAM time-division multiplexing? We shall now show that, in principle at least, the channel need not have a bandwidth larger than Nf_M.

The baseband signal, say $m_1(t)$, must be sampled at intervals not longer than $T_s = 1/2f_M$. Between successive samples of $m_1(t)$ will appear samples of the other $N - 1$ signals. Therefore the interval of separation between successive samples of different baseband signals is $1/2f_M N$. The composite signal, then, which is presented to the transmitting end of the communications channel, consists of a sequence of samples, that is, a sequence of *impulses*. If the bandwidth of the channel were arbitrarily great, the waveform at the receiving end would be the same as at the sending end and demultiplexing could be achieved in a straightforward manner.

If, however, the bandwidth of the channel is restricted, the channel response to an instantaneous sample will be a waveform which may well persist with significant amplitude long after the time of selection of the sample. In such a case, the signal at the receiving end at any particular sampling time may well have significant contributions resulting from previous samples of other signals. Consequently the signal which appears at any of the output terminals in Fig. 5.2-1 will not be a single baseband signal but will be instead a combination of many or even all the baseband signals. Such combining of baseband signals at a communication system output is called *crosstalk* and is to be avoided as far as possible.

Let us assume that our channel has the characteristics of an ideal low-pass filter with angular cutoff frequency $\omega_c = 2\pi f_c$, unity gain, and no delay. Let a sample be taken, say, of $m_1(t)$, at $t = 0$. Then at $t = 0$ there is presented at the transmitting end of the channel an impulse of strength $I_1 = m_1(0)\,dt$. The response at the receiving end is $s_{R1}(t)$ given by (see Prob. 5.3-1)

$$s_{R1}(t) = \frac{I_1 \omega_c}{\pi} \frac{\sin \omega_c t}{\omega_c t} \tag{5.3-1}$$

The normalized response $\pi s_R(t)/\omega_c$ is shown in Fig. 5.3-1 by the solid plot. At $t = 0$ the reponse attains a peak value proportional to the strength of the impulse

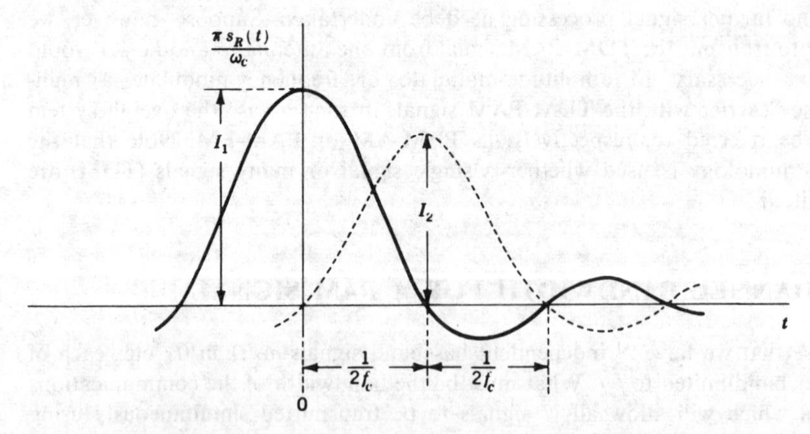

Figure 5.3-1 The response of an ideal low-pass filter to an instantaneous sample at $t = 0$ (solid plot). The response to a sample at $t = 1/2f_c$ (dashed plot).

$I_1 = m_1(0)\, dt$, which is in turn proportional to the value of the sample $m_1(0)$. This response persists indefinitely. Observe, however, that the response passes through zero at intervals which are multiples of $\pi/\omega_c = 1/2f_c$. Suppose, then, that a sample of $m_2(t)$ is taken and transmitted at $t = 1/2f_c$. If $I_2 = m_2(t = 1/2f_c)\, dt$,

$$s_{R2}(t) = \frac{I_2\, \omega_c}{\pi} \frac{\sin \omega_c(t - 1/2f_c)}{\omega_c(t - 1/2f_c)} \tag{5.3-2}$$

This response is shown by the dashed plot. Suppose, finally, that the demultiplexing is done also by instantaneous sampling at the receiving end of the channel, for $m_1(t)$ at $t = 0$ and for $m_2(t)$ at $t = 1/2f_c$. Then, in spite of the persistence of the channel response, there will be no crosstalk, and the signals $m_1(t)$ and $m_2(t)$ may be completely separated and individually recovered. Similarly, additional signals may be sampled and multiplexed, provided that each new sample is taken synchronously, every $1/2f_c$ s. The sequence must, of course, be continually repeated every $1/2f_M$ s, so that each signal is properly sampled.

We have then the result that with a channel of bandwidth f_c we need to separate samples by intervals $1/2f_c$. The sampling theorem requires that the samples of an individual baseband signal be separated by intervals not longer than $1/2f_M$. Hence the total number of signals which may be multiplexed is $N = f_c/f_M$, or $f_c = Nf_M$ as indicated earlier.

In principle then, multiplexing a number of signals by PAM time division requires no more bandwidth than would be required to multiplex these signals by frequency-division multiplexing using single-sideband transmission.

5.4 NATURAL SAMPLING

It was convenient, for the purpose of introducing some basic ideas, to begin our discussion of time multiplexing by assuming instantaneous commutation and decommutation. Such instantaneous sampling, however, is hardly feasible. Even if it were possible to construct switches which could operate in an arbitrarily short time, we would be disinclined to use them. The reason is that instantaneous samples at the transmitting end of the channel have infinitesimal energy, and when transmitted through a bandlimited channel give rise to signals having a peak value which is infinitesimally small. We recall that in Fig. 5.3-1 $I_1 = m_1(0)\, dt$. Such infinitesimal signals will inevitably be lost in background noise.

A much more reasonable manner of sampling, referred to as *natural sampling*, is shown in Fig. 5.4-1. Here the sampling waveform $S(t)$ consists of a train of pulses having duration τ and separated by the sampling time T_s. The baseband signal is $m(t)$, and the sampled signal $S(t)m(t)$ is shown in Fig. 5.4-1*c*. Observe that the sampled signal consists of a sequence of pulses of varying amplitude whose tops are not flat but follow the waveform of the signal $m(t)$.

With natural sampling, as with instantaneous sampling, a signal sampled at the Nyquist rate may be reconstructed exactly by passing the samples through an ideal low-pass filter with cutoff at the frequency f_M, where f_M is the highest-

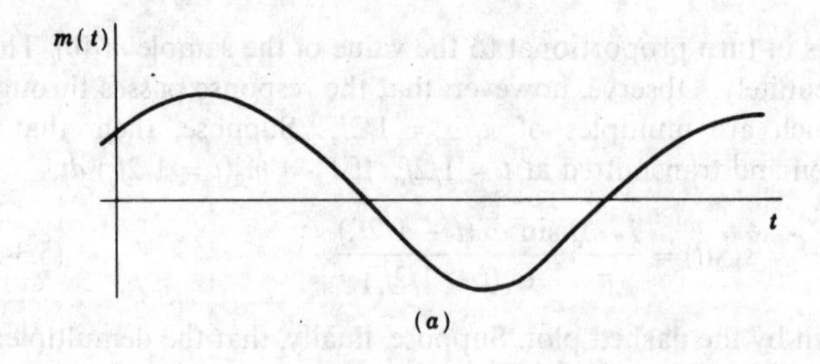

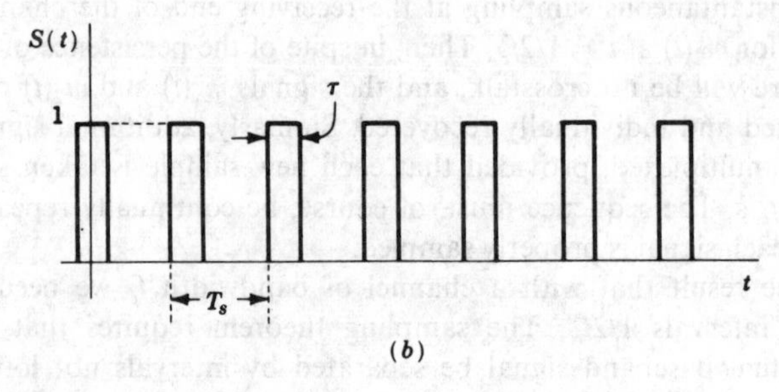

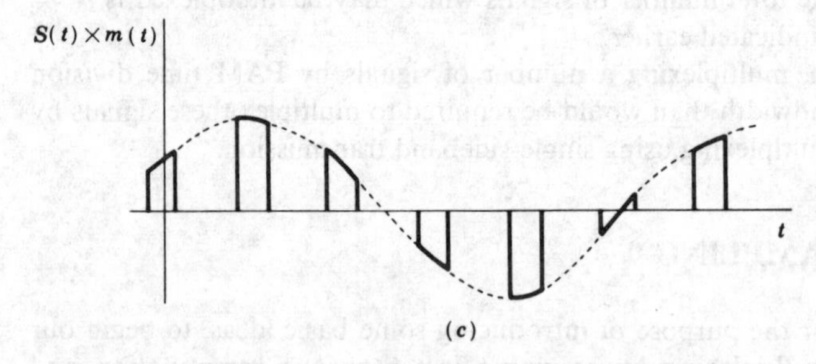

Figure 5.4-1 (*a*) A baseband signal *m*(*t*). (*b*) A sampling signal *S*(*t*) with pulses of finite duration. (*c*) The *naturally* sampled signal *S*(*t*)*m*(*t*).

frequency spectral component of the signal. To prove this, we note that the sampling waveform $S(t)$ shown in Fig. 5.4-1 is given by [see Eq. (1.3-12) with $A = 1$ and $T_0 = T_s$]

$$S(t) = \frac{\tau}{T_s} + \frac{2\tau}{T_s}\left(C_1 \cos 2\pi \frac{t}{T_s} + C_2 \cos 2 \times 2\pi \frac{t}{T_s} + \cdots\right) \tag{5.4-1}$$

with the constant C_n given by

$$C_n = \frac{\sin(n\pi\tau/T_s)}{n\pi\tau/T_s} \tag{5.4-2}$$

This sampling waveform differs from the sampling waveform of Eq. (5.1-1) for instantaneous sampling only in that dt is replaced by τ and by the fact that the amplitudes of the various harmonics are not the same. The sampled baseband signal $S(t)m(t)$ is, for $T_s = 1/2f_M$,

$$S(t)m(t) = \frac{\tau}{T_s} m(t) + \frac{2\tau}{T_s} [m(t)C_1 \cos 2\pi(2f_M)t + m(t)C_2 \cos 2\pi(4f_M)t + \cdots] \tag{5.4-3}$$

Therefore, as in instantaneous sampling, a low-pass filter with cutoff at f_M will deliver an output signal $s_o(t)$ given by

$$s_o(t) = \frac{\tau}{T_s} m(t) \tag{5.4-4}$$

which is the same as is given by the first term of Eq. (5.1-2) except with dt replaced by τ.

With samples of finite duration, it is not possible to completely eliminate the crosstalk generated in a channel, sharply bandlimited to a bandwidth f_c. If N signals are to be multiplexed, then the maximum sample duration is $\tau = T_s/N$. It is advantageous, for the purpose of increasing the level of the output signal, to make τ as large as possible. For, as is seen in Eq. (5.4-4), $s_o(t)$ increases with τ. However, to help suppress crosstalk, it is ordinarily required that the samples be limited to a duration much less than T_s/N. The result is a large *guard time* between the end of one sample and the beginning of the next.

5.5 FLAT-TOP SAMPLING

Pulses of the type shown in Fig. 5.4-1*c*, with tops contoured to follow the waveform of the signal, are actually not frequently employed. Instead *flat-topped* pulses are customarily used, as shown in Fig. 5.5-1*a*. A flat-topped pulse has a

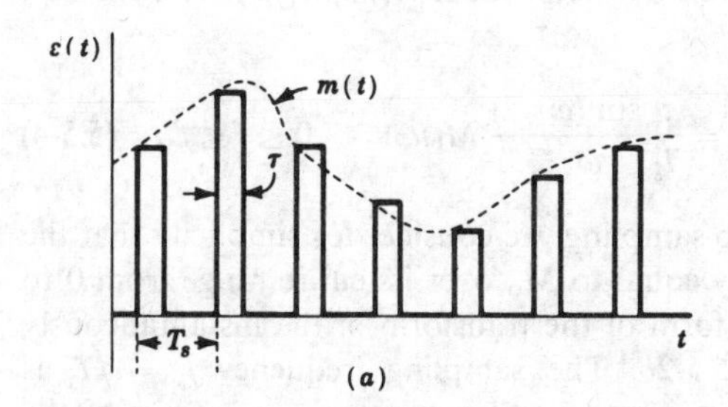

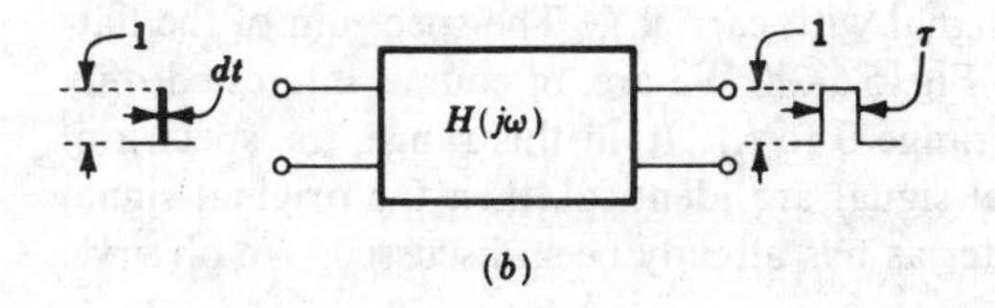

Figure 5.5-1 (*a*) Flat-topped sampling. (*b*) A network with transform $H(j\omega)$ which converts a pulse of width dt into a rectangular pulse of like amplitude but of duration τ.

constant amplitude established by the sample value of the signal at some point within the pulse interval. In Fig. 5.5-1*a* we have arbitrarily sampled the signal at the beginning of the pulse. In sampling of this type the baseband signal $m(t)$ cannot be recovered exactly by simply passing the samples through an ideal low-pass filter. However, the distortion need not be large. Flat-top sampling has the merit that it simplifies the design of the electronic circuitry used to perform the sampling operation.

To show the extent of the distortion, consider the signal $m(t)$ having a Fourier transform $M(j\omega)$. We have seen (see Figs. 5.1-2 and 5.1-3) how to deduce the transform of the sampled signal, when the sampling is instantaneous. The transform of the sampled signal for flat-top sampling is determined by considering that the flat-top pulse can be generated by passing the instantaneously sampled signal through a network which broadens a pulse of duration dt (an impulse) into a pulse of duration τ. The transform of a pulse of unit amplitude and width dt is

$$\mathscr{F}[\text{impulse of strength } dt \text{ at } t = 0] = dt \tag{5.5-1}$$

The transform of a pulse of unit amplitude and width τ is [see Eq. (1.10-20)]

$$\mathscr{F}\left[\text{pulse, amplitude} = 1, \text{ extending from } t = -\frac{\tau}{2} \text{ to } t = \frac{\tau}{2}\right] = \tau\,\frac{\sin(\omega\tau/2)}{\omega\tau/2} \tag{5.5-2}$$

Hence, the transfer function of the network shown in Fig. 5.5-1*b*, is required to be

$$H(j\omega) = \frac{\tau}{dt}\,\frac{\sin(\omega\tau/2)}{\omega\tau/2} \tag{5.5-3}$$

Let the signal $m(t)$, with transform $M(j\omega)$, be bandlimited to f_M and be sampled at the Nyquist rate or faster. Then in the range 0 to f_M the transform of the flat-topped sampled signal is given by the product $H(j\omega)M(j\omega)$ or, from Eqs. (5.5-1), (5.5-2), and (5.5-3)

$$\mathscr{F}[\text{flat-topped sampled } m(t)] = \frac{\tau}{T_s}\,\frac{\sin(\omega\tau/2)}{\omega\tau/2}\,M(j\omega) \qquad 0 \le f \le f_M \tag{5.5-4}$$

To illustrate the effect of flat-top sampling, we consider for simplicity that the signal $m(t)$ has a flat spectral density equal to M_0 over its entire range from 0 to f_M, as is shown in Fig. 5.5-2*a*. The form of the transform of the instantaneously sampled signal is shown in Fig. 5.5-2*b*. The sampling frequency $f_s = 1/T_s$ is assumed large enough to allow for a guard band between the spectrum of the baseband signal and the DSB-SC signal with carrier f_s. The spectrum of the flat-topped sampled signal is shown in Fig. 5.5-2*d*. We are, of course, interested only in the part of the spectrum in the range 0 to f_M. If, in this range, the spectra of the sampled signal and the original signal are identical, then the original signal may be recovered by a low-pass filter as has already been discussed. We observe,

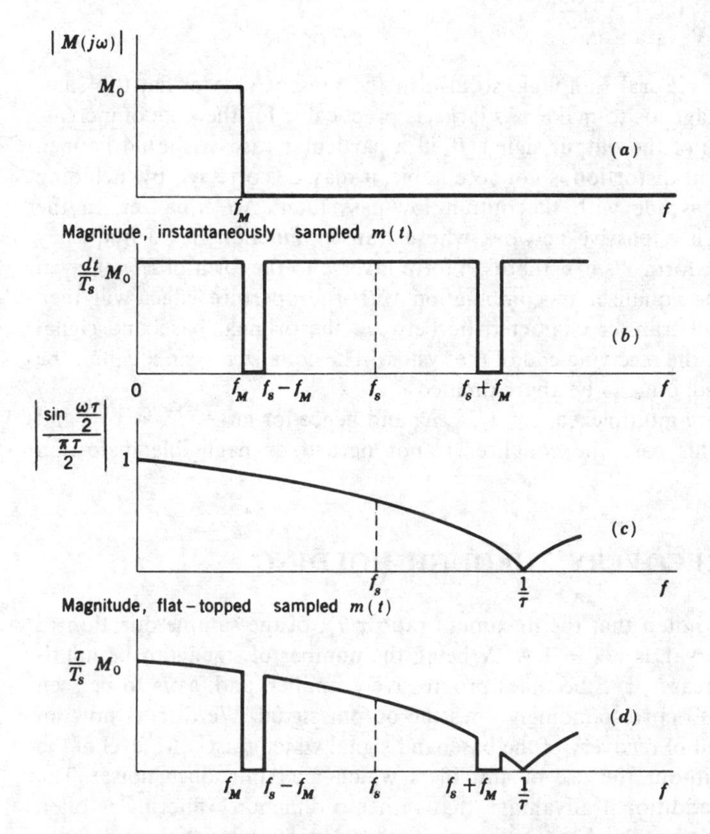

Figure 5.5-2 (*a*) An idealized spectrum of a baseband signal. (*b*) The spectrum of the signal with instantaneous sampling. (*c*) The form $[(\sin x)/x$, with $x = \omega\tau/2]$ of the distortion factor (aperture effect) introduced by flat-topped sampling. (*d*) The spectrum of the signal with flat-topped sampling.

however, that such is not the case and that, as a result, distortion will result. This distortion results from the fact that the original signal was "observed" through a finite rather than an infinitesimal time "aperture" and is hence referred to as *aperture effect* distortion.

The distortion results from the fact that the spectrum is multiplied by the sampling function $Sa(x) \equiv (\sin x)/x$ (with $x = \omega\tau/2$). The magnitude of the sampling function (see Sec. 1.4) falls off slowly with increasing x in the neighborhood of $x = 0$ and does not fall off sharply until we approach $x = \pi$, at which point $Sa(x) = 0$. To minimize the distortion due to the aperture effect, it is advantageous to arrange that $x = \pi$ correspond to a frequency very large in comparison with f_M. Since $x = \pi f\tau$, the frequency f_0 corresponding to $x = \pi$ is $f_0 = 1/\tau$. If $f_0 \gg f_M$, or, correspondingly, if $\tau \ll 1/f_M$, the aperture distortion will be small. The distortion becomes progressively smaller with decreasing τ. And, of course, as $\tau \to 0$ (instantaneous sampling), the distortion similarly approaches zero.

Equalization[1,2]

As in the case of natural sampling, so also in the present case of flat-top sampling, it is advantageous to make τ as large as practicable for the sake of increasing the amplitude of the output signal. If, in a particular case, it should happen that the consequent distortion is not acceptable, it may be corrected by including an *equalizer* in cascade with the output low-pass filter. An equalizer, in the present instance, is a passive network whose transfer function has a frequency dependence of the form $x/\sin x$, that is, a form inverse to the form of $H(j\omega)$ given in Eq. (5.5-3). The equalizer in combination with the aperture effect will then yield a flat overall transfer characteristic between the original baseband signal and the output at the receiving end of the system. The equalizer $x/\sin x$ cannot be exactly synthesized, but can be approximated.

If N signals are multiplexed, $\tau \leq 1/2f_M N$, and hence for large N $\tau \ll 1/f_M$ and $x/\sin x \simeq 1$. In this case the equalizer is not needed as negligible distortion results.

5.6 SIGNAL RECOVERY THROUGH HOLDING

We have already noted that the maximum ratio τ/T_s, of the sample duration to the sampling interval, is $\tau/T_s = 1/N$, N being the number of signals to be multiplexed. As N increases, τ/T_s becomes progressively smaller, and, as is to be seen from Eq. (5.4-4), so correspondingly does the output signal. We discuss now an alternative method of recovery of the baseband signal which raises the level of the output signal (without the use of amplifiers which may introduce noise). The method has the additional advantage that rather rudimentary filtering is often quite adequate, but has the disadvantage that some distortion must be accepted.

The method is illustrated in Fig. 5.6-1, where the baseband signal $m(t)$ and its flat-topped samples are shown. At the receiving end, and after *demultiplexing*, the sample pulses are extended; that is, the sample value of each individual baseband signal is *held* until the occurrence of the next sample of that same baseband signal. This operation is shown in Fig. 5.6-1 as the dashed extension of the sample pulses. The output waveform consists then, as shown, of an up and down staircase waveform with no blank intervals.

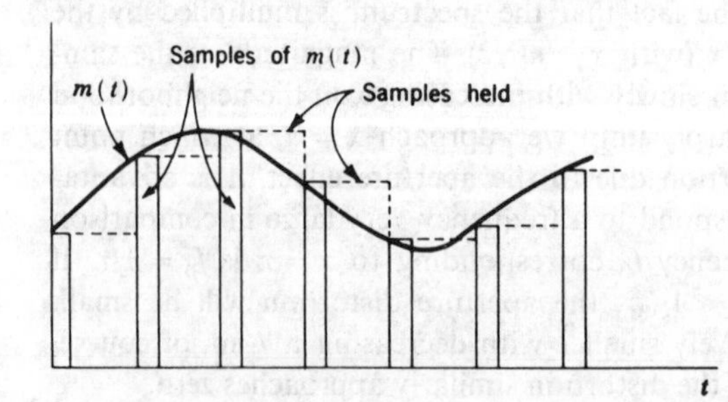

Figure 5.6-1 Illustrating the operation of holding.

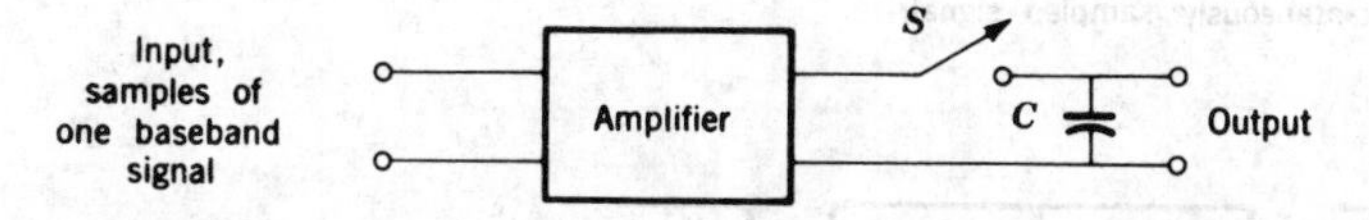

Figure 5.6-2 Illustrating a method of performing the operation of holding.

A method, in principle, by which this holding operation may be performed is shown in Fig. 5.6-2. The switch S operates in synchronism with the occurrence of input samples. This switch, ordinarily open, closes somewhat after the occurrence of the leading edge of a sample pulse and opens somewhat before the occurrence of the trailing edge. The amplifier, whose gain, if any, is incidental to the present discussion, has a low-output impedance. Hence, at the closing of the switch, the capacitor C charges abruptly to a voltage proportional to the sample value, and the capacitor holds this voltage until the operation is repeated for the next sample. In Fig. 5.6-1 we have idealized the situation somewhat by showing the output waveform maintaining a perfectly constant level throughout the sample pulse interval and its following holding interval. We have also indicated abrupt transitions in voltage level from one sample to the next. In practice, these voltage transitions will be somewhat rounded as the capacitor charges and discharges exponentially. Further, if the received sample pulses are natural samples rather than flat-topped samples, there will be some departure from a constant voltage level during the sample interval itself. As a matter of practice however, the sample interval is very small in comparison with the interval between samples, and the voltage variation of the baseband signal during the sampling interval is small enough to be neglected.

If the baseband signal is $m(t)$ with spectral density $M(j\omega) = \mathscr{F}[m(t)]$, we may deduce the spectral density of the sampled and held waveform in the manner of Sec. 5.5 and in connection with flat-topped sampling. We need but to consider that the flat tops have been stretched to encompass the entire interval between instantaneous samples. Hence the spectral density is given as in Eq. (5.5-4) except with τ replaced by the time interval between samples. We have, then,

$$\mathscr{F}[m(t), \text{sampled and held}] = \frac{\sin(\omega T_s/2)}{\omega T_s/2} M(j\omega) \qquad 0 \le f \le f_M \qquad (5.6\text{-}1)$$

In Fig. 5.6-3 we have again assumed for simplicity that the band-limited signal $m(t)$ has a flat spectral density of magnitude M_0. In Fig. 5.6-3*a* is shown the spectrum of the instantaneously sampled signal. In Fig. 5.6-3*b* has been drawn the magnitude of the aperture factor $(\sin x)/x$ (with $x = \omega T_s/2$), while in Fig. 5.6-3*c* is shown the magnitude of the spectrum of the sampled-and-held signal. These plots differ from the plots of Fig. 5.5-2 only in the location of the nulls of the factor $(\sin x)/x$. In Fig. 5.6-3 the first null occurs at the sampling frequency f_s. We observe that, as a consequence, the aperture effect, which is responsible for the $(\sin x)/x$ term, has accomplished most of the filtering which is required to suppress the part of the spectrum of the output signal above the

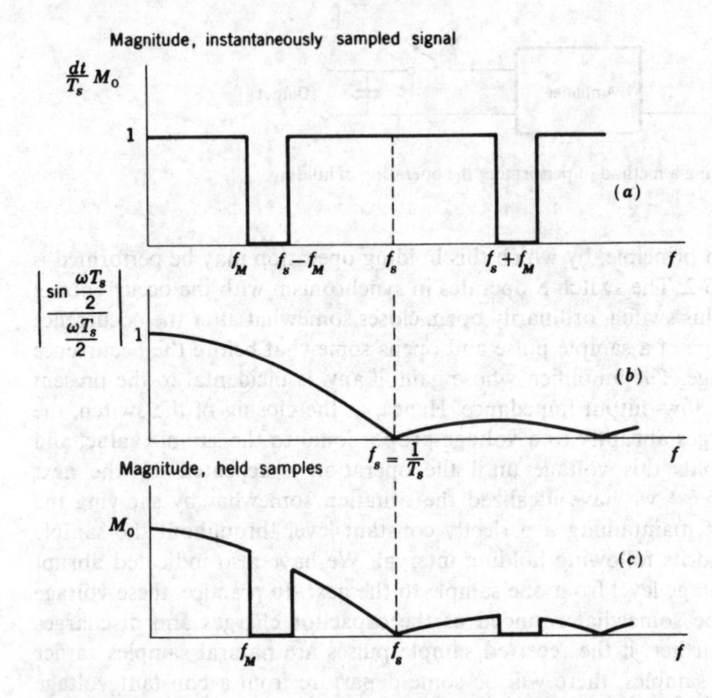

Figure 5.6-3 (*a*) Spectrum of instantaneously sampled signal $m(t)$ with $m(t)$ having idealized spectrum shown in Fig. 5.5-2*a*. (*b*) The magnitude of the aperture effect factor. (*c*) Spectrum of sampled-and-held signal.

bandlimit f_M. Of course, the filtering is not perfect, and some additional filtering may be required. We also note that, as in the case of flat-top sampling, there will be some distortion introduced by the unequal transmission of spectral components in the range 0 to f_M. If the distortion is not acceptable, then, as before, it may be corrected by an $x/\sin x$ equalizer.

Most importantly we note in comparing Eq. (5.6-1) with Eq. (5.5-4) that, aside from the relatively small effect of the $(\sin x)/x$ terms in the two cases, the sampled-and-held signal has a magnitude larger by the factor T_s/τ than the signal of sample duration τ. This increase in amplitude is, of course, intuitively to have been anticipated.

5.7 QUANTIZATION OF SIGNALS

The limitation of the system we have been describing for communicating over long channels is that once noise has been introduced any place along the channel, we are "stuck" with it. We now describe how the situation is modified by subjecting a signal to the operation of quantization. When quantizing a signal $m(t)$, we create a new signal $m_q(t)$ which is an approximation to $m(t)$. However, the quantized signal $m_q(t)$ has the great merit that it is, in large measure, separable from the additive noise.

The operation of quantization is represented in Fig. 5.7-1. Here we contemplate a signal $m(t)$ whose excursion is confined to the range from V_L to V_H. We have divided this total range into M equal intervals each of size S. Accordingly S, called the *step size*, is $S = (V_H - V_L)/M$. In Fig. 5.7-1 we show the specific example in which $M = 8$. In the center of each of these steps we locate *quantization levels* $m_0, m_1, \ldots, m_7$. The quantized signal $m_q(t)$ is generated in the following way: Whenever $m(t)$ is in the range Δ_0, the signal $m_q(t)$ maintains the constant level m_0; whenever $m(t)$ is in the range Δ_1, $m_q(t)$ maintains the constant level m_1; and so on. Thus the signal $m_q(t)$ will at all times be found at one of the levels m_0, $m_1, \ldots, m_7$. The transition in $m_q(t)$ from $m_q(t) = m_0$ to $m_q(t) = m_1$ is made abruptly when $m(t)$ passes the transition level L_{01} which is midway between m_0 and m_1 and so on. To state the matter in an alternative fashion, we say that, at every instant of time, $m_q(t)$ has the value of the quantization level to which $m(t)$ is closest. Thus the signal $m_q(t)$ does not change at all with time or it makes a "quantum" jump of step size S. Note the disposition of the quantization levels in the range from V_L to V_H. These levels are each separated by an amount S, but the separation of the extremes V_L and V_H each from its nearest quantization level is only $S/2$. Also, at every instant of time, the quantization error $m(t) - m_q(t)$ has a magnitude which is equal to or less than $S/2$.

We see, therefore, that the quantized signal is an approximation to the original signal. The quality of the approximation may be improved by reducing the size of the steps, thereby increasing the number of allowable levels. Eventually, with small enough steps, the human ear or the eye will not be able to distinguish the original from the quantized signal. To give the reader an idea of the number of quantization levels required in a practical system, we note that 256 levels can be used to obtain the quality of commercial color TV, while 64 levels gives only fairly good color TV performance. These results are also found to be valid when quantizing voice.

Now let us consider that our quantized signal has arrived at a repeater somewhat attenuated and corrupted by noise. This time our repeater consists of a quantizer and an amplifier. There is noise superimposed on the quantized levels of $m_q(t)$. But suppose that we have placed the repeater at a point on the communications channel where the instantaneous noise voltage is almost always less than half the separation between quantized levels. Then the output of the quantizer will consist of a succession of levels duplicating the original quantized signal and *with the noise removed.* In rare instances the noise results in an error in quantization level. A noisy quantized signal is shown in Fig. 5.7-2*a*. The allowable quantizer output levels are indicated by the dashed lines separated by amount S. The output of the quantizer is shown in Fig. 5.7-2*b*. The quantizer output is the level to which the input is closest. Therefore, as long as the noise has an instantaneous amplitude less than $S/2$, the noise will not appear at the output. One instance in which the noise does exceed $S/2$ is indicated in the figure, and, corre-

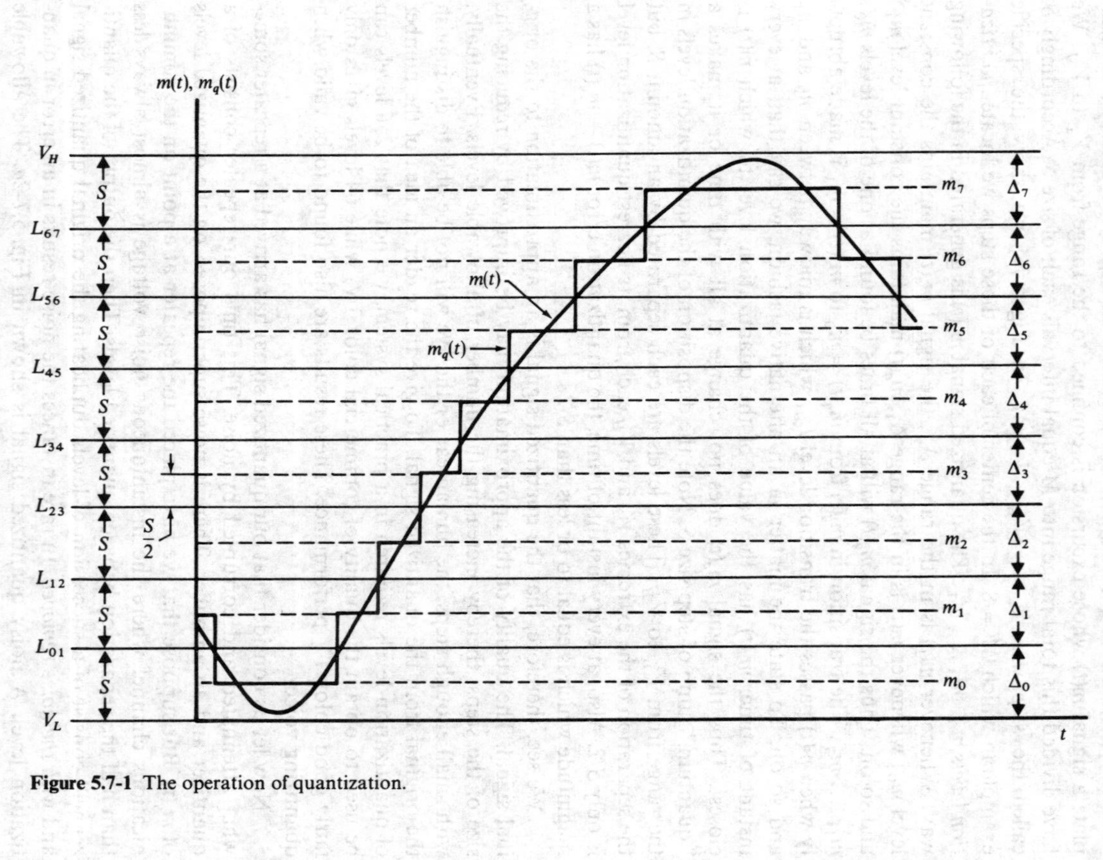

Figure 5.7-1 The operation of quantization.

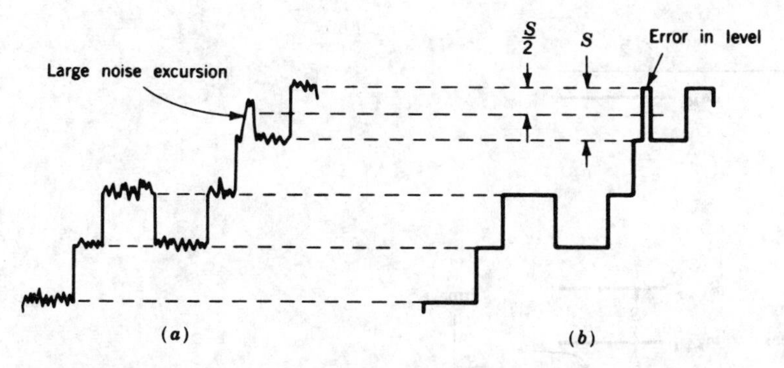

Figure 5.7-2 (*a*) A quantized signal with added noise. (*b*) The signal after requantization. One instance is recorded in which the noise level is so large that an error results.

spondingly, an error in level does occur. The statistical nature of noise is such that even if the average noise magnitude is much less than $S/2$, there is always a finite probability that from time to time, the noise magnitude will exceed $S/2$. Note that it is never possible to suppress completely level errors such as the one indicated in Fig. 5.7-2.

We have shown that through the method of signal quantization, the effect of additive noise can be significantly reduced. By decreasing the spacing of the repeaters, we decrease the attenuation suffered by $m_q(t)$. This effectively decreases the relative noise power and hence decreases the probability P_q of an error in level. P_q can also be reduced by increasing the step size S. However, increasing S results in an increased discrepancy between the true signal $m(t)$ and the quantized signal $m_q(t)$. This difference $m(t) - m_q(t)$ can be regarded as noise and is called *quantization noise*. Hence, the received signal is not a perfect replica of the transmitted signal $m(t)$. The difference between them is due to errors caused by additive noise and quantization noise. These noises are discussed further in Chap. 12.

5.8 QUANTIZATION ERROR

It has been pointed out that the quantized signal and the original signal from which it was derived differ from one another in a random manner. This difference or error may be viewed as a noise due to the quantization process and is called *quantization error*. We now calculate the mean-square quantization error $\overline{e^2}$, where e is the difference between the original and quantized signal voltages.

Let us divide total peak-to-peak range of the message signal $m(t)$ into M equal voltage intervals, each of magnitude S volts. At the center of each voltage interval we locate a quantization level $m_1, m_2, \ldots, m_M$ as shown in Fig. 5.8-1*a*. The dashed level represents the instantaneous value of the message signal $m(t)$ at a time t. Since, in this figure, $m(t)$ happens to be closest to the level m_k, the quantizer output will be m_k, the voltage corresponding to that level. The error is $e = m(t) - m_k$.

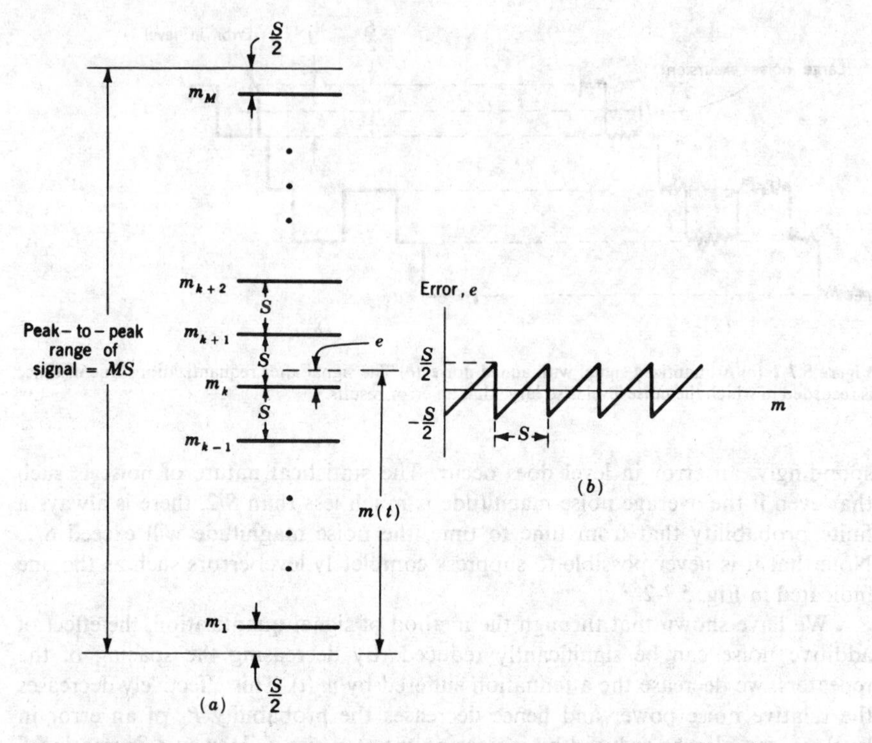

Figure 5.8-1 (*a*) A range of voltage over which a signal $m(t)$ makes excursions is divided into M quantization ranges each of size S. The quantization levels are located at the center of the range. (*b*) The error voltage $e(t)$ as a function of the instantaneous value of the signal $m(t)$.

Let $f(m)\,dm$ be the probability that $m(t)$ lies in the voltage range $m - dm/2$ to $m + dm/2$. Then the mean-square *quantization error* is

$$\overline{e^2} = \int_{m_1 - S/2}^{m_1 + S/2} f(m)(m - m_1)^2\,dm + \int_{m_2 - S/2}^{m_2 + S/2} f(m)(m - m_2)^2\,dm + \cdots \tag{5.8-1}$$

Now, ordinarily the probability density function $f(m)$ of the message signal $m(t)$ will certainly not be constant. However, suppose that the number M of quantization is large, so that the step size S is small in comparison with the peak-to-peak range of the message signal. In this case, it is certainly reasonable to make the approximation that $f(m)$ is constant within each quantization range. Then in the first term of Eq. (5.8-1) we set $f(m) = f^{(1)}$, a constant. In the second term $f(m) = f^{(2)}$, etc. We may now remove $f^{(1)}, f^{(2)}$, etc., from inside the integral sign. If we make the substitution $x \equiv m - m_k$, Eq. (5.8-1) becomes

$$\overline{e^2} = (f^{(1)} + f^{(2)} + \cdots) \int_{-S/2}^{S/2} x^2\,dx = (f^{(1)} + f^{(2)} + \cdots)\frac{S^3}{12} \tag{5.8-2a}$$

$$= (f^{(1)}S + f^{(2)}S + \cdots)\frac{S^2}{12} \tag{5.8-2b}$$

Now $f^{(1)}S$ is the probability that the signal voltage $m(t)$ will be in the first quantization range, $f^{(2)}S$ is the probability that m is in the second quantization range, etc. Hence the sum of terms in the parentheses in Eq. (5.8-2*b*) has a total value of unity. Therefore, the mean-square quantization error is

$$\overline{e^2} = \frac{S^2}{12} \tag{5.8-3}$$

5.9 PULSE-CODE MODULATION

A signal which is to be quantized prior to transmission is usually sampled as well. The quantization is used to reduce the effects of noise, and the sampling allows us to time-division multiplex a number of messages if we choose to do so. The combined operations of sampling and quantizing generate a quantized PAM waveform, that is, a train of pulses whose amplitudes are restricted to a number of discrete magnitudes.

We may, if we choose, transmit these quantized sample values directly. Alternatively we may represent each quantized level by a code number and transmit the code number rather than the sample value itself. The merit of so doing will be developed in the subsequent discussion. Most frequently the code number is converted, before transmission, into its representation in binary arithmetic, i.e., base-2 arithmetic. The digits of the binary representation of the code number are transmitted as pulses. Hence the system of transmission is called (binary) *pulse-code modulation* (PCM).

We review briefly some elementary points about binary arithmetic. The binary system uses only two digits, 0 and 1. An arbitrary number N is represented by the sequence $\ldots k_2 k_1 k_0$, in which the k's are determined from the equation

$$N = \cdots + k_2 2^2 + k_1 2^1 + k_0 2^0 \tag{5.9-1}$$

with the added constraint that each k has the value 0 or 1. The binary representations of the decimal numbers 0 to 15 are given in Table 5.9-1. Observe that to represent the four (decimal) numbers 0 to 3, we need only two binary digits k_1 and k_0. For the eight (decimal) numbers from 0 to 7 we require only three binary places, and so on. In general, if M numbers $0, 1, \ldots, M-1$ are to be represented, then an N binary digit sequence $k_{N-1} \cdots k_0$ is required, where $M = 2^N$.

The essential features of binary PCM are shown in Fig. 5.9-1. We assume that the analog message signal $m(t)$ is limited in its excursions to the range from -4 to $+4$ volts. We have set the step size between quantization levels at 1 volt. Eight quantization levels are employed, and these are located at $-3.5, -2.5, \ldots, +3.5$ volts. We assign the code number 0 to the level at -3.5 volts, the code number 1 to the level at -2.5 volts, etc., until the level at $+3.5$ volts, which is assigned the code number 7. Each code number has its representation in binary arithmetic ranging from 000 for code number 0 to 111 for code number 7.

In Fig. 5.9-1, in correspondence with each sample, we specify the sample value, the nearest quantization level, and the code number and its binary rep-

Table 5.9-1 Equivalent numbers in decimal and binary representation

Binary				Decimal
k_3	k_2	k_1	k_0	
0	0	0	0	0
0	0	0	1	1
0	0	1	0	2
0	0	1	1	3
0	1	0	0	4
0	1	0	1	5
0	1	1	0	6
0	1	1	1	7
1	0	0	0	8
1	0	0	1	9
1	0	1	0	10
1	0	1	1	11
1	1	0	0	12
1	1	0	1	13
1	1	1	0	14
1	1	1	1	15

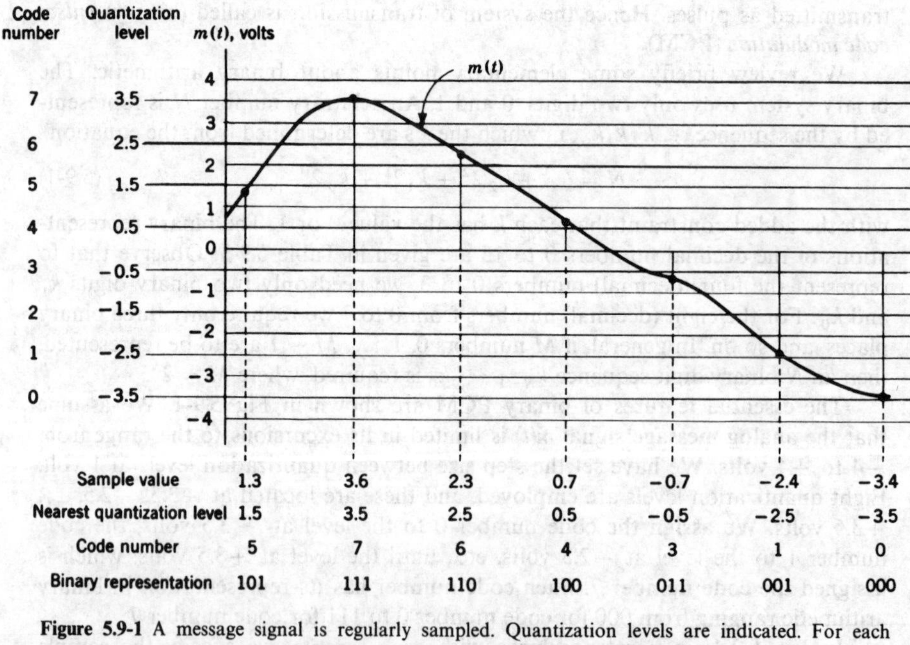

Figure 5.9-1 A message signal is regularly sampled. Quantization levels are indicated. For each sample the quantized value is given and its binary representation is indicated.

resentation. If we were transmitting the analog signal, we would transmit the sample values 1.3, 3.6, 2.3, etc. If we were transmitting the quantized signal, we would transmit the quantized sample values 1.5, 3.5, 2.5, etc. In binary PCM we transmit the binary representations 101, 111, 110, etc.

5.10 ELECTRICAL REPRESENTATIONS OF BINARY DIGITS

As intimated in the previous section, we may represent the binary digits by electrical pulses in order to transmit the code representations of each quantized level over a communication channel. Such a representation is shown in Fig. 5.10-1. Pulse time slots are indicated at the top of the figure, and, as shown in Fig. 5.10-1*a*, the binary digit 1 is represented by a pulse, while the binary digit 0 is represented by the absence of a pulse. The row of three-digit binary numbers given in Fig. 5.10-1 is the binary representation of the sequence of quantized samples in Fig. 5.9-1. Hence the pulse pattern in Fig. 5.10-1*a* is the (binary) PCM waveform that would be transmitted to convey to the receiver the sequence of quantized samples of the message signal $m(t)$ in Fig. 5.9-1. Each three-digit binary number that specifies a quantized sample value is called a *word*. The spaces between words allow for the multiplexing of other messages.

At the receiver, in order to reconstruct the quantized signal, all that is required is that a determination be made, within each pulse time slot, about whether a pulse is present or absent. The exact amplitude of the pulse is not important. There is an advantage in making the pulse width as wide as possible since the pulse energy is thereby increased and it becomes easier to recognize a pulse against the background noise. Suppose then that we eliminate the guard time τ_g between pulses. We would then have the waveform shown in Fig. 5.10-1*b*. We would be rather hard put to describe this waveform as either a sequence of positive pulses or of negative pulses. The waveform consists now of a sequence of transitions between two levels. When the waveform occupies the lower level in a

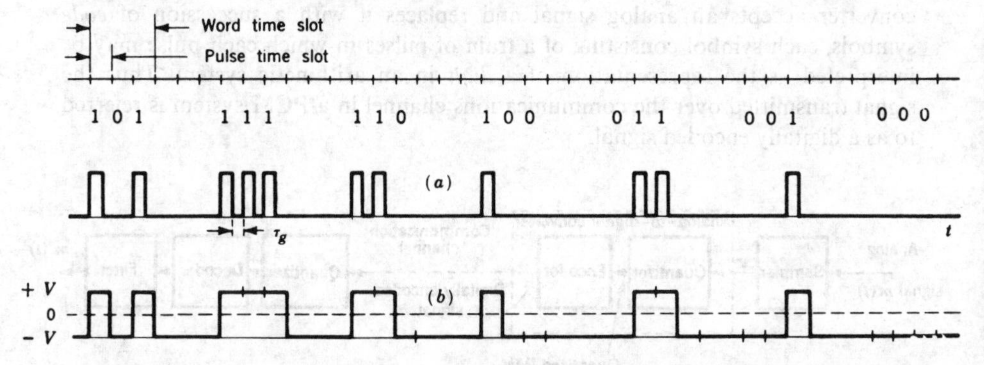

Figure 5.10-1 (*a*) Pulse representation of the binary numbers used to code the samples in Fig. 5.9-1. (*b*) Representation by voltage levels rather than pulses.

particular time slot, a binary 0 is represented, while the upper voltage level represents a binary 1.

Suppose that the voltage difference of $2V$ volts between the levels of the waveform of Fig. 5.10-1*b* is adequate to allow reliable determination at the receiver of which digit is being transmitted. We might then arrange, say, that the waveform make excursions between 0 and $2V$ volts or between $-V$ volts and $+V$ volts. The former waveform will have a dc component, the latter waveform will not. Since the dc component wastes power and contributes nothing to the reliability of transmission, the latter alternative is preferred and is indicated in Fig. 5.10-1*b*.

5.11 THE PCM SYSTEM

The Encoder

A PCM communication system is represented in Fig. 5.11-1. The analog signal $m(t)$ is sampled, and these samples are subjected to the operation of quantization. The quantized samples are applied to an *encoder*. The encoder responds to each such sample by the generation of a unique and identifiable binary pulse (or binary level) pattern. In the example of Figs. 5.9-1 and 5.10-1 the pulse pattern happens to have a numerical significance which is the same as the order assigned to the quantized levels. However, this feature is not essential. We could have assigned any pulse pattern to any level. At the receiver, however, we must be able to identify the level from the pulse pattern. Hence it is clear that not only does the encoder number the level, it also assigns to it an identification code.

The combination of the quantizer and encoder in the dashed box of Fig. 5.11-1 is called an *analog-to-digital converter*, usually abbreviated A/D converter. In commercially available A/D converters there is normally no sharp distinction between that portion of the electronic circuitry used to do the quantizing and that portion used to accomplish the encoding. In summary, then, the A/D converter accepts an analog signal and replaces it with a succession of *code* symbols, each symbol consisting of a train of pulses in which each pulse may be interpreted as the representation of a *digit* in an arithmetic system. Thus the signal transmitted over the communications channel in a PCM system is referred to as a digitally encoded signal.

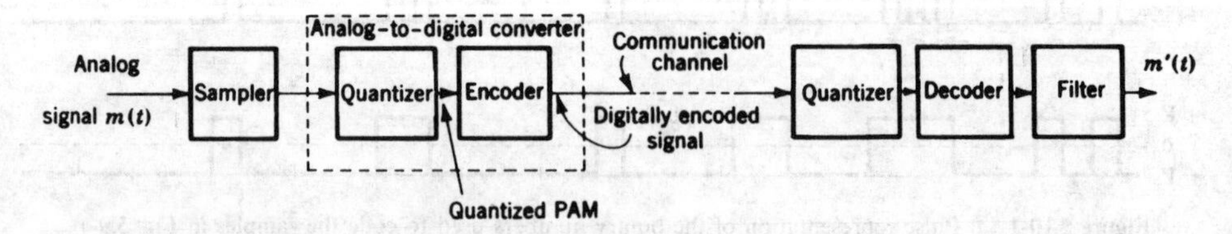

Figure 5.11-1 A PCM communication system.

The Decoder

When the digitally encoded signal arrives at the receiver (or repeater), the first operation to be performed is the separation of the signal from the noise which has been added during the transmission along the channel. As noted previously, separation of the signal from the noise is possible because of the quantization of the signal. Such an operation is again an operation of *requantization;* hence the first block in the receiver in Fig. 5.11-1 is termed a quantizer. A feature which eases the burden on this quantizer is that for each pulse interval it has only to make the relatively simple decision of whether a pulse has or has not been received or which of two voltage levels has occurred. Suppose the quantized sample pulses had been transmitted instead, rather than the binary-encoded codes for such samples. Then this quantizer would have had to have yielded, in each pulse interval, not a simple yes or no decision, but rather a more complicated determination about which of the many possible levels had been received. In the example of Fig. 5.10-1, if a quantized PAM signal had been transmitted, the receiver quantizer would have to decide which of the levels 0 to 7 was transmitted, while with a binary PCM signal the quantizer need only distinguish between two possible levels. The relative reliability of the yes or no decision in PCM over the multivalued decision required for quantized PAM constitutes an important advantage for PCM.

The receiver quantizer then, in each pulse slot, makes an educated and sophisticated estimate and then decides whether a positive pulse or a negative pulse was received and transmits its decisions, in the form of a reconstituted or regenerated pulse train, to the decoder. (If repeater operation is intended, the regenerated pulse train is simply raised in level and sent along the next section of the transmission channel.) The decoder, also called a *digital-to-analog* (*D/A*) *converter*, performs the inverse operation of the encoder. The decoder output is the sequence of quantized multilevel sample pulses. The quantized PAM signal is now reconstituted. It is then filtered to reject any frequency components lying outside of the baseband. The final output signal $m'(t)$ is identical with the input $m(t)$ except for quantization noise and the occasional error in yes-no decision making at the receiver due to the presence of channel noise.

5.12 COMPANDING

Referring again to Figs. 5.7-1 and 5.8-1 let us consider that we have established a quantization process employing M levels with step size S, the levels being established at voltages to accommodate a signal $m(t)$ which ranges from a low voltage V_L to a high voltage V_H. We can readily see that if the signal $m(t)$ should make excursions beyond the bounds V_H and V_L the system will operate at a disadvantage. For, within these bounds, the instantaneous quantization error never exceeds $\pm S/2$ while outside these bounds the error is larger.

Further, whenever $m(t)$ does not swing through the full available range the system is equally at a disadvantage. For, in order that $m_q(t)$ be a good approximation to $m(t)$ it is necessary that the step size S be small in comparison to the *range over which $m(t)$ swings.* As a very pointed example of this consideration consider a case in which $m(t)$ has a peak-to-peak voltage which is less than S and never crosses one of the transition levels in Fig. 5.7-1. In such a case $m_q(t)$ will be a fixed (dc) voltage and will bear no relationship to $m(t)$.

To explore this latter point somewhat more quantitatively let us consider that $m(t)$ is a signal, such as the sound signal output of a microphone, in which $V_H = -V_L = V$, i.e., a signal without dc components, and with (at least approximately) equal positive and negative peaks. Further, for simplicity, let us assume that in the range $\pm V$ the signal $m(t)$ is characterized by a uniform probability density. The probability density is then equal to $1/2V$ and the normalized average signal power of the applied input signal is

$$S_i = \overline{m^2(t)} = \int_{-V}^{+V} m^2(t) \frac{1}{2V} \, dm = \frac{V^2}{3} \tag{5.12-1}$$

The quantization noise, as given by Eq. (5.8-3) is

$$N_Q = \frac{S^2}{12} \tag{5.12-2}$$

If the number of quantization levels is M, then $MS = 2V$ so that

$$V = \frac{MS}{2} \tag{5.12-3}$$

Combining Eqs. (5.12-1), (5.12-2), and (5.12-3) we have that the input signal-to-quantization noise power ratio is

$$\frac{S_i}{N_Q} = M^2 \tag{5.12-4}$$

Eventually the received quantized signal will be smoothed out to generate an output signal with power S_o. If we have a useful communication system then presumably the effect of the quantization noise is not such as to cause an easily perceived difference between input and output signal. In such a case the output power S_o may be taken to be the same as the input power i.e., $S_o \simeq S_i$ so that finally we may replace Eq. (5.12-4) by

$$\frac{S_o}{N_Q} = M^2 \tag{5.12-5}$$

If there are M quantization levels then the code which singles out the closest quantization to the signal $m(t)$ will have to have N bits with $M = 2^N$. Hence Eq. (5.12-5) becomes

$$\frac{S_o}{N_Q} = (2^N)^2 = 2^{2N} \tag{5.12-6}$$

In decibels we have

$$\left[\frac{S_o}{N_Q}\right]_{dB} = 10 \log_{10} \left[\frac{S_o}{N_Q}\right] = 10 \log_{10} 2^{2N} = 6N \tag{5.12-7}$$

Equation (5.12-7) has the interpretation that, in a system where the signal is quantized using an N-bit code (i.e., the number of quantization levels is 2^N), and where the signal amplitude is capable of swinging through all available quantization regions without extending beyond the outermost ranges, the output *signal-to-quantization noise ratio* is $6N$ dB. In voice communication we use $N = 8$ corresponding to 256 quantization regions and $S_o/N_Q = (6)(8) = 48$ dB. If the signal is reduced in amplitude so that not all quantization ranges are used then S_o/N_Q becomes smaller, since N_Q, depending as it does only on step S is not affected by the amplitude reduction, while S_o is reduced. For example, if the amplitude were reduced by a factor of 2, the power is then reduced by a factor of 4 reducing the signal-to-noise ratio by 6 dB. It is interesting to observe that the effective number of quantization levels is also reduced by a factor of 2. Correspondingly, the number N of code bits is reduced by 1. In summary, the dependence of S_o/N_Q on the input signal power S_i is such that as the number of code bits needed decreases, S_o/N_Q decreases by 6 dB/bit.

It is generally required, for acceptable voice transmission, that the received signal have a ratio S_o/N_Q not less than 30 dB, and that this minimum 30 dB figure hold even though the signal power itself may vary by 40 dB. (The signal, in this case, is described as having a 40 dB *dynamic* range.) In an 8-bit system, at maximum signal level we have $S_o/N_Q \cong S_i/N_Q = 48$ dB. Now N_Q is fixed, and depends only on step size. If we are to allow S_i/N_Q to drop to no lower than 30 dB then the dynamic range would be restricted to $48 - 30 = 18$ dB.

The dynamic range can be materially improved by a process called *companding* (a word formed by combining the words *compressing* and *expanding*). As we have seen, to keep the signal-to-quantization noise ratio high we must use a signal which swings through a range which is large in comparison with the step size. This requirement is not satisfied when the signal is small. Accordingly before applying the signal to the quantizer we pass it through a network which has an input-output characteristic as shown in Fig. 5.12-1. Note that at low amplitudes the slope is larger than at large amplitudes. A signal transmitted through such a network will have the extremities of its waveform *compressed*. The peak signal which the system is intended to accommodate will, as before, range through all available quantization regions. But now, a small amplitude signal will range through *more* quantization regions than would be the case in the absence of compression. Of course, the compression produces signal distortion. To undo the distortion, at the receiver we pass the recovered signal through an *expander* network. An expander network has an input-output characteristic which is the inverse of the characteristic of the compressor. The inverse distortions of compressor and expander generate a final output signal without distortion.

The determination of the form of the compression plot of Fig. 5.12-1 is a somewhat subjective matter. In the United States, Canada, and Japan a *μ-law*

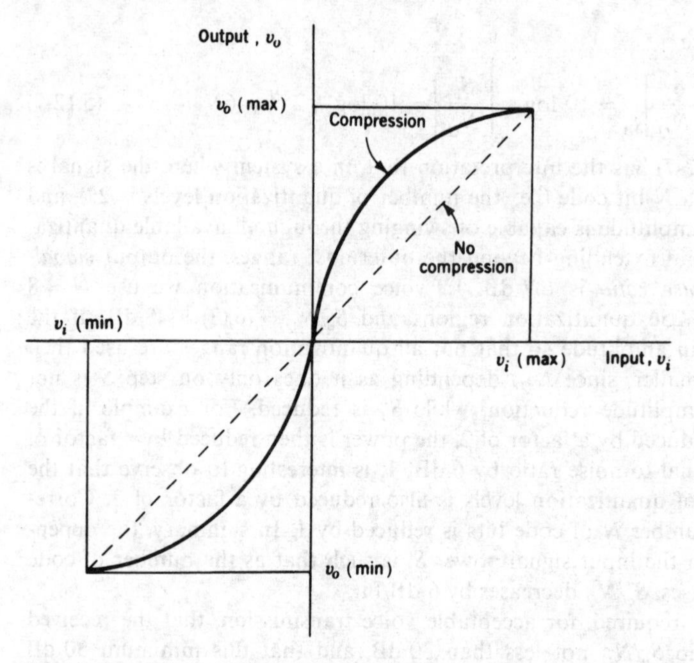

Figure 5.12-1 An input–output characteristic which provides compression.

compandor is used (see Prob. 5.12-1) and it differs somewhat from the *A-law* compander (see Prob. 5.12-2) used by the rest of the world. The "μ" and the "A" refer to parameters which appear in the equations for the compression and expansion characteristic.

In implementing the compression characteristic, the analog signal $m(t)$ is left unmodified and, instead, the step size is tapered so that the quantization levels are close together at low signal amplitudes and progressively further apart in regions attained as the signal increases in amplitude. To see one way in which the step sizes are altered consider again an 8-bit PCM system employing $2^8 = 256$ quantization levels. In such a system, the A/D converter shown in Fig. 5.11-1 would ordinarily be an 8-bit converter. Each input sample would generate an 8-bit output code identifying the closest quantization level. If the total range of the input was $\pm V$ then the step size would be $S = 2V/2^8$. Let us start however with a 12-bit A/D converter so that the step size is $2V/2^{12}$. This 12-bit PCM signal is not applied to the communication channel directly but is instead applied to the address pins of a read-only memory (ROM) whose content is to be described. The ROM has 12 address pins and 8 output data pins. The signal transmitted is the 8-bit data output of the ROM.

The content of the ROM is as follows: As shown in Fig. 5.12-2, in each successive memory location in the region where the step size is to be smallest (i.e., step size $= 2V/2^{12} = \Delta$) there are written successive 8-bit code words. Hence in this region a change in analog signal of step size Δ will change the code word presented for transmission by the amount Δ. This one-to-one correspondence between addresses and transmitted code applies as shown for the middle 64 addresses. For the next 64 addresses (32 on one side and 32 on the other side of the original 64) we arrange that at the memory locations specified by pairs of addresses, i.e., for *two* adjacent addresses the *same* code word is written into the memory. Hence the transmitted code word will change at every *second* addresses change and correspondingly the step size is 2Δ. Next we arrange that for the next

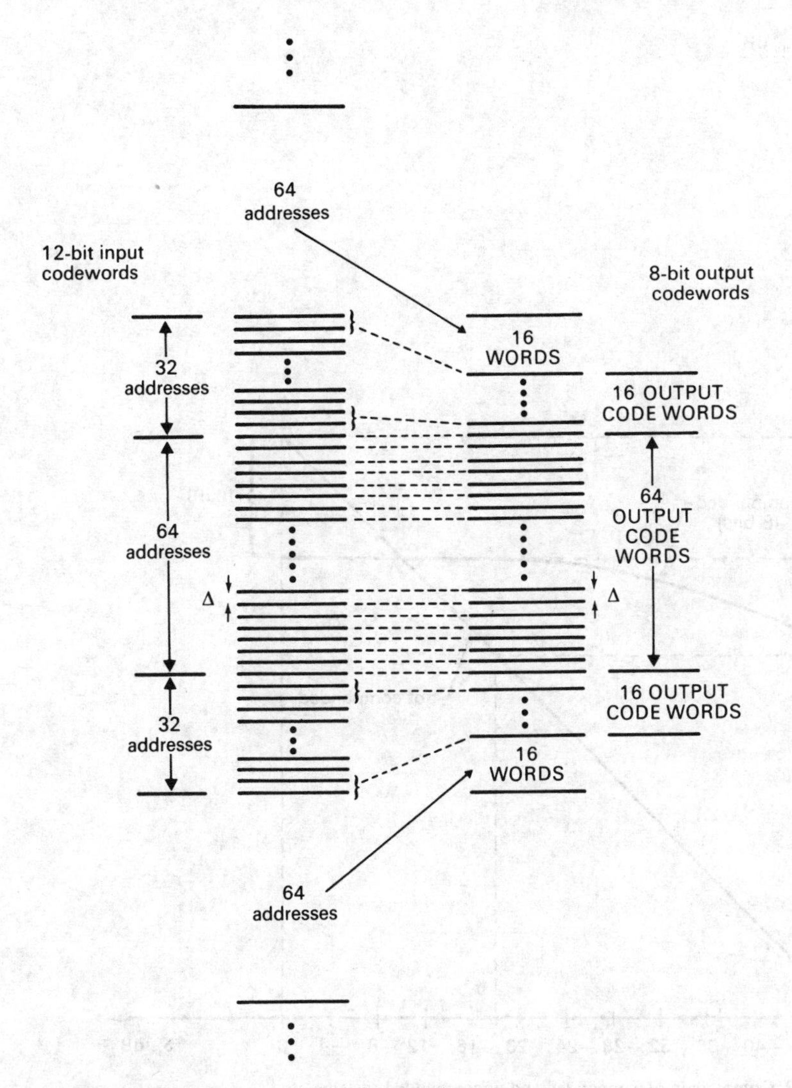

Figure 5.12-2 ROM characteristic for compression.

128 addresses (64 above + 64 below) the *same* code word is written into the locations of *four* successive memory locations so that the step size is 4Δ. We proceed in this manner, each time doubling the number of successive memory locations which contain the same code word. As can be verified, when we have used all $2^{12} = 4096$ addresses we shall have generated $2^8 = 256$ code words. At the receiver we have a mechanism for generating the quantization corresponding to each code word.

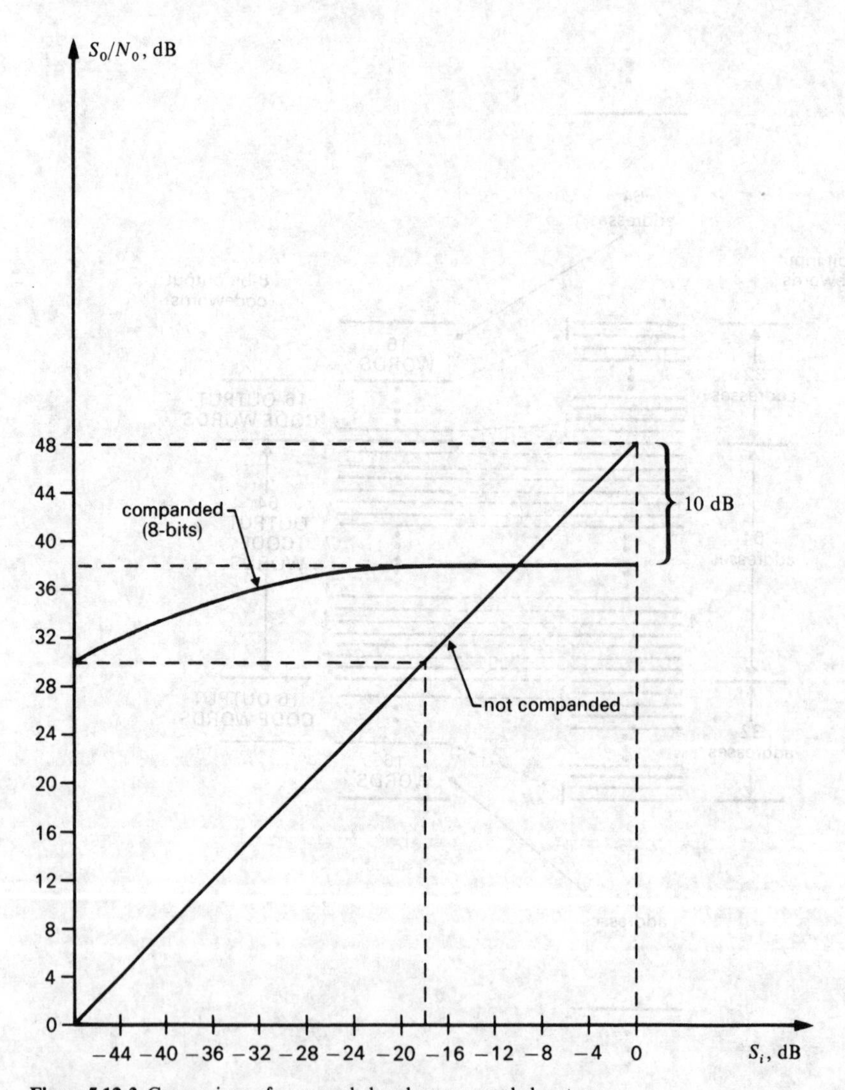

Figure 5.12-3 Comparison of companded and uncompanded systems.

It is interesting to observe that for the smallest 64 levels, the input and output signals are the same, i.e., small signals are *not* companded. One reason for this is that signals other than voice are often digitized using the same PCM system employed for voice. If a non-voice signal is subjected to the compression algorithm the result is usually a degradation of performance. Therefore, to avoid this possibility, such non-voice signals are kept 40 dB below the peak level of a voice signal.

In conclusion, we refer to Fig. 5.12-3 which compares the variation of output signal-to-noise ratio as a function of input signal power when companding is used, to the case of an uncompanded system. Note that the companded system has a far greater dynamic range than the uncompanded system and that theoretically the companded system has an output signal-to-noise ratio which exceeds 30 dB over a dynamic range of input signal power of 48 dB, while the uncompanded system has a dynamic range of 18 dB for the same conditions. It should be noted however, that the penalty paid using companding is approximately 10 dB. Thus an 8-bit uncompanded system, when operating at maximum amplitude, produces a signal-to-noise ratio of 48 dB. The same 8-bit system, using a compandor, yields a 38 dB SNR.

5.13 MULTIPLEXING PCM SIGNALS[3]

We have already noted the advantage of converting an analog signal into a PCM waveform when the signal must be transmitted over a noisy channel. When a large number of such PCM signals must be transmitted over a common channel, multiplexing of these PCM signals is required. In this section we discuss the multiplexing methods for PCM waveforms used in the United States by the common carriers (AT&T, GTE, etc.)

5.13-1 The $T1$ Digital System

Figure 5.13-1 shows the basic time division multiplexing scheme, called the $T1$ digital system, which is used to convey multiple signals over telephone lines using wideband coaxial cable. It accommodates 24 analog signals which we shall refer to as s_1 through s_{24}. Each signal is bandlimited to approximately 3.3 kHz and is sampled at the rate 8 kHz which exceeds, by a comfortable margin, the Nyquist rate of $2 \times 3.3 = 6.6$ kHz. Each of the time division multiplexed signals (still analog) is next A/D converted and companded as described in Sec. 5.12. The resulting digital waveform is transmitted over a coaxial cable, the cable serving to minimize signal distortion and serving also to suppress signal corruption due to noises from external sources. Periodically, at approximately 6000 ft intervals, the signal is regenerated by amplifiers called repeaters and then sent on toward its eventual destination. The repeater eliminates from each bit the effect of the distortion introduced by the channel. Also, the repeater removes from each bit any superimposed noise and thus, even having received a distorted and noisy signal, it

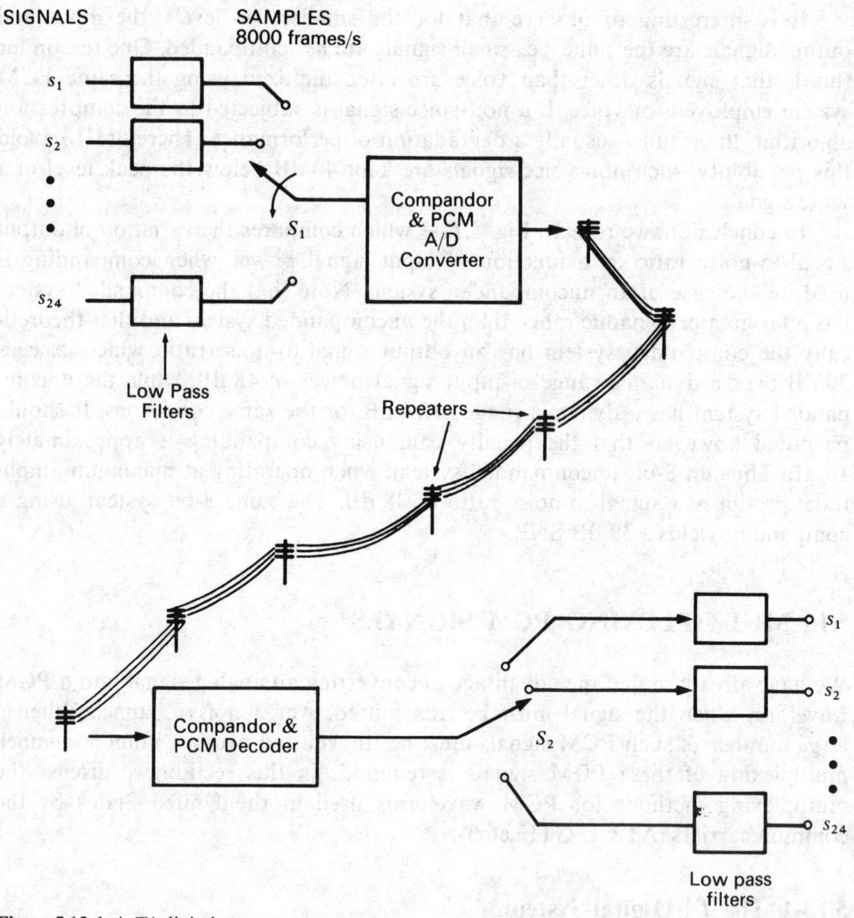

Figure 5.13-1 A $T1$ digital system.

retransmits a distortionless and noise-free duplicate of the signal originally sent. Such is, of course, the case for all bits except those infrequent bits which arrive so corrupted by noise that the bit is misread. At the destination the signal is companded, decoded, and demultiplexed, making available, individually, the 24 original signals.

Bits/Frame

The commutators sweep continuously from s_1 to s_{24} and back to s_1, etc., at the rate of 8000 revolutions per second thereby providing 8000 samples per second of each signal. Each sample is encoded into eight bits (corresponding to $2^8 = 256$ quantization levels). The digital signal generated during the course of one complete sweep of the commutator is therefore $24 \times 8 = 192$ bits.

Frame Synchronization

It is necessary to make available at the receiver not only the bits into which the signals have been encoded but also some synchronizing information. Without such synchronization (i.e., timing) information, the receiver cannot know which bits correspond to which of the original signals. To provide such synchronization an extra single bit is made available immediately preceding the 192 bits that carry the encoded signals. The 192 bit slots assigned to the encoded signal together with the one extra timing bit, for a total of 193 bits, is called a *frame.* The time slots for the 24 signals together with the extra frame-synchronizing bit F is shown in Fig. 5.13-2.

Twelve successive F slots are used to transmit a 12-bit code. The code happens to be 110111001000. This code is transmitted repetitively once every 12 frames and is used at the receiver to establish synchronization.

Bit Rate

Each signal is sampled 8000 times per second so that a complete frame occupies a time

$$T_p = 1/8000 = 125\ \mu s$$

This time T_p accommodates 193 bits so that the bit rate on a $T1$ channel is

$$f_b(T1) = \frac{193}{125}\ \text{Mb/s} = 1.544\ \text{Mb/s}$$

Bits in the frame synchronizing code occur once per frame, or every 125 μs. Hence the frame synchronization code repeats every 1.5 ms and the frame rate is 667 frames/s.

Signaling

A telephone system must be able to transmit not only the speech communication but also certain other signaling and supervisory information. Thus information

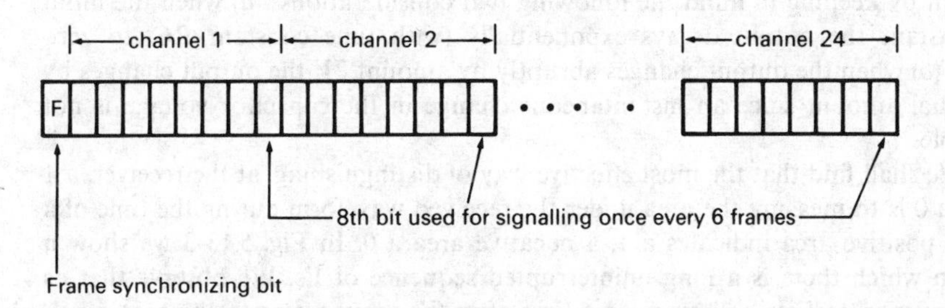

Figure 5.13-2 The PCM $T1$ frame using channel associated signalling.

needs to be conveyed that a call is being initiated, that a call is being terminated, specifying the address of the called party, etc. When analog transmission is employed, the signaling information is often conveyed over a channel separate from that which carries the voice. In the $T1$ digital system, a process of bit-slot sharing is used to allow the single channel to transmit both voice and signaling. In this sharing scheme, the eighth (least-significant) bit of each encoded sample is used for both voice transmission and also for signaling. During five successive samplings (and hence during five successive frames) each sample is encoded into eight bits. But during the sixth frame the samples are encoded into only seven bits, the eighth bit being used for signaling. This pattern is repeated every six frames. Thus in six frames the number of bits used for quantization encoding is $5 \times 8 + 1 \times 7 = 47$ so that samples are encoded on the average into $47/6 = 7\frac{5}{6}$ bits. The frequency of the bits used for signaling is $\frac{1}{6}$th of the frame bit rate, or

$$f_b(T1)_{(\text{signaling})} = \frac{1}{6}(8000) = 1{,}333 \text{ Hz}$$

This type of signaling is called *channel-associated* signaling.

Line Coding

We have frequently indicated that a digital baseband signal has the waveform shown in Fig. 5.13-3*a*. The signal makes excursions between $+V$ and $-V$ corresponding to logic-1 and logic-0 respectively. It is referred to as a *balanced* signal because of its symmetry with respect to zero volts or a *bipolar* signal to distinguish it from a *unipolar* signal which makes excursions between zero volts and $+2V$. It is also referred to as a NRZ (non-return-to-zero) signal since it jumps between $\pm V$ and never pauses at zero volts. As we shall now see, such a waveform is not suitable for use on a $T1$ channel.

For a variety of practical reasons the low-frequency response of a $T1$ channel is limited. This feature results from the fact that the signal needs, on occasion, to be transmitted through transformers, series capacitors, etc. To illustrate the influence of such frequency limiting we consider the effect on the bipolar signal of transmission through the rudimentary low-frequency limiting RC network of Fig. 5.13-3(*b*). The output response is shown in (*c*). The validity of this result can be seen by keeping in mind the following two considerations: (*a*) when the input is constant the output decays exponentially (with time constant RC) to zero volts; (*b*) when the output changes abruptly by amount $2V$ the output changes by an equal amount since an instantaneous change in the capacitor voltage is not possible.

We shall find that the most effective way of distinguishing, at the receiver, a 1 from a 0 is to measure the area under the received waveform during the time of a bit. A positive area indicates a 1, a negative area a 0. In Fig. 5.13-3 we show a case in which there is a long uninterrupted sequence of 1's. We observe that as the sequence persists and even for a time after the sequence ends, the area available to assure that a 1 has been transmitted is severely depressed. Of course, a

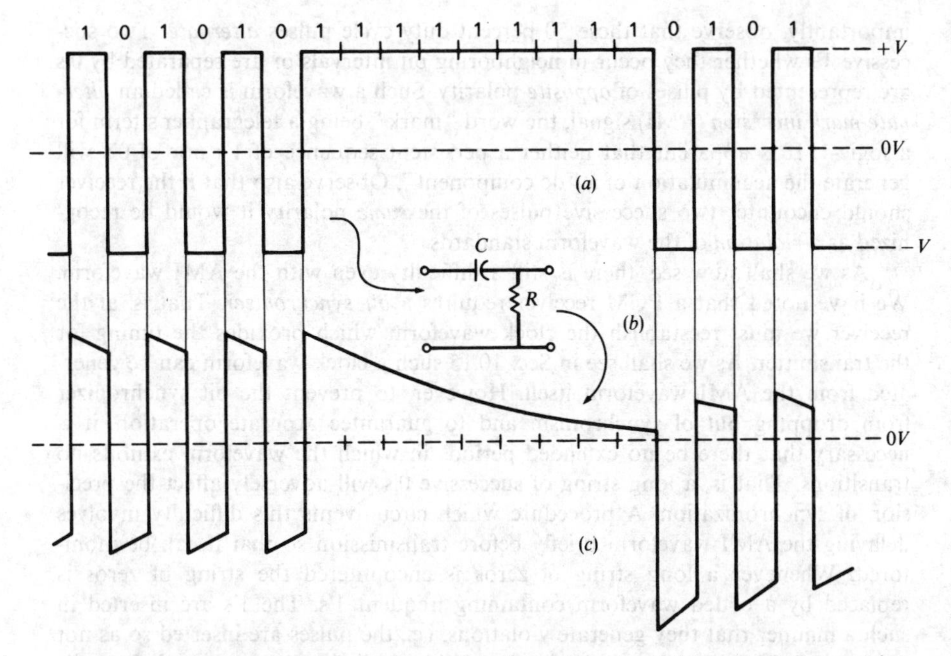

Figure 5.13-3 (*a*) An NRZ binary waveform. (*b*) A circuit whose transmission discriminates against low frequency. (*c*) The output which results when the waveform in (*a*) is transmitted through the circuit in (*b*).

corresponding difficulty results from a long string of 0's. In short, the difficulty with the NRZ waveform is that a long persistent sequence of 1's or 0's builds up a "dc component" which cannot be transmitted over a $T1$ channel.

In Fig. 5.13-4*a* we show again an NRZ waveform which we have just judged inadequate and in (*b*) we have shown an alternative more suitable waveform. A logic-0 is represented by zero volts over the whole bit interval. A logic-1 is represented by $+V$ or $-V$ which persists for a fraction of the bit interval. The fraction is referred to as the duty cycle and is often of the order of 50 percent. Most

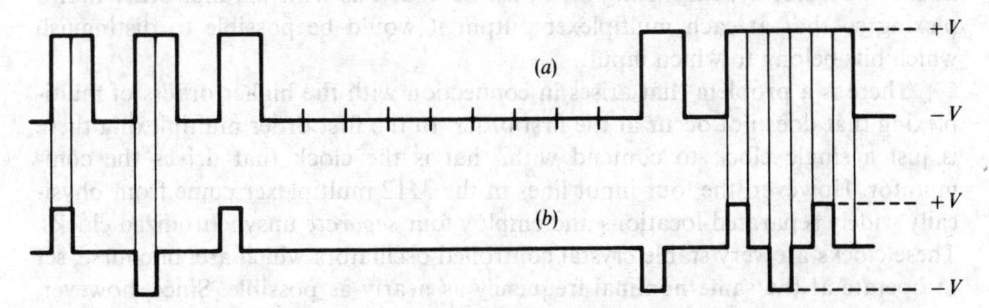

Figure 5.13-4 (*a*) An NRZ digital waveform. (*b*) An alternate Mark Inversion equivalent to the waveform in (*a*).

importantly, observe that these 50 percent duty cycle pulses *alternate*. Two successive 1's whether they occur in neighboring bit intervals or are separated by 0's are represented by pulses of *opposite* polarity. Such a waveform is called an *alternate mark inversion* (AMI) signal, the word "mark" being a telegrapher's term for a logic-1. It is apparent that neither a persistent sequence of 1's nor of 0's will generate the accumulation of a "dc component". Observe also that if the receiver should encounter two successive pulses of the *same* polarity it would be recognized as a *violation* of the waveform standards.

As we shall now see, there is still a difficulty even with the AMI waveform. We have noted that a PCM receiver requires a *bit synchronizer*. That is, at the receiver we must reestablish the clock waveform which provides the timing for the transmitter. As we shall see in Sec. 10.15 such a clock waveform can be generated from the AMI waveform itself. However, to prevent the bit synchronizer from dropping out of synchronism and to guarantee accurate operation, it is necessary that there be no extended periods in which the waveform exhibits no transitions. That is, a long string of successive 0's will adversely affect the precision of synchronization. A procedure which circumvents this difficulty involves delaying the AMI waveform briefly before transmission so that it can be monitored. Whenever a long string of zeros is encountered the string of zeros is replaced by a coded waveform containing frequent 1's. The 1's are inserted in such a manner that they generate violations, i.e., the pulses are inserted so as not to alternate. Hence at the receiver it is possible to distinguish proper 1's from 1's artificially added to keep the synchronizer going. In $T1$ transmission a simple procedure is employed: If eight or more consecutive zeros occur the least-significant bit (LSB) of a word in the string is changed to a 1 to break the pattern.

5.13-2 Multiplexing $T1$ Lines — The $T2$, $T3$, and $T4$ Lines

To take further advantage of the merits of TDM and digital transmission, the common carriers employ a hierarchy of further multiplexing as shown in Fig. 5.13-5. Four $T1$ lines are multiplexed in an $M12$ multiplexer to generate a $T2$ transmission system, seven $T2$ lines convert to a $T3$ line in an $M23$ multiplexer and six $T3$ lines convert to a $T4$ line in an $M34$ multiplexer. At each stage additional frame synchronizing bits must be added as with the first order multiplexing so that at each multiplexer output it would be possible to distinguish which bits belong to which input.

There is a problem that arises in connection with the higher orders of multiplexing that does not occur at the first order. In the first order multiplexing there is just a single clock to contend with, that is the clock that drives the commutator. However, the four input lines in the $M12$ multiplexer come from physically widely separated locations and employ four *separate* unsynchronized clocks. These clocks are very stable crystal controlled oscillators which are, of course, set to operate at the same nominal frequency as nearly as possible. Since, however, they do not have a means of communicating with one another, they are not synchronized and will hence experience a relative frequency drift. Design specifications allow a drift from the nominal set frequency of ± 130 parts/million. As

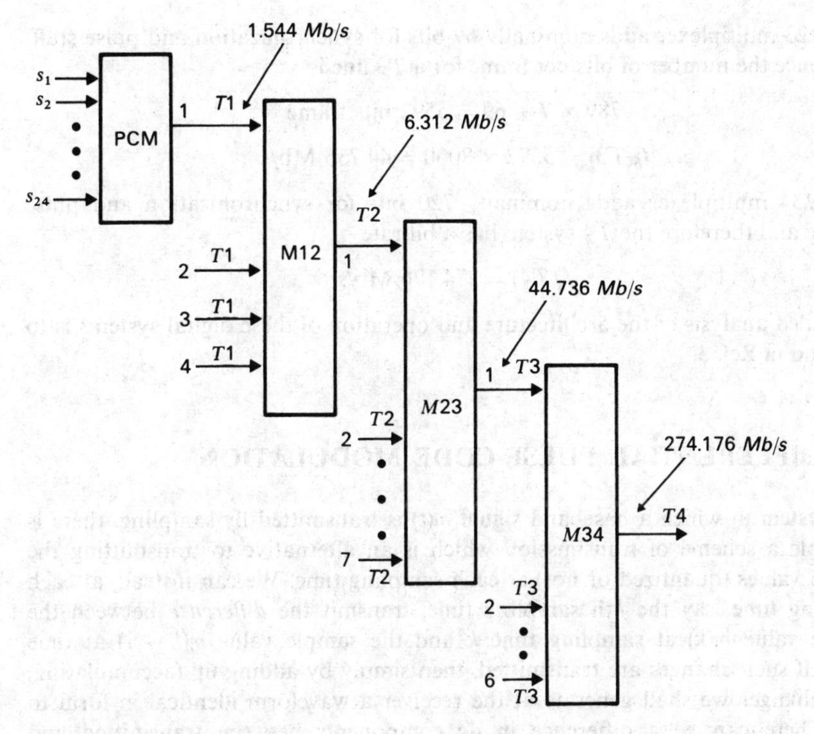

Figure 5.13-5 TDM hierarchy.

may be verified, if then two clocks have frequencies which differ by 2 × 130 = 260 parts/million, the faster clock will have generated one more time slot than the slower clock in the course of just 20 frames. For proper interleaving of bits at the *M*12 level it is necessary that all input bit streams have or be made to appear to have the same rate. To put the matter most simplistically, the process of adjusting bit rates to make them equal involves adding bits to the slower bit stream in an operation referred to as "pulse stuffing." Further bits must then be added to all bit streams to allow the receiver of the composite signal to distinguish time slots which carry information from slots which carry the "stuffed" bits.

Bit Rate

The *M*12 multiplexer adds nominally 17 bits for frame synchronization and pulse stuffing. Hence the number of bits per frame is

$$193 \times 4 + 17 = 789 \text{ bits/frame}$$

The *T*2 line bit rate is therefore

$$f_b(T2) = 789 \text{ bits/frame} \times 8000 \text{ frames/s}$$

$$= 6.312 \text{ Mb/s}$$

The *M*23 multiplexer adds nominally 69 bits for synchronization and pulse stuffing; hence the number of bits per frame for a *T*3 line is

$$789 \times 7 + 69 = 5592 \text{ bits/frame}$$

and
$$f_b(T3) = 5592 \times 8000 = 44.736 \text{ Mb/s}$$

The *M*34 multiplexer adds nominally 720 bits for synchronization and pulse stuffing and therefore the *T*4 system has a bit rate

$$f_b(T4) = 274.176 \text{ Mb/s}$$

A detailed analysis of the architecture and operation of these digital systems is to be found in Ref. 3.

5.14 DIFFERENTIAL PULSE-CODE MODULATION

In a system in which a baseband signal $m(t)$ is transmitted by sampling, there is available a scheme of transmission which is an alternative to transmitting the sample values (quantized or not) at each sampling time. We can instead, at each sampling time, say the kth sampling time, transmit the *difference* between the sample value $m(k)$ at sampling time k and the sample value $m(k-1)$ at time $k-1$. If such changes are transmitted, then simply by adding up (accumulating) these changes we shall generate at the receiver a waveform identical in form to $m(t)$. There can be a difference in dc components between transmitted and received signals but, almost invariably, such dc components are of no interest.

Such a differential scheme has special merit when these differences are to be transmitted by pulse code modulation. For we may well anticipate that the differences $m(k) - m(k-1)$ will be smaller than the sample values themselves. Hence fewer levels will be required to quantize the difference than are required to quantize $m(k)$ and correspondingly, fewer bits will be needed to encode the levels. For example, suppose that $m(k)$ extends over a range $V_H - V_L$, and using PCM, $m(k)$ is encoded using $2^8 = 256$ levels. Then the step size is $S = (V_H - V_L)/2^8$, that is $V_H - V_L = 256S$. If, however, the difference signal $m(k) - m(k-1)$ extends only over the range $\pm 2S$ then the quantized levels needed are at $\pm 0.5S$ and at $\pm 1.5S$. There are now only four levels and two bits per sample difference are adequate.

In an analog system, where we are able, at least in principle, to transmit the differences exactly, the differential system described above would operate in accordance with our description. In a digital (quantized) system we encounter the complication that the differences are not generally transmitted exactly because of the quantization. Further, we have the problem that the difference may be larger than the maximum that can be accommodated because of the restricted number of encoding bits we have provided. Hence it might well be that at some time there might be a large discrepancy between the original signal $m(t)$ and the signal $\hat{m}(t)$ generated at the receiver by accumulation. Suppose that over a number of samplings, while $m(t)$ is increasing, the transmitted differences were too small so that $\hat{m}(k)$ had fallen substantially short of keeping up with $m(t)$. Suppose, further, that in the interval sampling times k and $k+1$, $m(t)$ should decrease slightly. Clearly if we transmitted the negative change of $m(t)$ we would be giving the wrong signal.

In a digital differential system we circumvent the difficulty we have just described by making available at the transmitter a duplicate of the receiver accumulator so that at the transmitter we have available the same signal $\hat{m}(t)$. Then we arrange that the transmitted signal should not convey the most recent change in $m(t)$ but conveys instead the difference between $m(t)$ and $\hat{m}(t)$. In an analog system, this difference $m(t) - \hat{m}(t)$ is precisely the last change in $m(t)$. In a quantized system, as we have noted, such is not the case. In short, in a quantized system we add or subtract from $\hat{m}(t)$ a value which is appropriate to bring $\hat{m}(t)$ closer to $m(t)$. The waveform $\hat{m}(t)$ is generally referred to as the *approximation* to $m(t)$.

Altogether, then, the quantized differential transmission scheme is as shown in Fig. 5.14-1. (We ignore initially the "predictors" that appear in the figure.) The receiver consists of an acumulator which adds up the received quantized differences $\Delta_Q(k)$ and a filter which smooths out the quantization noise. The output of the accumulator is the signal approximation $\hat{m}(k)$ which becomes $\hat{m}(t)$ at the filter output. At the transmitter we need to know whether $\hat{m}(t)$ is larger or smaller than

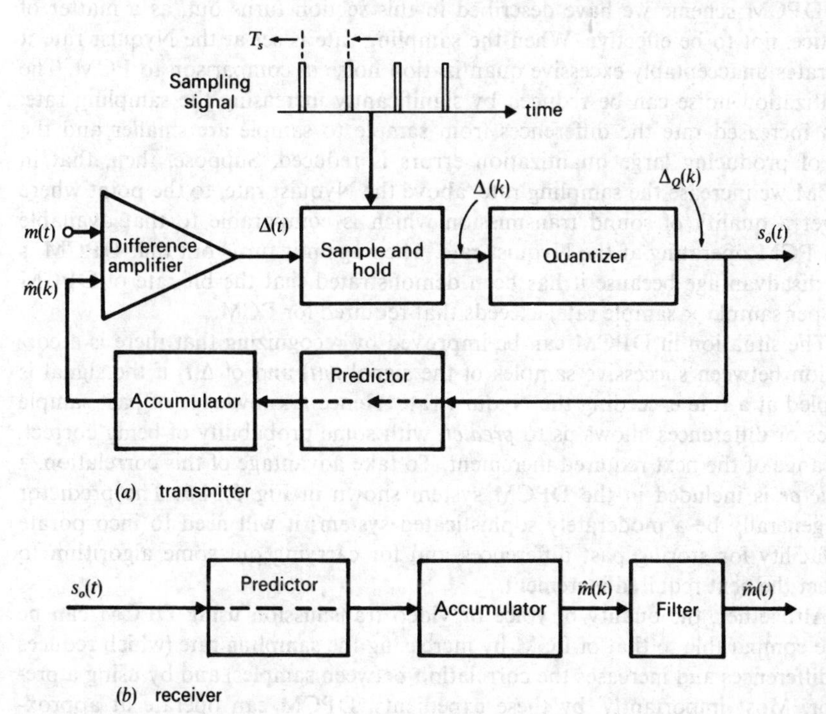

Figure 5.14-1 Representation of the basic principle of differential PCM.

$m(t)$, and by how much. We may then determine whether the next difference $\Delta_Q(k)$ needs to be positive or negative and of what amplitude in order to bring $\hat{m}(t)$ as close as possible to $m(t)$. For this reason we have a duplicate accumulator at the transmitter. At each sampling time, the transmitter difference amplifier compares $m(t)$ and $\hat{m}(t)$ and the sample and hold circuitry holds the result of that comparison $\Delta(t)$ for the duration of the interval between sampling times. The quantizer generates the signal $s_0(t) = \Delta_Q(k)$ both for transmission to the receiver and to provide input to the receiver accumulator in the transmitter. In a practical system the quantized differences would first be encoded into a binary bit stream before transmission and decoded at the receiver. For simplicity the encoder and decoder are not included in Fig. 5.14-1.

It needs to be emphasized that the basic limitation of the scheme we have just described is that the transmitted differences are quantized and are of limited maximum value. The quantization means that almost never will the increment $\Delta_Q(k)$ added to $m(k)$ make $\hat{m}(t)$ precisely equal to $m(t)$. The limitation on the maximum value of $\Delta_Q(k)$ means that when $m(t)$ changes monotonically at a rapid rate, $\hat{m}(t)$ may simply not be able to keep up.

Need for a Predictor

The DPCM scheme we have described in this section turns out, as a matter of practice, not to be effective. When the sampling rate is set at the Nyquist rate it generates unacceptably excessive quantization noise in comparison to PCM. The quantization noise can be reduced by significantly increasing the sampling rate. With increased rate the differences from sample to sample are smaller and the rate of producing large quantization errors is reduced. Suppose, then, that in DPCM we increase the sampling rate, above the Nyquist rate, to the point where we get a quality of sound transmission which is comparable to that available from PCM operating at the Nyquist rate. Then again it turns out that DPCM is at a disadvantage because it has been demonstrated that the bit rate of DPCM (bits per sample × sample rate) exceeds that required for PCM.

The situation in DPCM can be improved by recognizing that there is a correlation between successive samples of the signal $m(t)$ and of $\Delta(t)$ if the signal is sampled at a rate exceeding the Nyquist rate. Hence a knowledge of past sample values or differences allows us to *predict*, with some probability of being correct, the range of the next required increment. To take advantage of this correlation, a *predictor* is included in the DPCM system shown in Fig. 5.14-1. The predictor will generally be a moderately sophisticated system; it will need to incorporate the facility for storing past differences and for carrying out some algorithm to predict the next required increment.

Altogether, the quality of voice or video transmission using DPCM can be made comparable to that of PCM by increasing the sampling rate (which reduces the differences and increases the correlation between samples) and by using a predictor. Most importantly, by these expedients, DPCM can operate at approximately one-half of the bit rate of PCM with a consequent saving of spectrum space.

5.15 DELTA MODULATION[4]

Delta modulation (DM) is a DPCM scheme in which the difference signal $\Delta(t)$ is encoded into just a *single* bit. The single bit, providing for just two possibilities, is used to increase or decrease the estimate $\hat{m}(t)$.

One way in which a delta modulator can be assembled is shown in Fig. 5.15-1. This scheme is called linear delta modulation. The baseband signal $m(t)$ and its quantized approximation $\hat{m}(t)$ are applied as inputs to a comparator. A comparator, as its name suggests, simply makes a comparison between inputs. As indicated in the plot shown in Fig. 5.15-1, the comparator has one fixed output $V(H)$ when $m(t) > \hat{m}(t)$ and a different output $V(L)$ when $m(t) < \hat{m}(t)$. Ideally the transition between $V(H)$ and $V(L)$ is arbitrarily abrupt as $m(t) - \hat{m}(t)$ passes through zero. The comparator replaces the difference amplifier and quantizer of Fig. 5.14-1, since in the present special case we need to know only whether $m(t)$ is larger or smaller than $\hat{m}(t)$ and not the magnitude of the difference.

The up-down counter increments or decrements its count by 1 at each active edge of the clock waveform. (The active edge is either the rising or falling edge of the clock, depending on the hardware design of the counter. The sampling time is the time of occurrence of this active edge.) The count direction, i.e., incrementing or decrementing, is determined by the voltage levels at the "count direction command" input to the counter. When this binary input, which is also the transmitted output $s_o(t)$, is at the level $V(H)$, the counter counts-up and when it is at the level $V(L)$ the counter counts down. In the present case the counter serves as the accumulator shown in Fig. 5.14-1, since it adds or subtracts increments as

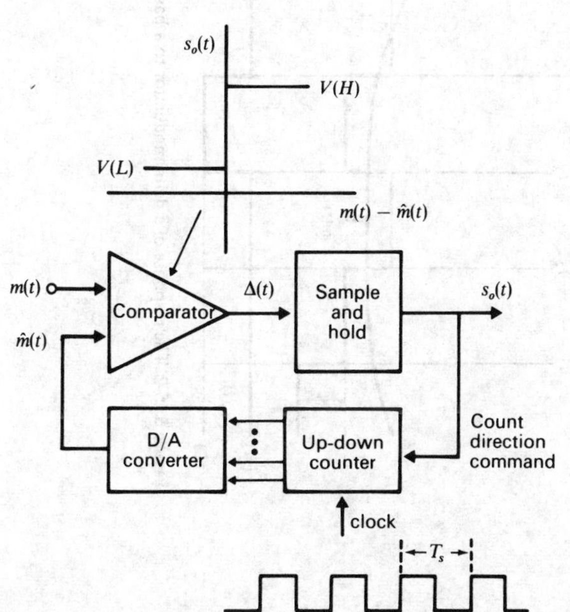

Figure 5.15-1 A delta modulator.

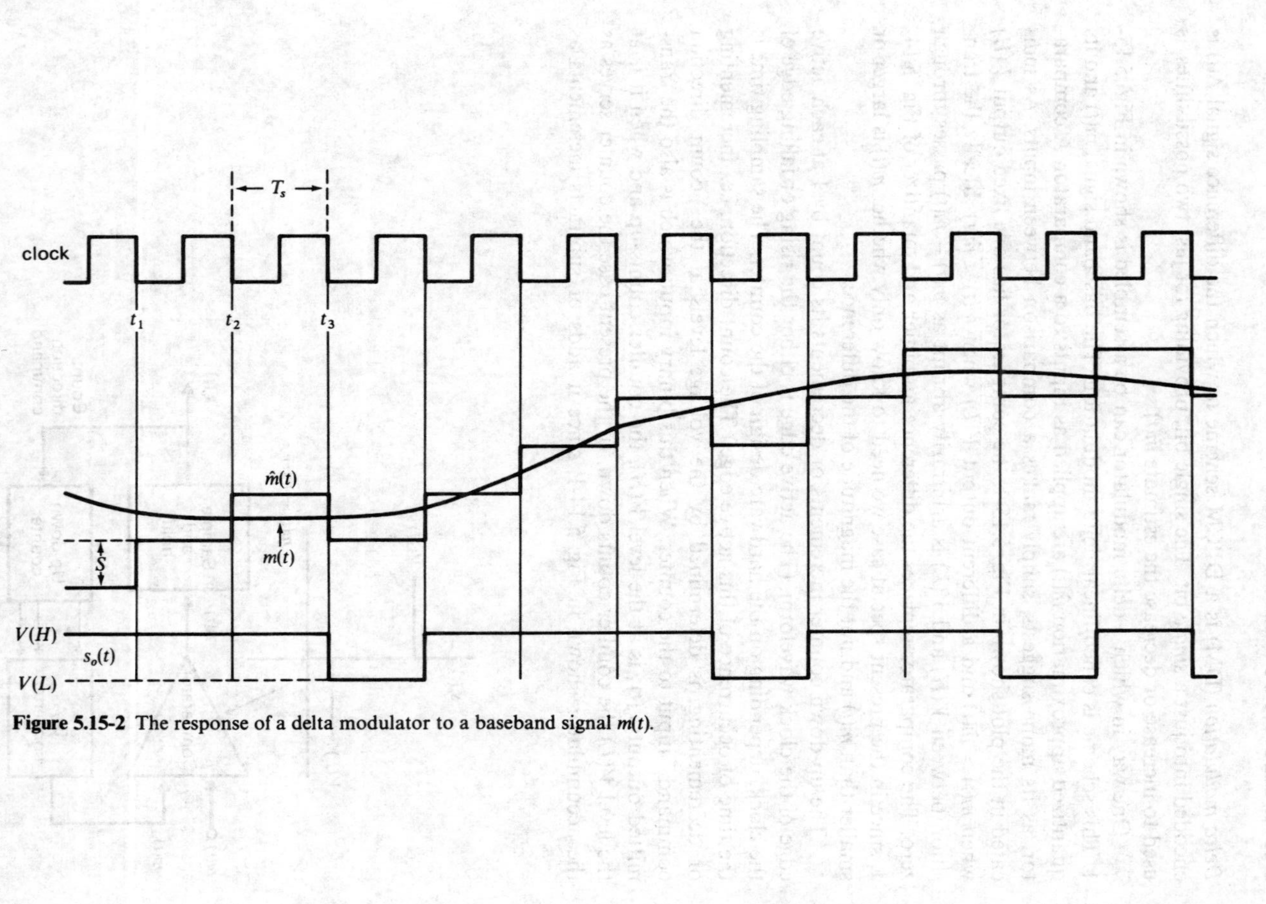

Figure 5.15-2 The response of a delta modulator to a baseband signal $m(t)$.

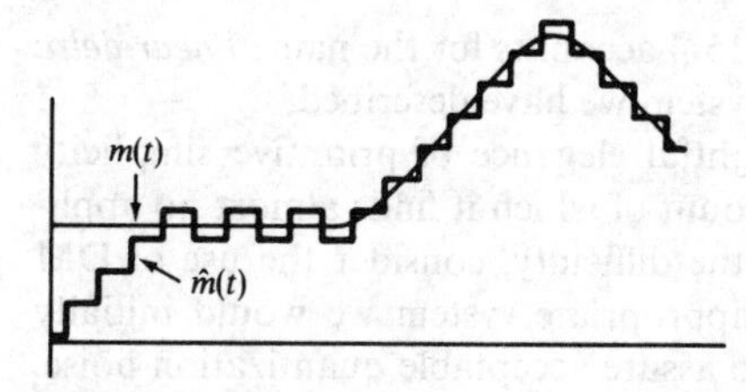

Figure 5.15-3 Illustrating the "start-up" response in delta modulation and the "hunting" of $\hat{m}(t)$ about $m(t)$.

directed and stores the accumulated result. The digital output of the counter is converted to the analog quantized approximation $\hat{m}(t)$ by the D/A converter. The waveforms of Fig. 5.15-2 are the waveforms for the system of Fig. 5.15-1, assuming that the active clock edge is the falling edge.

For a time preceding t_1 we find $m(t) > \hat{m}(t)$ so that $s_o(t) = V(H)$. At t_1, when the active clock edge appears, the counter is incremented and immediately (ignoring propagation delays in the counter and D/A converter) the signal $\hat{m}(t)$ jumps up by an amount equal to the step size S. At t_2 we still find $m(t) > \hat{m}(t)$ so $s_o(t)$ remains at $V(H)$ and there is another upward jump in $\hat{m}(t)$. At t_3, $m(t) < \hat{m}(t)$, so $s_o(t)$ becomes $s_o(t) = V(L)$, the counter decrements, and there is a consequent downward jump in $\hat{m}(t)$ by amount S and so on.

We note that at start-up, there may be a brief interval when $\hat{m}(t)$ may be a poor approximation to the baseband signal $m(t)$. This feature is illustrated in Fig. 5.15-3. Hence we note an initial large discrepancy between $m(t)$ and $\hat{m}(t)$ and the stepwise approach of $\hat{m}(t)$ to $m(t)$. We note further that even when $\hat{m}(t)$ has caught up to $m(t)$, and even though $m(t)$ remains unvarying, $\hat{m}(t)$ hunts, swinging up or down, above and below $m(t)$.

We have noted previously the limited ability of DPCM to accommodate to large increments in $m(t)$. For DM, this limitation is illustrated in Fig. 5.15-4. Here we have a signal $m(t)$, which, over an extended time, exhibits a slope which is so large that $\hat{m}(t)$ cannot keep up with it. As a consequence the error $m(t) - \hat{m}(t)$ becomes progressively larger, by far exceeding $S/2$. The excessive disparity between $m(t)$ and $\hat{m}(t)$ is described as a *slope-overload error* and occurs whenever $m(t)$ has a slope larger than the "slope" S/T_s which can be sustained by the wave-

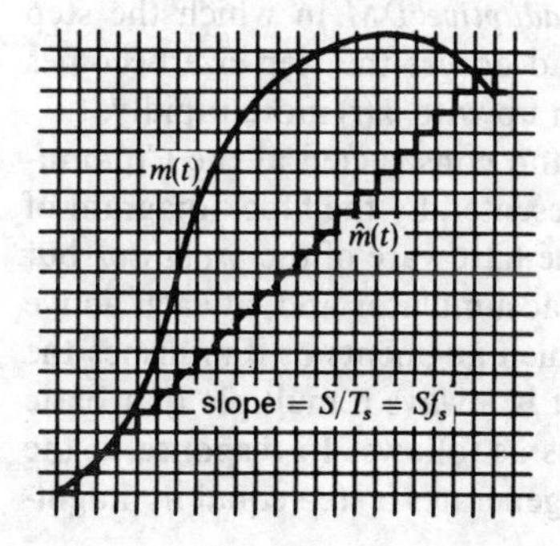

Figure 5.15-4 Slope-overload in the linear DM.

form $\hat{m}(t)$. The linear form of $\hat{m}(t)$ in Fig. 5.15-4 accounts for the name *linear delta modulation* being associated with the DM system we have described.

Linear delta modulation has the delightful elegance of primitive simplicity yet it suffers from severe limitations on account of which it finds almost no applications is real systems. As an example of the difficulty, consider the use of DM for speech transmission. In designing an appropriate system we would initially select a step size S which is small enough to assure acceptable quantization noise. As noted, $S = (V_H - V_L)/256$ is a reasonable selection. If now, because of the small size of S, slope overload should develop, we can hope to overcome the overload by increasing the sampling rate above the rate initially selected to satisfy the Nyquist criterion. Since we need only one bit per sample rather than eight, as in PCM, we can increase the sampling rate eightfold before we reach the rate required in PCM. Now it has been determined experimentally that DM will transmit speech without significant slope overload provided that the DM system is able to transmit a sinusoid of frequency $f = 800$ Hz whose amplitude is the same as the amplitude of the speech waveform. A sinusoid of amplitude A and frequency f has a maximum slope of $2\pi fA$ as it passes through zero. If slope overload is to be avoided then we require that the sampling rate $f_s (= 1/T_s)$ must satisfy the condition

$$Sf_s \geq 2\pi fA \tag{5.15-1}$$

or

$$f_s \geq \pi f \frac{2A}{S} = 256\pi f \cong 640 \text{ kHz} \tag{5.15-2}$$

where we have set $f = 800$ Hz. This frequency, which is also the bit rate, is to be compared with the bit rate in PCM. Assume that the highest frequency component of the speech signal is 3 kHz. Then the Nyquist rate is 6 kHz. To allow a margin of safety, let us use a sampling rate of 8 kHz. Then with 8 bits per sample the bit rate is 8 kHz × 8 = 64 kHz. Hence DM is not a practical alternative to PCM.

5.16 ADAPTIVE DELTA MODULATION (ADM)

We now consider a modification of DM called *adaptive DM* in which the step size is not kept fixed. Rather, when slope overload occurs the step size becomes progressively larger, thereby allowing $\hat{m}(t)$ to catch up with $m(t)$ more rapidly.

We describe now an ADM system designed and constructed at the Communications System Laboratory at CCNY. It is represented by the block diagram of Fig. 5.16-1. We shall not examine the details of the hardware of the *processor* but only describe what it does. The processor has an accumulator and, at each active edge of the clock waveform, generates a step S which augments or diminishes the accumulator. The step S is not of fixed size but it is always a multiple of a basic step S_0. The algorithm by which S is generated is as follows: In response to the kth active clock edge the processor, to start with, generates a step equal in magni-

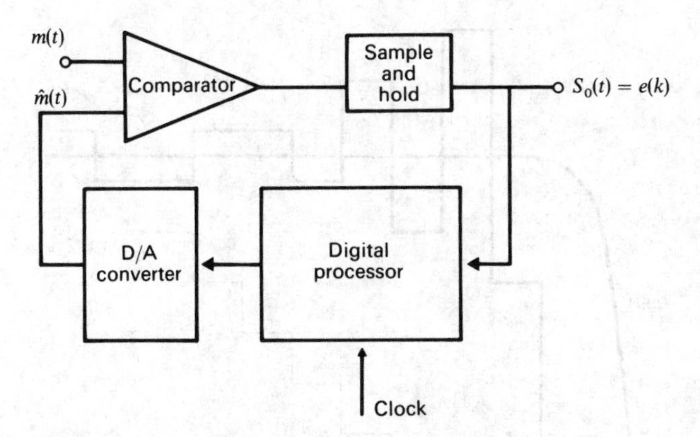

Figure 5.16-1 An adaptive DM.

tude to the step generated in response to the $(k-1)$st clock edge. This step is added to or subtracted from the accumulator, as required, to move $\hat{m}(t)$ toward $m(t)$. Further however, if the direction of the step at clock edge k is the *same* as at edge $k-1$ then the processor *increases* the magnitude of the step by amount S_0. If the directions are *opposite* then the processor *decreases* the magnitude of the step size by S_0. As the algorithm is carried out there are clock edges when the total step $S = 0$. In this case, at the next clock edge the step is S_0 in the direction again to move $\hat{m}(t)$ toward $m(t)$.

In Fig. 5.16-1 the output $S_0(t)$ is called $e(k)$. The symbol $e(k)$ represents the error, i.e., the discrepancy between $m(t)$ and $\hat{m}(t)$ and, as indicated in Fig. 5.15-1 is either $V(H)$ or $V(L)$. To express the algorithm by which the step size is determined it is convenient to arrange that:

$$e(k) = +1 \text{ if } m(t) > \hat{m}(t) \text{ immediately before the } k\text{th edge}$$

$$e(k) = -1 \text{ if } m(t) < \hat{m}(t) \text{ immediately before the } k\text{th edge}$$

We can now specify that at sampling time k the step size $S(k)$ is to be given by:

$$S(k) = |S(k-1)|\, e(k) + S_0\, e(k-1) \tag{5.16-1}$$

The important features of this adaptive scheme are shown in Fig. 5.16-2. Observe that as long as the condition $m(t) > \hat{m}(t)$ persists, the jumps in $\hat{m}(t)$ become progressively larger. The estimate $\hat{m}(t)$ catches up with $m(t)$ sooner than would be the case with linear delta modulation as shown by the waveform $\hat{m}'(t)$. On the other hand, when, in response to a large slope in $m(t)$, $\hat{m}(t)$ develops large jumps, it may require a large number of clock cycles for these jumps to decay in amplitude when they are no longer needed. Such a situation is to be seen in Fig. 5.16-2 in the neighborhood where $\hat{m}(t)$ first catches up with $m(t)$. Altogether, then, while this ADM system reduces slope error, it does so at the expense of increasing quantization error. The dashed waveform $\hat{m}'(t)$ corresponding to linear *DM* has small quantization error but extremely large slope error.

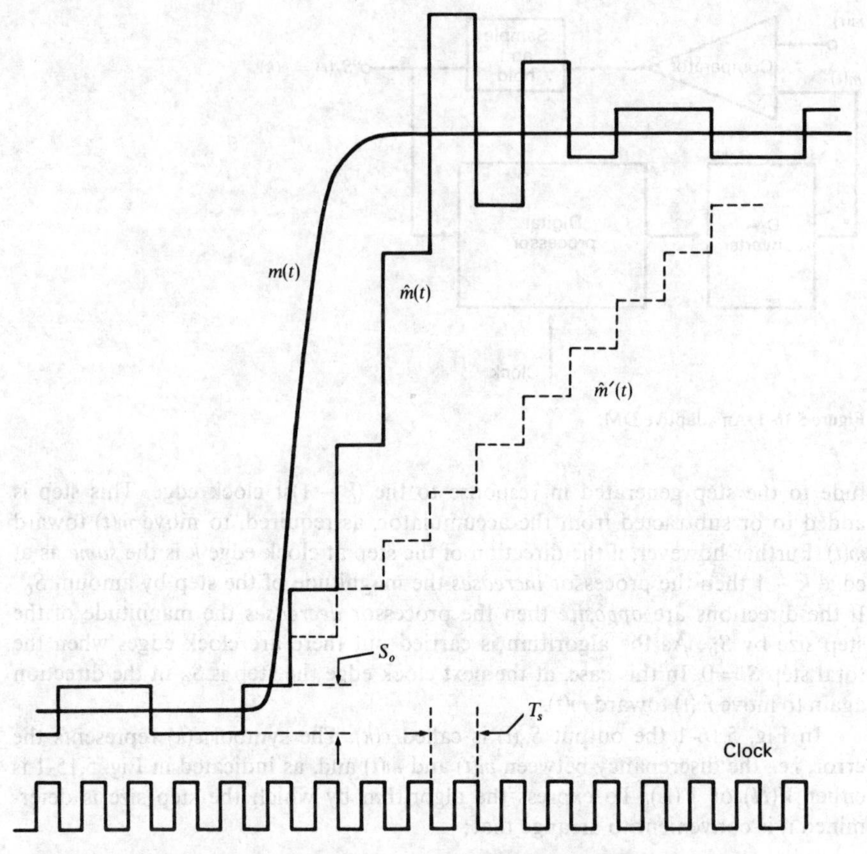

Figure 5.16-2 Waveforms comparing the response of the ADM and LDM.

Note, also, incidentally that when $m(t)$ remains constant (within an interval S_0) $\hat{m}(t)$ oscillates about $m(t)$ but that the oscillation frequency is one-half the clock frequency.

It turns out that in the matter of speech transmission the reduced slope error provides a net advantage in spite of the increased quantization error and that this adaptive delta modulator can operate at bit rates of 32 kb/s with performance comparable to that obtained when using PCM at 64 kb/s. Furthermore, the ADM can operate at 16 kb/s with only a slight degradation of performance.

It has been shown that more than 90 percent of the step sizes are less than or equal to $2S_0$ and that the probability of a step exceeding $15S_0$ is about 1 percent. In addition, the spurious frequency components introduced into the reconstructed signal by slope-overload error are principally in the low frequency range, while quantization error introduces principally high-frequency components. The power in speech is concentrated largely in the low-frequency components, and these low-frequency components are also the ones which must principally be reproduced without error in order to preserve intelligibility. Hence if we pass the reconstructed signal through a low-pass filter, we shall be able to discriminate against the high-frequency quantization error without materially degrading the voice signal. A variation of this algorithm is used on the Space Shuttle.

Continuously Variable Slope Delta Modulator (CVSD)

A DM system which adjusts its step size to any given value is shown in Fig. 5.16-3. The amplifier has a variable gain. That is, its gain is a function of the voltage applied at its gain-control terminal. We assume that the characteristics of the amplifier are such that when the gain-control voltage is zero, its gain is low, and that the gain increases with increasing positive gain-control voltage. The resistor-capacitor combination serves as an integrator, the voltage across C being proportional to the integral of the pulse signal $p_o(t)$. The voltage across C is used to control the gain of the amplifier. The square-law device ensures that whatever the polarity of the voltage across C, a positive voltage will be applied to the gain-control terminal of the amplifier.

Assume now that $m(t)$ is making only small excursions so that the modulator does not follow. The output $p_o(t)$ consists then of alternate polarity pulses. These pulses, when integrated, yield an average output of almost zero. The gain-control input is, hence, almost zero, the gain is low, and consequently the step size is reduced. Next consider the case of slope overload. If $m(t)$ increases positively or negatively at too rapid a rate, $\tilde{m}(t)$ cannot follow. The output $p_o(t)$ is then a train of all positive or all negative pulses. The integrator averages and provides a large voltage to increase the gain of the amplifier. Because of the squaring circuit the amplifier gain will increase no matter what the polarity of the integrator voltage.

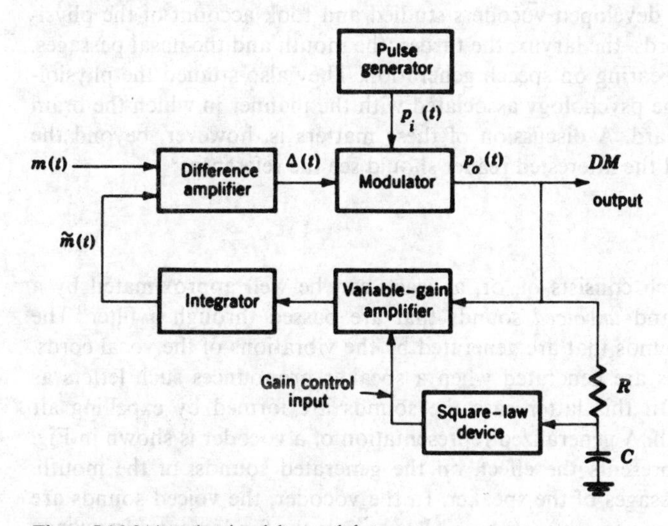

Figure 5.16-3 An adaptive delta modulator.

The end result is an increase in step size and a reduction in slope overload. It is of course, necessary that there be an adaptive adjustment of the step size at the receiver, as well. Thus, the receiver consists of the variable-gain amplifier, square-law device, RC filter and integrator shown in the feedback path of Fig. 5.16-3.

5.17 VOCODERS (VOICE CODERS)[5]

We listen, in turn, to one person after another pronounce a word or a sequence of words. One has a high-pitched voice, another a low pitch. One speaks with a foreign accent, another has no accent. One enunciates clearly, another slurs words and so on. Yet we understand each speaker. These considerations make it clear that there is a variability (within limits) that can be allowed in the waveform that must inpinge on a listener's ear before there appears a loss of recognition of the spoken word. It therefore occurs to us that to transmit speech we need not transmit the precise waveform generated by the speaker. Rather we can transmit information from which a waveform can be reconstructed at the receiver which is only similar to, rather than identical to, the waveform generated by the speaker. We anticipate that by allowing ourselves this lattitude we can operate a digital source coded transmission system at a lower bit rate. This expectation is indeed borne out. The source coders employed are called *vocoders* (voice coders) and they operate at a significantly lower bit rate than even ADM. Typically, vocoder bit rates are in the range 1.2 to 2.4 kb/s. However, the resulting reproduced voice has a synthetic-sounding and a somewhat artificial quality. As a result, vocoders are employed for special applications where it is acceptable to trade speech quality for the advantage of low bit rate. Applications are found in military communications, operator recorded messages, etc.

The people who developed vocoders studied and took account of the physiology of the vocal cords, the larynx, the throat, the mouth and the nasal passages, all of which have a bearing on speech generation. They also studied the physiology of the ear and the psychology associated with the manner in which the brain interprets sounds heard. A discussion of these matters is, however, beyond the scope of this text and the interested reader should see the references[5].

Voice Model

It appears that speech consists of, or, at least, can be well approximated by a sequence of *voiced* and *unvoiced* sounds that are passed through a filter. The voiced sounds are sounds that are generated by the vibrations of the vocal cords. The unvoiced sounds are generated when a speaker pronounces such letters as "s", "f", "p", etc. In this latter case the sounds are formed by expelling air through lips and teeth. A generalized representation of a vocoder is shown in Fig. 5.17-1. The filter represents the effect, on the generated sounds, of the mouth, throat, and nasal passages of the speaker. In the vocoder, the voiced sounds are simulated by an impulse generator whose frequency is the fundamental frequency

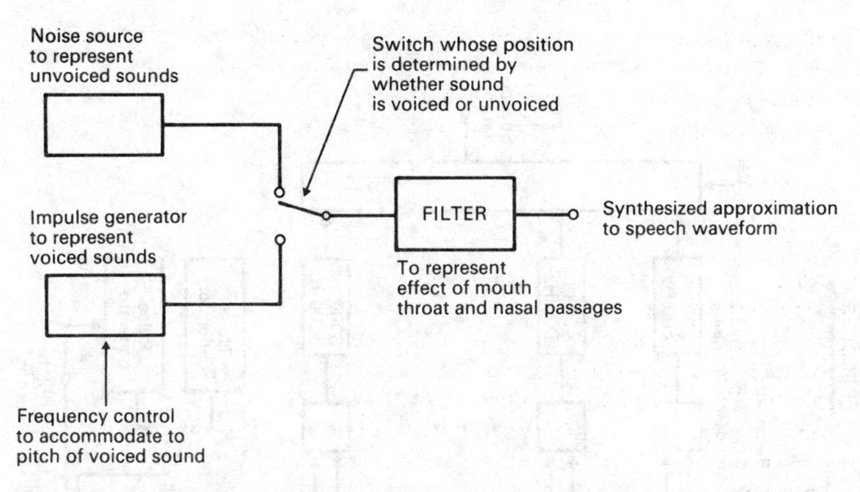

Figure 5.17-1 Speech model used in vocoders.

of vibration of the vocal cords. The unvoiced sounds are simulated by a noise source. Altogether, all vocoders employ the scheme shown in Fig. 5.17-1 to generate a synthesized approximation to a speech waveform. They differ only in the techniques employed to generate the voiced and unvoiced sounds and in the characteristics, and design of the filter.

5.18 CHANNEL VOCODER[5]

One of the many vocoder systems is the *channel vocoder* shown in Fig. 5.18-1. In this encoding system, the spectrum of the input speech is divided into 15 frequency ranges each of bandwidth 200 Hz. If the output of one of the bandpass filters were a sinusoidal waveform of fixed amplitude then the cascaded rectifier and 20 Hz low-pass filter would provide a dc voltage of magnitude proportional to that amplitude. In the real case represented in Fig. 5.18-1 each 20 Hz low-pass filter will provide instead a voltage which is proportional to the amplitude at the output of its associated 200 Hz bandpass filter.

Additionally, the input speech is applied to a frequency discriminator followed by a 20 Hz low-pass filter. When the signal is voiced, the output of the filter provides a voltage which is proportional to the voice frequency. This frequency is the *pitch* of the voice. The discriminator-filter combination is special in that when the speech is unvoiced the output of the filter is a smaller voltage than we encounter for voiced speech. Using a detector we can then determine, by noting the output of the discriminator-filter combination, whether the speech is voiced or unvoiced, and, if voiced, the voltage detected is determined by the pitch.

The outputs of the 16, 20 Hz low-pass filters are sampled, multiplexed and

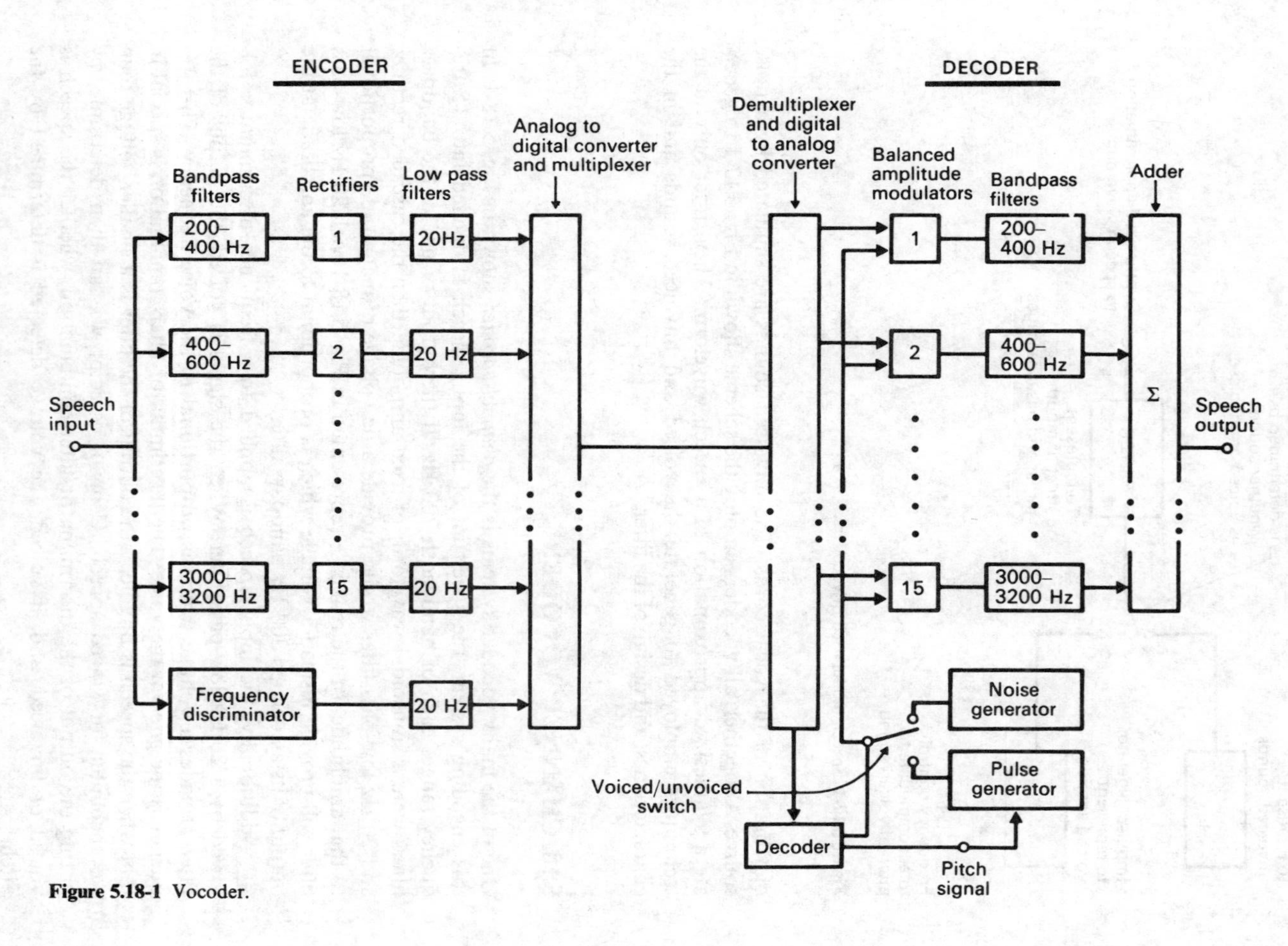

Figure 5.18-1 Vocoder.

A/D converted. If the sampling is at the Nyquist rate of 40 samples/s (corresponding to signals of 20 Hz bandwidth) and if we use 3 bits/sample to represent each voltage sample, the bit rate is

$$R = 40 \frac{\text{samples/s}}{\text{filter}} \times 16 \text{ filters} \times 3 \text{ bits/sample}$$

$$= 1.9 \times 10^3 \text{ bits/s} \qquad (5.18\text{-}1)$$

In any particular case the bit rate will depend on the number of low-pass filters, their passbands and the number of bits allowed per sample. Typical bit rates vary from 1.2×10^3 to 2.4×10^3 b/s.

At the vocoder receiver, as is to be seen in Fig. 5.18-1, the signal is demultiplexed and decoded, that is, converted back to analog form. Corresponding to each filter-rectifier combination at the encoder, there is provided, at the decoder, a balanced amplitude modulator and a bandpass filter with identical pass band. It will be recalled that a balanced modulator is a modulator that provides zero carrier output at zero modulation input. The carrier input to each modulator is the noise or pulse generator waveform. The modulation input is the amplitude signal (of each of the 15 rectifier-filters) provided by the encoder. At each sampling, the amplitude information is updated as is the information about whether the speech waveform is voiced or unvoiced and, if voiced, the pitch is provided by the discriminator.

Suppose, now, that in some sampling interval it happens that the speech is voiced and that the fundamental frequency of the generated sound is 450 Hz. This sound will generate an amplitude signal at the output associated with rectifier 2. The sound will have harmonics, so that there may be amplitude outputs at frequencies $2 \times 450 = 900$ Hz, $3 \times 450 = 1350$ Hz, etc., which will be seen at the outputs of rectifiers 4, 6, etc. The decoder will receive the information that the speech is voiced. Hence the decoder switch will connect to the pulse generator and the pulse generator frequency will be set to the pitch of the voice. The pulse generator waveform then constitutes a carrier input to all modulators. However, only modulators 2, 4, 6, etc. will generate outputs. These outputs are amplitude modulated reproductions of the pulse-generator waveform and hence are comprised of a fundamental and a succession of harmonics. However, the filter following modulator 2 will suppress all except the fundamental, the filter following modulator 4 will suppress all except the second harmonic, etc. All these waveforms are added and the resultant output waveform consists of a combination of a fundamental and harmonics with the same relative amplitudes as existed in the speech input. When the speech input is unvoiced the sound is noise-like and its spectrum extends through the entire speech-frequency range although not necessarily in a uniform manner. In this case we can expect that all paths in the encoder will provide an output as will all paths in the decoder.

It is interesting to note that a vocoder can also be designed which transmits only the largest envelope detector outputs, completely ignoring the remaining outputs. For example if the largest three outputs are sent, 4 bits/filter × 3 filters = 12 additional bits would be transmitted to tell the receiver which three filter outputs are sent. In this case, the bit rate becomes

$$R = 40\,\frac{\text{samples/s}}{\text{filter}} \times 3\text{ filters} \times 3\text{ bits/sample}$$

$$+\,40\text{ samples/s} \times \frac{12\text{ bits}}{\text{sample}}$$

$$= 840\text{ bits/s} \qquad (5.18\text{-}2)$$

Unfortunately, the quality of the signal is severely degraded. However, the bit rate is significantly reduced.

5.19 LINEAR PREDICTIVE CODER[5]

We observe, referring to Fig. 5.18-1, that the mechanism by which we regenerate speech at the decoder is the following: We make available at the decoder a noise generator and a pulse generator. We provide to the decoder information about whether the speech is voiced or unvoiced and, if voiced, the pitch. The outputs of these sound generators, i.e., the noise and pulse generator, are then passed through a *filter* with *adjustable parameters*. The filter is composed of the amplitude modulators and band pass filters. The adjusting signals are the modulating imputs to the modulators. These signals determine the relative proportions of fundamental and harmonics which are combined at the decoder output.

It occurs to us that we can improve the performance of the vocoder system in the following ways: First, let us employ a filter which has a greater versatility in its ability to be adjusted. Next, let us make this same filter available at the encoder. We shall then be able to regenerate the speech not only at the decoder but at the encoder as well. Thus the encoder will be able to "hear" how the speech will sound when regenerated at the decoder. Finally let us compare, at the encoder, the original speech and its regenerated version and, as in any feedback control system, use the differences, i.e., the error signal that appears, to adjust the filter parameters to minimize the error.

The *linear predictive coder* shown in Fig. 5.19-1 operates in this manner. As in the simpler encoder, a frequency discriminator and other hardware is provided to make available a pitch and a voiced-unvoiced signal. A single rectifier and filter is used to generate just one amplitude signal. We also provide the hardware necessary to regenerate the speech, that is a pulse generator, a noise generator, a switch, and a balanced modulator. The signal which is used to adjust the parameters of the adjustable filter is the difference (error) signal between the original speech and its regenerated form as it is formed at the filter output. This error signal is applied to a device which calculates and generates the signals which need to be applied to the adjustable filter to minimize the error signal. The decoder is shown in Fig. 5.19-2. It receives the amplitude, voiced-unvoiced and

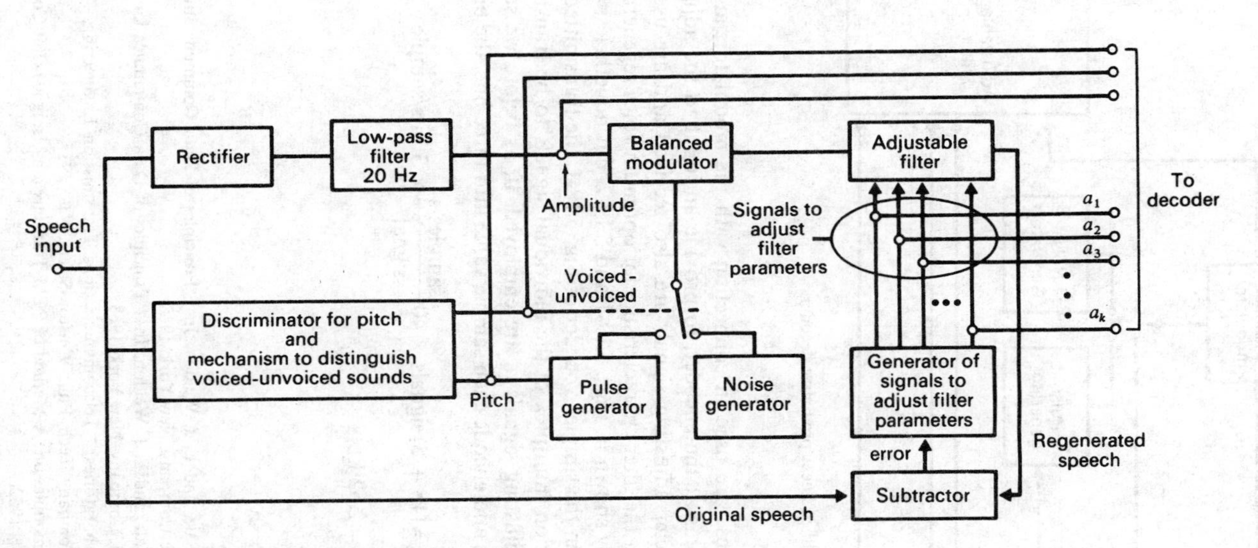

Figure 5.19-1 A simplified linear predictive encoder.

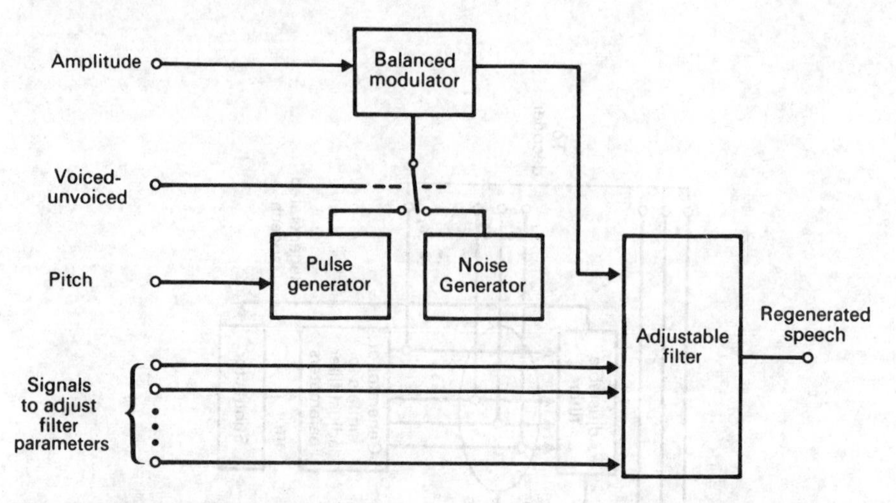

Figure 5.19-2 A simplified linear predictive decoder.

pitch signals. These are used in connection with the modulator, and pulse and noise generators, as at the encoder, to provide an input to the adjustable filter. The filter-parameter adjusting signals are also received and are used, as at the encoder, to adjust the filter characteristics for optimum voice regeneration.

Not explicitly shown in Figs. 5.19-1 and 5.19-2, but nonetheless to be understood is that, as in the simpler vocoder of Fig. 5.18-1, the transmitted signal must be the time-division multiplex of the individual signals to be transmitted. Typically, 18 filter-adjusting signals a_i are employed. If, as before we sample at the rate 40 samples/s and encode each sample value into three bits, the bit rate R is

$$R = (18 + 3)\,\text{signals} \times 40\,\frac{\text{samples/s}}{\text{signal}} \times 3\ \text{bits/sample}$$

$$= 2.52\ \text{kb/s} \tag{5.19-1}$$

REFERENCES

1. Lucky, R. W., J. Salz, and E. J. Weldon, Jr.: "Principles of Data Communication," pp. 61–87, McGraw-Hill Book Company, New York, 1968.
2. Lucky, R. W., J. Salz, and E. J. Weldon, Jr.: "Principles of Data Communication," pp. 128–165, McGraw-Hill Book Company, New York, 1968.
3. Bell Telephone Laboratories: "Telecommunications Transmission Engineering," Vol II, AT&T, Western Electric Company, Tech. Pub., Winston-Salem, N.C., 1977.
4. de Jager, F.: Deltamodulation: A Method of PCM Transmission Using a 1-unit Code, *Philips Res. Rept.* 7, pp. 442–446, 1952.
5. Jayant, N. S., and P. Noll: "Digital Coding of Waveforms," Prentice-Hall Inc., Englewood Cliffs, N.J., 1984.

PROBLEMS

5.1-1. It is required to transmit telephone messages across the United States, a 3000 mile run. The signal level is not to be allowed to drop below 1 millivolt before amplification and the signal is not to be allowed to be larger than 15 volts in order to avoid amplifier overload. Assuming that repeaters are to be located with equal spacings, how many repeaters will be required?

5.1-2. A bandpass signal has a spectral range that extends from 20 to 82 kHz. Find the acceptable range of the sampling frequency f_s.

5.1-3. A bandpass signal has a center frequency f_0 and extends from $f_0 - 5$ kHz to $f_0 + 5$ kHz. The signal is sampled at a rate $f_s = 25$ kHz. As the center frequency f_0 varies from $f_0 = 5$ kHz to $f_0 = 50$ kHz find the ranges of f_0 for which the sampling rate is adequate.

5.1-4. The signal $v(t) = \cos 5\pi t + 0.5 \cos 10\pi t$ is instantaneously sampled. The interval between samples is T_s.

(*a*) Find the maximum allowable value for T_s.

(*b*) If the sampling signal is $S(t) = 5 \sum_{k=-\infty}^{\infty} \delta(t - 0.1k)$, the sampled signal $v_s(t) = v(t)S(t)$ consists of a train of impulses, each with a different strength $v_s(t) = \sum_{k=-\infty}^{\infty} I_k\, \delta(t - 0.1k)$. Find I_0, I_1, and I_2, and show that $I_k = I_{4+k}$.

(*c*) To reconstruct the signal $v_s(t)$ is passed through a rectangular low-pass filter. Find the minimum filter bandwidth to reconstruct the signal without distortion.

5.1-5. We have the signal $v(t) = \cos 2\pi f_0 t + \cos 2 \times 2\pi f_0 t + \cos 3 \times 2\pi f_0 t$. Our interest extends, however, only to spectral components up to and including $2f_0$. We therefore sample at the rate $5f_0$ which is adequate for the $2f_0$ component of the signal.

(*a*) If sampling is asccomplished by multiplying $v(t)$ by an impulse train in which the impulses are of unit strength, write an expression for the sampled signal.

(*b*) To recover the part of the signal of interest, the sampled signal is passed through a rectangular low-pass filter with passband extending from 0 to slightly beyond $2f_0$. Write an expression for the filter output. Is the part of the signal of interest recovered exactly? If we want to reproduce the first two terms of $v(t)$ without distortion, what operation must be performed at the very outset?

5.1-6. The bandpass signal $v(t) = \cos 10\omega_0 t + \cos 11\omega_0 t + \cos 12\omega_0 t$ is sampled by an impulse train $S(t) = I \sum_{k=-\infty}^{\infty} \delta(t - kT_s)$.

(*a*) Find the maximum time between samples, T_s, to ensure reproduction without error.

(*b*) Using the result obtained in (*a*), obtain an expression for $v_s(t) = S(t)v(t)$.

(*c*) The sampled signal $v_s(t)$ is filtered by a rectangular low-pass filter with a bandwidth $B = 2f_0$. Obtain an expression for the filter output.

(*d*) The sampled signal $v_s(t)$ is filtered by a rectangular bandpass filter extending from $2f_0$ to $4f_0$. Obtain an expression for the filter output.

5.1-7. The bandpass signal $v(t) = \cos 10\omega_0 t + \cos 11\omega_0 t + \cos 12\omega_0 t$ is sampled by an impulse train, $S(t) = I \sum_{k=-\infty}^{\infty} \delta(t - k/8f_0)$. The sampled signal $v_s(t) = S(t)v(t)$ is then filtered by a rectangular low-pass filter having a bandwidth $B = 2f_0$. Obtain an expression for the filter output.

5.1-8. Let us view the waveform $v(t) = \cos \omega_0 t$ as a bandpass signal occupying an arbitrarily narrow frequency band. On this basis we find that the required sampling rate is $f_s = 0$. Discuss.

5.2-1. The TDM system shown in Fig. 5.2-1 is used to multiplex the four signals $m_1(t) = \cos \omega_0 t$, $m_2(t) = 0.5 \cos \omega_0 t$, $m_3(t) = 2 \cos 2\omega_0 t$, and $m_4(t) = \cos 4\omega_0 t$.

(*a*) If each signal is sampled at the same sampling rate, calculate the minimum sampling rate f_s.

(*b*) What is the commutator speed in revolutions per second.

(*c*) Design a commutator which will allow each of the four signals to be sampled at a rate faster than is required to satisfy the Nyquist criterion for the individual signal.

5.2-2. Three signals m_1, m_2, and m_3 are to be multiplexed. m_1 and m_2 have a 5-kHz bandwidth, and m_3 has a 10-kHz bandwidth. Design a commutator switching system so that each signal is sampled at its Nyquist rate.

5.3-1. Show that the response of a rectangular low-pass filter, with a bandwidth f_c, to the impulse function $I\,\delta(t - k/2f_c)$ is

$$S_R(t) = \frac{I\omega_c}{\pi} \frac{\sin \omega_c(t - k/2f_c)}{\omega_c(t - k/2f_c)}$$

Assume that in its passband the filter has $H(f) = 1$.

5.3-2. Four signals, $m_1(t) = 1 \cos \omega_0 t$, $m_2(t) = 1 \sin \omega_0 t$, $m_3(t) = -1 \cos \omega_0 t$, and $m_4(t) = -1 \sin \omega_0 t$ are sampled every $1/2f_0$ sec by the sampling function

$$S(t) = 1 \sum_{k=-\infty}^{\infty} \delta\left(t - \frac{k}{2f_0}\right)$$

The signals are then time-division multiplexed. The TDM signal is filtered by a rectangular low-pass filter having a bandwidth $f_c = 4f_0$ and then decommutated.

(*a*) Sketch the four outputs of the decommutator.

(*b*) Each of the four output signals is filtered by a rectangular low-pass filter having a bandwidth f_0. Show that the four signals are reconstructed without error.

5.3-3. The four signals of Prob. 5.3-2 are sampled, as indicated in that problem, and time-division multiplexed. The TDM signal is filtered by a rectangular low-pass filter having a bandwidth $f_c = 2f_0$ and then decommutated. Sketch the output at the decommutator switch segment where the samples of $m_1(t)$ should appear and show that $m_1(t)$ cannot be recovered.

5.4-1. The signal $v(t) = \cos \omega_0 t + \cos 8\omega_0 t$ is sampled by using *natural sampling.*

(*a*) Determine the minimum sampling rate f_s.

(*b*) Sketch $v_s(t) = S(t)v(t)$ if $S(t)$ is a train of pulses having unit height, occurring at the rate f_s; and $S(t) = 1$ for $nT - \tau/2 \le t \le nT + \tau/2$. The pulse duration is $\tau = 1/32f_0$.

(*c*) Repeat (*b*) if $\tau = 1/320f_0$.

5.5-1. Show that an impulse function $I\,\delta(t)$ can be stretched to have a width τ by passing the impulse function through a filter $(1 - e^{-j\omega\tau})/j\omega$. Show that this operation is identical with integrating the impulse for τ sec, that is, that the output $v_o(t)$ is given by

$$v_o(t) = \int_0^t I\,\delta(t)\,dt \qquad 0 \le t \le \tau$$
$$= 0 \qquad \text{otherwise}$$

5.8-1. For the quantizer characteristic shown in Fig. 5.8-1*b*.

(*a*) Plot the error characteristic $e = v_o - v_i$ versus v_i. Assume that $S = 1$, that is, $m_{k+1} - m_k =$ 1 volt.

(*b*) Since the error $e = v_o - v_i$ is periodic, it can be expanded in a Fourier series. Write the Fourier series for the error $e = e(v_i)$.

(*c*) If $v_i = S \sin \omega_0 t$, find the component of the error e at the angular frequency ω_0.

5.8-2. Show that if the signal is uniformly distributed, Eq. (5.8-3) results even if M is not large.

5.8-3. Consider a signal having a probability density

$$f(v) = \begin{cases} Ke^{-|v|} & -4 < v < 4 \\ 0 & \text{elsewhere} \end{cases}$$

(*a*) Find K.

(*b*) Determine the step size S if there are four quantization levels.

(*c*) Calculate the variance of the quantization error when there are four quantization levels. Do not assume that $f(v)$ is constant over each level. Compare your result with Eq. (5.8-3).

5.8-4. Consider a signal having a probability density

$$f(v) = K(1 - |v|) \qquad -1 \le v \le 1$$

Calculate (*a*) to (*c*) of Prob. 5.8-3.

5.9-1. Show that the numbers 0 to 7 can be written using 3 binary digits (bits). How many bits are required to write the numbers 0 to 5?

5.10-1. Consider that the signal $\cos 2\pi t$ is quantized into 16 levels. The sampling rate is 4 Hz. Assume that the sampling signal consists of pulses each having a unit height and duration dt. The pulses occur every $t = k/4$ sec, $-\infty < k < \infty$.

(*a*) Sketch the binary signal representing each sample voltage.

(*b*) How many bits are required per sample?

5.11-1. A D/A converter is shown in Fig. P5.11-1. Using the set-reset flip-flops shown explain the operation of the device.

5.11-2. An A/D converter is shown in Fig. P5.11-2. Using *trigger* flip-flops as indicated explain the operation of the device.

5.12-1. A μ-law compander uses a compressor which relates output to input by the relation

$$y = \pm \frac{\log(1 + \mu|x|)}{\log(1 + \mu)}$$

Here the $+$ sign applies when x is positive and the $-$ sign applies when x is negative. Also $x \equiv v_i/V$ and $y = v_o/V$ where v_i and v_0 are the input and output voltages and the range of allowable voltage is $-V$ to $+V$. The parameter μ determines the degree of compression.

(*a*) A commonly used value is $\mu = 255$. For this value make a plot of y versus x from $x = -1$ to $x = +1$.

(*b*) If $V = 40$ volts and 256 quantization levels are employed what is the voltage interval between levels when there is no compression? For $\mu = 255$ what is the minimum and what is the maximum effective separation between levels?

5.12-2. An A-law compander uses a compressor which relates output to input by the relations

$$y = \pm \frac{A|x|}{1 + \log A} \qquad \text{for} \qquad |x| \le \frac{1}{A}$$

$$y = \pm \frac{1 + \log A|x|}{1 + \log A} \qquad \text{for} \qquad \frac{1}{A} \le |x| \le 1$$

Here the $+$ sign applies when x is positive and the $-$ sign when x is negative. Also $x = v_i/V$ and $y = v_o/V$ where v_i and v_o are the input and output voltages and the range of allowable voltage is $-V$ to $+V$. The parameter A determines the degree of compression.

(*a*) A commonly used value is $A = 87.6$. For this value make a plot of y versus x from $x = -1$ to $x = +1$.

(*b*) If $\pm V = \pm 10$ volts and 256 quantization levels are employed, what is the voltage interval between levels when there is no compression? For $A = 87.6$ what is the minimum and the maximum effective separation between levels?

5.12-3. An analog signal $m(t)$ has a probability density function

$$f(x) = 1 - |x| \qquad -1 \le x \le +1$$

This signal is quantized by a quantizer that has eight quantization levels.

(*a*) Determine the voltages at which the quantization levels should be located so that there shall be equal probabilities that $m(t)$ is between any two adjacent levels or between the extreme levels and the $+1$ and -1 extreme voltages.

(*b*) If a quantizer is used with equal spacings between quantization levels, make an input–output plot of the compander that must precede the quantizer to satisfy the requirement of part (*a*).

5.12-4. An analog signal $m(t)$ has a probability density function

$$f(x) = 1 - |x| \qquad \text{for} \qquad -1 \le x \le -1$$

The signal is applied to a quantizer with two quantization levels at ± 0.5 volts. Calculate the mean square quantization error and compare with the result that would be given if Eq. (5.12-2) were applied.

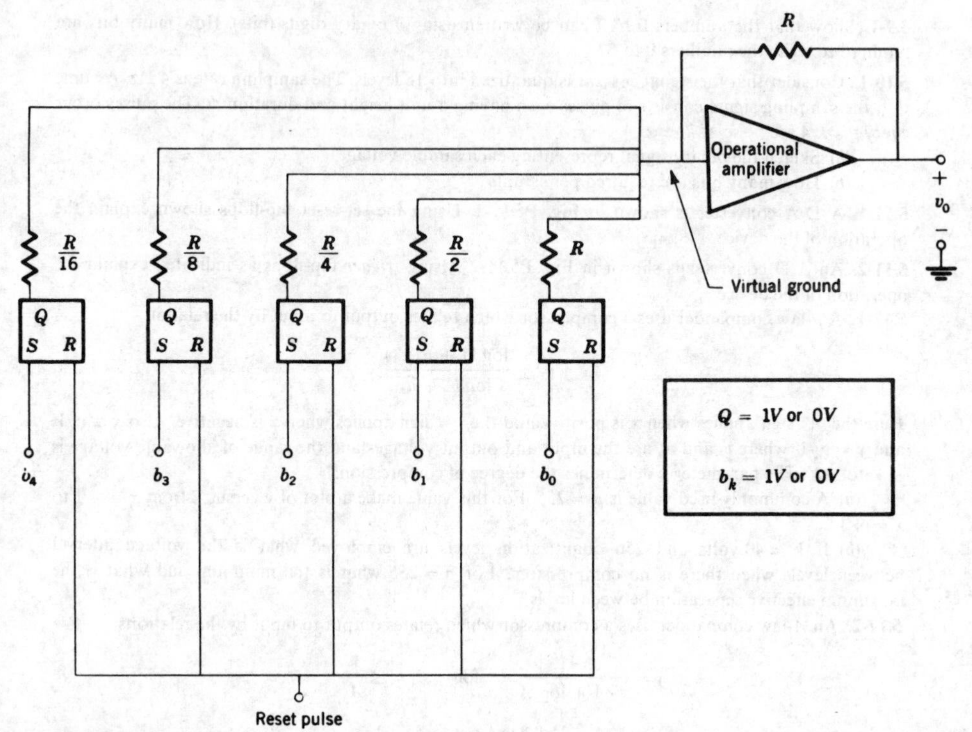

Figure P5.11-1

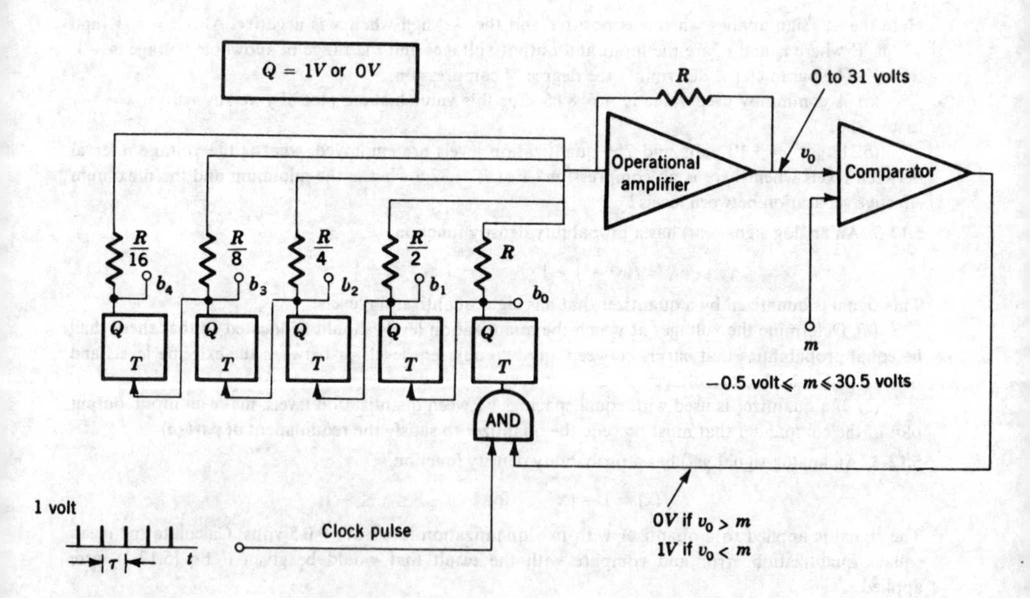

Figure P5.11-2

5.12-5. Verify that the procedure described in the text and represented in Fig. 5.12-2 for setting the content of the ROM yields $2^8 = 256$ code words for $2^{12} = 4096$ addresses. What is the ratio between the smallest quantization Δ and the largest quantization level?

5.13-1. (*a*) An NRZ waveform as in Fig. 5.13-3 consists of alternating 1's and 0's. The waveform swings between $+V$ and $-V$ and the bit duration is T_b. The waveform is passed through a RC network as in Fig. 5.13-3*b* whose time constant is $RC = T_b$. Calculate all the voltage levels of the output waveform for the circumstances that the NRZ waveform has persisted for a long time.

(*b*) For the circumstances of part (*a*) calculate the area under the waveform during the bit time T_b. Now consider that a sequence of 10 successive 1's appears. Calculate the area under the waveform during the 10th 1 bit and compare the area calculated for the case when 1's and 0's alternate.

5.13-2. Draw an AMI waveform corresponding to a binary bit stream 100110010101. The polarity of the first pulse in the waveform may be set arbitrarily.

5.13-3. Two of the $T1$ inputs to the $M12$ multiplexer of Fig. 5.13-5 are generated in systems whose clocks have frequencies which are different by 50 parts/million. In how long a time will the faster clock have generated one more time slot than the slower clock. In this time interval, how many frames will have been generated?

5.14-1. Consider the delta PCM system shown in Fig. P5.14-1.

(*a*) Explain its operation.

(*b*) Sketch the receiver.

(*c*) If $m(t) = 0.05 \sin 2\pi t$, find $\hat{m}(t)$ and $\Delta(t)$ graphically.

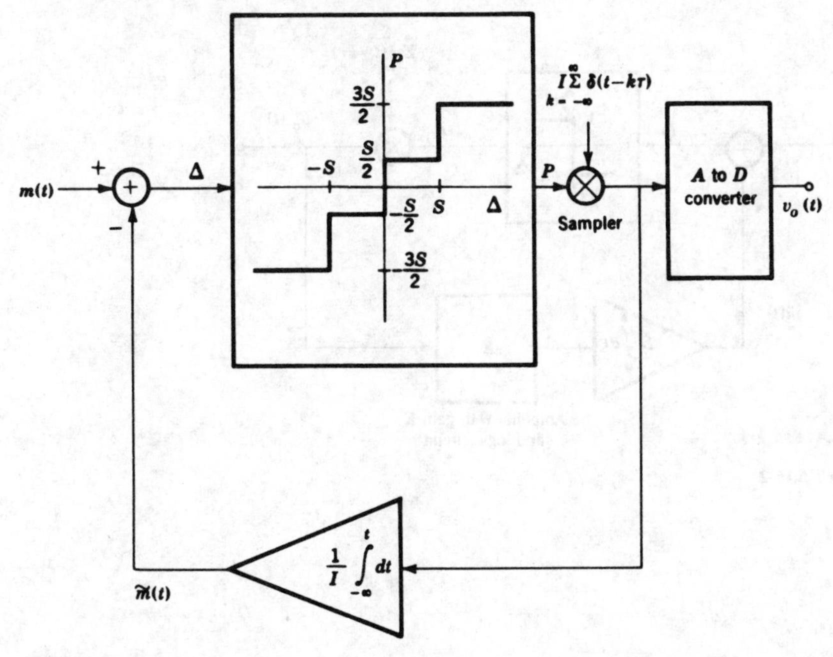

Figure P5.14-1

5.15-1. (*a*) Write the nonlinear difference equation of the delta modulator shown in Fig. 5.15-1. Let $m(t) = 18 \times 10^{-3} \sin 2\pi t$ and let the samples occur every 0.05 sec starting at $t = 0.01$ sec. The step size is 5 mV.

(*b*) Write and run a computer program to solve for $\Delta(t)$.

(*c*) Repeat (*b*) if the samples occur every 0.1 sec.

(*d*) Compare the results of (*b*) and (*c*).

5.15-2. The input to a DM is $m(t) = 0.01t$. The DM operates at a sampling frequency of 20 Hz and has a step size of 2 mV. Sketch the delta modulator output, $\Delta(t)$ and $\tilde{m}(t)$.

5.15-3. The input to a DM is $m(t) = kt$. Prove, by graphically determining $\tilde{m}(t)$, that slope overload occurs when k exceeds a specified value. What is this value in terms of the step size S and the sampling frequency f_s?

5.15-4. If the step size is S, the sampling frequency is $f_s^{(\Delta)}$, and $m(t) = M \sin \omega t$, explain what happens to $\tilde{m}(t)$ if $2M < S$. This is called *step-size limiting*.

5.15-5. The signal $m(t) = M \sin \omega_0 t$ is to be encoded by using a delta modulator. If the step size S and sampling frequency $f_s^{(\Delta)}$ are selected so as to ensure that neither slope overloads [Eq. (5.15-1)] nor step-size limiting (Prob. 5.15-4) occurs, show that $f_s^{(\Delta)} > 3f_0$.

5.16-1. The adaptive delta-modulation system described in Sec. 5.16 has an input which is $m(t) = 0$ until time $t = 0$ and thereafter $m(t) = 1250 \sin 2\pi t$. An inactive edge of the clock occurs at $t = 0$ and the clock period is 0.05 sec. On a single set of coordinate axes draw the clock waveform, the input $m(t)$ and the approximation $\hat{m}(t)$. Extend the plot through a full cycle of $m(t)$.

5.16-2. An adaptive delta modulator is shown in Fig. P5.16-2. The gain K is variable and is adjusted using the following logic. If $p_o(t)$ alternates between $+1$ and -1, $K = 1$; if a sequence of N positive or N negative pulses occurs, K increases by N; if after a sequence of N pulses the polarity changes, $|K|$ decreases by 2. Thus, consider the sequence 1, 1, 1. Then $K = 1, 2, 3, 1, 2$.

Find $\tilde{m}(t)$ if $m(t) = \sin 2\pi t$. Perform the analysis graphically. Consider a sampling time of 0.05 sec, and at $t = 0.01$ sec a sample occurs. At this time the step size $S = 1$ volt when $K = 1$.

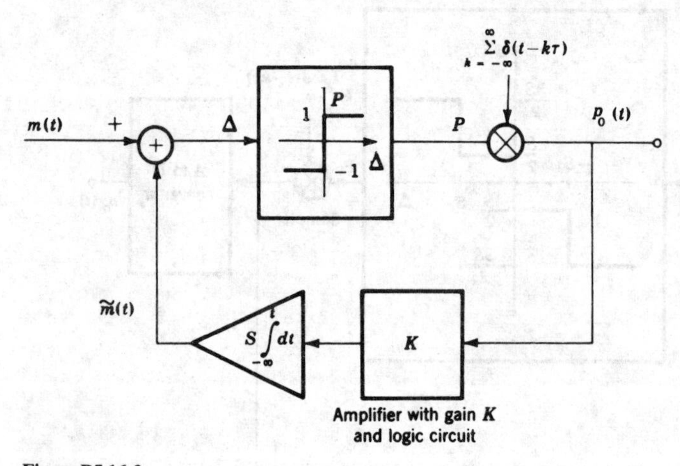

Figure P5.16-2

CHAPTER

SIX

DIGITAL MODULATION TECHNIQUES

6.1 INTRODUCTION

When it becomes necessary, for the purpose of transmission, to superimpose a binary waveform on a carrier, amplitude modulation (AM), phase modulation (PM), or frequency modulation (FM) may be used. Combination AM-PM systems such as quadrature amplitude modulation (QAM) are also commonly employed.

The selection of the particular modulation method used is determined by the application intended as well as by the channel characteristics such as available bandwidth and the susceptibility of the channel to fading. When radio communication is intended we must take account of antenna characteristics. As we have already noted (Sec. 3.1) an antenna is a narrow-band device whose operating frequency is related to its physical dimensions. When data transmission using a telephone channel is intended, we need to take account of the fact that often the channel may not transmit dc and low frequencies because of transformers which may be included in the transmission path. A fading channel is one in which the received signal amplitude varies with time because of variabilities in the transmission medium. When we must contend with such channels it is useful to use FM which is relatively insensitive to amplitude fluctuations.

In each of these situations we need a modulator at the transmitter and, at the receiver, a demodulator to recover the baseband signal. Such a modulator-demodulator combination is called a *MODEM*. In this chapter we present a description of many of the available modulation/demodulation techniques and compare them on the basis of their spectral occupancy. In Chap. 11 we shall complete our comparison and determine in each case the probability of error for each system as a function of signal-to-noise ratio and bandwidth available for transmission.

6.2 BINARY PHASE-SHIFT KEYING

In binary phase-shift keying (BPSK) the transmitted signal is a sinusoid of fixed amplitude. It has one fixed phase when the data is at one level and when the data is at the other level the *phase* is different by 180°. If the sinusoid is of amplitude A it has a power $P_s = \frac{1}{2}A^2$ so that $A = \sqrt{2P_s}$. Thus the transmitted signal is either

$$v_{\text{BPSK}}(t) = \sqrt{2P_s}\cos(\omega_0 t) \tag{6.2-1}$$

or

$$v_{\text{BPSK}}(t) = \sqrt{2P_s}\cos(\omega_0 t + \pi) \tag{6.2-2a}$$

$$= -\sqrt{2P_s}\cos(\omega_0 t) \tag{6.2-2b}$$

In BPSK the data $b(t)$ is a stream of binary digits with voltage levels which, as a matter of convenience, we take to be at $+1V$ and $-1V$. When $b(t) = 1V$ we say it is at logic level 1 and when $b(t) = -1V$ we say it is at logic level 0. Hence $v_{\text{BPSK}}(t)$ can be written, with no loss of generality, as

$$v_{\text{BPSK}}(t) = b(t)\sqrt{2P_s}\cos\omega_0 t \tag{6.2-3}$$

In practice, a BPSK signal is generated by applying the waveform $\cos\omega_0 t$, as a carrier, to a *balanced* modulator and applying the baseband signal $b(t)$ as the modulating waveform. In this sense BPSK can be thought of as an AM signal.

Reception of BPSK

The received signal has the form

$$v_{\text{BPSK}}(t) = b(t)\sqrt{2P_s}\cos(\omega_0 t + \theta) = b(t)\sqrt{2P_s}\cos\omega_0(t + \theta/\omega_0) \tag{6.2-4}$$

Here θ is a nominally fixed phase shift corresponding to the time delay θ/ω_0 which depends on the length of the path from transmitter to receiver and the phase shift produced by the amplifiers in the "front-end" of the receiver preceding the demodulator. The original data $b(t)$ is recovered in the demodulator. The demodulation technique usually employed is called synchronous demodulation and requires that there be available at the demodulator the waveform $\cos(\omega_0 t + \theta)$. A scheme for generating the carrier at the demodulator and for recovering the baseband signal is shown in Fig. 6.2-1.

The received signal is squared to generate the signal

$$\cos^2(\omega_0 t + \theta) = \tfrac{1}{2} + \tfrac{1}{2}\cos 2(\omega_0 t + \theta) \tag{6.2-5}$$

The dc component is removed by the bandpass filter whose passband is centered around $2f_0$ and we then have the signal whose waveform is that of $\cos 2(\omega_0 t + \theta)$. A frequency divider (composed of a flip-flop and narrow-band filter tuned to f_0) is used to regenerate the waveform $\cos(\omega_0 t + \theta)$. Only the waveforms of the signals at the outputs of the squarer, filter and divider are relevant to our discussion and not their amplitudes. Accordingly in Fig. 6.2-1 we have arbitrarily taken each amplitude to be unity. In practice, the amplitudes will

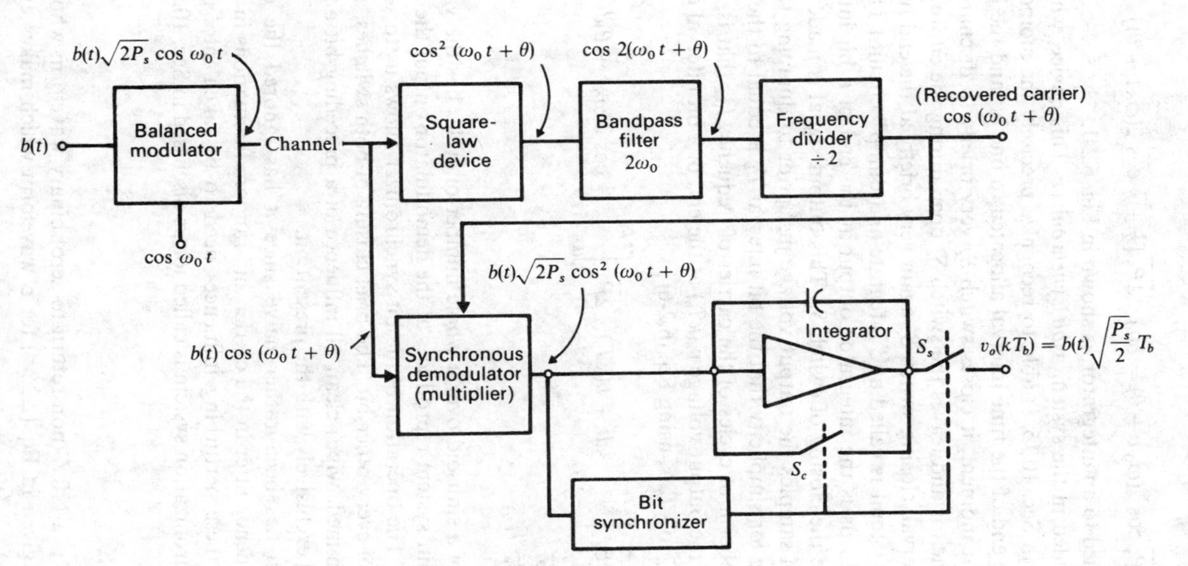

Figure 6.2-1 Scheme to recover the baseband signal in BPSK.

be determined by features of these devices which are of no present concern. In any event, the carrier having been recovered, it is multiplied with the received signal to generate

$$b(t)\sqrt{2P_s}\cos^2(\omega_0 t + \theta) = b(t)\sqrt{2P_s}\,[\tfrac{1}{2} + \tfrac{1}{2}\cos 2(\omega_0 t + \theta)] \tag{6.2-6}$$

which is then applied to an integrator as shown in Fig. 6.2-1.

We have included in the system a *bit synchronizer*. This device, whose operation is described in Sec. 10.15, is able to recognize precisely the moment which corresponds to the end of the time interval allocated to one bit and the beginning of the next. At that moment, it closes switch S_c very briefly to discharge (*dump*) the integrator capacitor and leaves the switch S_c open during the entire course of the ensuing bit interval, closing switch S_c again very briefly at the end of the next bit time, etc. (This circuit is called an "integrate-and-dump" circuit.) The output signal of interest to us is the integrator output at the end of a bit interval but immediately before the closing of switch S_c. This output signal is made available by switch S_s which samples the output voltage just prior to dumping the capacitor. Let us assume for simplicity that the bit interval T_b is equal to the duration of an integral number n of cycles of the carrier of frequency f_0, that is, $n \cdot 2\pi = \omega_0 T_b$. In this case the output voltage $v_o(kT_b)$ at the end of a bit interval extending from time $(k-1)T_b$ to kT_b is, using Eq. (6.2-6)

$$v_o(kT_b) = b(kT_b)\sqrt{2P_s}\int_{(k-1)T_b}^{kT_b} \tfrac{1}{2}dt + b(kT_b)\sqrt{2P_s}\int_{(k-1)T_b}^{kT_b} \tfrac{1}{2}\cos 2(\omega_0 t + \theta)dt \tag{6.2-7a}$$

$$= b(kT_b)\sqrt{\frac{P_s}{2}}\,T_b \tag{6.2-7b}$$

since the integral of a sinusoid over a whole number of cycles has the value zero. Thus we see that our system reproduces at the demodulator output the transmitted bit stream $b(t)$. The operation of the bit synchronizer allows us to sense each bit independently of every other bit. The brief closing of both switches, after each bit has been determined, wipes clean all influence of a preceding bit and allows the receiver to deal exclusively with the present bit.

Our discussion has been rather naive since it has ignored the effects of thermal noise, frequency jitter in the carrier and random fluctuations in propagation delay. When these perturbing influences need to be taken into account a phase-locked synchronization system is called for as discussed in Sec. 10.16.

Spectrum of BPSK

The waveform $b(t)$ is a NRZ (non-return-to-zero) binary waveform whose power spectral density is given in Eq. (2.25-4) for a waveform which makes excursions between $+\sqrt{P_s}$ and $-\sqrt{P_s}$. We have

$$G_b(f) = P_s T_b\left(\frac{\sin \pi f T_b}{\pi f T_b}\right)^2 \tag{6.2-8}$$

The BPSK waveform is the NRZ waveform multiplied by $\sqrt{2}\cos\omega_0 t$. Thus following the analysis of Sec. 3.2 we find that the power spectral density of the BPSK signal is

$$G_{\text{BPSK}}(f) = \frac{P_s T_b}{2}\left\{\left[\frac{\sin \pi(f-f_0)T_b}{\pi(f-f_0)T_b}\right]^2 + \left[\frac{\sin \pi(f+f_0)T_b}{\pi(f+f_0)T_b}\right]^2\right\} \tag{6.2-9}$$

Equations (6.2-8) and (6.2-9) are plotted in Fig. 6.2-2.

Note that, in principle at least, the spectrum of $G_b(f)$ extends over all frequencies and correspondingly so does $G_{\text{BPSK}}(f)$. Suppose then that we tried to multiplex signals using BPSK, using different carrier frequencies for different baseband signals. There would inevitably be overlap in the spectra of the various signals and correspondingly a receiver tuned to one carrier would also receive, albeit at a lower level, a signal in a different channel. This overlapping of spectra causes *interchannel* interference.

Since efficient spectrum utilization is extremely important in order to maximize the number of simultaneous users in a multi-user communication system, the FCC and CCITT require that the side-lobes produced in BPSK be reduced below certain specified levels. To accomplish this we employ a filter to restrict the bandwidth allowed to the NRZ baseband signal. For example, before modulation we might pass the bit stream $b(t)$ through a low-pass filter which suppresses (but does not completely eliminate) all the spectrum except the principal lobe. Since 90 percent of the power of the waveform is associated with this lobe the suggestion is not unreasonable. There is, however, the difficulty that such spectrum suppression distorts the signal and as a result, as we shall see, there is a partial overlap of a bit (symbol) and its adjacent bits in a single channel. This overlap is called *intersymbol* interference (ISI). Intersymbol interference can be somewhat alleviated by the use of *equalizers* at the receiver. Equalizers are filter-type structures used to undo the adverse effects of filters introduced, intentionally or unavoidably, at other places in a communications channel.

Geometrical Representation of BPSK Signals

Referring to Sec. 1.9 we see that a BPSK signal can be represented, in terms of one orthonormal signal $u_1(t) = \sqrt{(2/T_b)}\cos\omega_0 t$ as [see Eq. (6.2-1)]

$$v_{\text{BPSK}}(t) = \left[\sqrt{P_s T_b}\, b(t)\right]\sqrt{\frac{2}{T_b}}\cos\omega_0 t = \left[\sqrt{P_s T_b}\, b(t)\right]u_1(t) \tag{6.2-10}$$

The binary PSK signal can then be drawn as shown in Fig. 6.2-3. Note that the distance d between signals is

$$d = 2\sqrt{P_s T_b} = 2\sqrt{E_b} \tag{6.2-11}$$

where $E_b = P_s T_b$ is the energy contained in a bit duration. We show in Sec. 11.13

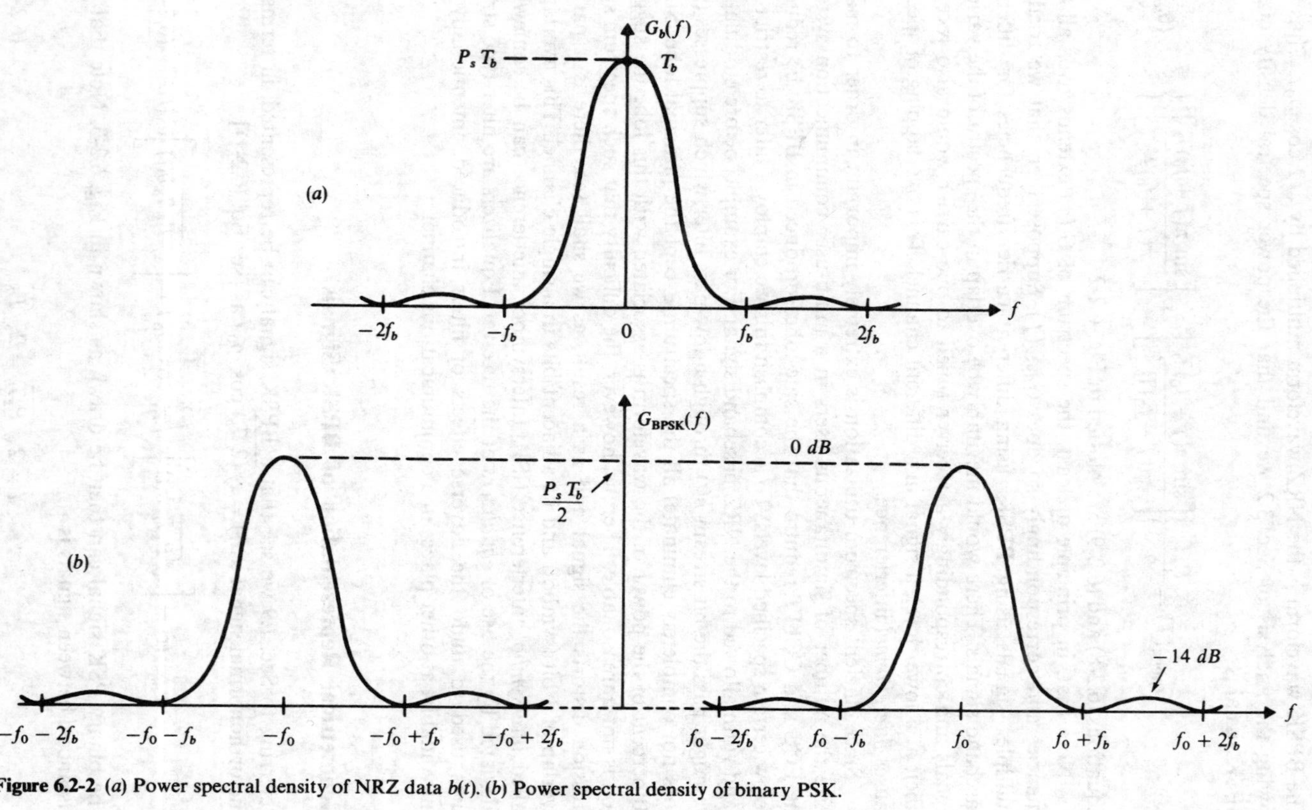

Figure 6.2-2 (*a*) Power spectral density of NRZ data $b(t)$. (*b*) Power spectral density of binary PSK.

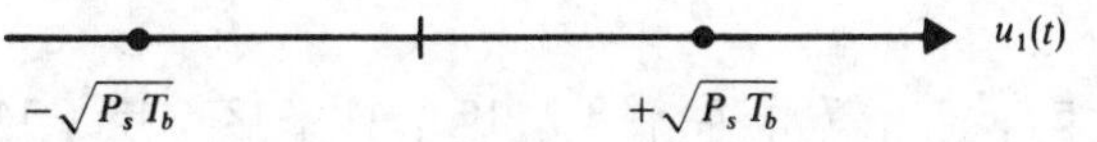

$-\sqrt{P_s T_b}$ $\quad$ $+\sqrt{P_s T_b}$ $\quad$ $u_1(t)$

Figure 6.2-3 Geometrical representation of BPSK signals.

that the distance d is inversely proportional to the probability that we make an error when, in the presence of noise, we try to determine which of the levels of $b(t)$ is being received.

6.3 DIFFERENTIAL PHASE-SHIFT KEYING

We observed in Fig. 6.2-1 that, in BPSK, to regenerate the carrier we start by squaring $b(t)\sqrt{2P_s}\cos\omega_0 t$. Accordingly, if the received signal were instead $-b(t)\sqrt{2P_s}\cos\omega_0 t$, the recovered carrier would remain as before. Therefore we shall not be able to determine whether the received baseband signal is the transmitted signal $b(t)$ or its negative $-b(t)$.

Differential phase-shift keying (DPSK) and differential encoded PSK (DEPSK) which is discussed in Sec. 6.4 are modifications of BPSK which have the merit that they eliminate the ambiquity about whether the demodulated data is or is not inverted. In addition DPSK avoids the need to provide the synchronous carrier required at the demodulator for detecting a BPSK signal.

A means for generating a DPSK signal is shown in Fig. 6.3-1. The data stream to be transmitted, $d(t)$, is applied to one input of an exclusive-OR logic gate. To the other gate input is applied the output of the exclusive or gate $b(t)$ delayed by the time T_b allocated to one bit. This second input is then $b(t - T_b)$. In Fig. 6.3-2 we have drawn logic waveforms to illustrate the response $b(t)$ to an input $d(t)$. The upper level of the waveforms corresponds to logic 1, the lower level to logic 0. The truth table for the exclusive-OR gate is given in Fig. 6.3-1

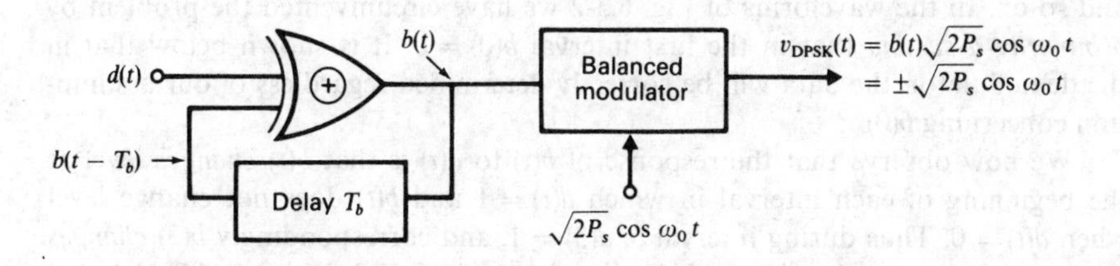

$d(t)$		$b(t - T_b)$		$b(t)$	
logic level	voltage	logic level	voltage	logic level	voltage
0	−1	0	−1	0	−1
0	−1	1	1	1	1
1	1	0	−1	1	1
1	1	1	1	0	−1

Figure 6.3-1 Means of generating a DPSK signal.

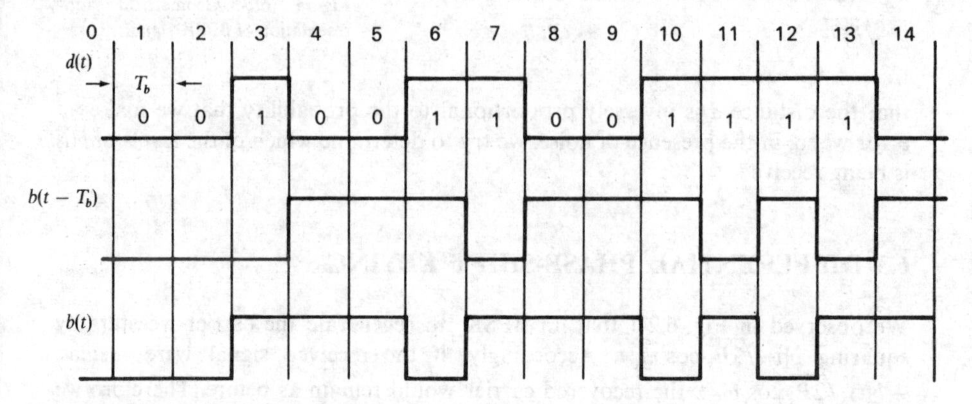

Figure 6.3-2 Logic waveforms to illustrate the response $b(t)$ to an input $d(t)$.

and with this table we can easily verify that the waveforms for $d(t)$, $b(t - T_b)$, and $b(t)$ are consistent with one another. We observe that, as required, $b(t - T_b)$ is indeed $b(t)$ delayed by one bit time and that in any bit interval the bit $b(t)$ is given by $b(t) = d(t) \oplus b(t - T_b)$. In the ensuing discussion we shall use the symbolism $d(k)$ and $b(k)$ to represent the logic levels of $d(t)$ and $b(t)$ during the kth interval.

Because of the feedback involved in the system of Fig. 6.3-2 there is a difficulty in determining the logic levels in the interval in which we start to draw the waveforms (interval 1 in Fig. 6.3-2). We cannot determine $b(t)$ in this first interval of our waveform unless we know $b(k = 0)$. But we cannot determine $b(0)$ unless we know both $d(0)$ and $b(-1)$, etc. Thus, to justify any set of logic levels in an initial bit interval we need to know the logic levels in the preceding interval. But such a determination requires information about the interval two bit times earlier and so on. In the waveforms of Fig. 6.3-2 we have circumvented the problem by *arbitrarily assuming* that in the first interval $b(0) = 0$. It is shown below that in the demodulator, the data will be correctly determined regardless of our assumption concerning $b(0)$.

We now observe that the response of $b(t)$ to $d(t)$ is that $b(t)$ *changes* level at the beginning of each interval in which $d(t) = 1$ and $b(t)$ does not change level when $d(t) = 0$. Thus during interval 3, $d(3) = 1$, and correspondingly $b(3)$ *changes* at the beginning at that interval. During intervals 6 and 7, $d(6) = d(7) = 1$ and there are *changes* in $b(t)$ at the beginnings of both intervals. During bits 10, 11, 12, and 13 $d(t) = 1$ and there are *changes* in $b(t)$ at the beginnings of each of these intervals. This behavior is to be anticipated from the truth table of the exclusive-OR gate. For we note that when $d(t) = 0$, $b(t) = b(t - T_b)$ so that, whatever the initial value of $b(t - T_b)$, it reproduces itself. On the other hand when $d(t) = 1$ then $b(t) = \overline{b(t - T_b)}$. Thus, in each successive bit interval $b(t)$ changes from its value in the previous interval. Note that in some intervals where $d(t) = 0$ we have $b(t) = 0$ and in other intervals when $d(t) = 0$ we have $b(t) = 1$. Similarly, when

$d(t) = 1$ sometimes $b(t) = 1$ and sometimes $b(t) = 0$. Thus there is no correspondence between the levels of $d(t)$ and $b(t)$, and the only invariant feature of the system is that a *change* (sometimes up and sometimes down) in $b(t)$ occurs whenever $d(t) = 1$, and that no change in $b(t)$ will occur whenever $d(t) = 0$.

Finally, we note that the waveforms of Fig. 6.3-2 are drawn on the assumption that, in interval 1, $b(0) = 0$. As is easily verified, if not intuitively apparent, if we had assumed $b(0) = 1$, the invariant feature by which we have characterized the system would continue to apply. Since $b(0)$ must be either $b(0) = 0$ or $b(0) = 1$, there being no other possibilities, our result is valid quite generally. If, however, we had started with $b(0) = 1$ the levels $b(1)$ and $b(0)$ would have been inverted.

As is seen in Fig. 6.3-1 $b(t)$ is applied to a balanced modulator to which is also applied the carrier $\sqrt{2P_s} \cos \omega_0 t$. The modulator output, which is the transmitted signal is

$$v_{\text{DPSK}}(t) = b(t) \sqrt{2P_s} \cos \omega_0 t$$

$$= \pm \sqrt{2P_s} \cos \omega_0 t \tag{6.3-1}$$

Thus altogether when $d(t) = 0$ the phase of the carrier does *not* change at the beginning of the bit interval, while when $d(t) = 1$ there is a phase change of magnitude π.

A method of recovering the data bit stream from the DPSK signal is shown in Fig. 6.3-3. Here the received signal and the received signal delayed by the bit time T_b are applied to a multiplier. The multiplier output is

$$b(t)b(t - T_b)(2P_s) \cos (\omega_0 t + \theta) \cos [\omega_0(t - T_b) + \theta]$$

$$= b(t)b(t - T_b)P_s\left\{\cos \omega_0 T_b + \cos \left[2\omega_0\left(t - \frac{T_b}{2}\right) + 2\theta\right]\right\} \tag{6.3-2}$$

and is applied to a bit synchronizer and integrator as shown in Fig. 6.2-1 for the BPSK demodulator. The first term on the right-hand side of Eq. (6.3-2) is, aside from a multiplicative constant, the waveform $b(t)b(t - T_b)$ which, as we shall see is precisely the signal we require. As noted previously in connection with BPSK, and so here, the output integrator will suppress the double frequency term. We should select $\omega_0 T_b$ so that $\omega_0 T_b = 2n\pi$ with n an integer. For, in this case we shall have $\cos \omega_0 T_b = +1$ and the signal output will be as large as possible.

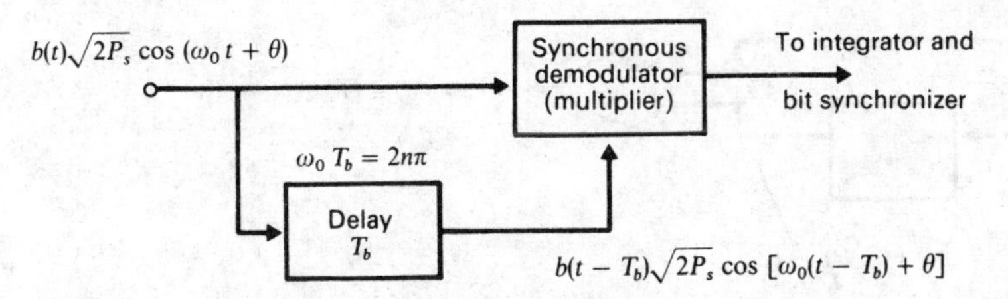

Figure 6.3-3 Method of recovering data from the DPSK signal.

Further, with this selection, the bit duration encompasses an integral number of clock cycles and the integral of the double-frequency term is exactly zero.

The transmitted data bit $d(t)$ can readily be determined from the product $b(t)b(t - T_b)$. If $d(t) = 0$ then there was no phase change and $b(t) = b(t - T_b)$ both being $+1V$ or both being $-1V$. In this case $b(t)b(t - T_b) = 1$. If however, $d(t) = 1$ then there was a phase change and either $b(t) = 1V$ with $b(t - T_b) = -1V$ or vice versa. In either case $b(t)b(t - T_b) = -1$.

The differentially coherent system, DPSK, which we have been describing has a clear advantage over the coherent BPSK system in that the former avoids the need for complicated circuitry used to generate a local carrier at the receiver. To see the relative disadvantage of DPSK in comparison with PSK, consider that during some bit interval the received signal is so contaminated by noise that in a PSK system an error would be made in the determination of whether the transmitted bit was a 1 or a 0. In DPSK a bit determination is made on the basis of the signal received in two successive bit intervals. Hence noise in one bit interval may cause errors to two bit determinations. The error rate in DPSK is therefore greater than in PSK, and, as a matter of fact, there is a tendency for bit errors to occur in pairs. It is not inevitable however that errors occur in pairs. Single errors are still possible. For consider a case in which the received signals in kth and $(k + 1)$st bit intervals are both somewhat noisy but that the signals in the $(k - 1)$st and $(k + 2)$nd intervals are noise free. Assume further that the kth interval signal is *not* so noisy that an error results from the comparison with the $(k - 1)$st interval signal and assume a similar situation prevails in connection with the $(k + 1)$st and the $(k + 2)$nd interval signals. Then it may be that only a single error will be generated, that error being the result of the comparison of the kth and $(k + 1)$st interval signals both of which are noisy.

6.4 DIFFERENTIALLY-ENCODED PSK (DEPSK)

As is noted in Fig. 6.3-3 the DPSK demodulator requires a device which operates at the carrier frequency and provides a delay of T_b. Differentially-encoded PSK eliminates the need for such a piece of hardware. In this system, synchronous demodulation recovers the signal $b(t)$, and the decoding of $b(t)$ to generate $d(t)$ is done at baseband.

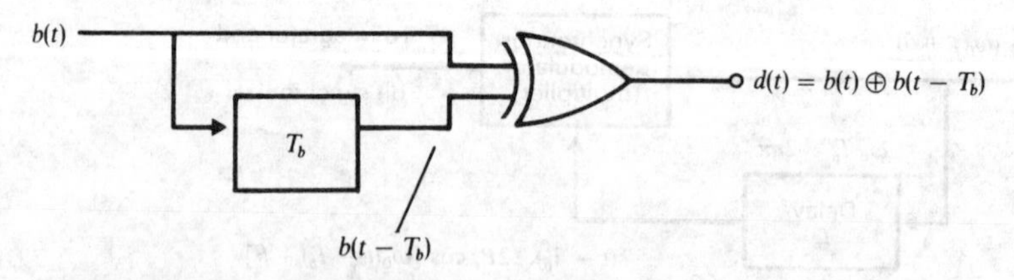

Figure 6.4-1 Baseband decoder to obtain $d(t)$ from $b(t)$.

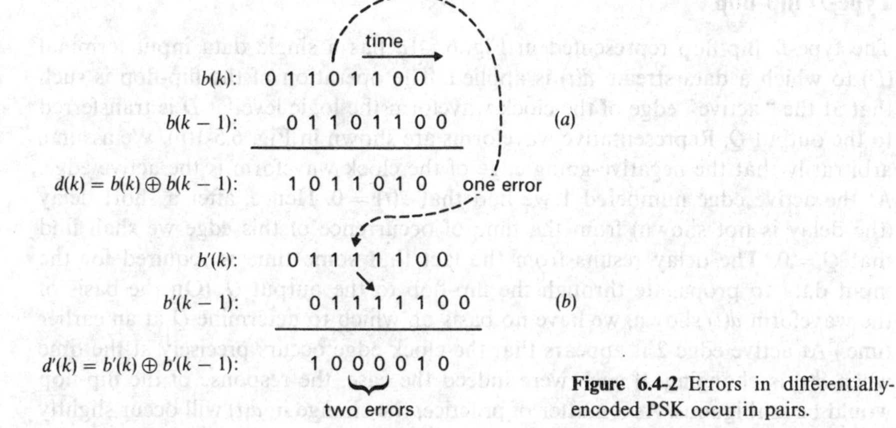

Figure 6.4-2 Errors in differentially-encoded PSK occur in pairs.

The transmitter of the DEPSK system is identical to the transmitter of the DPSK system shown in Fig. 6.3-1. The signal $b(t)$ is recovered in exactly the manner shown in Fig. 6.2-1 for a BPSK system. The recovered signal is then applied directly to one input of an exclusive-OR logic gate and to the other input is applied $b(t - T_b)$ (see Fig. 6.4-1). The gate output will be at one or the other of its levels depending on whether $b(t) = b(t - T_b)$ or $b(t) = \overline{b(t - T_b)}$. In the first case $b(t)$ did not change level and therefore the transmitted bit is $d(t) = 0$. In the second case $d(t) = 1$.

We have seen that in DPSK there is a tendency for bit errors to occur in pairs but that single bit errors are possible. In DEPSK errors *always* occur in pairs. The reason for the difference is that in DPSK we do not make a hard decision, in each bit interval about the phase of the received signal. We simply allow the received signal in one interval to compare itself with the signal in an adjoining interval and, as we have seen, a single error is not precluded. In DEPSK, a firm definite hard decision is made in each interval about the value of $b(t)$. If we make a mistake, then errors must result from a comparison with the preceding and succeeding bit. This result is illustrated in Fig. 6.4-2. In Fig. 6.4-2*a* is shown the error-free signals $b(k)$, $b(k - 1)$ and $d(k) = b(k) \oplus b(k - 1)$. In Fig. 6.4-2*b* we have assumed that $b'(k)$ has a single error. Then $b'(k - 1)$ must also have a single error. We note that the reconstructed waveform $d'(k)$ now has two errors.

6.5 QUADRATURE PHASE-SHIFT KEYING (QPSK)

We have seen that when a data stream whose bit duration is T_b is to be transmitted by BPSK the channel bandwidth must be nominally $2f_b$ where $f_b = 1/T_b$. Quadrature phase-shift keying, as we shall explain, allows bits to be transmitted using half the bandwidth. In describing the QPSK system we shall have occasion to use the type-*D* flip-flop as a one bit storage device. We therefore digress, very briefly, to remind the reader of the essential characteristics of this flip-flop.

Type-D flip-flop

The type-D flip-flop represented in Fig. 6.5-1a has a single data input terminal (D) to which a data stream $d(t)$ is applied. The operation of the flip-flop is such that at the "active" edge of the clock waveform the logic level at D is transferred to the output Q. Representative waveforms are shown in Fig. 6.5-1(b). We assume arbitrarily that the negative-going edge of the clock waveform is the active edge. At the active edge numbered 1 we find that $d(t) = 0$. Hence, after a short delay (the delay is not shown) from the time of occurrence of this edge we shall find that $Q = 0$. The delay results from the fact that some time is required for the input data to propagate through the flip-flop to the output Q. (On the basis of the waveform $d(t)$ shown, we have no basis on which to determine Q at an earlier time.) At active edge 2 it appears that the clock edge occurs precisely at the time when $d(t)$ is changing. If such were indeed the case, the response of the flip-flop would be ambiguous. As a matter of practice, the change in $d(t)$ will occur slightly after the active edge. Such is the case because, rather inevitably, the change in $d(t)$ is itself the response of some digital component (gate, flip-flop, etc.) to the very same clock waveform which is driving our type-D flip-flop. (The delays referred to are normally not indicated on a waveform diagram as in Fig. 6.5-1 because these delays are ordinarily very small in comparison with the bit time T_b.) In any event, the fact is that at active edge 2 we have $d(t) = 0$ and Q remains at $Q = 0$. The remainder of the waveforms are easily verified on the same basis. Observe that the Q waveform is the $d(t)$ waveform delayed by one bit interval T_b. The relevant point about the flip-flop in the matter of our present concern is the follow-

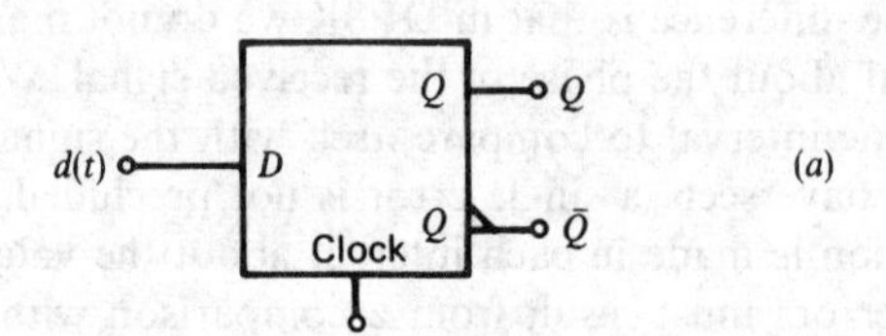

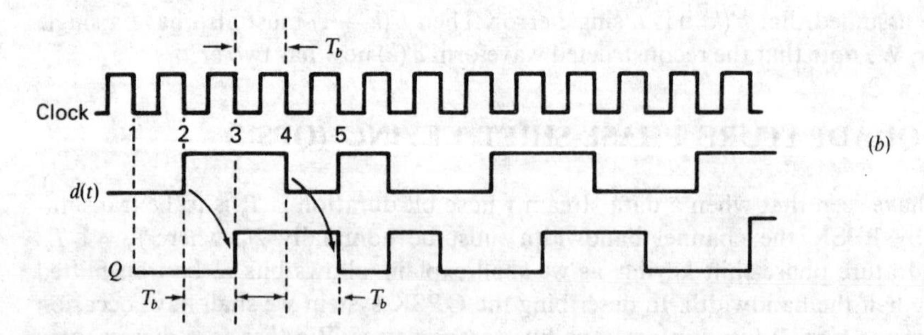

Figure 6.5-1 (a) Type-D flip-flop symbol. (b) Waveforms showing flip-flop characteristics.

ing: Once the flip-flop, in response to an active clock edge, has registered a data bit, it will hold that bit until updated by the occurrence of the next succeeding active edge.

QPSK Transmitter

The mechanism by which a bit stream $b(t)$ generates a QPSK signal for transmission is shown in Fig. 6.5-2 and relevant waveforms are shown in Fig. 6.5-3. In these waveforms we have arbitrarily assumed that in every case the active edge of the clock waveforms is the downward edge. The toggle flip-flop is driven by a clock waveform whose period is the bit time T_b. The toggle flip-flop generates an odd clock waveform and an even waveform. These clocks have periods $2T_b$. The active edge of one of the clocks and the active edge of the other are separated by the bit time T_b. The bit stream $b(t)$ is applied as the data input to both type-D flip-flops, one driven by the odd and one driven by the even clock waveform. The flip-flops register alternate bits in the stream $b(t)$ and hold each such registered bit for two bit intervals, that is for a time $2T_b$. In Fig. 6.5-3 we have numbered the bits in $b(t)$. Note that the bit stream $b_o(t)$ (which is the output of the flip-flop driven by the odd clock) registers bit 1 and holds that bit for time $2T_b$, then registers bit 3 for time $2T_b$, then bit 5 for $2T_b$, etc. The even bit stream $b_e(t)$ holds, for times $2T_b$ each, the alternate bits numbered 2, 4, 6, etc.

The bit stream $b_e(t)$ (which, as usual, we take to be $b_e(t) = \pm 1$ volt) is superimposed on a carrier $\sqrt{P_s} \cos \omega_0 t$ and the bit stream $b_o(t)$ (also ± 1 volt) is

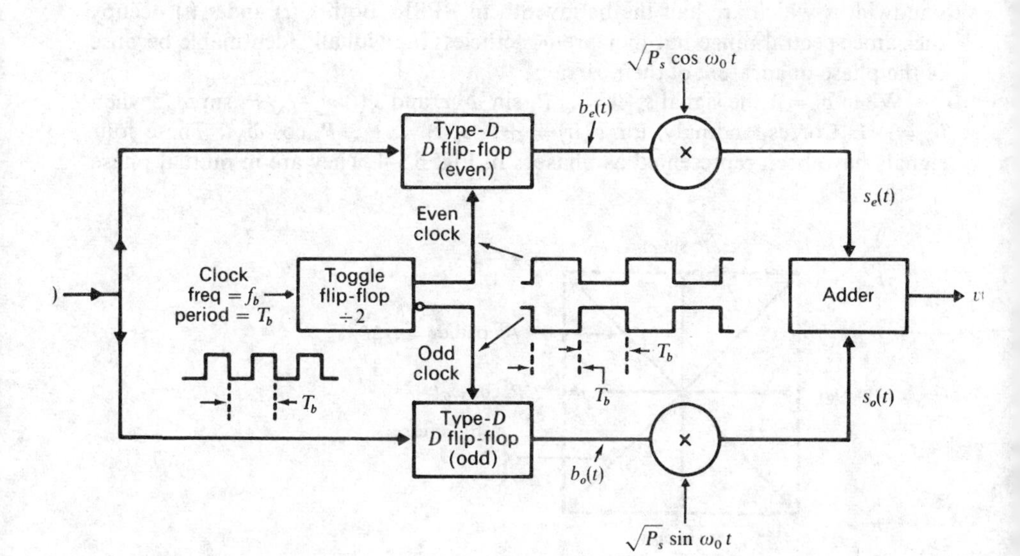

Figure 6.5-2 An offset QPSK transmitter.

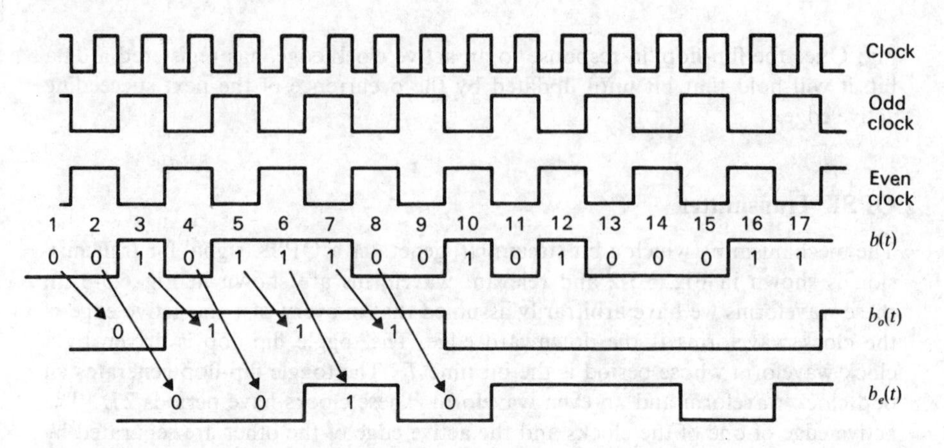

Figure 6.5-3 Waveforms for the QPSK transmitter of Fig. 6.5-2.

superimposed on a carrier $\sqrt{P_s} \sin \omega_0 t$ by the use of two multipliers (i.e., balanced modulators) as shown, to generate two signals $s_e(t)$ and $s_o(t)$. These signals are then added to generate the transmitted output signal $v_m(t)$ which is

$$v_m(t) = \sqrt{P_s}\, b_o(t) \sin \omega_0 t + \sqrt{P_s}\, b_e(t) \cos \omega_0 t \tag{6.5-1}$$

As may be verified, the total normalized power of $v_m(t)$ is P_s.

As we have noted in BPSK, a bit stream with bit time T_b multiplies a carrier, the generated signal has a nominal bandwidth $2 \times 1/T_b$. In the waveforms $b_o(t)$ and $b_e(t)$ the bit times are each $1/2T_b$, hence both $s_e(t)$ and $s_o(t)$ have nominal bandwidths which are half the bandwidth in BPSK. Both $s_e(t)$ and $s_o(t)$ occupy the same spectral range but they are nonetheless individually identifiable because of the phase quadrature of their carriers.

When $b_o = 1$ the signal $s_o(t) = \sqrt{P_s} \sin \omega_0 t$, and $s_o(t) = -\sqrt{P_s} \sin \omega_0 t$ when $b_o = -1$. Correspondingly, for $b_e(t) = \pm 1$, $s_e(t) = \pm\sqrt{P_s} \cos \omega_0 t$. These four signals have been represented as phasors in Fig. 6.5-4. They are in mutual phase

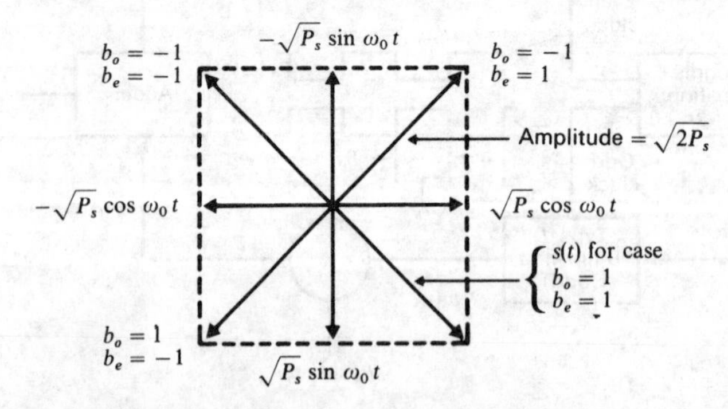

Figure 6.5-4 Phasor diagram for sinusoids in Fig. 6.5-2.

quadrature. Also drawn are the phasors representing the four possible output signals $v_m(t) = s_o(t) + s_e(t)$. These four possible output signals have equal amplitude $\sqrt{2P_s}$ and are in phase quadrature; they have been identified by their corresponding values of b_o and b_e. At the end of each bit interval (i.e., after each time T_b) either b_o or b_e can change, but both cannot change at the same time. Consequently, the QPSK system shown in Fig. 6.5-2 is called *offset* or *staggered* QPSK and abbreviated OQPSK. After each time T_b, the transmitted signal, if it changes, changes phase by 90° rather than by 180° as in BPSK.

Non-offset QPSK

Suppose that in Fig. 6.5-2 we introduce an additional flip-flop before either the odd or even flip-flop. Let this added flip-flop be driven by the clock which runs at the rate f_b. Then one or the other bit streams, odd or even, will be delayed by one bit interval. As a result, we shall find that two bits which occur in time sequence (i.e., serially) in the input bit stream $b(t)$ will appear at the same time (i.e., in parallel) at the outputs of the odd and even flip-flops. In this case $b_e(t)$ and $b_o(t)$ can change at the same time, after each time $2T_b$, and there can be a phase change of 180° in the output signal. There is no difference, in principle, between a staggered and non-staggered system.

In practice, there is often a significant difference between QPSK and OQPSK. At each transition time, T_b for OQPSK and $2T_b$ for QPSK, one bit for OQPSK and perhaps two bits for QPSK change from 1V to -1V or -1V to 1V. Now the bits $b_e(t)$ and $b_o(t)$ can, of course, not change instantaneously and, in changing, must pass through zero and dwell in that neighborhood at least briefly. Hence there will be brief variations in the *amplitude* of the transmitted waveform. These variations will be more pronounced in QPSK than in OQPSK since in the first case both $b_e(t)$ and $b_o(t)$ may be zero simultaneously so that the signal amplitude may actually be reduced to zero temporarily. There is a second mechanism through which amplitude variations are caused at the transmitter. In QPSK as in BPSK a filter is used to suppress sidebands. It turns out that when waveforms which exhibit abrupt phase changes, are filtered, the effect of the filter, at the time of the abrupt phase changes, is to cause substantial changes again in the *amplitude* of the waveform. Here too, we expect larger changes in QPSK where phase changes of 180° are possible than in OQPSK where the maximum phase change is 90°.

The amplitude variations can cause difficulty in QPSK communication systems which employ repeaters, i.e., stations which receive and rebroadcast signals, such as earth satellites. For such stations generally employ output power stages which operate nonlinearly, the nonlinearity being deliberately introduced because such nonlinear stages can operate with improved efficiency. However, precisely because of their nonlinearity, when presented with amplitude variations, they generate spectral components outside the range of the main lobe, thereby undoing the effect of the band limiting filter and causing interchannel interference. Further filtering to suppress the effect of amplitude variation has an effect on the phase of the signal and it is, of course, precisely the phase which conveys the signal message.

Symbol Versus Bit Transmission

In BPSK we deal individually with each bit of duration T_b. In QPSK we lump two bits together to form what is termed a *symbol.* The symbol can have any one of four possible values corresponding to the two-bit sequences 00, 01, 10, and 11. We therefore arrange to make available for transmission four distinct signals. At the receiver each signal represents *one symbol* and, correspondingly, *two bits.* When bits are transmitted, as in BPSK, the signal changes occur at the bit rate. When symbols are transmitted the changes occur at the symbol rate which is one-half the bit rate. Thus the symbol time is $T_s = 2T_b$.

The QPSK Receiver

A receiver for the QPSK signal is shown in Fig. 6.5-5. Synchronous detection is required and hence it is necessary to locally regenerate the carriers $\cos \omega_0 t$ and $\sin \omega_0 t$. The scheme for carrier regeneration is similar to that employed in BPSK. In that earlier case we squared the incoming signal, extracted a waveform

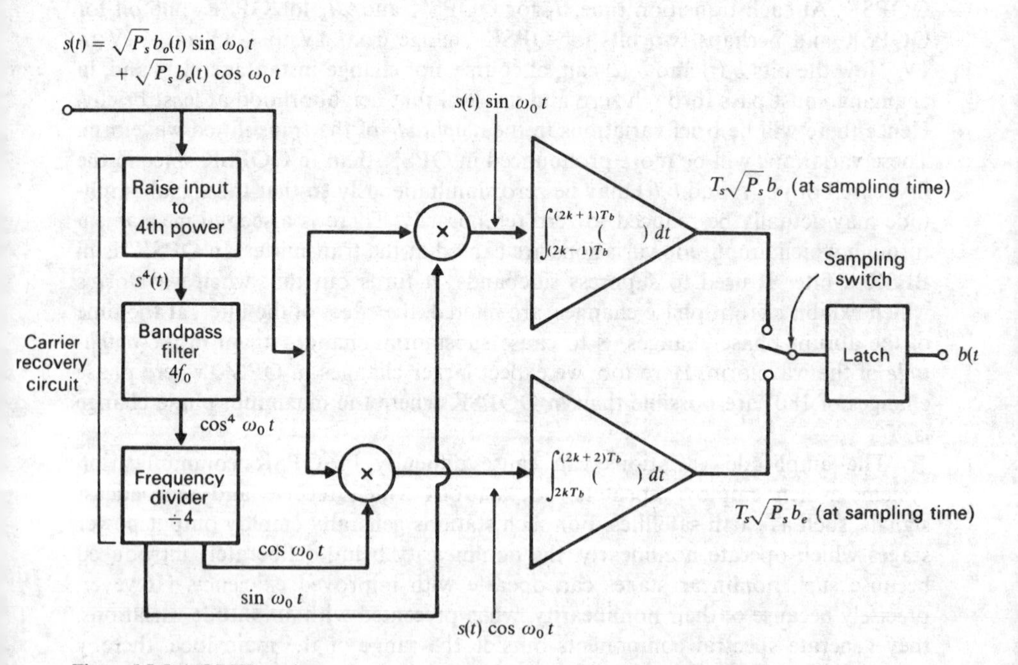

Figure 6.5-5 A QPSK receiver.

at twice the carrier frequency by filtering, and recovered the carrier by frequency dividing by two. In the present case, it is required that the incoming signal be raised to the fourth power after which filtering recovers a waveform at four times the carrier frequency and finally frequency division by four regenerates the carrier. In the present case, also, we require both sin $\omega_0 t$ and cos $\omega_0 t$. It is left as a problem (see Prob. 6.5-3) to verify that the scheme indicated in Fig. 6.5-5 does indeed yield the required waveforms sin $\omega_0 t$ and cos $\omega_0 t$.

The incoming signal is also applied to two synchronous demodulators consisting, as usual, of a multiplier (balanced modulator) followed by an integrator. The integrator integrates over a two-bit interval of duration $T_s = 2T_b$ and then dumps its accumulation. As noted previously, ideally the interval $2T_b = T_s$ should encompass an integral number of carrier cycles. One demodulator uses the carrier cos $\omega_0 t$ and the other the carrier sin $\omega_0 t$. We recall that when sinusoids in phase quadrature are multiplied, and the product is integrated over an integral number of cycles, the result is zero. Hence the demodulators will selectively respond to the parts of the incoming signal involving respectively $b_e(t)$ or $b_o(t)$.

Of course, as usual, a bit synchronizer is required to establish the beginnings and ends of the bit intervals of each bit stream so that the times of integration can be established. The bit synchronizer is needed as well to operate the sampling switch. At the end of each integration time for each individual integrator, and just before the accumulation is dumped, the integrator output is sampled. Samples are taken alternately from one and the other integrator output at the end of each bit time T_b and these samples are held in the latch for the bit time T_b. Each individual integrator output is sampled at intervals $2T_b$. The latch output is the recovered bit stream $b(t)$.

The voltages marked on Fig. 6.5-5 are intended to represent the waveforms of the signals only and not their amplitudes. Thus the actual value of the sample voltages at the integrator outputs depends on the amplitude of the local carrier, the gain, if any, in the modulators and the gain in the integrators. We have however indicated that the sample values depend on the normalized power P_s of the received signal and on the duration T_s of the symbol.

The mechanism used in Fig. 6.5-5 to regenerate the local carriers is a source of phase ambiguity of the type described in Sec. 6.4. That is, the carrier may be 180° out of phase with the carriers at the transmitter and as a result the demodulated signals may be complementary to the transmitted signal. This situation can be corrected, as before, by using differential encoding and decoding as in Figs. 6.4-1 and 6.4-2.

Signal Space Representation

In Sec. 1.9 we investigated four quadrature signals. Equation (1.9-19), repeated here, is

$$v_m(t) = \sqrt{2P_s} \cos \left[\omega_0 t + (2i + 1) \frac{\pi}{4} \right] \qquad m = 0, 1, 2, 3 \qquad (6.5\text{-}2)$$

These signals were then represented in terms of the two orthonormal signals $u_1(t) = \sqrt{(2/T)}\cos\omega_0 t$ and $u_2(t) = \sqrt{(2/T)}\sin\omega_0 t$. The result in Eq. (1.9-22), repeated here, is

$$v_m(t) = \left[\sqrt{P_s T}\cos(2i+1)\frac{\pi}{4}\right]\sqrt{\frac{2}{T}}\cos\omega_0 t$$

$$-\left[\sqrt{P_s T}\sin(2i+1)\frac{\pi}{4}\right]\sqrt{\frac{2}{T}}\sin\omega_0 t \qquad (6.5\text{-}3)$$

The QPSK signal $v_m(t)$ in Eq. (6.5-1) can be put in the form of Eq. (6.5-3) by setting

$$b_e = \sqrt{2}\cos(2i+1)\frac{\pi}{4} \qquad (6.5\text{-}4a)$$

and

$$b_o = -\sqrt{2}\sin(2i+1)\frac{\pi}{4} \qquad (6.5\text{-}4b)$$

Thus

$$v_m(t) = \sqrt{E_b}\, b_e(t)u_1(t) - \sqrt{E_b}\, b_o(t)u_2(t) \qquad (6.5\text{-}5)$$

where $T = 2T_b = T_s$. Now Fig. 1.9-9 can be redrawn as shown in Fig. 6.5-6, to show the geometrical representation of QPSK. The points in signal space corresponding to each of the four possible transmitted signals is indicated by dots. From each such signal we can recover two bits rather than one. The distance of a signal point from the origin is $\sqrt{E_s}$ which is the square root of the signal energy associated with the symbol, that is $E_s = P_s T_s = P_s(2T_b)$. As we have noted earlier

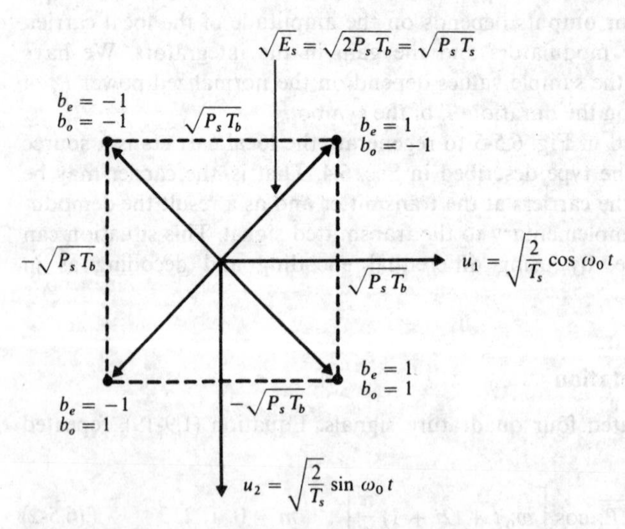

Figure 6.5-6 The four QPSK signals drawn in signal space.

and will verify in Sec. 11.14, our ability to determine a bit without error is measured by the distance in signal space between points corresponding to the different values of the bit. We note in Fig. 6.5-6 that points which differ in a single bit are separated by the distance

$$d = 2\sqrt{P_s T_b} = 2\sqrt{E_b} \tag{6.5-6}$$

where E_b is the energy contained in a bit transmitted for a time T_b. This distance for QPSK is the same as for BPSK (see Eq. (6.2-11)). Hence, altogether, we have the important result that, in spite of the reduction by a factor of two in the bandwidth required by QPSK in comparison with BPSK, the noise immunities of the two systems are the same.

6.6 *M*-ARY PSK

In BPSK we transmit each bit individually. Depending on whether $b(t)$ is logic 0 or logic 1, we transmit one or another of a sinusoid for the bit time T_b, the sinusoids differing in phase by $2\pi/2 = 180°$. In QPSK we lump together two bits. Depending on which of the four two-bit words develops, we transmit one or another of four sinusoids of duration $2T_b$, the sinusoids differing in phase by amount $2\pi/4 = 90°$. The scheme can be extended. Let us lump together N bits so that in this N-bit *symbol*, extending over the time NT_b, there are $2^N = M$ possible symbols. Now let us represent the symbols by sinusoids of duration $NT_b = T_s$ which differ from one another by the phase $2\pi/M$. Hardware to accomplish such M-ary communication is available.

Thus in M-ary PSK the waveforms used to identify the symbols are

$$v_m(t) = \sqrt{2P_s}\cos(\omega_0 t + \phi_m) \qquad (m = 0, 1, \ldots, M-1) \tag{6.6-1}$$

with the symbol phase angle given by

$$\phi_m = (2m + 1)\frac{\pi}{M} \tag{6.6-2}$$

The waveforms of Eq. (6.6-1) are represented by the dots in Fig. 6.6-1 in a signal space in which the coordinate axes are the orthonormal waveforms $u_1(t) = \sqrt{(2/T_s)}\cos\omega_0 t$ and $u_2(t) = \sqrt{(2/T_s)}\sin\omega_0 t$. The distance of each dot from the origin is $\sqrt{E_s} = \sqrt{P_s T_s}$.

From Eq. (6.6-1) we have

$$v_m(t) = (\sqrt{2P_s}\cos\phi_m)\cos\omega_0 t - (\sqrt{2P_s}\sin\phi_m)\sin\omega_0 t \tag{6.6-3}$$

Defining p_e and p_o by

$$p_e = \sqrt{2P_s}\cos\phi_m \tag{6.6-4a}$$

$$p_o = \sqrt{2P_s}\sin\phi_m \tag{6.6-4b}$$

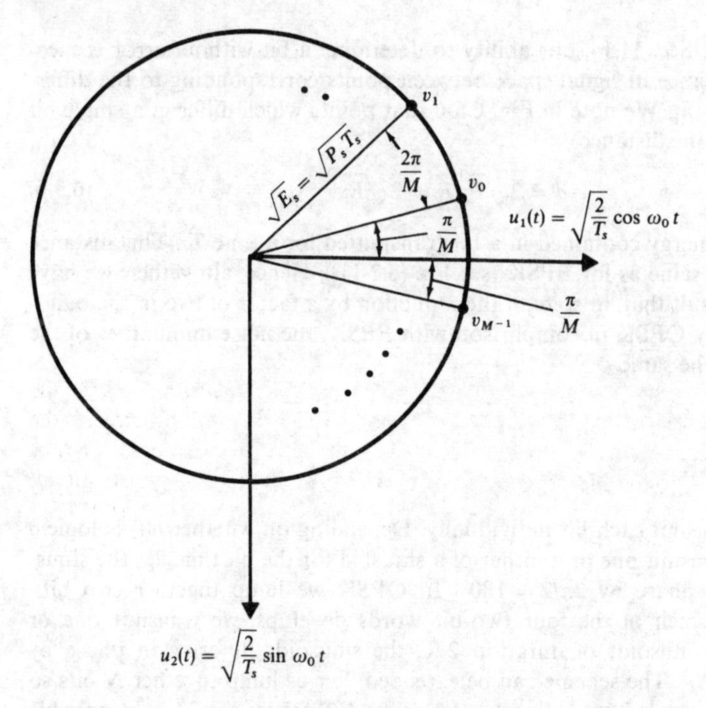

Figure 6.6-1 Geometrical representation of M-ary PSK signals.

Eq. (6.6-3) becomes

$$v_m(t) = p_e \cos \omega_0 t - p_o \sin \omega_0 t \tag{6.6-5}$$

Both p_e and p_o can change every $T_s = NT_b$ and can assume any of M possible values. The quantities p_e, p_o, and ϕ_m are random processes. The power spectral densities of p_e and p_o are given by Eq. (2.24-14) as

$$G_e(f) = \frac{\overline{|P_e(f)|^2}}{T_s} = 2P_s T_s \overline{\cos^2 \phi_m} \left(\frac{\sin \pi f T_s}{\pi f T_s}\right)^2 \tag{6.6-6}$$

and

$$G_o(f) = \frac{\overline{|P_o(f)|^2}}{T_s} = 2P_s T_s \overline{\sin^2 \phi_m} \left(\frac{\sin \pi f T_s}{\pi f T_s}\right)^2 \tag{6.6-7}$$

However, since ϕ_m is uniformly distributed

$$\overline{\cos^2 \phi_m} = \overline{\sin^2 \phi_m} = \tfrac{1}{2} \tag{6.6-8}$$

so that

$$G_e(f) = G_o(f) = P_s T_s \left(\frac{\sin \pi f T_s}{\pi f T_s}\right)^2 \tag{6.6-9}$$

As we have already noted, when signals with spectral density given by Eq. (6.6-9) are multiplied by a carrier, the resultant spectrum is centered at the carrier frequency and extends nominally over a bandwidth

$$B = \frac{2}{T_s} = 2f_s = 2\frac{f_b}{N} \tag{6.6-10}$$

We thus note that as we increase the number of bits N per symbol the bandwidth becomes progressively smaller. On the other hand as we can see from Fig. 6.6-1 the distance between symbol signal points becomes smaller. We readily calculate (using the law of cosines) that this distance is

$$d = \sqrt{4E_s \sin^2 (\pi/M)} = \sqrt{4NE_b \sin^2 (\pi/2^N)} \tag{6.6-11}$$

where E_s is the symbol energy $P_s \times (NT_b) = P_s T_s = NE_b$ and $E_b = P_s T_b$ is the energy associated with one bit. Thus as we increase N, i.e., as we increase the duration of the symbol, the bandwidth decreases, the distance d decreases and, as we shall see, the probability of error becomes higher. Such is the case for all increases in N except for the increase from $N = 1$ (BPSK) to $N = 2$ (QPSK).

M-ary Transmitter and Receiver

The physical implementation of an M-ary PSK transmission system is moderately elaborate. Such hardware is only of incidental concern to us in this text so we shall describe the M-ary transmitter-receiver somewhat superficially.

As shown in Fig. 6.6-2, at the transmitter, the bit stream $b(t)$ is applied to a *serial-to-parallel converter*. This converter has facility for storing the N bits of a symbol. The N bits have been presented serially, that is, in time sequence, one after another. These N bits, having been assembled, are then presented all at once on N output lines of the converter, that is they are presented in *parallel*. The converter output remains unchanging for the duration NT_b of a symbol during which time the converter is assembling a new group of N bits. Each symbol time the converter output is updated.

The converter output is applied to a D/A converter. This D/A converter generates an output voltage which assumes one of $2^N = M$ different values in a one-to-one correspondence to the M possible symbols applied to its input. That is,

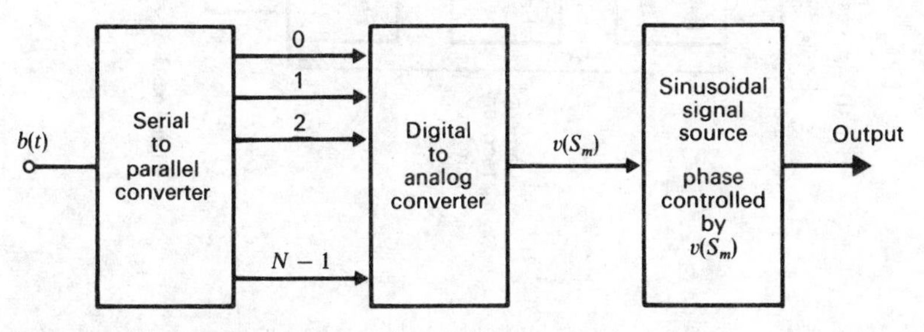

Figure 6.6-2 M-ary QPSK transmitter.

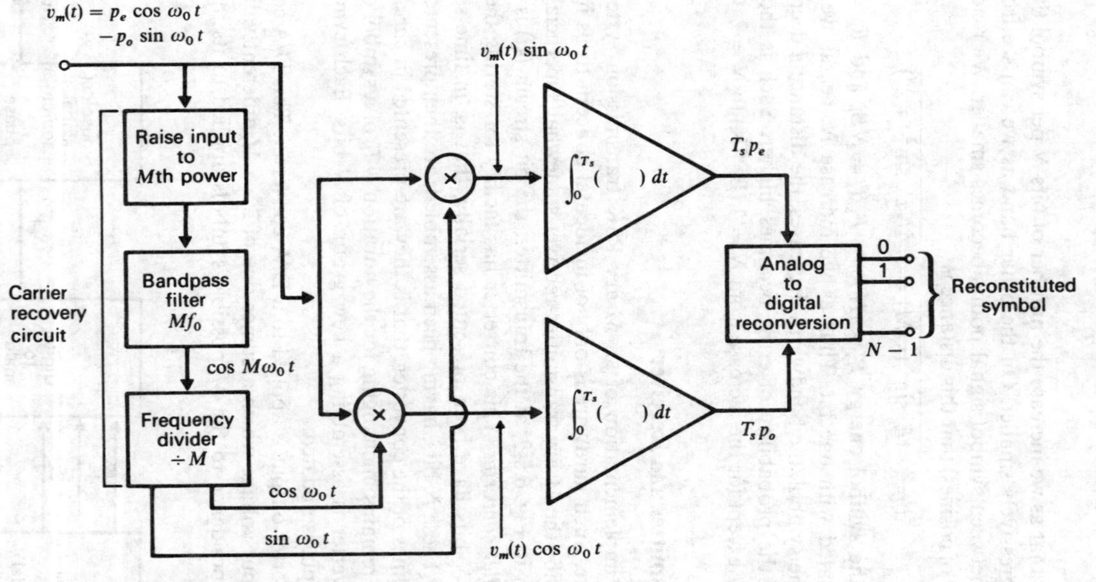

Figure 6.6-3 M-ary QPSK receiver.

the D/A output is a voltage $v(S_m)$ which depends on the symbol S_m ($m = 0, 1, \ldots, M - 1$). Finally $v(S_m)$ is applied as a control input to a special type of constant-amplitude sinusoidal signal source whose phase ϕ_m is determined by $v(S_m)$. Altogether, then, the output is a fixed amplitude, sinusoidal waveform, whose phase has a one-to-one correspondence to the assembled N-bit symbol. The phase can change once per symbol time.

The receiver, shown in Fig. 6.6-3 is similar to the nonoffset QPSK receiver. The carrier recovery system requires, in the present case a device to raise the received signal to the Mth power, filter to extract the Mf_0 component and then divide by M. As was the case in Fig. 6.5-5 the signals indicated in Fig. 6.5-5 are intended to represent the waveforms only and not the amplitude. Since there is no staggering of parts of the symbol, the integrators extend their integration over the same time interval. Of course, again, a bit synchronizer is needed. The integrator outputs are voltages whose amplitudes are proportional to $T_s p_e$ and $T_s p_o$ respectively and change at the symbol rate. These voltages measure the components of the received signal in the directions of the quadrature phasors $\sin \omega_0 t$ and $\cos \omega_0 t$. Finally the signals $T_s p_e$ and $T_s p_o$ are applied to a device which reconstructs the digital N-bit signal which constitutes the transmitted signal.

There may or may not be a need to regenerate the bit stream. As a matter of fact, the idea of transmitting information one bit at a time by a bit stream $b(t)$ arises when we have a system, like BPSK which can handle only one bit at a time. If, on the other hand, our system handles M-bit symbols, then the data may originate as M-bit words. In such a case the serial to parallel converter shown in Fig. 6.6-2 is not needed.

Current operating systems are common in which $M = 16$. In this case the bandwidth is $B = 2f_b/4 = f_b/2$ in comparison to $B = f_b$ for QPSK. PSK systems transmit information through signal phase and not through signal amplitude. Hence such systems have great merit in situations where, on account of the vagaries of the transmission medium, the received signal varies in amplitude (i.e., fading channels).

6.7 QUADRATURE AMPLITUDE SHIFT KEYING (QASK)

In BPSK, QPSK, and M-ary PSK we transmit, in any symbol interval, one signal or another which are distinguished from one another in phase but are all of the *same amplitude*. In each of these individual systems the end points of the signal vectors in signal space falls on the circumference of a circle. Now we have noted that our ability to distinguish one signal vector from another in the presence of noise will depend on the distance between the vector end points. It is hence rather apparent that we shall be able to improve the noise immunity of a system by allowing the signal vectors to differ, not only in their phase but also in amplitude. We now describe such an *amplitude and phase shift keying* system. Like QPSK it involves direct (balanced) modulation of carriers in *quadrature* (i.e.,

$\cos \omega_0 t$ and $\sin \omega_0 t$) and hence might be abbreviated QAPSK. However, the accepted abbreviation is simply QASK.

As an example of a QASK system, let us consider that we propose to transmit a symbol for every 4-bits. There are then $2^4 = 16$ different possible symbols and we shall have to be able to generate 16 distinguishable signals. One possible geometrical representation of 16 signals is shown in Fig. 6.7-1. In this configuration each signal point is equally distant from its nearest neighbors, the distance being $d = 2a$. We have placed the points symmetrically about the origin of the signal space to simplify the hardware design of the system while keeping the energy per signal near a minimum.

Let us assume (quite reasonably as a matter of practice) that all 16 signals are equally likely. Because of the symmetry displayed in Fig. 6.7-1 we can determine the average energy associated with a signal, from the four signals in the first quadrant. The average normalized energy of a signal is

$$E_s = \tfrac{1}{4}[(a^2 + a^2) + (9a^2 + a^2) + (a^2 + 9a^2) + (9a^2 + 9a^2)]$$
$$= 10a^2 \tag{6.7-1}$$

so that

$$a = \sqrt{0.1E_s} \tag{6.7-2}$$

and

$$d = 2\sqrt{0.1E_s} \tag{6.7-3}$$

In the present case since each symbol represents 4 bits, the normalized symbol energy is $E_s = 4E_b$ where E_b is the normalized bit energy. Hence

$$a = \sqrt{0.1E_s} = \sqrt{0.4E_b} \quad \text{and} \quad d = 2\sqrt{0.4E_b} \tag{6.7-4}$$

This distance is significantly less than the distance between adjacent QPSK signals where, from Eq. (6.5-6) $d = 2\sqrt{E_b}$; however the distance is greater for 16 QASK than for 16 MPSK where, from Eq. (6.6-11),

$$d = \sqrt{16E_b \sin^2 \frac{\pi}{16}} = 2\sqrt{0.15E_b} \tag{6.7-5}$$

Thus, 16 QASK will be shown to have a lower error rate than 16 MPSK, but a higher error rate than QPSK.

A typical signal in Fig. 6.7-1 is

$$v_{\text{QASK}} = k_1 a u_1(t) + k_2 a u_2(t) \tag{6.7-6}$$

in which k_1 and k_2 are each equal to ± 1 or ± 3. Since we have that $u_1(t) = \sqrt{(2/T_s)} \cos \omega_0 t$, and that $u_2(t) = \sqrt{(2/T_s)} \sin \omega_0 t$ and $a = \sqrt{0.1E_s}$ we can write Eq. (6.7-6) as

$$v_{\text{QASK}} = k_1 \sqrt{0.2\frac{E_s}{T_s}} \cos \omega_0 t + k_2 \sqrt{0.2\frac{E_s}{T_s}} \sin \omega_0 t \tag{6.7-7}$$

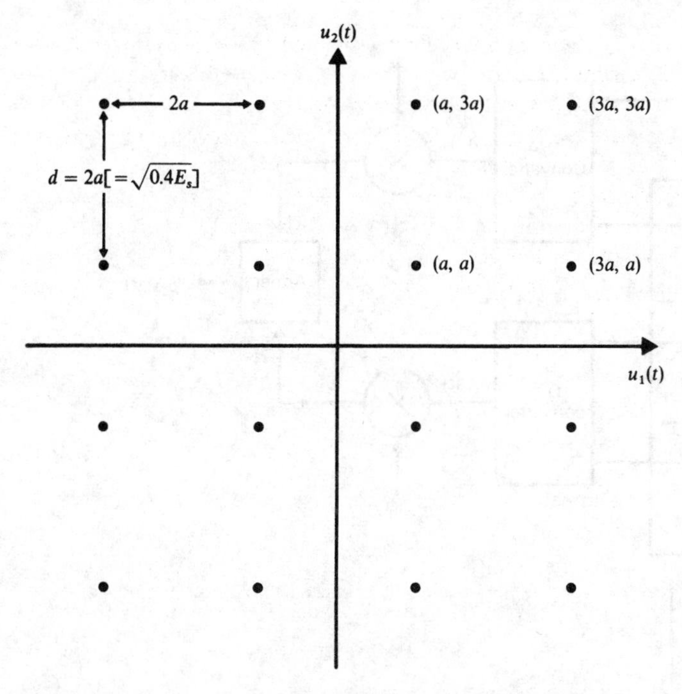

Figure 6.7-1 Geometrical representation of 16 signals in a QASK system.

and since $E_s/T_s = P_s$ we have

$$v_{\text{QASK}} = k_1\sqrt{0.2P_s}\cos\omega_0 t + k_2\sqrt{0.2P_s}\sin\omega_0 t \tag{6.7-8}$$

A generator of a QASK signal for 4-bit symbol is shown in Fig. 6.7-2. The 4-bit symbol $b_{k+3}\,b_{k+2}\,b_{k+1}b_k$ is stored in the 4-bit register made up of four flip-flops. A new symbol is presented once per interval $T_s = 4T_b$ and the content of the register is correspondingly updated at each active edge of the clock which also have a period T_s. Two bits are presented to one D/A converter and two bits to a second converter. The converter output $A_e(t)$ modulates the balanced modulator whose input carrier is the even function $\sqrt{P_s}\cos\omega_0 t$ and $A_o(t)$ modulates the modulator with odd-function carrier. The transmitted signal is then

$$v_{\text{QASK}}(t) = A_e(t)\sqrt{P_s}\cos\omega_0 t + A_o(t)\sqrt{P_s}\sin\omega_0 t \tag{6.7-9}$$

Comparing Eq. (6.7-8) with Eq. (6.7-9) we find that

$$A_e,\ A_o = \pm\sqrt{0.2} \quad \text{or} \quad \pm 3\sqrt{0.2} \tag{6.7-10}$$

Also, since all four values of A_e and A_o are equally likely, we readily verify that

$$\overline{A_e^2} = \overline{A_o^2} = 1 \tag{6.7-11}$$

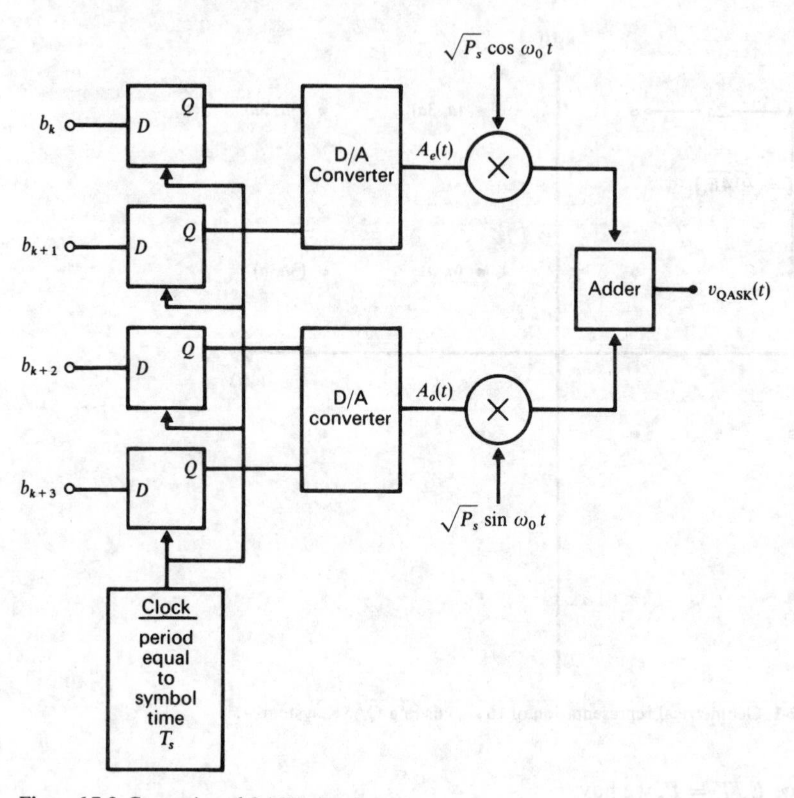

Figure 6.7-2 Generation of QASK signal.

Thus each of the quadrature terms in Eq. (6.7-8) conveys on the average, one half of the average total transmitted power.

Bandwidth of a QASK Signal

The power spectral density and bandwidth of a QASK signal can be calculated by the procedure applied in the case of M-ary PSK, since Eq. (6.7-8) is similar to Eq. (6.6-3). Thus, we have that the power spectral density is (see Prob. 6.7-5)

$$G_{\text{QASK}}(f) = \frac{P_s T_s}{2}\left[\frac{\sin \pi(f - f_0)T_s}{\pi(f - f_0)T_s}\right]^2 + \frac{P_s T_s}{2}\left[\frac{\sin \pi(f + f_0)T_s}{\pi(f + f_0)T_s}\right]^2 \tag{6.7-12}$$

where $T_s = NT_b$. The bandwidth of the QASK signal is

$$B = 2f_b/N \tag{6.7-13}$$

which is the same as in the case of M-ary PSK. For the present case of QASK with $N = 4$ corresponding to 16 possible distinguishable signals we have $B_{\text{QASK}(16)} = f_b/2$ which is one-fourth of the bandwidth required for binary PSK.

The QASK Receiver

The QASK receiver shown in Fig. 6.7-3 is similar to the QPSK receiver of Fig. 6.5-5. As in QPSK, a local set of quadrature carriers for synchronous demodulation is generated by raising the received signal to the fourth power, extracting the component at frequency $4f_0$ and then dividing the frequency by 4. In the present case, since the coefficients A_e and A_o are not of fixed value, it behooves us to inquire whether there is still a recoverable carrier. We have

$$v_{\text{QASK}}^4(t) = P_s^2(A_e(t) \cos \omega_0 t + A_o(t) \sin \omega_0 t)^4 \tag{6.7-14}$$

Neglecting all terms not at the frequency $4f_0$ we are left with

$$\frac{v_{\text{QASK}}^4(t)}{P_s} = \left[\frac{A_e^4(t) + A_o^4(t) - 6A_e^2(t)A_o^2(t)}{8}\right] \cos 4\omega_0 t + \left[\frac{A_e(t)A_o(t)[A_e^2(t) - A_o^2(t)]}{2}\right] \sin 4\omega_0 t \tag{6.7.15}$$

The average value of the coefficient of $\cos 4\omega_0 t$ is not zero while the average value of the coefficient of $\sin 4\omega_0 t$ is zero. Thus, a narrow bandwidth filter centered at $4f_0$ will recover a signal at frequency $4f_0$.

The quadrature carriers being available, two balanced modulators are used together with two integrators as shown to recover the signals $A_e(t)$ and $A_o(t)$. The

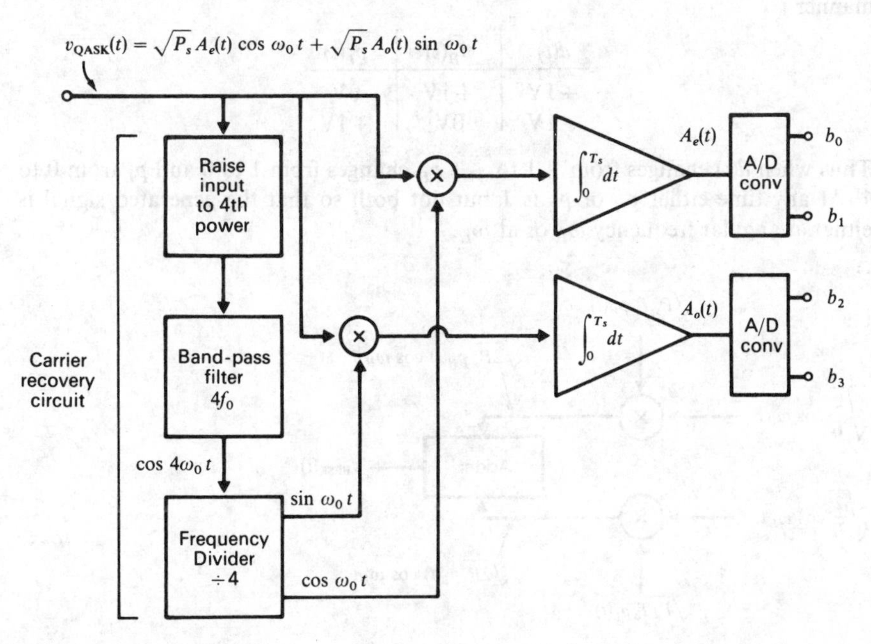

Figure 6.7-3 The QASK receiver.

integrators have an integration time equal to the symbol time T_s and, of course, as usual, symbol time synchronizers (not shown) are required. Finally, the original input bits are recovered by using A/D converters.

6.8 BINARY FREQUENCY-SHIFT KEYING

In binary frequency-shift keying (BFSK) the binary data waveform $d(t)$ generates a binary signal

$$v_{\text{BFSK}}(t) = \sqrt{2P_s} \cos [\omega_0 t + d(t)\Omega t] \tag{6.8-1}$$

Here $d(t) = +1$ or -1 corresponding to the logic levels 1 and 0 of the data waveform. The transmitted signal is of amplitude $\sqrt{2P_s}$ and is either

$$v_{\text{BFSK}}(t) = s_H(t) = \sqrt{2P_s} \cos (\omega_0 + \Omega)t \tag{6.8-2}$$

or

$$v_{\text{BFSK}}(t) = s_L(t) = \sqrt{2P_s} \cos (\omega_0 - \Omega)t \tag{6.8-3}$$

and thus has an angular frequency $\omega_0 + \Omega$ or $\omega_0 - \Omega$ with Ω a constant offset from the nominal carrier frequency ω_0. We shall call the higher frequency $\omega_H(= \omega_0 + \Omega)$ and the lower frequency $\omega_L(= \omega_0 - \Omega)$. We may conceive that the BFSK signal is generated in the manner indicated in Fig. 6.8-1. Two balanced modulators are used, one with carrier ω_H and one with carrier ω_L. The voltage values of $p_H(t)$ and of $p_L(t)$ are related to the voltage values of $d(t)$ in the following manner

$d(t)$	$p_H(t)$	$p_L(t)$
+1V	+1V	0V
−1V	0V	+1V

Thus when $d(t)$ changes from $+1$ to -1 p_H changes from 1 to 0 and p_L from 0 to 1. At any time either p_H or p_L is 1 but not both so that the generated signal is either at angular frequency ω_H or at ω_L.

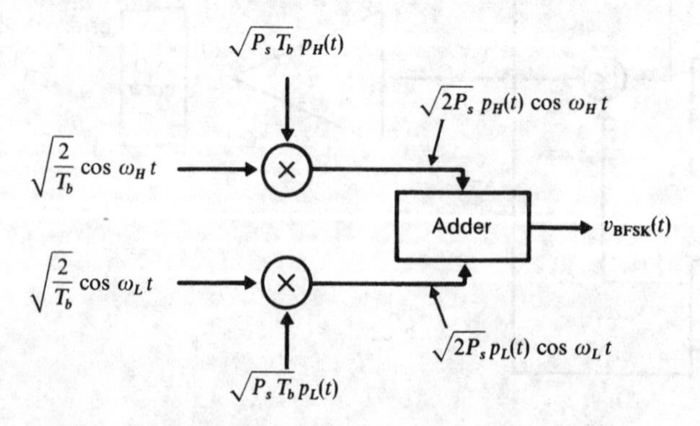

Figure 6.8-1 A representation of a manner in which a BFSK signal can be generated.

Spectrum of BFSK

In terms of the variables p_H and p_L the BFSK signal is

$$v_{\text{BFSK}}(t) = \sqrt{2P_s}\, p_H \cos(\omega_H t + \theta_H) + \sqrt{2P_s}\, p_L \cos(\omega_L t + \theta_L) \qquad (6.8\text{-}4)$$

where we have assumed that each of the two signals are of independent and random, uniformly distributed phase. Each of the terms in Eq. (6.8-4) looks like the signal $\sqrt{2P_s}\, b(t) \cos \omega_0 t$ which we encountered in BPSK [see Eq. (6.2-3)] and for which we have already deduced the spectrum, but there is an important difference. In the BPSK case, $b(t)$ is bipolar, i.e., it alternates between $+1$ and -1 while in the present case p_H and p_L are unipolar, alternating between $+1$ and 0. We may, however, rewrite p_H and p_L as the sums of a constant and a bipolar variable, that is

$$p_H(t) = \tfrac{1}{2} + \tfrac{1}{2}\, p'_H(t) \qquad (6.8\text{-}5a)$$

$$p_L(t) = \tfrac{1}{2} + \tfrac{1}{2}\, p'_L(t) \qquad (6.8\text{-}5b)$$

In Eq. (6.8-5) p'_H and p'_L are bipolar, alternating between $+1$ and -1 and are complementary. When p'_H is $+1$, $p'_L = -1$ and vice versa. We have then

$$v_{\text{BFSK}}(t) = \sqrt{\frac{P_s}{2}} \cos(\omega_H t + \theta_H) + \sqrt{\frac{P_s}{2}} \cos(\omega_L t + \theta_L)$$
$$+ \sqrt{\frac{P_s}{2}}\, p'_H \cos(\omega_H t + \theta_H) + \sqrt{\frac{P_s}{2}}\, p'_L \cos(\omega_L t + \theta_L) \qquad (6.8\text{-}6)$$

The first two terms in Eq. (6.8-6) produce a power spectral density which consists of two impulses, one at f_H and one at f_L. The last two terms produce the spectrum of two binary PSK signals (see Fig. 6.2-2a) one centered about f_H and one about f_L. The individual power spectral density patterns of the last two terms in Eq. (6.8-6) are shown in Fig. 6.8-2 for the case $f_H - f_L = 2f_b$. For this separation between f_H and f_L we observe that the overlapping between the two parts of the spectra is not large and we may expect to be able, without excessive difficulty, to distinguish the levels of the binary waveform $d(t)$. In any event, with this separation the bandwidth of BFSK is

$$BW(\text{BFSK}) = 4f_b \qquad (6.8\text{-}7)$$

which is twice the bandwidth of BPSK.

Receiver for BFSK Signal

A BFSK signal is typically demodulated by a receiver system as in Fig. 6.8-3. The signal is applied to two bandpass filters one with center frequency at f_H the other at f_L. Here we have assumed, as above, that $f_H - f_L = 2(\Omega/2\pi) = 2f_b$. The filter frequency ranges selected do not overlap and each filter has a passband wide enough to encompass a main lobe in the spectrum of Fig. 6.8-2. Hence one filter

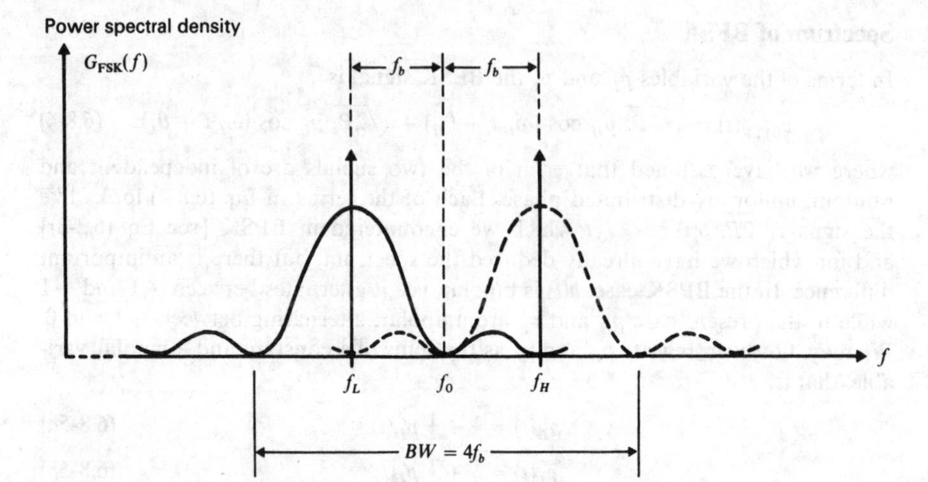

Figure 6.8-2 The power spectral densities of the individual terms in Eq. (6.8-6).

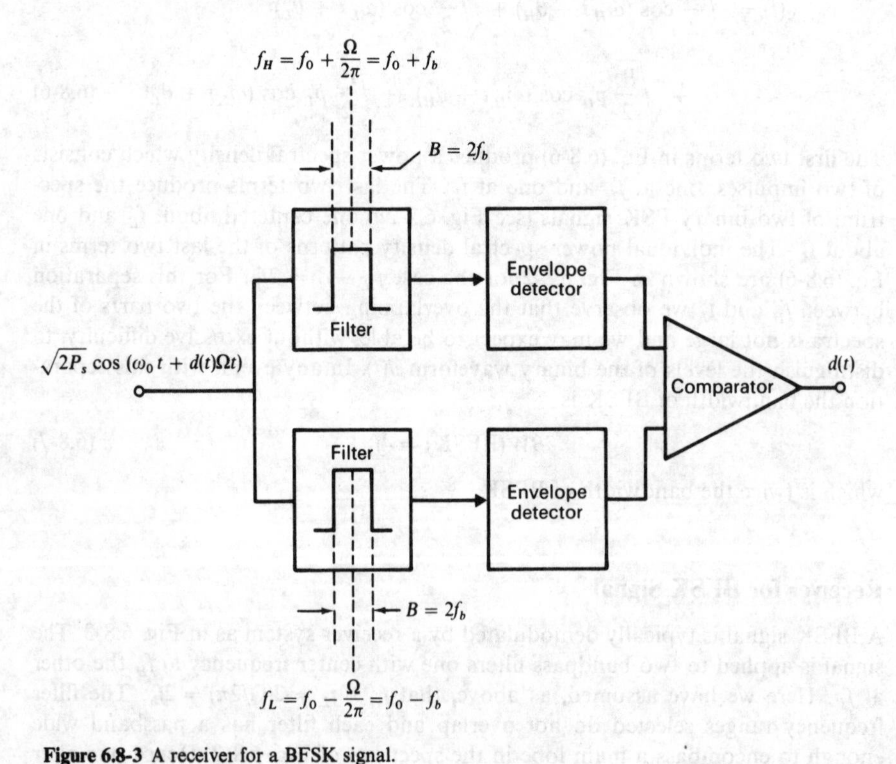

Figure 6.8-3 A receiver for a BFSK signal.

will pass nearly all the energy in the transmission at f_H the other will perform similarly for the transmission at f_L. The filter outputs are applied to envelope detectors and finally the envelope detector outputs are compared by a comparator. A comparator is a circuit that accepts two input signals. It generates a binary output which is at one level or the other depending on which input is larger. Thus at the comparator output the data $d(t)$ will be reproduced.

When noise is present, the output of the comparator may vary due to the systems response to the signal and noise. Thus, practical systems use a bit synchronizer and an integrator and sample the comparator output only once at the end of each time interval T_b.

Geometrical Representation of Orthogonal BFSK

We noted, in M-ary phase-shift keying and in quadrature-amplitude shift keying, that any signal could be represented as $C_1 u_1(t) + C_2 u_2(t)$. There $u_1(t)$ and $u_2(t)$ are the orthonormal vectors in signal space, that is, $u_1(t) = \sqrt{2/T_s} \cos \omega_0 t$ and $u_2(t) = \sqrt{2/T_s} \sin \omega_0 t$. The functions u_1 and u_2 are orthonormal over the symbol interval T_s and, if the symbol is a single bit, $T_s = T_b$. The coefficients C_1 and C_2 are constants. The normalized energies associated with $C_1 u_1(t)$ and with $C_2 u_2(t)$ are respectively C_1^2 and C_2^2 and the total signal energy is $C_1^2 + C_2^2$. In M-ary PSK and QASK the orthogonality of the vectors u_1 and u_2 results from their *phase quadrature.* In the present case of BFSK it is appropriate that the orthogonality should result from a special *selection of the frequencies* of the unit vectors. Accordingly, with m and n integers, let us establish unit vectors

$$u_1(t) = \sqrt{\frac{2}{T_b}} \cos 2\pi m f_b t \tag{6.8-8}$$

and

$$u_2(t) = \sqrt{\frac{2}{T_b}} \cos 2\pi n f_b t \tag{6.8-9}$$

in which, as usual, $f_b = 1/T_b$. The vectors u_1 and u_2 are the mth and nth harmonics of the (fundamental) frequency f_b. As we are aware, from the principles of Fourier analysis, different harmonics ($m \pm n$) are orthogonal over the interval of the fundamental period $T_b = 1/f_b$.

If now the frequencies f_H and f_L in a BFSK system are selected to be (assuming $m > n$)

$$f_H = m f_b \tag{6.8-10a}$$

$$f_L = n f_b \tag{6.8-10b}$$

then the corresponding signal vectors are

$$s_H(t) = \sqrt{E_b}\, u_1(t) \tag{6.8-11a}$$

$$s_L(t) = \sqrt{E_b}\, u_2(t) \tag{6.8-11b}$$

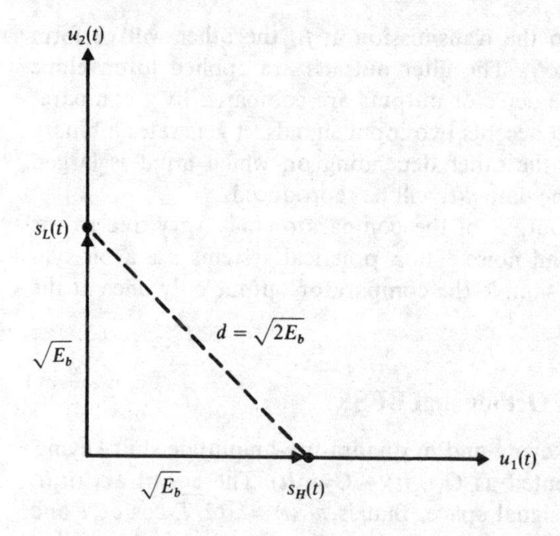

Figure 6.8-4 Signal space representation of orthogonal BFSK.

The signal space representation of these signals is shown in Fig. 6.8-4. The signals, like the unit vectors are orthogonal. The distance between signal end points is therefore

$$d = \sqrt{2E_b} \tag{6.8-12}$$

Note that this distance is considerably smaller than the distance separating end points of BPSK signals, which are *antipodal.*

Geometrical Representation of Non-Orthogonal BFSK

When the two FSK signals $s_H(t)$ and $s_L(t)$ are not orthogonal, the Gram–Schmidt procedure can still be used to represent the signals of Eqs. (6.8-2) and (6.8-3).

Let us represent the higher frequency signal $s_H(t)$ as:

$$s_H(t) = \sqrt{2P_s} \cos \omega_H t = s_{11} u_1(t) \qquad 0 \le t \le T_b \tag{6.8-13a}$$

and the lower frequency signal $s_L(t)$ as:

$$s_L(t) = \sqrt{2P_s} \cos \omega_L t = s_{12} u_1(t) + s_{22} u_2(t) \qquad 0 \le t \le T_b \tag{6.8-13b}$$

The representation of these two signals in signal space is shown in Fig. 6.8-5. Referring to this figure we see that the distance separating s_H and s_L is:

$$d^2_{\text{BFSK}} = (s_{11} - s_{12})^2 + s^2_{22} = s^2_{11} - 2s_{11}s_{12} + s^2_{12} + s^2_{22} \tag{6.8-14}$$

In order to determine d^2_{BFSK} when the two signals are not orthogonal we must evaluate s_{11}, s_{12}, and s_{22} using Eqs. (6.8-13). From Eq. (6.8-13*a*) we have:

$$s^2_{11} = 2P_s \int_0^{T_b} \cos^2 \omega_H t \, dt = E_b \left[1 + \frac{\sin 2\omega_H T_b}{2\omega_H T_b} \right] \tag{6-8-15}$$

Using Eq. (6.8-13*b*) we first determine s_{12} by multiplying both sides of the equation by $u_1(t)$ and integrating from $0 \le t \le T_b$. The result is:

$$s_{12} = \sqrt{2P_s} \int_0^{T_b} u_1(t) \cos \omega_L t \, dt$$

$$= \frac{E_b}{s_{11}} \left[\frac{\sin (\omega_H - \omega_L)T_b}{(\omega_H - \omega_L)T_b} + \frac{\sin (\omega_H + \omega_L)T_b}{(\omega_H + \omega_L)T_b} \right] \tag{6.8-16a}$$

To arrive at this result we have used Eq. (6.8-13*a*) where

$$u_1(t) = (\sqrt{2P_s}/s_{11}) \cos \omega_H t \tag{6.8-16b}$$

Finally, s_{22} is found from Eq. (6.8-13*b*) by squaring both sides of the equation and then integrating from 0 to T_b. Since u_1 and u_2 are orthogonal, the result is:

$$\int_0^{T_b} s_L^2(t) \, dt = 2P_s \int_0^{T_b} \cos^2 \omega_L t \, dt = s_{12}^2 + s_{12}^2 \tag{6.8-17a}$$

Hence

$$s_{12}^2 + s_{22}^2 = E_b \left[1 + \frac{\sin 2\omega_L T_b}{2\omega_L T_b} \right] \tag{6.8-17b}$$

The distance d between s_H and s_L given in Eq. (6.8-14) can now be determined by substituting Eqs. (6.8-15), (6.8-16*a*), and (6.8-17*b*) into Eq. (6.8-14). The result is:

$$d^2 = E_b \left[1 + \frac{\sin 2\omega_H T_b}{2\omega_H T_b} \right] - 2E_b \left[\frac{\sin (\omega_H - \omega_L)T_b}{(\omega_H - \omega_L)T_b} + \frac{\sin (\omega_H + \omega_L)T_b}{(\omega_H + \omega_L)T_b} \right] + E_b \left[1 + \frac{\sin 2\omega_L T_b}{2\omega_L T_b} \right] \tag{6.8-18}$$

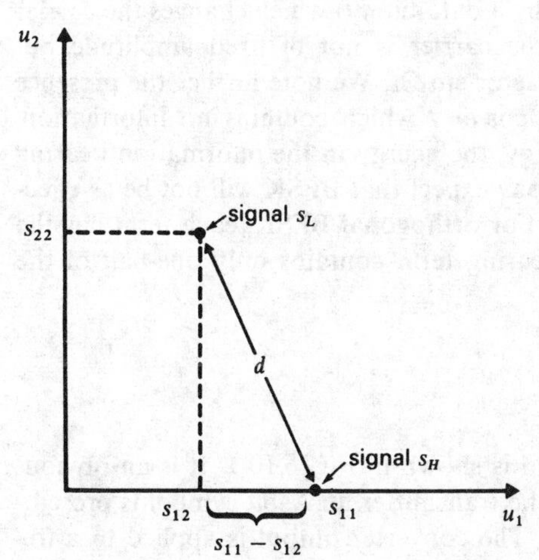

Figure 6.8-5 Signal space representation of BFSK when $s_H(t)$ and $s_L(t)$ are not orthogonal.

Equation (6.8-18) can be simplified by recognizing that:

$$\left|\frac{\sin 2\omega_H T_b}{2\omega_H T_b}\right| \ll 1$$

$$\left|\frac{\sin 2\omega_L T_b}{2\omega_L T_b}\right| \ll 1$$

and

$$\left|\frac{\sin (\omega_H + \omega_L)T_b}{(\omega_H + \omega_L)T_b}\right| \ll \left|\frac{\sin (\omega_H - \omega_L)T_b}{(\omega_H - \omega_L)T_b}\right|$$

The final result is then

$$d^2 \cong 2E_b\left[1 - \frac{\sin (\omega_H - \omega_L)T_b}{(\omega_H - \omega_L)T_b}\right] \tag{6.8-19}$$

Note that when $s_H(t)$ and $s_L(t)$ are orthogonal $(\omega_H - \omega_L)T_b = 2\pi(m - n)f_b T_b = 2\pi(m - n)$ and Eq. (6.8-19) gives $d = \sqrt{2E_b}$ as expected. Note also that if $(\omega_H - \omega_L)T_b = 3\pi/2$, the distance d is increased and becomes

$$d_{\text{opt}} = \left[2E_b\left(1 + \frac{2}{3\pi}\right)\right]^{1/2} \simeq \sqrt{2.4E_b} \tag{6.8-20}$$

an increase in d^2 by 20 percent.

6.9 COMPARISON OF BFSK AND BPSK

Let us start with the BFSK signal of Eq. (6.8-1). Using the trigonometric identity for the cosine of the sum of two angles and recalling that $\cos\theta = \cos(-\theta)$ while $\sin\theta = -\sin(-\theta)$ we are led to the alternate equivalent expression

$$v_{\text{BFSK}}(t) = \sqrt{2P_s}\cos\Omega t\cos\omega_0 t - \sqrt{2P_s}\, d(t)\sin\Omega t\sin\omega_0 t \tag{6.9-1}$$

Note that the second term in Eq. (6.9-1) looks like the signal encountered in BPSK i.e., a carrier $\sin\omega_0 t$ multiplied by a data bit $d(t)$ which changes the carrier phase. In the present case however, the carrier is not of fixed amplitude but rather the amplitude is shaped by the factor $\sin\Omega t$. We note further the presence of a quadrature reference term $\cos\Omega t\cos\omega_0 t$ which contains no information. Since this quadrature term carries energy, the energy in the information bearing term is thereby diminished. Hence we may expect that BFSK will not be as effective as BPSK in the presence of noise. For orthogonal BFSK, each term has the same energy, hence the information bearing term contains only one-half of the total transmitted energy.

6.10 *M*-ARY FSK

An M-ary FSK communications system is shown in Fig. 6.10-1. It is an obvious extension of a binary FSK system. At the transmitter an N-bit symbol is presented each T_s to an N-bit D/A converter. The converter output is applied to a fre-

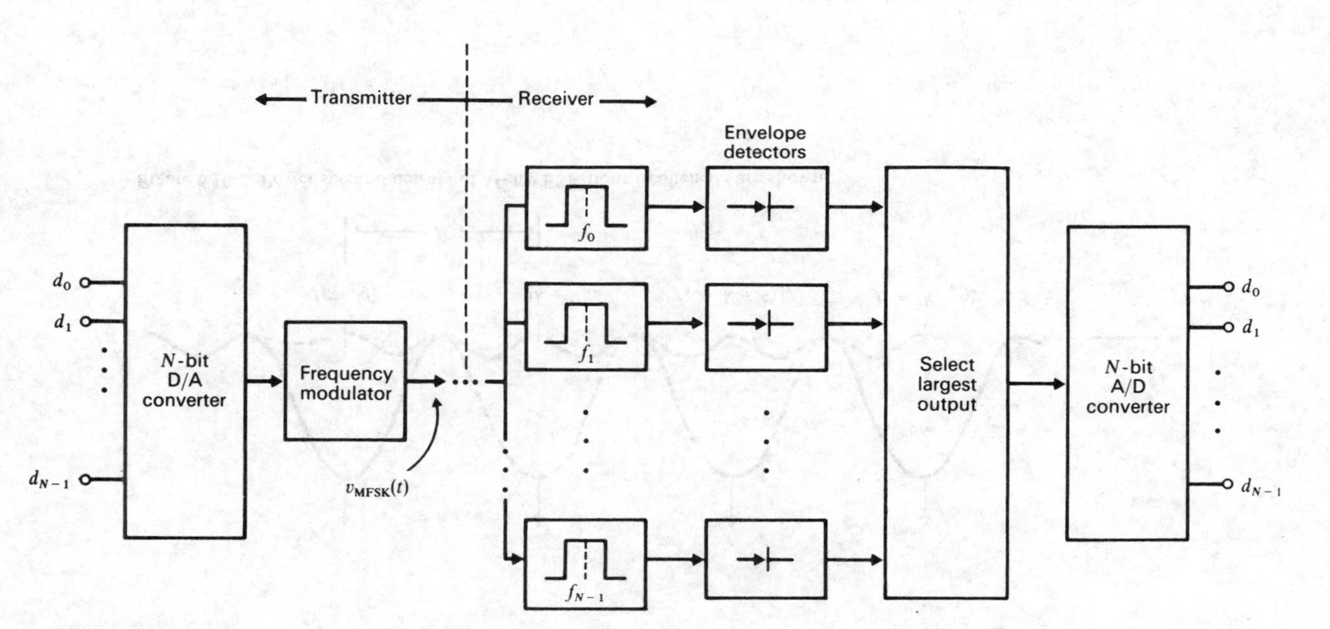

Figure 6.10-1 An M-ary communications system.

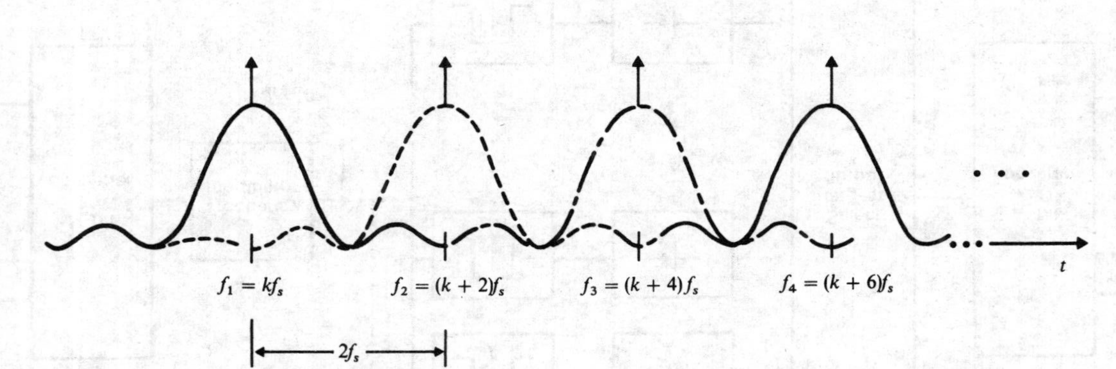

Figure 6.10-2 Power spectral density of M-ary FSK (four frequencies are shown).

quency modulator, i.e., a piece of hardware which generates a carrier waveform whose frequency is determined by the modulating waveform. The transmitted signal, for the duration of the symbol interval, is of frequency f_0 or f_1 ... or f_{M-1} with $M = 2^N$. At the receiver, the incoming signal is applied to M paralleled bandpass filters each followed by an envelope detector. The bandpass filters have center frequencies $f_0, f_1, \ldots, f_{M-1}$. The envelope detectors apply their outputs to a device which determines which of the detector indications is the largest and transmits that envelope output to an N-bit A/D converter.

As we shall see (Sec. 11.15) the probability of error is minimized by selecting frequencies $f_0, f_1, \ldots, f_{M-1}$ so that the M signals are mutually orthogonal. One commonly employed arrangement simply provides that the carrier frequency be successive even harmonics of the symbol frequency $f_s = 1/T_s$. Thus the lowest frequency, say f_0, is $f_0 = kf_s$, while $f_1 = (k+2)f_s, f_2 = (k+4)f_s$, etc. In this case, the spectral density patterns of the individual possible transmitted signals overlap in the manner shown in Fig. 6.10-2, which is an extension to M-ary FSK of the pattern of Fig. 6.8-2 which applies to binary FSK. We observe that to pass M-ary FSK the required spectral range is

$$B = 2Mf_s \tag{6.10-1}$$

Since $f_s = f_b/N$ and $M = 2^N$ we have

$$B = 2^{N+1} f_b/N \tag{6.10-2}$$

Note that M-ary FSK requires a considerably increased bandwidth in comparison with M-ary PSK. However, as we shall see, the probability of error for M-ary FSK *decreases* as M increases, while for M-ary PSK, the probability of error increases with M.

Geometrical Representation of M-ary FSK

In Fig. 6.8-4, we provided a signal space representation for the case of orthogonal binary FSK. The case of M-ary orthogonal FSK signals is clearly an extension of this figure. We simply conceive of a coordinate system with M mutually orthogonal coordinate axes. The signal vectors are then parallel to these axes. The best we can do pictorially is the three-dimensional case shown in Fig. 6.10-3. As usual, and as is indicated in the figure, the square of the length of the signal vector is the normalized signal energy. Note that, as in Fig. 6.8-4, the distance between signal points is

$$d = \sqrt{2E_s} = \sqrt{2NE_b} \tag{6.10-3}$$

Note that this value of d is greater than the values of d calculated for M-ary PSK with the exception of the cases $M = 2$ and $M = 4$. It is also greater than d in the case of 16-QASK.

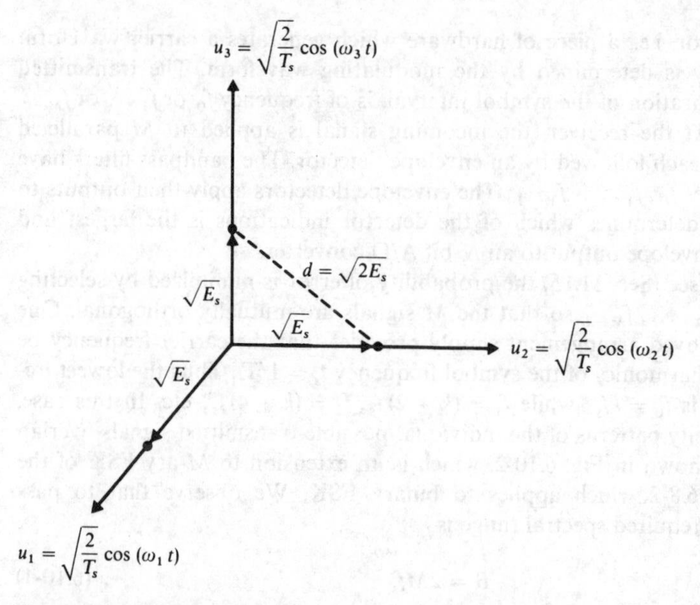

Figure 6.10-3 Geometrical representation of orthogonal M-ary FSK ($M = 3$) when the frequencies are selected to generate orthogonal signals.

6.11 MINIMUM SHIFT KEYING (MSK)

In discussing *minimum shift keying*, we shall want to make a number of comparisons between MSK and QPSK. One of these comparisons will concern the spectra of the two systems. For this reason we review briefly some matters concerning the spectrum of QPSK. In offset QPSK the transmitted signal is [see Eq. (6.5-1)]

$$v_{\text{OQPSK}}(t) = \sqrt{P_s}\, b_e(t) \cos \omega_0 t + \sqrt{P_s}\, b_o(t) \sin \omega_0 t \tag{6.11-1}$$

To find the power spectral density of this signal we start with the power spectral density of the baseband waveform $b_e(t)$. The waveform $p(t) \equiv \sqrt{P_s}\, b_e(t)$ is a random sequence of rectangular waveforms, flat topped for symbol duration $T_s = 2T_b$, and having an amplitude $\pm\sqrt{P_s}$. Its power spectral density $G(f)$ is given by the general formula (Eq. (2.24-14))

$$G_p(f) = \frac{\overline{|P(f)|^2}}{T_s} \tag{6.11-2}$$

where $P(f)$ is the Fourier transform $p(t)$. We readily calculate that the two-sided power spectral density of $p(t)$ is, with $f_s = 1/T_s = f_b/2$ and $P_s = E_b/T_b$

$$G_p(f) = 2E_b \left(\frac{\sin 2\pi f/f_b}{2\pi f/f_b} \right)^2 \tag{6.11-3}$$

The spectral density of $\sqrt{P_s}\, b_e(t) \cos \omega_0 t$ is generated by translating the two-sided pattern of Eq. (6.11-3) to $+f_0$ and also to $-f_0$, each such translated pattern being reduced in magnitude by 4 because the multiplication of $\sqrt{P_s}\, b_e(t)$ by $\cos \omega_0 t$ translates one-half of the voltage spectrum to f_0 and the other half to $-f_0$. (See Sec. 3.2.) Thus the voltage of each term is reduced by 2 and the power by 4. The power spectral density of the second term in Eq. (6.11-1), that is the term $\sqrt{P_s}\, b_o(t) \sin \omega_0 t$ is identical to the density of the first. Finally we note that the two terms are not correlated since $b_e(t)$ and $b_o(t)$ are quite independent of one another. Hence the total power density is twice the density generated by either term. Altogether then we find

$$G_{\text{OQPSK}}(f) = E_b \left\{ \left[\frac{\sin 2\pi(f - f_0)/f_b}{2\pi(f - f_0)/f_b} \right]^2 + \left[\frac{\sin 2\pi(f + f_0)f_b}{2\pi(f + f_0)/f_b} \right]^2 \right\} \tag{6.11-4}$$

Earlier, upon examining the pattern for the power spectral density (see Fig. 6.2-2) we took the attitude that the bandwidth of QPSK (QPSK and OQPSK yield the same result) is $B = f_b$ because such a bandwidth is adequate to encompass the main lobe. The main lobe contains 90 percent of the signal energy. Still, the not inconsiderable power outside the main lobe is a source of trouble when QPSK is to be used for multichannel communication on adjacent carriers. If, say, we establish additional channels at carrier frequencies $f_0' = f_0 \pm f_b$, then the side lobe associated with the first channel, having a peak value at frequency $f_0 + 3f_b/4$, will be a source of serious interchannel interference. These side lobes, as is to be noted in Fig. 6.2-2, are smaller than the main lobe by only 14 dB.

This difficulty, i.e., the wide spectrum of QPSK, is due to the character of the baseband signal. This signal consists of *abrupt changes*, and abrupt changes give rise to *spectral components at high frequencies*. In short, the baseband spectral range is very large and multiplication by a carrier translates the spectral pattern without changing its form. We might try to alleviate the difficulty by passing the baseband signal through a low-pass filter to suppress the many side lobes. Such filtering will cause intersymbol interference. The problem of interchannel interference in QPSK is so serious that regulatory and standardization agencies such as the FCC and CCIR will not permit these systems to be used except with bandpass filtering at the carrier frequency (i.e., at the transmitter output) to suppress the side lobes.

The filtering which we have just described does not, in certain situations, necessarily resolve the problem of interchannel interference. We shall discuss the matter qualitatively: We recall that QPSK (staggered or not) is a system in which the signal is of constant amplitude, the information content being borne by phase changes. In both QPSK and OQPSK there are abrupt phase changes in the signal. In QPSK these changes can occur at the symbol rate $1/T_s = 1/2T_b$ and can be as large as 180°. In OQPSK phase changes of 90° can occur at the bit rate. Now it turns out that when such waveforms with abrupt phase changes, are filtered to suppress sidebands, the effect of the filter, at the times of the abrupt phase changes, is to cause substantial changes in the *amplitude* of the waveform. Such amplitude variations can cause problems in QPSK communication systems which employ repeaters, i.e., stations which receive and rebroadcast signals such as earth satellites. For such stations often employ nonlinear power output stages in their transmitters. These nonlinear stages suppress the amplitude variations. However, precisely because of their nonlinearity, they generate spectral components outside the range of the main lobe thereby undoing the effect of the bandlimiting filtering and causing interchannel interference. In this matter QPSK with its 180° phase changes is a substantially worse offender than OQPSK with only 90° phase changes.

With these preliminary comments we turn now to MSK. There are two important differences between QPSK and MSK:

1. In MSK the baseband waveform, that multiplies the quadrature carrier, is much "smoother" than the abrupt rectangular waveform of QPSK. While the spectrum of MSK has a main center lobe which is 1.5 times as wide as the main lobe of QPSK, the side lobes in MSK are relatively much smaller in comparison to the main lobe, making filtering much easier.
2. The waveform of MSK exhibits *phase continuity*, that is, there are not abrupt phase changes as in QPSK. As a result we avoid the intersymbol interference caused by nonlinear amplifiers.

The waveforms of MSK are shown in Fig. 6.11-1. In (*a*) we start with a typical data bit stream $b(t)$. This bit stream is divided into an odd an even bit stream in (*b*) and (*c*), as in the manner of OQPSK. The odd stream $b_o(t)$ consists of the alternate bits b_1, b_3, etc., and the even stream $b_e(t)$ consists of b_2, b_4, etc. Each bit in both streams is held for two bit intervals $2T_b = T_s$, the symbol time. The staggering, which is optional in QPSK, is essential in MSK as we shall see. Observe again, as we noted earlier, the effect of the staggering is that the changes in the odd and even stream do not occur at the same time.

Also generated at the MSK transmitter are the waveforms $\sin 2\pi(t/4T_b)$ and $\cos 2\pi(t/4T_b)$ as in (*d*). These waveforms, and their phases with respect to the bit streams $b_o(t)$ and $b_e(t)$, meet the essential requirements that $\sin 2\pi(t/4T_b)$ passes through zero precisely at the end of the symbol time in $b_e(t)$ and $\cos 2\pi(t/4T_b)$ passes through zero at the end of the symbol time in $b_o(t)$. We now generate the products $b_e(t) \sin 2\pi(t/4T_b)$ and $b_o(t) \cos 2\pi(t/4T_b)$ which are shown in (*e*) and (*f*).

In MSK the transmitted signal is

$$v_{\text{MSK}}(t) = \sqrt{2P_s}\left[b_e(t) \sin 2\pi\left(\frac{t}{4T_b}\right)\right] \cos \omega_0 t + \sqrt{2P_s}\left[b_o(t) \cos 2\pi\left(\frac{t}{4T_b}\right)\right] \sin \omega_0 t \tag{6.11-5}$$

(As is readily verified, the coefficient $\sqrt{P_s}$ in Eq. (6.11-1) needs to appear as $\sqrt{2P_s}$ in Eq. (6.11-5) in order that P_s shall continue to represent the total signal power.) Thus, in comparing Eq. (6.11-1) and Eq. (6.11-5) we observe that in OQPSK the quadrature carriers $\cos \omega_0 t$ and $\sin \omega_0 t$ are multiplied by the rectangular

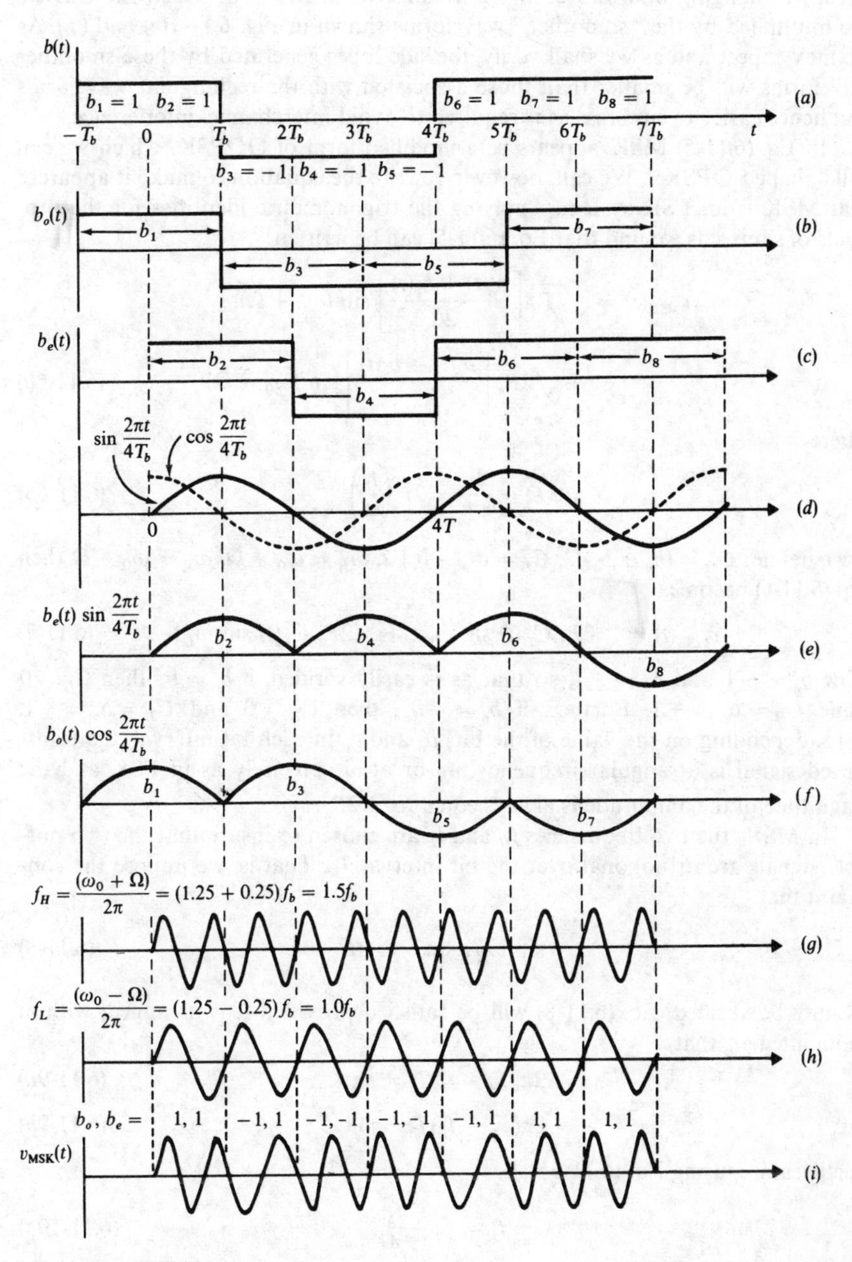

Figure 6.11-1 MSK waveforms.

abruptly changing odd and even bit streams. In MSK, in contrast, the carriers are multiplied by the "smoother" waveforms shown in Fig. 6.11-1(*e*) and (*f*). As we may expect and as we shall verify, the side lobes generated by these smoother waveforms will be smaller than those associated with the rectangular waveforms and hence easier to suppress as is required to avoid interchannel interference.

In Eq. (6.11-5) MSK appears as a modified form of OQPSK, which we can call "shaped QPSK". We can, however, rewrite the equation to make it apparent that MSK is an FSK system. Applying the trigonometric identities for the products of sinusoids we find that Eq. (6.11-5) can be written:

$$v_{\text{MSK}}(t) = \sqrt{2P_s}\left[\frac{b_o(t) + b_e(t)}{2}\right] \sin(\omega_0 + \Omega)t + \sqrt{2P_s}\left[\frac{b_o(t) - b_e(t)}{2}\right] \sin(\omega_0 - \Omega)t \tag{6.11-6a}$$

where

$$\Omega = \frac{2\pi}{4T_b} = 2\pi\left(\frac{f_b}{4}\right) \tag{6.11-6b}$$

If we define: $C_H = (b_o + b_e)/2$, $C_L = (b_o - b_e)/2$, $\omega_H = \omega_0 + \Omega$, $\omega_L = \omega_0 - \Omega$ then Eq. (6.11-6) becomes

$$v_{\text{MSK}}(t) = \sqrt{2P_s}\, C_H(t) \sin \omega_H t + \sqrt{2P_s}\, C_L(t) \sin \omega_L t \tag{6.11-7}$$

Now $b_o = \pm 1$ and $b_e = \pm 1$, so that as is easily verified, if $b_o = b_e$ then $C_L = 0$ while $C_H = b_o = \pm 1$. Further, if $b_o = -b_e$, then $C_H = 0$ and $C_L = b_o = \pm 1$. Thus, depending on the value of the bits b_o and b_e in each bit interval, the transmitted signal is at angular frequency ω_H or at ω_L precisely as in FSK and the magnitude of the amplitude is always equal to $\sqrt{2P_s}$.

In MSK, the two frequencies f_H and f_L are chosen to insure that the two possible signals are orthogonal over the bit interval T_b. That is, we impose the constraint that

$$\int_0^{T_b} \sin \omega_H t \sin \omega_L t \, dt = 0 \tag{6.11-8}$$

As may be verified, Eq. (6.11-8) will be satisfied provided it is arranged, with m and n integers, that

$$2\pi(f_H - f_L)T_b = n\pi \tag{6.11-9a}$$

and

$$2\pi(f_H + f_L)T_b = m\pi \tag{6.11-9b}$$

Furthermore, using Eq. (6.11-6*b*):

$$f_H = f_0 + \frac{f_b}{4} \tag{6.11-10a}$$

and

$$f_L = f_0 - \frac{f_b}{4} \tag{6.11-10b}$$

Using Eq. (6.11-10) and Eq. (6.11-9) we find that

$$f_b T_b = f_b \cdot \frac{1}{f_b} = 1 = n \tag{6.11-11a}$$

and

$$f_0 = \frac{m}{4} f_b \tag{6.11-11b}$$

Equation (6.11-11*a*) shows that since $n = 1$, f_H and f_L are as *close together as possible for orthogonality to prevail.* It is for this reason that the present system is called "minimum shift keying." Equation (6.11-11*b*) shows that the carrier frequency f_0 is an integral multiple of $f_b/4$. Thus

$$f_H = (m + 1)\frac{f_b}{4} \tag{6.11-12a}$$

and

$$f_L = (m - 1)\frac{f_b}{4} \tag{6.11-12b}$$

Signal Space Representation of MSK

The signal space representation of MSK is shown in Fig. 6.11-2. The orthonormal unit vectors of the coordinate system are given by $u_H(t) = \sqrt{2/T_s}\sin\omega_H t$ and $u_L(t) = \sqrt{2/T_s}\sin\omega_L t$. The end points of the four possible signal vectors are indicated by dots. The smallest distance between signal points is

$$d = \sqrt{2E_s} = \sqrt{4E_b} \tag{6.11-13}$$

just as for the case of QPSK.

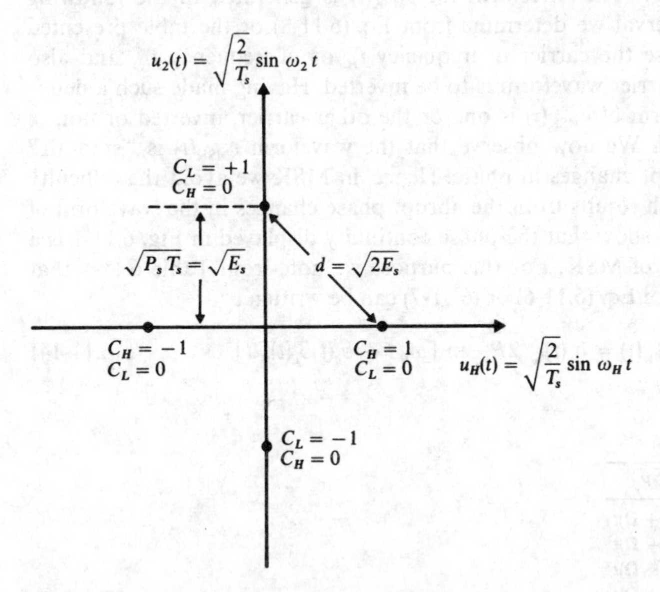

Figure 6.11-2 Signal space representation of MSK.

We recall that QPSK generates two BPSK signals which are orthogonal to one another by virtue of the fact that the respective carriers are in phase quadrature. Such phase quadrature can also be characterized as *time* quadrature since, at a carrier frequency f_0 a phase shift of $\pi/2$ is accomplished by a time shift in amount $1/4f_0$, that is $\sin 2\pi f_0(t + 1/4f_0) = \sin (2\pi f_0 t + \pi/2) = \cos (2\pi f_0 t)$. It is of interest to note, in contrast, that in MSK we have again two BPSK signals [i.e., the two individual terms in Eq. (6.11-7)]. Here, however, the respective carriers are orthogonal to one another by virtue of the fact that they are in *frequency* quadrature.

Phase Continuity in MSK

A most important and useful feature of MSK is its *phase continuity*. This matter is illustrated in Fig. 6.11-1 in waveforms in (*g*), (*h*), and (*i*). Here we have assumed that $f_0 = 5f_b/4$ so that

$$f_H = f_0 + f_b/4 = 5f_b/4 + f_b/4 = 1.5f_b$$

and also $f_L = f_0 - f_b/4 = 1.0f_b$. Carriers of frequencies f_H and f_L are shown in (*g*) and (*h*). We also find, from Eq. (6.11-6*a*), that for the various combinations of b_o and b_e, $v_{MSK}(t)/\sqrt{2P_s}$ is given in Table 6.11-1.

For the bit sequences b_o and b_e in Fig. 6.11-1*b* and *c* the corresponding waveform $v_{MSK}(t)$ is shown in (*i*). The values of b_o and b_e are tabulated in (*i*) for each bit intervals. We note, of course, that because of the staggering, b_o and b_e do not change simultaneously. The waveform for $v_{MSK}(t)$ is generated in the following way: In each bit interval we determine from Eq. (6.11-8) or the table presented below, whether to use the carrier of frequency f_H or of frequency f_L, and also whether or not the carrier waveform is to be inverted. Having made such a determination, the waveform of $v_{MSK}(t)$ is one or the other carrier, inverted or not, *in that same bit interval*. We now observe that the waveform $v_{MSK}(t)$ is "smooth" and exhibits no abrupt changes in phase. Hence, in MSK we avoid the difficulty described above which results from the abrupt phase changes in the waveform of QPSK. We shall now show that the phase continuity displayed in Fig. 6.11-1 is a general characteristic of MSK. For this purpose we note from Table 6.11-1 that the $v_{MSK}(t)$ waveform of Eq. (6.11-6) or (6.11-7) can be written as

$$v_{MSK}(t) = b_o(t)\sqrt{2P_s} \sin [\omega_0 t + b_o(t)b_e(t)\Omega t] \tag{6.11-14}$$

Table 6.11-1

b_e	b_o	$v_{MSK}(t)/\sqrt{2P_s}$
−1	−1	$-\sin (\omega_0 + \Omega)t$
−1	1	$\sin (\omega_0 - \Omega)t$
1	−1	$-\sin (\omega_0 - \Omega)t$
1	1	$\sin (\omega_0 + \Omega)t$

The equivalence of Eq. (6.11-6) or (6.11-7) and Eq. (6.11-14) can be established by comparing the two in each of the four cases corresponding to $b_o(t) = \pm 1$ and $b_e(t) = \pm 1$. Now let us turn our attention to the instantaneous phase $\phi(t)$ of the sinusoid in Eq. (6.11-14) where

$$\phi(t) = \omega_0 t + b_o(t)b_e(t)\Omega t \tag{6.11-15}$$

For convenience we represent the phase as $\phi_+(t)$ or $\phi_-(t)$ where

$$\phi_+(t) = (\omega_0 + \Omega)t; \qquad b_o(t) \cdot b_e(t) = +1 \tag{6.11-16a}$$

and

$$\phi_-(t) = (\omega_0 - \Omega)t; \qquad b_o(t) \cdot b_e(t) = -1 \tag{6.11-16b}$$

In Fig. 6.11-3 we have plotted $\phi_+(t)$ and $\phi_-(t)$.

The term $b_o(t) \cdot b_e(t)$ in Eq. (6.11-14) can change at times kT_b (k an integer). Whenever $b_o(t)\, b_e(t)$ changes from $+1$ to -1 or vice versa we see from Fig. 6.11-3 that there is an abrupt phase change in $\phi(t)$ of magnitude $2 \cdot \Omega \cdot kT_b = k\pi$. The occurrence and the magnitudes of these abrupt changes are shown in the figure. Now we recall that $b_o(t)$ and $b_e(t)$ do not change at the same time and that, if $b_o(t)$ does change, it changes at times $t = kT_b$ with k odd, while $b_e(t)$ can only change when k is even. Consider, then, first a change in $b_e(t)$. Such a change will cause a phase change which is a multiple of 2π, which is equivalent to no change at all. Next consider the effect of a change in $b_o(t)$. In this case the phase change in $\phi(t)$ will be an odd multiple of π, that is, a phase change equivalent to π. But now, looking back at Eq. (6.11-14) we take note of the coefficient $b_o(t)$, which multiplies $\sqrt{2P_s}\sin[\phi(t)]$. Whenever there is a change in $b_o(t)$ to change the phase $\phi(t)$ by π, the coefficient $b_o(t)$ will also change sign, yielding an additional π phase change. Hence, a change in $b_o(t)$ produces no net phase discontinuity.

Thus, we have established that Eq. (6.11-5) assures the phase continuity that we have been seeking.

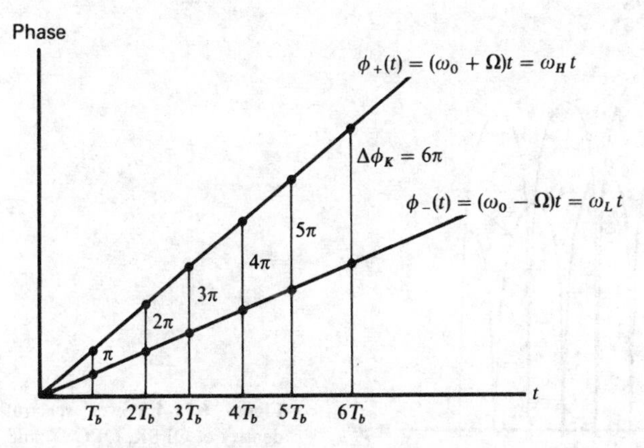

Figure 6.11-3 Illustrating why the phase in MSK is continuous.

Power Spectral Density of MSK

Referring to Eq. 6.11-5 we see that the baseband waveform which multiplies the $\sin \omega_0 t$ in MSK is

$$p(t) = \sqrt{2P_s}\, b_o(t) \cos \frac{\pi}{2} f_0 t \qquad -T_b \leq t \leq T_b \tag{6.11-17}$$

As can be verified (Prob. 6.11-5), the waveform $p(t)$ has a power spectral density $G_p(f)$ given by (since again $P_s = E_b/T_b$)

$$G_p(f) = \frac{32E_b}{\pi^2}\left[\frac{\cos 2\pi f/f_b}{1-\left(\frac{4f}{f_b}\right)^2}\right]^2 \tag{6.11-18}$$

Proceeding as we did above in the case of QPSK, we find that the power spectral density for the total MSK signal of Eq. (6.11-5) is

$$G_{\text{MSK}}(f) = \frac{8}{\pi^2} E_b \left(\left\{\frac{\cos 2\pi(f-f_0)/f_b}{1-\left[\frac{4(f-f_0)}{f_b}\right]^2}\right\}^2 + \left\{\frac{\cos 2\pi(f+f_0)f_b}{1-\left[\frac{4(f+f_0)}{f_b}\right]^2}\right\}^2\right) \tag{6.11-19}$$

A comparison of the baseband spectral densities of QPSK (which is the same as OQPSK) and MSK is shown in the plots of Fig. 6.11-4. Observe that the main lobe in MSK is wider than the main lobe in QPSK. In MSK the bandwidth required to accommodate this lobe is $2 \times 3/4f_b = 1.5f_b$ while it is only $1.0f_b$ in QPSK. However in MSK the side lobes are very greatly suppressed in comparison to QPSK. This result is to be anticipated since, except near $f = 0$ (or $f = f_0$),

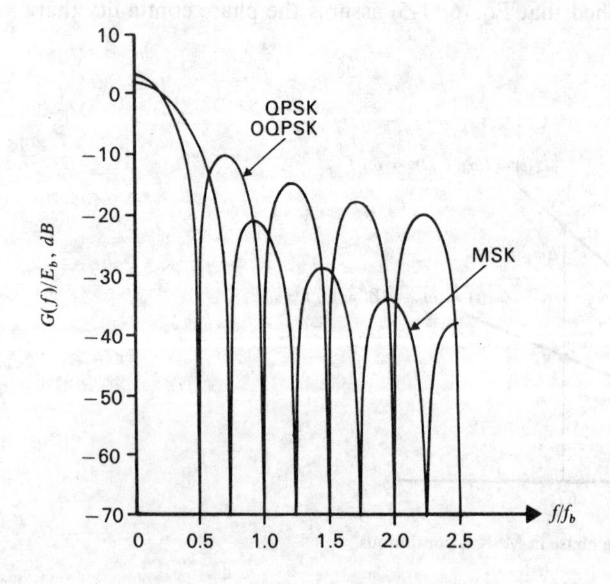

Figure 6.11-4 Power spectral density of QPSK, OPQSK and MSK.

in QPSK $G(f)$ falls off as $1/f^2$, while in MSK $G(f)$ falls off as $1/f^4$. It turns out that in MSK 99 percent of the signal power is contained in a bandwith of about $1.2f_b$ while in QPSK the corresponding bandwidth is about $8f_b$.

Generation and Reception of MSK

One way to generate an MSK signal is the following: We start with $\sin \Omega t$ and $\sin \omega_0 t$, and use 90° phase shifters to generate $\sin (\Omega t + \pi/2) = \cos \Omega t$ and $\sin (\omega_0 t + \pi/2) = \cos \omega_0 t$. We then use multipliers to form the products $\sin \Omega t \cos \omega_0 t$ and $\cos \Omega t \sin \omega_0 t$. Additional multipliers generate $\sqrt{2P_s}\, b_e(t) \sin \Omega t \cos \omega_0 t$ and $\sqrt{2P_s}\, b_o(t) \cos \Omega t \sin \omega_0 t$. Finally an adder is used to form the sum as called for in Eq. (6.11-5). An alternative and more favored scheme is shown in Fig. 6.11-5*a*. This technique has the merit that it avoids the need for precise 90° phase shifters at angular frequencies ω_0 and Ω. It is left as a student exercise (Prob 6.11-6) to verify that, in Fig. 6.11-5*a*, the waveforms named $x(t)$ and $y(t)$ are as given and that the output is indeed identical to $v_{\text{MSK}}(t)$ given in Eq. (6.11-5).

The MSK receiver is shown in Fig. 6.11-5*b*. Detection is performed synchronously, i.e., by determining the correlation of the received signal with the waveform $x(t) = \cos \Omega t \sin \omega_0 t$ to determine bit $b_o(t)$, and with $y(t) = \sin \Omega t \cos \omega_0 t$ to determine bit $b_e(t)$. As is customary, such correlation is performed by multiplication and integration over the symbol interval. As is to be noted, the integrators integrate over staggered overlapping intervals of symbol time $T_s = 2T_b$. At the end of each integration time the integrator output is stored and then the integrator output is dumped. (The dumping is not explicitly indicated in Fig. 6.11-5*b*.) The switch at the output swings back and forth at the bit rate so that finally the output waveform is the original data bit stream $d_k(t)$.

Of course, to make the receiver effective we need to reconstruct the waveforms $x(t)$ and $y(t)$. A method for locally regenerating $x(t)$ and $y(t)$ is shown in Fig. 6.11-6. From Eq. (6.11-7) we see that MSK consists of transmitting one of two possible BPSK signals, the first at frequency $\omega_0 - \Omega$ and the second at frequency $\omega_0 + \Omega$. Thus as in BPSK detection, we first square and filter the incoming signal. The output of the squarer has spectral components at the frequency $2\omega_H = 2(\omega_0 + \Omega)$ and at $2\omega_L = 2(\omega_0 - \Omega)$. These are separated out by bandpass filters. Division by 2 yields waveforms $\frac{1}{2} \sin \omega_H t$ and $\frac{1}{2} \sin \omega_L t$ from which, as indicated, $x(t)$ and $y(t)$ are regenerated by addition and subtraction respectively. Further, the multiplier and low-pass filter shown regenerate a waveform at the symbol rate $f_s = f_b/2$ which can be used to operate the sampling switches in Fig. 6.11-4*b*.

The reader will, at this point, want to know how the bandpass filters and divide by 2 circuits work, if, for a long period of time, say, $C_L = 0$, that is, only the high frequency signal is transmitted. The answer to this question is given in detail in Sec. 10.16 which is concerned with Phase Locked Loop synchronization. For the present let us just note that the bandpass filter—divide by 2 circuits are actually oscillators built to operate at frequencies $2\omega_H$ and $2\omega_L$. The incoming

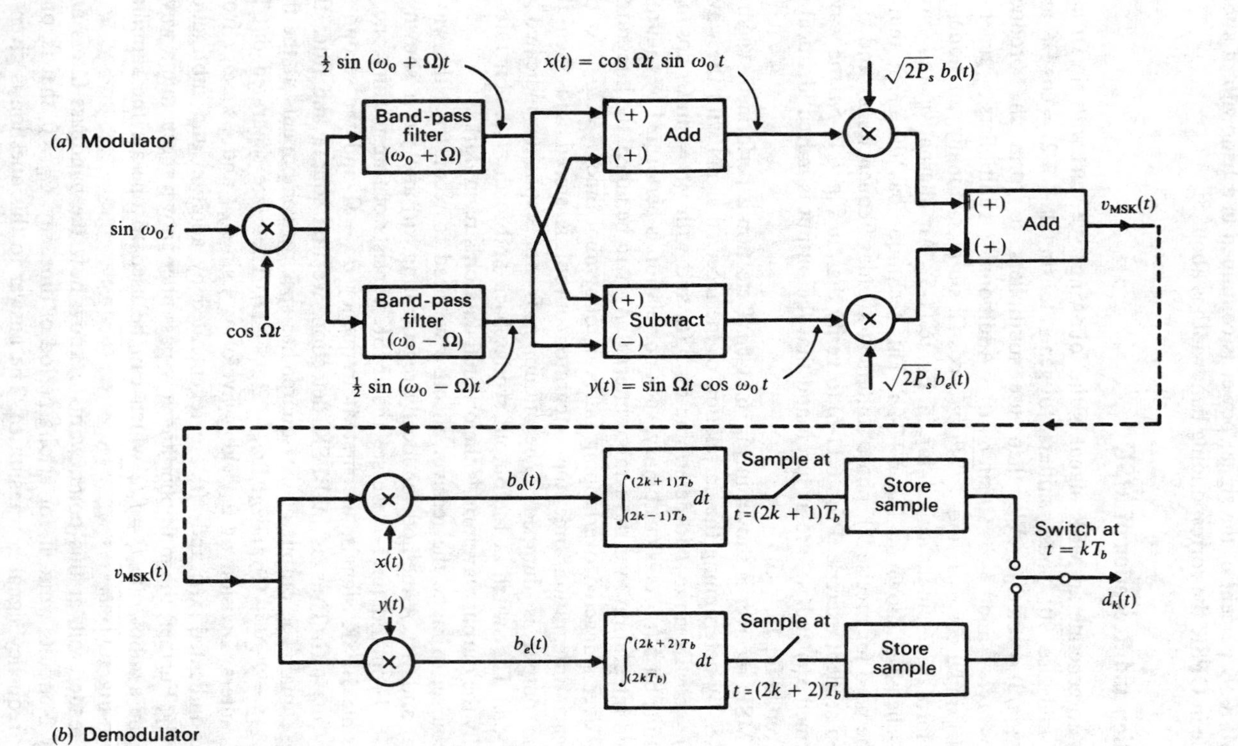

Figure 6.11-5 Method of modulating and demodulating MSK.

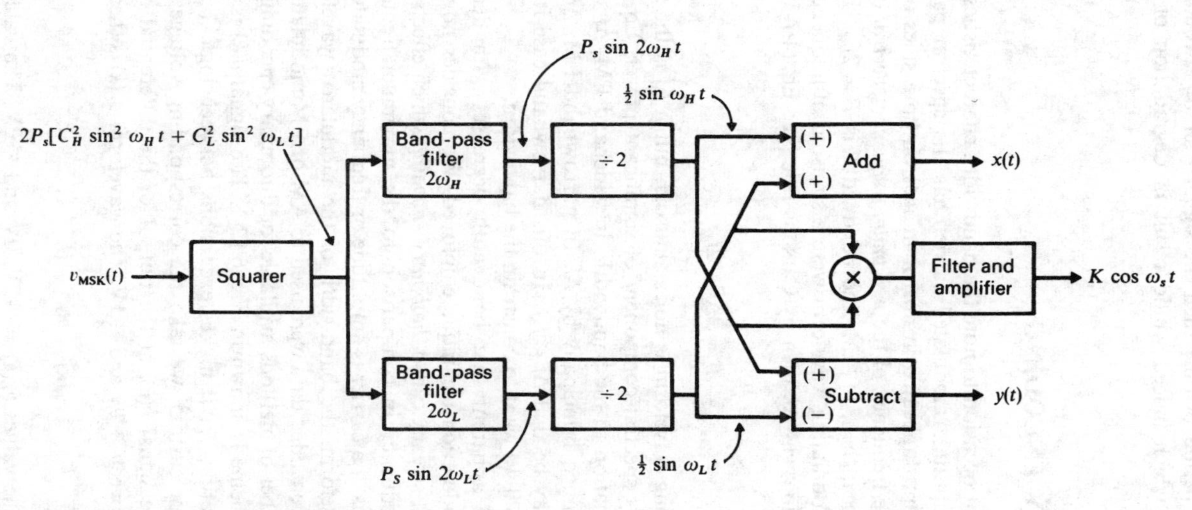

Figure 6.11-6 Technique to regenerate $x(t)$ and $y(t)$.

signal at each frequency merely provides phase synchronizing information and does not effect the oscillator frequency or amplitude. If, say, only ω_H is transmitted for an extended period of time then the signal $\frac{1}{2} \sin \omega_L t$ would have some phase jitter present which indeed would result in degradation of system performance.

6.12 DUOBINARY ENCODING

We have had frequent occasion to note the bandwidth characteristics of a modulated carrier. An amplitude modulated carrier has the spectral pattern of the baseband signal except duplicated in mirror images on both sides of the carrier frequency. If f_M is the frequency of the *maximum frequency spectral component of the baseband waveform*, then, in AM, the bandwidth is $B = 2f_M$. In frequency modulation, if the modulating waveform were a sinusoid of frequency f_M, and if the frequency deviation was Δf, then by Carsons rule [see Eq. (4.7-3)] the bandwidth would be:

$$B = 2\Delta f + 2f_M \tag{6.12-1}$$

Even if the modulating waveform is not a sinusoid, but if f_M is the frequency of the highest frequency spectral component of significant power, Eq. (6.12-1) is a reasonable approximation to the bandwidth. In wideband FM, $\Delta f \gg f_M$ and the bandwidth is determined principally by Δf. In narrowband FM, Δf and f_M are comparable, or it may be that $\Delta f \ll f_M$. In this narrowband case it is advantageous to minimize f_M if we want to constrain the bandwidth.

Altogether, it is apparent that bandwidth decreases with decreasing f_M, regardless of the modulation technique employed. We consider now a mode of encoding a binary bit stream, called *duobinary encoding* which effects a reduction of the maximum frequency in comparison to the maximum frequency of the unencoded data. Thus, if a carrier is amplitude or frequency modulated by a duobinary encoded waveform, the bandwidth of the modulated waveform will be smaller than if the unencoded data were used to AM or FM modulate the carrier.

There are a number of methods available for duobinary encoding and decoding. One popular scheme is shown in Fig. 6.12-1. The signal $d(k)$ is the data bit stream with bit duration T_b. It makes excursions between logic 1 and logic 0, and, as had been our custom, we take the corresponding voltage levels to be $+1$V and -1V. The signal $b(k)$, at the output of the differential encoder also makes excursions between $+1$V and -1V. The waveform $v_D(k)$ is therefore

$$v_D(k) = b(k) + (k - 1) \tag{6.12-2}$$

which can take on the values $v_D(k) = +2$V, 0V and -2V. The value of $v_D(k)$ in any interval k depends on both $b(k)$ and $b(k - 1)$. Hence there is a correlation between the values of $v_D(k)$ in any two successive intervals. For this reason the coding of Fig. 6.12-1 is referred to as *correlative* coding.

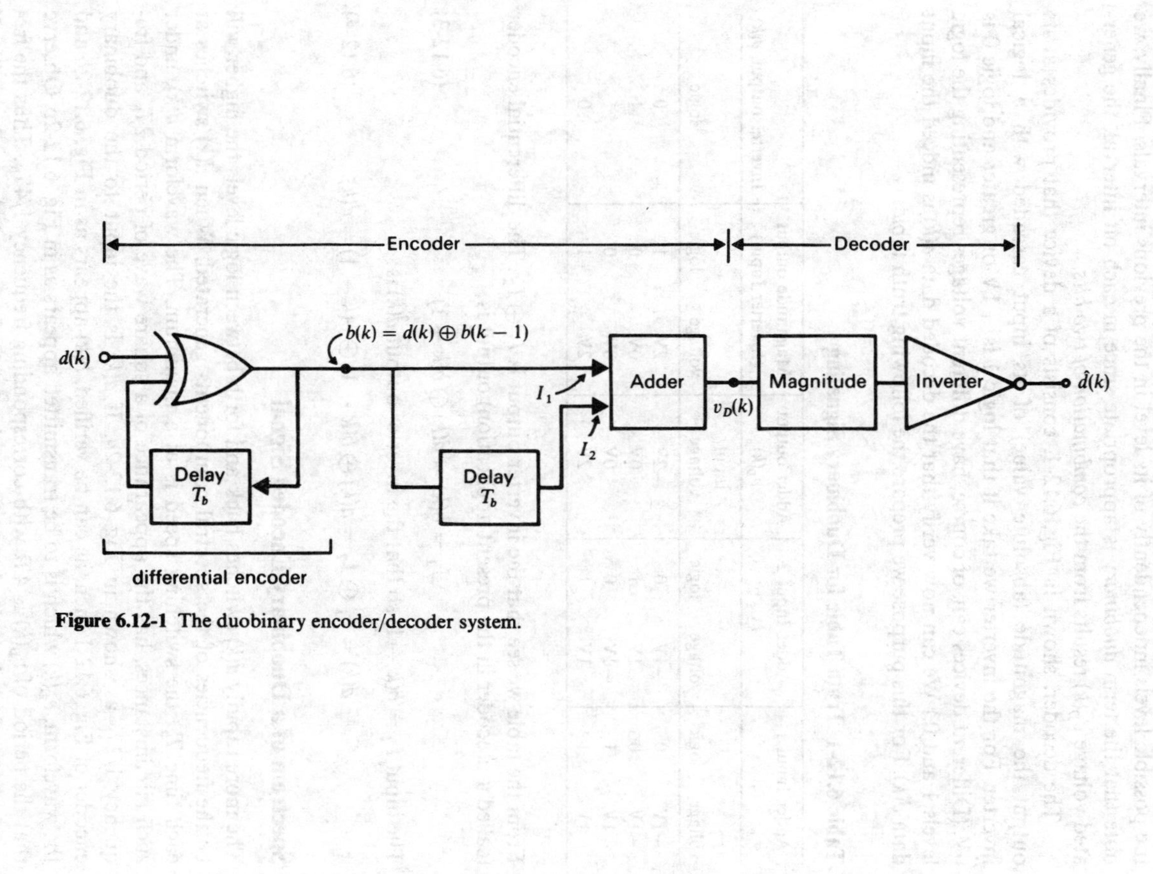

Figure 6.12-1 The duobinary encoder/decoder system.

The correlation can be made apparent in another way. When the transition is made from one interval to the next, *it is not possible for* $v_D(k)$ *to change from* $+2V$ *to* $-2V$ *or vice versa.* In short, in any interval, $v_D(k)$ cannot always assume any of the possible levels independently of its level in the previous intervals. Finally, we note that the term *duobinary* is appropriate since in each bit interval, the generated voltage $v_D(k)$ results from the *combination of two bits.*

The decoder, shown in Fig. 6.12-1, consists of a device that provides at its output the magnitude (absolute value) of its input cascaded with a logical inverter. For the inverter we take it that logic 1 is +1V or greater and logic 0 is 0V. (Different devices can, of course, have different voltages representing the logic levels 1 and 0.) We can now verify that the decoded data $\hat{d}(k)$ is indeed the input data $d(k)$. For this purpose we prepare the following truth table:

Table 6.12-1 Truth Table for Duobinary Signaling

Adder Input 1 I_1		Adder Input 2 I_2		Adder output $v_D(k)$	Magnitude output (Inverter Input)		Inverter output, $\hat{d}(k)$
voltage	logic	voltage	logic	input voltage	voltage	logic	logic
−1V	0	−1V	0	−2V	2V	1	0
−1V	0	1V	1	0V	0V	0	1
1V	1	−1V	0	0V	0V	0	1
1V	1	1V	1	2V	2V	1	0

From the table we see that the inverter output is $I_1 \oplus I_2$. The differential encoder (called a precoder in the present application) output is:

$$I_1 = b(k) = d(k) \oplus b(k-1) \tag{6.12-3}$$

The input $I_2 = b(k-1)$ so that the inverter output $\hat{d}(k)$ is:

$$\hat{d}(k) = I_1 \oplus I_2 = d(k) \oplus b(k-1) \oplus b(k-1) = d(k) \tag{6.12-4}$$

Spectrum of a Duobinary Encoded Signal

The more rapidly $d(k)$ switches back and forth between logic levels the higher will be the frequencies of the spectral components generated. When $d(k)$ switches at each time T_b, the switching speed is at a maximum. The waveform $d(k)$, under such circumstances, has the appearance of a square wave of period $2T_b$ and frequency $1/2T_b$ as shown in Fig. 6.12-2*a*. If $d(k)$ is the input to the duobinary encoder of Fig. 6.12-1 then, as can be verified, $b(k)$ appears as in Fig. 6.12-2*b* and the waveform, $v_D(k)$ which is to be transmitted appears as in Fig. 6.12-2*c*. Observe that the period of $v_D(k)$ is $4T_b$ with corresponding frequency $1/4T_b$. Thus the frequency of $v_D(k)$ is half the frequency of the original unencoded waveform $d(k)$. If we are inclined to make a moderately crude approximation, we might even think of the waveform $d(k)$ as a sinusoid of frequency $1/2T_b$ and of the waveform $v_D(t)$ as a sinusoid of frequency $1/4T_b$. If we were free to select either $d(k)$ or $v_D(k)$ as a

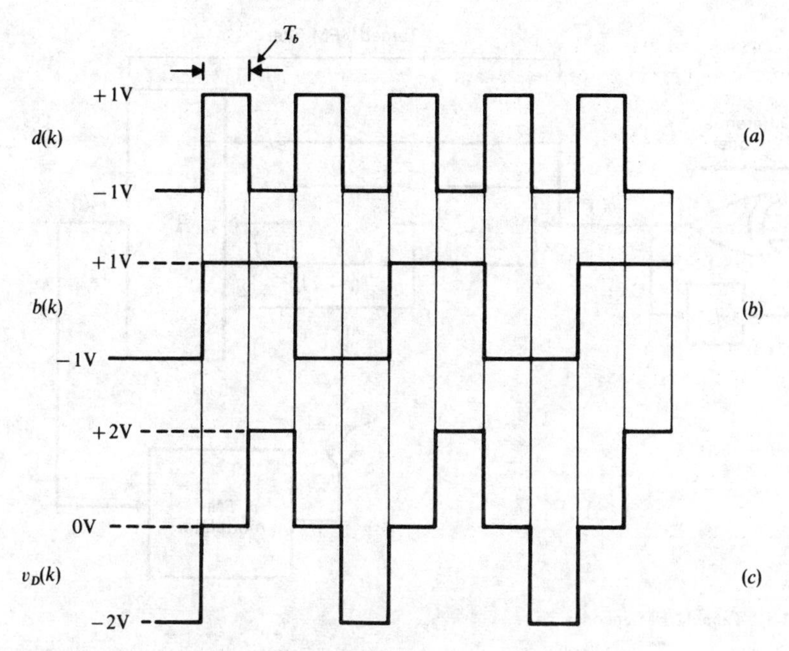

Figure 6.12-2 Waveforms of $d(k)$, $b(k)$ and $v_D(k)$.

modulating waveform for a carrier, and if we were interested in conserving bandwidth, we would choose $v_D(k)$. For, persisting in our crude approximation, if amplitude modulation were involved, the bandwidth of the modulated waveform would be $2(1/4T_b) = f_b/2$ using $v_D(k)$ since the modulating frequency is $f_M = 1/4T_b$, and would be $2(1/2T_b) = f_b$ using $d(k)$. With frequency modulation, if the peak-to-peak carrier frequency deviation were $2\Delta f$, then, using Carson's rule, the modulated carrier would have a bandwidth $2(\Delta f) + 2(1/2T_b)$ with $d(k)$ as the modulating signal, as in BFSK; and $2(\Delta f) + 2(1/4T_b)$ with $v_D(k)$ as the modulating signal.

Tamed FM

A great deal of effort has been expended to find an arrangement, following the pattern of Fig. 6.12-1 which will yield a $v_D(k)$ with maximum compression of bandwidth in comparison to $d(k)$. Although *the* optimum arrangement is yet to be found, the arrangement of Fig. 6.12-3 does offer a significant improvement over the duobinary arrangement of Fig. 6.12-1. The arrangement (i.e., filter) of Fig. 6.12-3 is often used to encode a baseband waveform intended to FM modulate a carrier. When so employed it restrains (i.e., tames) the frequency excursions of the FM signal and the technique is referred to as "tamed FM".

To compare the schemes of Figs. 6.12-1 and 6.12-3 we shall calculate the transfer function $H_T(f)$ of the taming filter and $H_D(f)$ of the duobinary filter. We

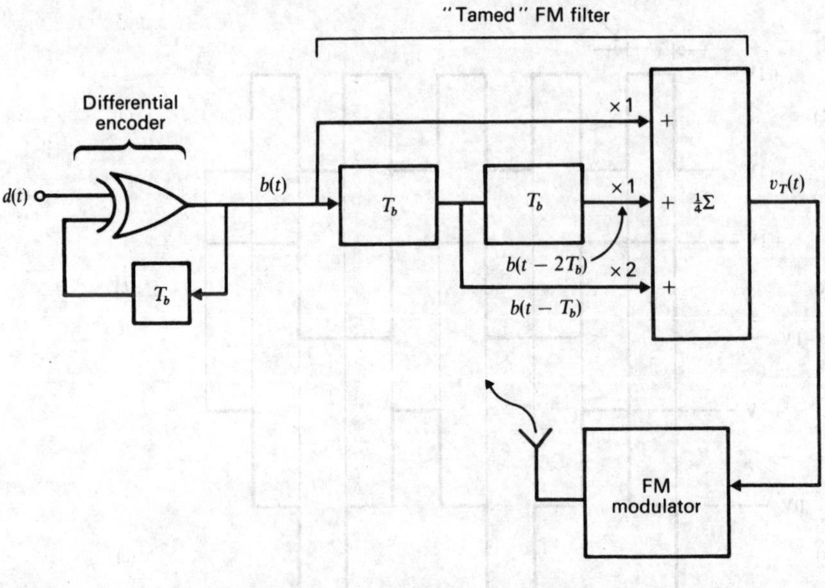

Figure 6.12-3 "Tamed" FM encoder.

keep in mind that if $B(f)$ is the Fourier transform $\mathscr{F}(b(t))$ then the Fourier transform of $b(t - kT_b)$ is $B(f)e^{-j2\pi k T_b}$. We have then in Fig. 6.12-3,

$$v_T(t) = \tfrac{1}{4}[b(t) + 2b(t - T_b) + b(t - 2T_b)] \tag{6.12-5}$$

With $V_T(f) \equiv \mathscr{F}(v_T(t))$, we then have that the transform of the *taming* filter is:

$$H_T(f) = \frac{V_T(f)}{B(f)} = \tfrac{1}{4}[1 + 2e^{-j2\pi f T_b} + e^{-j4\pi f T_b}]$$

$$= \left(\frac{1 + e^{-j2\pi f T_b}}{2}\right)^2 = \tfrac{1}{2}e^{j2\pi f T_b} \cos^2 \pi f T_b \tag{6.12-6}$$

and

$$|H_T(f)|^2 = \cos^4 \pi f T_b \tag{6.12-7}$$

In the same way we can calculate $H_D(f)^2$ for the duobinary filter of Fig. 6.12-1. We find

$$|H_D(f)|^2 = \cos^2 \pi f T_b \tag{6.12-8}$$

The power spectral density of the bit stream $b(t)$ is [see Eq. (6.2-8)]

$$G_b(f) = V_b^2 T_b \left(\frac{\sin \pi f T_b}{\pi f T_b}\right)^2 \tag{6.12-9}$$

in which V_b is the amplitude of a bit. Thus altogether the power spectral densities of the duobinary encoded waveform $v_D(t)$ and the tamed FM encoded waveform $v_T(t)$ are then:

$$G_D(f) = H_D(f)^2 G_b(f) = V_b^2 T_b \left(\frac{\sin \pi f T_b}{\pi f T_b}\right)^2 \cos^2 \pi f T_b \qquad (6.12\text{-}10)$$

$$G_T(f) = H_T(f)^2 G_b(f) = V_b^2 T_b \left(\frac{\sin \pi f T_b}{\pi f T_b}\right)^2 \cos^4 \pi f T_b \qquad (6.12\text{-}11)$$

It can be shown (see Prob. 6.12-4) that in the frequency range $0 \leq f < f_b/2$ a larger fraction of the total power is contained in the TFM waveform than in the duobinary waveform.

6.13 A COMPARISON OF NARROWBAND FM SYSTEMS

The great merit of FM over AM is that FM allows us to suppress the effects of noise albeit at the expense of bandwidth. There are, however, frequent occasions when FM is used in preference to AM for an entirely different purpose.

We have already noted that as a matter of practice, the amplitude variations in an AM signal may be a source of difficulty. Such signals when amplified by nonlinear amplifiers (as is generally required for high power efficiency) generate spurious out-of-band spectral components which are filtered out only with difficulty. Even AM systems such as BPSK and QPSK which do not, in principle, generate signals with amplitude variation, do, as a matter of practice, generally have such variation. Further, BPSK and QPSK generate discontinuities in the carrier phase, which are a further source of difficulty. When it is necessary to avoid such amplitude variations and phase discontinuities, FM is the solution, for the FM waveform

$$v(t) = A \cos (\omega_0 t + k \int m(t)\, dt) = A \cos \Phi(t) \qquad (6.13\text{-}1)$$

has a constant amplitude. Also, no matter how discontinuous may be the modulating waveform, no matter how abruptly $m(t)$ may jump, the phase $\Phi(t)$ is absolutely continuous.

When FM is used principally to avoid amplitude variation and phase discontinuity, narrowband FM is usually employed. For example we have seen that MSK is an FSK (frequency shift keying) system and hence is an FM system in which $f_M = 1/2T_b = f_b/2$ and $\Delta f = f_b/4$. Hence the bandwidth, using Eq. (6.12-1), is

$$B = 2\left(\frac{f_b}{4}\right) + 2(f_b/2) = 1.5 f_b \qquad (6.13\text{-}2)$$

If data, $d(k)$, of bit duration T_b is encoded by a duobinary filter before being used to frequency modulate a carrier then, as we have seen, it is reasonable to take f_M in Eq. (6.12-1), to be $f_M = 1/4T_b$. If also we take $\Delta f = f_M/4$ then we have

$$B = 2\Delta f + 2f_M = 2(f_b/4) + 2(1/4T_b) = f_b \qquad (6.13\text{-}3)$$

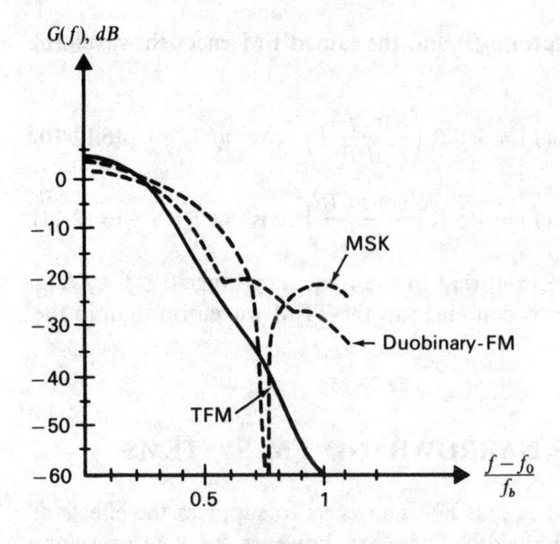

Figure 6.13-1 Spectrum of MSK, duobinary FM and TFM.

If we were to draw, for the taming filter of Fig. 6.12-3, waveforms of $d(k)$, $b(k)$, and $v_T(k)$ comparable to Fig. 6.12-2 we would again come to the conclusion that $f_M = 1/4T_b$ just as in the duobinary case. On this basis, for the taming filter as for the duobinary filter again with $\Delta f + f_b/4$,

$$B = \frac{1}{T_b} = f_b \tag{6.13-4}$$

The two cases yield the same estimated bandwidth because of the crudeness of the estimate. Actually, in the taming filter case, the sidelobes outside the bandwidth range are substantially reduced in comparison with the duobinary filter.

The power spectral density for duobinary—FM, tamed FM, and MSK are compared in Fig. 6.13-1.

6.14 PARTIAL RESPONSE SIGNALING

We have observed that if we are interested in constraining the bandwidth of a modulated carrier, especially in AM, we must look to the baseband signal. Now the power spectral density of the baseband waveform is determined by the waveform associated with an *individual bit* (or symbol). For example in QPSK this waveform is rectangular, in MSK it is a half cycle of a sinusoid, etc. We observe that this waveform for an individual bit, or symbol, is a waveform of *finite extension of time*; its duration being the bit or symbol time. Any such waveform of limited time duration has a spectrum which, in principle at least, extends *throughout the frequency range*, i.e., from $f = -\infty$ to $f = +\infty$. It appears, accordingly, that while some waveforms may be better than others, there is no waveform that will completely eliminate interchannel interference.

It occurs to us that we may be able to minimize the spectrum of a signal by passing the baseband waveform through a low-pass filter to remove the offending spectral range of the baseband waveform. However, any such filtering will increase the time extension of the baseband bit waveform so that it will overlap neighboring bit intervals causing *intersymbol interference*. In short, any unfiltered baseband waveform must cause some interchannel interference and if we filter it, we can reduce the interchannel interference but only at the expense of causing intersymbol interference.

There turns out to be a way out of this dilemma by taking advantage of a result which we already encountered. We have already noted, in connection with our discussion of the sampling theorem (Sec. 5.4) that an impulse of unit strength when applied to a "brickwall filter" with transfer characteristic $H_B(f)$

$$H_B(f) = \begin{cases} 1 & |f| \leq f_c \\ 0 & \text{elsewhere} \end{cases} \tag{6.14-1a}$$

as shown in Fig. 6.14-1*a*, produces a response which is

$$h_B(t) = 2f_c \frac{\sin 2\pi f_c t}{2\pi f_c t} \tag{6.14-1b}$$

This response displays a maximum at $t = 0$ and is zero at multiples of the time $1/2f_c$. Hence if impulses are generated at time $t = k/2f_c$ (k an integer) then, at the time of peak response to any particular impulse, the response to all other impulses will be zero. Hence, even if all the individual responses are superimposed, we shall be able, by sampling at times $t = k/2f_c$, to observe a response proportional to the kth impulse, without interference from the waveforms due to other impulses.

We can apply these considerations to our present concern in the following way: Let us represent the bits in our data stream by impulses, a positive impulse for logic 1, a negative impulse for logic 0, the impulses being separated by the bit time T_b. Let us pass the bit stream through a brickwall filter of cut off frequency $f_c = 1/2T_b = f_b/2$. The filtering will allow us to suppress interchannel interference. While the bit waveform will now be of infinite extent we will still be able to avoid intersymbol interference by sampling at $t = k/f_b$.

The use of a brickwall filter which cuts off at $f_c = f_b/2$ has associated with it a number of awkward features. First of all, it is very difficult to construct a good approximation to such a filter (the ideal filter itself is, of course, unrealizable). Secondly, we observe, from Eq. (6.14-1*b*) that the amplitude of each of the successive peaks falls off slowly, i.e., only linearly with time. Hence, intersymbol interference, even between bits which are greatly separated in time, can result if the sampling operation is not timed with great precision. That is, the intersymbol interference is very sensitive to timing errors.

Accordingly, we look to other filters. A filter of interest in this regard is the "cosine" filter with transfer characteristics shown in Fig. 6.14-1*b*,

$$H_C(f) = \begin{cases} 2 \cos \pi f T_b & \text{for } f \leq f_b/2 \\ 0 & \text{for } f > f_b/2 \end{cases} \tag{6.14-2}$$

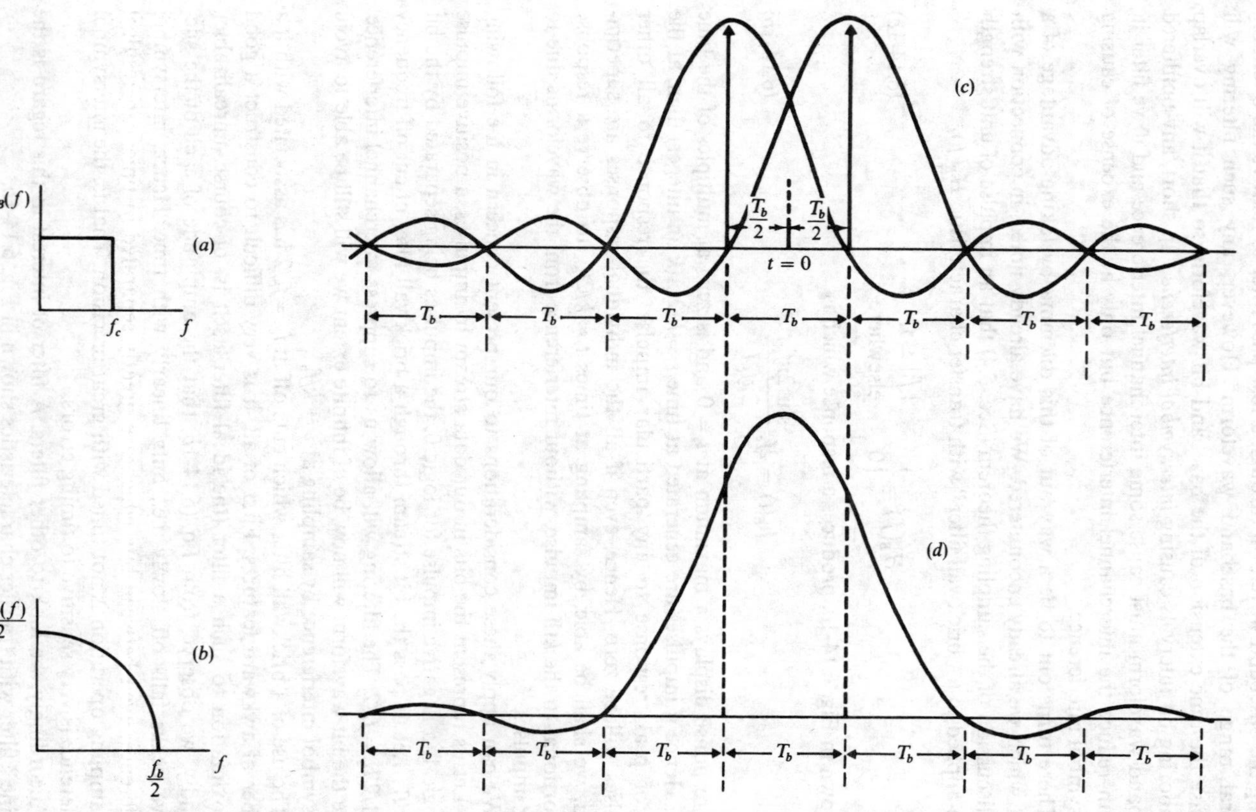

Figure 6.14-1 Duobinary filters and waveforms (*a*) brickwall filter; (*b*) cosine filter; (*c*) impulse response of a cosine filter; (*d*) impulse response of a brickwall filter.

or equivalently

$$H_C(f) = [H_B(f)][2 \cos \pi f T_B] \tag{6.14-3}$$

Equation (6.14-3) can be rewritten as

$$H_C(f) = H_B(f)(e^{j\pi f T_b} + e^{-j\pi f T_b}) \tag{6.14-4a}$$

Accordingly, the response of the cosine filter is the response of a cascade of two filters, one with transfer characteristic $H_B(f)$ (with $f_c = f_b/2$) and a second with transfer characteristic given by $2 \cos \pi f T_b$. We recall further that if a time function $q(t)$ has a Fourier transform $Q(f)$ then $Q(f)e^{-j2\pi f T}$ is the transform of $q(t - T)$. Thus, altogether we have that

$$h_C(t) = h_B(t + T_b/2) + h_B(t - T_b/2) \tag{6.14-4b}$$

The two parts of $h_C(t)$ are drawn in Fig. 6.14-1*c* and the total response is shown in Fig. 6.14-1*d*. The response is still non-causal, since, as we note, there is a response before $t = 0$. This non-causal feature is a result of the fact that, while we have not assumed that the filter response falls from $H_C(f) = 1$ to $H_C(f) = 0$ precipitously at cut off, we have persisted in assuming that the filter cuts off completely at $f_b/2$. Nonetheless, a real filter can be constructed which approximates the unrealizable filter of Eq. (6.14-2).

Observe that the side peaks in the complete response $h_C(t)$ are smaller than the side peaks of the response $h_B(t)$. Such is the case because the side peaks in $h_C(t)$ are the *differences* of the side peaks in $h_B(t)$. This point is also brought out when it is shown (Prob. 6.14-1) that the response $h_C(t)$ of the cosine filter is:

$$h_C(t) = \frac{4f_b \cos \pi f_b t}{\pi(1 - 4f_b^2 t^2)} \tag{6.14-5}$$

From Eq. (6.14-5) we observe that the sidelobes fall off as the square of the time rather than linearly with time as does $h_B(t)$ given in Eq. (6.14-1*b*). This smaller peak will reduce the sensitivity to errors in the time at which the sampling operation is performed. We observe further, however, that the central peak of $h_C(t)$, being of duration $3T_b$, is 1.5 times as wide as the central peak of $h_B(t)$ which has a duration $2T_b$.

Now suppose, that corresponding to each bit of duration T_b, of a data stream we generate a positive impulse of strength $+I$ whenever the bit is at logic 1 and a negative impulse $-I$ whenever the bit is at logic 0. Suppose, further, that these impulses are applied to the input of the cosine filter. In Fig. 6.14-2 we have drawn the filter responses individually to five successive positive impulses. For simplicity, we have in each case drawn only the central lobe and we have indicated by dots all the places where the individual response waveforms pass through zero. Where there is no dot, the waveforms has a finite value. The peaks of the responses are separated by times T_b and the widths of the central lobes are $3T_b$. The total response is, of course, simply the sum of the individual responses.

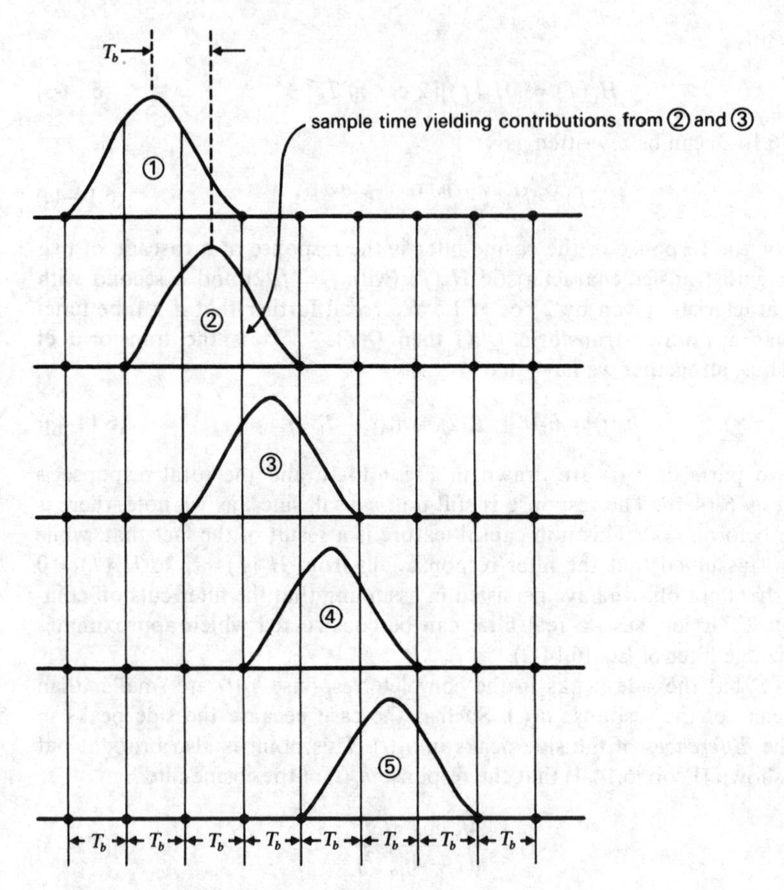

Figure 6.14-2 Filter responses to five separate impulses.

We can make the following observations from Fig. 6.14-2:

1. If we sample the total response at a time when an individual response is at its peak, the sample will have contributions from *all* the individual responses.
2. There is *no* possible time at which a sample of the total response is due only to a single individual response.
3. Importantly, if we sample the total response midway between times when the individual responses are at peak value, i.e., at $t = (2k - 1)T_b/2$, then the sample value will have contributions in equal amount from *only the two individual responses* that straddle the sampling time. These sampling times are indicated in Fig. 6.14-2 by the light vertical lines. One such sampling time, yielding contributions from individual responses 2 and 3 is explicitly marked. It can be calculated from Eq. (6.14-1*b*) that at the sampling time the contribution from each of the straddling individual responses will be a voltage If_b. Note that in

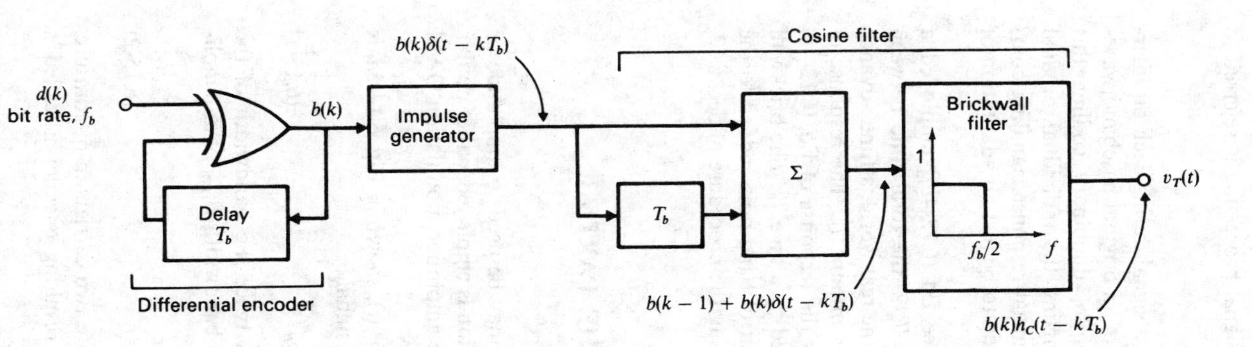

(a) Encoder

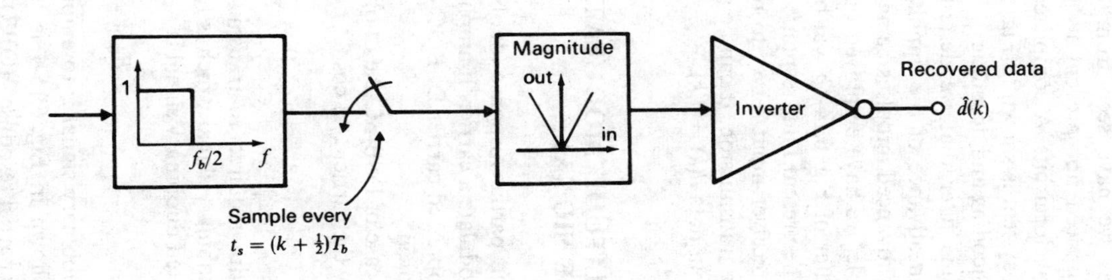

(b) Decoder

Figure 6.14-3 Duobinary encoder and decoder using cosine filter.

sampling at the indicated times, we sample when the individual responses are *not* at peak value. For this reason, the present signal processing is referred to as *partial-responses signaling.*

In partial-response signaling, we shall transmit a signal during each bit interval that has contributions from two successive bits of an original baseband waveform. But as we have seen from our discussion of duobinary coding, this superposition need not prevent us from disentangling the individual original baseband waveform bits. A complete (baseband) partial-response signaling communications system is shown in Fig. 6.14-3. It is seen to be just an adaptation of duobinary encoding and decoding.

The cosine filter that we described above [see Eq. (6.12-4*b*)] employed a delay and an advance of the impulse by amount $T_b/2$, the total time between delayed and advanced impulses being T_b. Since, in the real world, a time advance is not possible, we have employed only a delay by amount T_b. The response of the cosine filter of Fig. 6.14-1*b* will be the same as the response of Eq. (6.12-4*b*) except that the present filter response will be delayed by a time T_b. The brickwall filter at the receiver input serves to remove any out of band noise added to the signal during transmission. It can be shown, following the procedure of Sec. 6.12, that the output data $\hat{d}(k) = d(k)$.

6.15 AMPLITUDE MODULATION OF THE PARTIAL RESPONSE SIGNAL

The baseband partial response (duobinary) signal may be used to amplitude or frequency modulate a carrier. If amplitude modulation is employed, either double sideband suppressed carrier DSB/SC or quadrature amplitude modulation QAM can be employed.

For the case of DSB/SC the duobinary signal, $v_T(t)$, shown in Fig. 6.14-3*a*, is multiplied by the carrier $\sqrt{2}\cos\omega_0 t$. The resulting signal is

$$v_{DSB}(t) = \sqrt{2}\, v_T(t) \cos \omega_0 t \tag{6.15-1}$$

The bandwidth required to transmit the signal is twice the bandwidth of the baseband duobinary signal which is $f_b/2$. Hence the bandwidth B_{DSB} of an amplitude modulated duobinary signal is

$$B_{DSB} = 2(f_b/2) = f_b \tag{6.15-2}$$

If the duobinary signal is to amplitude modulate two carriers in quadrature, the circuit shown in Fig. 6.15-1 is used and the resulting encoder is called a "quadrature partial response" (QPR) encoder.

Figure 6.15-1 shows that the data $d(t)$ at the bit rate f_b is first separated into an even and an odd bit stream $d_e(t)$ and $d_o(t)$ each operating with the bit rate $f_b/2$. Both $d_e(t)$ and $d_o(t)$ are then separately duobinary encoded into signals $v_{T_e}(t)$ and

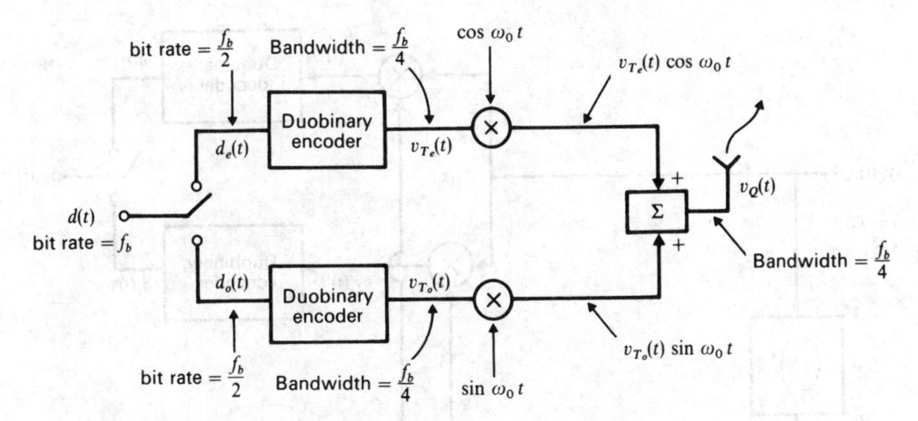

Figure 6.15-1 QPR encoder.

$v_{T_o}(t)$. Each duobinary encoder is similar to the encoder shown in Fig. 6.14-3*a* except that each delay is now $2T_b$ rather than T_b, the data rate of the input is $f_b/2$ rather than f_b and the bandwidth of the brickwall filter is now $\frac{1}{2}(f_b/2) = f_b/4$ rather than $f_b/2$. Thus the bandwidth required to pass $v_{T_e}(t)$ or $v_{T_o}(t)$ is $f_b/4$. Each duobinary signal is then modulated using the quadrature carrier signals $\cos \omega_0 t$ and $\sin \omega_0 t$.

The bandwidth of each of the quadrature amplitude modulated signals is

$$B_{QPR} = 2(f_b/4) = f_b/2 \tag{6.15-3}$$

Hence the total bandwidth required to pass a QPR signal is also B_{QPR}, since the two quadrature components occupy the same frequency band.

It should be noted that if QPSK, rather than QPR, were used to encode the data $d(t)$, the bandwidth required would be $B_{QPSK} = f_b$. However, if 16 QAM or 16 PSK were used to encode the data the required bandwidth would be $B_{16QAM} = B_{16PSK} = f_b/2$. Thus the spectrum required to pass a QPR signal is similar to that required to pass 16 QAM or 16 PSK. However, the QPR signal displays no (or in practice very small) sidelobes which makes QPR the encoding system of choice when spectrum width is the major problem. The drawback in using QPR is that the transmitted signal envelope is not a constant but varies with time.

QPR Decoder

A QPR decoder is shown in Fig. 6.15-2. As in 16-QAM and 16-PSK to decode (see Secs. 6.6 and 6.7) the input signal, $v_Q(t)$ is first raised to the fourth power, filtered and then frequency divided by 4. The result yields the two quadrature carriers: $\cos \omega_0 t$ and $\sin \omega_0 t$. The detailed derivation of this result is left to Prob. 6.15-1. Using the two quadrature carriers we demodulate $v_Q(t)$ and obtain the two baseband duobinary signals $v_{T_e}(t)$ and $v_{T_o}(t)$. Duobinary decoding then

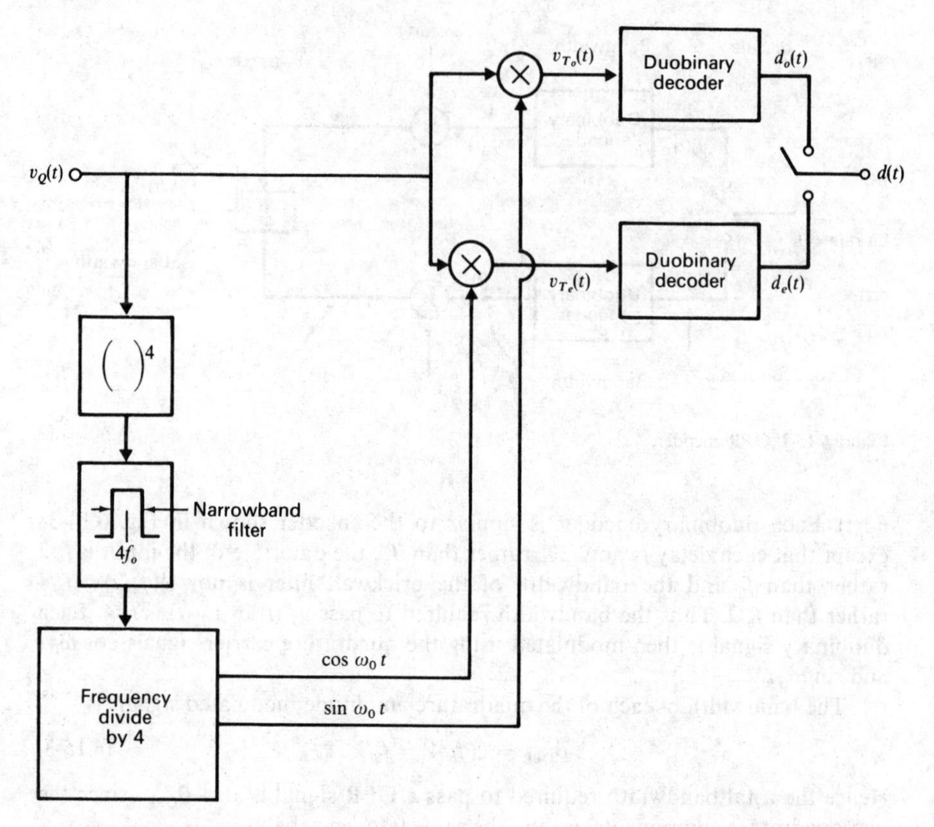

Figure 6.15-2 QPR decoder.

takes place; each duobinary decoder being similar to the decoder shown in Fig. 6.14-3b except that they operate at $f_b/2$ rather than at f_b. The reconstructed data $d_o(t)$ and $d_e(t)$ is then combined to yield the data $d(t)$.

PROBLEMS

6.2-1 The data $b(t)$ consists of the bit stream 001010011010. Assume that the bit rate f_b is equal to the carrier frequency f_0 and sketch $v_{\text{BPSK}}(t)$.

6.2-2. Calculate and plot the power P_f contained in the power spectral density $G_b(f)$ in Eq. (6.2-8) as a function of the frequency range 0 to f. Note that when $f = 0$, $P_f = 0$ and when f is infinite $P_f = P_s$.

6.3-1. The bit stream $d(t)$ is to be transmitted using DPSK. If $d(t)$ is 001010011010, determine $b(t)$. Show that $b(t)b(t - T_b)$ yields the original data.

6.4-1. The bit stream $d(t)$ is to be transmitted using DEPSK. If $d(t)$ is 001010011010, determine $b(t)$. Show that after decoding, using the circuit of Fig. 6.4-1, the data $d(t)$ is recovered. Show that if the fourth bit in $b(t)$ is in error, then the fourth and fifth data bits, $d(t)$, will also be in error.

6.5-1. The bit stream $b(t)$ is to be transmitted using OQPSK. If $b(t)$ is 001010011010 sketch $v_m(t)$ (see Eq. 6.5-1). Assume $f_s = f_b/2 = f_0$.

6.5-2. Repeat Prob. 6.5-1 if the odd bit stream is delayed by T_b so that QPSK is generated.

6.5-3. Verify that the circuit shown in Fig. 6.5-5 yields $\sin \omega_0 t$ and $\cos \omega_0 t$, each with the same phase as the received signal.

6.5-4. Verify Eq. (6.5-5). Show that the minimum distance between signals is given by Eq. (6.5-6).

6.6-1. 8-ary PSK is used to modulate the bit stream 001010011010. Sketch the transmitted waveform $v_{8PSK}(t)$. Assume $f_s = f_b/3 = f_0$.

6.6-2. Repeat Prob. 6.6-1 if 16-ary PSK is used. Assume $f_s = f_b/4 = f_0$.

6.6-3. Verify Eq. (6.6-9) taking note of the fact that ϕ_m has the discrete values given by Eq. (6.6-2) and does not take on all values from $-\pi$ to π.

6.6-4. Verify Eq. (6.6-11).

6.6-5. If M becomes large, in Eq. (6.6-11), show that d approaches zero.

6.7-1. Verify Eqs. (6.7-2) and (6.7-3).

6.7-2. The bit stream 001010011010 is to be transmitted using 16-QASK. Sketch the transmitted waveform. Assume $f_s = f_b/4 = f_0$.

6.7-3. Verify Eq. (6.7-5).

6.7-4. Verify Eq. (6.7-11).

6.7-5. Verify Eq. (6.7-12).

6.7-6. Verify Eq. (6.7-15). Calculate the average power contained in the signal $v_{QASK}^4(t)$ at the frequency $4f_0$.

6.8-1. The bit stream 001010011010 is to be transmitted using BFSK. Sketch the transmitted waveform. Assume $f_L = f_b$ and $f_H = 2f_b$.

6.8-2. One way to obtain the power spectral density of a waveform is to first obtain the autocorrelation function $R(\tau)$. The power spectral density is the Fourier transform of $R(\tau)$. Since $R(\tau) = E[v(t)v(t + \tau)]$,

(*a*) Find $R(\tau)$ for the BFSK signal of Eq. (6.8-6) and verify Fig. (6.8-2).

(*b*) If $\theta_H = \theta_L$ find $R(\tau)$ and the power spectral density.

6.8-3. Verify Eq. (6.8-12).

6.8-4. Verify Eqs. (6.8-15).

6.8-5. Verify Eqs. (6.8-16*a*) and (6.8-17*b*).

6.8-6. Verify Eq. (6.8-19). What is the relationship between f_H, f_L and f_b in order that Eq. (6.8-19) reduce to Eq. (6.8-12).

6.10-1. Consider using 4-ary FSK.

(*a*) Show that if the frequencies are separated by f_s, they are each orthogonal.

(*b*) Calculate the bandwidth B under the condition of (*a*). Show that the bandwidth is five-eighths of the value required by Eq. (6.10-1). Explain the difference.

(*c*) Determine the ratio of bandwidths of 2-ary FSK and 4-ary FSK when the frequencies are separated by f_s. Repeat for $2f_s$.

6.11-1. Verify Eq. (6.11-6*a*).

6.11-2. Verify Eq. (6.11-8) using Eq. (6.11-12).

6.11-3. If $b(t)$ is 001010011010 sketch $v_{MSK}(t)$. Assume in Eq. (6.11-12) that $m = 5$.

6.11-4. Repeat Prob. 6.11-3 if $m = 3$.

6.11-5. Verify Eq. (6.11-18).

6.11-6. Verify that in Fig. 6.11-5*a* the waveforms $x(t)$ and $y(t)$ are as shown and that the output is $v_{MSK}(t)$.

6.12-1. Verify Table 6.12-1.

6.12-2. Verify Eqs. (6.12-7) and (6.12-8).

6.12-3. Verify Eqs. (6.12-10) and (6.12-11).

6.12-4. Plot $G_D(f)$ and $G_T(f)$ as a function of frequency using Eqs. (6.12-10) and (6.12-11).

6.14-1. Verify Eq. (6.14-5).

6.15-1. Show that if two duobinary signals $v_{T_e} = b_e(k) \cdot b_e(k-1)$ and $v_{T_o} = b_o(k)b_o(k-1)$ are modulated using carrier signals which are in time quadrature so that:

$$v_Q(t) = \sqrt{P_s}\, V_{T_e}(t) \cos \omega_0 t + \sqrt{P_s}\, V_{T_o}(t) \sin \omega_0 t$$

then the carrier signals $\cos \omega_0 t$ and $\sin \omega_0 t$ can be recovered in the receiver by filtering the signal $V_Q^4(t)$.

6.15-2. Repeat 6.15-1 if $V_{T_e}(t)$ and $V_{T_o}(t)$ are each filtered by a "brick wall" low-pass-filter prior to being amplitude modulated.

CHAPTER

SEVEN

MATHEMATICAL REPRESENTATION OF NOISE

All the communication systems discussed in the preceding chapters accomplish the same end. They allow us to reproduce the signal, impressed on the communication channel at the transmitter, at the demodulator output. Our only basis for comparison between systems, up to this point, has been bandwidth occupancy, convenience of multiplexing, and ease of implementation of the physical hardware. We have neglected, however, in our preceding discussions, the very important and fundamental fact that, in any real physical system, when the signal voltage arrives at the demodulator, it will be accompanied by a voltage waveform which varies with time in an entirely *unpredictable* manner. This unpredictable voltage waveform is a random process called *noise*. A signal accompanied by such a waveform is described as being *contaminated* or *corrupted* by noise. We now find that we have a new basis for system comparison, that is, the extent to which a communication system is able to distinguish the signal from the noise and thereby yield a low-*distortion* and low-*error* reproduction of the original signal.

In the present chapter we shall make only a brief reference to the sources of noise which are discussed more extensively in Chap. 14. Here we shall be concerned principally with a discussion of the mathematical representation and statistical characterizations of noise.

7.1 SOME SOURCES OF NOISE

One source of noise is the constant agitation which prevails throughout the universe at the molecular level. Thus, a piece of solid metal may appear to our gross view to be completely at rest. We know, however, that the individual molecules are vibrating about their positions of equilibrium in a crystal lattice, and that the conduction electrons of the metal are wandering randomly throughout the volume of the metal. Similarly the molecules of an enclosed gas are in constant motion, colliding with one another and colliding also with the walls of the container. These agitations of molecules are called *thermal* agitations because they increase with temperature.

Let us consider a simple resistor. It is a resistor, or rather a conductor, because there are within it conduction electrons which are free to wander randomly through the entire volume of the resistor. On the average these electrons will be uniformly distributed through the volume, as will positive ions, and the entire structure will be electrically neutral. However, because of the random and erratic wanderings of the electrons, there will be *statistical fluctuations* away from neutrality. Thus at one time or another the distribution of charge may not be uniform, and a voltage difference will appear between the resistor terminals. The random, erratic, unpredictable voltage which so appears is referred to as *thermal resistor noise*. As is to be expected, thermal resistor noise increases with temperature. Resistor noise also increases with the resistance value of the resistor, being zero in a perfect conductor.

A second type of noise results from a phenomenon associated with the flow of current across semiconductor junctions. The charge carriers, electrons or holes, enter the junction region from one side, drift or are accelerated across the junction, and are collected on the other side. The average junction current determines the average interval that elapses between the times when two successive carriers enter the junction. However, the exact interval that elapses is subject to random statistical fluctuations. This randomness gives rise to a type of noise which is referred to as *shot noise*. Shot noise is also encountered as a result of the randomness of emission of electrons from a heated surface and is consequently also associated with thermionic devices.

When a signal reaches a receiver it may well arrive very greatly attenuated. It is therefore necessary to provide amplification. This amplification is accomplished in circuits using active devices (transistors, etc.) and resistors. Hence the signal becomes corrupted by *thermal* and *shot* noise. Even more, the signal may have been contaminated by noise as a result of many types of random disturbances superimposed on the signal during the course of its transfer over the communication channel. The contamination of the signal may take several forms. The noise may be added to the signal, in which case it is called *additive* noise, or the noise may multiply the signal, in which case the effect is called *fading*.

We shall confine our interest, for the most part, albeit not exclusively, to noise which may be described as an ergodic random process. The characteristic of ergodicity of interest here is that an ergodic process is also stationary, that is, statistical averages taken over an ensemble representing the processes yield a result that is independent of the time at which the averages are evaluated. We shall further assume, except where specifically noted, that the probability density of the noise is gaussian. In very many communication systems and in a wide variety of circumstances the assumption of a gaussian density is justifiable. On the other hand, it needs to be noted that such an assumption is hardly universally valid. For example, if gaussian noise is applied to the input of a rectifier circuit, the output is not gaussian. Similarly, it may well be that the noise encountered on a telephone line or on other channels consists of short, pulse-type disturbances whose amplitude distribution is decidedly not gaussian.

7.2 A FREQUENCY-DOMAIN REPRESENTATION OF NOISE

In a communication system noise is often passed through filters. These filters are usually described in terms of their characteristics in the frequency domain. Hence, to determine the influence of these filters on the noise, it is convenient to have a *frequency-domain characterization* of the noise.[1] We shall now establish such a frequency-domain characterization. On the basis of this representation we shall be able to define a *power spectral density* for a noise waveform that has characteristics similar to those of the power spectral density of a deterministic waveform. Our discussion will be somewhat heuristic.

Let us select a particular sample function of the noise, and select from that sample function an interval of duration T extending, say, from $t = -T/2$ to $t = T/2$. Such a noise sample function $n^{(s)}(t)$ is shown in Fig. 7.2-1*a*. Let us generate, as in Fig. 7.2-1*b*, a periodic waveform in which the waveform in the selected interval is repeated every T sec. This periodic waveform $n_T^{(s)}(t)$ can be expanded in a Fourier series, and such a series will properly represent $n^{(s)}(t)$ in the interval

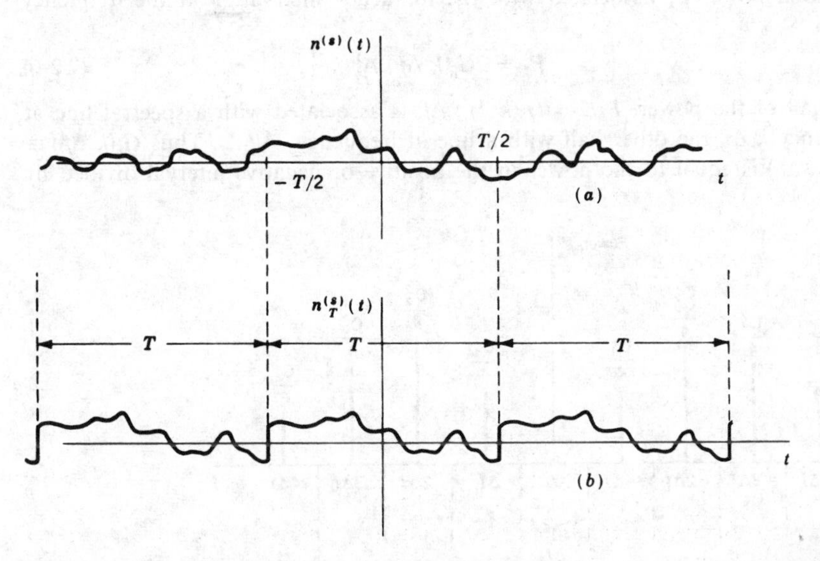

Figure 7.2-1 (*a*) A sample noise waveform. (*b*) A periodic waveform is generated by repeating the interval in (*a*) from $-T/2$ to $T/2$.

$-T/2$ to $T/2$. The fundamental frequency of the expansion is $\Delta f = 1/T$, and, assuming no dc component, we have

$$n_T^{(s)}(t) = \sum_{k=1}^{\infty} (a_k \cos 2\pi k \, \Delta f t + b_k \sin 2\pi k \, \Delta f t) \tag{7.2-1}$$

or alternately

$$n_T^{(s)}(t) = \sum_{k=1}^{\infty} c_k \cos (2\pi k \, \Delta f t + \theta_k) \tag{7.2-2}$$

in which a_k, b_k, and c_k are the constant coefficients of the spectral terms and θ_k is a phase angle. Of course,

$$c_k^2 = a_k^2 + b_k^2 \tag{7.2-3}$$

and

$$\theta_k = -\tan^{-1} \frac{b_k}{a_k} \tag{7.2-4}$$

The power spectrum of the expansion is shown in Fig. 7.2-2. The power associated with each spectral term is $c_k^2/2 = (a_k^2 + b_k^2)/2$. Since a two-sided spectrum is shown, each power spectral line is of height $c_k^2/4$. The frequency axis has been marked off in intervals Δf, and a power spectral line is located at the center of each interval. We now define the power spectral density at the frequency $k \, \Delta f$ as the quantity

$$G_n(k \, \Delta f) \equiv G_n(-k \, \Delta f) \equiv \frac{c_k^2}{4 \, \Delta f} = \frac{a_k^2 + b_k^2}{4 \, \Delta f} \tag{7.2-5}$$

The total power P_k associated with the frequency interval Δf at the frequency $k \, \Delta f$ is

$$P_k = 2G_n(k \, \Delta f) \, \Delta f \tag{7.2-6}$$

One-half of the power, $P_k/2 = G_n(k \, \Delta f) \, \Delta f$, is associated with a spectral line at frequency $k \, \Delta f$, the other half with a line at frequency $-k \, \Delta f$. Thus $G_n(k \, \Delta f) \equiv G_n(-k \, \Delta f)$ is equal to the power in the positive or negative interval divided by

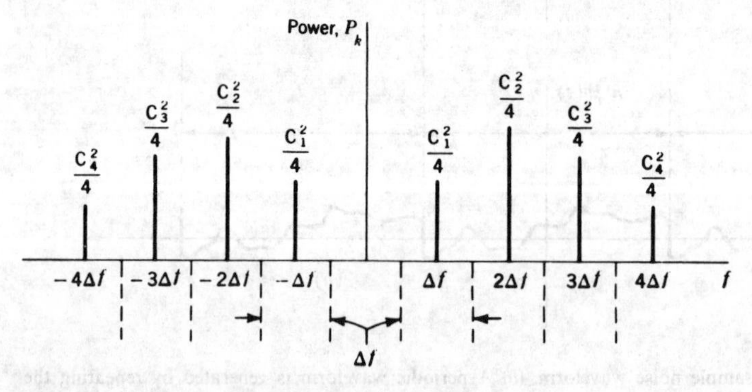

Figure 7.2-2 The power spectrum of the waveform $n_T^{(s)}$.

the size of the interval. Hence $G_n(k\,\Delta f)$ is the (two-sided) *mean power spectral density* within each interval.

For a particular sample function of the noise one obtains specific values of coefficients a_k and b_k in Eq. (7.2-1) [or coefficients c_k and angles θ_k in Eq. (7.2-2)]. Different sample functions will result in different coefficients. If we propose that the representations of Eq. (7.2-1) or (7.2-2) are to represent generally the random process under discussion, i.e., the periodic noise, we need but consider that the a_k's, b_k's, c_k's, and θ_k's are not fixed numbers but are instead *random variables.*† Finally, let us allow that $T \to \infty$ $(\Delta f \to 0)$, so that the periodic sample functions of the noise revert to the actual noise sample functions. We then have that the noise $n(t)$ is to be represented as

$$n(t) = \lim_{\Delta f \to 0} \sum_{k=1}^{\infty} (a_k \cos 2\pi k\,\Delta f t + b_k \sin 2\pi k\,\Delta f t) \tag{7.2-7}$$

or as

$$n(t) = \lim_{\Delta f \to 0} \sum_{k=1}^{\infty} c_k \cos (2\pi k\,\Delta f t + \theta_k) \tag{7.2-8}$$

We continue to accept Eq. (7.2-5) as the definition of the power spectral density of the noise $n(t)$, except that we now replace c_k^2 by $\overline{c_k^2}$, that is, by the expected or ensemble average value of the square of the random variable c_k. Further, as $\Delta f \to 0$, the discrete spectral lines get closer and closer, finally forming a continuous spectrum. In Eq. (7.2-5) we therefore replace $k\,\Delta f$ by the continuous frequency variable f. From Eq. (7.2-3) we also have that

$$\overline{c_k^2} = \overline{a_k^2} + \overline{b_k^2} \tag{7.2-9}$$

so that finally Eq. (7.2-5) becomes

$$G_n(f) = \lim_{\Delta f \to 0} \frac{\overline{c_k^2}}{4\,\Delta f} = \lim_{\Delta f \to 0} \frac{\overline{a_k^2} + \overline{b_k^2}}{4\,\Delta f} \tag{7.2-10}$$

Note that the power in the frequency range from f_1 to f_2 is

$$P(f_1 \to f_2) = \int_{-f_2}^{-f_1} G_n(f)\,df + \int_{f_1}^{f_2} G_n(f)\,df = 2\int_{f_1}^{f_2} G_n(f)\,df \tag{7.2-11}$$

while the total power P_T is

$$P_T = \int_{-\infty}^{\infty} G_n(f)\,df = 2\int_{0}^{\infty} G_n(f)\,df \tag{7.2-12}$$

† In Chap. 2, following accepted practice, we used capital symbols to represent random variables (e.g., X, Y, etc.) and lowercase symbols (x, y, etc.) to represent possible particular values attainable by the random variables. Since it is inconvenient to do so, we shall no longer persist in this precision of notation. No confusion will result thereby since the text will always make clear the significance of a symbol.

7.3 THE EFFECT OF FILTERING ON THE PROBABILITY DENSITY OF GAUSSIAN NOISE

We shall now show that if, as in Fig. 7.3-1*a*, gaussian noise $n_i(t)$ is applied to the input of a filter, the output noise $n_o(t)$ is also gaussian. Let the impulse response of the filter be $h(t)$. Then, applying Eq. (1.12-8) to the present situation, we have that

$$n_o(t) = \int_{-\infty}^{\infty} n_i(\tau)h(t-\tau)\,d\tau = \int_{-\infty}^{t} n_i(\tau)h(t-\tau)\,d\tau \tag{7.3-1}$$

In Eq. (7.3-1) we have taken cognizance of the fact that, in general, the upper limit may actually be set at $\tau = t$, since the variable of integration is τ and $h(t-\tau) = 0$ for $\tau > t$. Equation (7.3-1) expresses the output $n_o(t)$ as the superposition of a succession of impulses of strength $n_i(\tau)\,d\tau$ applied at the filter input. This point is emphasized by rewriting Eq. (7.3-1) in the form

$$n_o(t) = \lim_{\Delta\tau \to 0} \sum_{k=-\infty}^{k=t/\Delta\tau} n_i(k\,\Delta\tau)h(t - k\,\Delta\tau)\,\Delta\tau \tag{7.3-2}$$

in which k ranges over integral values. In the limit $k\,\Delta\tau$ reverts to the continuous variable τ, and the summation reverts to the integral. In Eq. (7.3-2), $n_i(k\,\Delta\tau)$ is a random variable. That is, the noise is represented by an ensemble of sample functions, and the value of $n_i(k\,\Delta\tau)$ depends on the sample function we consider. On the other hand, the quantities $h(t - k\,\Delta\tau)$ are fixed, deterministic numbers, regulated by the nature of the filter.

Let us assume that the input noise is white and gaussian. Then, as discussed in Sec. 2.24, the past history of the noise waveform provides no information about its future behavior. That is to say, on any one sample noise waveform of the ensemble, successive noise voltage values are independent of one another no matter how small the interval $\Delta\tau$ between voltage determinations. Hence, the random variables $n_i(k\,\Delta\tau)$ and $n_i(l\,\Delta\tau)$ are independent except for the case where $k = l$. Thus, at any time, with $t = t_0$, $n_o(t) = n_o(t_0)$, as given by Eq. (7.3-2), is a

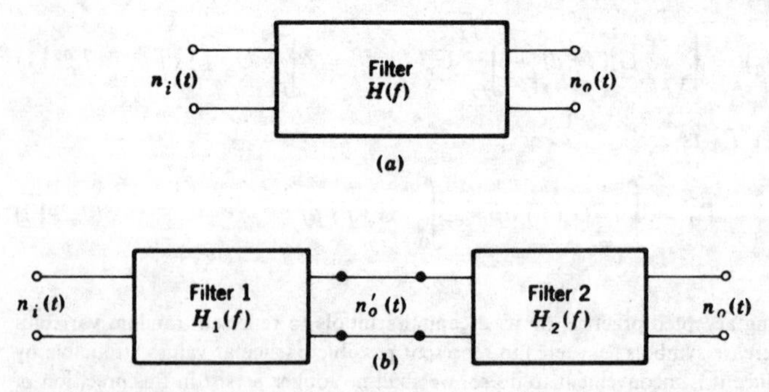

Figure 7.3-1 (*a*) Gaussian noise $n_i(t)$ is applied to a linear filter whose output is $n_o(t)$. (*b*) The filter in (*a*) is split into two parts.

linear superposition of independent gaussian random variables. Hence, as discussed in Sec. 2.18, $n_o(t_0)$ is a gaussian random variable, and $n_o(t)$ is a gaussian random process.

Now suppose that the filter of transfer function $H(f)$ is split into two parts, as shown in Fig. 7.3-1*b*, with $H(f) = H_1(f)H_2(f)$. Then if $n_i(t)$ is white and gaussian, both $n_o'(t)$ and $n_o(t)$ are, on the basis of the above discussion, gaussian, albeit, in general, not white. Further, since both $n_o'(t)$ and $n_o(t)$ are gaussian, we have the result that nonwhite gaussian noise, when applied to a linear filter, yields an output noise which is again gaussian. Strictly, we have established this last result only for the case where the nonwhite gaussian noise input is itself derived by filtering white noise. But this result is quite general enough for our purposes.

Equation (7.3-2) may be applied directly to filter 2 in Fig. 7.3-1*b* by simply replacing the impulses $n_i(k\ \Delta\tau)$ by the impulses $n_o'(k\ \Delta\tau)$. However, since $n_o'(t)$ is not white noise, successive noise voltage values are *not* independent of one another; that is, the random variable $n_o'(k\ \Delta\tau)$ and $n_o'(l\ \Delta\tau)$ are gaussian but not independent. Nonetheless, we found that $n_o(t)$ is gaussian. Hence we deduce that a linear superposition of *dependent* gaussian random variables is still gaussian.

7.4 SPECTRAL COMPONENTS OF NOISE

We have represented noise $n(t)$ as a superposition of noise spectral components. The spectral component associated with the kth frequency interval is given, in the limit as $\Delta f \to 0$, by $n_k(t)$ written as

$$n_k(t) = a_k \cos 2\pi k\ \Delta f t + b_k \sin 2\pi k\ \Delta f t \tag{7.4-1}$$

or as

$$n_k(t) = c_k \cos (2\pi k\ \Delta f t + \theta_k) \tag{7.4-2}$$

The spectral components which compose a deterministic waveform are themselves deterministic. The spectral components, as in the present case, which compose noise, are themselves random processes. Thus, in Eqs. (7.4-1) and (7.4-2), a_k, b_k, c_k, and θ_k are random variables, and $n_k(t)$ represents an ensemble of sample functions, one sample function for each possible set of values of a_k and b_k or of c_k and θ_k. The sample functions are each deterministic waveforms. They are, as a matter of fact, pure sinusoids differing from one another in phase and amplitude, depending on the value of θ_k and c_k. The random process $n_k(t)$ is stationary; that is, its statistical properties do not change with time. However, it is not ergodic; that is, the time averages of the individual sample function of the ensemble are different from one another.

We look now at some of the properties of the random variables a_k and b_k. The normalized power P_k (variance) of $n_k(t)$ is determined by taking the average over the *ensemble* of $[n_k(t)]^2$. We find, from Eq. (7.4-1), that

$$P_k = \overline{[n_k(t)]^2} = \overline{a_k^2} \cos^2 2\pi k\ \Delta f t + \overline{b_k^2} \sin^2 2\pi k\ \Delta f t + \overline{2a_k b_k} \sin 2\pi k\ \Delta f t \cos 2\pi k\ \Delta f t \tag{7.4-3}$$

As noted, $n_k(t)$ is a stationary process, so that $\overline{[n_k(t)]^2}$ does not depend on the time selected for its evaluation. We are therefore at liberty to evaluate P_k by substituting in Eq. (7.4-3) a value $t = t_1$ for which $\cos 2\pi k\, \Delta f t_1 = 1$, in which case $\sin 2\pi k\, \Delta f t_1 = 0$. We then have

$$P_k = \overline{a_k^2} \tag{7.4-4}$$

Similarly we may show that $P_k = \overline{b_k^2}$ so that

$$\overline{a_k^2} = \overline{b_k^2} \tag{7.4-5}$$

From Eqs. (7.2-6), (7.2-9), (7.4-4), and (7.4-5) we have

$$P_k = 2G_n(k\,\Delta f)\,\Delta f = 2G_n(-k\,\Delta f)\,\Delta f = \overline{a_k^2} = \overline{b_k^2} = \frac{\overline{a_k^2}}{2} + \frac{\overline{b_k^2}}{2} = \frac{\overline{c_k^2}}{2} \tag{7.4-6}$$

Since $\overline{a_k^2} = \overline{b_k^2}$, Eq. (7.4-3) may be written

$$P_k = \overline{a_k^2}(\cos^2 2\pi k\,\Delta f t + \sin^2 2\pi k\,\Delta f t) + \overline{2a_k b_k} \sin 2\pi k\,\Delta f t \cos 2\pi k\,\Delta f t \tag{7.4-7a}$$

Thus

$$P_k = \overline{a_k^2} + \overline{2a_k b_k} \sin 2\pi k\,\Delta f t \cos 2\pi k\,\Delta f t \tag{7.4-7b}$$

We note, however, from Eq. (7.4-4) that $P_k = \overline{a_k^2}$ independently of time. In order that Eq. (7.4-7b) be consistent with this result, we require that

$$\overline{a_k b_k} = 0 \tag{7.4-8}$$

Thus, the coefficients a_k and b_k are *uncorrelated*.

We may also establish that the coefficients a_k and b_k are gaussian. For this purpose we use Eq. (7.4-1) and substitute for t a value $t = t_1$, for which $\cos 2\pi k\,\Delta f t_1 = 1$ and $\sin 2\pi k\,\Delta f t_1 = 0$. Then

$$n_k(t_1) = a_k \tag{7.4-9}$$

Now $n_k(t_1)$ is a gaussian random variable. We have this result from the consideration that $n_k(t)$ may be viewed as the output of a very narrowband filter whose input is gaussian noise. As discussed in Sec. 7.3 such a noise output is gaussian, and the noise voltage at any time, as at $t = t_1$ above, has a gaussian probability density. Hence, Eq. (7.4-9) says that a_k is also a gaussian random variable. It is, of course, similarly established that b_k is gaussian.

We note also that since $n_k(t)$ is the output of a narrowband filter at frequency $k\,\Delta f$, it has no dc component. Hence from Eq. (7.4-9) a_k has no dc component; that is, $\overline{a_k} = 0$, and, of course, similarly $\overline{b_k} = 0$.

Finally, let us consider two spectral components of noise, one at frequency $k\,\Delta f$ and the other at frequency $l\,\Delta f$. Then

$$n_k(t) = a_k \cos 2\pi k\,\Delta f t + b_k \sin 2\pi k\,\Delta f t \tag{7.4-10}$$

and

$$n_l(t) = a_l \cos 2\pi l\,\Delta f t + b_l \sin 2\pi l\,\Delta f t \tag{7.4-11}$$

If we form the product $n_k(t)n_l(t)$ from Eqs. (7.4-10) and (7.4-11), we find that the product has four terms (products of sinusoids), all of which are time-dependent, provided that $k \neq l$. The coefficients of these terms are $a_k a_l$, $a_k b_l$, $b_k a_l$, and $b_k b_l$. Now let us take the ensemble average of the product, that is, $\overline{n_k(t)n_l(t)}$. Then, again, because of the stationary character of the random processes involved, this ensemble average must be time-independent. Hence we have that

$$\overline{a_k a_l} = \overline{a_k b_l} = \overline{b_k a_l} = \overline{b_k b_l} = 0 \tag{7.4-12}$$

That is, each of the coefficients a_k and b_k is uncorrelated with each of the coefficients a_l and b_l.

In summary, we find that we have described noise in the following manner. Noise $n(t)$ is a gaussian, ergodic, random process. It may be represented as a linear superposition of spectral components of the form given in Eq. (7.4-1) with the understanding, of course, that the description becomes more precise as $\Delta f \to 0$. The coefficients a_k and b_k are gaussian random variables of average value zero and equal variance (normalized power) related to the two-sided power spectral density as in Eq. (7.4-6). The coefficients a_k and b_k are uncorrelated with one another and are uncorrelated also with the coefficients of a spectral component at a different frequency.

We may now deduce some statistical characteristics of interest concerning the c_k's and θ_k's in the noise spectral component representation of Eq. (7.4-2). We note from Eqs. (7.2-3) and (7.2-4) that the c_k's and θ_k's are related to the a_k's and b_k's in precisely the manner in which the random variables R and θ of Sec. 2.16 are related to the gaussian variables X and Y. From the results of Sec. 2.16 we then find (Prob. 7.4-2) that the c_k's have a Rayleigh probability density

$$f(c_k) = \frac{c_k}{P_k} e^{-c_k^2/2P_k} \qquad c_k \geq 0 \tag{7.4-13}$$

where P_k [Eq. (7.4-6)] is the normalized power in the spectral range Δf at the frequency $k\,\Delta f$. Similarly, the angle θ_k has a uniform probability density

$$f(\theta_k) = \frac{1}{2\pi} \qquad -\pi \leq \theta_k \leq \pi \tag{7.4-14}$$

Furthermore, the amplitude c_k and phase θ_k are independent of one another and are independent as well of the amplitude and phase of a spectral component at a different frequency.

7.5 RESPONSE OF A NARROWBAND FILTER TO NOISE

If the representation of noise as a superposition of spectral components is a reasonable one, we should expect that when noise is passed through a narrowband filter the output of the filter should look rather like a sinusoid. We find that such

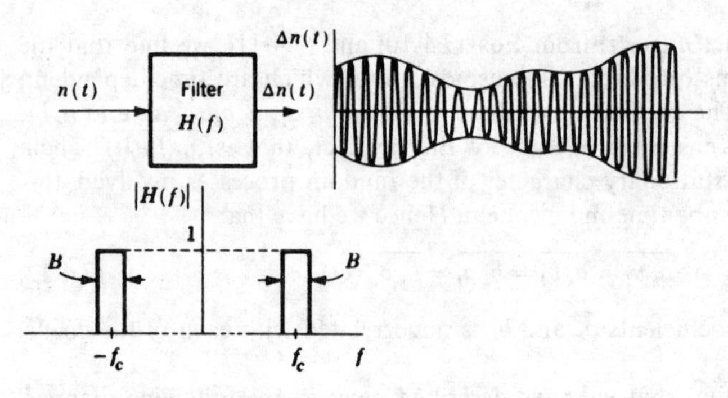

Figure 7.5-1 Response of a narrowband filter to noise.

is indeed the case, for we find that the output of a narrowband filter with noise input has the appearance shown in Fig. 7.5-1. The output waveform looks like a sinusoid except that, as expected, the amplitude varies randomly. The spectral range of the *envelope* of the filter output encompasses the spectral range from $-B/2$ to $B/2$, where B is the filter bandwidth. The average frequency of the waveform is the center frequency f_c of the filter. If $B \ll f_c$, the envelope changes very "slowly" and makes an appreciable change only over very many cycles. Thus, while the spacings of the *zero crossings* of the waveform are not precisely constant, the change from cycle to cycle is small, and when averaged over many cycles is quite constant at the value $1/2f_c$. Finally, we may note that as B becomes progressively smaller, so also does the average amplitude, and the waveform becomes more and more sinusoidal.

7.6 EFFECT OF A FILTER ON THE POWER SPECTRAL DENSITY OF NOISE

Let a spectral component of noise $n_{k_i}(t)$ given by Eq. (7.4-1) be the input to a filter whose transfer function at the frequency $k\,\Delta f$ is

$$H(k\,\Delta f) = |H(k\,\Delta f)|\,e^{j\varphi_k} = |H(k\,\Delta f)|\underline{/\varphi_k} \tag{7.6-1}$$

The corresponding output spectral component of noise will be $n_{k_o}(t)$

$$n_{k_o}(t) = |H(k\,\Delta f)|\,a_k \cos\,(2\pi k\,\Delta f t + \varphi_k) + |H(k\,\Delta f)|\,b_k \sin\,(2\pi k\,\Delta f t + \varphi_k) \tag{7.6-2}$$

The power P_{k_i} associated with $n_{k_i}(t)$ is, from Eq. (7.4-6),

$$P_{k_i} = \frac{\overline{a_k^2 + b_k^2}}{2} \tag{7.6-3}$$

Since $|H(k\,\Delta f)|$ is a deterministic function, $\overline{[|H(k\,\Delta f)|\,a_k]^2} = |H(k\,\Delta f)|^2\overline{a_k^2}$ and $\overline{[|H(k\,\Delta f)|\,b_k]^2} = |H(k\,\Delta f)|^2\overline{b_k^2}$. Hence, comparing Eq. (7.6-2) with Eq. (7.4-1), we find that the power P_{k_o} associated with $n_{k_o}(t)$ is

$$P_{k_o} = |H(k\,\Delta f)|^2 \frac{\overline{a_k^2} + \overline{b_k^2}}{2} \tag{7.6-4}$$

Finally, then, from Eqs. (7.6-3) and (7.6-4), using also Eq. (7.4-6), we have that the power spectral densities at input and output, $G_{n_i}(k\,\Delta f)$ and $G_{n_o}(k\,\Delta f)$, are related by

$$G_{n_o}(k\,\Delta f) = |H(k\,\Delta f)|^2 G_{n_i}(k\,\Delta f) \tag{7.6-5}$$

In the limit as $\Delta f \to 0$ and $k\,\Delta f$ is replaced by a continuous variable f, Eq. (7.6-5) becomes

$$G_{n_o}(f) = |H(f)|^2 G_{n_i}(f) \tag{7.6-6}$$

Note the similarity between this result and Eq. (1.9-7), which applies to a deterministic waveform.

7.7 SUPERPOSITION OF NOISES

The concept of a power spectrum is useful because it allows us to resolve a deterministic waveform, or a random process, $f(t)$ into a sum

$$f(t) = f_1(t) + f_2(t) + \cdots \tag{7.7-1}$$

in such manner that *superposition of power* applies, i.e., the power of $f(t)$ is the sum of the powers of $f_1(t), f_2(t), \ldots$. When a deterministic waveform is resolved into a series of spectral components, superposition of power applies because of the orthogonality of spectral components of different frequencies. This point was discussed in Sec. 1.7.

We have also represented a noise waveform as a superposition of spectral components, all of which are harmonics of some fundamental frequency Δf which, in the limit, approaches zero. Hence, up to the present point, this feature alone would have been enough to justify superposition of noise power as expressed in Eqs. (7.2-11) and (7.2-12). But suppose that we have two noise processes $n_1(t)$ and $n_2(t)$, whose spectral ranges overlap in part or in their entirety. Then the power P_{12} of the sum $n_1(t) + n_2(t)$ would be

$$P_{12} = E\{[n_1(t) + n_2(t)]^2\} = E[n_1^2(t)] + E[n_2^2(t)] + 2E[n_1(t)n_2(t)] \tag{7.7-2a}$$

$$= P_1 + P_2 + 2E[n_1(t)n_2(t)] \tag{7.7-2b}$$

where P_1 and P_2 are the powers, respectively, of the noise processes $n_1(t)$ and $n_2(t)$, and $E[n_1(t)n_2(t)]$, which is the expected value of the product, is the *cross correlation* of the processes. Thus superposition of power, $P_{12} = P_1 + P_2$, continues to apply, provided the processes are *uncorrelated*. Such would be the case, for example, if $n_1(t)$ and $n_2(t)$ were the thermal noises of two different resistors even in the same frequency bands.

7.8 MIXING INVOLVING NOISE

Noise with Sinusoid

A situation often encountered in communication systems is one in which noise is mixed with (i.e., multiplied by) a deterministic sinusoidal waveform. Let the sinusoidal waveform be $\cos 2\pi f_0 t$. Then the product of this waveform with a spectral noise component, as given by Eq. (7.4-1), yields

$$n_k(t) \cos 2\pi f_0 t = \frac{a_k}{2} \cos 2\pi(k\,\Delta f + f_0)t + \frac{b_k}{2} \sin 2\pi(k\,\Delta f + f_0)t + \frac{a_k}{2} \cos 2\pi(k\,\Delta f - f_0)t + \frac{b_k}{2} \sin 2\pi(k\,\Delta f - f_0)t \qquad (7.8\text{-}1)$$

Thus the mixing gives rise to two noise spectral components, one at the sum frequency $f_0 + k\,\Delta f$ and one at the difference frequency $f_0 - k\,\Delta f$. In addition, the amplitudes of each of the two noise spectral components generated by mixing has been reduced by a factor of 2 with respect to the original noise spectral component. Hence the variances (normalized power) of the two new noise components are smaller by a factor of 4. Accordingly, if the power spectral density of the original noise component at frequency $k\,\Delta f$ is $G_n(k\,\Delta f)$, then, from Eq. (7.8-1), the new components have spectral densities

$$G_n(k\,\Delta f + f_0) = G_n(k\,\Delta f - f_0) = \frac{G_n(k\,\Delta f)}{4} \qquad (7.8\text{-}2)$$

In the limit as $\Delta f \to 0$, we replace $k\,\Delta f$ by the continuous variable f, and Eq. (7.8-2) becomes

$$G_n(f + f_0) = G_n(f - f_0) = \frac{G_n(f)}{4} \qquad (7.8\text{-}3)$$

In words: given the power spectral density plot $G_n(f)$ of a noise waveform $n(t)$, the power spectral density of $n(t) \cos 2\pi f_0 t$ is arrived at as follows: divide $G_n(f)$ by 4, shift the divided plot to the left by amount f_0, to the right by amount f_0, and add the two shifted plots.

Now consider the following situation: We have noise $n(t)$ from which we single out two spectral components, one at frequency $k\,\Delta f$ and one at frequency $l\,\Delta f$. We mix with a sinusoid at frequency f_0, with f_0 selected to be midway between $k\,\Delta f$ and $l\,\Delta f$; that is, $f_0 = \frac{1}{2}(k + l)\,\Delta f$. Then the mixing will give rise to *four* spectral components, two difference-frequency components and two sum-frequency components. The two difference-frequency components will be at the same frequency $p\,\Delta f = f_0 - k\,\Delta f = l\,\Delta f - f_0$. However, we now show that these difference-frequency components are uncorrelated. Representing the spectral components at $k\,\Delta f$ and at $l\,\Delta f$ as in Eqs. (7.4-10) and (7.4-11), we find that the difference-frequency components are

$$n_{p_1}(t) = \frac{a_k}{2} \cos 2\pi p\,\Delta f t - \frac{b_k}{2} \sin 2\pi p\,\Delta f t \tag{7.8-4}$$

and

$$n_{p_2}(t) = \frac{a_l}{2} \cos 2\pi p\,\Delta f t + \frac{b_l}{2} \sin 2\pi p\,\Delta f t \tag{7.8-5}$$

where $n_{p_1}(t)$ is the difference component due to the mixing of frequencies f_0 and $k\,\Delta f$, while $n_{p_2}(t)$ is the difference component due to the mixing of frequencies f_0 and $l\,\Delta f$. If we now take into account, as we have established in Sec. 7.4, that $\overline{a_k a_l} = \overline{a_k b_l} = \overline{b_k a_l} = \overline{b_k b_l} = 0$, then we find from Eqs. (7.8-4) and (7.8-5) that

$$E[n_{p_1}(t) n_{p_2}(t)] = 0 \tag{7.8-6}$$

Thus, as discussed in connection with Eq. (7.7-2), superposition of power applies, and the power at the difference frequency due to the superposition of $n_{p_1}(t)$ and $n_{p_2}(t)$ is

$$E\{[n_{p_1}(t) + n_{p_2}(t)]^2\} = E\{[n_{p_1}(t)]^2\} + E\{[n_{p_2}(t)]^2\} \tag{7.8-7}$$

Thus, mixing noise with a sinusoidal signal results in a frequency shifting of the original noise by f_0. The variance of the shifted noise is found by adding the variance of each new noise component. Thus we see that the principle stated immediately after Eq. (7.8-3) applies even when there is overlap in the two shifted power spectral density plots.

Noise-noise Mixing

We consider now the result of multiplying two spectral components of noise. Representing the spectral components as in Eq. (7.4-2), we find that the product of two components, one at the frequency $k\,\Delta f$, the other at the frequency $l\,\Delta f$ is

$$\begin{aligned} n_k(t) n_l(t) = {} & \tfrac{1}{2} c_k c_l \cos [2\pi(k + l)\,\Delta f t + \theta_k + \theta_l] \\ & + \tfrac{1}{2} c_k c_l \cos [2\pi(k - l)\,\Delta f t + \theta_k - \theta_l] \end{aligned} \tag{7.8-8}$$

The multiplication thus gives rise to two new spectral components of noise, one at the sum frequency $(k + l)\,\Delta f$ and one at the difference frequency $(k - l)\,\Delta f$. The terms in Eq. (7.8-8) are of the form of Eq. (7.4-2), except that $\frac{1}{2} c_k c_l$ replaces c_k, and θ_k in Eq. (7.4-2) is replaced by $\theta_k + \theta_l$ in one case and $\theta_k - \theta_l$ in the other. Since θ_k and θ_l have uniform probability densities, it is intuitively apparent from the principle of *minimum astonishment* that $\theta_k + \theta_l$ and $\theta_k - \theta_l$ also have uniform densities. (A more formal proof is possible; see Prob. 7.8-3.) Hence the normalized power associated with each of the terms in Eq. (7.8-8) may be deduced as for the spectral component of Eq. (7.4-2). We then have, using Eq. (7.4-6),

$$P_{k+l} = P_{k-l} = \tfrac{1}{2}\,\overline{(\tfrac{1}{2}c_k c_l)^2} \tag{7.8-9}$$

Since c_k and c_l are independent random variables,

$$P_{k+l} = P_{k-l} = \tfrac{1}{8}\,\overline{c_k^2}\;\overline{c_l^2} = \tfrac{1}{2}P_k P_l \tag{7.8-10}$$

7.9 LINEAR FILTERING

Thermal noise has a power spectral density which is quite uniform up to frequencies of the order of 10^{13} Hz. Shot noise has a power spectral density which is reasonably constant up to frequencies which are of the order of the reciprocal of the transit time of charge carriers across the junction. Other noise sources similarly have very wide spectral ranges. We shall assume, in discussing the effect of noise on communication systems, that we have to contend with *white* noise. White noise is noise whose power spectral density is uniform over the entire frequency range of interest. The term *white* is used in analogy with white light, which is a superposition of all visible spectral components. Thus we assume, as shown in Fig. 7.9-1, that over the entire spectrum, including positive and negative frequencies,

$$G_n(f) = \frac{\eta}{2} \tag{7.9-1}$$

in which η is a constant (see Sec. 14.5).

In order to minimize the noise power that is presented to the demodulator of a receiving system, we introduce a filter before the demodulator as indicated in Fig. 7.9-2. The bandwidth B of the filter is made as narrow as possible so as to avoid transmitting any unnecessary noise to the demodulator. For example, in an AM system in which the baseband extends to a frequency of f_M, the bandwidth $B = 2f_M$. In a wideband FM system the bandwidth is proportional to twice the frequency deviation.

It is useful to consider the effect of certain types of filters on the noise. One of the filters most often used is the simple *RC* low-pass filter.

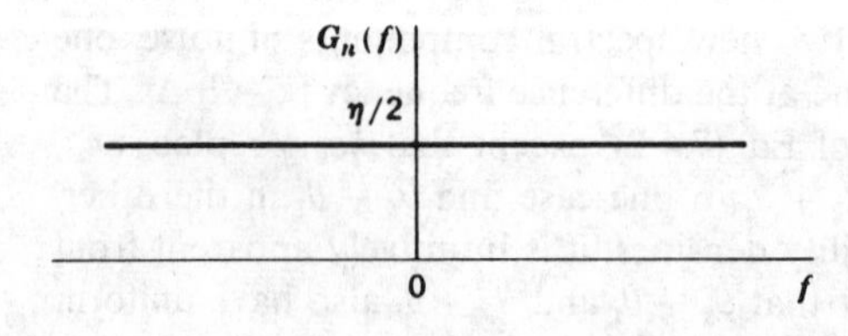

Figure 7.9-1 Power spectral density of white noise.

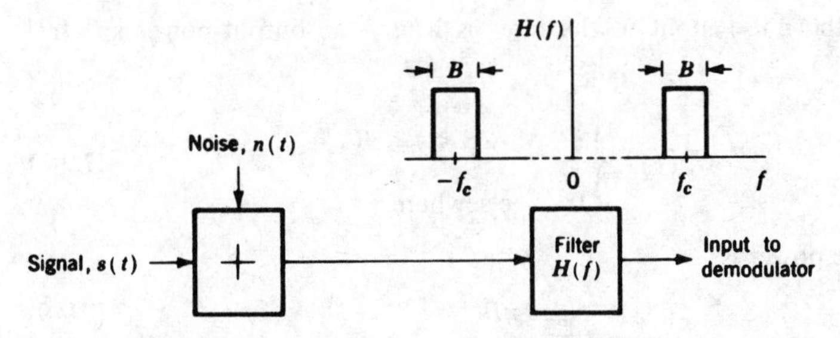

Figure 7.9-2 A filter is placed before a demodulator to limit the noise power input to the demodulator.

The RC Low-pass Filter

An *RC* low-pass filter with a 3-dB frequency f_c has the transfer function

$$H(f) = \frac{1}{1 + jf/f_c} \tag{7.9-2}$$

If the input noise to his filter has a power spectral density $G_{n_i}(f)$ and the power spectral density of the output noise is $G_{n_o}(f)$, then, using Eq. (7.6-6), we have

$$G_{n_o}(f) = G_{n_i}(f)|H(f)|^2 \tag{7.9-3}$$

If the noise is white, $G_{n_i}(f) = \eta/2$ for all frequencies, Eq. (7.9-3) becomes

$$G_{n_o}(f) = \frac{\eta}{2} \frac{1}{1 + (f/f_c)^2} \tag{7.9-4}$$

The noise power at the filter output N_o is

$$N_o = \int_{-\infty}^{\infty} G_{n_o}(f)\, df = \frac{\eta}{2} \int_{-\infty}^{\infty} \frac{df}{1 + (f/f_c)^2} \tag{7.9-5}$$

Changing variables to $x \equiv f/f_c$, and noting that $\int_{-\infty}^{\infty} dx/(1 + x^2) = \pi$, we have

$$N_o = \frac{\pi}{2} \eta f_c \tag{7.9-6}$$

The Rectangular (Ideal) Low-pass Filter

A *rectangular* low-pass filter has the transfer function

$$H(f) = \begin{cases} 1 & |f| \leq B \\ 0 & \text{elsewhere} \end{cases} \tag{7.9-7}$$

Assuming that the noise input to the filter is *white*, the output-power spectral density is

$$G_{n_o}(f) = \begin{cases} \dfrac{\eta}{2} & -B \leq f \leq B \\ 0 & \text{elsewhere} \end{cases} \tag{7.9-8}$$

The output noise power is

$$N_o = \eta B \tag{7.9-9}$$

A Rectangular Bandpass Filter

A rectangular bandpass filter is shown in Fig. 7.9-3. The bandwidth of the filter is $f_2 - f_1$. Then, with a *white* noise input, the output-noise power is

$$N_o = 2\frac{\eta}{2}(f_2 - f_1) = \eta(f_2 - f_1) \tag{7.9-10}$$

A Differentiating Filter

A differentiating filter is a network which yields at its output a waveform which is proportional to the time derivative of the input waveform. As discussed in Sec. 1.9, such a network has a transfer function $H(f)$ which is proportional to the frequency; that is,

$$H(f) = j2\pi\tau f \tag{7.9-11}$$

where τ is a constant factor of proportionality. If white noise with $G_{n_i}(f) = \eta/2$ is passed through such a filter, then the output-noise-power spectral density is

$$G_{n_o}(f) = |H(f)|^2 G_{n_i}(f) = 4\pi^2\tau^2 f^2 \frac{\eta}{2} \tag{7.9-12}$$

If the differentiator is followed by a rectangular low-pass filter having a band-

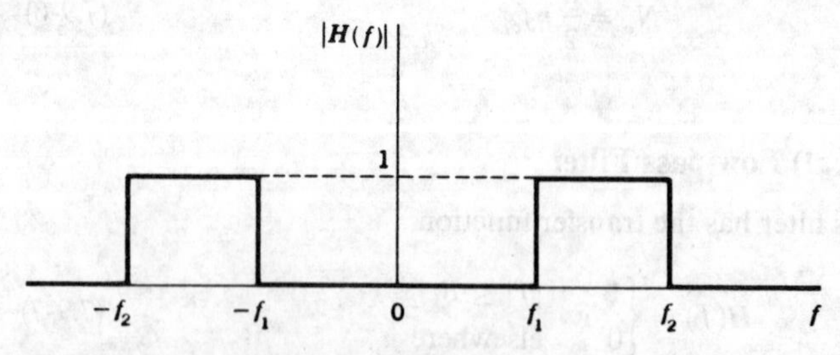

Figure 7.9-3 A rectangular bandpass filter.

width B, as described by Eq. (7.9-7), the noise power at the output of the low-pass filter is

$$N_o = \int_{-B}^{B} 4\pi^2\tau^2 f^2 \frac{\eta}{2}\, df = \frac{4\pi^2}{3}\eta\tau^2 B^3 \tag{7.9-13}$$

An Integrator

Let noise $n(t)$ be applied to the input of an integrator at time $t = 0$. We calculate the noise power at the integrator output at a time $t = T$. The result will be of interest in connection with the discussion of the *matched filter* in Chap. 11.

A network which performs the operation of integration has a transfer function $1/j\omega\tau$. A delay by an interval T is represented by a factor $e^{-j\omega T}$. Hence a network which performs an integration over an interval T may be represented by a network whose transfer function is

$$H(f) = \frac{1}{j\omega\tau} - \frac{e^{-j\omega T}}{j\omega\tau} = \frac{1 - e^{-j\omega T}}{j\omega\tau} \tag{7.9-14}$$

where τ is a constant. We find, with $\omega = 2\pi f$, that

$$|H(f)|^2 = \left(\frac{T}{\tau}\right)^2 \left(\frac{\sin \pi T f}{\pi T f}\right)^2 \tag{7.9-15}$$

The noise power output of such a filter with white input noise of power spectral density $\eta/2$ is (using $x \equiv \pi\tau f$)

$$N_o = \int_{-\infty}^{\infty} \frac{\eta}{2} |H(f)|^2\, df = \frac{\eta}{2}\left(\frac{T}{\tau}\right)^2 \int_{-\infty}^{\infty} \left(\frac{\sin \pi T f}{\pi T f}\right)^2 df \tag{7.9-16a}$$

$$= \frac{\eta T}{2\pi\tau^2} \int_{-\infty}^{\infty} \left(\frac{\sin x}{x}\right)^2 dx \tag{7.9-16b}$$

The definite integral in Eq. (7.9-16b) has the value π, so that finally

$$N_o = \frac{\eta T}{2\tau^2} \tag{7.9-17}$$

It is instructive to obtain Eq. (7.9-17) by a calculation in the time domain. If an input noise *sample* to the integrator is $n_i(t)$ and the corresponding output of the integrator is $n_o(T)$ then

$$n_o(T) = \frac{1}{\tau} \int_0^T n_i(t)\, dt \tag{7.9-18}$$

Note that $n_o(T)$ is a random variable since T is a constant.

The expected value (ensemble average) of the random variable $n_o(T)$ is

$$m_{n_o} = E[n_o(T)] = E\left\{\frac{1}{\tau}\int_0^T n_i(t)\, dt\right\} \tag{7.9-19a}$$

In general, as discussed in Sec. 2.11, if $f(x)$ is the probability density of a random variable x, and $u(x)$ is some function of x, then

$$E[u(x)] = \int_{-\infty}^{\infty} u(x)f(x)\,dx \tag{7.9-19b}$$

Correspondingly, we have in the present case, that

$$m_{n_o} = \int_{-\infty}^{\infty} dn_i\, f(n_i)\left\{\frac{1}{\tau}\int_0^T n_i(t)\,dt\right\} \tag{7.9-19c}$$

Interchanging the order of integration (see Prob. 7.9-3) yields

$$m_{n_o} = \frac{1}{\tau}\int_0^T dt\left\{\int_{-\infty}^{\infty} n_i\, f(n_i)\,dn_i\right\} \tag{7.9-20}$$

But the integral in brackets is the mean value of n_i, i.e.,

$$m_{n_i} = E(n_i) = \int_{-\infty}^{\infty} n_i\, f(n_i)\,dn_i \tag{7.9-21}$$

Assuming, as before, that the average value of the input noise $m_{n_i} = 0$, we see that the average value of the output noise $m_{n_o} = 0$ as well.

The normalized noise power N_o corresponding to the random variable $n_o(T)$ is equal to the variance of $n_o(T)$ (see Sec. 2.12). Thus

$$N_o = \sigma_{n_o}^2 = E\{[n_o(T)]^2\} = E\left\{\frac{1}{\tau^2}\int_0^T n_i(t)\,dt \cdot \int_0^T n_i(\lambda)\,d\lambda\right\} \tag{7.9-22}$$

where t and λ are dummy variables of integration. N_o can be rewritten as

$$N_o = E\left\{\frac{1}{\tau^2}\int_0^T\int_0^T n_i(t)n_i(\lambda)\,dt\,d\lambda\right\} \tag{7.9-23}$$

If we define

$$\alpha \equiv n_i(t) \qquad \text{and} \qquad \beta \equiv n_i(\lambda)$$

Equation (7.9-23) can be written as

$$N_o = \frac{1}{\tau^2}\int_{-\infty}^{\infty} f(\alpha, \beta)\left\{\int_0^T dt \int_0^T d\lambda\alpha\beta\right\} d\alpha\,d\beta \tag{7.9-24}$$

Interchanging the order of integration (see Prob. 7.9-4) yields

$$N_o = \frac{1}{\tau^2}\int_0^T dt \int_0^T d\lambda \int_{-\infty}^{\infty} \alpha\beta f(\alpha, \beta)\,d\alpha\,d\beta \tag{7.9-25}$$

But

$$R_{n_i}(t - \lambda) = E[n_i(t)n_i(\lambda)] = E(\alpha\beta) = N_o \tag{7.9-26}$$

where $R_{n_i}(t - \lambda)$ is the autocorrelation function of $n_i(t)$.

We recall (see Eq. 2.24-5) that the correlation and the power spectral density are Fourier transform pairs. Since for white noise the power spectral density is $G(f) = \eta/2$ we have

$$E[n_i(t)n_i(\lambda)] = R_{n_i}(t-\lambda) = \mathscr{F}\{G(f)\} = \int_{\infty}^{\infty} \eta/2 \; e^{-j2\pi f(t-\lambda)} \, df = (\eta/2)\, \delta(t-\lambda) \tag{7.9-27}$$

Altogether, then we have, from Eq. (7.9-25) and (7.9-26) that

$$N_o = \frac{1}{\tau^2}\int_0^T dt \int_0^T d\lambda(\eta/2)\, \delta(t-\lambda) \tag{7.9-28a}$$

$$= \frac{1}{\tau^2}\int_0^T dt \cdot \eta/2 = \frac{\eta T}{2\tau^2} \tag{7.9-28b}$$

in agreement with Eq. (7.9-17).

7.10 NOISE BANDWIDTH

Consider that white noise is present at the input to a receiver. Suppose also that a filter with transfer function $H(f)$ centered at f_0, such as is indicated by the solid plot of Fig. 7.10-1, is being used to restrict the noise power actually passed on to the receiver. Now contemplate a rectangular filter as shown by the dotted plot in Fig. 7.10-1. This filter is also centered at f_0. Let the rectangular filter bandwidth B_N be adjusted so that the real filter and the rectangular filter transmit the same noise power. Then the bandwidth B_N is called the *noise bandwidth* of the real filter. The noise bandwidth, then, is the bandwidth of an idealized (rectangular) filter which passes the same noise power as does the real filter.

We illustrate the concept of noise bandwidth by considering the case of the low-pass RC filter with transfer function as given by Eq. (7.9-2). For this filter $H(f)$ attains its maximum value $H(f) = 1$ at $f = 0$. As given by Eq. (7.9-6), with white-noise input of power spectral density $\eta/2$, the noise output of the filter is

$$N_o(RC) = \frac{\pi}{2}\eta f_c \tag{7.10-1}$$

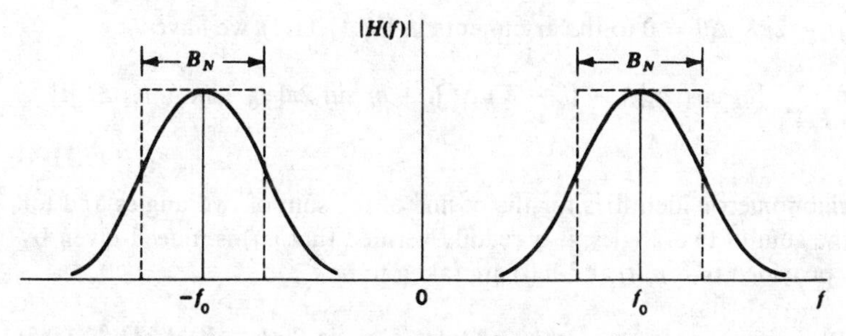

Figure 7.10-1 Illustration of the noise bandwidth of a filter.

In the presence of such noise, a rectangular low-pass filter with $H(f) = 1$ over its bandpass B_N would yield an output-noise power

$$N_o(\text{rectangular}) = \frac{\eta}{2} 2B_N = \eta B_N \tag{7.10-2}$$

Setting $N_o(RC) = N_o$ (rectangular), we find the noise bandwidth to be

$$B_N = \frac{\pi}{2} f_c \tag{7.10-3}$$

Thus, the noise bandwidth of the RC filter is $\pi/2$ (= 1.57) times its 3-dB bandwidth f_c.

7.11 QUADRATURE COMPONENTS OF NOISE

We have represented noise $n(t)$, as in Eq. (7.2-7), as the superposition of spectral components of noise in the expression

$$n(t) = \lim_{\Delta f \to 0} \sum_{k=1}^{\infty} (a_k \cos 2\pi k \, \Delta f t + b_k \sin 2\pi k \, \Delta f t) \tag{7.11-1}$$

It is sometimes more advantageous to represent the noise through an alternative representation given by

$$n(t) = n_c(t) \cos 2\pi f_0 t - n_s(t) \sin 2\pi f_0 t \tag{7.11-2}$$

in which f_0 is an arbitrary frequency. The representation of Eq. (7.11-2) is frequently used with great convenience in dealing with noise confined to a relatively narrow frequency band in the neighborhood of f_0. For this reason Eq. (7.11-2) is often referred to as the *narrowband* representation. The term *quadrature component* representation is also often used because of the appearance in the equation of sinusoids in quadrature.

We may readily transform Eq. (7.11-1) into Eq. (7.11-2) and, in so doing, arrive at explicit expressions for $n_c(t)$ and $n_s(t)$. Let us select f_0 to correspond to $k = K$; that is, we set

$$f_0 = K \, \Delta f \tag{7.11-3}$$

Adding $2\pi f_0 t - 2\pi K \, \Delta f t = 0$ to the arguments in Eq. (7.11-1), we have

$$n(t) = \lim_{\Delta f \to 0} \sum_{k=1}^{\infty} \{a_k \cos 2\pi[f_0 + (k - K) \, \Delta f]t + b_k \sin 2\pi[f_0 + (k - K) \, \Delta f]t\} \tag{7.11-4}$$

Using the trigonometric identities for the cosine of the sum of two angles and for the sine of the sum of two angles, it is readily verified that $n(t)$ is indeed given by Eq. (7.11-2), provided that $n_c(t)$ and $n_s(t)$ are taken to be

$$n_c(t) = \lim_{\Delta f \to 0} \sum_{k=1}^{\infty} [a_k \cos 2\pi(k - K) \, \Delta f t + b_k \sin 2\pi(k - K) \, \Delta f t] \tag{7.11-5}$$

and

$$n_s(t) = \lim_{\Delta f \to 0} \sum_{k=1}^{\infty} [a_k \sin 2\pi(k - K)\, \Delta f t - b_k \cos 2\pi(k - K)\, \Delta f t] \quad (7.11\text{-}6)$$

Like $n(t)$, so also $n_c(t)$ and $n_s(t)$ are stationary random processes which are represented as linear superpositions of spectral components. We recall, from Sec. 7.4, that the a_k's and b_k's are gaussian random variables of zero mean and equal variance and that, further, the a_k's and b_k's are uncorrelated. We may then readily establish (Probs. 7.11-1 to 7.11-4) that $n_c(t)$ and $n_s(t)$ are gaussian random processes of zero mean value and of equal variance and that, further, $n_c(t)$ and $n_s(t)$ are uncorrelated.

To see the significance of the quadrature representation of noise, let us use it in connection with narrowband noise. We observe in Eqs. (7.11-5) and (7.11-6) that a noise spectral component in $n(t)$ of frequency $f = k\, \Delta f$ gives rise in $n_c(t)$ and in $n_s(t)$ to a spectral component of frequency $(k - K)\, \Delta f = f - f_0$. Suppose then that the noise $n(t)$ is narrowband, extending over a bandwidth B. And suppose that f_0 is selected midway in the frequency range of the noise. Then the spectrum of the noise $n(t)$ extends over the range $f_0 - B/2$ to $f_0 + B/2$. On the other hand, the spectrum of $n_c(t)$ and of $n_s(t)$ extends over only the range from $-B/2$ to $B/2$. By way of example, if the noise $n(t)$ is confined to a frequency band of only 10 kHz centered around $f_0 = 10$ MHz, then while $n(t)$ is a superposition of spectral components around the 10-MHz frequency, $n_c(t)$ and $n_s(t)$ change only insignificantly during the time the sinusoid of frequency f_0 executes a full cycle.

In view of the slow variations of $n_c(t)$ and $n_s(t)$ relative to the sinusoid of frequency f_0, it is reasonable and useful to give the quadrature representation of noise an interpretation in terms of phasors and a phasor diagram. Thus, in Eq. (7.11-2) the term $n_c(t) \cos 2\pi f_0 t$ is of frequency f_0 and of relatively slowly varying amplitude $n_c(t)$. Similarly, the term $-n_s(t) \sin 2\pi f_0 t$ is in quadrature with the first term and has a relatively slowly varying amplitude $n_s(t)$. In a coordinate system rotating counterclockwise with angular velocity $2\pi f_0$, these phasors are as represented in Fig. 7.11-1. These two phasors of varying amplitude give rise to a resultant phasor of amplitude $r(t) = [n_c^2(t) + n_s^2(t)]^{1/2}$ which makes an angle

$$\theta(t) = \tan^{-1} [n_s(t)/n_c(t)] \quad (7.11\text{-}7)$$

with the horizontal. With the passage of time, the end point of this resultant phasor wanders about randomly over the phasor diagram.

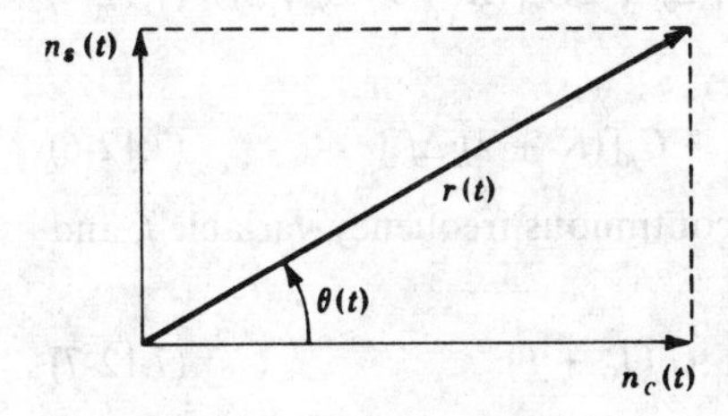

Figure 7.11-1 A phasor diagram of the quadrature representation of noise.

We shall find the quadrature representation useful generally in the analysis of noise and shall find the phasor interpretation especially useful in discussing angle modulation communications systems.

7.12 POWER SPECTRAL DENSITY OF $n_c(t)$ AND $n_s(t)$

To determine the power spectral density of $n_c(t)$, let us select from the original noise $n(t)$ those spectral components corresponding to $k = K + \lambda$ and $k = K - \lambda$, where λ, like k and K, is an integer. Since $k = K$ corresponds to the frequency f_0, the selected components correspond to frequencies $f_0 + \lambda\,\Delta f$ and $f_0 - \lambda\,\Delta f$. These two frequencies give rise to four power spectral lines in a two-sided power spectral density plot as shown in Fig. 7.12-1. In this figure we have assumed band-limited noise. However, for the sake of generality we have not assumed that the power spectral density is uniform in the band, nor have we assumed that the frequency $f_0 = K\,\Delta f$ is located at the center of the band.

We now select from $n_c(t)$, as given by Eq. (7.11-5), that part, $\Delta n_c(t)$, corresponding to our selection of frequencies, $f_0 \pm \lambda\,\Delta f$, from $n(t)$. We find

$$\Delta n_c(t) = a_{K-\lambda} \cos 2\pi\lambda\,\Delta f t - b_{K-\lambda} \sin 2\pi\lambda\,\Delta f t + a_{K+\lambda} \cos 2\pi\lambda\,\Delta f t + b_{K+\lambda} \sin 2\pi\lambda\,\Delta f t \tag{7.12-1}$$

Note that all four terms in Eq. (7.12-1) are of the same frequency $\lambda\,\Delta f$. These four terms represent four uncorrelated random processes, since the a's and b's are uncorrelated random variables. Hence we may find the power P_λ of $\Delta n_c(t)$ by determining the ensemble average of $[\Delta n_c(t)]^2$. Since $\Delta n_c(t)$ is a stationary random process, the ensemble average can be calculated at any time $t = t_1$. Following the procedure employed in Sec. 7.4, we choose t_1 so that $\lambda\,\Delta f t_1$ is an integer. Then

$$\Delta n_c(t_1) = a_{K-\lambda} + a_{K+\lambda} \tag{7.12-2}$$

and

$$P_\lambda = E\{[\Delta n_c(t_1)]^2\} = E[(a_{K-\lambda} + a_{K+\lambda})^2] \tag{7.12-3}$$

Using Eq. (7.4-12), which says that $E(a_{K-\lambda}\,a_{K+\lambda}) = 0$, we have, from Eq. (7.12-3), that

$$P_\lambda = \overline{a_{K-\lambda}^2} + \overline{a_{K+\lambda}^2} \tag{7.12-4}$$

From Eqs. (7.4-6) and (7.12-4) we then find

$$P_\lambda = 2G_{n_c}(\lambda\,\Delta f)\,\Delta f = 2G_n[(K-\lambda)\,\Delta f]\,\Delta f + 2G_n[(K+\lambda)\,\Delta f]\,\Delta f \tag{7.12-5}$$

Hence

$$G_{n_c}(\lambda\,\Delta f) = G_n[(K-\lambda)\,\Delta f] + G_n[(K+\lambda)\,\Delta f] \tag{7.12-6}$$

We now set $K\,\Delta f = f_0$ and replace $\lambda\,\Delta f$ by a continuous frequency variable f, and we have, from Eq. (7.12-6),

$$G_{n_c}(f) = G_n(f_0 - f) + G_n(f_0 + f) \tag{7.12-7}$$

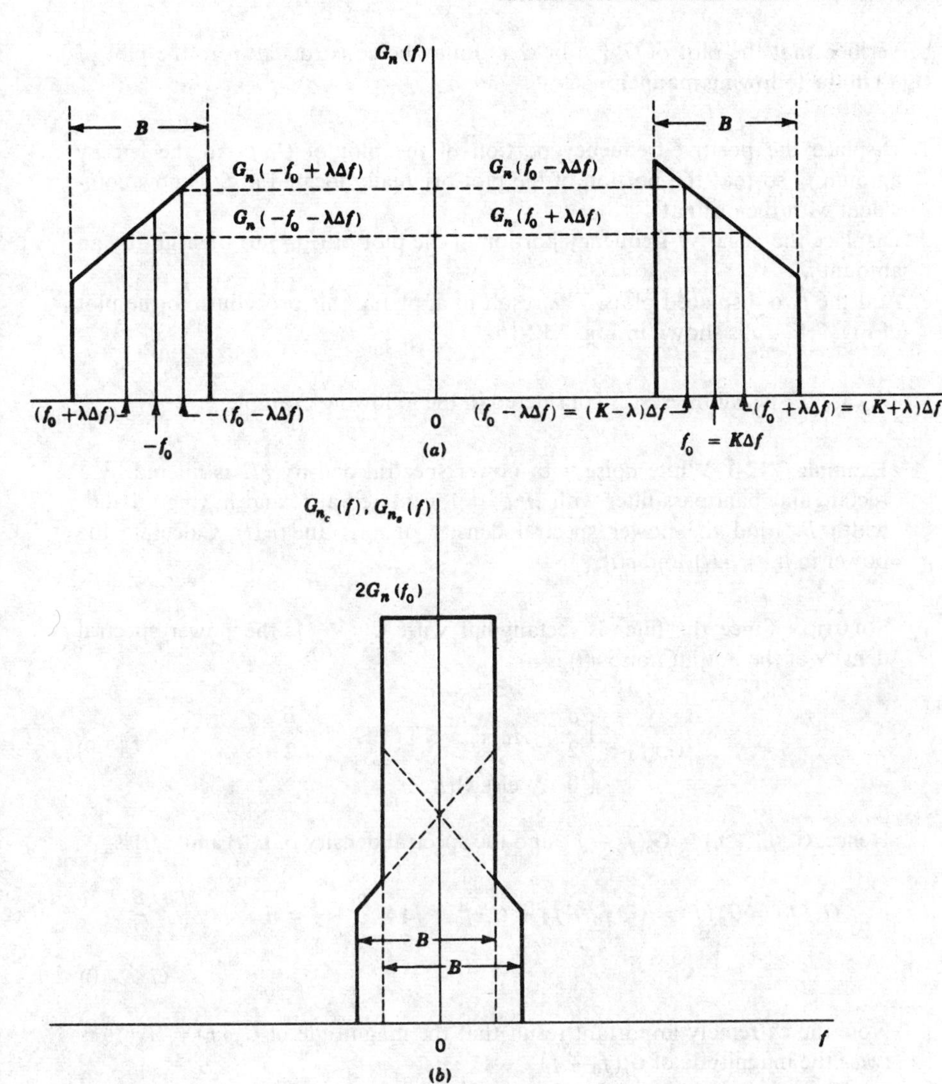

Figure 7.12-1 (*a*) Power spectrum of bandlimited noise. (*b*) Power spectral density of n_c and n_s.

In a similar manner we may deduce an identical result for $G_{n_s}(f)$, namely,

$$G_{n_s}(f) = G_n(f_0 - f) + G_n(f_0 + f) \tag{7.12-8}$$

Expressed in words, Eqs. (7.12-7) and (7.12-8) say, that to find the power spectral density of $n_c(t)$ or of $n_s(t)$ at a frequency f, add the power spectral densities of $n(t)$ at the frequencies $f_0 - f$ and $f_0 + f$. In view of this result, and in view of the symmetry of a two-sided power spectral density plot as in Fig. 7.12-1, it may readily be verified that the plot of $G_{n_c}(f)$ or $G_{n_s}(f)$ may be constructed from the plot of $G_n(f)$ in the following manner:

1. Displace the positive-frequency portion of the plot of $G_n(f)$ to the left by amount f_0 so that the portion of the plot originally located at f_0 is now coincident with the ordinate.
2. Displace the negative-frequency portion of the plot of $G_n(f)$ to the right by an amount f_0.
3. Add the two displaced plots. The result of applying this procedure to the plot of Fig. 7.12-1*a* is shown in Fig. 7.12-1*b*.

A case of special interest is considered in the following example.

Example 7.12-1 White noise with power spectral density $\eta/2$ is filtered by a rectangular bandpass filter with $H(f) = 1$, centered at f_0 and having a bandwidth B. Find the power spectral density of $n_c(t)$ and $n_s(t)$. Calculate the power in $n_c(t)$, $n_s(t)$, and $n(t)$.

Solution Since the filter is rectangular with $H(f) = 1$, the power spectral density of the output noise $n(t)$ is

$$G_n(f) = \begin{cases} \dfrac{\eta}{2} & f_0 - \dfrac{B}{2} \le |f| \le f_0 + \dfrac{B}{2} \\ 0 & \text{elsewhere} \end{cases} \tag{7.12-9}$$

Hence, $G_n(f_0 + f) = G_n(f_0 - f)$, and the spectral density of $n_c(t)$ and $n_s(t)$ is

$$G_{n_c}(f) = G_{n_s}(f) = G_n(f_0 - f) + G_n(f_0 + f) = \frac{\eta}{2} + \frac{\eta}{2} = \eta \qquad |f| \le \frac{B}{2} \tag{7.12-10}$$

Note the extremely important result that the magnitude of $G_{n_c}(f) = G_{n_s}(f)$ is *twice* the magnitude of $G_n(f_0 + f)$.

The power (variance) of $n_c(t)$, and of $n_s(t)$, is

$$\sigma_{n_c}^2 = \sigma_{n_s}^2 = \int_{-B/2}^{B/2} G_{n_c}(f)\, df = \eta B \tag{7.12-11}$$

The power (variance) of $n(t)$ is

$$\sigma_n^2 = \int_{-f_0 - B/2}^{-f_0 + B/2} G_n(f)\, df + \int_{f_0 - B/2}^{f_0 + B/2} G_n(f)\, df = 2\,\frac{\eta}{2}\,B = \eta B \tag{7.12-12}$$

Thus, the power of $n_c(t)$, $n_s(t)$, and $n(t)$ are each equal.

7.13 PROBABILITY DENSITY OF $n_c(t)$, $n_s(t)$, AND THEIR TIME DERIVATIVES

We have noted that $n_c(t)$ and $n_s(t)$ are gaussian random processes with mean values of zero. If the noise $n(t)$ has a power spectral density $\eta/2$ over a bandwidth B, then, as noted in the preceding example, $\sigma_{n_c}^2 = \sigma_{n_s}^2 = \eta B$. Using Eq. (2.12-1) with $m = 0$, we find that the probability densities of the random variables n_c and n_s [that is, $n_c(t)$ and $n_s(t)$ at any fixed time] are given by

$$f(n_c) = \frac{1}{\sqrt{2\pi\eta B}} e^{-n_c{}^2/2\eta B} \tag{7.13-1a}$$

$$f(n_s) = \frac{1}{\sqrt{2\pi\eta B}} e^{-n_s{}^2/2\eta B} \tag{7.13-1b}$$

Since $n_c(t)$ and $n_s(t)$ are gaussian, the time derivatives $\dot{n}_c(t)$ and $\dot{n}_s(t)$ are also gaussian, because the operation of differentiation is an operation performed by a linear filter (Eq. 7.9-11), and from Sec. 7.3 we know that filtering gaussian noise does not change its probability density. To write the probability densities of $\dot{n}_c(t)$ and $\dot{n}_s(t)$, we need first to evaluate their variances $\sigma_{\dot{n}_c}^2$ and $\sigma_{\dot{n}_s}^2$. Noting that differentiation is equivalent to multiplying each spectral component by $j\omega$, we find

$$G_{\dot{n}_c}(f) = |j\omega|^2 \, G_{n_c}(f) = 4\pi^2 f^2 G_{n_c}(f) \tag{7.13-2}$$

so that, using Eq. (7.12-10), we find

$$\sigma_{\dot{n}_c}^2 = \int_{-B/2}^{B/2} G_{\dot{n}_c}(f)\, df = \int_{-B/2}^{B/2} 4\pi^2 f^2 \eta \, df = \frac{\pi^2}{3} \eta B^3 \tag{7.13-3}$$

with an identical result for $\sigma_{\dot{n}_s}^2$. Hence we find

$$f(\dot{n}_c) = \frac{\exp\{-\dot{n}_c^2/[(2\pi^2/3)\eta B^3]\}}{\sqrt{(2\pi^3/3)\eta B^3}} \tag{7.13-4}$$

with a similar expression for $f(\dot{n}_s)$. Assuming that the four random variables n_c, n_s, $\dot{n}_c$, and $\dot{n}_s$ are independent, the joint distribution function for the four variables is the product of the individual densities. Hence from Eqs. (7.13-1) and (7.13-3) we find

$$f(n_c, n_s, \dot{n}_c, \dot{n}_s) = \frac{\exp[-(n_c^2 + n_s^2)/2\eta B] \exp[-(\dot{n}_c^2 + \dot{n}_s^2)/(2\pi^2/3)\eta B^3]}{[(2\pi^2/\sqrt{3})\eta B^2]^2} \tag{7.13-5}$$

We shall have occasion to use Eq. (7.13-5) in Chap. 10 in connection with an analysis of threshold effects in frequency modulation.

In arriving at Eq. (7.13-5), we assumed that the four random variables involved were independent. That such is indeed the case may be verified in the manner indicated in Prob. 7.13-1.

7.14 REPRESENTATION OF NOISE USING ORTHONORMAL COORDINATES

In our discussion of the frequency domain representation of noise we saw that a noise process can be represented as a sum of orthonormal functions. These orthonormal functions are the sines and cosines. In our discussion of the Gram-Schmitt technique, where our interest concerned a waveform defined over a time interval T, we pointed out that limitless other functions, orthonormal over T, are also possible. We consider here some features of the representation of noise in terms of orthonormal functions.

If $u_i(t)$ are a set of orthonormal functions in the interval T then in that interval, noise $n(t)$ is

$$n(t) = \sum_{i=0}^{\infty} n_i u_i(t) \tag{7.14-1}$$

in which n_i is the coefficient of the ith component and is evaluated in the usual manner, that is,

$$n_i = \int_0^T n(t)u_i(t)\, dt \tag{7.14-2}$$

If the noise $n(t)$ is a gaussian random process with a mean value of zero then n_i is a gaussian random variable with a zero mean value.

We shall now determine the correlation between coefficients say n_i and n_j. We have

$$n_i n_j = \int_0^T n(t)u_i(t)\, dt \int_0^T n(\lambda)u_j(\lambda)\, d\lambda \tag{7.14-3a}$$

$$= \int_0^T dt \int_0^T d\lambda n(t)n(\lambda)u_i(t)u_j(\lambda) \tag{7.14-3b}$$

where t and λ are dummy variable of integration. We now take the ensemble average of both sides of Eq. (7.14-3*b*). Interchanging the order of averaging and integrating as in Sec. 7.9 we have

$$E(n_i n_j) = \int_0^T dt \int_0^T d\lambda E[n(t)n(\lambda)]u_i(t)u_j(\lambda) \tag{7.14-4}$$

Since the noise process is ergodic, Eq. (2.24-5) applies and we have that the autocorrelation of the process is

$$R(t - \lambda) = E[n(t)n(\lambda)] \tag{7.14-5}$$

Further, assuming white noise of power spectral density $G(f) = \eta/2$ we have, as in Eq. (7.9-27) that

$$R(t - \lambda) = \eta/2\ \delta(t - \lambda) \tag{7.14-6}$$

From Eqs. (7.14-3), (7.14-4), and (7.14-5) we have that

$$E(n_i n_j) = \int_0^T dt \int_0^T d\lambda\ \eta/2\ \delta(t - \lambda) u_i(t) u_j(\lambda) \tag{7.14-7}$$

$$= \eta/2 \int_0^T u_i(t) u_j(t)\, dt = \begin{cases} \eta/2 & \text{if } i = j \\ 0 & \text{if } i \neq j \end{cases} \tag{7.14-8}$$

We shall have occasion in Sec. 11.2 in our study of the probability of error to make use of the fact that the additive white gaussian noise $n(t)$ can be represented as Eq. 7.14-1 where the n_i form a set of statistically independent random variables, each with a mean of zero and a variance of $\eta/2$.

7.15 IRRELEVANT NOISE COMPONENTS

In Sec. 1.22 we considered the situation where we have M signals s_l ($l = 1, 2, \ldots, M$) defined over an interval T. We showed that each of these signals could be represented as a superposition of N orthonormal waveforms u_i ($i = 1, 2, \ldots, N$). The signals s_l being given, the orthonormal waveforms u_i were generated through the Gram-Schmitt technique. We have then

$$s_l(t) = \sum_{i=1}^{N} s_{il}\, u_i(t) \qquad l = 1, 2, \ldots, M \tag{7.15-1}$$

the s_{il} being time-independent coefficients. For example, in QPSK $M = 4$ and $N = 2$.

In the absence of noise, it is a rather trivial task to determine at a receiver, which signal $s_l(t)$ was transmitted. In the presence of noise, the receiver system of Fig. 7.15-1 is optimum in the sense that it maximizes the probability of being correct. In this system we make available at the receiver, either in fact or in principle, the very same orthonormal signals $u_i(t)$ referred to above. The receiver determines the correlations of the received signal with each of the orthonormal waveforms. On the basis of this examination we assume that the transmitted signal is the one which yields the greatest correlation.

We noted in the previous section that white noise can also be represented as a superposition of orthonormal components. In the case of white noise an infinite number of orthonormal components are required. Let us now consider that we select an array of orthonormal components, N of which (say the first N) are identical to the N waveforms u_i ($i = 1, 2, \ldots, N$) used to represent the possible signals $s_l(t)$. Then the noise is

$$n(t) = \sum_{i=1}^{N} n_i\, u_i(t) + \sum_{i=N+1}^{\infty} n_i\, u_i(t) \tag{7.15-2}$$

We note that the noise due to the first summation is uncorrelated to the noise terms contained in the second summation. For this reason it is clear that the

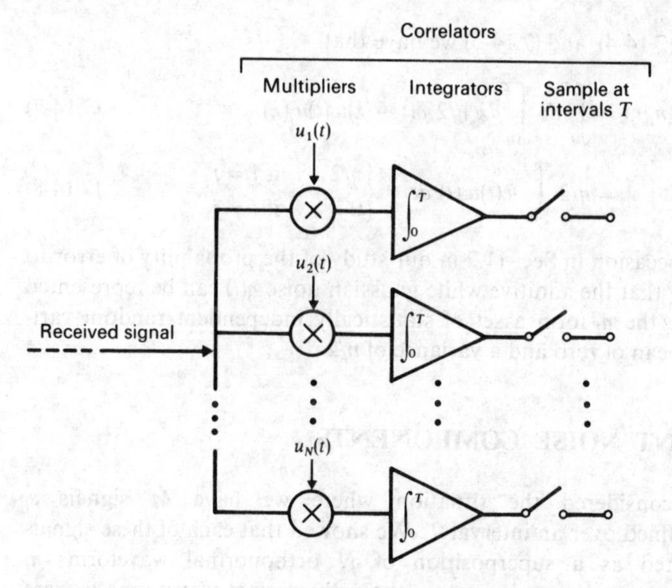

Figure 7.15-1 An optimum receiver consists of a bank of correlators.

noise components in the second term of Eq. (7.15-2) will yield no output at any of the correlators of the receiver of Fig. 7.15-1. Hence these components are reasonably referred to as "irrelevant."

In succeeding sections we shall make use of this result by representing the noise only in terms of the orthonormal components of the signal and we shall ignore the "irrelevant" noise terms.

REFERENCE

1. Davenport, W., and W. Root: "Random Signals and Noise," McGraw-Hill Book Company, New York, 1958.

Papoulis, A.: "Probability, Random Variables, and Stochastic Processes," McGraw-Hill Book Company, New York, 1965.

PROBLEMS

7.2-1. (*a*) A symmetrical square wave makes excursions between $+V$ and $-V$ volt, where V is a random variable uniformly distributed between 1 V and 2 V. It has a fundamental frequency 10^3 Hz. Make a plot of the two-sided power spectral density of the waveform. Assume that the delay τ between $t = 0$ and a transition is a random variable uniformly distributed between 0 and 1 ms.

(*b*) What fraction of the normalized power of the waveform is contained in the frequency range -3×10^3 to $+3 \times 10^3$ Hz?

7.2-2. Noise $n(t)$ has the power spectral density shown.

(*a*) Find the normalized power P of the noise in terms of η and f_M. Find P for $\eta = 1\ \mu\text{V}^2/\text{Hz}$ and $f_M = 10$ kHz.

(*b*) Find and plot the autocorrelation function $R_n(\tau)$ of the noise.

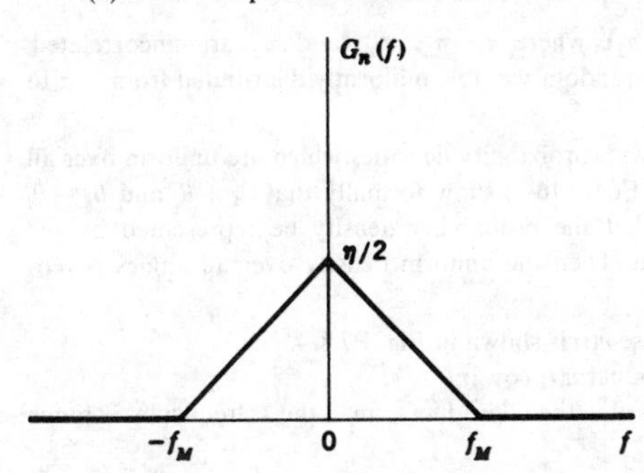

Figure P7.2-2

7.3-1. Gaussian noise $n(t)$ of zero mean has a power spectral density

$$G_n(f) = 2\ \mu\text{V}^2/\text{Hz} \qquad |f| \leq 1\ \text{kHz}$$
$$= 0 \qquad \text{elsewhere}$$

(*a*) What is the normalized power of the noise?

(*b*) Write the probability density function $f(n)$ of the noise.

(*c*) The noise $n(t)$ is passed through a filter. The power output of the filter is one-half the power of $n(t)$. Write the probability density function for the output noise of the filter.

7.4-1. (*a*) Two gaussian noise spectral components $n_k(t)$ and $n_l(t)$ are approximated as in Eqs. (7.4-10) and (7.4-11). The variance of a_k is $\sigma_k^2 = 1$, and the variance of a_l is $\sigma_l^2 = 2$. What is the normalized power of the sum $n_s(t) = n_k(t) + n_l(t)$?

Write the probability density function for the noise waveform $n_s(t)$.

7.4-2. (*a*) Gaussian noise in a very narrow spectral range is represented approximately by Eq. (7.4-1). The normalized power of the noise is 0.01 μW. Write the expression for the probability density functions of the coefficients a_k and b_k.

(*b*) The narrowband noise in (*a*) is approximated by the representation of Eq. (7.4-2). Write the expression for the probability density functions of c_k and of θ_k.

7.6-1. White noise with two-sided power spectral density $\eta/2$ is passed through a low-pass *RC* network with time constant $\tau = RC$ and thereafter through an ideal amplifier with voltage gain of 10.

(*a*) Write the expression for the autocorrelation function $R_n(\tau)$ of the white noise.

(*b*) Write the expression for the power spectral density of the noise at the output of the amplifier.

(*c*) Write the expression for the autocorrelation of the output noise in (*b*).

7.7-1. Consider the noise waveforms $n_1(t), n_2(t), n_3(t), \ldots, n_N(t)$ where

$$E[n_j(t)] = 0 \qquad j = 1, 2, \ldots, N$$

and

$$E[n_j(t)n_k(t)] = \begin{cases} 1 & j = k \\ \frac{1}{2} & |j - k| = 1 \\ 0 & |j - k| > 1 \end{cases}$$

(*a*) Calculate the power dissipated in a 1-ohm resistor by the noise

$$n(t) = \sum_{j=1}^{N} n_j(t)$$

(*b*) Assuming that each of the $n_j(t)$ has a gaussian probability density, write the probability density for $n(t)$.

7.8-1. Noise $n(t)$ amplitude modulates a carrier having a random phase $v(t) = n(t) \cos(\omega_0 t + \varphi)$. Consider that $E[n^2(t)] = \sigma^2$ and that the probability density of the random variable φ is $1/2\pi$ from $-\pi$ to π. Assume that $n(t)$ and φ are independent. Show that the power dissipated by $v(t)$ is $E(v^2) = \sigma^2/2$. Is $v(t)$ stationary?

7.8-2. $n(t) = n_1 \cos(\omega_1 t + \varphi_1) + n_2 \cos(k\omega_1 t + \varphi_2)$, where n_1, n_2, φ_1, and φ_2 are uncorrelated. Show that if $v(t) = n(t) \cos(\omega_0 t + \theta)$, where θ is a random variable uniformly distributed from $-\pi$ to π, then $E(v^2) = E(n^2)/2 = \frac{1}{4}[E(n_1^2) + E(n_2^2)]$.

7.8-3. The angles θ_k and θ_l are random variables with probability densities which are uniform over all angles. Starting with the convolution integral of Eq. (2.16-4) show formally that $\theta_k + \theta_l$ and $\theta_k - \theta_l$ also have uniform probability densities. (*Hint:* Let the probability density be represented by the radius vector r in a cylindrical coordinate system. Then the uniform density over all angles is represented by the radius of a circle of length $1/2\pi$.)

7.8-4. The two-sided power spectral density of noise $n(t)$ is shown in Fig. P7.8-4.

(*a*) Plot the power spectral density of the product $n(t) \cos 2\pi f_1 t$.

(*b*) Calculate the normalized power of the product in the frequency range $-(f_2 - f_1)$ to $(f_2 - f_1)$.

(*c*) Repeat parts (*a*) and (*b*) for the product $n(t) \cos 2\pi[(f_2 + f_1)/2]t$.

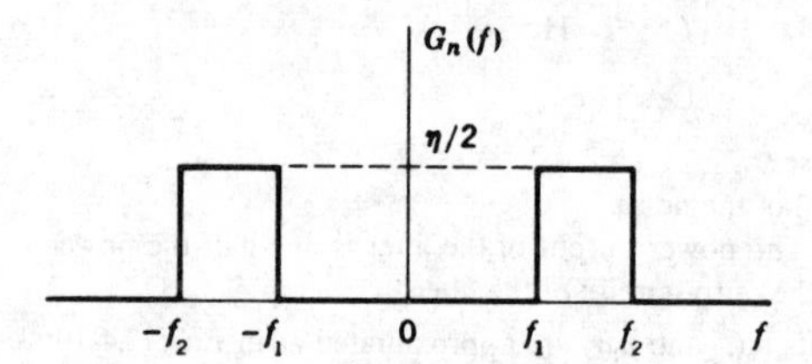

Figure P7.8-4

7.8-5. A noise waveform $n(t)$ has the bandlimited power spectral density shown in Fig. P7.8-5, with $a = 10^{-6}$ V^2/Hz and $f_M = 10^4$ Hz. Plot the power spectral density of $n(t) \sin 2\pi \times 10^6 t$ and find its normalized power over all frequencies.

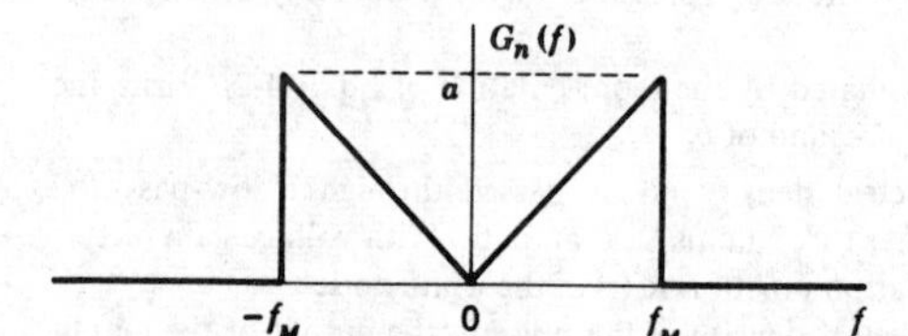

Figure P7.8-5

7.9-1. A white-noise current source having a power spectral density $G_n(f) = \eta/2$ is filtered by a narrowband, parallel, single-tuned circuit having a resonant frequency at f_0 and a 3-dB bandwidth $B \ll f_0$.

(*a*) Find the output voltage noise power.

(*b*) Compare your result with Eq. (7.9-6).

7.9-2. A random process $v(t) = A \cos(\omega_0 t + \Theta)$ where Θ is a random variable uniformly distributed between $-\pi$ and π, is integrated for a time T. Calculate, using Eq. (7.9-23) the power, N_0, in the integrator output. Is this result expected?

7.9-3. Show that Eq. (7.9-20) follows from Eq. (7.9-19*c*).

7.9-4. Show that Eq. (7.9-25) can be obtained from Eq. (7.9-24).

7.10-1. Calculate the noise bandwidth of a single-tuned *RLC* filter centered at frequency f_0 and having a 3-dB bandwidth $B \ll f_0$.

7.10-2. Calculate the noise bandwidth of the gaussian filter $|H(\omega)|^2 = e^{-\omega^2}$.

7.11-1. Show that if a_n and b_n are gaussian, $n_c(t)$ and $n_s(t)$ are gaussian.

7.11-2. Show that $E(n_c^2) = E(n_s^2)$.

7.11-3. Show that the autocorrelation functions of $n_c(t)$ and $n_s(t)$ are the same.

7.11-4. Show that $E[n_c(t)n_s(t)] = 0$.

7.11-5. Show that $E[n_c(t)n_s(t+\tau)]$ is not always equal to zero. Prove that it is equal to zero when $G_{n_c}(f) = G_{n_s}(f)$ is an *even* function with respect to f_0.

7.11-6. Verify that, with $n_c(t)$ and $n_s(t)$ given as in Eqs. (7.11-5) and (7.11-6), $n(t)$ given by Eq. (7.11-2) is identical with $n(t)$ given by Eq. (7.11-1).

7.12-1. Using Eq. (7.11-2), show that when $E[n_c(t_1)n_s(t_2)] = 0$ [that is, $G_n(f)$ is an even function of f, with respect to f_0], $R_{nn}(\tau) = R_{n_c n_c}(\tau) \cos \omega_0 \tau = R_{n_s n_s}(\tau) \cos \omega_0 \tau$.

7.12-2. Noise $n(t)$ has the power spectral density shown. We write

$$n(t) = n_c(t) \cos 2\pi f_0 t - n_s(t) \sin 2\pi f_0 t.$$

Make plots of the power spectral densities of $n_c(t)$ and $n_s(t)$ for the cases

(a) $f_0 = f_1$
(b) $f_0 = f_2$
(c) $f_0 = \frac{1}{2}(f_2 + f_1)$

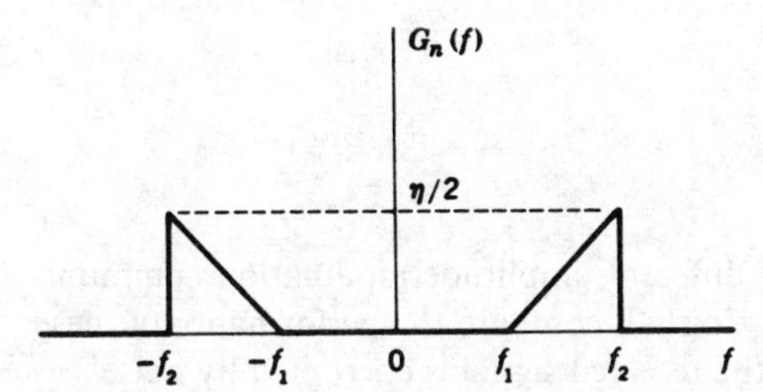

Figure P7.12-2

7.12-3. (a) If $G_n(f) = \alpha^2/(\alpha^2 + \omega^2)$, show that $R_n(\tau) = Ke^{-\alpha|\tau|}$. Find K.
(b) If

$$G_n(f) = \frac{\alpha^2/2}{\alpha^2 + (\omega - \omega_0)^2} + \frac{\alpha^2/2}{\alpha^2 + (\omega + \omega_0)^2}$$

show that $R_n(\tau) = Ke^{-\alpha|\tau|} \cos \omega_0 \tau$. Here $G_n(f)$ is the spectral density of noise which has been filtered by a narrowband (*symmetrical*) single-tuned filter.

7.13-1. Verify that $n_c(t)$, $n_s(t)$, $\hat{n}_c(t)$, and $\hat{n}_s(t)$ are uncorrelated.

7.14-1. Verify Eq. (7.14-8).

7.15-1. The signal $\sqrt{2P_s}\, d(t) \cos (\omega_0 t)$ is received embedded in white gaussian noise. In order to reduce the probability of error in determining the data $d(t)$ it was proposed that a second receiver be employed using an antenna facing away from the signal, so that only the white gaussian noise is received.

Can the noise received by this second system help in reducing the error rate by cancelling the noise received with the signal?

CHAPTER

EIGHT

NOISE IN AMPLITUDE-MODULATION SYSTEMS

In Chap. 3 we described a number of different amplitude-modulation communication systems. In the present chapter we shall compare the performance of these systems under the circumstances that the received signal is corrupted by noise.

8.1 AMPLITUDE-MODULATION RECEIVER

A system for processing an amplitude-modulated carrier and recovering the baseband modulating system is shown in Fig. 8.1-1. We assume that the signal has suffered great attenuation during the course of its transmission over the communication channel and hence is in need of amplification. The input to the system might be a signal furnished by a receiving antenna which receives its signal from a transmitting antenna. The carrier of the received signal is called a *radio-frequency* (RF) carrier, and its frequency is the *radio* frequency f_{rf}. The input signal is amplified in an RF amplifier and then passed on to a *mixer*. In the mixer the modulated RF carrier is mixed (i.e., multiplied) with a sinusoidal waveform generated by a local oscillator which operates at a frequency f_{osc}. The process of mixing is also called *heterodyning*, and since, as is to be explained, the heterodyning local-oscillator frequency f_{osc} is selected to be *above* the radio frequency f_{rf}, the system is often referred to as a *superheterodyne* system.

The process of mixing generates sum and difference frequencies. Thus the mixer output consists of a carrier of frequency $f_{osc} + f_{rf}$ and a carrier $f_{osc} - f_{rf}$. Each carrier is modulated by the baseband signal to the same extent as was the input RF carrier. The sum frequency is rejected by a filter. This filter is not shown

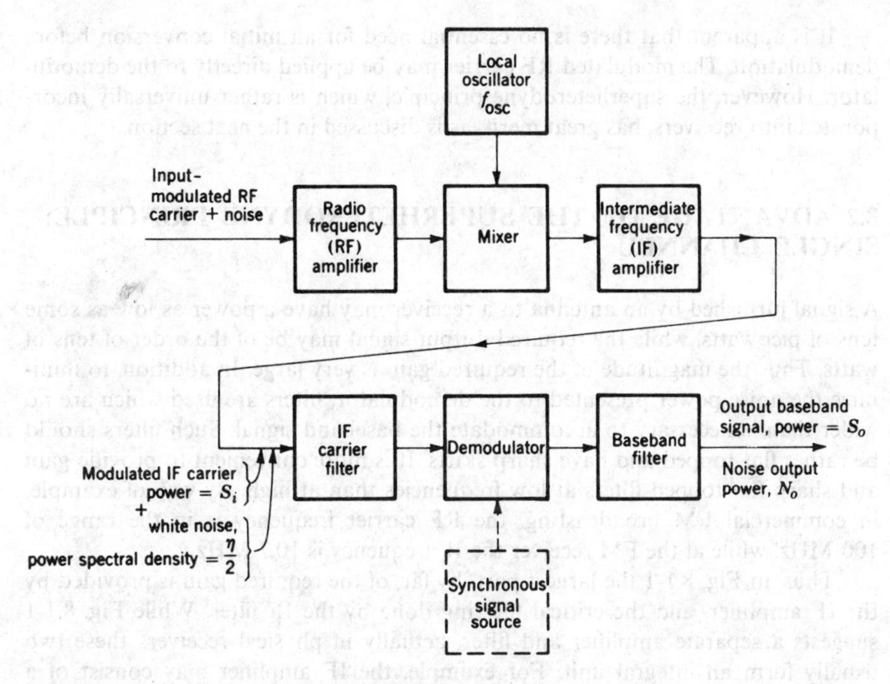

Figure 8.1-1 A receiving system for an amplitude-modulated signal.

explicitly in Fig. 8.1-1 and may be considered to be part of the mixer. The difference-frequency carrier is called the *intermediate frequency* (IF) carrier, that is, $f_{if} = f_{osc} - f_{rf}$. The modulated IF carrier is applied to an IF amplifier. The process just described, in which a modulated RF carrier is replaced by a modulated IF carrier, is called *conversion.* The combination of the mixer and local oscillator is called a *converter*.

The IF amplifier output is passed, through an IF carrier filter, to the demodulator in which the baseband signal is recovered, and finally through a baseband filter. The baseband filter may include an amplifier, not explicitly indicated in Fig. 8.1-1. If synchronous demodulation is used, a synchronous signal source will be required.

The only absolutely essential operation performed by the receiver is the process of frequency translation back to baseband. This process is, of course, the inverse of the operation of modulation in which the baseband signal is frequency-translated to a carrier frequency. The process of frequency translation is performed in the system of Fig. 8.1-1 in part by the converter and in part by the demodulator. For this reason the converter is sometimes referred to as the *first detector*, while the demodulator is then called the *second detector*. The only other components of the system are the linear amplifiers and filters, none of which would be essential if the signal were strong enough and there were no need for multiplexing.

It is apparent that there is no essential need for an initial conversion before demodulation. The modulated RF carrier may be applied directly to the demodulator. However, the superheterodyne principle, which is rather universally incorporated into receivers, has great merit, as is discussed in the next section.

8.2 ADVANTAGE OF THE SUPERHETERODYNE PRINCIPLE: SINGLE CHANNEL

A signal furnished by an antenna to a receiver may have a power as low as some tens of picowatts, while the required output signal may be of the order of tens of watts. Thus the magnitude of the required gain is very large. In addition, to minimize the noise power presented to the demodulator, filters are used which are no wider than is necessary to accommodate the baseband signal. Such filters should be rather flat-topped and have sharp skirts. It is more convenient to provide gain and sharp flat-topped filters at low frequencies than at high. By way of example, in commercial FM broadcasting, the RF carrier frequency is in the range of 100 MHz, while at the FM receiver the IF frequency is 10.7 MHz.

Thus, in Fig. 8.1-1 the largest part, by far, of the required gain is provided by the IF amplifier, and the critical filtering done by the IF filter. While Fig. 8.1-1 suggests a separate amplifier and filter, actually in physical receivers these two usually form an integral unit. For example, the IF amplifier may consist of a number of amplifier stages, each one contributing to the filtering. Some filtering will also be incorporated in the RF amplifier. But this filtering is not critical. It serves principally to limit the total noise power input to the mixer and thereby avoids *overloading* the mixer with a noise waveform of excessive amplitude.

RF amplification is employed whenever the incoming signal is very small. This is because of the fact that RF amplifiers, such as masers, are *low-noise* devices; i.e., an RF amplifier can be designed to provide relatively high gain while generating relatively little noise. When RF amplification is not employed, the signal is applied directly to the mixer. The mixer provides relatively little gain and generates a relatively large noise power. Calculations showing typical values of gain and noise power generation in RF, mixer, and IF amplifiers are presented in Sec. 14.14.

Multiplexing

An even greater merit of the superheterodyne principle becomes apparent when we consider that we shall want to tune the receiver to one or another of a number of different signals, each using a different RF carrier. If we were not to take advantage of the superheterodyne principle, we would require a receiver in which many stages of RF amplification were employed, each stage requiring tuning. Such tuned-radio-frequency (TRF) receivers were, as a matter of fact, commonly employed during the early days of radio communication. It is difficult enough to operate at the higher radio frequencies; it is even more difficult to *gang-tune* the individual stages over a wide band, maintaining at the same time a reasonably sharp flat-topped filter characteristic of constant bandwidth.

In a *superhet* receiver, however, we need but change the frequency of the local oscillator to go from one RF carrier frequency to another. Whenever f_{osc} is set so that $f_{osc} - f_{rf} = f_{if}$, the mixer will convert the input modulated RF carrier to a modulated carrier at the IF frequency, and the signal will proceed through the demodulator to the output. Of course, it is necessary to gang the tuning of the RF amplifier to the frequency control of the local oscillator. But again this ganging is not critical, since only one or two RF amplifiers and filters are employed.

Finally, we may note the reason for selecting f_{osc} higher than f_{rf}. With this higher selection the fractional change in f_{osc} required to accommodate a given range of RF frequencies is smaller than would be the case for the alternative selection.

8.3 SINGLE-SIDEBAND SUPPRESSED CARRIER (SSB-SC)

The receiver of Fig. 8.1-1 is suitable for the reception and demodulation of all types of amplitude-modulated signals, single sideband or double sideband, with and without carrier. The only essential changes required to accommodate one type of signal or another are in the demodulator and in the bandwidth of the IF carrier filter. Hence, from this point, our interest will focus on the section of receiver beginning with the IF filter and through to the output.

The signal input to the IF filter is an amplitude-modulated IF carrier. The normalized power (power dissipated in a 1-ohm resistor) of this signal is S_i. The signal arrives with noise. Added, is the noise generated in the RF amplifier and amplified in the RF amplifier and IF amplifiers. The IF amplifiers and mixer are also sources of noise, i.e., thermal noise, shot noise, etc., but this noise, lacking the gain of the RF amplifier, represents a second-order effect. (See Sec. 14.11.) We shall assume that the noise is gaussian, white, and of two-sided power spectral density $\eta/2$. The IF filter is assumed rectangular and of bandwidth no wider than is necessary to accommodate the signal. The output baseband signal has a power S_o and is accompanied by noise of total power N_o.

Calculation of Signal Power

With a single-sideband suppressed-carrier signal, the demodulator is a multiplier as shown in Fig. 8.3-1*a*. The carrier is $A \cos 2\pi f_c t$. For synchronous demodulation the demodulator must be furnished with a synchronous locally generated carrier $\cos 2\pi f_c t$. We assume that the upper sideband is being used; hence the carrier filter has a bandpass, as shown in Fig. 8.3-1*b*, that extends from f_c to $f_c + f_M$, where f_M is the baseband bandwidth. The bandwidth of the baseband filter extends from zero to f_M as shown in Fig. 8.3-1*c*.

Let us assume that the baseband signal is a sinusoid of angular frequency

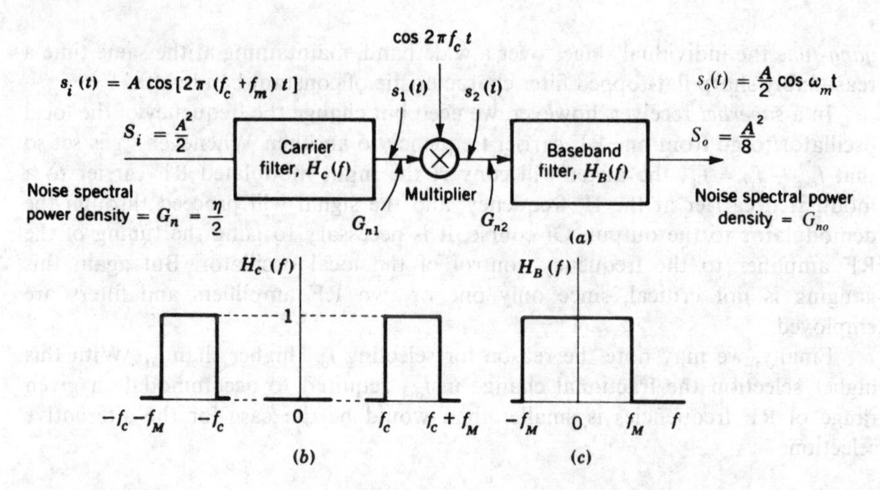

Figure 8.3-1 (*a*) A synchronous demodulator operating on a single-sideband single-tone signal. (*b*) The bandpass range of the carrier filter. (*c*) The passband of the lowpass baseband filter.

$f_m(f_m \leq f_M)$. The carrier frequency is f_c, and, since we have assumed that the upper sideband is being used, the received signal is

$$s_i(t) = A \cos [2\pi(f_c + f_m)t] \tag{8.3-1}$$

The output of the multiplier is

$$s_2(t) = s_1(t) \cos \omega_c t = \frac{A}{2} \cos [2\pi(2f_c + f_m)t] + \frac{A}{2} \cos 2\pi f_m t \tag{8.3-2}$$

Only the difference-frequency term will pass through the baseband filter. Therefore the output signal is

$$s_o(t) = \frac{A}{2} \cos 2\pi f_m t \tag{8.3-3}$$

which is the modulating signal amplified by $\frac{1}{2}$.

The input signal power is

$$S_i = \frac{A^2}{2} \tag{8.3-4}$$

while the output signal power is

$$S_o = \frac{1}{2}\left(\frac{A}{2}\right)^2 = \frac{A^2}{8} = \frac{S_i}{4} \tag{8.3-5}$$

Thus

$$\frac{S_o}{S_i} = \frac{1}{4} \tag{8.3-6}$$

We may readily see that, even though Eq. (8.3-6) was deduced on the assumption of a sinusoidal baseband signal, the result is entirely general. For suppose that the baseband signal were quite arbitrary in waveshape. Then the single-sideband signal generated by this baseband signal may be resolved into a series of harmonically related spectral components. The input power is the sum of the powers in these individual components. Next, we note, as was discussed in Chap. 3, that superposition applies to the multiplication process being used for demodulation. Therefore, the output signal power generated by the simultaneous application at the input of many spectral components is simply the sum of the output powers that would result from each spectral component individually. Hence S_i and S_o in Eqs. (8.3-4) and (8.3-5) are properly the *total* powers, independently of whether a single or many spectral components are involved.

Calculation of Noise Power

We now calculate the output noise N_o. For this purpose, we recall from Sec. 7.8 that when a noise spectral component at a frequency f is multiplied by $\cos 2\pi f_c t$, the original noise component is replaced by two components, one at frequency $f_c + f$ and one at frequency $f_c - f$, each new component having one-fourth the power of the original.

The input noise is white and of spectral density $\eta/2$. The noise input to the multiplier has a spectral density G_{n1} as shown in Fig. 8.3-2*a*. The density of the noise after multiplication by $\cos 2\pi f_c t$ is G_{n2} as is shown in Fig. 8.3-2*b*. Finally the noise transmitted by the baseband filter is of density G_{no} as in Fig. 8.3-2*c*. The total noise output is the area under the plot in Fig. 8.3-2*c*. We have, then, that

$$N_o = 2f_M \frac{\eta}{8} = \frac{\eta f_M}{4} \tag{8.3-7}$$

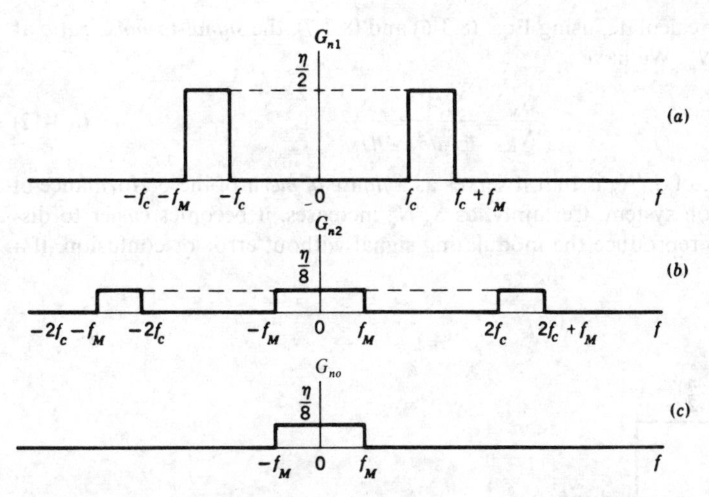

Figure 8.3-2 Spectral densities of noises in SSB demodulator. (*a*) Density G_{n1} of noise input to multiplier. (*b*) Density G_{n2} of noise output of multiplier. (*c*) Density G_{no} of noise output of baseband filter.

Use of Quadrature Noise Components

It is of interest to calculate the output noise power N_o in an alternative manner using the transformation of Eq. (7.11-2):

$$n(t) = n_c(t) \cos 2\pi f_c t - n_s(t) \sin 2\pi f_c t \tag{8.3-8}$$

We now apply Eq. (8.3-8) to the noise output of the IF filter so that $n(t)$ has the spectral density G_{n1} as in Fig. 8.3-2*a*. The spectral densities of $n_c(t)$ and $n_s(t)$ are [see Eqs. (7.12-7) and (7.12-8)]:

$$G_{n_c}(f) = G_{n1}(f) = G_{n1}(f_c - f) + G_{n1}(f_c + f) \tag{8.3-9}$$

We observe that for $0 \leq f \leq f_M$, $G_{n1}(f_c + f) = \eta/2$, while $G_{n1}(f_c - f) = 0$, so that $G_{n_c}(f)$ and $G_{n_s}(f)$ are as shown in Fig. 8.3-3.

Multiplying $n(t)$ by $\cos 2\pi f_c t$ yields

$$\begin{aligned} n(t) \cos 2\pi f_c t &= n_c(t) \cos^2 2\pi f_c t - n_s(t) \sin 2\pi f_c t \cos 2\pi f_c t \\ &= \tfrac{1}{2} n_c(t) + \tfrac{1}{2} n_c(t) \cos 4\pi f_c t - \tfrac{1}{2} n_s(t) \sin 4\pi f_c t \end{aligned} \tag{8.3-10}$$

The spectra of the second and third terms in Eq. (8.3-10) extend over the range $2f_c - f_M$ to $2f_c + f_M$ and are outside the baseband filter. The output noise is, therefore,

$$n_o(t) = \tfrac{1}{2} n_c(t) \tag{8.3-11}$$

The spectral density of $n_o(t)$ is then $G_{no} = \frac{1}{4} G_{n_c} = \frac{1}{4}(\eta/2) = \eta/8$. Hence as before, as shown in Fig. 8.3-2*c*, the spectral density G_{no} is $\eta/8$ over the range $-f_M$ to f_M, and the total noise is again $N_o = \eta f_M/4$.

Calculation of Signal-to-Noise Ratio (SNR)

Finally we may calculate, using Eqs. (8.3-6) and (8.3-7), the *signal-to-noise ratio* at the output, S_o/N_o. We have

$$\frac{S_o}{N_o} = \frac{S_i/4}{\eta f_M/4} = \frac{S_i}{\eta f_M} \tag{8.3-12}$$

The importance of S_o/N_o is that it serves as a *figure of merit* of the performance of a communication system. Certainly, as S_o/N_o increases, it becomes easier to distinguish and to reproduce the modulating signal without error or confusion. If a

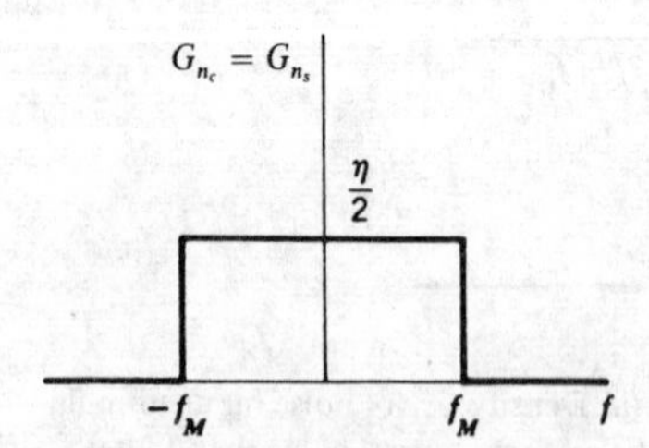

Figure 8.3-3 Power spectral densities of G_{n_c} and G_{n_s}.

system of communication allows the use of more than a single type of demodulator (say, synchronous or nonsynchronous), that ratio S_o/N_o will serve as a figure of merit with which to compare demodulators.

We observe from Eq. (8.3-12) that to increase the output signal-to-noise power ratio, we can increase the transmitted signal power, restrict the baseband frequency range, or make the receiver quieter.

8.4 DOUBLE-SIDEBAND SUPPRESSED CARRIER (DSB-SC)

When a baseband signal of frequency range f_M is transmitted over a DSB-SC system, the bandwidth of the carrier filter must be $2f_M$ rather than f_M. Thus, input noise in the frequency range $f_c - f_M$ to $f_c + f_M$ will contribute to the output noise, rather than only in the range f_c to $f_c + f_M$ as in the SSB case.

Calculation of Noise Power

This situation is illustrated in Fig. 8.4-1*a*, which shows the spectral density $G_{n1}(f)$ of the white input noise after the IF filter. This noise is multiplied by $\cos \omega_c t$. The multiplication results in a frequency shift by $\pm f_c$ and a reduction of power in the power spectral density of the noise by a factor of 4. Thus, the noise in region *d* of Fig. 8.4-1*a* shifts to regions *d* shown in Fig. 8.4-1*b*. Similarly regions *a*, *b*, and *c* of Fig. 8.4-1*a* are translated by $\pm f_c$ and are also attenuated by 4 as shown in

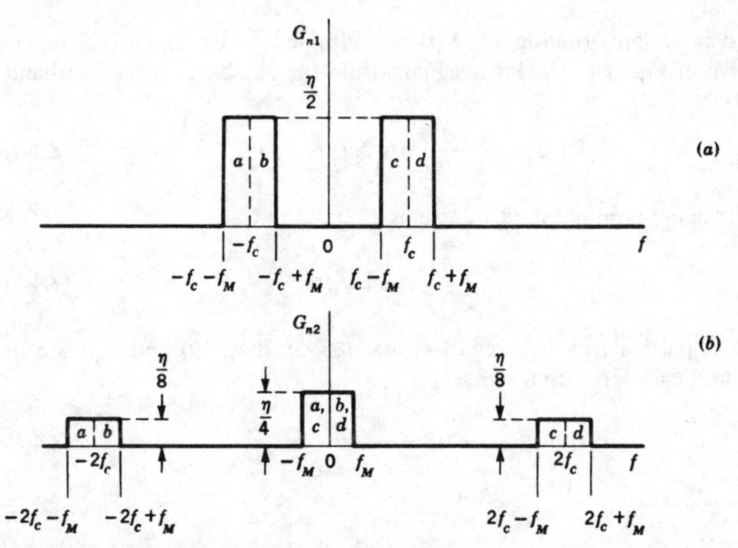

Figure 8.4-1 Spectral densities of noise in DSB demodulation. (*a*) Density G_{n1} of noise at output of IF filter. (*b*) Density G_{n2} of noise output of baseband filter.

Fig. 8.4-1*b*. Note that the noise-power spectral density in the region between $-f_M$ and $+f_M$ is $\eta/4$, while the noise density in the SSB case, as shown in Fig. 8.3-2*c* is only $\eta/8$. Hence the output noise power in twice as large as the output noise power for SSB given in Eq. (8.3-5). The output noise for DSB after baseband filtering is therefore

$$N_o = \frac{\eta}{4}(2f_M) = \frac{\eta f_M}{2} \tag{8.4-1}$$

Calculation of Signal Power

We might imagine that for equal received powers, the ratio S_o/N_o for DSB would be only half the corresponding ratio for SSB. We shall now see that such is not the case, and that the ratio S_o/N_o is the *same* in the two cases. Again, without loss in generality, let us assume a sinusoidal baseband signal of frequency $f_m \leq f_M$. To keep the received power the same as in the SSB case, that is, $S_i = A^2/2$, we write

$$\begin{aligned} s_i(t) &= \sqrt{2}\, A \cos 2\pi f_m t \cos 2\pi f_c t \\ &= \frac{A}{\sqrt{2}} \cos [2\pi(f_c + f_m)t] + \frac{A}{\sqrt{2}} \cos [2\pi(f_c - f_m)t] \end{aligned} \tag{8.4-2}$$

The received power is then

$$S_i = \frac{1}{2}\left(\frac{A}{\sqrt{2}}\right)^2 + \frac{1}{2}\left(\frac{A}{\sqrt{2}}\right)^2 = \frac{A^2}{2} \tag{8.4-3}$$

as in Eq. (8.3-4).

In the demodulator (multiplier), $s_i(t)$ is multiplied by $\cos \omega_c t$. The upper-sideband term in Eq. (8.4-2) yields a signal within the passband of the baseband filter given by

$$s_o'(t) = \frac{A}{2\sqrt{2}} \cos 2\pi f_m t \tag{8.4-4}$$

The lower-sideband term of Eq. (8.4-2) yields

$$s_o''(t) = \frac{A}{2\sqrt{2}} \cos 2\pi f_m t \tag{8.4-5}$$

Observe, most particularly in Eqs. (8.4-4) and (8.4-5), that $s_o'(t)$ and $s_o''(t)$ are in phase and that hence the output signal is

$$s_o(t) = s'(t) + s''(t) = \frac{A}{\sqrt{2}} \cos 2\pi f_m t \tag{8.4-6}$$

which has a power

$$S_o = \frac{A^2}{4} = \frac{S_i}{2} \tag{8.4-7}$$

rather than $S_o = A^2/8 = S_i/4$ as in Eq. (8.3-5) for the SSB case. Thus we see that when a received signal of fixed power is split into two sideband components each of half power, as in DSB, rather than being left in a single sideband, the output signal power increases by a factor of 2. This increase results from the fact that the contributions from each sideband yield output signals which are in phase. A doubling in amplitude causes a fourfold increase in power. This fourfold increase, due to the inphase addition of s'_o and s''_o, is in part undone by the need to split the input power into two half-power sidebands. Thus the overall improvement in output signal power is by a factor of 2.

On the other hand, the noise outputs due to noise spectral components symmetrically located with respect to the carrier are uncorrelated with one another. The two resultant noise spectral components in the output, although of the same frequency, are uncorrelated. Hence the combination of the two yields a power which is the sum of the two powers individually, not larger than the sum, as is the case with the signal.

Calculation of Signal-to-Noise Ratio

Returning now to the calculation of signal-to-noise ratio for DSB-SC, we find from Eqs. (8.4-1) and (8.4-7) that

$$\frac{S_o}{N_o} = \frac{S_i}{\eta f_M} \tag{8.4-8}$$

exactly as for SSB-SC.

Arbitrary Modulating Signal

In the discussion in the present section concerning DSB we have assumed that the baseband signal waveform is sinusoidal. As pointed out in Sec. 8.3, this assumption causes no loss of generality because of the linearity of the demodulation. Nonetheless it is often convenient to have an expression for the power of a DSB signal in terms of the arbitrary waveform $m(t)$ of the baseband modulating signal. Hence let the received signal be

$$s_i(t) = m(t) \cos 2\pi f_c t \tag{8.4-9}$$

The power of $s_i(t)$ is

$$S_i \equiv \overline{s_i^2(t)} = \overline{m^2(t) \cos^2 2\pi f_c t} = \tfrac{1}{2}\overline{m^2(t)} + \tfrac{1}{2}\overline{m^2(t) \cos (4\pi f_c t)} \tag{8.4-10}$$

Now $m(t)$ can always be represented as a sum of sinusoidal spectral components. [Of interest, albeit of no special relevance in the present discussion, is the fact that if $m(t)$ is bandlimited to f_M, $m^2(t)$ is bandlimited to $2f_M$. See Prob. 8.4-1.] Hence $m^2(t) \cos 4\pi f_c t$ consists of a sum of sinusoidal waveforms in the frequency range $2f_c \pm 2f_M$. The average value of such a sum is zero, and we therefore have

$$S_i \equiv \overline{s_i^2(t)} = \tfrac{1}{2}\overline{m^2(t)} \tag{8.4-11}$$

When the signal $s_i(t)$ in Eq. (8.4-9) is demodulated by multiplication by $\cos 2\pi f_c t$, and the product passed through the baseband filter, the output is $s_o(t) = m(t)/2$. The output signal power is

$$S_o = \frac{\overline{m^2(t)}}{4} \tag{8.4-12}$$

so that, from Eqs. (8.4-11) and (8.4-12),

$$S_o = \frac{S_i}{2} \tag{8.4-13}$$

that is, the same result as given in Eq. (8.4-7) for an assumed single sinusoidal modulating signal.

Use of Quadrature Noise Components to Calculate N_o

It is again interesting to calculate N_o using the transformation of Eq. (7.11-2):

$$n(t) = n_c(t) \cos 2\pi f_c t - n_s(t) \sin 2\pi f_c t \tag{8.4-14}$$

The power spectral density of $n_c(t)$ and $n_s(t)$ are [see Eqs. (7.12-7) and (7.12-8)]

$$G_{n_c}(f) = G_{n_s}(f) = G_{n1}(f_c + f) + G_{n1}(f_c - f) \tag{8.4-15}$$

In the frequency range $|f| \leq f_M$, $G_{n1}(f_c + f) = G_{n1}(f_c - f) = \eta/2$. Thus

$$G_{n_c}(f) = G_{n_s}(f) = \eta \qquad |f| \leq f_M \tag{8.4-16}$$

(This result was also derived in Example 7.12-1.)

The result of multiplying $n(t)$ by $\cos 2\pi f_c t$ yields

$$n(t) \cos 2\pi f_c t = \tfrac{1}{2}n_c(t) + \tfrac{1}{2}n_c(t) \cos 4\pi f_c t - \tfrac{1}{2}n_s(t) \sin 4\pi f_c t \tag{8.4-17}$$

Baseband filtering eliminates the second and third terms, leaving

$$n_o(t) = \tfrac{1}{2}n_c(t) \tag{8.4-18}$$

The power spectral density of $n_o(t)$ is then

$$G_{no}(f) = \frac{1}{4} G_{n_c}(f) = \frac{\eta}{4} \qquad -f_M \leq f \leq f_M \tag{8.4-19}$$

The output noise power N_o is, therefore,

$$N_o = \frac{\eta}{4} 2f_M = \frac{\eta f_M}{2} \tag{8.4-20}$$

This result is, of course, identical with Eq. (8.4-1), which was obtained by considering directly the effect on a noise spectral component of a multiplication by $\cos 2\pi f_c t$.

8.5 DOUBLE SIDEBAND WITH CARRIER

Let us now consider the case where a carrier accompanies the double-sideband signal. Demodulation is achieved synchronously as in SSB-SC and DSB-SC. The carrier is used as a *transmitted reference* to obtain the reference signal $\cos \omega_c t$ (see Prob. 8.5-1). We note that the carrier increases the total input-signal power but makes no contribution to the output-signal power. Equation (8.4-8) applies directly to this case, provided only that we replace S_i by $S_i^{(SB)}$, where $S_i^{(SB)}$ is the power in the sidebands alone. Then

$$\frac{S_o}{N_o} = \frac{S_i^{(SB)}}{\eta f_M} \tag{8.5-1}$$

Suppose that the received signal is

$$\begin{aligned} s_i(t) &= A[1 + m(t)] \cos 2\pi f_c t \\ &= A \cos 2\pi f_c t + Am(t) \cos 2\pi f_c t \end{aligned} \tag{8.5-2}$$

where $m(t)$ is the baseband signal which amplitude-modulates the carrier $A \cos 2\pi f_c t$. The carrier power is $A^2/2$. The sidebands are contained in the term $Am(t) \cos 2\pi f_c t$. The power associated with this term is $(A^2/2)\overline{m^2(t)}$, where $\overline{m^2(t)}$ is the time average of the square of the modulating waveform. We then have that the total input power S_i is given by

$$S_i = \frac{A^2}{2} + S_i^{(SB)} = \frac{A^2}{2} [1 + \overline{m^2(t)}] \tag{8.5-3}$$

Eliminating A^2, we have

$$S_i^{(SB)} = \frac{\overline{m^2(t)}}{1 + \overline{m^2(t)}} S_i \tag{8.5-4}$$

or, with Eq. (8.5-1),

$$\frac{S_o}{N_o} = \frac{\overline{m^2(t)}}{1 + \overline{m^2(t)}} \frac{S_i}{\eta f_M} \tag{8.5-5}$$

In terms of the carrier power $P_c \equiv A^2/2$, we get, from Eqs. (8.5-3) and (8.5-5), that

$$\frac{S_o}{N_o} = \overline{m^2(t)} \frac{P_c}{\eta f_M} \tag{8.5-6}$$

If the modulation is sinusoidal, with $m(t) = m \cos 2\pi f_m t$ (m a constant), then

$$s_i(t) = A(1 + m \cos 2\pi f_m t) \cos 2\pi f_c t \tag{8.5-7}$$

In this case $\overline{m^2(t)} = m^2/2$ and

$$\frac{S_o}{N_o} = \frac{m^2}{2 + m^2} \frac{S_i}{\eta f_M} \tag{8.5-8}$$

When the carrier is transmitted only to synchronize the local demodulator waveform $\cos 2\pi f_c t$, relatively little carrier power need be transmitted. In this case $m \gg 1$, $m^2/(2 + m^2) \cong 1$, and the signal-to-noise ratio is not greatly reduced by the presence of the carrier. On the other hand, when envelope demodulation is used (Sec. 3.4), it is required that $m \leq 1$. When $m = 1$, the carrier is 100 percent modulated. In this case $m^2/(2 + m^2) = \frac{1}{3}$, so that of the power transmitted, only one-third is in the sidebands which contribute to signal power output.

A Figure of Merit

We observe that in each demodulation system considered so far, the ratio $S_i/\eta f_M$ appeared in the expression for output SNR [see Eqs. (8.3-12), (8.4-8), and (8.5-5)]. This ratio is the output signal power S_i divided by the product ηf_M. To give the product ηf_M some physical significance, we consider it to be the noise power N_M at the input, measured in a frequency band equal to the *baseband frequency*. Thus

$$N_M \equiv \frac{\eta}{2} 2f_M = \eta f_M \tag{8.5-9}$$

The ratio $S_i/\eta f_M$ is, therefore, often referred to as the *input signal-to-noise ratio* S_i/N_M. It needs to be kept in mind that N_M is the noise power transmitted through the IF filter only when the IF filter bandwidth is f_M. Thus N_M is the true input noise power only in the case of single sideband.

For the purpose of comparing systems, we introduce the *figure of merit* γ, defined by

$$\gamma \equiv \frac{S_o/N_o}{S_i/N_M} \tag{8.5-10}$$

The results given above may now be summarized as follows:

$$\gamma = \begin{cases} 1 & \text{SSB-SC} \quad (8.5\text{-}11) \\ 1 & \text{DSB-SC} \quad (8.5\text{-}12) \\ \dfrac{\overline{m^2(t)}}{1 + \overline{m^2(t)}} & \text{DSB} \quad (8.5\text{-}13) \\ \dfrac{m^2}{2 + m^2} & \text{DSB with sinusoidal modulation} \quad (8.5\text{-}14) \end{cases}$$

A point of interest in connection with *double-sideband* synchronous demodulation is that, for the purpose of suppressing output-noise power, the carrier filter of Fig. 8.3-1 is not necessary. A noise spectral component at the input which lies outside the range $f_c \pm f_M$ will, after multiplication in the demodulator, lie outside the passband of the baseband filter. On the other hand, if the carrier filter is eliminated, the magnitude of the noise signal which reaches the modulator may be large enough to overload the active devices used in the demodulator. Hence, such carrier filters are normally included, but the purpose is *overload suppression* rather than *noise suppression.* In single sideband, of course, the situation is different, and the carrier filter does indeed suppress noise.

8.6 SQUARE-LAW DEMODULATOR

We saw in Sec. 3.6 that a double-sideband signal with carrier may be demodulated by passing the signal through a network whose input-output characteristic is not linear. Such nonlinear demodulation has the advantage, over the linear synchronous demodulation methods, that a synchronous local carrier need not be obtained. This eliminates the rather costly synchronizing circuits. In this section we discuss the performance and determine the output SNR of a nonlinear demodulator which uses a network whose output signal y (voltage or current) is related to the input signal x (voltage or current) by

$$y = \lambda x^2 \tag{8.6-1}$$

in which λ is a constant. As shown in Fig. 8.6-1, this nonlinear network, which constitutes the demodulator, is preceded by a bandpass IF filter of bandwidth $2f_M$ and is followed by a baseband low-pass filter of bandwidth f_M.

We discuss this square-law demodulator in part for its own intrinsic interest, but also because it exhibits an important characteristic which is not displayed by the linear synchronous demodulators. We have previously adopted the quantity $\gamma \equiv (S_o/N_o)/(S_i/N_M)$ as a figure of merit for the performance of demodulators in the presence of noise. We observed [Eqs. (8.5-11) to (8.5-14)] that this figure of merit is not a function of the input signal-to-noise ratio S_i/N_M. Therefore, if the input S_i/N_M decreases, say, by a factor α, the output S_o/N_o will also decrease by α. The nonlinear demodulator also has a range where the figure of merit γ is inde-

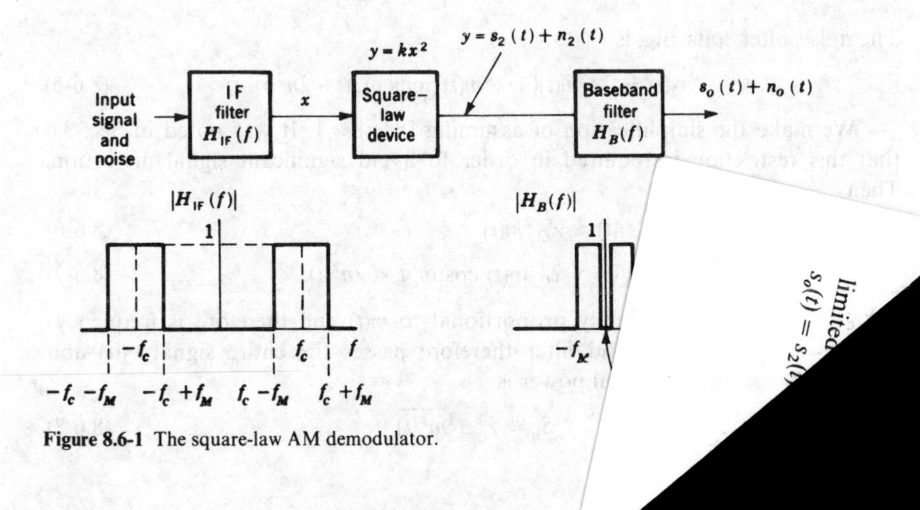

Figure 8.6-1 The square-law AM demodulator.

pendent of S_i/N_M. However, as S_i/N_M decreases, there is a point, a *threshold*, at which the output S_o/N_o decreases more rapidly than does the input S_i/N_M. This threshold often marks the limit of usefulness of the demodulator.

Analysis

We assume that we have a carrier of amplitude A and angular frequency ω_c, amplitude-modulated by a baseband signal $m(t)$. Then the received signal is $s_i(t) = A[1 + m(t)] \cos \omega_c t$. We assume that input noise $n(t)$, having a power spectral density $\eta/2$ for $f_c - f_M < |f| < f_c + f_M$ after IF filtering, has been added to the signal. The input to the demodulator is therefore

$$x(t) = A[1 + m(t)] \cos \omega_c t + n(t) \tag{8.6-2}$$

and the output is, from Eq. (8.6-1),

$$y(t) = \lambda\{A[1 + m(t)] \cos \omega_c t + n(t)\}^2 \tag{8.6-3}$$

We now expand Eq. (8.6-3) and drop those terms whose power spectral density falls outside the baseband filter. The baseband filter is designed to pass the modulation and suppress as much noise as possible. Thus, the maximum frequency passed is f_M. On the low-frequency end, the filter extends only low enough to pass the minimum signal frequency, often 100 to 300 Hz. However, we assume for simplicity that the filter passes all low-frequency components with the exception of dc (zero frequency). Thus we drop dc terms.

The spectrum of the baseband signal $m(t)$ extends to f_M. Then, as shown in Prob. 8.4-1, the spectrum of $m^2(t)$ extends to $2f_M$. We therefore drop terms of the form $m(t) \cos 2\omega_c t$ and $m^2(t) \cos 2\omega_c t$, since the spectrum of these terms extends, respectively, over the range $2f_c \pm f_M$ and $2f_c \pm 2f_M$. The signal that then remains, at the input to the baseband filter, is

$$s_2(t) = \lambda A^2 m(t) \left[1 + \frac{m(t)}{2}\right] \tag{8.6-4}$$

The noise, after squaring, is

$$n_2(t) = 2\lambda A n(t)[1 + m(t)] \cos \omega_c t + \lambda n^2(t) \tag{8.6-5}$$

We make the simplification of assuming $|m(t)| \ll 1$. It was noted in Sec. 3.6 that this restriction is required in order to avoid significant signal distortion. Then

$$s_2(t) \approx \lambda A^2 m(t) \tag{8.6-6}$$

and

$$n_2(t) \approx 2\lambda A n(t) \cos \omega_c t + \lambda n^2(t) \tag{8.6-7}$$

The signal $s_2(t)$ is directly proportional to $m(t)$ and therefore is frequency-[illegible] to f_M. The baseband filter therefore passes the entire signal $s_2(t)$ and [illegible]. The output signal power is

$$S_o = \lambda^2 A^4 \overline{m^2(t)} \tag{8.6-8}$$

Since $n(t)$ has a power spectral density $\eta/2$, the spectral density of $n(t) \cos \omega_c t$ is $\eta/4$ between $-f_M$ and $+f_M$. (See Fig. 8.4-1*b*.) Therefore the noise power N'_o, due to the term $2\lambda A n(t) \cos \omega_c t$ in Eq. (8.6-7), is

$$N'_o = 4\lambda^2 A^2 \frac{\eta}{4} 2f_M = 2\lambda^2 A^2 \eta f_M \tag{8.6-9}$$

Calculation of N''_o

We turn now to the calculation of the noise power N''_o which results from the term $\lambda n^2(t)$. For this purpose, as shown in Fig. 8.6-2, we divide the spectral range of the noise ($f_c - f_M$ to $f_c + f_M$) into $2K + 1$ intervals, each Δf wide. Then, as discussed in Sec. 7.2, we represent the power in each interval by a spectral line of power $\eta \Delta f/2$ located at the center of the interval. We represent the noise $n(t)$ in the manner of Eq. (7.2-2) so that

$$n(t) = \sum_{k=-K}^{+K} c_k \cos [(2\pi f_c + k \Delta f)t + \theta_k] \tag{8.6-10}$$

From Eq. (7.4-6), with $G_n(k \Delta f) = \eta/2$, we have

$$\overline{c_k^2} = 2\eta \Delta f \tag{8.6-11}$$

Let us now single out two *particular* spectral components in Eq. (8.6-10), separated by a frequency $\rho \Delta f$. Calling the sum of this specific set of spectral components $n_{k,\rho}(t)$, we have

$$n_{k,\rho}(t) = c_k \cos [(2\pi f_c + k \Delta f)t + \theta_k] + c_{k+\rho} \cos \{[2\pi f_c + (k+\rho) \Delta f]t + \theta_{k+\rho}\} \tag{8.6-12}$$

In the course of forming the product $n(t) \times n(t) = n^2(t)$, we shall have to form the product $n_{k,\rho}(t) \times n_{k,\rho}(t) = n^2_{k,\rho}(t)$. Such a product will give rise to a spectral component at the sum frequency $2f_c + (2k + \rho) \Delta f$ and to a component at the differ-

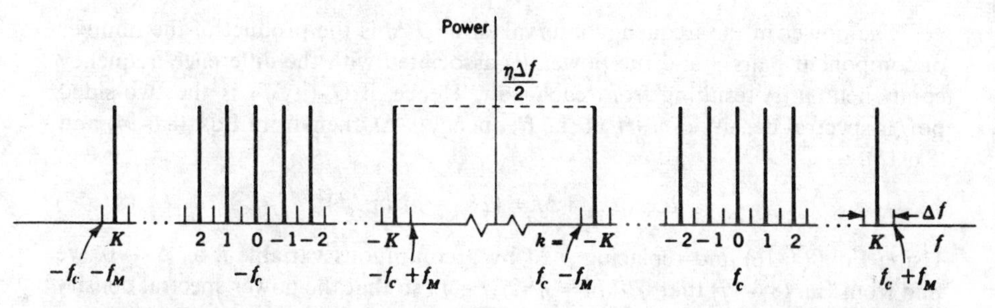

Figure 8.6-2 The spectral range $|f - f_c| \leq f_M$ of the noise $n(t)$ of power spectral density $\eta/2$ is divided into intervals Δf. The power in each interval is represented approximately by a single spectral line of power $\eta \Delta f/2$.

ence frequency $\rho\ \Delta f$. The sum frequency component is of no interest to us since it lies outside the baseband frequency range 0 to f_M. The difference frequency component, as is readily verified from Eq. (8.6-12), is

$$n_\rho(t) = c_k c_{k+\rho} \cos(2\pi\rho\ \Delta f t + \theta_{k+\rho} - \theta_k) \tag{8.6-13}$$

The power P_ρ associated with $n_\rho(t)$ may readily be deduced on the basis of the discussion in Sec. 7.8 concerning noise-noise mixing. We note that $n_\rho(t)$ in Eq. (8.6-13) is larger by a factor of 2 than the difference frequency term in Eq. (7.8-8). This factor of 2 results from the fact that $n_\rho(t)$ arises from the multiplication of the *sum* of two noise spectral components by itself, while the difference term in Eq. (7.8-8) arises from the product of one noise spectral component with another. Applying the result given in Eq. (7.8-8) to the present case, we find, since $\overline{c_k^2} = \overline{c_{k+\rho}^2}$ and using Eq. (8.6-11), that

$$P_\rho \equiv \overline{n_\rho^2(t)} = \tfrac{1}{2}\overline{c_k^2}\ \overline{c_{k+\rho}^2} = 2(\eta\ \Delta f)^2 \tag{8.6-14}$$

Different particular sets of spectral components in Eq. (8.6-10) are separated by the frequency difference $\rho\ \Delta f$ as in Eq. (8.6-13) and have power given by Eq. (8.6-14). These spectral components, albeit at the same frequency, are uncorrelated. Hence the total power at frequency $\rho\ \Delta f$ is simply the sum of the individual powers. Therefore, to find the total power associated with the frequency $\rho\ \Delta f$, we need now only calculate the number of pairs of spectral components in Eq. (8.6-10) which are separated by $\rho\ \Delta f$.

Now there are $2K + 1$ spectral components in $n(t)$, as given by Eq. (8.6-10), spaced by intervals Δf. Hence the number of pairs p of components separated by a frequency $\rho\ \Delta f$ is $p = 2K + 1 - \rho$. Since, in the limit, as $\Delta f \to 0$, $K \to \infty$, we shall ignore the unity in comparison with K and write

$$p \approx 2K - \rho \tag{8.6-15}$$

We have further, as is apparent in Fig. 8.6-2, that $f_M = (K + \frac{1}{2})\ \Delta f$. For the reason just indicated we write

$$f_M \approx K\ \Delta f \tag{8.6-16}$$

The power in the frequency interval Δf at $\rho\ \Delta f$ is the product of the number of component pairs p and the power P_ρ associated with the difference-frequency component $n_\rho(t)$ resulting from each pair. Hence, if $G_{n^2}(\rho\ \Delta f)$ is the two-sided power spectral density of $n^2(t)$, at the frequency $\rho\ \Delta f$, then, from Eqs. (8.6-14) and (8.6-15),

$$2G_{n^2}(\rho\ \Delta f)\ \Delta f = (2K - \rho)2(\eta\ \Delta f)^2 \tag{8.6-17}$$

Using Eq. (8.6-16) and replacing $\rho\ \Delta f$ by a continuous variable f, as $\Delta f \to 0$, we find from Eq. (8.6-17) that $G_{n^2}(f) = \eta^2(2f_M - f)$ so that the power spectral density of $\lambda n^2(t)$ is

$$G_{\lambda n^2}(f) = \lambda^2\eta^2(2f_M - f) \tag{8.6-18}$$

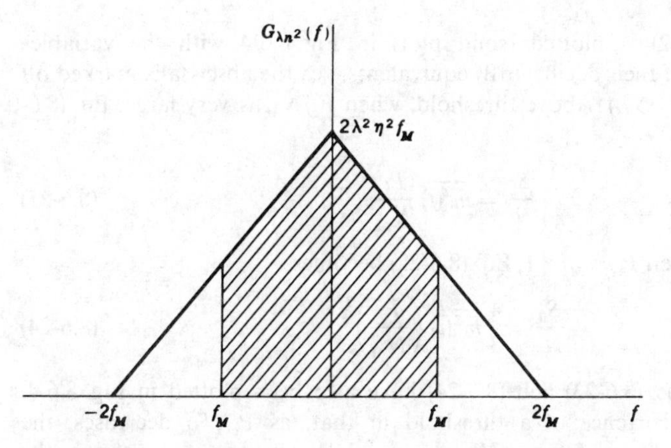

Figure 8.6-3 Plot of power spectral density $G_{\lambda n^2}(f)$ in baseband region.

as plotted in Fig. 8.6-3. The baseband filter has a bandwidth that extends only to f_M. Therefore the noise power output N''_o is given by the area of the shaded region in the triangular plot of Fig. 8.6-3. We find that

$$N''_o = 3\lambda^2\eta^2 f_M^2 \tag{8.6-19}$$

Signal-to-Noise Ratio; Threshold

The total output-noise power is, from Eqs. (8.6-9) and (8.6-19),

$$N_o = N'_o + N''_o = 2\lambda^2\eta f_M A^2 + 3\lambda^2\eta^2 f_M^2 \tag{8.6-20}$$

The output signal-to-noise ratio is, from Eqs. (8.6-8) and (8.6-20),

$$\frac{S_o}{N_o} = \frac{A^4\overline{m^2(t)}}{2\eta f_M A^2 + 3\eta^2 f_M^2} \tag{8.6-21}$$

We now rewrite Eq. (8.6-21) using as before the symbols $P_c \equiv A^2/2$ for the carrier power and $N_M \equiv \eta f_M$ for the input-noise power in the frequency range f_M. We find that

$$\frac{S_o}{N_o} = \overline{m^2(t)}\,\frac{P_c}{N_M}\,\frac{1}{1 + \frac{3}{4}(N_M/P_c)} \tag{8.6-22}$$

A comparison of Eqs. (8.6-22) and (8.5-6) is of interest. In both cases the input signal is an amplitude-modulated carrier with both sidebands present. The equations are identical except for the additional factor $1/[1 + \frac{3}{4}(N_M/P_c)]$ which appears in the case of the square-law demodulator. This factor has its origin in the noise-times-noise term $\lambda n^2(t)$ in Eq. (8.6-5). When the carrier power P_c is very much larger than N_M, the extra factor may be ignored, and the square-law demodulator performs as well as the linear synchronous demodulator. Otherwise, however, the square-law demodulator is at a disadvantage.

Equation (8.6-22) is plotted (solid plot) in Fig. 8.6-4 with the variables expressed in terms of their decibel (dB) equivalents, i.e., the abscissa is marked off in units of 10 log (P_c/N_M). Above threshold, when P_c/N_M is very large, Eq. (8.6-22) becomes

$$\frac{S_o}{N_o} = \overline{m^2(t)}\,\frac{P_c}{N_M} \tag{8.6-23}$$

Below threshold, when $P_c/N_M \ll 1$, Eq. (8.6-22) becomes

$$\frac{S_o}{N_o} = \frac{4}{3}\,\overline{m^2(t)}\left(\frac{P_c}{N_M}\right)^2 \tag{8.6-24}$$

For comparison, Eqs. (8.6-23) and (8.6-24) have also been plotted in Fig. 8.6-4. We observe the occurrence of a threshold in that, as P_c/N_M decreases, the demodulator performance curve falls progressively further away from the straight-line plot corresponding to P_c/N_M very large. The *threshold* point is

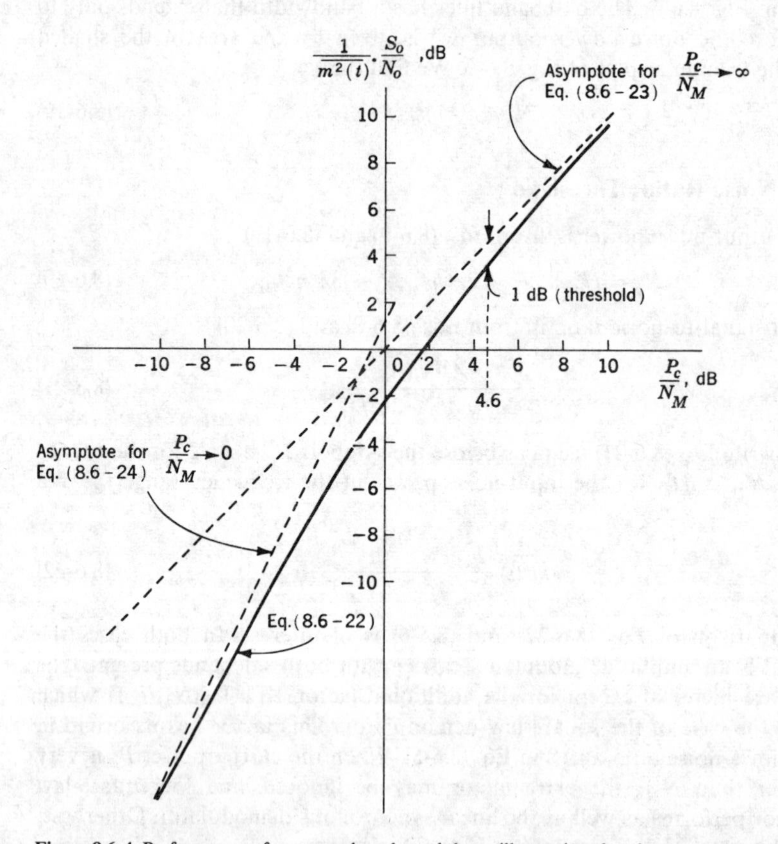

Figure 8.6-4 Performance of a square-law demodulator illustrating the phenomenon of threshold.

chosen arbitrarily to be the point at which the performance curve falls away by 1 dB as shown. On this basis it turns out that the threshold occurs when $P_c/N_M = 4.6$ dB or when $P_c = 2.9N_M$.

8.7 THE ENVELOPE DEMODULATOR

We again consider an AM signal with modulation $|m(t)| < 1$. To demodulate this DSB signal we shall use a network which accepts the modulated carrier and provides an output which follows the waveform of the *envelope* of the carrier. The diode demodulator of Sec. 3.6 is a physical circuit which performs the required operation to a good approximation. As usual, as in Fig. 8.3-1, the demodulator is preceded by a bandpass filter with center frequency f_c and bandwidth $2f_M$, and is followed by a low-pass baseband filter of bandwidth f_M.

It is convenient in the present discussion to use the noise representation given in Eq. (7.11-2):

$$n(t) = n_c(t) \cos \omega_c t - n_s(t) \sin \omega_c t \tag{8.7-1}$$

If the noise $n(t)$ has a power spectral density $\eta/2$ in the range $|f - f_c| \le f_M$ and is zero elsewhere as shown in Fig. 8.4-1, then, as explained in Sec. 7.12, both $n_c(t)$ and $n_s(t)$ have the spectral density η in the frequency range $-f_M$ to f_M.

At the demodulator input, the input signal plus noise is

$$s_1(t) + n_1(t) = A[1 + m(t)] \cos \omega_c t + n_c(t) \cos \omega_c t - n_s(t) \sin \omega_c t \tag{8.7-2a}$$

$$= \{A[1 + m(t)] + n_c(t)\} \cos \omega_c t - n_s(t) \sin \omega_c t \tag{8.7-2b}$$

where A is the carrier amplitude and $m(t)$ the modulation. In a phasor diagram, the first term of Eq. (8.7-2b) would be represented by a phasor of amplitude $A[1 + m(t)] + n_c(t)$, while the second term would be represented by a phasor perpendicular to the first and of amplitude $n_s(t)$. The phasor sum of the two terms is then represented by a phasor of amplitude equal to the square root of the sum of the squares of the amplitudes of the two terms. Thus, the output signal plus noise just prior to baseband filtering is the envelope (phasor sum):

$$s_2(t) + n_2(t) = \{(A[1 + m(t)] + n_c(t))^2 + n_s^2(t)\}^{1/2} \tag{8.7-3a}$$

$$= \{A^2[1 + m(t)]^2 + 2A[1 + m(t)]n_c(t) + n_c^2(t) + n_s^2(t)\}^{1/2} \tag{8.7-3b}$$

We should now like to make the simplification in Eq. (8.7-3b) that would be allowed if we might assume that both $|n_c(t)|$ and $|n_s(t)|$ were much smaller than the carrier amplitude A. The difficulty is that n_c and n_s are noise "waveforms" for which an explicit time function may not be written and which are described only in terms of the statistical distributions of their instantaneous amplitudes. No matter how large A and how small the values of the standard deviation of $n_c(t)$ or $n_s(t)$, there is always a finite probability that $|n_c(t)|$, $|n_s(t)|$, or both, will be comparable to, or even larger than, A. On the other hand, if the standard deviations of $n_c(t)$ and $n_s(t)$ are much smaller than A, the *likelihood* that n_c or n_s will approach or exceed A is rather small. For example, since n_c and n_s have gaussian distributions, the probability that $n_c(t)$ is greater than twice the standard deviation is only 0.045, and only 0.00006 that it exceeds 4 times its standard deviation. Hence, if $\sqrt{\overline{n_c^2(t)}} \ll A$ and we assume that $|n_c(t)| \ll A$, the assumption is *usually* valid.

Assuming then that $|n_c(t)| \ll A$ and $|n_s(t)| \ll A$, the "noise-noise" terms $n_c^2(t)$ and $n_s^2(t)$ may be dropped, leaving us with the approximation

$$s_2(t) + n_2(t) \approx \{A^2[1 + m(t)]^2 + 2A[1 + m(t)]n_c(t)\}^{1/2} \tag{8.7-4a}$$

$$= A[1 + m(t)]\left\{1 + \frac{2n_c(t)}{A[1 + m(t)]}\right\}^{1/2} \tag{8.7-4b}$$

Using now the further approximation that $(1 + x)^{1/2} \approx 1 + x/2$ for small x we have finally that

$$s_2(t) + n_2(t) \approx A[1 + m(t)] + n_c(t) \tag{8.7-5}$$

The output-signal power measured after the baseband filter, and neglecting dc terms, is $S_o = A^2\overline{m^2(t)}$. Since the spectral density of $n_c(t) = \eta$, the output-noise power after baseband filtering is $N_o = 2\eta f_M$. Again using the symbol $N_M (\equiv \eta f_M)$ to stand for the noise power at the input in the baseband range f_M, and using Eq. (8.5-3), we find that

$$\gamma \equiv \frac{S_o/N_o}{S_i/N_M} = \frac{\overline{m^2(t)}}{1 + \overline{m^2(t)}} \tag{8.7-6}$$

The result is the same as given in Eq. (8.5-13) for synchronous demodulation. To make a comparison with the square-law demodulator, we assume $\overline{m^2(t)} \ll 1$. In this case, as before, $S_i \cong P_c$, and Eq. (8.7-6) reduces to Eq. (8.5-6). Hence we have the important result that *above threshold* the synchronous demodulator, the square-law demodulator, and the envelope demodulator all perform equally well, provided $\overline{m^2(t)} \ll 1$.

Threshold

Like the square-law demodulator, the envelope demodulator exhibits a threshold. As the input signal-to-noise ratio decreases, a point is reached where the signal-to-noise ratio at the output decreases more rapidly than at the input. The calculation of signal-to-noise ratio is quite complex, and we shall therefore be content to simply state the result[1] that for $S_i/N_M \ll 1$, and $\overline{m^2(t)} \ll 1$

$$\frac{S_o}{N_o} = \frac{\overline{m^2(t)}}{1.1}\left(\frac{S_i}{N_M}\right)^2 \tag{8.7-7}$$

Equation (8.7-7) obviously indicates a poorer performance than indicated by Eq. (8.7-6), which applies above threshold.

Since both square-law demodulation and envelope demodulation exhibit a threshold, a comparison is of interest. We had assumed in square-law demodulation that $\overline{m^2(t)} \ll 1$. Then, as noted above, $S_i \cong A^2/2 = P_c$ the carrier power, and Eq. (8.7-7) becomes

$$\frac{S_o}{N_o} = \frac{\overline{m^2(t)}}{1.1}\left(\frac{P_c}{N_M}\right)^2 \tag{8.7-8}$$

which is to be compared with Eq. (8.6-24) giving S_o/N_o below threshold for the square-law demodulator.

The comparison indicates that, below threshold, the square-law demodulator performs better than the envelope detector. Actually this advantage of the square-law demodulator is of dubious value. Generally, when a demodulator is operated below threshold to any appreciable extent, the performance may be so poor as to be nearly useless. What is of greater significance is that the comparison suggests that the threshold in square-law demodulation is lower than the threshold in envelope demodulation. Therefore a square-law demodulator will operate above threshold on a weaker signal than will an envelope demodulator.

In summary, on strong signals all demodulators work equally well except that the square-law demodulator requires that $\overline{m^2(t)} \ll 1$ to avoid baseband-signal distortion. On weak signals, synchronous demodulation does best since it exhibits no threshold. When synchronous demodulation is not feasible, square-law demodulation does better than envelope demodulation. It is also interesting to note that voice signals require a 40-dB output signal-to-noise ratio for high quality. In this case both the linear-envelope detector and the square-law detector operate above threshold.

REFERENCE

1. Davenport, W., and W. Root: "Random Signals and Noise," McGraw-Hill Book Company, New York, 1958.

PROBLEMS

8.1-1. (*a*) A superheterodyne receiver using an IF frequency of 455 kHz is tuned to 650 kHz. It is found that the receiver picks up a transmission from a transmitter whose carrier frequency is 1560 kHz. Suggest a reason for this undesired reception and suggest a remedy. (These frequencies, 650 kHz and 1560 kHz, are referred to as *image frequencies*. Why?)

8.3-1. Let $g(t)$ be a waveform characterized by a power spectral density $G(f)$. Assume $G(f) = 0$ for $|f| \geq f_1$. Show that the time-average value of $g(t) \cos 2\pi f_c t$ is zero if $f_c > f_1$.

8.3-2. As noted in Sec. 3.10, if $m(t)$ is an arbitrary baseband waveform, a received SSB signal may be written $s_i(t) = m(t) \cos 2\pi f_c t + \hat{m}(t) \sin 2\pi f_c t$. Here $\hat{m}(t)$ is derived from $m(t)$ by shifting by 90° the phase of every spectral component in $m(t)$.

(*a*) Show that $m(t)$ and $\hat{m}(t)$ have the same power spectral densities and that $\overline{m^2(t)} = \overline{\hat{m}^2(t)}$.

(*b*) Show that if $m(t)$ has a spectrum which extends from zero frequency to a maximum frequency f_M, then $m^2(t)$, $\hat{m}^2(t)$, and $m(t)\hat{m}(t)$ all have spectra which extend from zero frequency to $2f_M$.

(*c*) Show that the normalized power $\bar{S}_i$ of $s_i(t)$ is $\overline{m^2(t)} = \overline{\hat{m}^2(t)}$. (*Hint*: Use the results of Prob. 8.3-1.)

(*d*) Calculate the normalized power S_o of the demodulated SSB signal, i.e., the signal $s_i(t)$ multiplied by $\cos 2\pi f_c t$ and then passed through a baseband filter. Show that $S_o = \overline{m^2(t)}/4$ and hence that $S_o/S_i = \frac{1}{4}$. [*Note*: This problem establishes more generally the result given in Eq. (8.3-6) which was derived on the basis of the assumption that the modulating waveform is a sinusoid.]

8.3-3. Prove that Eq. (8.3-11) is correct by sketching the power spectral density of Eq. (8.3-10).

8.3-4. A baseband signal $m(t)$ is transmitted using SSB as in Prob. 8.3-2. Assume that the power spectral density of $m(t)$ is

$$G_m(f) = \begin{cases} \dfrac{\eta_m}{2} \dfrac{|f|}{f_M} & |f| < f_M \\ 0 & |f| > f_M \end{cases}$$

Find:

(*a*) The input signal power.

(*b*) The output signal power.

(*c*) If white gaussian noise with power spectral density $\eta/2$ is added to the SSB signal, find the output SNR. The baseband filter cuts off at $f = f_M$.

8.3-5. A received SSB signal has a power spectrum which extends over the frequency range from $f_c =$ 1 MHz to $f_c + f_M = 1.003$ MHz. The signal is accompanied by noise with uniform power spectral density 10^{-9} watt/Hz.

(*a*) The noise $n(t)$ is expressed as $n(t) = n_c(t) \cos 2\pi f_c t - n_s(t) \sin 2\pi f_c t$. Find the power spectral densities of the quadrature components $n_c(t)$ and $n_s(t)$ of the noise in the spectral range specified as $f_c \leq f \leq f_c + f_M$.

(*b*) The signal plus its accompanying noise is multiplied by a local carrier $\cos 2\pi f_c t$. Plot the power spectral density of the noise at the output of the multiplier.

(*c*) The signal plus noise, after multiplication, is passed through a baseband filter and an amplifier which provides a voltage gain of 10. Plot the power spectral density of the noise at the amplifier output, and calculate the total noise output power.

8.4-1. Show that $m^2(t)$ in Eq. (8.4-10) is bandlimited to $2f_M$.

8.4-2. Repeat Prob. 8.3-4 if DSB rather than SSB modulation is employed.

8.4-3. A carrier of amplitude 10 mV at f_c is 50 percent amplitude-modulated by a sinusoidal waveform of frequency 750 Hz. It is accompanied by thermal noise of two-sided power spectral density

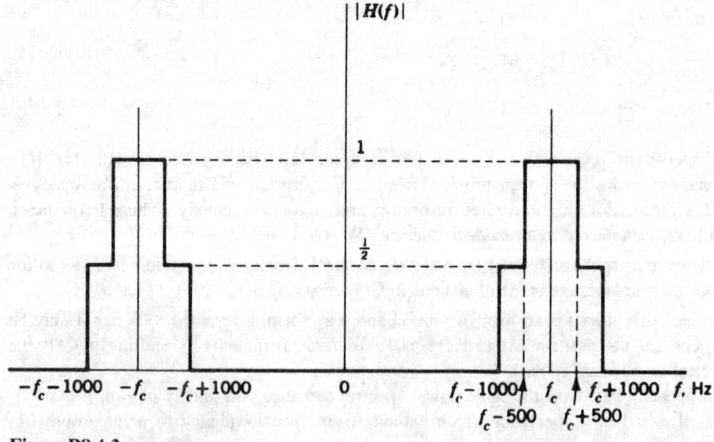

Figure P8.4-3

$\eta/2 = 10^{-3}$ watt/Hz. The signal plus noise is passed through the filter shown. The signal is demodulated by multiplication with a local carrier of amplitude 1 volt.

(*a*) Find the output signal power.

(*b*) Find the output noise power.

8.5-1. The signal $[\varepsilon + m(t)] \cos \omega_c t$ is synchronously detected. The reference signal $\cos \omega_c t$ used to multiply the incoming signal, is obtained by passing the input signal through a narrowband filter of bandwidth B, as shown in Fig. P8.5-1.

(*a*) Calculate S_i.

(*b*) Calculate $v_R(t)$ due to the input signal alone. Calculate the noise power accompanying $v_R(t)$.

(*c*) Comment on the effect of the noisy reference on the output SNR.

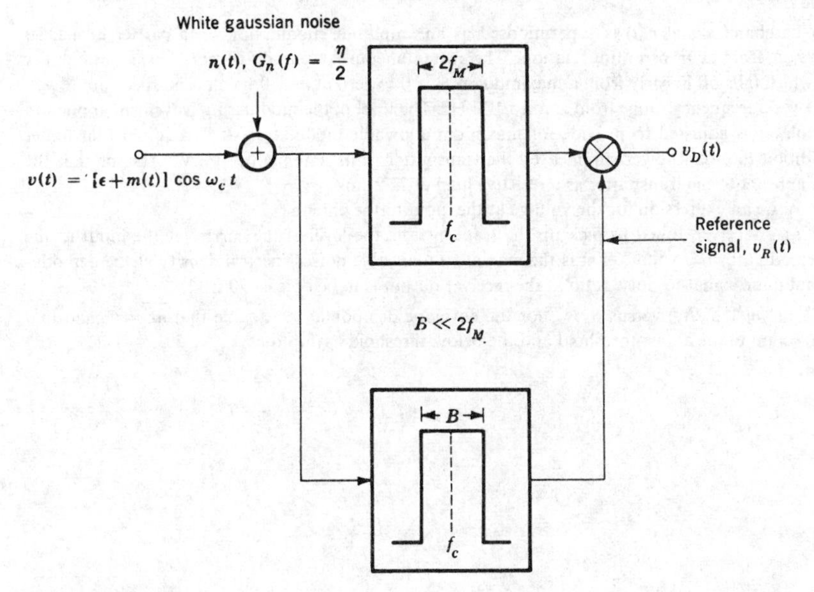

Figure P8.5-1

8.5-2. Verify Eqs. (8.5-5) and (8.5-8).

8.5-3. (*a*) Show that the output SNR of a DSB-SC signal which is synchronously detected is independent of the IF bandwidth; i.e., Eq. (8.4-8) is independent of the IF bandwidth.

(*b*) Show that the output SNR of an SSB signal which is synchronously detected is dependent on the IF bandwidth. To do this, consider an IF bandwidth which extends from $f_0 - B$ to $f_0 + f_M$, where $f_M > B > 0$. Calculate the output SNR using this bandwidth.

8.5-4. In the received amplitude-modulated signal $s_i(t) = A[1 + m(t)] \cos 2\pi f_c t$, $m(t)$ has the power spectral density $G_m(f)$ specified in Prob. 8.3-4. The received signal is accompanied by noise of power spectral density $\eta/2$. Calculate the output signal-to-noise ratio.

8.6-1. Verify Eq. (8.6-4) by showing graphically that all the neglected terms have spectra falling outside the range $|f| \leq f_M$.

8.6-2. Given $2K + 1$ spectral components of a noise waveform spaced by intervals Δf. Show that the number of pairs of components separated by a frequency $\rho\, \Delta f$ is $p = 2K - 1 - \rho$ [i.e., verify the discussion leading to Eq. (8.6-15)].

8.6-3. In a DSB transmission, a carrier of frequency 1 MHz and of amplitude 2 volts is amplitude-modulated to the extent of 10 percent by a sinusoidal baseband signal of frequency 5 kHz. The signal is corrupted by white noise of two-sided power spectral density 10^{-6} watt/Hz. The demodulator is a device whose input/output characteristic is given by $v_o = 3v_i^2$, where v_o and v_i are respectively the output and input voltage. The IF filter, before the demodulator, has a rectangular transfer characteristic of unity gain and 10-kHz bandwidth. By error, the IF filter is tuned so that its center frequency is at 1.002 MHz.

(*a*) Calculate the signal waveform at the demodulator output and calculate its normalized power.

(*b*) Calculate the noise power at the demodulator output and the signal-to-noise ratio.

8.6-4. Plot $(1/\overline{m^2})(S_o/N_o)$ versus P_c/N_M in Eq. (8.6-22) and show that the 1-dB threshold occurs when $P_c/N_M = 4.6$ dB.

8.6-5. Assume that $\overline{m^2(t)} = 0.1$ and that $S_o/N_o = 30$ dB. Find P_c/N_M in Eq. (8.6-22). Are we above threshold?

8.7-1. A baseband signal $m(t)$ is superimposed as an amplitude modulation on a carrier in a DSB transmission from a transmitting station. The instantaneous amplitude of $m(t)$ has a probability density which falls off linearly from a maximum at $m = 0$ to zero at $m = 0.1$ volt. The spectrum of $m(t)$ extends over a frequency range from zero to 10 kHz. The level of the modulating waveform applied to the modulator is adjusted to provide for maximum allowable modulation. It is required that under this condition the total power supplied by the transmitter to its antenna be 10 kW. Assume that the antenna appears to the transmitter as a resistive load of 72 ohms.

(*a*) Write an expression for the voltage at the input to the antenna.

(*b*) At a receiver, tuned to pick up the transmission, the level of the carrier at the input to the diode demodulator is 3 volts. What is the maximum allowable power spectral density at the demodulator input if the signal-to-noise ratio at the receiver output is to be at least 20 dB?

8.7-2. Plot $(1/\overline{m^2})(S_o/N_o)$ versus S_i/N_M for the envelope demodulator. Assume that $\overline{m^2} \ll 1$, and find the intersection of the above-threshold and the below-threshold asymptotes.

CHAPTER

NINE

NOISE IN FREQUENCY-MODULATION SYSTEMS

In this chapter we discuss the performance of frequency-modulation systems in the presence of additive noise. We shall show how, in an FM system, an improvement in output signal-to-noise power ratio can be made through the sacrifice of bandwidth.

9.1 AN FM DEMODULATOR

The receiving system of Fig. 8.1-1 may be used with an AM or FM signal. When used to recover a frequency-modulated signal, the AM demodulator is replaced by an FM demodulator such as the limiter-discriminator shown in Fig. 9.1-1.

The Limiter

In an FM system, the baseband signal varies only the frequency of the carrier. Hence any amplitude variation of the carrier must be due to noise alone. The *limiter* is used to suppress such amplitude-variation noise. In a limiter, diodes, transistors, or other devices are used to construct a circuit in which the output voltage v_1 is related to the input voltage v_i in the manner shown in Fig. 9.1-2*a*. The output follows the input only over a limited range. A cycle of the carrier is shown in Fig. 9.1-2*b* and the output waveform in Fig. 9.1-2*c*. In limiter operation

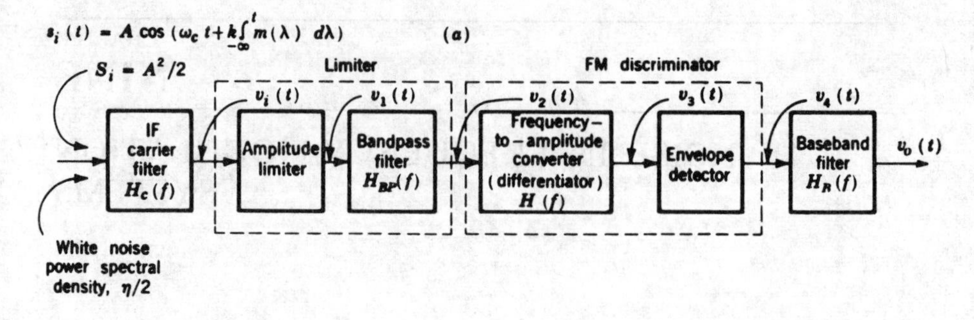

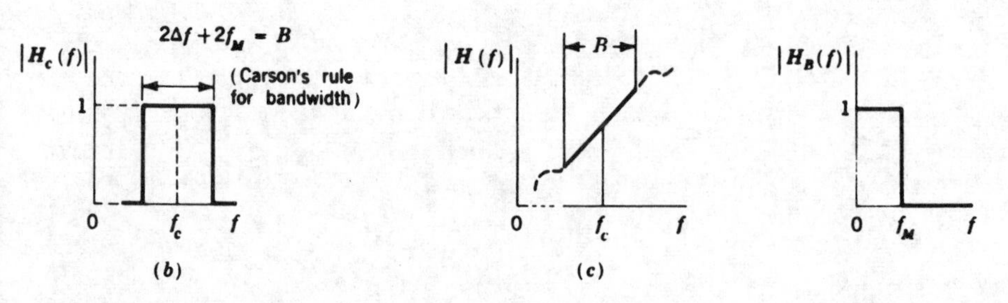

Figure 9.1-1 A limiter-discriminator used to demodulate an FM signal.

the carrier amplitude is very large in comparison with the limited range of the limiter, actually much larger than is indicated in Fig. 9.1-2. As a consequence, the output waveform is a *square* wave. Thus, the output has a waveshape which is nearly entirely independent of modest changes in carrier amplitude. The bandpass filter, following the limiter, selects the fundamental frequency component of the square wave. Therefore the filter output is again sinusoidal. It has an amplitude which is very nearly independent of the input-carrier amplitude, but does, of course, have the same instantaneous frequency as does the input carrier.

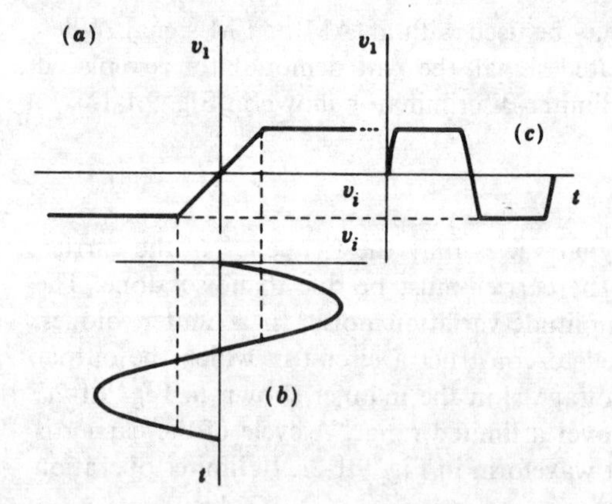

Figure 9.1-2 (*a*) A limiter input-output characteristic. (*b*) A cycle of the input carrier. (*c*) The output waveform.

In a physical circuit the limiter and filter generally form an integral unit so that actually there is no point in the limiter-filter combination where the square-wave waveform may be observed.

In this text we shall assume that the discriminator is always preceded by an ideal *hard* limiter, i.e., that the limiter output is a perfect square wave, no matter how small $v_i(t)$ may be.

The Discriminator

The discriminator consists also of two component parts. The first of these is a network which, over the range of excursion of the instantaneous frequency, exhibits a transfer characteristic $H(f)$ such that $|H(f)|$ varies linearly with frequency. When the constant-amplitude FM signal passes through this network, it will appear at the output with an amplitude variation (i.e., an envelope) which varies with time precisely as does the instantaneous frequency of the carrier. The baseband signal is now recovered by passing this amplitude-modulated waveform through an envelope demodulator such as the diode detector of Fig. 3.4-2. The input to the envelope detector is frequency-modulated as well as amplitude-modulated, but the detector does not respond to the frequency modulation.

Mathematical Representation of the Operation of the Limiter-Discriminator

The frequency-to-amplitude converter necessary to obtain frequency demodulation need have an $|H(f)|$ which varies linearly with ω over only a limited range, and its slope may be positive or negative. Further, the phase variation of $H(f)$ is of no consequence. However, as a matter of mathematical convenience we shall assume that $H(j\omega)$ is given by

$$H(j\omega) = j\sigma\omega \tag{9.1-1}$$

where σ is a constant. The advantage of such a selection of $H(j\omega)$ is that (see Fig. 9.1-1) the output of the converter $v_3(t)$ is related to the input $v_2(t)$ by the equation

$$v_3(t) = \sigma \frac{d}{dt} v_2(t) \tag{9.1-2}$$

Equation (9.1-2) results from the fact that a multiplication by $j\omega$ in the frequency domain is equivalent to differentiation in the time domain; i.e.,

$$\sigma \frac{d}{dt} \Leftrightarrow j\sigma\omega \tag{9.1-3}$$

Now suppose that the voltage $v_2(t)$ applied to the converter is

$$v_2(t) = A_L \cos\left[\omega_c t + \phi(t)\right] \tag{9.1-4}$$

Here A_L is the *limited* amplitude of the carrier so that A_L is fixed and independent of the input amplitude, and $\omega_c t + \phi(t)$ is the instantaneous phase. Then from Eq. (9.1-2)

$$v_3(t) = -\sigma A_L \left[\omega_c + \frac{d}{dt} \phi(t) \right] \sin [\omega_c t + \phi(t)] \tag{9.1-5}$$

The output of the envelope detector is, using $\alpha \equiv \sigma A_L$,

$$v_4(t) = \sigma A_L \left[\omega_c + \frac{d}{dt} \phi(t) \right] = \alpha \omega_c + \alpha \frac{d}{dt} \phi(t) \tag{9.1-6}$$

Thus, in summary, we see that if the input waveform to the discriminator is given by Eq. (9.1-4), the discriminator output is calculated from Eq. (9.1-6). Note that the discriminator output is proportional to the frequency of the input, $d\phi/dt$ (see Sec. 4.2).

9.2 CALCULATION OF OUTPUT SIGNAL AND NOISE POWERS

Let us now consider that the input signal to the IF carrier filter of Fig. 9.1-1 is

$$s_i(t) = A \cos \left[\omega_c t + k \int_{-\infty}^{t} m(\lambda)\, d\lambda \right] \tag{9.2-1}$$

where $m(t)$ is the frequency-modulating baseband waveform. We assume that the signal is embedded in additive white gaussian noise of power spectral density $\eta/2$. The IF carrier filter has a bandwidth $B = 2\,\Delta f + 2f_M$ (Carson's rule for bandwidth; see Sec. 4.7). This filter passes the signal with negligible distortion and eliminates all noise outside the bandwidth B. The signal with its accompanying noise is ideally limited, discriminated, and after passing through the baseband filter, appears at the output as a signal $s_o(t)$ and a noise waveform $n_o(t)$.

It can be shown that, when the signal-to-noise ratio is high, the noise does not affect the output-signal power. We shall accept this result without giving a proof.[1] In calculating the output-signal power we shall therefore ignore the noise. When the signal $s_i(t)$ in Eq. (9.2-1) arrives at the output of the limiter shown in Fig. 9.1-1, the signal is $s_2(t)$ [corresponding to $v_2(t)$] given by

$$s_2(t) = A_L \cos \left[\omega_c t + k \int_{-\infty}^{t} m(\lambda)\, d\lambda \right] \tag{9.2-2}$$

Using the result of Eq. (9.1-6) and setting

$$\phi(t) = k \int_{-\infty}^{t} m(\lambda)\, d\lambda \tag{9.2-3}$$

we find for the output of the discriminator

$$s_4(t) = \alpha \omega_c + \alpha k m(t) \tag{9.2-4}$$

The baseband filter rejects the dc component and passes the signal component without distortion. Thus, the output signal is $s_o(t) = \alpha k m(t)$, and the output-signal power is

$$S_o = \alpha^2 k^2 \overline{m^2(t)} \tag{9.2-5}$$

Output-Noise Power

Let us now calculate the noise output of the FM discriminator which results from the presence at the input of white noise having a power spectral density $\eta/2$. To facilitate the computation, we set the modulation $m(t) = 0$. It can be shown, although the proof is complex and will not be undertaken here, that the noise output is approximately independent of $m(t)$.[1] The carrier and noise pass through the IF filter $H_c(\omega)$, which filters the noise. The resulting noise is represented as in Eq. (7.11-2). Thus the carrier and noise at the limiter input are

$$\begin{aligned} v_i(t) &= A \cos \omega_c t + n_c(t) \cos \omega_c t - n_s(t) \sin \omega_c t \\ &= [A + n_c(t)] \cos \omega_c t - n_s(t) \sin \omega_c t \end{aligned} \tag{9.2-6}$$

A phasor diagram of the signal and noise is shown in Fig. 9.2-1. Note that the phasor representing $n_c(t) \cos \omega_c t$ is in phase with the carrier phasor $A \cos \omega_c t$. The phasor representing $n_s(t) \sin \omega_c t$ has an amplitude $n_s(t)$ and is in phase-quadrature with the other two terms. The envelope $R(t)$ is easily computed using Fig. 9.2-1 and is

$$R(t) = \sqrt{[A + n_c(t)]^2 + [n_s(t)]^2} \tag{9.2-7}$$

Similarly, the phase $\theta(t)$ is

$$\theta(t) = \tan^{-1} \frac{n_s(t)}{A + n_c(t)} \tag{9.2-8}$$

Thus, the signal and noise forming $v_i(t)$ can be written as

$$v_i(t) = R(t) \cos [\omega_c t + \theta(t)] \tag{9.2-9}$$

We ignore the time-varying envelope $R(t)$, since all time variations are removed by the limiter. The output of the limiter-bandpass filter is therefore

$$v_2(t) = A_L \cos [\omega_c t + \theta(t)] \tag{9.2-10}$$

where again A_L is determined by the limiter, and is a constant.

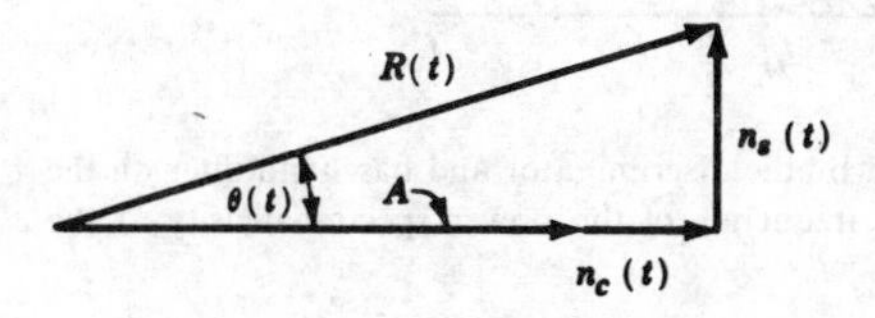

Figure 9.2-1 A phasor diagram of the terms in Eq. (9.2-6).

Let us assume that we are operating under the condition of high-input SNR. Then the noise power is much smaller than the carrier power. In this case we assume that most of the time $|n_c(t)| \ll A$ and $|n_s(t)| \ll A$, on the basis of the justification presented in Sec. 8.7. With these assumptions and using the approximation $\tan\theta \approx \theta$ for small θ, we have, from Eq. (9.2-8),

$$\theta(t) \approx \frac{n_s(t)}{A} \tag{9.2-11}$$

Thus, $v_2(t)$ is approximately

$$v_2(t) = A_L \cos\left[\omega_c t + \frac{n_s(t)}{A}\right] \tag{9.2-12}$$

Comparing Eq. (9.2-12) with Eq. (9.1-4) we see that $n_s(t)/A$ is $\phi(t)$. Then, from Eq. (9.1-6) we find

$$v_4(t) = \alpha\left[\omega_c + \frac{1}{A}\frac{d}{dt}n_s(t)\right] \tag{9.2-13}$$

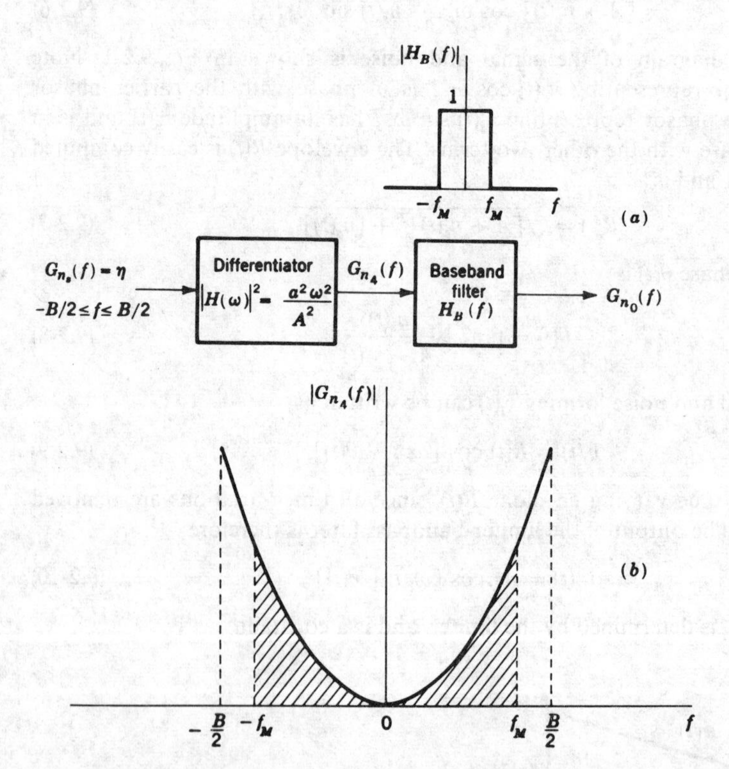

Figure 9.2-2 (*a*) Indicating the operations performed by the discriminator and baseband filter on the noise output of the limiter. (*b*) The variation with frequency of the power spectral density at the output of an FM demodulator.

If we drop the dc term in Eq. (9.2-13), the noise $n_4(t)$ at the input to the baseband filter is

$$v_4(t) = \frac{\alpha}{A} \frac{d}{dt} n_s(t) \tag{9.2-14}$$

The spectral density of $n_s(t)$ is η over the frequency range $|f| \leq B/2$ (see Sec. 7.12). The differentiation is equivalent to passing $n_s(t)$ through a network whose transfer function is $H(j\omega) = j\omega$. Hence, as shown in Fig. 9.2-2a, the operation performed on $n_s(t)$ is equivalent to passing $n_s(t)$ through a network with $H(j\omega) = j\alpha\omega/A$. Then $|H(j\omega)|^2 = \alpha^2\omega^2/A^2$. Therefore the spectral density of $n_4(t)$ is $G_{n_4}(f)$ given by

$$G_{n_4}(f) = \frac{\alpha^2\omega^2}{A^2}\eta \qquad |f| \leq \frac{B}{2} \tag{9.2-15}$$

This spectral density is plotted in Fig. 9.2-2b.

Since the baseband filter passes frequencies just up to f_M, only the shaded area in Fig. 9.2-2b contributes to the output-noise power. This output power can therefore be calculated by computing the shaded area. The output-noise power N_o is

$$N_o = \int_{-f_M}^{f_M} G_{n_4}(f)\, df = \frac{\alpha^2\eta}{A^2}\int_{-f_M}^{f_M} 4\pi^2 f^2\, df = \frac{8\pi^2}{3}\frac{\alpha^2\eta}{A^2} f_M^3 \tag{9.2-16}$$

Output SNR

The output signal-to-noise ratio can now be computed using Eqs. (9.2-5) and (9.2-16). We find

$$\frac{S_o}{N_o} = \frac{\alpha^2 k^2 \overline{m^2(t)}}{(8\pi^2/3)(\alpha^2\eta/A^2)f_M^3} = \frac{3}{4\pi^2}\frac{k^2\overline{m^2(t)}}{f_M^2}\frac{A^2/2}{\eta f_M} \tag{9.2-17}$$

Now let us consider that the modulating signal $m(t)$ is sinusoidal and produces a frequency deviation Δf. Then the input signal $s_i(t)$ given in Eq. (9.2-1) may be written [see Eq. (4.4-4)]

$$s_i(t) = A\cos\left(\omega_c t + \frac{\Delta f}{f_m}\sin 2\pi f_m t\right) \tag{9.2-18}$$

where f_m is the modulating frequency. Comparing Eq. (9.2-1) with (9.2-18), we have, after differentiating the argument, that

$$km(t) = 2\pi\,\Delta f \cos 2\pi f_m t \tag{9.2-19}$$

Hence

$$k^2\overline{m^2(t)} = \frac{4\pi^2(\Delta f)^2}{2} = 2\pi^2(\Delta f)^2 \tag{9.2-20}$$

Substituting this result into Eq. (9.2-17) yields

$$\frac{S_o}{N_o} = \frac{3}{2}\left(\frac{\Delta f}{f_M}\right)^2 \frac{A^2/2}{\eta f_M} = \frac{3}{2}\beta^2\frac{S_i}{N_M} \tag{9.2-21}$$

where $\beta \equiv \Delta f/f_M$ is the modulation index, $S_i = A^2/2$ is the input-signal power, and $N_M \equiv \eta f_M$ is the noise power at the input in the baseband bandwidth f_M. We also have that

$$\gamma_{\text{FM}} \equiv \frac{S_o/N_o}{S_i/N_M} \equiv \frac{3}{2}\beta^2 \tag{9.2-22}$$

which is to be compared with values of γ given in Eqs. (8.5-11) and (8.5-12) for AM systems.

It is to be noted that in Eqs. (9.2-21) and (9.2-22) the symbol β is being used in a sense which is somewhat different from the sense in which we have used it previously in Chap. 4. Previously, we used β to represent the modulation index associated with a sinusoidal modulating waveform of frequency f_m, that is, $\beta = \Delta f/f_m$, where Δf is the frequency deviation produced by the sinusoidal waveform. In the present instance, $\beta(\equiv \Delta f/f_M)$ characterizes not a particular sinusoidal modulating waveform but rather a particular set of specifications associated with an FM system. Thus Eq. (9.2-22) has the following interpretation. Suppose that in an FM system whose baseband frequency range is f_M we consider only sinusoidal modulation. Suppose we assume that independently of the frequency of the modulation its amplitude (and, hence, the frequency deviation Δf) is kept fixed. Then, independently of the frequency of the modulation (up to f_M) the performance criterion γ will be fixed at $3\beta^2/2$.

9.3 COMPARISON OF FM AND AM

A comparison of the performances of FM and conventional AM (double sideband with carrier) is of interest. Equation (9.2-22) applies for sinusoidal frequency modulation of the carrier with modulation index β. Let us compare this result with the corresponding result for sinusoidal amplitude modulation with 100 percent modulation. We find for γ_{AM} from Eq. (8.5-14) that $\gamma_{\text{AM}} = \frac{1}{3}$, so that from Eq. (9.2-22)

$$\frac{\gamma_{\text{FM}}}{\gamma_{\text{AM}}} = \frac{9}{2}\beta^2 \tag{9.3-1}$$

In the comparison of FM and AM leading to Eq. (9.3-1), we have assumed in the two cases equal input-noise-power spectral density $\eta/2$, equal baseband bandwidth f_M, and equal input-signal power S_i.

Several authors prefer to make the comparison not on the basis of equal signal power but rather on the basis of *equal signal power measured when the modulation* $m(t) = 0$. In this case, as is easily verified (Prob. 9.3-2), we find that Eq. (9.3-1) is replaced by

$$\frac{\gamma_{FM}}{\gamma_{AM}} = 3\beta^2 \tag{9.3-2}$$

A comparison on this basis is not completely fair, however, since the total *transmitted* powers are unequal when the unmodulated carrier powers are the same.

It is clear from either Eq. (9.3-1) or Eq. (9.3-2) that FM offers the possibility of improved signal-to-noise ratio over AM. The improvement begins when $9\beta^2/2 \approx 1$ (Eq. 9.3-1) or when $3\beta^2 \approx 1$ (Eq. 9.3-2) corresponding to $\beta \cong \sqrt{2}/3 \cong 0.5$ or $\beta \cong 1/\sqrt{3} \cong 0.6$. As β increases, the improvement becomes more pronounced, but this improvement is achieved at the expense of requiring greater bandwidth. To see the relationship between improvement and bandwidth sacrifice, let us assume that β is large enough so that Carson's rule for bandwidth formula [Eq. (4.7-2)]

$$B_{FM} = 2(\beta + 1)f_M \tag{9.3-3}$$

may be approximated by $B_{FM} \approx 2\beta f_M$. The bandwidth of the AM system is $B_{AM} = 2f_M$ so that Eq. (9.3-1) may be rewritten

$$\frac{\gamma_{FM}}{\gamma_{AM}} = \frac{9}{2}\left(\frac{B_{FM}}{B_{AM}}\right)^2 \tag{9.3-4}$$

Accordingly, each increase in bandwidth by a factor of 2 increases γ_{FM}/γ_{AM} by a factor of 4 (6 dB).

We observe then a characteristic of FM which is not shared by AM. The FM system allows us to sacrifice bandwidth for the sake of improving signal-to-noise ratio. The improvement begins to make itself apparent when $\beta \cong 0.5$ or 0.6. This value of β is roughly the value of β which establishes the demarcation between "narrowband" and "wideband" FM. Thus, signal-to-noise improvement is a feature of wideband FM not shared by narrowband FM. Equations (9.3-1) and (9.3-2) suggest a continuous improvement in performance with increased β (and correspondingly increased bandwidth). Such is indeed the case as long as the noise power admitted by the carrier filter continues to be small in comparison with the signal power. It will be recalled that in deriving Eq. (9.2-22) we made this assumption of relatively small noise (see Eq. 9.2-11). When the bandwidth becomes so large that the noise power is not relatively small, the performance of the FM system degrades rapidly, i.e., the system exhibits a *threshold*. We shall see in Sec. 10.1 that when the input-noise power is not small in comparison with the input-signal power, the system performance may be improved by *restricting* the bandwidth, by *reducing* the modulation index.

We shall now comment briefly on what might appear to be an anomalous situation. We refer to the fact that the performance of the FM system is improved as the bandwidth increases. We might, on intuitive grounds, have expected the opposite effect, since widening the bandwidth admits more noise into the system.

In the AM system the carrier filter bandwidth extends over the range $f_c \pm f_M$, while in FM the bandwidth extends over the range $f_c \pm \beta f_M$. The fact is, however, that noise in the range $f_c \pm \beta f_M$ but outside the range $f_c \pm f_M$ does not appear at the output of the FM system. To see this clearly, we refer to Eq. (9.2-14), which relates the output noise to the quadrature component of the input noise. This equation indicates that each spectral component of input noise gives rise to an output spectral component of the same frequency. Consider, for example, the pair of noise spectral components at frequencies $f_c \pm f_n$ with $f_M < f_n < \beta f_M$. Such noise components give rise to a baseband noise component at a frequency f_n which will not pass through the baseband filter whose cutoff is at f_M. Each such pair of noise spectral components gives rise to a baseband noise component which is independent of all other noise components. That is, the noise spectral components are *not correlated*.

On the other hand, let the carrier be FM-modulated by a sinusoidal baseband signal of frequency $f_s > f_M/2$. Then, as discussed in Sec. 4.5, the carrier is accompanied by sidebands separated from the carrier by $\pm f_s$, $\pm 2f_s$, etc. Except for the sidebands at $\pm f_s$, all other sidebands are separated from the carrier by more than the baseband bandwidth. However, these sidebands *are correlated* in such a way that the sidebands at, say, $\pm 2f_s$, do not give rise to a baseband signal at frequency $2f_s$ but serve rather to increase the amplitude of the baseband signal at f_s, as do the higher-order sidebands. Thus, the apparent anomaly is resolved by recognizing that noise outside the range $f_c \pm f_M$ does not pass through the system, while signal sidebands outside this range do contribute to the eventual output signal of the demodulator.

9.4 PREEMPHASIS AND DEEMPHASIS, SINGLE CHANNEL

Suppose that we undertake to transmit a baseband signal using FM modulation and naturally require the best possible signal-to-noise ratio for a given carrier power, noise spectral density, IF bandwidth, and baseband bandwidth. Then clearly, before applying the baseband signal to the FM modulator, we shall raise the level of the modulating baseband signal to the maximum extent possible in order to modulate the carrier as vigorously as possible. How shall we know that we have reached the maximum allowable level of the modulating signal? One way of making such a determination is to demodulate the modulated signal and measure the distortion. The distortion occurs because eventually the frequency deviation exceeds the specified IF bandwidth. The modulating signal level may be raised only until the distortion exceeds a specified value.

When, however, the baseband signal happens to be an audio signal, it turns out that something further can be done. An audio signal usually has the characteristic that its power spectral density is relatively high in the low-frequency range and falls off rapidly at higher frequencies. For example, speech has little power spectral density above about 3 kHz. And while music, of course, extends farther into the high-frequency range, the feature still persists that most of its power is in the low-frequency region. As a consequence, when we examine the spectrum of the sidebands associated with a carrier which is frequency-modulated by an audio signal, we find that the power spectral density of the sidebands is greatest near the carrier and relatively small near the limits of the allowable frequency band allocated to the transmission. The manner in which we may take advantage of these spectral features, which are characteristic of audio signals, in order to improve the performance of an FM system, is shown in Fig. 9.4-1.

We observe in Fig. 9.4-1 that, at the transmitting end, the baseband signal $m(t)$ is not applied directly to the FM modulator but is first passed through a filter of transfer characteristic $H_p(\omega)$, so that the modulating signal is $m_p(t)$. The modulated carrier is transmitted across a communication channel, during which process, as usual, noise is added to the signal. The receiver is a conventional discriminator except that a filter has been introduced before the baseband filter. The transfer characteristic of this filter is the reciprocal of the characteristic of the transmitter filter. This receiver filter of transfer characteristic $1/H_p(\omega)$ may equally well be placed after the baseband filter since both filters are linear. We observe that any modification introduced into the baseband signal by the first filter, prior to modulation, is presently undone by the second filter which follows the discriminator. Hence, the output signal at the receiver is exactly the same as it would be if the filters had been omitted entirely. The noise, however, passes through only the receiver filter, and this filter may then be used to suppress the noise to some extent.

The selection of the transfer characteristic $H_p(\omega)$ is based on the following considerations. We note that at the output of the demodulator the spectral density of the noise, given by $G_{n4}(f)$ in Eq. (9.2-15) and shown in Fig. 9.2-2*b*, increases with the square of the frequency. Hence the receiver filter will be most effective in suppressing noise if the response of the filter falls off with increasing frequency, that is, if the filter transmission is lowest where the spectral density of the noise is highest. In such a case the transmitter filter must exhibit a rising response with increasing frequency. Let us assume initially that we design the transmitter filter so that it serves only to increase the spectral density of the higher-frequency components of the signal $m(t)$. Such a filter must necessarily increase the power in the modulating signal, thereby increasing the distortion above its specified maximum value. We have, however, noted above that the spectral density of the modulated carrier is relatively small near the edges of the allowed frequency band. We may then expect that such a filter may possibly raise

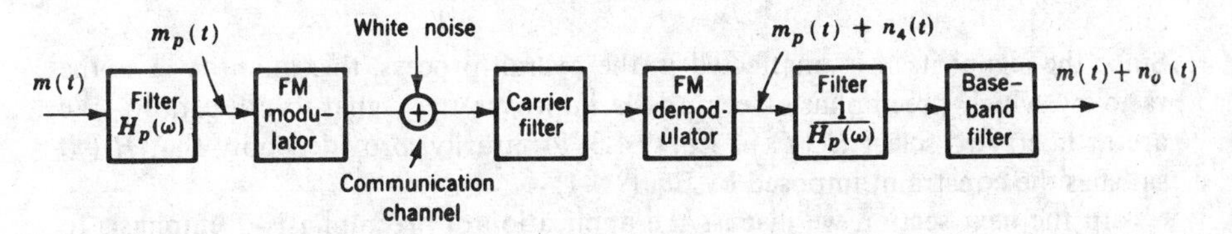

Figure 9.4-1 Preemphasis and deemphasis in an FM system.

the signal spectral density only near the edges of the allowed frequency band and cause only a small increase in distortion. In this case we might expect that if the modulating signal power is lowered to decrease the distortion to the allowed value, we end up with a net advantage, i.e., the improvement due to raising the spectral density near the edges of the allowable band outweighs the disadvantage due to the need to lower the level of the modulating signal. We shall see that such is indeed the case. The premodulation filtering in the transmitter, to raise the power spectral density of the baseband signal in its upper-frequency range, is called *preemphasis* (or *predistortion*). The filtering at the receiver to undo the signal preemphasis and to suppress noise is called *deemphasis*.

SNR Improvement Using Preemphasis

We recall from Sec. 4.13 that the bandwidth occupied by the output of an FM modulator is fixed if the normalized power (or mean square frequency deviation) of the modulating signal is kept fixed (see Prob. 9.4-3). Hence, referring to Fig. 9.4-1, we require that the normalized power of the baseband signal $m(t)$ must be the same as the normalized power of the preemphasized signal $m_p(t)$. We begin by expressing the normalized power of a signal in terms of its spectral density. If then $G_m(f)$ is the power spectral density of $m(t)$, then the power spectral density of $m_p(t)$ is $|H_p(f)|^2 G_m(f)$, and we require that

$$P_m = \int_{-f_M}^{f_M} G_m(f)\, df = \int_{-f_M}^{f_M} |H_p(f)|^2 G_m(f)\, df \tag{9.4-1}$$

where f_M is the maximum frequency of the modulating signal.

In the absence of deemphasis the output noise is given by N_o in Eq. (9.2-16). With the deemphasis filter, the output noise is

$$N_{od} = \left(\frac{\alpha}{A}\right)^2 4\pi^2\eta \int_{-f_M}^{f_M} f^2 \left|\frac{1}{H_p(f)}\right|^2 df \tag{9.4-2}$$

The ratio of the noise output without deemphasis to the noise output with deemphasis is $N_o/N_{od} \equiv \mathscr{R}$. From Eqs. (9.2-16) and (9.4-2) we have

$$\mathscr{R} = \frac{(\alpha/A)^2(4\pi^2\eta)\displaystyle\int_{-f_M}^{f_M} f^2\, df}{(\alpha/A)^2(4\pi^2\eta)\displaystyle\int_{-f_M}^{f_M} f^2/|H_p(f)|^2\, df} = \frac{f_M^3/3}{\displaystyle\int_0^{f_M} f^2\, df/|H_p(f)|^2} \tag{9.4-3}$$

Since the signal itself is unaffected in the overall process, this quantity $\mathscr{R}$ is the ratio by which preemphasis-deemphasis improves the signal-to-noise ratio. We are at liberty to select $H_p(f)$ in Eq. (9.4-3) arbitrarily, provided only that $H_p(f)$ satisfies the constraint imposed by Eq. (9.4-1).

In the next section we discuss the application of preemphasis-deemphasis to commercial FM broadcasting.

9.5 PREEMPHASIS AND DEEMPHASIS IN COMMERCIAL FM BROADCASTING

We have noted that the spectral density of the noise at the output of an FM demodulator increases with the square of the frequency. Hence a deemphasis network at the receiver will be most effective in suppressing noise if its response falls with increasing frequency. In commercial FM the deemphasis is performed by the simple low-pass resistance-capacitance network of Fig. 9.5-1*a*. This network has a transfer function $H_d(f)$ given by

$$H_d(f) = \frac{1}{1 + jf/f_1} \tag{9.5-1}$$

where $f_1 = 1/2\pi RC$. At the transmitter we require an inverse network. A simple network which may be adjusted to provide the required response is shown in Fig. 9.5-1*b*. Let us assume that $r \ll R$, and that in the baseband audio-frequency range r is also very small in comparison with the reactance of the capacitance C. In this case we can approximately compute the current $I(f)$ by neglecting the presence of r. We find, for the audio range,

$$I(f) = V_i(f)\left(\frac{1}{R} + j\omega C\right) \tag{9.5-2}$$

The output voltage $V_o(f) = rI(f)$, and the transfer function $H_p(f) \equiv V_o(f)/I(f)$ is

$$H_p(f) = \frac{r}{R}(1 + j\omega CR) = \frac{r}{R}\left(1 + j\frac{f}{f_1}\right) \tag{9.5-3}$$

where, as before, $f_1 = 1/2\pi RC$. Hence $H_p(f)$ has a frequency dependence inverse to $H_d(f)$ as required, in order that no net distortion be introduced into the signal. Thus, $H_p(f)H_d(f) = r/R =$ constant.

Normalized logarithmic plots of H_p and H_d are shown in Fig. 9.5-2. The transfer function H_d actually has a second breakpoint, as shown, at $f_2 = 1/2\pi rC$. Since $r \ll R, f_2 \gg f_1$, and it is easily arranged that this second breakpoint lie well outside the baseband spectral range and hence be irrelevant.

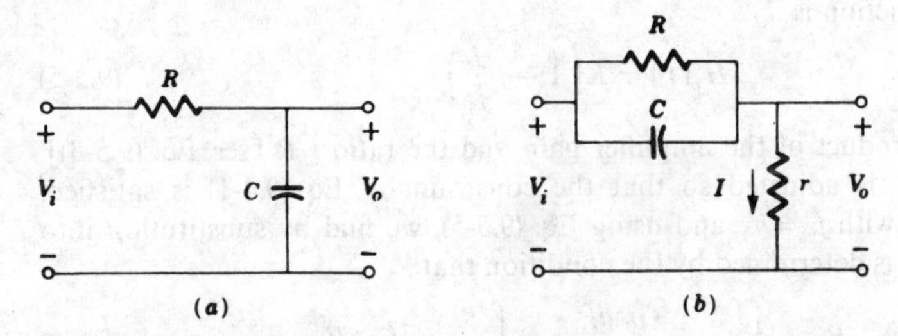

Figure 9.5-1 (*a*) Deemphasis network and (*b*) preemphasis network used in commercial radio.

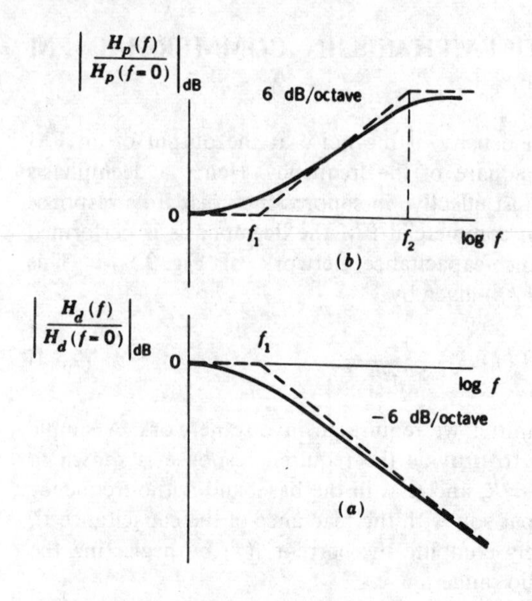

Figure 9.5-2 Normalized logarithmic plots of the frequency characteristics of (*a*) the deemphasis network and (*b*) the preemphasis network.

The improvement in signal-to-noise ratio which results from preemphasis depends on the frequency dependence of the power spectral density of the baseband signal. Let us assume that the spectral density of a typical audio signal, say music, may reasonably be represented as having a frequency dependence given by

$$G_m(f) = \begin{cases} G_0 \dfrac{1}{1 + (f/f')^2} & |f| \leq f_M \\ 0 & \text{elsewhere} \end{cases} \tag{9.5-4}$$

where G_0 is the spectral density at low frequencies, while f' is the frequency at which $G(f)$ has fallen by 3 dB from its low-frequency value. Let us further assume that we have adjusted the preemphasis network so that $f_1 = f'$.

The baseband signal to be preemphasized is transmitted through the network of Fig. 9.5-1*b* and also through an adjustable gain amplifier. The amplifier will make up for the attenuation produced by the preemphasis network, and its gain is adjusted to the point where the full allowable bandwidth is occupied by the modulated carrier. Hence, the baseband signal $m(t)$ passes through a network whose transfer function is

$$H_p(f) = K\left(1 + j\frac{f}{f_1}\right) \tag{9.5-5}$$

where K is the product of the amplifier gain and the ratio r/R [see Eq. (9.5-3)]. The coefficient K is adjusted so that the constraint of Eq. (9.4-1) is satisfied. Using Eq. (9.5-4) with $f' = f_1$ and using Eq. (9.5-5), we find by substitution into Eq. (9.4-1) that K is determined by the condition that

$$P_m = \int_{-f_M}^{f_M} \frac{G_0\, df}{1 + (f/f_1)^2} = \int_{-f_M}^{f_M} K^2 G_0\, df \tag{9.5-6}$$

Integrating and solving for K^2, we find

$$K^2 = \frac{f_1}{f_M} \tan^{-1} \frac{f_M}{f_1} \tag{9.5-7}$$

Using this value of K, and substituting $H_p(f)$ in Eq. (9.5-5) into Eq. (9.4-3), we compute $\mathscr{R}$, the improvement which results from preemphasis and subsequent deemphasis. We find that

$$\mathscr{R} = \frac{\tan^{-1}(f_M/f_1)}{3(f_1/f_M)[1 - (f_1/f_M)\tan^{-1}(f_M/f_1)]} \tag{9.5-8}$$

If the power in the baseband signal is principally confined to the lower frequencies, so that $f_M/f_1 \gg 1$, then $\mathscr{R}$ becomes approximately

$$\mathscr{R} \cong \frac{\pi}{6} \frac{f_M}{f_1} \tag{9.5-9}$$

In commercial FM broadcasting $f_1 = 2.1$ kHz, while f_M may reasonably be taken as $f_M = 15$ kHz. In this case $f_M/f_1 = 7.5$, and from Eq. (9.5-9) $\mathscr{R} \cong 4.7$ corresponding to 6.7 dB improvement in the output SNR when using preemphasis. Since output signal-to-noise ratios are typically 40 to 50 dB, this represents an improvement of approximately 15 percent; a significant improvement.

Preemphasis is particularly effective in FM systems which are used for transmission of audio signals. This effectiveness results from the fact that the spectral density of the audio signal is smallest precisely where the spectral density of the noise is greatest. For, as noted, the audio signal spectral density falls off with increasing frequency while the noise spectral density increases with the square of the frequency. The advantage of using preemphasis is less pronounced in an AM system, such as DSB with carrier, which is employed in commercial AM broadcasting. For in AM the noise spectral density is constant and does not increase with frequency. Preemphasis is also frequently employed in connection with phonographic recording to suppress needle scratch noise.

Stereo FM

As noted in Sec. 9.2, and as shown in Fig. 9.2-2, the noise output of the discriminator in an FM receiver has a parabolic power spectral density in the range $-B/2$ to $B/2$. In commercial FM the bandwidth is $B \simeq 200$ kHz so that $B/2 \simeq 100$ kHz. In monophonic FM, only the noise in the spectral range of the baseband filter is passed on to the output. In stereo (Sec. 4.14), the situation is different. The difference signal $L(t) - R(t)$ as received occupies the frequency range from 23 to 53 kHz. When this difference signal is translated to baseband, so also is the noise in this difference-signal frequency interval. The noise in this difference-signal frequency interval is substantially larger than in the baseband frequency range, because the difference-signal frequency range is twice as large as the baseband range and also because of the parabolic form of the noise power spectral density. As a result, commercial stereo FM is noisier than monophonic FM.

For the sake of the advantage gained thereby, as well as for the sake of compatibility, preemphasis is incorporated in stereo FM just as in monophonic FM. In stereo FM, at the transmitter, the sum signal $L(t) + R(t)$ and the difference signal $L(t) - R(t)$ are passed through identical preemphasis filters. At the receiver the sum and difference signals are passed through identical deemphasis filters before the sum and difference signals are combined to generate the individual signals $L(t)$ and $R(t)$. The preemphasis and deemphasis filters are as shown in Fig. 9.5-1, with transfer characteristics as indicated in Fig. 9.5-2. As in monophonic FM, $f_1 = 2.1$ kHz (corresponding to a time constant of 75 μs).

The overall result, taking the preemphasis and deemphasis into account, and assuming a baseband frequency range extending from nominally zero to 15 kHz, is that commercial stereo FM yields a signal-to-noise ratio about 22 dB poorer than monophonic FM. (See Prob. 9.5-2.) This disadvantage turns out to be tolerable simply because of the high power with which commercial broadcasts are transmitted.

9.6 PHASE MODULATION IN MULTIPLEXING

One method which is used to multiplex voice channels prior to long distance transmission is shown in Fig. 9.6-1. The speech signals $m_1(t)$, $m_2(t)$, etc., are each bandlimited to 3 kHz and are the baseband signals used to generate a sequence of SSB-SC signals, whose carrier frequencies f_1, f_2, etc., are separated by 4 kHz. The outputs of the SSB modulators are added, and this sum of signals forms a composite baseband signal that is used to angle-modulate (FM, PM, or some

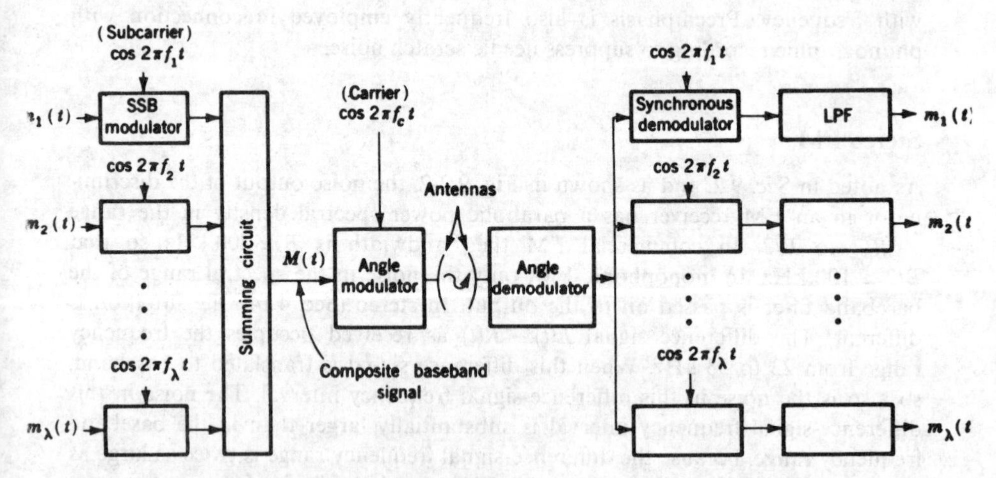

Figure 9.6-1 A system of frequency division multiplexing.

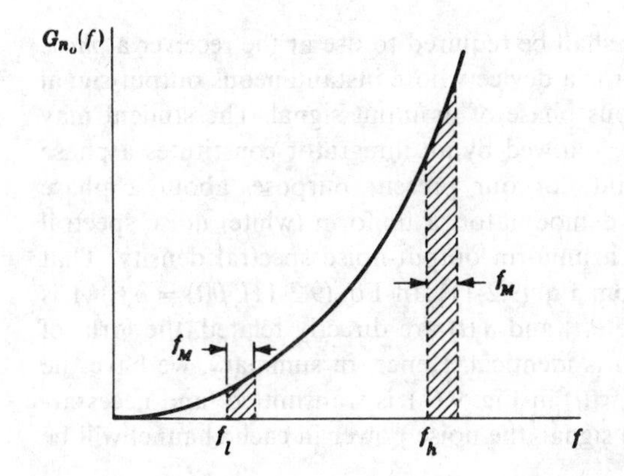

Figure 9.6-2 To illustrate that in the multiplex system of Fig. 9.6-1, using FM, channels associated with high carrier frequencies are noisier than those associated with lower frequencies.

combination of the two) a carrier of frequency f_c. If 1000 speech signals are multiplexed in this way, the baseband signal nominally extends over a bandwidth 1000×4 kHz $= 4$ MHz. The angle modulator excites an antenna whose signal is directed at a receiving antenna. Not indicated in the figure are the repeater stations which may be interposed between initial transmitter and final receiver in a system which connects very distant points. For example, much of the telephone communication between the United States and Europe presently employs a system as shown in Fig. 9.6-1, and uses a satellite repeater.

At the receiver, an angle demodulator recovers the composite baseband signal. The individual baseband signals $m_1(t)$, $m_2(t)$, etc., are recovered by synchronous demodulation and by passing the output of each synchronous demodulator through a low-pass filter of bandwidth 3 kHz.

Suppose now that the angle modulator at the transmitter is an FM modulator, that is, the instantaneous frequency of the modulator is directly proportional to the instantaneous magnitude of the composite baseband signal $M(t)$. Then the angle demodulator would be a frequency demodulator, such as the discriminator discussed in Sec. 9.1. However, we have noted in that section that, with white noise present at the input to the discriminator, the spectral density of the noise output is quadratic. As a consequence a channel associated with a higher subcarrier frequency will be *noisier* than a channel associated with a lower subcarrier frequency. That such is the case is easily seen in Fig. 9.6-2. Here we have plotted the positive half of the parabolic spectral density $G_{n_o}(f)$ of the output noise of the discriminator. Two subcarrier frequencies f_l and f_h are indicated, together with the frequency band occupied by the associated SSB signals (the upper sidebands are arbitrarily chosen). The noise power in each of the individual channels is proportional to its respective shaded area. If then the transmitter power is raised to a level where the signal-to-noise ratio is acceptable in the highest channel, the signal-to-noise ratio in the lower channels will be better than is required.

This variation in noise power from channel to channel may be corrected, and the entire system rendered thereby more efficient, by transmitting a carrier which is *phase-modulated* by the composite signal, rather than frequency-modulated.

For under these circumstances we shall be required to use at the receiver a *phase demodulator*. A phase demodulator is a device whose instantaneous output signal is proportional to the instantaneous phase of its input signal. The student may readily verify that a discriminator followed by an integrator constitutes a phase demodulator. The important point, for our present purpose, about a phase demodulator is that, with such a demodulator, a uniform (white) noise spectral density at the input gives rise to a uniform output-noise spectral density. That such is the case is readily seen from Eq. (9.2-11). In Eq. (9.2-11), $\theta(t) = n_s(t)/A$ is the phase-modulation noise. Since $\theta(t)$ and $n_s(t)$ are directly related, the form of the power spectral density of each is identical. Hence, in summary, we have the result that if the composite signal $M(t)$ in Fig. 9.6-1 is transmitted (and necessarily received) as a phase-modulated signal, the noise power in each channel will be the same.

9.7 COMPARISON BETWEEN FM AND PM IN MULTIPLEXING

We noted in the preceding section that phase modulation offers some advantage over frequency modulation in a multichannel FDM system. We compare the two systems now, to see more quantitatively the advantage that accrues from the use of phase modulation.

A frequency-modulation system is shown in Fig. 9.7-1*a* and a phase-modulation system is indicated in Fig. 9.7-1*b*. We have chosen to construct the phase modulator as a frequency modulator preceded by a differentiator. That such an arrangement does indeed constitute a phase modulator was pointed out

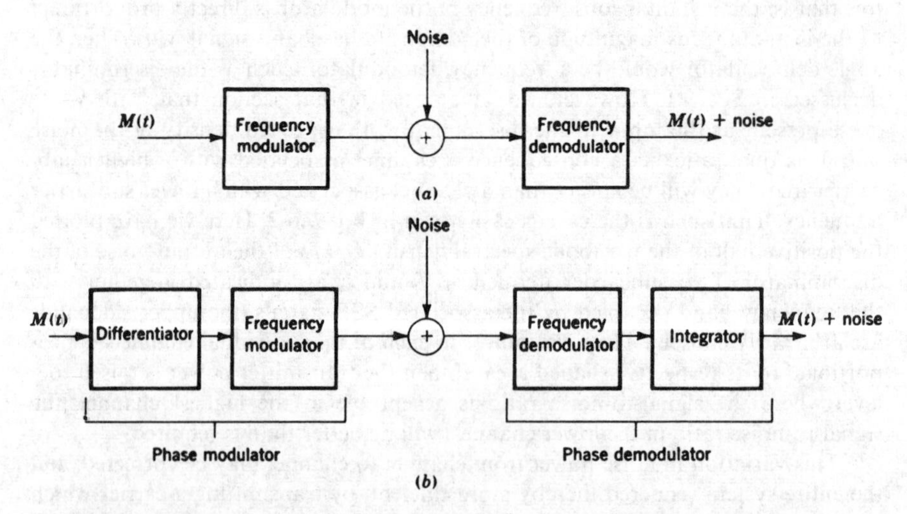

Figure 9.7-1 Comparison of an FM system in (*a*) with a phase-modulation system in (*b*).

in Sec. 4.3. Similarly we have constructed a phase demodulator as a discriminator followed by an integrator. We may, if we please, view the system of Fig. 9.7-1*b* as a frequency-modulation system in which preemphasis and deemphasis have been incorporated. The differentiator is the preemphasis network, and the integrator is the deemphasis network. The combination of preemphasis and deemphasis leaves the signal unaltered; hence the output signal is the same in Fig. 9.7-1*a* and *b*. However, the noise power per channel is the same for each channel in (*b*), while in (*a*) the noise is most pronounced in the top channel. Assuming that both channels (*a*) and (*b*), are constrained to use the same bandwidth, we shall now compute the improvement afforded by channel (*b*) in the signal-to-noise ratio of the top channel.

If there are N channels, the frequency range of the topmost channel of the composite signal $M(t)$ extends from $(N-1)f_M$ to Nf_M, where f_M is the frequency range of an individual component signal. Applying Eq. (9.2-16) to the present case, we find, that, in the absence of deemphasis, the noise output of the top channel is

$$N_{o,\text{top}} = 2\frac{\alpha^2\eta}{A^2}\int_{(N-1)f_M}^{Nf_M} 4\pi^2 f^2\,df$$

$$\simeq \frac{8\pi^2\alpha^2\eta N^2 f_M^3}{A^2} \tag{9.7-1}$$

since $N \gg 1$. The preemphasis (differentiator) circuit has a transfer function whose magnitude squared is

$$|H_p(f)|^2 = 4\pi^2\tau^2 f^2 \tag{9.7-2}$$

in which τ^2 is a constant. For the deemphasis filter we have

$$|H_d(f)|^2 = \frac{1}{|H_p(f)|^2} = \frac{1}{4\pi^2\tau^2 f^2} \tag{9.7-3}$$

Applying Eq. (9.4-1) to the composite signal $M(t)$, we find that the condition of equal bandwidth requires that

$$\int_{-Nf_M}^{+Nf_M} G_M(f)\,df = \int_{-Nf_M}^{+Nf_M} |H_p(f)|^2 G_M(f)\,df \tag{9.7-4}$$

In Eq. (9.7-4) $G_M(f)$ is the power spectral density of the composite signal. Assuming that $G_M(f)$ is constant, we find by substituting Eq. (9.7-2) into Eq. (9.7-4) that

$$\tau^2 = \frac{3}{4\pi^2 N^2 f_M^2} \tag{9.7-5}$$

In the presence of a deemphasis filter with transfer function $H_d(f)$ the output-noise power of the top channel becomes, from Eq. (9.7-1),

$$N_{od,\text{top}} = 2\frac{\alpha^2\eta}{A^2}\int_{(N-1)f_M}^{Nf_M} \frac{4\pi^2 f^2\,df}{|H_p(f)|^2} \tag{9.7-6}$$

The improvement factor $\mathscr{R}$ for the top channel is calculated from Eqs. (9.7-1), (9.7-3), (9.7-5), and (9.7-6). We find

$$\mathscr{R}_{\text{top}} \equiv \frac{N_{o,\text{top}}}{N_{od,\text{top}}} = 3(= 4.8 \text{ dB}) \tag{9.7-7}$$

9.8 EFFECT OF TRANSMITTER NOISE

The discussion of the preceding section is somewhat unrealistic in that it was assumed that the only noise with which we have to contend is noise introduced in the communications channel. Actually, some noise is introduced by the transmitter itself. For example, the frequency modulator may have some "jitter." That is, there is some random variation of frequency even in the absence of a modulating signal. Let us represent this jitter as due to a noise voltage present at the input to the frequency modulator, and let us assume for explicitness that the spectral density of this noise voltage is uniform. If we now also assume that the spectral density of the composite baseband signal is also uniform, then the situation with the PM system can be represented as shown in Fig. 9.8-1*a*. In this figure the composite input signal is $M(t)$ of spectral density $G_M(f)$, and the output of the amplifier-differentiator is $M'(t)$ of spectral density $G_{M'}(f)$. The transmitter noise $n(t)$ with spectral density $G_n(f)$, is placed at the junction between the amplifier-differentiator and the frequency modulator. The positive-frequency portion of the assumed *flat* power spectral density of $G_M(f)$ is shown in Fig. 9.8-1*b*, while the

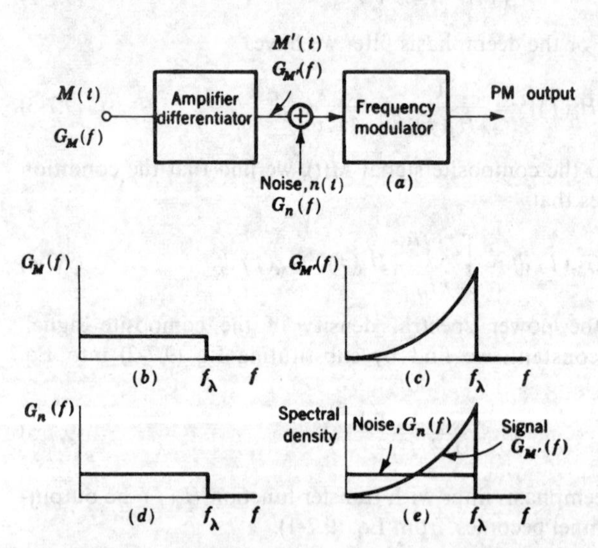

Figure 9.8-1 (*a*) A PM system in which noise is introduced before transmission. (*b*) The spectral density of the system. (*c*) The spectral density of the signal after differentiation. (*d*) The spectral density of the noise. (*e*) Comparison of spectral densities of signal and noise at input to modulator.

parabolic density of $G_{M'}(f)$ appears in Fig. 9.8-1*c*. The noise spectral density is shown in Fig. 9.8-1*d*, while finally in Fig. 9.8-1*e* we compare the spectral densities of the signal and noise at the input to the frequency modulator.

Observe that at the lower frequencies the noise exceeds the signal. To keep the signal level above the noise at low frequencies, it is necessary to modify the transfer characteristic of the network which precedes the frequency modulator. A suitable network is one whose transfer characteristic increases with frequency to emphasize the high frequencies without suppressing the low frequencies. A network similar to the preemphasis circuit of Fig. 9.5-1*b* is suitable. In practice when such a preemphasis circuit is employed, together with its reciprocal deemphasis circuit, the 4.8-dB advantage quoted above for PM over FM is not realized. In a 1000-channel voice multiplex system the advantage is more nearly 3 or 4 dB.

REFERENCE

1. Sakrison, D.: "Communication Theory," John Wiley & Sons, Inc., New York, 1968.

PROBLEMS

9.1-1. In an FM receiver we find it advantageous to use an amplitude limiter. In the "amplitude-modulation" receivers discussed in Chap. 8 we did not find it advantageous to use a "frequency limiter." Discuss.

9.1-2. The AM-FM signal $[1 + m(t)] \cos (\omega_c t + \beta \sin \omega_M t)$, $|m(t)| < 1$ is ideally limited and then bandpass-filtered. If the bandwidth of the bandpass filter is selected to pass the fundamental frequency and to reject frequencies of the order of $2\omega_c$, $3\omega_c$, etc., and, in addition, if the filter bandwidth is also chosen to pass 98 percent of the signal energy contained in the fundamental and its sidebands, find:

(*a*) The bandwidth of the filter.

(*b*) The minimum ratio of ω_c/ω_M for a given β.

9.1-3. (*a*) The signal $v(t) = 2(1 + 0.1 \cos 2\pi \times 10^3 t) \cos 2\pi \times 10^6 t$ is passed through a filter which removes the lower sideband. Write an expression for the filter output $v_o(t)$ in a form which makes explicitly evident that $v_o(t)$ is an angle-modulated waveform with varying amplitude, that is

$$v_o(t) = A(t) \cos [2\pi \times 10^6 t + \theta(t)]$$

Find $A(t)$ and $\theta(t)$ and make reasonable approximations to simplify $v_o(t)$.

(*b*) The waveform $v_o(t)$ is applied to a network whose transfer function is $H(f) = j\sigma\omega$ (with $\sigma = 1/2\pi$ sec) in the frequency range $f_c = 10^6 \pm 100$ Hz and $H(f) = 0$ otherwise. The output of this network is then applied to an envelope demodulator. Write an expression for the output of the demodulator.

9.1-4. We note from Eq. (9.1-6) that, since $\alpha = \sigma A_L$, the output of the limiter-discriminator of Fig. 9.1-1 is $v_4 = \sigma A_L \, d\phi/dt$. Hence the more severe the limiting, i.e., the smaller A_L, the smaller will be the signal output of the limiter-discriminator. A student commented that limiting should be just barely adequate to remove amplitude variations in order to keep A_L as large as possible. Is this comment valid? Discuss.

9.2-1. For Eq. (9.2-21) make a plot of $(S_o/N_o)_{\text{dB}}$ against $(S_i/N_M)_{\text{dB}}$ for values of $\beta = 1, 5, 10, 100$. On the same set of axes include the corresponding plot for synchronous demodulation in a linear system (i.e., SSB or DSB-SC).

9.2-2. Commercial FM transmission is allocated a bandwidth $B = 200$ kHz. Assume a receiver with a rectangular IF filter of corresponding bandwidth. If the discriminator output is applied directly to a baseband filter with $f_M = 15$ kHz, what fraction of the noise output of the discriminator passes through the filter?

9.2-3. Consider two frequency-modulated signals, the first where $km(t) = \beta\omega_M \cos \omega_M t$ and the second where $km(t)$ is equal to a gaussian signal, which is white with power spectral density $\eta_m/2$, $|f| \leq f_M$, and zero elsewhere. If the output SNR of each system is the same, calculate the ratio of the IF bandwidths needed to pass 98 percent of the signal energy in each case.

9.2-4. If the input SNR $S_i/N_M = 30$ dB, calculate S_o/N_o when $km(t) = \beta\omega_M \cos \omega_M t$ and $\beta = 1, 5, 10, 100$.

9.2-5. In an FM system the baseband bandwidth is f_M. The modulation is sinusoidal and the bandwidth B is to be kept constant. Write an expression for γ_{FM}, defined in Eq. (9.2-22) as a function of the modulation index, in terms of B and f_M.

9.2-6. The input SNR $S_i/N_M = 30$ dB, and an $S_o/N_o = 48$ dB is required. Find the mean-squared frequency deviation in terms of the frequency range, f_M.

9.3-1. Verify Eq. (9.3-1).

9.3-2. Derive Eq. (9.3-2).

9.3-3. A baseband signal $m(t)$ whose spectrum extends to $f_M = 5$ kHz is transmitted by FM. The rms frequency deviation produced by $m(t)$ is 100 kHz. The same signal $m(t)$ is also transmitted by SSB. If the output signal-to-noise ratios for FM and SSB are to be the same, compare the input signal-to-noise ratios for the two types of transmission.

9.4-1. The power spectral density of a modulating signal $m(t)$ is given by

$$G_m(f) = \frac{G_0}{1 + (f/f_1)^2}$$

where $f_1 \ll f_M$. If a preemphasis circuit is to be used, where $|H_p(f)|^2 = K^2 f^2$, find K^2 if the preemphasis is not to increase the bandwidth.

9.4-2. If $G_m(f) = \eta_m/2$, $|f| \leq f_M$ and zero elsewhere,

(*a*) Find $\int_{-f_M}^{f_M} |H_p(f)|^2 \, df$.

(*b*) If $|H_p(f)|^2 = k_1^2 + k_2^2 f^2$, find k_1 and k_2 to maximize $\mathscr{R}$, subject to the constraint of part *a* of this problem.

9.4-3. Consider that a carrier is FM-modulated by a baseband waveform $m(t)$. Let the bandwidth occupied by the FM waveform be defined as in Eq. (4.14-1):

$$B = 2\left[\frac{\int_{-\infty}^{\infty} v^2 G(v)\, dv}{\int_{-\infty}^{\infty} G(v)\, dv}\right]^{1/2}$$

(*a*) Show that, in this case, the bandwidth is proportional to the normalized power of $m(t)$, i.e., $B = c^2 \overline{m^2(t)}$, with c^2 a constant.

(*b*) Let $m(t)$ have a normalized power of 2 volts^2 and suppose that the sensitivity of the FM modulator is such that a change by 1 volt in $m(t)$ produces a frequency change of 10 kHz. Find the bandwidth B.

9.5-1. Preemphasis is to be used in conjunction with a DSB-SC system. The spectral density of the baseband signal, of spectral range f_M, is as given in Eq. (9.5-4), and the preemphasis filter has the transfer function given in Eq. (9.5-5). Assume $f_1 = f'$. Impose the constraint that the preemphasis is not to increase the transmitted power. Calculate the preemphasis improvement $\mathscr{R}$ and compare with Eq. (9.5-9).

9.5-2. (*a*) In a stereo FM system the baseband spectral range is $f_M = 15$ kHz. Assume that preemphasis is not used. Find the ratio of the noise output power to the noise power for monophonic transmission. See Fig. 4.24-1.

(*b*) Taking preemphasis into account as described in Sec. 9.5, show that stereo FM is 22 dB noisier than monophonic FM.

9.5-3. Plot Eqs. (9.5-8) and (9.5-9) to determine when Eq. (9.5-9) can be employed.

9.5-4. If, in Eq. (9.5-8), $f_1 = 2.1$ kHz and $f_M = 15$ kHz, calculate $\mathscr{R}$. Compare your result with that given in the text which was obtained using Eq. (9.5-9).

9.7-1. (*a*) One thousand signals are multiplexed by the system of Fig. 9.6-1. The angle modulator is a frequency modulator, i.e, the composite baseband signal $M(t)$ frequency modulates the carrier. The lowest SSB carrier frequency is 10 kHz and each baseband signal is allocated 4 kHz. Assume that each baseband signal has the same normalized power. The channel with the least noise turns out to have a signal-to-noise ratio of 80 dB. If a signal-to-noise ratio lower than 20 dB were not acceptable, how many channels of the system would be usable?

(*b*) Assume instead that $M(t)$ phase-modulates the carrier. Assume also that the total transmitted power is the same as in part (*a*). Calculate the signal-to-noise ratio in any channel.

9.7-2. A 4-MHz TV signal, and one thousand 4-kHz audio signals, are multiplexed onto a single FM carrier (the audio signals are SSB-modulated to obtain this goal, the TV signal is left at baseband and is therefore channel 1). The power spectral density of the composite signal is constant over its entire spectral range.

(*a*) Find the spectral range of the composite signal.

(*b*) Calculate the output SNR for channel 1, the TV signal, in terms of the input SNR.

(*c*) Calculate the output SNR for the top channel.

CHAPTER

TEN

THRESHOLD IN FREQUENCY MODULATION

We have seen (Secs. 8.6 and 8.7) that the nonlinear AM demodulators, such as the square-law demodulator and the envelope demodulator, exhibit a threshold. The FM discriminator also exhibits a threshold, and in this chapter we shall discuss the mechanism which is responsible for such an FM threshold. We shall also discuss two demodulator circuits which have lower thresholds than the discriminator. These circuits are the *phase-locked loop* (PLL) and the *frequency demodulator with feedback* (FMFB).

10.1 THRESHOLD IN FREQUENCY MODULATION

In a communication system in which the modulation is linear and demodulation is accomplished by coherent detection (for example, SSB and DSB-SC), we have the result that [see Eqs. (8.5-11) and (8.5-12)]

$$\frac{S_o}{N_o} = \frac{S_i}{N_M} \tag{10.1-1}$$

or equivalently

$$10 \log \frac{S_o}{N_o} \equiv \left[\frac{S_o}{N_o}\right]_{\text{dB}} = \left[\frac{S_i}{N_M}\right]_{\text{dB}} \equiv 10 \log \frac{S_i}{N_M} \tag{10.1-2}$$

In FM we have [see Eq. (9.2-21)]

$$\frac{S_o}{N_o} = \frac{3}{2} \beta^2 \frac{S_i}{N_M} \tag{10.1-3}$$

or equivalently

$$\left[\frac{S_o}{N_o}\right]_{\text{dB}} = \left[\frac{S_i}{N_M}\right]_{\text{dB}} + 10 \log \frac{3}{2} \beta^2 \tag{10.1-4}$$

In Fig. 10.1-1 we have plotted Eq. (10.1-2) in a coordinate system in which the coordinate axes are marked off in decibels. The plot is a straight line passing through the origin. We have also plotted Eq. (10.1-4) for two values of $\beta(> \sqrt{\frac{2}{3}})$. These plots, as indicated by the dashed extensions, are also straight lines. In the FM case a plot for a value of β is raised by the amount $10 \log (3\beta^2/2)$ above the plot for a linear coherent modulation system. The quantity $10 \log (3\beta^2/2)$ expresses in decibels precisely the improvement afforded by the FM system in return for a sacrifice of bandwidth.

Experimentally it is determined, however, that the FM system exhibits a threshold. Thus, as indicated by the solid-line plots, for each value of β, as S_i/N_M decreases, a point is reached where S_o/N_o falls off much more sharply than S_i/N_M. The threshold value of S_i/N_M is arbitrarily taken to be the value at which S_o/N_o falls 1 dB below the dashed extension. We note that for larger β the threshold S_i/N_M is also higher. Suppose, then, that we are operating with a modulation index β_1 above the threshold for β_1 but below the threshold for β_2. Suppose,

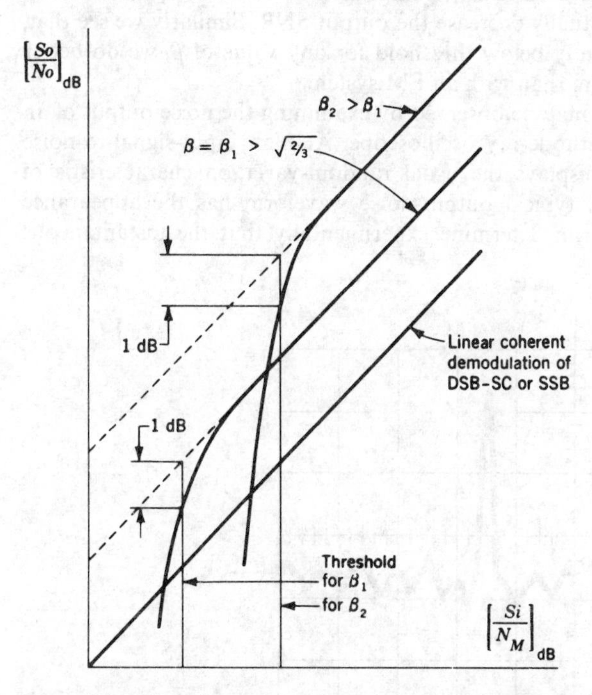

Figure 10.1-1 Plots of output signal-to-noise ratio against input signal-to-noise ratio for linear modulation and demodulation and also for an FM system. Illustrating the phenomenon of threshold in FM.

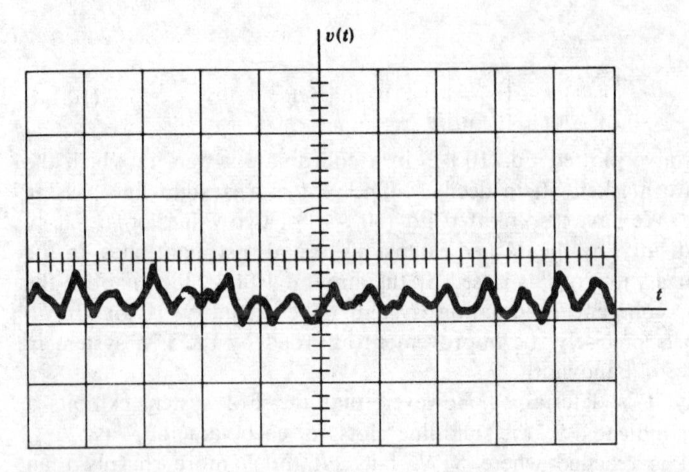

Figure 10.1-2 Thermal noise at discriminator output.

further, that at this input S_i/N_M we should increase β from β_1 to β_2 hoping thereby to improve the output-signal-to-noise ratio S_o/N_o by a sacrifice of bandwidth. We would find, however, as is apparent from Fig. 10.1-1, that such a sacrifice of bandwidth would actually decrease the output SNR. Similarly we see that, if we are operating sufficiently below threshold for any value of β, we do better with a linear coherent system than with an FM system.

The onset of threshold may be observed by examining the noise output of an FM discriminator on a cathode-ray oscilloscope. At high input-signal-to-noise ratios (S_i/N_M), the noise displays the usual random-variation characteristic of thermal noise generally. A typical output-noise waveform has the appearance shown in Fig. 10.1-2. We can determine experimentally that the instantaneous

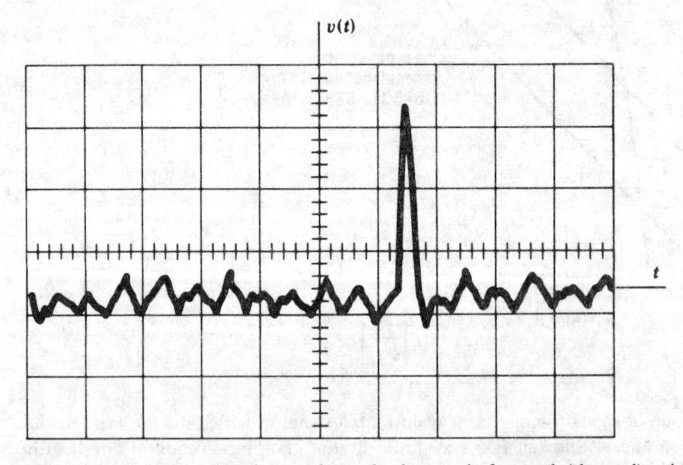

Figure 10.1-3 A spike superimposed on a background of smooth (thermal) noise.

noise amplitude has a gaussian probability density as is to be anticipated from the discussion of Sec. 7.3. As S_i/N_M decreases, a point is reached where the character of the noise waveform changes markedly. The noise now has the appearance indicated in Fig. 10.1-3. Here we observe, superimposed on the background thermal-type noise (usually referred to as *smooth* noise), a pulse-type noise waveform. Because of its appearance this new component of noise is often referred to as *spike* noise or *impulse* noise. If we were to listen to the discriminator, we would hear a clicking sound on each occasion that we observed a spike.

The appearance of these spikes denotes the onset of threshold. In succeeding sections we shall discuss the origin of these spikes. We shall show that although the frequency of occurrence of a spike is small, the noise energy associated with a spike is very large compared with the energy of the smooth noise occurring during a comparable time interval. Hence the *spike* noise greatly increases the total noise output and thereby causes a threshold.

10.2 OCCURRENCE OF SPIKES

An FM demodulator consisting of an IF filter, limiter-discriminator, and a baseband filter is shown in Fig. 10.2-1. The IF filter has a bandwidth B, centered at the IF frequency f_c. The baseband filter extends from $-f_M$ to $+f_M$ but does not pass dc. In order to simplify our discussion, we assume that the input signal is an unmodulated carrier; that is, $m(t) = 0$, accompanied by white noise. The white noise is filtered and shaped by the IF filter.

In the present and in succeeding sections we shall have occasion to use the quadrature component representation of noise as expressed in Eq. (7.11-2). It will be convenient to use a notation different from the notation of Eq. (7.11-2). We

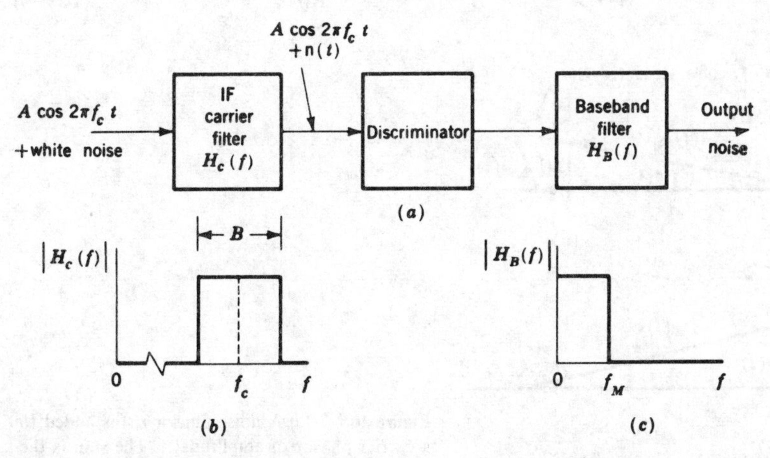

Figure 10.2-1 (*a*) An FM discriminator and associated filters. (*b*) The bandpass range of the carrier filter. (*c*) The passband of the baseband filter.

shall replace $n_c(t)$ by $x(t)$ and $n_s(t)$ by $y(t)$. In this alternative notation, which is as commonly found in the literature as is the original notation, Eq. (7.11-2) reads, with $2\pi f_c$ replaced by ω_c,

$$n(t) = x(t) \cos \omega_c t - y(t) \sin \omega_c t \tag{10.2-1}$$

The notation of Eq. (10.2-1) is especially appropriate when $n(t)$ is to be represented as a combination of phasors in a coordinate system rotating counterclockwise with angular frequency ω_c. In such a coordinate system $n(t)$ is represented by the phasor sum of $x(t)$ in the horizontal direction (i.e., the x direction) and $y(t)$ in the vertical direction (i.e., the y direction). In terms of this new notation, the carrier and noise output of the IF filter, which is the input $v_i(t)$ to the demodulator, is given by

$$v_i(t) = A \cos \omega_c t + x(t) \cos \omega_c t - y(t) \sin \omega_c t \tag{10.2-2}$$

This equation can be rewritten in phasor notation as

$$\begin{aligned} v_i(t) &= \text{Re}\,\{[A + x(t)]e^{j\omega_c t} + y(t)e^{j[\omega_c t + (\pi/2)]}\} \\ &= \text{Re}\,\{e^{j\omega_c t}\underbrace{[A + x(t) + jy(t)]}_{\text{phasors}}\} \end{aligned} \tag{10.2-3}$$

The phasor $A + x(t)$ lies along the horizontal axis and $y(t)$ lies along the vertical axis.

It is convenient, when discussing the occurrence of spikes, to talk about the *amplitude* of the noise $r(t) = \sqrt{x^2 + y^2}$ and its random phase angle $\phi(t) = \tan^{-1} (y(t)/x(t))$. [It was shown in Sec. 2.16 that the density of $r(t)$ is Rayleigh and that the density of $\phi(t)$ is $1/2\pi$ from $-\pi$ to $+\pi$.]

The phasor diagram for $v_i(t)$ is shown in Fig. 10.2-2. Here we see that the phasor sum of signal A and noise $r(t)$ is defined as $R(t)$. The angle that $R(t)$ makes with the horizontal axis is called $\theta(t)$. Then

$$A + x + jy = A + r(t)e^{j\phi(t)} = R(t)e^{j\theta(t)} \tag{10.2-4}$$

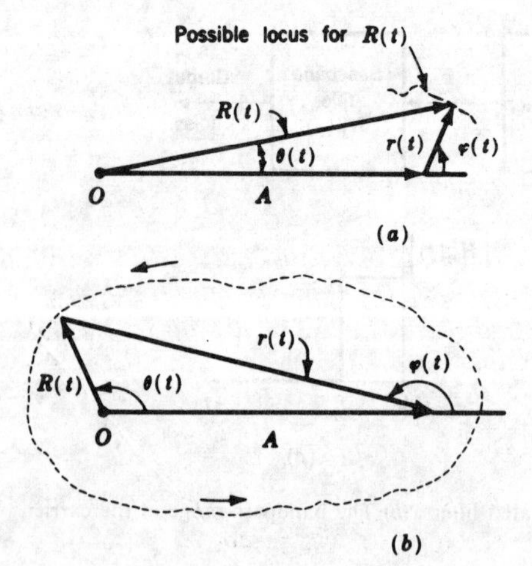

Figure 10.2-2 (*a*) A noise phasor $r(t)$ is added to a carrier phasor of amplitude A. The sum is the resultant phasor $R(t)$. (*b*) A locus for the end point of $R(t)$ which will give rise to a spike.

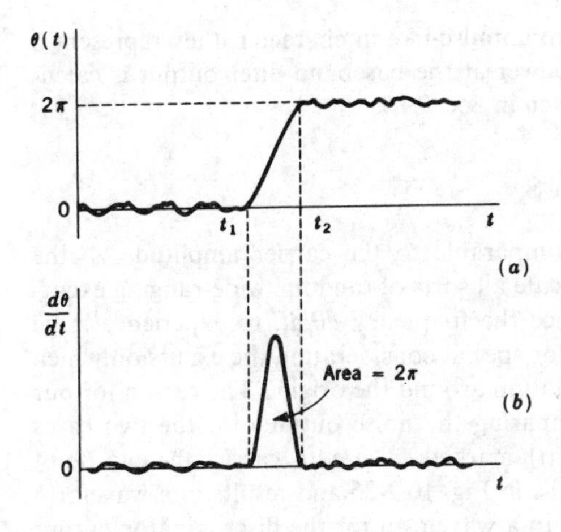

Figure 10.2-3 (*a*) A plot of $\theta(t)$ for a case in which the end point of $R(t)$ in Fig. 10.2-2*b* executes a rotation around the origin. (*b*) A plot of $d\theta/dt$ as a function of time.

and that $v_i(t)$ in Eq. (10.2-2) is

$$v_i(t) = \text{Re}\,[e^{j\omega_c t}R(t)e^{j\theta(t)}] = R(t)\cos\,[\omega_c t + \theta(t)] \tag{10.2-5}$$

As was discussed in Sec. 9.2, the output of the discriminator is proportional to $\dot{\theta}(t) = d\theta/dt$.

Let us now consider Fig. 10.2-2*a* to see how $\theta(t)$, and hence $\dot{\theta}(t)$, are affected as the noise $r(t)$ and $\phi(t)$ vary. If the noise power is small in comparison with the carrier power, we expect that $r(t) \ll A$ most of the time, and that the end point of the resultant phasor $R(t)$ will never wander far from the end point of the carrier phasor. Thus, as $\phi(t)$ changes, the angle $\theta(t)$ remains small.

If, however, the ratio $S_i/\eta f_M$ decreases, the likelihood of $r(t)$ being much less than A also decreases. When $r(t)$ becomes comparable in magnitude to the carrier amplitude A, the locus of the end point of the resultant phasor $R(t)$ moves away from the end point of the carrier phasor and may, as shown in Fig. 10.2-2*b*, even rotate about the origin. The variation of the phase angle $\theta(t)$, at and near the time of occurrence of this event, is shown in Fig. 10.2-3*a*. If the rotation of the end point of $R(t)$ around the origin occurs in the interval between t_1 and t_2, the angle $\theta(t)$ changes by 2π rad during this time interval. Preceding t_1 and following t_2, $r(t) \ll A$, and the usual small random variations of $\theta(t)$ occur.

We are interested in the discriminator output, which is proportional to the instantaneous frequency $d\theta/dt$, and we have therefore plotted $d\theta/dt$ in Fig. 10.2-3*b*. Notice that when $\theta(t)$ changes by 2π rad, $d\theta/dt$ appears as a sharp *spike* or *impulse* with area 2π. To show that the area under the spike waveform is indeed 2π, we simply integrate $d\theta/dt$ over the time interval, t_1 to t_2 during which $\theta(t)$ changes by 2π. Thus

$$\text{Area} = \int_{t_1}^{t_2} \frac{d\theta}{dt}\,dt = \theta\Big|_{t_1}^{t_2} = 2\pi \tag{10.2-6}$$

The waveform shown in Fig. 10.2-3*b* is the *spike* noise referred to in Sec. 10.1 and is the waveform which is presented to the input of the baseband filter. Since these spikes occur only rarely and are impulse-like in character, they represent a *shot* noise phenomenon. The noise power at the baseband filter output is calculated in Sec. 10.4, using the results given in Sec. 2.25.

10.3 SPIKE CHARACTERISTICS

When the noise amplitude $r(t)$ is comparable to the carrier amplitude A, the resultant phasor $R(t)$ may clearly execute all sorts of random, wide-ranging excursions which will cause $\theta(t)$, and hence the frequency $d\theta/dt$, to experience large changes. Why then do we single out for special consideration the excursion which carries the resultant in a complete rotation around the origin? The reason for our special concern may be seen by comparing the noise outputs for the two cases shown in Fig. 10.3-1*a*. The path of $R(t)$, marked *spike path*, carries the end point of $R(t)$ completely around the origin as in Fig. 10.2-2*b* and results in a waveform for $\theta(t)$ as shown in Fig. 10.2-3*a* and in a waveform for the discriminator output as shown in Fig. 10.2-3*b*. The second path, marked *triplet path*, departs from the spike path only slightly, but, most importantly, it does *not* encircle the origin. For

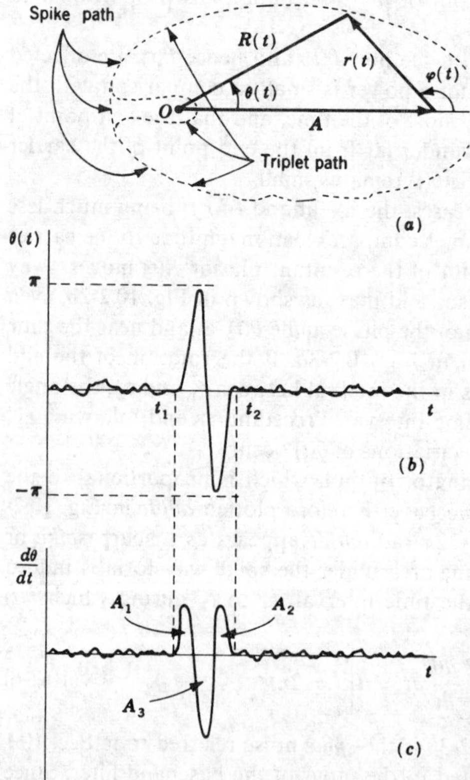

Figure 10.3-1 (*a*) Comparing two nearly identical paths for the end point of $R(t)$. One path encircles the origin, the other does not. (*b*) and (*c*) $\theta(t)$ and $d\theta/dt$ for the path which does not encircle the origin.

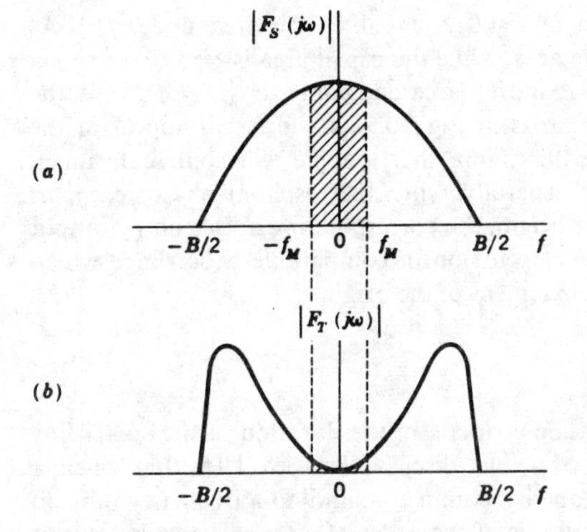

Figure 10.3-2 Qualitative plots of the Fourier transform of a pulse type waveform with spectral components limited to range $-B/2$ to $+B/2$. (*a*) For case where waveform has dc component. (*b*) For case where dc component is zero.

this path, $\theta(t)$ and the discriminator output appear as shown in Fig. 10.3-1*b* and *c*. Notice that the angle $\theta(t)$ increases nearly to π, reverses to nearly $-\pi$, and then returns to a small value. The waveform of $\theta(t)$ displays, not a simple jump of 2π, but rather a pulse doublet. The waveform of $d\theta/dt$ displays not a single pulse but rather three pulses, two of positive and one of negative polarity. It is most important to recognize that the total algebraic area under the waveform in Fig. 10.3-1*c* is zero. That is, the sum of the positive areas $A_1 + A_2$ is equal to the negative area A_3. This can be seen by applying the calculation of Eq. (10.2-6) to the determination of the total area and noting that the total change in θ between t_1 and t_2 is zero.

We can now illustrate how, in an FM system, the noise energy present at the output of the baseband filter shown in Fig. 10.2-1 is much larger when a spike is generated than when a triplet occurs. We have already noted that if a carrier, accompanied by noise, is passed through a bandpass filter of bandwidth B, the spectral components of the noise output of the discriminator will extend over the frequency range from $-B/2$ to $B/2$. Of this noise, only the part which lies in the range $-f_M$ to f_M will appear at the output of the baseband filter. In Fig. 10.3-2*a* and *b* we have sketched the magnitudes of the Fourier transforms: $F_S(j\omega)$, the spike waveform shown in Fig. 10.2-3*b*, and $F_T(j\omega)$, the triplet waveform shown in Fig. 10-3.1*c*. We have confined the spectral range to the interval $-B/2$ to $+B/2$ as required, and have indicated that for the triplet $F_T(j\omega) = 0$ at $f = 0$, while for the spike $F_S(j\omega) \neq 0$ at $f = 0$. These features of the transforms, at $f = 0$, are verified in the following way. The Fourier transform of a function $f(t)$ is

$$F(j\omega) = \int_{-\infty}^{\infty} f(t)e^{-j\omega t}\, dt \tag{10.3-1}$$

so that

$$[F(j\omega)]_{\omega=0} = \int_{-\infty}^{\infty} f(t)\, dt = \text{area under } f(t) \tag{10.3-2}$$

Thus, the value of the transform at $f = 0$ is equal to the area under $f(t)$. As already noted, the spike has a finite area, while the triplet area is zero.

Since the passband of the baseband filter extends only to $f_M \ll B/2$, only the energy represented by the shaded areas in Fig. 10.3-2*a* and *b* will appear at the filter output. It is apparent from this figure that a spike will contribute much more noise than the triplet. More generally, when the resultant phasor executes any path that carries it around the origin, the corresponding noise energy output will be much larger than any other excursion of comparable wide range which does not succeed in causing an encirclement of the origin.

Spike Duration

We may make an order-of-magnitude estimate of the duration of the spike. For this purpose we use the general principle, discussed in Sec. 1.15, that when a pulse is made up of spectral components, confined mainly to a frequency range 0 to f_{max}, the time duration of the pulse is of the order of $1/f_{max}$. The spike output of the discriminator has a spectrum that extends over the range 0 to $B/2$, where B is the bandpass of the carrier filter. Hence, we estimate that the spike duration is approximately $2/B$ sec.

10.4 CALCULATION OF THRESHOLD IN AN FM DISCRIMINATOR

To calculate the spike noise contribution at the output of an FM discriminator, we need to know the power spectral density of the spike noise. The spike noise is similar to the shot noise discussed in Sec. 2.25. We may therefore use the result of Eq. (2.25-7), which states that the power spectral density is given by

$$G_S(f) = \frac{1}{T_s} |P(j\omega)|^2 \tag{10.4-1}$$

where $P(j\omega)$ is the Fourier transform of the spike, and T_s is the mean time interval between spikes. From the discussion of Sec. 10.3 we know that $|P(j\omega)|$ will have the general form shown in Fig. 10.3-2*a*. For the purpose of calculating noise power at the output of the baseband filter, we need know $G_S(f)$ only in the range $-f_M$ to $+f_M$. In this range $|P(j\omega)|$ does not change very much, provided that $B/2 \gg f_M$. We may therefore assume that $|P(j\omega)|$ is constant and equal to $|P(0)|$, that is, to the value of the Fourier transform at $f = 0$.

Let $\theta(t)$ in Fig. 10.3-1 make a rotation by 2π so that a spike is formed. Then the waveform of the output-noise voltage spike from the discriminator is

$$[n(t)]_{spike} = \alpha \frac{d\theta}{dt} \tag{10.4-2}$$

where α is a constant, i.e., the discriminator constant. From Eq. (10.3-2), $|P(0)|$ is equal to the area of $[n(t)]_{\text{spike}}$. Since α is a constant and the area under the waveform $d\theta/dt$ is 2π, we have

$$|P(0)| = 2\pi\alpha \tag{10.4-3}$$

Since each spike has the same area, 2π, the average value of $|P(0)|$ is also given by Eq. (10.4-3), and from Eq. (10.4-1)

$$G_S(f) = \frac{4\pi^2\alpha^2}{T_s} \tag{10.4-4}$$

Thus the total output-noise power due to the spikes is simply

$$N_S = \frac{4\pi^2\alpha^2}{T_s} 2f_M \tag{10.4-5}$$

The expression for N_S given in Eq. (10.4-5) is not very useful until we learn how T_s depends on the parameters of the system. It is proven in Sec. 10.5 that, for an unmodulated carrier,

$$\frac{1}{T_s} = \frac{B}{2\sqrt{3}} \operatorname{erfc} \sqrt{\frac{f_M}{B} \frac{S_i}{N_M}} \tag{10.4-6}$$

where erfc is the complimentary error function, i.e.,

$$\operatorname{erfc} x \equiv \frac{2}{\sqrt{\pi}} \int_x^\infty e^{-\lambda^2}\, d\lambda = 1 - \frac{2}{\sqrt{\pi}} \int_0^x e^{-\lambda^2}\, d\lambda \tag{10.4-7}$$

From Eqs. (10.4-5) and (10.4-6) we have

$$N_S = \frac{4\pi^2\alpha^2 B f_M}{\sqrt{3}} \operatorname{erfc} \sqrt{\frac{f_M}{B} \frac{S_i}{N_M}} \tag{10.4-8}$$

The total output-noise power is $N_o = N_G + N_S$, where N_G is the gaussian (smooth) noise given by Eq. (9.2-16), and the noise N_S is the spike noise given by Eq. (10.4-8). The signal power is given by Eq. (9.2-5). Combining these equations, we have, with $S_i = A^2/2$, and $N_M = \eta f_M$,

$$\frac{S_o}{N_o} = \frac{[3k^2\overline{m^2(t)}/4\pi^2 f_M^2](S_i/N_M)}{1 + (\sqrt{3}B/f_M)(S_i/N_M) \operatorname{erfc} \sqrt{(f_M/B)(S_i/N_M)}} \tag{10.4-9}$$

The threshold effect is evident in Eq. (10.4-9). At very high S_i/N_M the complementary error function approaches zero. The right-hand member of the denominator of Eq. (10.4-9) becomes zero, and S_o/N_o varies linearly with S_i/N_M. As S_i/N_M decreases, the denominator increases above unity, and S_o/N_o decreases at a more rapid rate than does S_i/N_M.

We shall see in Sec. 10.6 that the spike output of a discriminator is very much larger when the input carrier is modulated than when the carrier is not modulated. In deriving Eq. (10.4-8), we have used Eq. (10.4-6) which applies only in the absence of modulation, or when the carrier is frequency-modulated to a very slight extent. Thus Eq. (10.4-9) is valid only when the bandwidth occupied by the modulated signal is very small in comparison with the bandwidth B of the IF carrier filter.

10.5 CALCULATION OF MEAN TIME BETWEEN SPIKES

In this section we shall calculate the mean time between spikes. We shall derive Eq. (10.4-6), which was used in the preceding section in our calculation of S_o/N_o near threshold.

A phasor diagram of the carrier and noise at the input to the discriminator is shown in Fig. 10.5-1. The input noise is expressed in terms of the quadrature components $x(t)$ and $y(t)$. In Secs. 10.2 and 10.3 we showed that a spike resulted in the discriminator output when $\theta(t)$ rotated 2π rad. If the rotation was counterclockwise so that $\theta(t)$ *increased* by 2π the spike was positive, and if the rotation was clockwise so that $\theta(t)$ *decreased* by 2π the spike was negative.

We shall assume that if the end point of $R(t)$ has crossed the horizontal axis so that $\theta(t)$ *passes through* π rad, then $\theta(t)$ will continue to increase (or decrease) giving rise to a positive (or negative) spike. Thus we assume that to ensure a spike output, it is not actually necessary to observe a complete rotation of $R(t)$, but it is adequate that there be guaranteed a rotation of at least π rad. If, then, at an instant of time $t = t_1$,

$$x(t_1) < -A \tag{10.5-1a}$$

$$y(t_1) = 0 \tag{10.5-1b}$$

and

$$\dot{y}(t_1) < 0 \tag{10.5-1c}$$

then $\theta(t)$ has *increased* through π rad, and a positive spike will result. Similarly if, at an instant of time $t = t_2$,

$$x(t_2) < -A \tag{10.5-2a}$$

$$y(t_2) = 0 \tag{10.5-2b}$$

and

$$\dot{y}(t_2) > 0 \tag{10.5-2c}$$

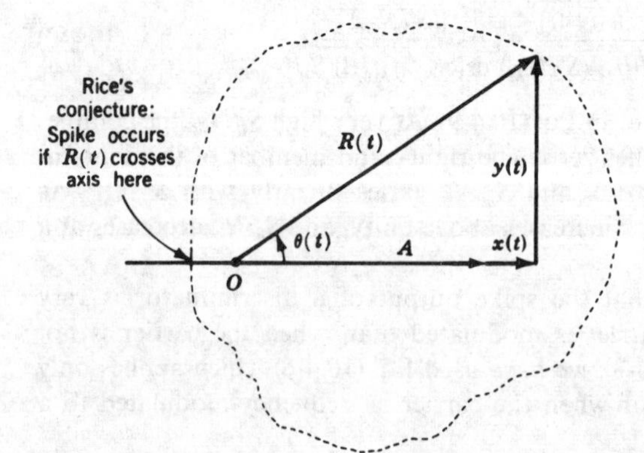

Figure 10.5-1 Phasor diagram using quadrature noise components.

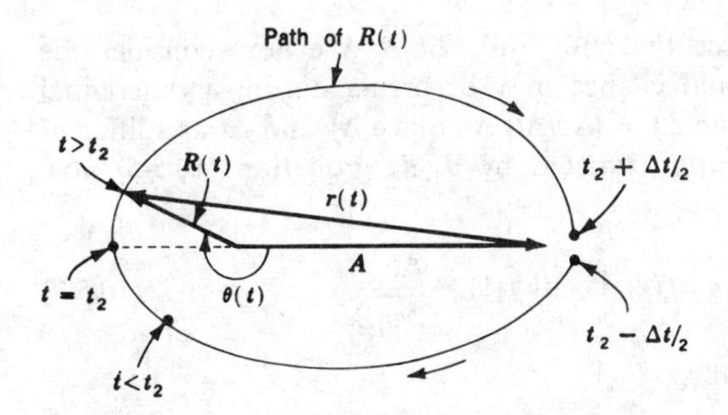

Figure 10.5-2 Locus of $R(t)$ and $\theta(t)$ to cause a negative spike.

then $\theta(t)$ has *decreased* through π rad, and a negative spike results. This hypothesis is due to Rice[1] and has been verified experimentally in a variety of different circumstances.[2] We shall employ these results, Eqs. (10.5-1) and (10.5-2), to determine the probability of occurrence of a spike.

Let us consider an interval of time Δt and calculate the probability P_- of a negative spike occurring during this time interval. This probability is the probability that the conditions given by Eq. (10.5-2) are satisfied at some instant t_2 within the time interval Δt. These conditions are illustrated in Fig. 10.5-2. In this figure we see $R(t)$ and $\theta(t)$ just after $\theta(t)$ has *decreased* through π, which is equivalent to $y(t)$ *increasing* through zero. It is this *increase* which results in $\dot{y}(t_2) > 0$.

Thus, we write

$$P_- = P\left[x(t_2) < -A,\ y(t_2) = 0,\ \left.\frac{dy}{dt}\right|_{t_2} > 0 \right] \tag{10.5-3}$$

In Sec. 7.13 we determined the joint probability density of x, $\dot{x}$, y, and $\dot{y}$. In the present instance we need only the joint density of x, y, and $\dot{y}$. Using Eq. (7.13-5), we have, after replacing n_c and n_s by x and y, respectively

$$f(x, y, \dot{y}) = \overbrace{\frac{e^{-x^2/2\eta B}}{\sqrt{2\pi\eta B}}}^{f(x)} \overbrace{\frac{e^{-y^2/2\eta B}}{\sqrt{2\pi\eta B}}}^{f(y)} \overbrace{\frac{e^{-\dot{y}^2/(2\pi^2 B^3\eta/3)}}{\sqrt{2\pi^3 B^3 \eta/3}}}^{f(\dot{y})} \tag{10.5-4}$$

and

$$P_- = \int_{-\infty}^{-A} dx \int_0^{\infty} d\dot{y} \int_{-\Delta y/2}^{\Delta y/2} dy\, f(x, y, \dot{y}) \tag{10.5-5}$$

Before integrating Eq. (10.5-5), let us investigate the limits on the integrals to verify that Eq. (10.5-5) really represents Eq. (10.5-3). First we see that $\int_{-\infty}^{-A} dx\, f(x)$ will yield the probability that $x(t_2) < -A$. Next, we note that $\int_{-\Delta y/2}^{\Delta y/2} dy\, f(y) = f(0)\,\Delta y$ gives the probability of $y(t_2)$ passing through 0. Finally we have $\int_0^{\infty} d\dot{y}\, f(\dot{y})$, which is the probability of $\dot{y}(t_2)$ being greater than zero. These are exactly the conditions of Eq. (10.5-3).

We now evaluate Eq. (10.5-5). The first integral that we determine is

$$\int_{-\Delta y/2}^{\Delta y/2} dy\, f(y) = f(0)\,\Delta y \tag{10.5-6}$$

Referring to Eq. (10.5-4), we see that $f(0) = 1/\sqrt{2\pi\eta B}$. We next consider the factor Δy. This is the differential change in y occurring during a differential change in time Δt. Thus, we write $\Delta y = (\Delta y/\Delta t)\,\Delta t$. Since Δy and Δt are differential quantities, $\Delta y/\Delta t$ can be approximated by dy/dt. Equation (10.5-5) now becomes

$$P_- = \int_{-\infty}^{-A} dx\, f(x) \int_0^{\infty} d\dot{y}\, f(\dot{y}) \frac{\Delta t}{\sqrt{2\pi\eta B}}\, \dot{y} \tag{10.5-7}$$

The next integral to be evaluated is

$$\int_0^{\infty} d\dot{y}\, f(\dot{y})\dot{y} = \int_0^{\infty} d\dot{y}\, \dot{y}\, \frac{e^{-\dot{y}^2/(2\pi^2 B^3\eta/3)}}{\sqrt{2\pi^3 B^3\eta/3}} \tag{10.5-8}$$

This integral is easily evaluated after changing the variable of integration to

$$\lambda = \frac{\dot{y}^2}{2\pi^2 B^3\eta/3} \tag{10.5-9}$$

Then

$$d\lambda = \frac{\dot{y}\, d\dot{y}}{\pi^2 B^3\eta/3} \tag{10.5-10}$$

Thus, Eq. (10.5-8) becomes

$$\sqrt{\frac{\pi B^3\eta}{6}} \int_0^{\infty} d\lambda\, e^{-\lambda} = \sqrt{\frac{\pi B^3\eta}{6}} \tag{10.5-11}$$

Equation (10.5-7) now becomes

$$P_- = \int_{-\infty}^{-A} dx\, f(x) \left(\sqrt{\frac{\pi B^3\eta}{6}} \frac{\Delta t}{\sqrt{2\pi\eta B}} \right)$$

$$= \frac{B\,\Delta t}{\sqrt{12}} \int_{-\infty}^{-A} dx\, \frac{e^{-x^2/2\eta B}}{\sqrt{2\pi\eta B}} \tag{10.5-12}$$

Again, we make a change in variable of integration. This time we set

$$\lambda = \frac{-x}{\sqrt{2\eta B}} \tag{10.5-13}$$

Then

$$P_- = \frac{B\,\Delta t}{4\sqrt{3}} \int_{\frac{A}{\sqrt{2\eta B}}}^{\infty} \frac{2}{\sqrt{\pi}}\, d\lambda\, e^{-\lambda^2} = \frac{B\,\Delta t}{4\sqrt{3}} \operatorname{erfc} \sqrt{\frac{A^2}{2\eta B}} \tag{10.5-14}$$

The input-signal power is $S_i = A^2/2$, while the input noise in the baseband bandwidth is $N_M = \eta f_M$. Hence we may rewrite Eq. (10.5-14) in the form

$$P_- = \left(\frac{B}{4\sqrt{3}} \operatorname{erfc} \sqrt{\frac{f_M}{B}\frac{S_i}{N_M}} \right) \Delta t \tag{10.5-15}$$

In one second there are $1/\Delta t$ intervals in which a spike might occur. Since P_- is the probability of occurrence of a negative spike within any such individual interval, the expected number of pulses in a second, N_-, is

$$N_- = \frac{P_-}{\Delta t} \tag{10.5-16}$$

Hence, N_- is found from Eqs. (10.5-15) and (10.5-16) to be

$$N_- = \frac{B}{4\sqrt{3}} \operatorname{erfc} \sqrt{\frac{f_M}{B} \frac{S_i}{N_M}} \tag{10.5-17}$$

From the symmetry displayed in Fig. 10.5-2 we see that the average number of positive spikes N_+ is the same as the average number of negative spikes N_-. Thus, the total number of spikes occurring per second in the presence of a carrier alone (no modulation) is, on the average,

$$N_c = N_- + N_+ = 2N_- = 2N_+ \tag{10.5-18}$$

The average time between spikes T_s is as anticipated in Eq. (10.4-6),

$$T_s = \frac{1}{N_c} = \frac{2\sqrt{3}}{B \operatorname{erfc} \sqrt{(f_M/B)(S_i/N_M)}} \tag{10.5-19}$$

10.6 EFFECT OF MODULATION

When the noise spectral components swing the resultant phasor in Fig. 10.2-2 around the origin in the counterclockwise direction, a positive spike occurs. Similarly, a clockwise rotation produces a negative spike. When the carrier frequency is at the center frequency of the carrier filter, the noise spectral components are symmetrically located in frequency above and below the carrier. In this case, because of the symmetry, positive and negative spikes are equally likely.

Now, however, let us assume that we have *offset* the carrier so that it lies just within the passband of the carrier filter at the lower-frequency limit of the passband. Let us consider a new coordinate system in which the phasor for this offset carrier is at rest. This offset phasor is accompanied by noise spectral components almost all of which are at higher frequency than the carrier and extend over a frequency range, with respect to the carrier, of B rather than $B/2$ as was the case when the carrier was centrally located. In our new coordinate system all the noise spectral components rotate counterclockwise. Hence only positive-frequency spikes can occur. Further, since the noise causing the spike has a spectral range B rather than $B/2$ (see Prob. 10.6-1), the spikes will be narrower than in the case of the symmetrically located carrier. Since the spike area is fixed, the spike amplitude must be larger. Finally, because of the higher relative frequencies of the noise components with respect to the the offset carrier, whatever is going to happen will happen more frequently. Hence the rate of occurrence of spikes will increase.

Suppose we use a cathode-ray oscilloscope to examine the output waveform of the discriminator of Fig. 10.2-1. Let us assume that polarities have been adjusted so that a frequency increase in the carrier produces a positive deflection on the scope. In the case where the carrier is centrally located in the carrier filter passband, the scope trace will display the output noise with an equal frequency of occurrence of positive and negative spikes. As the carrier frequency is lowered, the trace will move downward, and the number of positive and negative spikes will become asymmetrical. The positive spikes will become more frequent and larger in amplitude, while the negative spikes will become smaller and occur less frequently. Eventually, only positive spikes will appear, and the frequency of occurrence of these positive spikes will be greater than the frequency of positive and negative spikes together in the symmetrical case.

When the carrier is modulated, we may view the modulation as simply a continuously varying frequency offset. When the baseband signal appears at the output of the discriminator, negative spikes are encountered most frequently at the positive extremity of the signal, and positive spikes at the negative extremity. The characteristic appearance of a recovered sinusoidal modulation is shown in Fig. 10.6-1. The important fact to note is not so much the spike polarity; rather, it is the unfortunate fact that in the region of threshold the noise power increases further when the carrier is modulated.

When the frequency of the input is offset by an amount $\delta f = km(t)/2\pi$ from the frequency f_c carrier, the number of spikes increases by an amount δN. This increase δN (in the presence of a carrier which is frequency-modulated) is related to δf by

$$\delta N = |\,\delta f\,| e^{-(f_M/B)(S_i/N_M)} \tag{10.6-1}$$

The polarity of these *additional* spikes as observed at the discriminator output is always in the direction *opposite* to the polarity of the output signal produced by the offset δf. The derivation of Eq. (10.6-1) is an extension of the derivation of Eq. (10.5-18) and will not be given here. (See Prob. 10.6-2.)

Suppose that the input signal is sinusoidally modulated at a frequency $f_m(\leq f_M)$. Then the input is

$$v(t) = A \cos\left(\omega_c t + \frac{\Delta f}{f_m} \sin 2\pi f_m t\right) \tag{10.6-2}$$

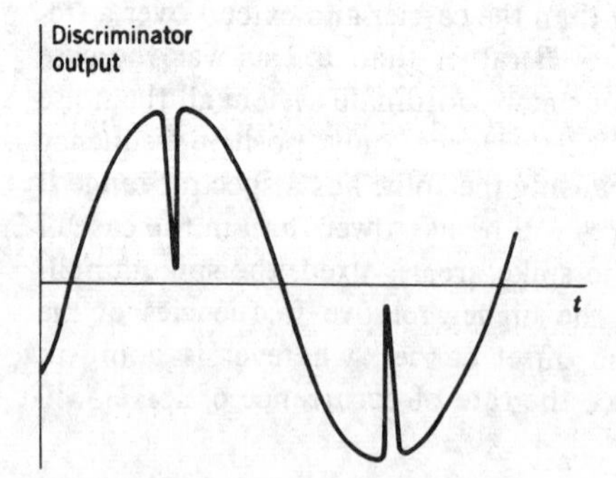

Figure 10.6-1 Spikes present on sinusoidal output signal.

where Δf is the frequency deviation. Then

$$\delta f = \Delta f \cos 2\pi f_m t \tag{10.6-3}$$

and the average $|\overline{\delta f}| = (2/\pi)\,\Delta f$. The average value of δN is, therefore,

$$\overline{\delta N} = \frac{2\Delta f}{\pi} e^{-(f_M/B)(S_i/N_M)} \tag{10.6-4}$$

The total number of spikes per second is $N = N_c + \overline{\delta N}$, the latter two terms given, respectively, by Eqs. (10.5-19) and (10.6-4). It may, however, be verified (Prob. 10.6-4) that near and below threshold $\overline{\delta N} \gg N_c$. Therefore the total average number of spikes is $N \cong \overline{\delta N}$. The average time between spikes T_s is, then,

$$T_s = \frac{1}{N} \approx \frac{1}{\overline{\delta N}} \tag{10.6-5}$$

We may now calculate the output SNR of the FM discriminator when demodulating a sinusoidally modulated carrier where

$$\delta f = km(t)/2\pi = \Delta f \cos \omega_m t$$

We use Eqs. (9.2-16), (9.2-21), (10.4-5), (10.6-4), and (10.6-5) to find

$$\frac{S_o}{N_o} = \frac{(\frac{3}{2})\beta^2(S_i/N_M)}{1 + (12\beta/\pi)(S_i/N_M)\exp\{-\frac{1}{2}[1/(\beta+1)](S_i/N_M)\}} \tag{10.6-6}$$

where $\beta = \Delta f/f_M$.

Equation (10.6-6) is plotted in Fig. 10.6-2 for $\beta = 3$ and $\beta = 12$. Note that, when we are operating at a value of S_i/N_M which is above threshold for both values of β, the higher β, corresponding to a wider bandwidth, results in an increase in output SNR. On the other hand, at $S_i/N_M = 20$ dB, the output SNR is higher for $\beta = 3$ than for $\beta = 12$.

If the input signal is modulated by a sample function of a gaussian random process, we have

$$v(t) = A\cos\left[\omega_c t + k\int_{-\infty}^{t} m(\lambda)\,d\lambda\right] + n(t) \tag{10.6-7}$$

where $km(t)$ is the instantaneous frequency deviation. Hence the rms frequency deviation produced is

$$(\Delta f)_{\text{rms}} = \sqrt{k^2\overline{m^2(t)}}/2\pi \tag{10.6-8}$$

We begin our calculation of the output SNR by obtaining an expression for the output-signal power. The output-signal power is found from Eqs. (9.2-5) and (10.6-8):

$$S_o = \alpha^2 k^2 \overline{m^2(t)} = \alpha^2 4\pi^2 (\Delta f_{\text{rms}})^2 \tag{10.6-9}$$

The FM noise, neglecting spikes, is given by Eq. (9.2-16):

$$N_G = \alpha^2 \frac{8\pi^2}{3}\frac{\eta f_M^3}{A^2} \tag{10.6-10}$$

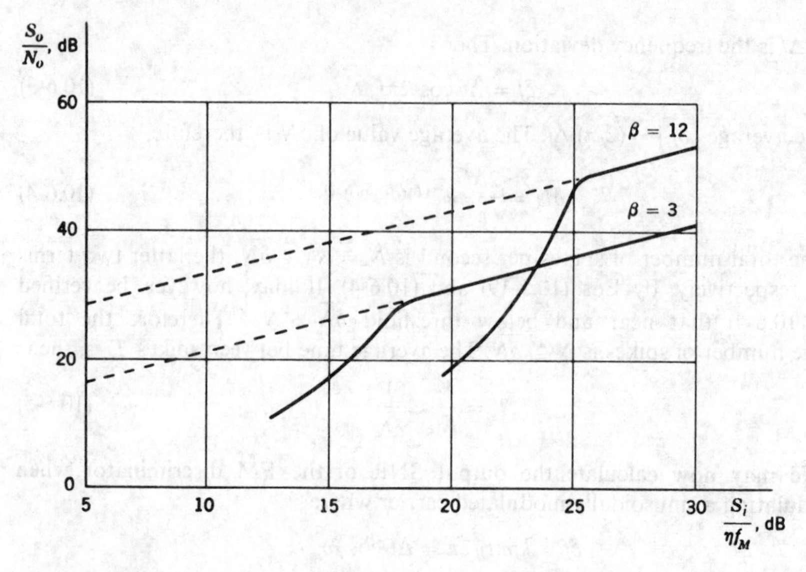

Figure 10.6-2 Output SNR of an FM discriminator when demodulating an FM signal which is sinusoidally modulated.

The spike noise is found by combining Eqs. (10.4-5), (10.6-1), and (10.6-5), with the result

$$N_S = \alpha^2(8\pi^2 f_M)\,\overline{|\delta f|}\,\exp\left[-(f_m/B)(S_i/\eta f_M)\right] \tag{10.6-11}$$

Here $\delta f = km(t)/2\pi$ is the instantaneous frequency deviation and thus $\overline{|\delta f|}$ is the average value of the magnitude of the instantaneous frequency deviation. Noting that $m(t)$ is gaussian, we have that δf is gaussian with an average value of 0 and a standard deviation $(\Delta f)_{\text{rms}}$. Hence

$$\overline{|\delta f|} = \int_{-\infty}^{\infty} \frac{|\delta f|\exp\left[-(\delta f)^2/2(\Delta f_{\text{rms}})^2\right]}{\sqrt{2\pi}(\Delta f_{\text{rms}})}\, d\,(\delta f) \tag{10.6-12}$$

This integral can be evaluated by integrating from 0 to infinity and doubling the result. Thus

$$\overline{|\delta f|} = \sqrt{\frac{2}{\pi}}\,(\Delta f_{\text{rms}}) \int_0^{\infty} \frac{\delta f}{(\Delta f_{\text{rms}})^2} \exp\left[-(\delta f)^2/2(\Delta f_{\text{rms}})^2\right] d\,(\delta f) \tag{10.6-13}$$

This integral is directly integrable [change variables to $x = (\delta f)^2/2(\Delta f_{\text{rms}})^2$]. The result is

$$\overline{|\delta f|} = \sqrt{\frac{2}{\pi}}\,(\Delta f_{\text{rms}}) \tag{10.6-14}$$

Substituting Eq. (10.6-14) into Eq. (10.6-11) yields

$$N_S = \alpha^2(8\pi^2 f_M)\left(\sqrt{\frac{2}{\pi}}\,(\Delta f_{\text{rms}})\right)\exp\left[-(f_M/B)(S_i/\eta f_M)\right] \qquad (10.6\text{-}15)$$

The output SNR is found by combining Eqs. (10.6-9) to (10.6-11). The final result is

$$\frac{S_o}{N_o} = \frac{3(\Delta f_{\text{rms}}/f_M)^2 S_i/\eta f_M}{1 + 6\sqrt{2/\pi}(\Delta f_{\text{rms}}/f_M)(S_i/\eta f_M)\exp\left[-(f_M/B)(S_i/\eta f_M)\right]} \qquad (10.6\text{-}16)$$

Equation (10.6-13) is a function of the input SNR, $S_i/\eta f_M$, and the rms modulation index $\Delta f_{\text{rms}}/f_M$ [note that using Eq. (4.13-5), we have $B/f_M = 4.6\Delta f_{\text{rms}}/f_M$].

10.7 THE PHASE-LOCKED LOOP

When it is necessary to modulate a carrier, we usually employ either amplitude modulation or frequency modulation. When noise is a problem (as is often the case), we would be inclined to use FM and allow a sacrifice of bandwidth for the sake of improved output signal-to-noise ratio. In particular we should like to be able to avail ourselves of the advantage of FM when the input signal-to-noise ratio is low. It is therefore most disconcerting to find that there is an FM threshold which precludes such use of FM. For this reason a great deal of effort has gone into studies of methods to lower the FM threshold.

In the following sections we discuss several types of FM demodulators which, while they offer no improvement over the conventional discriminator above threshold, do provide threshold improvement. One such demodulator is the *phase-locked loop.*[3]

The phase-locked loop (PLL) is a feedback system which may be used to extract a baseband signal from a frequency-modulated carrier. The basic component building blocks of a phase-locked loop are a *phase comparator* and a *voltage-controllable oscillator* (abbreviated VCO). Before discussing the PLL, we shall discuss the phase comparator and the VCO.

Phase Comparator

A phase comparator is a device with two input ports and a single output port. If periodic signals of identical frequency but with a timing difference (i.e., a phase difference if the signals are sinusoidal) are applied to the inputs, the output is a voltage which depends on the timing difference. One way in which a phase comparator may be constructed is by the combination of a multiplier and a filter as shown in Fig. 10.7-1. Here we assume two sinusoidal input voltages of amplitudes A and B, frequency ω, and with time-varying phases $\theta_1(t)$ and $\theta_2(t)$. The

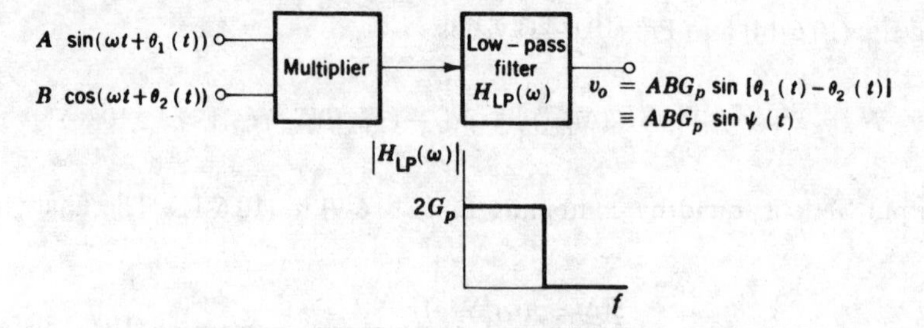

Figure 10.7-1 A phase comparator consisting of a multiplier and a filter.

student may easily verify (Prob. 10.7-1) that the output of the multiplier consists of the term

$$v_o = \frac{AB}{2} \sin\left[\theta_1(t) - \theta_2(t)\right] \equiv \frac{AB}{2} \sin \psi(t) \tag{10.7-1}$$

plus other terms whose spectral components cluster around 2ω. Hence, if $\theta_1(t)$ and $\theta_2(t)$ have bandwidths less than 2ω, a low-pass filter may separate and pass only the term $(AB/2) \sin \psi(t)$. In Fig. 10.7-1 we have assumed an amplifier with a gain $2G_p$ in the filter stage of the phase comparator so that the output voltage is $v_o = ABG_p \sin \psi(t)$. The comparator output is plotted in Fig. 10.7-2*a* as a function of ψ.

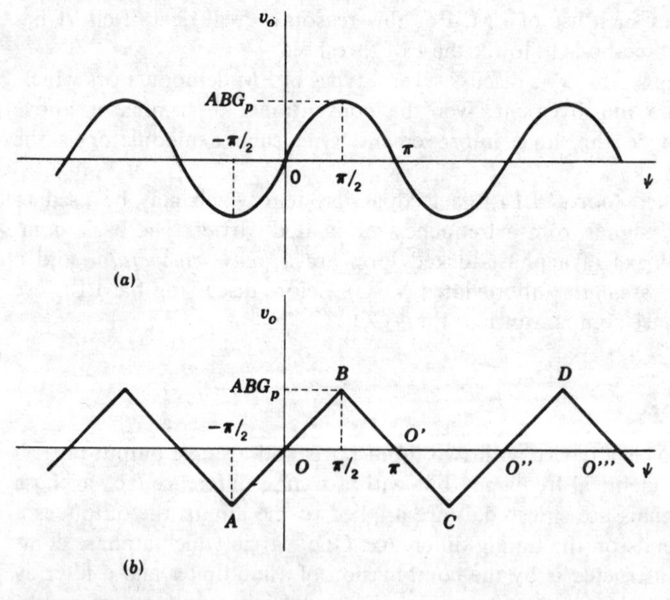

Figure 10.7-2 (*a*) Phase characteristic of the comparator of Fig. 10.7-1 with sinusoidal inputs. (*b*) Phase characteristic with square-wave inputs.

Suppose, however, that we convert the input sinusoids to square waves before application to the multiplier. Such conversion may be achieved by hard limiting and amplifying the input signals. The student may easily verify that if the square waves are clipped so that they have peak amplitudes A and B, the comparator output varies with ψ in the manner shown in Fig. 10.7-2*b*. The plot in Fig. 10.7-2*b* displays the same periodicity as does the plot in Fig. 10.7-2*a* except that the sinusoidal variation has been replaced by a piecewise-linear variation. Thus

$$v_o = \begin{cases} ABG_p \dfrac{\psi}{\pi/2} & |\psi| \leq \dfrac{\pi}{2} \\ -ABG_p \dfrac{\psi - \pi}{\pi/2} & \dfrac{\pi}{2} \leq \psi \leq \dfrac{3\pi}{2} \\ \text{etc.} \end{cases} \tag{10.7-2}$$

Voltage-Controlled Oscillator

A voltage-controlled oscillator is a source of a periodic signal whose frequency may be determined by a voltage applied to the VCO from an external source. Any frequency modulator may serve as a VCO. In practice, for simplicity, the VCO's employed in phase-locked loops are of the parameter-variation type described in Sec. 4.15.

Suppose that a VCO generates a sinusoidal waveform of amplitude B and, in the absence of a frequency-controlling voltage, operates at an angular frequency ω_c. Let the frequency sensitivity of the VCO be G_0 rad/(s)(volt). G_0 is the change in the instantaneous angular frequency ω_i produced by a change in the frequency-controlling voltage v; that is, $G_0 = d\omega_i/dv$. In the present discussion, G_0 plays the same role as the constant k in, for example, Eq. (9.2-1). Thus the oscillator signal v_{osc} is

$$v_{\text{osc}} = B \cos \left(\omega_c t + G_0 \int_{-\infty}^{t} v(\lambda)\, d\lambda \right) \tag{10.7-3}$$

In this case, the instantaneous angular frequency is

$$\omega_i(t) = \omega_c + G_0 v(t) \tag{10.7-4}$$

as required.

Phase-Locked Loop Demodulator

The manner in which a phase comparator and VCO are connected to form a PLL is shown in Fig. 10.7-3. The controlling voltage for the VCO is v_o, taken from the output of the comparator, and the output of the VCO furnishes one input of the comparator. The PLL of Fig. 10.7-3 is called a *first-order* loop for reasons that are explained below.

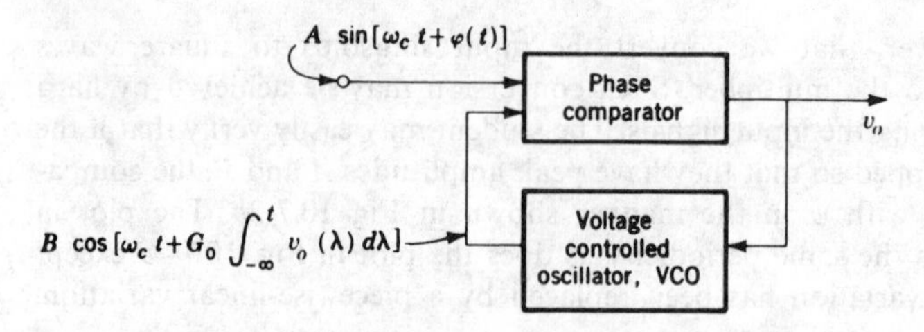

Figure 10.7-3 A (first-order) phase-locked loop.

We discuss qualitatively how the PLL may be used to recover the baseband signal from an FM modulated carrier. Figure 10.7-3 shows a frequency-modulated carrier $A \sin [\omega_c t + \phi(t)]$ applied to one input of the phase comparator. The carrier frequency is ω_c, and if the modulating baseband signal is $m(t)$, then $\phi(t) = k \int_{-\infty}^{t} m(\lambda)\, d\lambda$, with k a constant. Now let us assume that initially $\phi(t) = 0$, and that we have adjusted the VCO so that when its input voltage $v_o = 0$, its frequency is precisely ω_c, the carrier frequency. Let us further adjust the VCO output signal to have a 90° phase shift relative to the carrier. This phase shift is required so that the comparator output shall be zero when $v_o(t) = 0$. Then the situation we have established is certainly a state of equilibrium. The two inputs to the comparator differ in phase by 90°; the comparator output, which is the VCO input, is zero. Therefore this initial setting of the VCO will not be disturbed.

Now let the frequency of the input signal make an abrupt change ω at time $t = 0$. Then, beginning at $t = 0$, $\phi(t) = \omega t$, since $d\phi/dt = \omega$. That is, the abrupt frequency change causes the phase $\phi(t)$ to begin to increase linearly with time. The phase difference at the comparator input will generate a positive output v_o, which will in turn increase the frequency of the VCO. A new equilibrium point will eventually be established in the system, when the frequency of the VCO has been increased to equal the frequency of the input signal. When equilibrium is established, the input signal and the VCO output will be of identical frequency but no longer different in phase by 90°. For if the VCO is to operate at a frequency other than its initial frequency ω_c, there must be an output v_o, and hence a departure from the 90° phase difference at the input which yields $v_o = 0$.

The VCO output given in Fig. 10.7-3 is the same as Eq. (10.7-3) with v replaced by v_o. If the input and VCO frequencies are to be the same at equilibrium, we require that

$$\frac{d\phi(t)}{dt} = \frac{d}{dt} G_0 \int_{-\infty}^{t} v_o(\lambda)\, d\lambda \tag{10.7-5}$$

Letting $d\phi/dt = \omega$, we have

$$v_o(t) = \frac{\omega}{G_0} \tag{10.7-6}$$

Thus the output voltage is proportional to the frequency change as required in an FM demodulator. We see then that if the input-carrier frequency changes continuously, and at a rate which is slow in comparison with the time required for the PLL to establish a new equilibrium operating point, the PLL output is continuously proportional to the *frequency variation* of the carrier.

We readily note that the PLL is a feedback control system in which the error signal is the phase difference between the modulated carrier and the VCO signal. Initially, with an unmodulated carrier, the operating point of the phase comparator may be adjusted to be at the origin in Fig. 10.7-2*a* or *b*. In the presence of modulation the operating point will move up and down along the portion of the plot between $-\pi/2$ and $\pi/2$.

10.8 ANALYSIS OF THE PHASE-LOCKED LOOP

We now present an elementary analysis of the PLL of Fig. 10.7-3. We assume for simplicity that the phase characteristic of the phase comparator is *piecewise-linear* as shown in Fig. 10.7-2*b*. The results will apply equally well to a PLL using a comparator with the characteristic shown in Fig. 10.7-2*a*. From Fig. 10.7-3 we have that the phase angle difference $\psi(t)$ is

$$\psi(t) \equiv \phi(t) - G_0 \int_{-\infty}^{t} v_o(\lambda)\, d\lambda \tag{10.8-1}$$

Differentiating Eq. (10.8-1) and transposing, we have

$$\frac{d\psi}{dt} + G_0 v_o = \frac{d\phi}{dt} \tag{10.8-2}$$

The output voltage can be eliminated using Eq. (10.7-2). The result is

$$\frac{d\psi}{dt} + \frac{\psi}{\tau} = \frac{d\varphi}{dt} \qquad \text{for } |\psi| \leq \frac{\pi}{2} \tag{10.8-3}$$

where

$$\tau \equiv \frac{\pi}{2ABG_p G_0} \tag{10.8-4}$$

Since ψ and v_o are directly related by Eq. (10.7-2), Eq. (10.8-3) can also be written as

$$\frac{dv_o}{dt} + \frac{v_o}{\tau} = \frac{1}{G_0 \tau} \frac{d\phi}{dt} \qquad \text{for } |\psi| \leq \frac{\pi}{2} \tag{10.8-5}$$

Equation (10.8-5) shows explicitly the relationship between the frequency offset $d\varphi/dt$ and the output voltage $v_o(t)$. Equation (10.8-3) or (10.8-5) is the differential equation of the PLL. Its solution, subject to the appropriate initial conditions, describes the PLL performance. Parenthetically, we note that with the phase comparator having the sinusoidal characteristic shown in Fig. 10.7-2*a*, the equation corresponding to Eq. (10.8-3) is

$$\frac{d\psi}{dt} + \frac{\sin\psi}{\tau'} = \frac{d\phi(t)}{dt} \tag{10.8-6}$$

with $\tau' = 1/ABG_p G_0$. For small ψ, where $\sin\psi \approx \psi$, Eq. (10.8-6) would be identical with Eq. (10.8-3), except that the time constants would be different, this difference reflecting only the fact that in Fig. 10.7-2 the slopes at the origin are different in parts (*a*) and (*b*).

Suppose that the PLL is operating initially with $\psi = 0$, that is, the VCO and input-carrier frequency are identical and exactly 90° out of phase. Now let us again ask the question raised before; that is, what happens if the angular frequency of the carrier is abruptly changed by an amount ω ($d\varphi/dt = \omega$)? Subject to these initial conditions, the solution of Eq. (10.8-3) is

$$\psi = \omega\tau(1 - e^{-t/\tau}) \qquad |\psi| \leq \frac{\pi}{2} \tag{10.8-7}$$

Then if $\omega\tau \leq \pi/2$, ψ will approach its new steady-state value with a time constant τ. The new equilibrium steady state is at

$$\psi = \psi_e = \omega\tau \leq \frac{\pi}{2} \tag{10.8-8}$$

or correspondingly, using Eqs. (10.7-2) and (10.8-4), the steady-state output voltage v_o is given by

$$v_o = \frac{\omega}{G_0} \tag{10.8-9}$$

Operating Range

We note that as long as the phase comparator operates in the range *AOB* in Fig. 10.7-2*b*, that is, $|\psi| \leq \pi/2$, the steady-state operating point is given by $\psi_e = \omega\tau$. The range R of the PLL is the angular frequency change ω of the input carrier which will just carry the steady-state operating point from O to B or from O to A. This range R is given by

$$R = \omega_{\max} = \frac{\psi_{\max}}{\tau} = \frac{\pi/2}{\tau} = \frac{\pi}{2\tau} \tag{10.8-10}$$

For the phase comparator of Fig. 10.7-2*a* we find R by using Eq. (10.8-6). Set $d\phi(t)/dt = \omega$ and $d\psi/dt = 0$ (steady state). Then, when $\psi = \pi/2$, $\sin\psi = 1$. Thus $R = \omega_{\max} = 1/\tau'$. We observe that the behavior of the PLL is determined entirely by the single parameter τ (or τ'). This parameter depends on the gains G_0 and G_p and also on the amplitudes A and B of the input and VCO signals.

As τ becomes progressively smaller, the range of the PLL becomes progressively larger, and the speed with which the PLL responds to a frequency change ω

increases. The reason the range becomes larger is, as a matter of fact, precisely because the loop does respond faster. A *fast-acting* loop will not permit the phase $\phi(t)$ to depart appreciably from the phase of the VCO. In the limit, when $\tau \to 0$, ψ will remain close to $\psi = 0$ and hence close to the origin 0 in Fig. 10.7-2*b*, no matter how large ω should become.

Bandwidth

Since the PLL response is given by a linear differential equation, it has the characteristics of a filter. To see that such is the case, let $\Psi(s) \equiv \mathscr{L}[\psi(t)]$ and $\Omega(s) \equiv \mathscr{L}(d\phi/dt) \equiv \mathscr{L}[\omega(t)]$. Then from Eq. (10.8-3) we have

$$\Psi(s) = \frac{\Omega(s)}{s + 1/\tau} \tag{10.8-11}$$

The transfer function of the PLL, $\psi(s)/\Omega(s)$, which relates the output signal to the input signal, i.e., the change in the input angular frequency, has a transfer function

$$H(s) = \frac{\Psi(s)}{\Omega(s)} = \frac{1}{s + 1/\tau} \tag{10.8-12}$$

The transfer function has a single pole at $s = -1/\tau$, and the PLL is therefore referred to as a *first-order* PLL.

Suppose that the angular frequency variation of the input signal were sinusoidal: $\omega(t) = \Delta\omega_m \cos \omega_m t$, with $\Delta\omega_m$ the amplitude, and ω_m the angular frequency of the sinusoidal variation of the instantaneous frequency. Then for a fixed frequency deviation $\Delta\omega_m$, an increase in the modulating frequency ω_m results in a decrease in $\psi(t)$ and hence $v_o(t)$. The angular frequency $(\omega_m)_{3\text{ dB}}$ at which the response will have fallen by 3 dB is seen from Eq. (10.8-12) to be

$$(\omega_m)_{3\text{ dB}} = \frac{1}{\tau} \tag{10.8-13}$$

A PLL will consequently introduce distortion between the original modulating signal and the signal recovered at the PLL output. The distortion will be the same as would be introduced if the modulating signal had been passed through a low-pass resistance-capacitance network of time constant τ. The distortion may be decreased by making the PLL 3-dB cutoff frequency much higher than the highest-frequency spectral component of the modulating signal.

A point worthy of note is to be seen in Eqs. (10.8-10) and (10.8-13). We observe that a single time constant τ determines both the range of the PLL and its frequency response. We are therefore not at liberty, in this first-order loop, to vary these parameters independently. We shall see in Sec. 10.11 how this limitation can be remedied.

As long as the operating point of the comparator remains in the range *AOB* in Fig. 10.7-2*b*, the PLL demodulator accomplishes no function which is not performed at least equally well by a conventional discriminator. The PLL displays its special merit, however, precisely when it is operated in such a manner that its operating point ranges outside the limits A and B in Fig. 10.7-2b. In the next section we discuss such operation.

10.9 STABLE AND UNSTABLE OPERATING POINTS

We have seen that in the range AOB in Fig. 10.7-2b, the differential equation for the PLL is

$$\frac{d\psi}{dt} + \frac{\psi}{\tau} = \frac{d\phi}{dt} \tag{10.9-1}$$

This equation applies equally well for operation over any positive-slope region in Fig. 10.7-2b, provided that ψ measures the departure of the phase from the phase at which the phase characteristic crosses the axis $v_o = 0$. Thus, for example, the equation applies to the region $CO''D$ if ψ is measured from O''. Similarly, by retracing the derivation, we find that in a negative-slope region, as, say, $BO'C$, if ψ is measured from O', the equation is (see Prob. 10.9-1)

$$\frac{d\psi}{dt} - \frac{\psi}{\tau} = \frac{d\phi}{dt} \tag{10.9-2}$$

In Fig. 10.9-1a we have drawn a portion of the phase-comparator characteristic in which the slope is positive. Let us assume that the input-carrier angular frequency has been offset by an amount ω. Then, as given in Eq. (10.8-9), the new *steady-state* equilibrium is, as shown, at $v_o = \omega/G_0$, and the corresponding equilibrium value of ψ is $\psi_e = \omega\tau$ as in Eq. (10.8-8). Suppose, however, that because

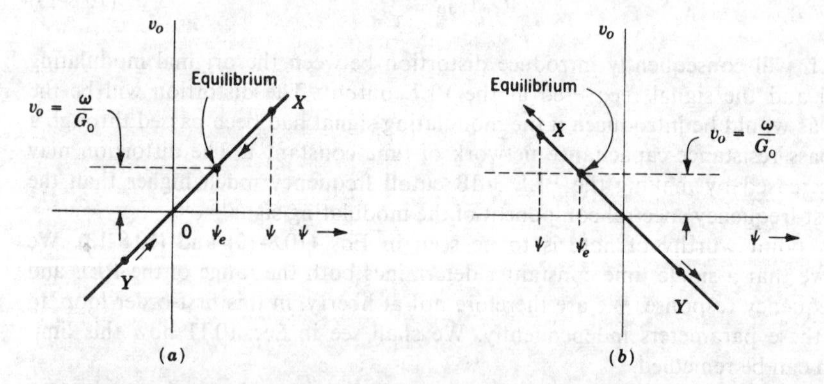

Figure 10.9-1 (a) Illustrating that an equilibrium point is stable if the operating point of a PLL is on a positive-slope portion of the phase comparator. (b) Illustrating the instability that results when the operating point is on a negative-slope portion of the comparator.

of the past history of the PLL, the operating point happens to find itself initially at point X where the phase angle is ψ. We have then from Eq. (10.9-1) that

$$\frac{d\psi}{dt} = \frac{d\phi}{dt} - \frac{\psi}{\tau} \tag{10.9-3a}$$

$$= \omega - \frac{\psi}{\tau} \tag{10.9-3b}$$

$$= \frac{\psi_e}{\tau} - \frac{\psi}{\tau} = \frac{1}{\tau}(\psi_e - \psi) \tag{10.9-3c}$$

Now if, as in the case of point X, $\psi > \psi_e$, then $d\psi/dt$ is negative. Hence the instantaneous operating point X moves to the left along the comparator-phase characteristic. The point X will, as we have already seen, approach the equilibrium point asymptotically, with time constant τ. Similarly if the operating point is at Y, then $\psi < \psi_e$, and $d\psi/dt$ is positive so that again Y approaches the equilibrium point. The equilibrium point is a *stable* point.

Let us now consider the negative-slope portion of the phase characteristic such as $BO'C$ in Fig. 10.7-2b. Here we use Eq. (10.9-2) instead of Eq. (10.9-1). Repeating the above discussion using Fig. 10.9-1b, we find that initial operating points such as X and Y result in ψ moving *away* from, not *toward*, the equilibrium point ψ_e. Thus in Fig. 10.9-1b the equilibrium point is *unstable*. In the next section we apply these simple notions to explain how the PLL acts to suppress spikes.

10.10 SPIKE SUPPRESSION

Let us assume that the input to a PLL is an unmodulated carrier, offset from ω_c by an amount ω. Let the steady-state output voltage of the PLL corresponding to this offset be $v_o = V_o = \omega/G_0$ as indicated by the dashed horizontal line in Fig. 10.10-1. That is, the offset ω is so large that the horizontal line ZZ' does not intersect the phase characteristic. There are no *actual* equilibrium points. However, two extended equilibrium points have been indicated at points P_2 and P_3. The point P_2 is the intersection of AOB with ZZ'. An initial operating point P_1 on AOB will move as though headed for equilibrium at P_2 for, until point B is reached, the PLL does not know that AOB does not continue to P_2. Similarly the *equilibrium* point for operation along $BO'C$ is at P_3.

If the PLL is initially at the origin with $\psi = 0$, and at time $t = 0$ a phase offset ω is added to ω_c, the operating point will move from 0 to B. The operating point cannot remain at B, since B is not an equilibrium point. However, the point B is also on the negative-slope segment $BO'C$. Hence it will recede from the equilibrium point P_3 and proceed toward C. Continuing in the same way, we see that from C the operating point will go to D, i.e., toward the equilibrium point, and so on. If the output voltage corresponding to point B is $v_o = v_{oB}$, then the output

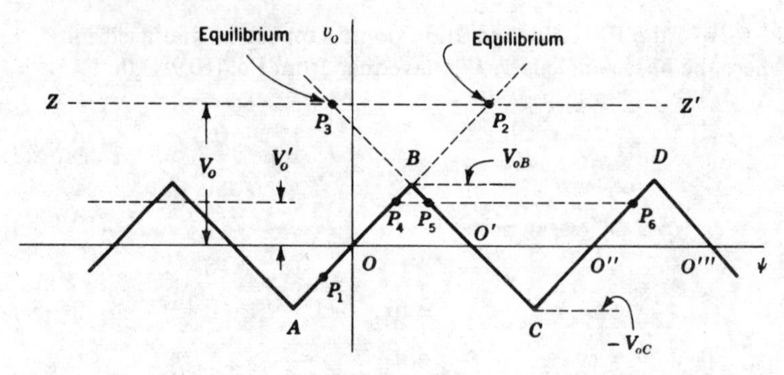

Figure 10.10-1 Illustrating the PLL response for various assumed fixed-frequency offsets of the carrier signal.

voltage will oscillate with excursion from $+v_{oB}$ to $-v_{oB}$. The waveform of this oscillation is easily calculated. Note that ψ constantly increases and does not oscillate.

Suppose that the offset were such to establish equilibrium at the level V_o' as indicated in Fig. 10.10-1; then if the operating point is initially on AOB, it will settle at P_4. If the operating point is initially on $BO'C$, it will eventually settle at P_4 or P_6, depending on whether the initial point is above or below P_5. If the operating point is initially constrained to be precisely at P_5, it will remain there only as long as no disturbance displaces it, even ever so slightly. If a disturbance does move the operating point from P_5, the operating point will end up again at P_4 or P_6, depending on the direction of the disturbance.

After these preliminaries we may turn our attention to the matter of spikes. For this purpose we shall inquire into the response of the PLL to an "artificial" spike. Thus we consider that the input carrier is at the carrier angular frequency ω_c except for a short time T during which the angular frequency is offset by an amount ω. Such artificial spikes are shown by the solid-line plot in Fig. 10.10-2 in which the offset angular frequency ω is plotted as a function of time. In a physical situation the duration T of the spike will be determined by the bandwidth of the IF carrier filter. The spike amplitude ω_S will then be determined by the condition that the area under the spike be 2π, that is, $\omega_S T = 2\pi$.

We have seen [Eq. (10.8-10)] that the angular frequency corresponding to a steady-state operating point at B in Fig. 10.10-1 is $\omega_{\text{max}} = \pi/2\tau$. The dashed waveforms in Fig. 10.10-2 indicate the PLL responses for three relative values of ω_{max} and ω_S.

Case 1 The steady-state output voltage v_o corresponding to ω is $v_o = \omega/G_0$ [see Eq. (10.8-9)]. In Fig. 10.10-2*a* we assume $\omega_{\text{max}} > \omega_S$. The output voltage of the PLL responds to the spike by rising asymptotically with time constant τ to the steady-state level $v_o = \omega_S/G_0$. At this level the operating point of the PLL has not reached the point B in Fig. 10.10-1 since we assume that $\omega_{\text{max}} > \omega_S$. At the termination of the spike the PLL returns to its initial level. The output-voltage waveform v_o has the form of the dashed plot. This output waveform is a replica of the input-spike waveform except rounded because of the time constant of the PLL. In this case, a spike input produces a spike output.

Case 2 Here in Fig. 10.10-2*b* $\omega_{max} < \omega_S$. Hence the operating point reaches the limit of its range (point *B* in Fig. 10.10-1) before it has been able to respond fully to the spike amplitude. After reaching point *B*, the operating point starts down along $BO'C$. However, the spike ends before point O' is reached. When the spike has terminated, ω is again equal to zero, and O' is an *unstable* equilibrium point. Thus, the operating point retraces its path back over *B* and into the positive-slope region, finally returning to the starting point at *O*. An input spike now yields two spikes at the output.

Note that the change from case 1 to case 2 is made by increasing τ, thereby lowering ω_{max} and also slowing the response of the PLL.

Case 3 Here in Fig. 10.10-2*c* ω_{max} has been reduced still further and has been set at such a level that the operating point reaches the limit of the range in a time slightly less than one-half the spike duration. Therefore when the spike terminates, the operating point has passed O' and hence continues to *C* and finally

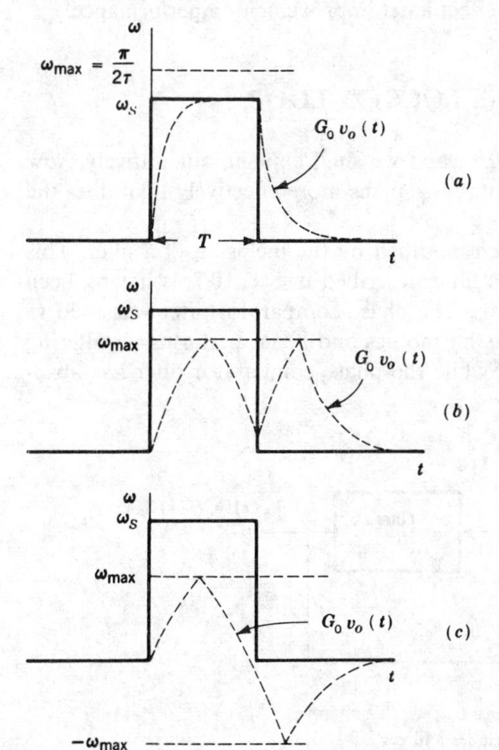

Figure 10.10-2 Various possible responses of a PLL to an "artificial" spike. The different cases correspond to different selections of the time constant of the PLL.

settles at O''. In this case a spike input has yielded an output which consists of a positive spike followed by a negative spike. Such an output waveform as indicated by the dashed plot in Fig. 10.10-2*c* is called a *doublet*. We may reasonably expect that the total area under a doublet will be near zero. Therefore the doublet will yield very little energy in the baseband in comparison with the energy yielded by the spike itself. Here, then, qualitatively is the mechanism by which the PLL suppresses the noise of spikes. The PLL changes spikes into doublets of relatively small energy.

If all the spikes encountered in a physical situation were of precisely the same waveform, then, as the preceding discussion indicates, it would be possible to adjust the time constant τ of the PLL so that each spike might be replaced by a doublet. Observe that only a small range of τ is suitable. The time constant τ must be large enough to avoid the responses indicated in Fig. 10.10-2*a* and *b*. On the other hand, increasing τ results in decreasing the PLL bandwidth [Eq. (10.8-13)]. If the PLL is being used to demodulate a carrier frequency modulated by a baseband signal of bandwidth f_M, then we require that $(1/\tau) \gg f_M$.

Unfortunately all spikes are not identical. Even in the absence of modulation there will be some variation in spike waveform, and with modulation the spikes will vary considerably during the course of the modulation cycle. Still, as a matter of practice it turns out to be possible to adjust a PLL to convert many of the spikes into doublets and thereby effect a net improvement in performance.

10.11 SECOND-ORDER PHASE-LOCKED LOOP

We now discuss the second-order PLL, and we shall explain, qualitatively, how the second-order loop manages to suppress spikes more effectively than does the first-order loop.

In Fig. 10.11-1 the PLL has been modified by the inclusion of a filter. This filter is not to be compared with the filter described in Sec. 10.7 as having been incorporated in the phase comparator. The phase-comparator filter was used to suppress the carrier frequency and its harmonics and hence performs no filtering in the passband of the PLL. Hence, while the phase-comparator filter has abso-

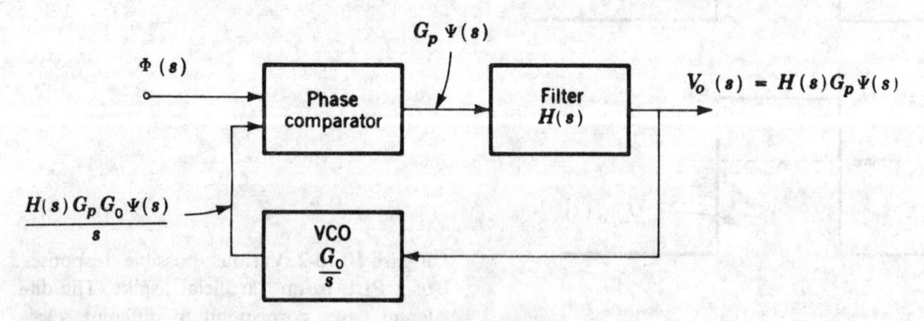

Figure 10.11-1 A phase-locked loop which includes a filter.

lutely no influence on the PLL performance, the filter introduced in Fig. 10.11-1 is deliberately introduced to have such influence.

It will be more convenient, in our present discussion, to refer not to the angle-modulated carrier waveforms encountered in the PLL but rather directly to the phases of these waveforms. Thus, in Fig. 10.11-1 the signal input is $\phi(t)$ with Laplace transform $\Phi(s)$. The output voltage of the phase comparator is $G_p \Psi(s)$, where $\psi(t)$ is the phase-angle difference at the phase-comparator input, and G_p is a constant of proportionality. The VCO provides a carrier having a phase proportional to the *integral* of its input voltage $v_o(t)$. Hence the VCO has been characterized as having a transform G_0/s.

With $H(s) \equiv 1$, the PLL of Fig. 10.11-1 is identical with the loop of Fig. 10.7-3. We have, in this case

$$\Psi(s) \equiv \Phi(s) - \frac{G_p G_0 \Psi(s)}{s} \tag{10.11-1}$$

Using $s\Phi(s) = \mathscr{L}[d\phi(t)/dt] \equiv \Omega(s)$ and $1/\tau = G_0 G_p$, we find from Eq. (10.11-1) that

$$\Psi(s) = \frac{s\Phi(s)}{s + G_p G_0} = \frac{\Omega(s)}{s + G_0 G_p} = \frac{\Omega(s)}{s + 1/\tau} \qquad |\psi(t)| \leq \frac{\pi}{2} \tag{10.11-2}$$

exactly as in Eq. (10.8-11).

Now let us assume a *proportional-plus-integral* filter with $H(s)$ given by

$$H(s) = 1 + \frac{K}{s} \tag{10.11-3}$$

where K is a constant. We now find that the phase-angle difference $\psi(t)$ has a transform

$$\Psi(s) = \frac{s^2\Phi(s)}{s^2 + s/\tau + K/\tau} = \frac{s\Omega(s)}{s^2 + s/\tau + K/\tau} \qquad |\psi(t)| \leq \frac{\pi}{2} \tag{10.11-4}$$

while the output voltage $v_o(t)$ has a transform

$$V_o(s) = G_p H(s)\Psi(s) = \frac{s(s + K)G_p \Phi(s)}{s^2 + s/\tau + K/\tau} = \frac{(s + K)G_p \Omega(s)}{s^2 + s/\tau + K/\tau} \tag{10.11-5}$$

We observe that the expressions for $\Psi(s)$ and $V_o(s)$ have denominators which, being quadratic in s, give rise to two poles. Hence the term *second-order* PLL.

We inquire now, as we did for the first-order loop, about the steady-state response of the second-order loop to an abrupt change of magnitude ω in the angular frequency of the input carrier. To make this calculation, we use the *final-valve theorem* which states that if $F(s) = \mathscr{L}[f(t)]$, then

$$\lim_{s \to 0} [sF(s)] = \lim_{t \to \infty} f(t) \tag{10.11-6}$$

Applying this theorem to $\Psi(s)$ and $V_o(s)$ in Eqs. (10.11-4) and (10.11-5) with $\Omega(s) = \omega/s$, we find

$$\psi(\infty) = 0 \tag{10.11-7}$$

and

$$v_o(\infty) = G_p \omega \tau \tag{10.11-8}$$

Thus, we observe from Eq. (10.11-8) that the steady-state response at the output of the second-order loop is the same as in the first-order loop. However, we see from Eq. (10.11-7) that the phase difference $\psi(t)$, unlike the situation obtained in the first-order loop, does not respond with a steady-state displacement from its initial position of equilibrium. Instead, while it may make a transient excursion, it thereafter settles right back to its starting point. Another way of emphasizing this same point is to recognize that the transfer function that relates $\Psi(s)$ to the input signal $\Omega(s)$ is Eq. (10.11-4) is the transfer function of a bandpass filter, while the transfer function that relates $V_o(s)$ to $\Omega(s)$ in Eq. (10.11-5) is the transfer function of a low-pass filter. Therefore, by appropriately selecting the cutoff frequencies of these filters, it is possible to arrange that the output of the PLL respond properly to the modulating signal, while the operating point of the phase comparator responds hardly at all and always hovers close to its initial point of equilibrium.

The relevance of these considerations to spike suppression may now be seen. Consider the first-order loop and let the input be a step in frequency which carries the phase-comparator operating point toward A in Fig. 10.10-1 and holds it there at point P_1. If, now, a spike develops because of noise, this spike will be in the direction to drive the operating point in the opposite direction toward B. The spike will not be suppressed unless the operating point does indeed get to point B and beyond toward C, as explained in Sec. 10.10. However, the initial displacement of the operating point away from O toward A due to the input signal makes such spike suppression less likely than if the initial operating point were at point O. On the other hand, in the second-order loop, after some initial response to the input-frequency step, the phase comparator settles back at point O. Thus if an input spike occurs, that is, $\phi(t)$ changes by 2π rad, the distance moved to point B and hence the time required to reach point B is less than in the first-order loop. As consequence such a spike has a greater likelihood of suppression.

10.12 OUTPUT SNR OF A PHASE-LOCKED LOOP[4]

The input signal to an FM demodulator is

$$v_i(t) = R(t) \cos [\omega_c t + \phi(t)] \tag{10.12-1}$$

Here $R(t)$ is the envelope of the waveform which results from the superposition of the carrier of amplitude A and the noise $n(t)$. Similarly $\phi(t) = \phi_s(t) + \phi_n(t)$, where $\phi_s(t)$ is the angular modulation due to the signal, and $\phi_n(t)$ due to the noise. Let us ignore, for the present, the effect of the noise on the envelope and assume that $R(t) = A$.

With this assumption, an FM discriminator, when presented with the waveform of Eq. (10.12-1), will yield an output voltage

$$v_o(t) = \alpha \frac{d\varphi(t)}{dt} \tag{10.12-2}$$

where α is a constant. The student may verify from Eq. (10.8-5) that the output voltage of a first-order PLL is related to $d\phi(t)/dt$ as indicated in Fig. 10.12-1 (see Prob. 10.12-1). The time constant of the *RC* circuit is equal to the time constant τ of the PLL. If we neglect the effect of this low-pass *RC* circuit, then

$$v_o(t) = \frac{1}{G_0} \frac{d\varphi(t)}{dt} \tag{10.12-3}$$

just as for the discriminator except with α replaced by $1/G_0$. Now both the discriminator and the PLL will be followed by a low-pass baseband filter of cutoff frequency f_M. If, as is always the case, we select the 3-dB frequency of the *RC* circuit, which represents the phase-locked loop, to be much larger than f_M, then as far as the baseband filter output is concerned, the *RC* circuit is indeed of no consequence. Above threshold, where spikes are extremely rare, the discriminator and the PLL, when combined with the baseband filter, process the input waveform, signal plus noise, in exactly the same way. The output SNR does not depend on the constant α nor on the constant $1/G_0$. Therefore, above threshold, both discriminator and PLL yield the same output SNR.

The Effect of Limiting

Now let us take account of the fact that the envelope of $v_i(t)$ in Eq. (10.12-1) actually varies because of noise. A discriminator is always accompanied in practice by a limiter. Hence the result stated in Eq. (10.12-2) continues to apply, since the limiter removes all amplitude variation. In the case of the PLL we note from Fig. 10.12-1 that the carrier amplitude enters only through the time constant τ, for as is to be seen in Eq. (10.8-4) τ depends on *A*. Still, if the 3-dB frequency of the *RC* filter in Fig. 10.12-1 is very much larger than the baseband filter cutoff frequency, the small variations in τ resulting from the noise can have little effect. At least such will be the case above threshold where the noise power is very small in comparison with the carrier power. Hence, we have the result that above threshold the PLL does not require a limiter, and the result of Eq. (10.12-3) is valid.

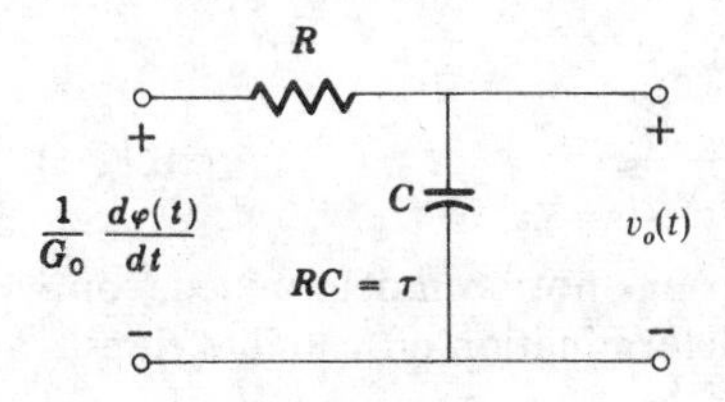

Figure 10.12-1 A representation of the relationship between the output voltage v_o of a first-order PLL and the frequency modulation $d\phi/dt$ of an input carrier.

Above threshold it is immaterial whether we use a limiter. We shall now show qualitatively that below threshold it is advantageous *not* to use a limiter. As explained in Sec. 10.10 in connection with Fig. 10.10-2, it is necessary to restrict the range $\omega_{\max}$ ($= \pi/2\tau$) of the PLL in order to achieve spike suppression. Such restriction of the range has the disadvantage that it also restricts the ability of the PLL to follow the frequency deviations of the carrier due to signal modulation. It would therefore be of advantage if we could keep the range large in the absence of a spike and yet be able to restrict the range during the course of the formation of a spike. We may now see that if the PLL is operated without a limiter, the PLL will automatically restrict its own range during a spike.

Referring to Fig. 10.2-2*b*, we see that when the noise $r(t)$ is comparable to A, that is, when a spike occurs, then as the resultant $R(t)$ executes a rotation by 2π, its average magnitude will be smaller than A. [It is of course possible to conceive that $r(t)$ is so large that $R(t)$ is larger than A, but such a situation would have a small probability of occurrence and is therefore of no interest in the present discussion.] Furthermore, since the range $\omega_{\max} = \pi/2\tau$ and τ is inversely proportional to the carrier envelope [see Eq. (10.8-4)], the range of the PLL automatically reduces during a spike. This advantageous operating characteristic of the PLL would be lost if the envelope amplitude were kept constant by the use of a limiter.

Output SNR

We have seen that the number of spikes present at the PLL output is fewer than at the output of the FM discriminator, and threshold *extension* results. The output-noise power due to spikes present at the PLL output is N_S and is found by replacing $1/T_s$ by $N \equiv N_c + \delta N$, the average number of spikes per second occurring at the PLL output, in Eq. (10.4-5). The result is

$$N_S = \frac{8\pi^2 f_M N}{G_0^2} \tag{10.12-4}$$

Note that α is replaced by $1/G_0$ in this equation. Since the PLL suppresses spikes, N is smaller than for the discriminator. Thus, N_S is smaller in the PLL.

From Eqs. (9.2-5) and (9.2-16) with α replaced by $1/G_0$, and using Eq. (10.12-4), we have

$$\frac{S_o}{N_o} = \frac{k^2\overline{m^2(t)}}{(8\pi^2/3)\eta f_M^3/A^2 + 8\pi^2 f_M N} \tag{10.12-5}$$

If we simplify and assume sinusoidal modulation, $km(t) = \Delta\omega \cos \omega_m t$, then Eq. (10.12-5) becomes

$$\frac{S_o}{N_o} = \frac{\frac{3}{2}\beta^2 S_i/N_M}{1 + 6(N/f_M)(S_i/N_M)} \tag{10.12-6}$$

While an analytic expression for N of a PLL has not been determined, computer simulation of the PLL has resulted in the determination of N and of thresh-

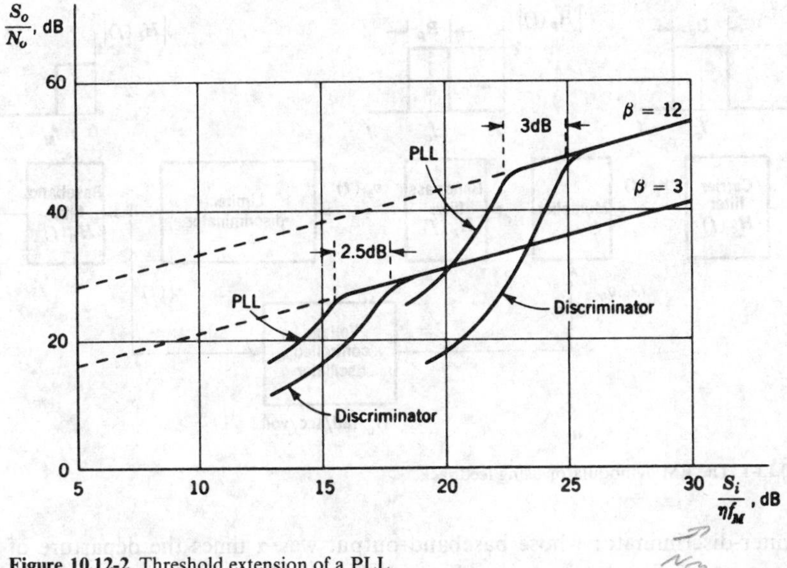

Figure 10.12-2 Threshold extension of a PLL.

old.[5] The results obtained are shown in Fig. 10.12-2. We see from this figure that with $\beta = 12$, the PLL extends threshold by 3 dB, while with $\beta = 3$, threshold extension of 2.5 dB results.

10.13 THE FM DEMODULATOR USING FEEDBACK

An FM demodulator using feedback (FMFB) is shown in Fig. 10.13-1. This demodulator, like the PLL, decreases the S_i/N_M at threshold. The construction of the FMFB and the PLL are similar inasmuch as a VCO is frequency-modulated by the output signal $v_o(t)$ of the FMFB demodulator. Also, as in the PLL, the input $v_i(t)$ of carrier frequency $f_c(= \omega_c/2\pi)$ is multiplied with the output v_{osc} of the VCO. In the present case, however, the frequency of the VCO is offset from f_c by an amount f_0. The FMFB includes in its forward transmission path a bandpass filter and also a limiter-discriminator, neither of which is encountered in the PLL. The bandpass filter, following the multiplier, is centered at the offset frequency f_0, and hence passes the difference-frequency output of the multiplier.

We show now that the FMFB recovers the baseband signal from an FM modulated carrier, and when operating above threshold yields the same signal-to-noise output as does a simple limiter-discriminator. Let the input signal and noise $v_i(t)$ be

$$v_i(t) = R(t) \sin [\omega_c t + \phi_s(t) + \phi_n(t)] \tag{10.13-1}$$

where $R(t)$ is the envelope of the carrier signal and the noise, $\phi_s(t)$ is the angular modulation due to the signal, and $\phi_n(t)$ is due to the noise. If $v_i(t)$ were the input

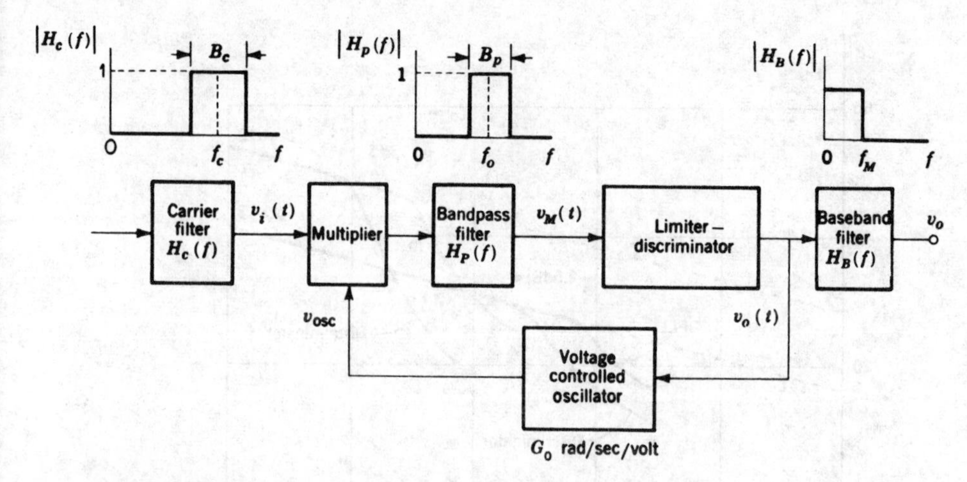

Figure 10.13-1 The FM demodulator using feedback.

to a limiter-discriminator whose baseband output was α times the departure of the instantaneous angular frequency from the carrier frequency ω_c, the output voltage of the discriminator $v_o(t)$ would be

$$v_o(t) = \alpha \frac{d}{dt} [\phi_s(t) + \phi_n(t)] \tag{10.13-2}$$

Returning to the FMFB shown in Fig. 10.13-1, we represent the VCO output as

$$v_{osc} = B \cos \left[(\omega_c - \omega_0)t + G_0 \int_{-\infty}^{t} v_o(\lambda)\, d\lambda \right] \tag{10.13-3}$$

where B is the VCO amplitude and G_0, as introduced in Eq. (10.7-3), is the change in angular frequency of the VCO per unit change in $v_o(t)$. We neglect temporarily the effect of the bandpass filter following the multiplier except to take account of the fact that it passes only the difference-frequency component of the multiplier. On this basis, the product signal $v_M(t)$, which is equal to the low-pass component of $v_i v_{osc}$, is

$$v_M(t) = \frac{AB}{2} \cos \left[\omega_0 t + \phi_s(t) + \phi_n(t) - G_0 \int_{-\infty}^{t} v_o(\lambda)\, d\lambda \right] \tag{10.13-4}$$

This product signal is applied to the limiter-discriminator. The output of the discriminator $v_o(t)$ is then

$$v_o(t) = \alpha \left[\frac{d\phi_s(t)}{dt} + \frac{d\phi_n(t)}{dt} - G_0 v_o(t) \right] \tag{10.13-5}$$

Solving for $v_o(t)$ yields

$$v_o(t) = \frac{\alpha}{1 + \alpha G_0} \frac{d}{dt} (\phi_s + \phi_n) \tag{10.13-6}$$

We observe from Eq. (10.13-6), and from comparing this equation with Eq. (10.13-2), that the FMFB does indeed demodulate. We see further that the only difference between the output of the discriminator and the FMFB is that the amplitude of output of the FMFB is smaller by the factor $(1 + \alpha G_0)$. Since both signal and noise have been reduced by the same factor $1/(1 + \alpha G_0)$, the signal-to-noise ratio is the same for the FMFB and the FM discriminator.

We turn our attention now to the bandpass filter and ask how narrow it can be made to pass the signal without undue distortion. Using Eq. (10.13-6), we may rewrite Eq. (10.13-4) as

$$v_M(t) = \frac{AB}{2} \cos\left[\omega_0 t + \frac{1}{1 + \alpha G_0}(\phi_s + \phi_n)\right] \tag{10.13-7}$$

We observe, in comparing Eq. (10.13-7) with Eq. (10-13.1), that the feedback has suppressed the frequency deviation, produced by the signal ϕ_s, by the factor $1/(1 + \alpha G_0)$. Consider, for example, that the modulation is sinusoidal. Then $\phi_s(t) = \beta \sin \omega_m t$, and the phase of the signal present in the multiplied signal is $\phi_s(t)/(1 + \alpha G_0) = [\beta/(1 + \alpha G_0)] \sin \omega_m t$. Hence, if the bandwidth of the carrier filter is B_c and the bandwidth of the bandpass filter preceding the discriminator is B_p, then

$$B_p = 2\left(\frac{\beta}{1 + \alpha G_0} + 1\right) f_m \tag{10.13-8}$$

and

$$B_c = 2(\beta + 1) f_m \tag{10.13-9}$$

so that

$$B_p = \frac{[\beta/(1 + \alpha G_0)] + 1}{\beta + 1} B_c \tag{10.13-10}$$

If $\beta = 9$ and $\alpha G_0 = 8$, then $B_p = \frac{1}{5} B_c$.

Note that the bandwidths B_p and B_c given by Eqs. (10.13-8) and (10.13-9) employ Carson's rule for the bandwidth of *rectangular* filters. In practice the carrier filter with bandwidth B_c is *almost* rectangular. However, the bandpass filter with bandwidth B_p is usually a single-tuned filter, and to pass 98 percent of the signal energy, B_p must be greater than the value given in Eq. (10.13-8). The reason for employing the single-tuned circuit is related to the fact that the FMFB is a feedback system. If a single-tuned filter is used, the system is stable; if, however, a more sophisticated filter containing many poles is employed, the FMFB may oscillate.

10.14 THRESHOLD EXTENSION USING THE FMFB[6]

Let us consider the operation of the FMFB under two conditions of G_0. First let the VCO sensitivity $G_0 = 0$. Then, referring to Fig. 10.13-1, we find there is no feedback. The modulation index is not reduced, and $B_p = B_c$ [see Eq. (10.13-10)]. The VCO now serves only to shift the carrier frequency ω_c to an IF frequency ω_0. Therefore no threshold extension occurs.

Our second condition for G_0 is $G_0 \to \infty$. Let the output voltage of the VCO be written as

$$v_{osc} = B \cos [(\omega_c - \omega_0)t + \phi_{osc}(t)] \tag{10.14-1}$$

where, comparing Eq. (10.14-1) with Eq. (10.13-3), we see that

$$\phi_{osc}(t) = G_0 \int_{-\infty}^{t} v_o(\lambda)\, d\lambda \tag{10.14-2}$$

Since G_0 is extremely large, $v_o(t)$ must be infinitesimally small for $\phi_{osc}(t)$ to remain finite.

Refer now to Eq. (10.13-5) and replace $G_0 v_o(t)$ by $\dot{\phi}_{osc}(t)$. We have then

$$v_o(t) = \frac{1}{G_0}\frac{d\phi_{osc}(t)}{dt} = \alpha\left(\frac{d\phi_s}{dt} + \frac{d\phi_n}{dt} - \frac{d\phi_{osc}}{dt}\right) \tag{10.14-3}$$

For very large G_0 we can assume $v_o(t) = 0$. Equation (10.14-3) now becomes

$$\frac{d\phi_{osc}}{dt} \approx \frac{d\phi_s}{dt} + \frac{d\phi_n}{dt} \tag{10.14-4}$$

Equation (10.14-4) shows that when the amount of feedback becomes very large, the frequency of the VCO approaches the sum of the frequency of the input signal and the frequency of the input noise. Note that the output of the multiplier becomes a very narrowband FM signal.

Thus, whenever $\phi_n(t)$ rotates by 2π, resulting in a discriminator spike, we see from Eq. (10.14-4) that ϕ_{osc} also rotates by 2π, resulting in a spike in the FMFB. Thus there is no threshold extension when $G_0 \to \infty$.

We now show that if G_0 is neither zero nor very large, threshold extension results. We assume first that the modulation is sinusoidal and that the demodulation is performed by a conventional limiter-discriminator circuit. From Eqs. (10.5-18), (10.5-19), and (10.6-4) we find that the total average number of spikes per second N is

$$\begin{aligned} N_{discr} &= N_c + \overline{\delta N} \\ &= \frac{B}{2\sqrt{3}} \operatorname{erfc} \sqrt{\frac{f_M}{B}\frac{S_i}{N_M}} + \frac{2\Delta f}{\pi} \exp\left[-(f_M/B)(S_i/N_M)\right] \end{aligned} \tag{10.14-5}$$

In Eq. (10.14-5), B is the bandwidth of the IF filter and Δf is the frequency deviation of the sinusoidally modulated carrier.

In the case of the FMFB, N is also given by Eq. (10.14-5), provided that B is replaced by B_p (see Fig. 10.13-1) and Δf is multiplied by the factor $1/(1 + \alpha G_0)$. Hence,

$$N_{FMFB} = \frac{B_p}{2\sqrt{3}} \operatorname{erfc} \sqrt{\frac{f_M}{B_p}\frac{S_i}{N_M}} + \frac{2\Delta f}{\pi(1 + \alpha G_0)} \exp\left[-(f_M/B_p)(S_i/N_M)\right] \tag{10.14-6}$$

Both N_{discr} and N_{FMFB} depend on the ratio S_i/N_M. It can, however, be shown that S_i/N_M is the same whether measured at the output of the IF filter or at the output of the filter of bandwidth B_p. However, since $B_p < B$ and in addition, since $1 + \alpha G_0 > 1$, we see in comparing Eqs. (10.14-5) and (10.14-6) that $N_{\text{FMFB}} < N_{\text{discr}}$ and that hence the threshold is extended.

10.15 BIT SYNCHRONIZER

A bit synchronizer is a circuit which establishes at a receiver the times at which a bit interval ends and the succeeding bit interval begins. The synchronizer thus regenerates, at the receiver, a clock waveform which is synchronous with the original clock waveform, at the transmitter. Bit synchronizers can be extremely sophisticated devices especially when the timing information has to be extracted from a noisy received signal. In this section we shall describe a basic bit synchronizer which is effective with signals having a signal to noise ratio of 6 dB or better. The bit synchronizer circuit is shown in Fig. 10.15-1. It is a phase locked loop in which the timing (phase) comparison is done by a flip-flop.

The flip-flop is designed to respond, not to the logic levels at its S (set) and R (reset) input terminals, but rather to positive-going *transitions*. Thus, when the data $d(t)$ makes a positive transition from logic 0 to logic 1, the flip-flop will go to the set state with $Q = 1$. A positive-going transition on R (reset) will transfer the flip-flop back to the reset state with $Q = 0$. Waveforms for the bit synchronizer are shown in Fig. 10.15-2. The original timing clock of the transmitter is shown in (*a*) while a typical data stream, synchronous with the clock, is shown in (*b*). Now let us consider that we have disconnected the VCO input from the "time-to-voltage converter" and that we have applied at that input the dc voltage required to cause the VCO to oscillate at exactly the clock frequency. The VCO waveform is shown in Fig. 10.15-2*c*. (The VCO output need not be sinusoidal. It is easily arranged that it be a square wave as shown.) In the general case when

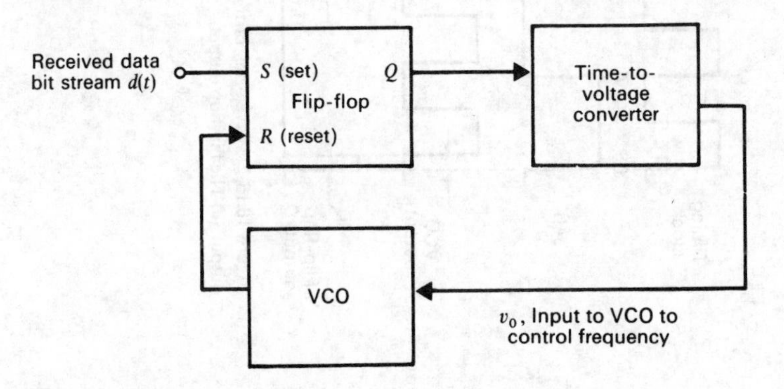

Figure 10.15-1 A bit synchronizer.

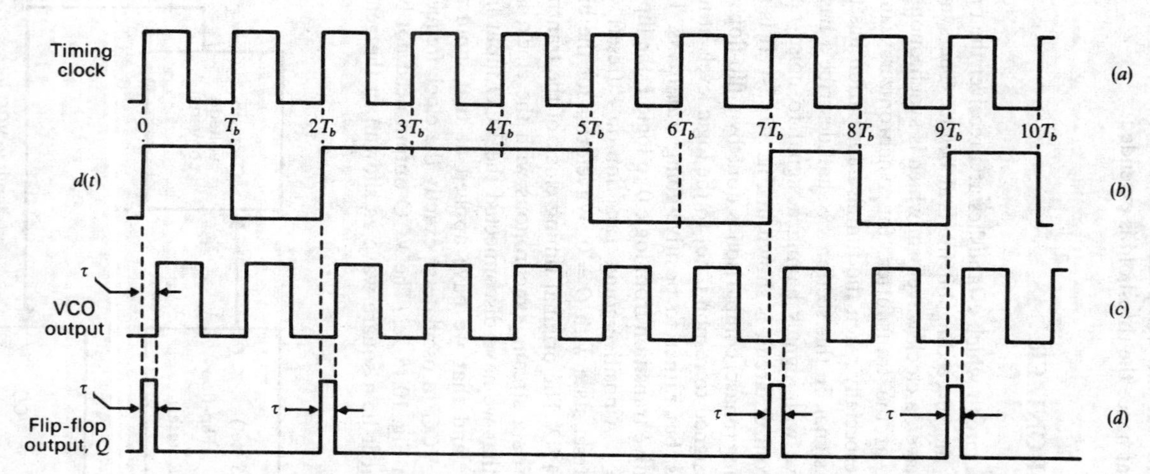

Figure 10.15-2 Waveforms of the bit synchronizer. (*a*) The transmitter timing clock. (*b*) A binary data bit stream. (*c*) The VCO output. (*d*) The flip-flop output.

the VCO frequency is equal to the clock frequency we shall be left with a timing offset τ between corresponding edges of clock and VCO waveforms. It is now easily verified that the output Q of the flip-flop is as appears in Fig. 10.15-2*d*. Whenever $d(t)$ makes an upward transition Q goes to logic 1 and Q is brought back to logic 0 by the next following upward transition in the VCO waveform. When a pulse appears at the flip-flop output it has a duration τ_Q which is equal to the offset τ. A plot of τ_Q as a function of τ is shown in Fig. 10.15-3.

The "time-to-voltage converter" shown in Fig. 10.15-1 is a circuit which, on receiving an input pulse from the flip-flop, generates a dc voltage at its output which is proportional to the pulse duration τ_Q and then (except as noticed below) holds its output v_o constant, until updated on the occasion of the next received pulse. Now let us arrange that when $\tau_Q = \tau = T_b/2$, the converter output v_o is exactly the voltage which causes the VCO to oscillate at the correct clock frequency, i.e., at the data bit rate $f_b = 1/T_b$. Under these circumstances, when the loop of Fig. 10.15-1 is closed it will find itself in a situation of equilibrium with an operating point as indicated on Fig. 10.15-3. For, if indeed, the VCO is oscillating at frequency f_b and if indeed $\tau = T_b/2$ then the converter output will provide exactly the voltage to keep the VCO frequency at f_b and, if the VCO frequency does not change there be no change in $\tau_Q = \tau$. Further, we may see that this equilibrium operating condition is one of *stable equilibrium.* For suppose, say, because of some parameter variation in a component of the VCO, or because of noise, causing an apparent change in the rate of the transmitter clock, the rate of the VCO should increase momentarily relative to the rate of the transmitter clock. Then successive edges of the VCO clock would occur progressively earlier in comparison with the edges of the original timing clock and $\tau_Q = \tau$ would become progressively smaller. As τ_Q decreases so also does v_o and correspondingly the frequency of the VCO would decrease momentarily thereby increasing τ. The correction would continue until the frequency of the VCO were returned to its original value. The PLL would then return to its equilibrium point. The operating point in Fig. 10.15-3 would therefore move downward. In the same manner, an increase in the VCO frequency would momentarily move the operating point upward. As the sensitivity of the VCO frequency to the offset τ is increased,

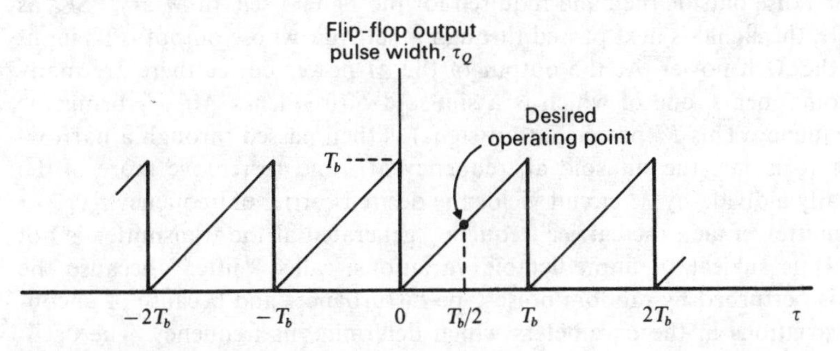

Figure 10.15-3 Plot of the VCO input as a function of τ.

the momentary shift of the operating point will become smaller. Note that with $\tau = T_b/2$, there is an edge of the VCO waveform which occurs at the time that marks the transition from one bit interval to the next. Setting $\tau = T_b/2$ also has the advantage that thereby we allow maximum latitude for shifts of the operating point in either direction.

If the pulses which are input to the converter occurred quite regularly we might well be able to replace the converter with a simple low-pass averaging filter. If, however, there is a long sequence of successive 1's or 0's in the data $d(t)$ then, correspondingly, during these sequences there will be no pulses. It is for this reason that we have provided to the converter the ability to hold its output, updating as each new pulse is received. Actually, such a scheme leaves something to be desired. For if we update at each pulse, the entire system will be excessively responsive to noise-induced fluctuations in the timing of the edges of the received data waveform. For this reason, the practice is to arrange that the converter generate and hold a voltage which is proportional to the weighted average of some number of the most recent pulses, the heavier weighting being given to the latest.

A discussion of the details of the circuitry of the "time-to-voltage converter" is not appropriate for this text. We shall note, however, that it generally involves, initially, an analog-to-digital converter to change the analog pulse duration data to digital form. Thereafter, the weighted averaging and holding is done by a digital filter. Finally the digital filter output is often processed by a digital-to-analog converter to make available an analog signal for application to the VCO. (It should be noted that many VCO's are capable of receiving a digital word rather than an analog voltage.)

10.16 CARRIER RECOVERY

As we have noted, in many communication schemes it is necessary to regenerate at the receiver a waveform which is synchronous with the transmitter carrier. The method we have described for achieving this carrier recovery is shown in Fig. 10.16-1. (We disregard temporarily the phase-locked loop and the dashed connections to it.) The receiver signal is initially passed through a bandpass filter to remove the noise outside the band required for the signal itself. In M-ary PSK, as an example, the signal is next passed through a network whose output is its input raised to the Mth power. At the output of the M-power device there are many spectral components, one of which is a sinusoid of frequency Mf_c, f_c being the carrier frequency. This M-power output signal is then passed through a narrowband filter to isolate the sinusoid at frequency Mf_c and to remove more of the noise. Finally a divide-by-M circuit yields the desired carrier of frequency f_c.

As a matter of fact, the carrier frequency generated at the transmitter is not constant. It is subject to unpredictable variations, called "jitter" because the oscillator is perturbed by random noise-type disturbances and because of uncontrollable variations in the parameters which determine its frequency. The oscillator frequency drifts back and forth through a frequency deviation $\pm\Delta f$. The

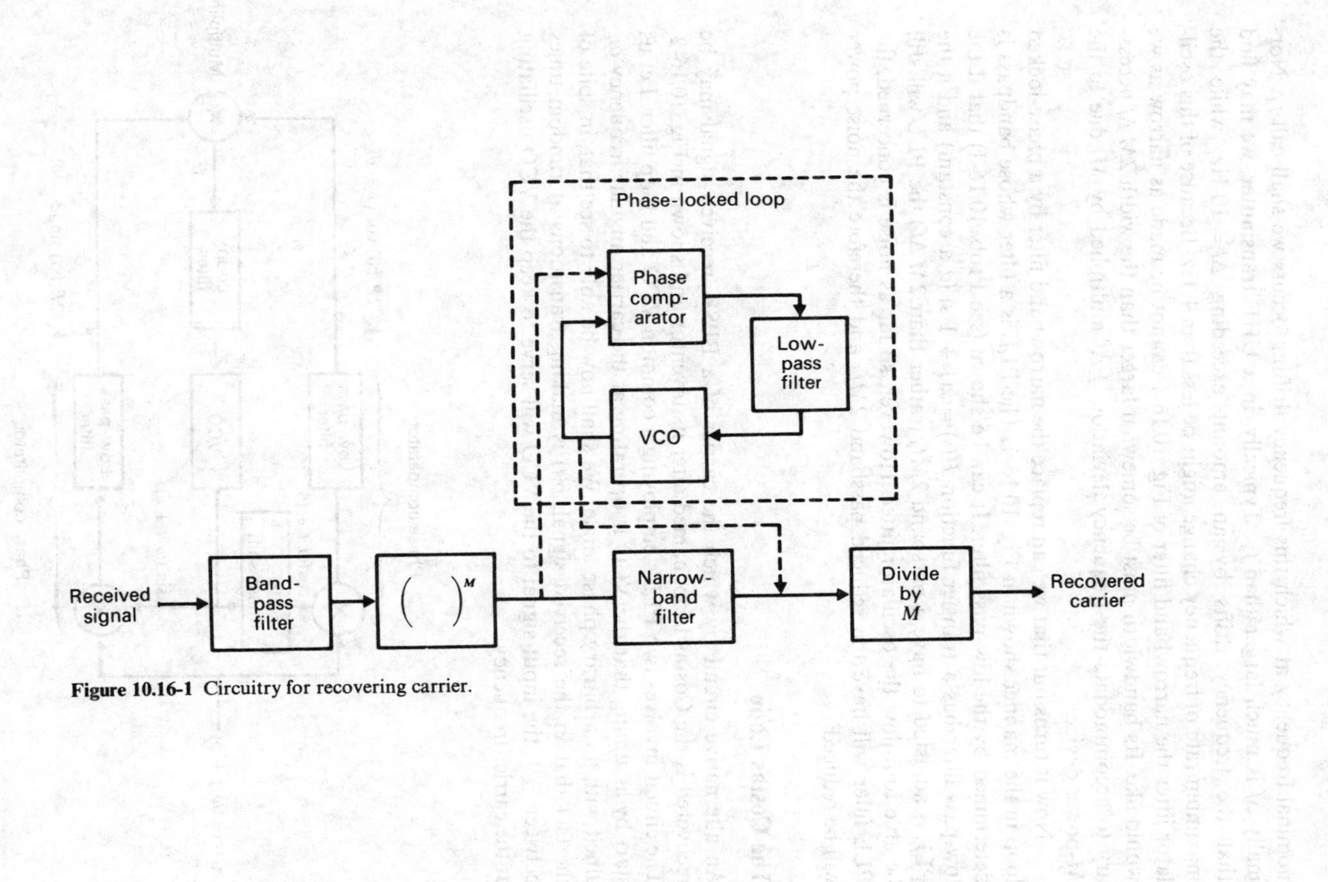

Figure 10.16-1 Circuitry for recovering carrier.

nominal frequency at which this frequency drifting occurs we shall call f_d. Normally Δf is much larger than f_d. Typically, in a UHF transmitter, we may find that the frequency drifts by an amount exceeding $\Delta f = 10$ Hz, while the maximum rate of frequency change might be less than 1 Hz. Because of this oscillator jitter the narrowband filter of Fig. 10.16-1 cannot be made as narrow as we would like. Its bandwidth must be somewhat larger than the width $2M\,\Delta f$ necessary to accommodate the frequency deviation $\pm\Delta f$ multiplied by M, due to the M-power device.

Now it turns out that we can replace the narrowband filter by a phase-locked loop in the manner shown in Fig. 10.16-1. The PLL is a filter whose bandpass is determined by the low-pass filter. It can be shown (see Prob. 10.16-1) that if the low-pass filter has a transfer function $H(s) = \alpha(1 + 1/s)$ (α a constant) and if the PLL is adjusted to have a passband $2Mf_d$ (rather than $2M\,\Delta f$) the PLL will still be able to follow the oscillator jitter. However, so far as noise is concerned, the PLL filter will have the reduced passband $2Mf_d$ and therefore the noise power will be reduced.

The Costas Loop

An alternative circuit by which the carrier of a BPSK received signal may be recovered is the Costas loop, named after its inventor and shown in Fig. 10.16-2. The circuit involves two PLL's employing a common VCO and loop filter. Let us start by assuming that the VCO is operating at the carrier angular frequency ω_c albeit with an arbitrary phase angle. We shall now be able to see that, in spite of the fact that in the received signal, $b(t)$ is making transitions at random times between ± 1, the input signal to the VCO will serve to keep the VCO oscillating at the carrier frequency.

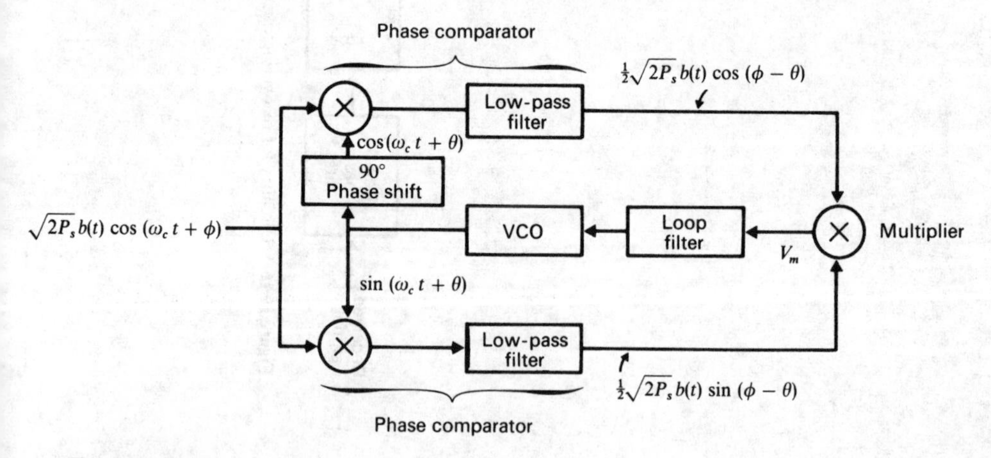

Figure 10.16-2 The Costas loop.

The low-pass filters remove the double frequency terms generated in the phase comparators and make available the waveforms $\frac{1}{2}\sqrt{2P_s}\, b(t) \cos(\phi - \theta)$ and $\frac{1}{2}\sqrt{2P_s}\, b(t) \sin(\phi - \theta)$ as shown. The multiplier output signal is then:

$$V_m = \tfrac{1}{4}(2P_s)b^2(t) \sin(\phi - \theta) \cos(\phi - \theta) \tag{10.16-1a}$$

$$= \frac{P_s}{4} \sin 2(\phi - \theta) \tag{10.16-1b}$$

Observe in Eq. (10.16-1) that, because of the multiplication, the $b(t)$ coefficient has dropped out. Further, we note that if some perturbation should cause the VCO frequency to differ from the carrier frequency, this frequency difference would manifest itself by a progressive change in the phase difference $\phi - \theta$. But a change in $(\phi - \theta)$ will cause a change in V_m which can now serve to increase or decrease the VCO frequency to maintain sychronism. Finally, we note that, as in the case in Fig. 10.16-1, when the PLL is employed, so in the present case, the required bandwidth required of the Costas loop to accommodate oscillator jitter is determined by f_d and not Δf.

REFERENCES

1. Rice, S. O.: "Time-series Analysis," chap. 25, John Wiley & Sons, Inc., New York, 1963.
2. Schilling, D. L., E. Nelson, and K. Clarke: Discriminator Response to an FM Signal in a Fading Channel, *IEEE Trans. Commun. Tech.*, April, 1967.
   Schilling, D. L., E. Hoffman, and E. Nelson: Error Rates For Digital Signals Demodulated by an FM Discriminator, *IEEE Trans. Commun. Tech.*, August, 1967.
   Nelson, E., and D. L. Schilling: The Response of an FM Discriminator to a Digital FM Signal in Randomly Fading Channels, *IEEE Trans. Commun. Tech.*, August, 1968.
3. Schilling, D. L.: The Response of an APC System to FM Signals and Noise, *Proc. IEEE*, October, 1963.
4. Schilling, D. L., and J. Billig: Threshold Extension Capability of the PLL and the FMFB, *Proc. IEEE*, May, 1964.
5. Osborne, P., and D. L. Schilling: Threshold Response of a Phase Locked Loop, *Proc. Int. Conf. Commun.*, 1968.
6. Hoffman, E., and D. L. Schilling: Threshold of the FMFB, *Proc. Int. Conf. Commun.*, 1969.

PROBLEMS

10.2-1. (*a*) A carrier, $A \cos 2\pi f_c t$, is accompanied by noise having quadrature components $x(t)$ and $y(t)$. Assume that for an interval from $t = 0$ to $t = 1/f$, $x(t)$ and $y(t)$ may be approximated by

$$x(t) = \frac{A}{2} [\sin 2\pi(f_c + f)t]$$

$$y(t) = \frac{A}{2} [\cos 2\pi(f_c + f)t - 1]$$

Draw the path of the resultant phasor $R(t)$ in the phasor diagram of Fig. 10.2-2. Make a qualitative plot of $\theta(t)$ and of $d\theta/dt$.

(b) Repeat (a) if f is replaced by $2f$ in the expression for $y(t)$.

(c) Repeat (a) if the functions describing $x(t)$ and $y(t)$ are interchanged.

(d) Repeat (c) if the amplitude A in the expression for $x(t)$ and $y(t)$ is replaced by $3A$.

10.2-2. Consider a random pulse train where each pulse has a duration $\tau = 2/B$ and an amplitude πB. If the average time between pulses is T, and if the pulse train is filtered by a rectangular low-pass filter of bandwidth f_M,

(a) Show that the power measured at the filter output is $(4\pi^2/T)(2f_M)$, if $f_M \ll B/2$.

(b) Obtain an expression for the power measured at the filter output when f_M is not much less than B.

10.3-1. A random pulse train consists of pulses given by the equation $p(t) = \sin Bt$, $0 \le t \le 2\pi/B$, separated by the average time interval T.

(a) Find $|P(f)|^2$.

(b) If the pulse train is filtered by a low-pass filter with a cutoff frequency $f_M \ll B$, show that the power measured at the filter output is very small in comparison with the power of the pulse train.

10.4-1. Threshold is defined as the value of $S_i/\eta f_M$ which results in S_o/N_o decreasing by 1 dB from the value

$$\frac{S_o}{N_o} = \frac{3k^2\overline{m^2}}{4\pi^2 f_M^2}\frac{S_i}{\eta f_M}$$

(a) Show that when the carrier is unmodulated, threshold is reached when

$$0.26 = \sqrt{3}\,\frac{B}{f_M}\frac{S_i}{\eta f_M}\operatorname{erfc}\sqrt{\frac{f_M}{B}\frac{S_i}{\eta f_M}}$$

(b) Plot $S_i/\eta f_M$ at threshold as a function of B/f_M.

10.6-1. Consider that the signal and noise at the input to an FM system are

$$v_i(t) = \cos(\omega_c t + \Omega t) + x(t)\cos\omega_0 t - y(t)\sin\omega_0 t$$

and that

$$G_x(f) = G_y(f) = \eta \qquad -\frac{B}{2} \le f \le +\frac{B}{2}$$

(a) Show that $v_i(t)$ can also be written as

$$v_i(t) = \cos(\omega_c t + \Omega t) + x'(t)\cos(\omega_c t + \Omega t) + y'(t)\sin(\omega_c t + \Omega t)$$

Find $x'(t)$ and $y'(t)$.

(b) Find $G_{x'}(f) = G_{y'}(f)$. Show that if $\Omega = B/2$, the spectral density of x' (and y') extends to B Hz. Comment on the significance of this result with regard to the duration of a spike.

(c) Find $E[x'(t)y'(t)]$, i.e., show that even though x and y are uncorrelated, x' and y' are correlated.

10.6-2. Using the results of Prob. 10.6-1 and Eqs. (10.5-1) and (10.5-2),

(a) Show that a negative spike occurs when $x' < 1$, y' goes through 0, and $\dot{y}' > 0$.

(b) Show also that a positive spike occurs when $x' < -1$, y' goes through 0, and $\dot{y}' < 0$.

(c) Using Eq. (10.5-4) with x replaced by x', and y by y', derive Eq. (10.6-1), where $\delta f = \Omega/2\pi$.

10.6-3. An FM carrier is modulated by the signal $m(t)$, which is gaussian, and bandlimited to f_M Hz. The instantaneous frequency deviation produced by $m(t)$ is $km(t)$. If the mean value of the square of $km(t)$ is $\overline{(\Delta\omega)^2}$, find $\overline{|\delta f|}$.

10.6-4. An FM carrier is sinusoidally modulated as in Eq. (10.6-2). Show, using Eqs. (10.6-4) and (10.5-19), that $\overline{\delta N} \gg N_c$.

10.6-5. Verify Eq. (10.6-6).

10.6-6. (*a*) Plot the ratio $S_i/\eta f_M$ at threshold as a function of β [use Eq. (10.6-6)].

(*b*) Compare your results with those found in Prob. 10.4-1*b*.

10.6-7. The FM signal $v(t) = \cos [\omega_c t + k \int_{-\infty}^{t} m(\lambda)\, d\lambda]$ is demodulated using an FM discriminator. If the signal $m(t)$ is gaussian and is bandlimited to f_M Hz, and $S_i/\eta f_M = 20$ dB, find the maximum rms deviation possible so that the discriminator is operating at threshold.

10.7-1. Show that if the two inputs to a phase comparator are $V_1(t) = A \cos [\omega_c t + \theta_1(t)]$ and $V_2(t) = B \cos [\omega_c t + \theta_2(t)]$, the output of the phase comparator is $v_o(t) = (AB/2) \cos [\theta_1(t) - \theta_2(t)] + (AB/2) \cos [2\omega_c t + \theta_1(t) + \theta_2(t)]$.

10.7-2. Show that Eq. (10.7-2) describes the output of a phase comparator when its input consists of "square" waves rather than sine waves.

10.8-1. A first-order PLL, with a phase-comparator characteristic as shown in Fig. 10.7-2*b*, is initially adjusted so that in the presence of a carrier alone $\psi = 0$. The VCO has the property that when its input is changed by 1V, its frequency changes by 10^5 Hz. The time constant of the PLL is $\tau = 10^{-4}$ s.

(*a*) The carrier is frequency-modulated by a sinusoidal waveform. What is the maximum allowable peak frequency deviation of the carrier if the PLL is to recover the modulating waveform without distortion?

(*b*) With the frequency deviation at its peak as in (*a*), make a plot of the PLL output voltage v_0 as a function of the frequency of the modulating waveform.

10.8-2. Derive Eq. (10.8-6).

10.9-1. Derive Eq. (10.9-2).

10.9-2. A first-order PLL has a phase-comparator characteristic as in Fig. 10.7-2*b*. The input carrier is unmodulated and the PLL is adjusted so that it is in equilibrium with $\psi = 0$. By the application of external constraints, the operating point is forced to $\psi = 3\pi/4$. At the time $t = 0$, these constraints are removed.

(*a*) Write the differential equation for the phase angle ψ.

(*b*) If $\tau = 10^{-3}$ s, calculate the time at which the operating point reaches *B* in Fig. 10.7-2*b*. Calculate $d\psi/dt$ at the time the constraints are released and at the time point *B* is reached.

10.10-1. A first-order PLL has a phase-comparator characteristic as in Fig. 10.7-2*b*. The PLL is adjusted so that, in the presence of a carrier alone, $\psi = 0$. The time constant of the PLL is $\tau = 10^{-4}$ s. At time $t = 0$ the carrier is abruptly offset by 5 kHz. Plot ψ as a function of time up to the point where ψ attains the value $\psi = 3\pi/2$.

10.10-2. A first-order PLL has a phase characteristic as in Fig. 10.7-2*b*. The PLL is adjusted so that in the presence of a carrier alone $\psi = -\pi/4$. The time constant of the PLL is $\tau = 5 \times 10^{-5}$ s and the sensitivity of the VCO is such that a 1V input changes its frequency by 10 kHz. At time $t = 0$ the carrier is abruptly offset by 7.5 kHz (in the direction to increase ψ) for the interval $0 \le t \le 10^{-4}$ s. Draw a plot of the output voltage waveform of the PLL.

10.10-3. A first-order PLL has a phase characteristic as in Fig. 10.7-2*b*. In the presence of a carrier alone $\psi = 0$. The time constant is $\tau = 10^{-3}$ s and $G_0 = 1$ kHz/volt. At time $t = 0$ the carrier is offset by 0.5 kHz for the interval $0 \le t \le \tau$, i.e., the frequency of the carrier is modulated by a pulse of duration τ.

(*a*) How long must the pulse last (say $\tau = \tau_0$) if at the end of the pulse $\psi = \pi$ (in which case the subsequent behavior of the PLL is indeterminate).

(*b*) Assume $\tau = 0.9\tau_0$. Draw the waveform at the output of the PLL.

(*c*) Assume $\tau = 1.1\tau_0$. Draw the waveform at the output of the PLL.

10.10-4. A first-order PLL is used to demodulate the signal $\cos [\omega_c t + \Omega t - \phi_s(t)]$. The VCO is initially operating at a frequency ω_c. Assume $\Omega = \pi/4\tau$.

(*a*) Find $v_0(t)$ when $\phi_s(t) = 0$.

(*b*) If, after the PLL has reached equilibrium, $\phi_s(t) = -\lambda t$, $0 \le t \le 2\pi/\lambda$ and zero elsewhere, find the minimum value of the product $\lambda\tau$ to *avoid* a spike.

(*c*) If $\phi_s(t) = +\lambda t$, $0 \le t \le 2\pi/\lambda$ and zero elsewhere, find the minimum value of the $\lambda\tau$ product to *avoid* a spike.

10.11-1. Find the differential equation of the piecewise-linear second-order PLL in the region in which $\pi/2 \leq \psi \leq 3\pi/2$.

10.11-2. Find the differential equation of the second-order PLL if the phase comparator has a sinusoidal characteristic.

10.12-1. Show that Fig. 10.12-1 represents the first-order PLL.

10.12-2. Plot N/f_M versus $S_i/\eta f_M$ at threshold. This is called the *threshold hyperbola*. If $S_i/\eta f_M = 20$ dB and $f_M = 5$ kHz, find N to produce threshold.

10.14-1. Show that the ratio $S_i/\eta f_M$ is unchanged when measured after the carrier filter or the bandpass filter shown in Fig. 10.13-1.

10.14-2. Show that for the FMFB, $\delta N \gg N_c$. Use Eq. (10.14-6) and plot $N_c/\delta N$ as a function of $S_i/\eta B_p$ with $[(1 + \alpha G_0)B_p]/\Delta f = 4$ (this corresponds to choosing $1 + \alpha G_0 = \beta$, which represents reasonably good design).

10.14-3. Find the threshold extension possible for $\beta = 3$ and 5, if $1 + \alpha G_0 = \beta$.

10.15-1. Refer to Fig. 10.15-3. Show that if, due to noise, τ_Q is momentarily displaced to $\tau = 3T_b/4$ or $T_b/4$, then when the disturbance subsides, τ_Q will return to $\tau_Q = T_b/2$.

10.15-2. Design a *time-to-voltage* converter having the characteristics that 20 past samples are used to compute v_o and that all samples receive equal weight.

10.16-1. Show that if $H(s) = \alpha(1 + 1/s)$ and the PLL is adjusted to have a bandwidth $2Mf_d$, the PLL will be able to follow jitter occurring at the rate f_d with peak deviations less than Δf.

10.16-2. Repeat Prob. 10.16-1 for the Costas loop shown in Fig. 10.16-2.

CHAPTER

ELEVEN

DATA TRANSMISSION

A data transmission system using binary encoding transmits a sequence of binary digits, that is, 1's and 0's. These digits may be represented in a number of ways. For example, a 1 may be represented by a voltage V held for a time T, while a zero is represented by a voltage $-V$ held for an equal time. In general the binary digits are encoded so that a 1 is represented by a signal $s_1(t)$ and a 0 by a signal $s_2(t)$, where $s_1(t)$ and $s_2(t)$ each have a duration T. The resulting signal may be transmitted directly or, as is more usually the case, used to modulate a carrier. The received signal is corrupted by noise, and hence there is a finite probability that the receiver will make an error in determining, within each time interval, whether a 1 or a 0 was transmitted.

In this chapter we make calculations of such error probabilities and discuss methods to minimize them. The discussion will lead us to the concept of the *matched filter* and *correlator*.

11.1 A BASEBAND SIGNAL RECEIVER

Consider that a binary-encoded signal consists of a time sequence of voltage levels $+V$ or $-V$. If there is a guard interval between the bits, the signal forms a sequence of positive and negative pulses. In either case there is no particular interest in preserving the waveform of the signal after reception. We are interested only in knowing within each bit interval whether the transmitted voltage was $+V$ or $-V$. With noise present, the received signal and noise together will yield sample values generally different from $\pm V$. In this case, what deduction shall we make from the sample value concerning the transmitted bit?

Suppose that the noise is gaussian and therefore the noise voltage has a probability density which is entirely symmetrical with respect to zero volts. Then the probability that the noise has increased the sample value is the same as the probability that the noise has decreased the sample value. It then seems entirely reasonable that we can do no better than to assume that if the sample value is positive the transmitted level was $+V$, and if the sample value is negative the transmitted level was $-V$. It is, of course, possible that at the sampling time the noise voltage may be of magnitude larger than V and of a polarity opposite to the polarity assigned to the transmitted bit. In this case an error will be made as indicated in Fig. 11.1-1. Here the transmitted bit is represented by the voltage $+V$ which is sustained over an interval T from t_1 to t_2. Noise has been superimposed on the level $+V$ so that the voltage v represents the received signal and noise. If now the sampling should happen to take place at a time $t = t_1 + \Delta t$, an error will have been made.

We can reduce the probability of error by processing the received signal plus noise in such a manner that we are then able to find a sample time where the sample voltage due to the signal is emphasized relative to the sample voltage due to the noise. Such a processer (receiver) is shown in Fig. 11.1-2. The signal input during a bit interval is indicated. As a matter of convenience we have set $t = 0$ at the beginning of the interval. The waveform of the signal $s(t)$ before $t = 0$ and after $t = T$ has not been indicated since, as will appear, the operation of the receiver during each bit interval is independent of the waveform during past and future bit intervals.

The signal $s(t)$ with added white gaussian noise $n(t)$ of power spectral density $\eta/2$ is presented to an integrator. At time $t = 0+$ we require that capacitor C be uncharged. Such a discharged condition may be ensured by a brief closing of switch SW_1 at time $t = 0-$, thus relieving C of any charge it may have acquired during the previous interval. The sample is taken at the output of the integrator by closing this sampling switch SW_2. This sample is taken at the end of the bit interval, at $t = T$. The signal processing indicated in Fig. 11.1-2 is described by the phrase *integrate and dump*, the term *dump* referring to the abrupt discharge of the capacitor after each sampling.

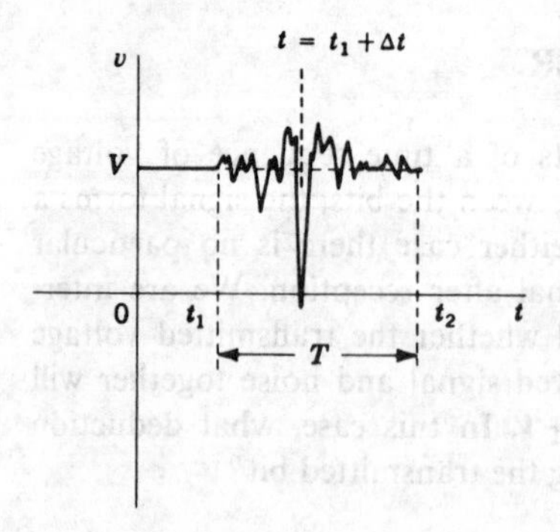

Figure 11.1-1 Illustration that noise may cause an error in the determination of a transmitted voltage level.

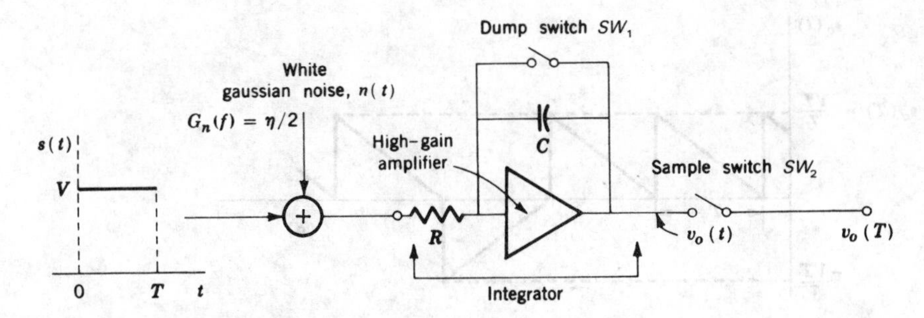

Figure 11.1-2 A receiver for a binary coded signal.

Peak Signal to RMS Noise Output Voltage Ratio

The integrator yields an output which is the integral of its input multiplied by $1/RC$. Using $\tau = RC$, we have

$$v_o(T) = \frac{1}{\tau}\int_0^T [s(t) + n(t)]\,dt = \frac{1}{\tau}\int_0^T s(t)\,dt + \frac{1}{\tau}\int_0^T n(t)\,dt \tag{11.1-1}$$

The sample voltage due to the signal is

$$s_o(T) = \frac{1}{\tau}\int_0^T V\,dt = \frac{VT}{\tau} \tag{11.1-2}$$

The sample voltage due to the noise is

$$n_o(T) = \frac{1}{\tau}\int_0^T n(t)\,dt \tag{11.1-3}$$

This noise-sampling voltage $n_o(T)$ is a gaussian random variable in contrast with $n(t)$, which is a gaussian random process.

The variance of $n_o(T)$ was found in Sec. 7.9 [see Eq. (7.9-17)] to be

$$\sigma_o^2 = \overline{n_o^2(T)} = \frac{\eta T}{2\tau^2} \tag{11.1-4}$$

and, as noted in Sec. 7.3, $n_o(T)$ has a gaussian probability density.

The output of the integrator, before the sampling switch, is $v_o(t) = s_o(t) + n_o(t)$. As shown in Fig. 11.1-3*a*, the signal output $s_o(t)$ is a ramp, in each bit interval, of duration T. At the end of the interval the ramp attains the voltage $s_o(T)$ which is $+VT/\tau$ or $-VT/\tau$, depending on whether the bit is a 1 or a 0. At the end of each interval the switch SW_1 in Fig. 11.1-2 closes momentarily to discharge the capacitor so that $s_o(t)$ drops to zero. The noise $n_o(t)$, shown in Fig. 11.1-3*b*, also starts each interval with $n_o(0) = 0$ and has the random value $n_o(T)$ at the end of each interval. The sampling switch SW_2 closes briefly just before the closing of SW_1 and hence reads the voltage

$$v_o(T) = s_o(T) + n_o(T) \tag{11.1-5}$$

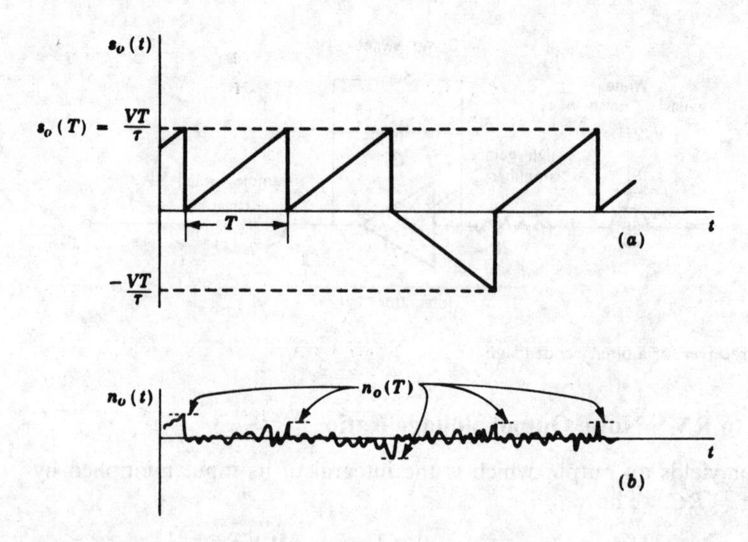

Figure 11.1-3 (*a*) The signal output and (*b*) the noise output of the integrator of Fig. 11.1-2.

We would naturally like the output signal voltage to be as large as possible in comparison with the noise voltage. Hence a figure of merit of interest is the signal-to-noise ratio

$$\frac{[s_o(T)]^2}{[n_o(T)]^2} = \frac{2}{\eta} V^2 T \tag{11.1-6}$$

This result is calculated from Eqs. (11.1-2) and (11.1-4). Note that the signal-to-noise ratio increases with increasing bit duration T and that it depends on V^2T which is the normalized energy of the bit signal. Therefore, a bit represented by a narrow, high amplitude signal and one by a wide, low amplitude signal are equally effective, provided V^2T is kept constant.

It is instructive to note that the integrator filters the signal and the noise such that the signal voltage increases linearly with time, while the standard deviation (rms value) of the noise increases more slowly, as $\sqrt{T}$. Thus, the integrator enhances the signal relative to the noise, and this enhancement increases with time as shown in Eq. (11.1-6).

11.2 PROBABILITY OF ERROR

Since the function of a receiver of a data transmission is to distinguish the bit 1 from the bit 0 in the presence of noise, a most important characteristic is the probability that an error will be made in such a determination. We now calculate this error probability P_e for the integrate-and-dump receiver of Fig. 11.1-2.

We have seen that the probability density of the noise sample $n_o(T)$ is gaussian and hence appears as in Fig. 11.2-1. The density is therefore given by

$$f[n_o(T)] = \frac{e^{-n_o^2(T)/2\sigma_o^2}}{\sqrt{2\pi\sigma_o^2}} \tag{11.2-1}$$

where σ_o^2, the variance, is $\sigma_o^2 \equiv \overline{n_o^2(T)}$ given by Eq. (11.1-4). Suppose, then, that during some bit interval the input-signal voltage is held at, say, $-V$. Then, at the sample time, the signal sample voltage is $s_o(T) = -VT/\tau$, while the noise sample is $n_o(T)$. If $n_o(T)$ is positive and larger in magnitude than VT/τ, the total sample voltage $v_o(T) = s_o(T) + n_o(T)$ will be positive. Such a positive sample voltage will result in an error, since as noted earlier, we have instructed the receiver to interpret such a positive sample voltage to mean that the signal voltage was $+V$ during the bit interval. The probability of such a misinterpretation, that is, the probability that $n_o(T) > VT/\tau$, is given by the area of the shaded region in Fig. 11.2-1. The probability of error is, using Eq. (11.2-1).

$$P_e = \int_{VT/\tau}^{\infty} f[n_o(T)]\, dn_o(T) = \int_{VT/\tau}^{\infty} \frac{e^{-n_o^2(T)/2\sigma_o^2}}{\sqrt{2\pi\sigma_o^2}}\, dn_o(T) \tag{11.2-2}$$

Defining $x \equiv n_o(T)/\sqrt{2}\sigma_o$, and using Eq. (11.1-4), Eq. (11.2-2) may be rewritten as

$$P_e = \frac{1}{2}\frac{2}{\sqrt{\pi}}\int_{x=V\sqrt{T/\eta}}^{\infty} e^{-x^2}\, dx$$

$$= \frac{1}{2}\operatorname{erfc}\left(V\sqrt{\frac{T}{\eta}}\right) = \frac{1}{2}\operatorname{erfc}\left(\frac{V^2T}{\eta}\right)^{1/2} = \frac{1}{2}\operatorname{erfc}\left(\frac{E_s}{\eta}\right)^{1/2} \tag{11.2-3}$$

in which $E_s = V^2T$ is the signal energy of a bit.

If the signal voltage were held instead at $+V$ during some bit interval, then it is clear from the symmetry of the situation that the probability of error would again be given by P_e in Eq. (11.2-3). Hence Eq. (11.2-3) gives P_e quite generally.

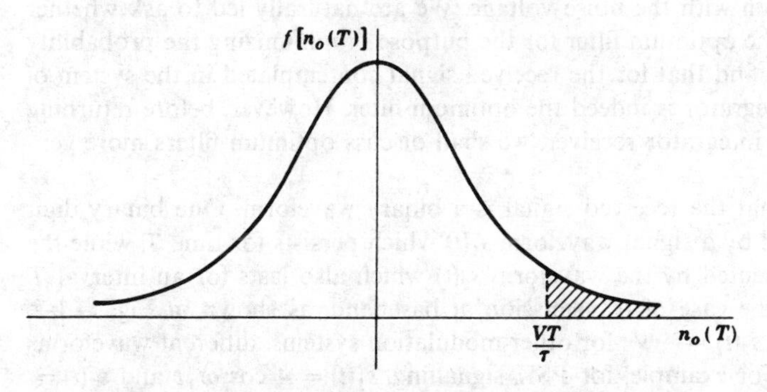

Figure 11.2-1 The gaussian probability density of the noise sample $n_o(T)$.

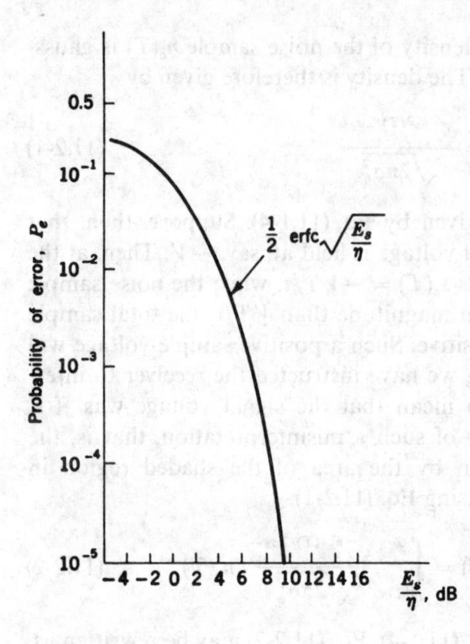

Figure 11.2-2 Variation of P_e versus E_s/η.

The probability of error P_e, as given in Eq. (11.2-3), is plotted in Fig. 11.2-2. Note that P_e decreases rapidly as E_s/η increases. The maximum value of P_e is $\frac{1}{2}$. Thus, even if the signal is entirely lost in the noise so that any determination of the receiver is a sheer guess, the receiver cannot be wrong more than half the time on the average.

11.3 THE OPTIMUM FILTER

In the receiver system of Fig. 11.1-2, the signal was passed through a filter (i.e., the integrator), so that at the sampling time the signal voltage might be emphasized in comparison with the noise voltage. We are naturally led to ask whether the integrator is the optimum filter for the purpose of minimizing the probability of error. We shall find that for the received signal contemplated in the system of Fig. 11.1-2 the integrator is indeed the optimum filter. However, before returning specifically to the integrator receiver, we shall discuss optimum filters more generally.

We assume that the received signal is a binary waveform. One binary digit (bit) is represented by a signal waveform $s_1(t)$ which persists for time T, while the other bit is represented by the waveform $s_2(t)$ which also lasts for an interval T. For example, in the case of transmission at baseband, as shown in Fig. 11.1-2, $s_1(t) = +V$, while $s_2(t) = -V$; for other modulation systems, different waveforms are transmitted. For example, for PSK signalling, $s_1(t) = A \cos \omega_0 t$ and $s_2(t) = -A \cos \omega_0 t$; while for FSK, $s_1(t) = A \cos (\omega_0 + \Omega)t$ and $s_2(t) = A \cos (\omega_0 - \Omega)t$.

As shown in Fig. 11.3-1 the input, which is $s_1(t)$ or $s_2(t)$, is corrupted by the addition of noise $n(t)$. The noise is gaussian and has a spectral density $G(f)$. [In most cases of interest the noise is white, so that $G(f) = \eta/2$. However, we shall assume the more general possibility, since it introduces no complication to do so.] The signal and noise are filtered and then sampled at the end of each bit interval. The output sample is either $v_o(T) = s_{o1}(T) + n_o(T)$ or $v_o(T) = s_{o2}(T) + n_o(T)$. We assume that immediately after each sample, every energy-storing element in the filter has been discharged.

We have already considered in Sec. 2.22, the matter of signal determination in the presence of noise. Thus, we note that in the absence of noise the output sample would be $v_o(T) = s_{o1}(T)$ or $s_{o2}(T)$. When noise is present we have shown that to minimize the probability of error one should assume that $s_1(t)$ has been transmitted if $v_o(T)$ is *closer* to $s_{o1}(T)$ than to $s_{o2}(T)$. Similarly, we assume $s_2(t)$ has been transmitted if $v_o(T)$ is *closer* to $s_{o2}(T)$. The decision boundary is therefore midway between $s_{o1}(T)$ and $s_{o2}(T)$. For example, in the baseband system of Fig. 11.1-2, where $s_{o1}(T) = VT/\tau$ and $s_{o2}(T) = -VT/\tau$, the decision boundary is $v_o(T) = 0$. In general, we shall take the decision boundary to be

$$v_o(T) = \frac{s_{o1}(T) + s_{o2}(T)}{2} \tag{11.3-1}$$

The probability of error for this general case may be deduced as an extension of the considerations used in the baseband case. Suppose that $s_{o1}(T) > s_{o2}(T)$ and that $s_2(t)$ was transmitted. If, at the sampling time, the noise $n_o(T)$ is positive and larger in magnitude than the voltage difference $\frac{1}{2}[s_{o1}(T) + s_{o2}(T)] - s_{o2}(T)$, an error will have been made. That is, an error [we decide that $s_1(t)$ is transmitted rather than $s_2(t)$] will result if

$$n_o(T) \geq \frac{s_{o1}(T) - s_{o2}(T)}{2} \tag{11.3-2}$$

Hence the probability of error is

$$P_e = \int_{[s_{o1}(T) - s_{o2}(T)]/2}^{\infty} \frac{e^{-n_o^2(T)/2\sigma_o^2}}{\sqrt{2\pi\sigma_o^2}}\, dn_o(T) \tag{11.3-3}$$

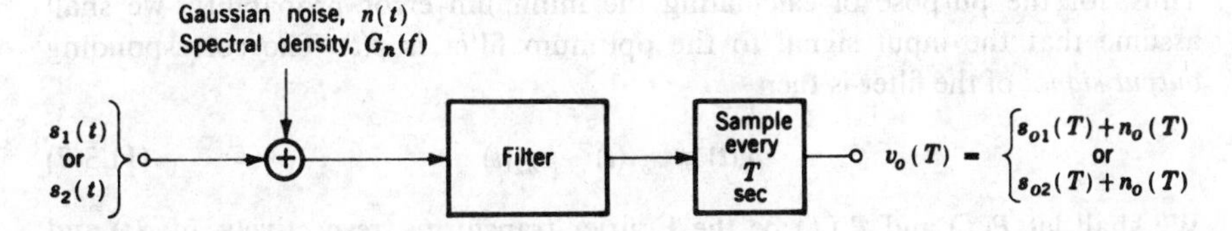

Figure 11.3-1 A receiver for binary coded signalling.

If we make the substitution $x \equiv n_o(T)/\sqrt{2}\sigma_o$, Eq. (11.3-3) becomes

$$P_e = \frac{1}{2}\frac{2}{\sqrt{\pi}}\int_{[s_{o1}(T)-s_{o2}(T)]/2\sqrt{2}\sigma_o}^{\infty} e^{-x^2}\,dx \tag{11.3-4a}$$

$$P_e = \frac{1}{2}\operatorname{erfc}\left[\frac{s_{o1}(T)-s_{o2}(T)}{2\sqrt{2}\sigma_o}\right] \tag{11.3-4b}$$

Note that for the case $s_{o1}(T) = VT/\tau$ and $s_{o2}(T) = -VT/\tau$, and, using Eq. (11.1-4), Eq. (11.3-4*b*) reduces to Eq. (11.2-3) as expected.

The complementary error function is a monotonically decreasing function of its argument. (See Fig. 11.2-2.) Hence, as is to be anticipated, P_e decreases as the difference $s_{o1}(T) - s_{o2}(T)$ becomes larger and as the rms noise voltage σ_o becomes smaller. The optimum filter, then, is the filter which maximizes the ratio

$$\gamma = \frac{s_{o1}(T)-s_{o2}(T)}{\sigma_o} \tag{11.3-5}$$

We now calculate the transfer function $H(f)$ of this optimum filter. As a matter of mathematical convenience we shall actually maximize γ^2 rather than γ.

Calculation of the Optimum-Filter Transfer Function $H(f)$

The fundamental requirement we make of a binary encoded data receiver is that it distinguishes the voltages $s_1(t) + n(t)$ and $s_2(t) + n(t)$. We have seen that the ability of the receiver to do so depends on how large a particular receiver can make γ. It is important to note that γ is proportional not to $s_1(t)$ nor to $s_2(t)$, but rather to the *difference* between them. For example, in the baseband system we represented the signals by voltage levels $+V$ and $-V$. But clearly, if our only interest was in distinguishing levels, we would do just as well to use $+2$ volts and 0 volt, or $+8$ volts and $+6$ volts, etc. (The $+V$ and $-V$ levels, however, have the advantage of requiring the least average power to be transmitted.) Hence, while $s_1(t)$ or $s_2(t)$ is the received signal, the signal which is to be compared with the noise, i.e., the signal which is relevant in all our error-probability calculations, is the difference signal

$$p(t) \equiv s_1(t) - s_2(t) \tag{11.3-6}$$

Thus, for the purpose of calculating the minimum error probability, we shall assume that the input signal to the optimum filter is $p(t)$. The corresponding *output signal* of the filter is then

$$p_o(t) \equiv s_{o1}(t) - s_{o2}(t) \tag{11.3-7}$$

We shall let $P(f)$ and $P_o(f)$ be the Fourier transforms, respectively, of $p(t)$ and $p_o(t)$.

If $H(f)$ is the transfer function of the filter,

$$P_o(f) = H(f)P(f) \tag{11.3-8}$$

and

$$p_o(T) = \int_{-\infty}^{\infty} P_o(f)e^{j2\pi fT}\, df = \int_{-\infty}^{\infty} H(f)P(f)e^{j2\pi fT}\, df \tag{11.3-9}$$

The input noise to the optimum filter is $n(t)$. The output noise is $n_o(t)$ which has a power spectral density $G_{n_o}(f)$ and is related to the power spectral density of the input noise $G_n(f)$ by

$$G_{n_o}(f) = |H(f)|^2 G_n(f) \tag{11.3-10}$$

Using Parseval's theorem (Eq. 1.13-5), we find that the normalized output noise power, i.e., the noise variance σ_o^2, is

$$\sigma_o^2 = \int_{-\infty}^{\infty} G_{n_o}(f)\, df = \int_{-\infty}^{\infty} |H(f)|^2 G_n(f)\, df \tag{11.3-11}$$

From Eqs. (11.3-9) and (11.3-11) we now find that

$$\gamma^2 = \frac{p_o^2(T)}{\sigma_o^2} = \frac{|\int_{-\infty}^{\infty} H(f)P(f)e^{j2\pi Tf}\, df|^2}{\int_{-\infty}^{\infty} |H(f)|^2 G_n(f)\, df} \tag{11.3-12}$$

Equation (11.3-12) is unaltered by the inclusion or deletion of the absolute value sign in the numerator since the quantity within the magnitude sign $p_o(T)$ is a positive real number. The sign has been included, however, in order to allow further development of the equation through the use of the *Schwarz inequality*.

The *Schwarz inequality* states that given arbitrary complex functions $X(f)$ and $Y(f)$ of a common variable f, then

$$\left|\int_{-\infty}^{\infty} X(f)Y(f)\, df\right|^2 \le \int_{-\infty}^{\infty} |X(f)|^2\, df \int_{-\infty}^{\infty} |Y(f)|^2\, df \tag{11.3-13}$$

The equal sign applies when

$$X(f) = KY^*(f) \tag{11.3-14}$$

where K is an arbitrary constant and $Y^*(f)$ is the complex conjugate of $Y(f)$.

We now apply the Schwarz inequality to Eq. (11.3-12) by making the identification

$$X(f) \equiv \sqrt{G_n(f)}\, H(f) \tag{11.3-15}$$

and

$$Y(f) \equiv \frac{1}{\sqrt{G_n(f)}} P(f)e^{j2\pi Tf} \tag{11.3-16}$$

Using Eqs. (11.3-15) and (11.3-16) and using the Schwarz inequality, Eq. (11.3-13), we may rewrite Eq. (11.3-12) as

$$\frac{p_o^2(T)}{\sigma_o^2} = \frac{|\int_{-\infty}^{\infty} X(f)Y(f)\, df|^2}{\int_{-\infty}^{\infty} |X(f)|^2\, df} \le \int_{-\infty}^{\infty} |Y(f)|^2\, df \tag{11.3-17}$$

or, using Eq. (11.3-16),

$$\frac{p_o^2(T)}{\sigma_o^2} \le \int_{-\infty}^{\infty} |Y(f)|^2\, df = \int_{-\infty}^{\infty} \frac{|P(f)|^2}{G_n(f)}\, df \tag{11.3-18}$$

The ratio $p_o^2(T)/\sigma_o^2$ will attain its maximum value when the equal sign in Eq. (11.3-18) may be employed as is the case when $X(f) = KY^*(f)$. We then find from Eqs. (11.3-15) and (11.3-16) that the optimum filter which yields such a maximum ratio $p_o^2(T)/\sigma_o^2$ has a transfer function

$$H(f) = K\frac{P^*(f)}{G_n(f)}e^{-j2\pi fT} \tag{11.3-19}$$

Correspondingly, the maximum ratio is, from Eq. (11.3-18),

$$\left[\frac{p_o^2(T)}{\sigma_o^2}\right]_{\max} = \int_{-\infty}^{\infty} \frac{|P(f)|^2}{G_n(f)}\, df \tag{11.3-20}$$

In succeeding sections we shall have occasion to apply Eqs. (11.13-19) and (11.13-20) to a number of cases of interest.

11.4 WHITE NOISE: THE MATCHED FILTER

An optimum filter which yields a maximum ratio $p_o^2(T)/\sigma_o^2$ is called a *matched filter* when the input noise is *white*. In this case $G_n(f) = \eta/2$, and Eq. (11.3-19) becomes

$$H(f) = K\frac{P^*(f)}{\eta/2}e^{-j2\pi fT} \tag{11.4-1}$$

The impulsive response of this filter, i.e., the response of the filter to a unit strength impulse applied at $t = 0$, is

$$h(t) = \mathscr{F}^{-1}[H(f)] = \frac{2K}{\eta}\int_{-\infty}^{\infty} P^*(f)e^{-j2\pi fT}e^{j2\pi ft}\, df \tag{11.4-2a}$$

$$= \frac{2K}{\eta}\int_{-\infty}^{\infty} P^*(f)e^{j2\pi f(t-T)}\, df \tag{11.4-2b}$$

A physically realizable filter will have an impulse response which is real, i.e., not complex. Therefore $h(t) = h^*(t)$. Replacing the right-hand member of Eq. (11.4-2*b*) by its complex conjugate, an operation which leaves the equation unaltered, we have

$$h(t) = \frac{2K}{\eta}\int_{-\infty}^{\infty} P(f)e^{j2\pi f(T-t)}\, df \tag{11.4-3a}$$

$$= \frac{2K}{\eta}\, p(T-t) \tag{11.4-3b}$$

Finally, since $p(t) \equiv s_1(t) - s_2(t)$ [see Eq. (11.3-6)], we have

$$h(t) = \frac{2K}{\eta}[s_1(T-t) - s_2(T-t)] \tag{11.4-4}$$

The significance of these results for the matched filter may be more readily appreciated by applying them to a specific example. Consider then, as in Fig. 11.4-1*a*, that $s_1(t)$ is a triangular waveform of duration T, while $s_2(t)$, as shown in Fig. 11.4-1*b*, is of identical form except of reversed polarity. Then $p(t)$ is as shown in Fig. 11.4-1*c*, and $p(-t)$ appears in Fig. 11.4-1*d*. The waveform $p(-t)$ is the waveform $p(t)$ rotated around the axis $t = 0$. Finally, the waveform $p(T - t)$ called for as the impulse response of the filter in Eq. (11.4-3*b*) is this rotated waveform $p(-t)$ translated in the positive t direction by amount T. This last translation ensures that $h(t) = 0$ for $t < 0$ as is required for a *causal* filter.

In general, the impulsive response of the matched filter consists of $p(t)$ rotated about $t = 0$ and then delayed long enough (i.e., a time T) to make the filter realiz-

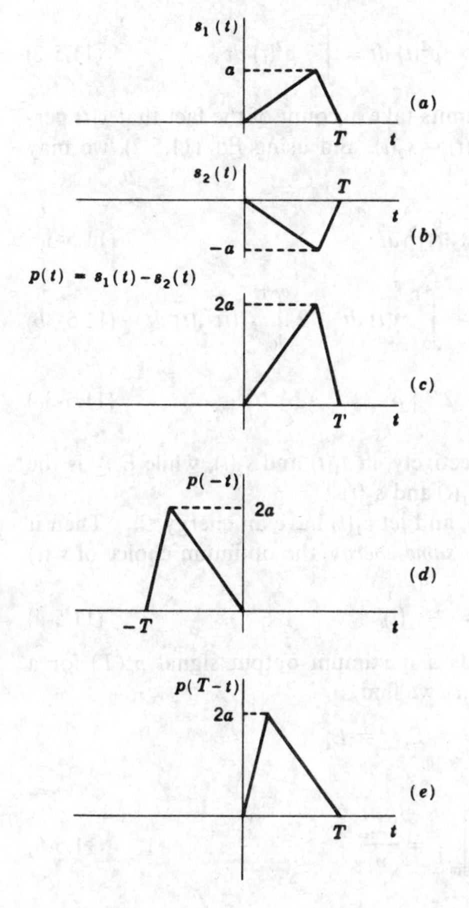

Figure 11.4-1 The signals (*a*) $s_1(t)$, (*b*) $s_2(t)$, and (*c*) $p(t) = s_1(t) - s_2(t)$. (*d*) $p(t)$ rotated about the axis $t = 0$. (*e*) The waveform in (*d*) translated to the right by amount T.

able. We may note in passing, that any additional delay that a filter might introduce would in no way interfere with the performance of the filter, for both signal and noise would be delayed by the same amount, and at the sampling time (which would need similarly to be delayed) the ratio of signal to noise would remain unaltered.

11.5 PROBABILITY OF ERROR OF THE MATCHED FILTER

The probability of error which results when employing a matched filter, may be found by evaluating the maximum signal-to-noise ratio $[p_o^2(T)/\sigma_o^2]_{max}$ given by Eq. (11.3-20). With $G_n(f) = \eta/2$, Eq. (11.3-20) becomes

$$\left[\frac{p_o^2(T)}{\sigma_o^2}\right]_{max} = \frac{2}{\eta}\int_{-\infty}^{\infty} |P(f)|^2\, df \tag{11.5-1}$$

From Parseval's theorem we have

$$\int_{-\infty}^{\infty} |P(f)|^2\, df = \int_{-\infty}^{\infty} p^2(t)\, dt = \int_0^T p^2(t)\, dt \tag{11.5-2}$$

In the last integral in Eq. (11.5-2), the limits take account of the fact that $p(t)$ persists for only a time T. With $p(t) = s_1(t) - s_2(t)$, and using Eq. (11.5-2), we may write Eq. (11.5-1) as

$$\left[\frac{p_o^2(T)}{\sigma_o^2}\right]_{max} = \frac{2}{\eta}\int_0^T [s_1(t) - s_2(t)]^2\, dt \tag{11.5-3a}$$

$$= \frac{2}{\eta}\left[\int_0^T s_1^2(t)\, dt + \int_0^T s_2^2(t)\, dt - 2\int_0^T s_1(t)s_2(t)\, dt\right] \tag{11.5-3b}$$

$$= \frac{2}{\eta}(E_{s1} + E_{s2} - 2E_{s12}) \tag{11.5-3c}$$

Here E_{s1} and E_{s2} are the energies, respectively, in $s_1(t)$ and $s_2(t)$, while E_{s12} is the energy due to the correlation between $s_1(t)$ and $s_2(t)$.

Suppose that we have selected $s_1(t)$, and let $s_1(t)$ have an energy E_{s1}. Then it can be shown that if $s_2(t)$ is to have the *same energy*, the optimum choice of $s_2(t)$ is

$$s_2(t) = -s_1(t) \tag{11.5-4}$$

The choice is optimum in that it yields a maximum output signal $p_o^2(T)$ for a given signal energy. Letting $s_2(t) = -s_1(t)$, we find

$$E_{s1} = E_{s2} = -E_{s12} \equiv E_s$$

and Eq. (11.5-3c) becomes

$$\left[\frac{p_o^2(T)}{\sigma_o^2}\right]_{max} = \frac{8E_s}{\eta} \tag{11.5-5}$$

Rewriting Eq. (11.3-4*b*) using $p_o(T) = s_{o1}(T) - s_{o2}(T)$, we have

$$P_e = \frac{1}{2}\,\text{erfc}\left[\frac{p_o(T)}{2\sqrt{2}\,\sigma_o}\right] = \frac{1}{2}\,\text{erfc}\left[\frac{p_o^2(T)}{8\sigma_o^2}\right]^{1/2} \tag{11.5-6}$$

Combining Eq. (11.5-6) with (11.5-5), we find that the minimum error probability $(P_e)_{\min}$ corresponding to a maximum value of $p_o^2(T)/\sigma_o^2$ is

$$(P_e)_{\min} = \frac{1}{2}\,\text{erfc}\left\{\frac{1}{8}\left[\frac{p_o^2(T)}{\sigma_o^2}\right]_{\max}\right\}^{1/2} \tag{11.5-7}$$

$$= \frac{1}{2}\,\text{erfc}\left(\frac{E_s}{\eta}\right)^{1/2} \tag{11.5-8}$$

We note that Eq. (11.5-8) establishes more generally the idea that the error probability depends only on the signal energy and not on the signal waveshape. Previously we had established this point only for signals which had constant voltage levels.

We note also that Eq. (11.5-8) gives $(P_e)_{\min}$ for the case of the matched filter and when $s_1(t) = -s_2(t)$. In Sec. 11.2 we considered the case when $s_1(t) = +V$ and $s_2(t) = -V$ and the filter employed was an integrator. There we found [Eq. (11.2-3)] that the result for P_e was identical with $(P_e)_{\min}$ given in Eq. (11.5-8). This agreement leads us to suspect that for an input signal where $s_1(t) = +V$ and $s_2(t) = -V$, the integrator is the matched filter. Such is indeed the case. For when we have

$$s_1(t) = V \qquad 0 \le t \le T \tag{11.5-9a}$$

and

$$s_2(t) = -V \qquad 0 \le t \le T \tag{11.5-9b}$$

the impulse response of the matched filter is, from Eq. (11.4-4),

$$h(t) = \frac{2K}{\eta}\,[s_1(T-t) - s_2(T-t)] \tag{11.5-10}$$

The quantity $s_1(T-t) - s_2(T-t)$ is a pulse of amplitude $2V$ extending from $t = 0$ to $t = T$ and may be rewritten, with $u(t)$ the unit step,

$$h(t) = \frac{2K}{\eta}\,(2V)[u(t) - u(t-T)] \tag{11.5-11}$$

The constant factor of proportionality $4KV/\eta$ in the expression for $h(t)$ (that is, the gain of the filter) has no effect on the probability of error since the gain affects signal and noise alike. We may therefore select the coefficient K in Eq. (11.5-11) so that $4KV/\eta = 1$. Then the inverse transform of $h(t)$, that is, the transfer function of the filter, becomes, with s the Laplace transform variable,

$$H(s) = \frac{1}{s} - \frac{e^{-sT}}{s} \tag{11.5-12}$$

The first term in Eq. (11.5-12) represents an integration beginning at $t = 0$, while the second term represents an integration with reversed polarity beginning at $t = T$. The overall response of the matched filter is an integration from $t = 0$ to $t = T$ and a zero response thereafter. In a physical system, as already described, we achieve the effect of a zero response after $t = T$ by sampling at $t = T$, so that so far as the determination of one bit is concerned we ignore the response after $t = T$.

11.6 COHERENT RECEPTION: CORRELATION

We discuss now an alternative type of receiving system which, as we shall see, is identical in performance with the matched filter receiver. Again, as shown in Fig. 11.6-1, the input is a binary data waveform $s_1(t)$ or $s_2(t)$ corrupted by noise $n(t)$. The bit length is T. The received signal plus noise $v_i(t)$ is multiplied by a locally generated waveform $s_1(t) - s_2(t)$. The output of the multiplier is passed through an integrator whose output is sampled at $t = T$. As before, immediately after each sampling, at the beginning of each new bit interval, all energy-storing elements in the integrator are discharged. This type of receiver is called a *correlator*, since we are *correlating* the received signal and noise with the waveform $s_1(t) - s_2(t)$.

The output signal and noise of the correlator shown in Fig. 11.6-1 are

$$s_o(T) = \frac{1}{\tau} \int_0^T s_i(t)[s_1(t) - s_2(t)]\, dt \tag{11.6-1}$$

$$n_o(T) = \frac{1}{\tau} \int_0^T n(t)[s_1(t) - s_2(t)]\, dt \tag{11.6-2}$$

where $s_i(t)$ is either $s_1(t)$ or $s_2(t)$, and where τ is the constant of the integrator (i.e., the integrator output is $1/\tau$ times the integral of its input). We now compare these outputs with the matched filter outputs.

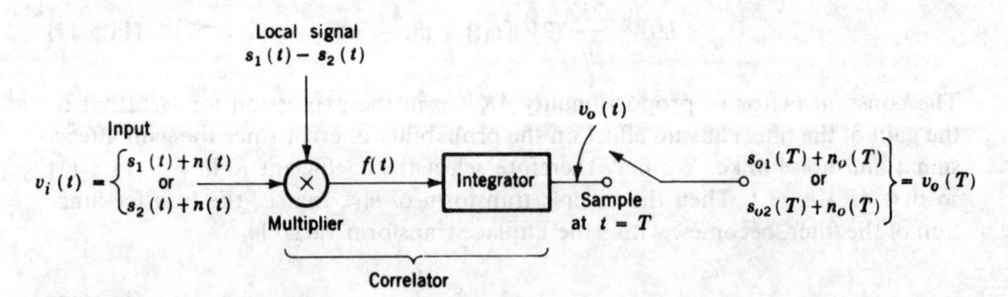

Figure 11.6-1 A coherent system of signal reception.

If $h(t)$ is the impulsive response of the matched filter, then the output of the matched filter $v_o(t)$ can be found using the convolution integral (see Sec. 1.12). We have

$$v_o(t) = \int_{-\infty}^{\infty} v_i(\lambda)h(t-\lambda)\,d\lambda = \int_0^T v_i(\lambda)h(t-\lambda)\,d\lambda \tag{11.6-3}$$

The limits on the integral have been changed to 0 and T since we are interested in the filter response to a bit which extends only over that interval. Using Eq. (11.4-4) which gives $h(t)$ for the matched filter, we have

$$h(t) = \frac{2K}{\eta}[s_1(T-t) - s_2(T-t)] \tag{11.6-4}$$

so that $$h(t-\lambda) = \frac{2K}{\eta}[s_1(T-t+\lambda) - s_2(T-t+\lambda)] \tag{11.6-5}$$

Substituting Eq. (11.6-5) into (11.6-3), we have

$$v_o(t) = \frac{2K}{\eta}\int_0^T v_i(\lambda)[s_1(T-t+\lambda) - s_2(T-t+\lambda)]\,d\lambda \tag{11.6-6}$$

Since $v_i(\lambda) = s_i(\lambda) + n(\lambda)$, and $v_o(t) = s_o(t) + n_o(t)$, setting $t = T$ yields

$$s_o(T) = \frac{2K}{\eta}\int_0^T s_i(\lambda)[s_1(\lambda) - s_2(\lambda)]\,d\lambda \tag{11.6-7}$$

where $s_i(\lambda)$ is equal to $s_1(\lambda)$ or $s_2(\lambda)$. Similarly we find that

$$n_o(T) = \frac{2K}{\eta}\int_0^T n(\lambda)[s_1(\lambda) - s_2(\lambda)]\,d\lambda \tag{11.6-8}$$

Thus $s_o(T)$ and $n_o(T)$, as calculated from Eqs. (11.6-1) and (11.6-2) for the correlation receiver, and as calculated from Eqs. (11.6-7) and (11.6-8) for the matched filter receiver, are identical. Hence the performances of the two systems are identical.

The *matched filter* and the *correlator* are not simply two distinct, independent techniques which happen to yield the same result. In fact they are two techniques of synthesizing the optimum filter $h(t)$.

11.7 PHASE-SHIFT KEYING

An important application of the coherent reception system of Sec. 11.6 is its use in phase-shift keying (PSK). Here the input signal is

$$s_1(t) = A\cos\omega_0 t \tag{11.7-1}$$

or $$s_2(t) = -A\cos\omega_0 t \tag{11.7-2}$$

At the receiver a coherent local signal $s_1(t) - s_2(t) = 2A \cos \omega_0 t$ needs to be provided for the multiplier (see Fig. 11.6-1).

Since, in PSK, $s_1(t) = -s_2(t)$, Eq. (11.5-8) gives the error probability. Then, in PSK, as in baseband transmission.

$$P_e = \frac{1}{2} \operatorname{erfc} \sqrt{\frac{E_s}{\eta}} \tag{11.7-3}$$

If a bit duration extends for a time T, which encompasses a whole number of cycles, then the signal energy is $E_s = A^2T/2$ so that from Eq. (11.7-3) the error probability is

$$P_e = \frac{1}{2} \operatorname{erfc} \sqrt{\frac{A^2T}{2\eta}} \tag{11.7-4}$$

Imperfect Phase Synchronization

In PSK, where the signals $s_1(t)$ and $s_2(t)$ are as given in Eqs. (11.7-1) and (11.7-2), the required local waveform for the correlator which is shown in Fig. 11.6-1 is given by $s_1(t) - s_2(t) = 2A \cos \omega_0 t$. Then, when $s_1(t)$ is received, the output of the correlator at the sampling time $t = T$ is $s_{o1}(T) = cA^2T$, where c is some constant depending on the gain of the integrator. Similarly, $s_{o2}(T) = -cA^2T$.

Suppose now that the local signal used at the correlator were not $2A \cos \omega_0 t$ as required, but rather $2A \cos (\omega_0 t + \phi)$, where ϕ is some fixed phase offset. Then, as is easily verified at the sampling instant, the correlator output would become $s_o(T) = \pm 2cA^2T \cos \phi$, that is, the output signal is reduced, being multiplied by the factor $\cos \phi$. In this case the energy becomes $E_s \cos^2 \phi$ and Eq. (11.7-3) would be replaced by

$$P_e = \frac{1}{2} \operatorname{erfc} \sqrt{\frac{E_s}{\eta} \cos^2 \phi} \tag{11.7-5}$$

while Eq. (11.7-4) would read

$$P_e = \frac{1}{2} \operatorname{erfc} \sqrt{\frac{A^2T \cos^2 \phi}{2\eta}} \tag{11.7-6}$$

The phase shift ϕ increases the probability of error P_e. Typical error probabilities in a communication system range from 10^{-4} to 10^{-7}. In this range, if $\phi = 25°$, the probability of error is increased by a factor of 10 as compared with the result obtained for $\phi = 0$.

Imperfect Bit Synchronization

The PSK system shown in Fig. 11.6-1 and characterized by Eq. 11.6-3 assumes that the *bit synchronizer* at the receiver operates with perfect precision. The bit synchronizer (Sec. 10.15), it will be recalled, is used to assure that the integration starts at $t = 0$ and ends at $t = T$, that is, that the integration extend exactly over the duration of the bit. As a matter of practice, either because of limitations in the synchronizer or because of the influence of noise, the integration may extend not from 0 to T but rather from τ to $T + \tau$. We note that the output noise voltage $n_o(T + \tau)$ will, of course, not be affected by this overlap since the integration time T remains unchanged. Further, if the two bits overlapped have the same logic value then again the overlap has no affect on the signal voltage $s_o(T + \tau)$. But, if the overlapped bits are different, the overlap will cause a reduction in $s_o(T + \tau)$. To see that such is the case, consider that the transmitted signal is $A \cos \omega_0 t$ in the interval 0 to T and $-A \cos \omega_0 t$ for T to $2T$. If there is an overlap of amount τ then from Eqs. 11.6-7, 11.7-1, and 11.7-2,

$$s_o(T + \tau) = \frac{2K}{\eta} \int_{\tau}^{T} A \cos \omega_0 t[2A \cos \omega_0 t]\, dt$$

$$- \frac{2K}{\eta} \int_{T}^{T+\tau} A \cos \omega_0 t[2A \cos \omega_0 t] \tag{11.7-7}$$

$$= \frac{2K}{\eta} [A^2(T - \tau) - A^2\tau] = \frac{2K}{\eta} [A^2T]\left[1 - \frac{2\tau}{T}\right] \tag{11.7-8}$$

It is readily verified that if the overlap is in the other direction, i.e., the integration extends from $-\tau$ to $T - \tau$, the result is the same, so that more generally Eq. (11.7-8) becomes

$$s_o(T + \tau) = \frac{2K}{\eta} [A^2T]\left[1 - \frac{2|\tau|}{T}\right] \tag{11.7-9}$$

Correspondingly Eq. (11.7-3) should be corrected to read

$$P_e = \frac{1}{2} \operatorname{erfc} \sqrt{\left(\frac{E_s}{\eta}\right)\left(1 - \frac{2|\tau|}{T}\right)^2} \tag{11.7-10}$$

In the range of probabilities of error that are of interest to us, we note that if $\tau = 0.05T$ the probability of error is increased by a factor of 10.

If both phase error and timing error are present then Eq. (11.7-3) becomes

$$P_e = \frac{1}{2} \operatorname{erfc} \left[\left(\frac{E_s}{\eta}\right)(\cos^2 \phi)\left(1 - \frac{2|\tau|}{T}\right)^2\right]^{1/2} \tag{11.7-11}$$

11.8 FREQUENCY-SHIFT KEYING

In frequency-shift keying (FSK) the received signal is either

$$s_1(t) = A \cos (\omega_0 + \Omega)t \tag{11.8-1}$$

or

$$s_2(t) = A \cos (\omega_0 - \Omega)t \tag{11.8-2}$$

As was explained in Sec. 11.6, one way of synthesizing the matched filter is to construct the correlation receiver system shown in Fig. 11.6-1. This receiver will give precisely the same performance as a matched filter, provided that the local waveform is $s_1(t) - s_2(t)$. In FSK the required local waveform is

$$s_1(t) - s_2(t) = A \cos(\omega_0 + \Omega)t - A \cos(\omega_0 - \Omega)t \tag{11.8-3}$$

Probability of Error

In Sec. 11.5 we calculated the probability of error for the matched filter and arrived at the result $P_e = \frac{1}{2} \operatorname{erfc} \sqrt{E_s/\eta}$ given in Eq. (11.5-8). The derivation was general, and would apply in the present case, except for the fact that we had assumed there that $s_1(t) = -s_2(t)$. This assumption is obviously not valid for FSK.

To calculate the probability of error for FSK, we return to a point in the derivation of Sec. 11.5 just before the introduction of the assumption $s_1(t) = -s_2(t)$. We start with Eq. (11.5-3*a*), which reads

$$\left[\frac{p_o^2(T)}{\sigma_o^2}\right]_{\max} = \frac{2}{\eta} \int_0^T [s_1(t) - s_2(t)]^2 \, dt \tag{11.8-4}$$

Substituting $s_1(t)$ and $s_2(t)$ as given in Eqs. (11.8-1) and (11.8-2) into Eq. (11.8-4) and performing the indicated integration, we find that

$$\left[\frac{p_o^2(T)}{\sigma_o^2}\right]_{\max} = \frac{2A^2T}{\eta}\left[1 - \frac{\sin 2\Omega T}{2\Omega T} + \frac{1}{2}\frac{\sin[2(\omega_0 + \Omega)T]}{2(\omega_0 + \Omega)T} - \frac{1}{2}\frac{\sin[2(\omega_0 - \Omega)T]}{2(\omega_0 - \Omega)T} - \frac{\sin 2\omega_0 T}{2\omega_0 T}\right] \tag{11.8-5}$$

Note that in Sec. 6.8 we arrived at this same result using the Gram-Schmitt procedure.

If we assume that the offset angular frequency Ω is very small in comparison with the carrier angular frequency ω_0 (a situation usually encountered in physical systems), then the last three terms in Eq. (11.8-5) each have the form $(\sin 2\omega_0 T)/2\omega_0 T$. This ratio approaches zero as $\omega_0 T$ increases. We further assume, as is generally the case, that $\omega_0 T \gg 1$. We may therefore neglect these last three terms. We are left with

$$\left[\frac{p_o^2(T)}{\sigma_o^2}\right]_{\max} = \frac{2A^2T}{\eta}\left(1 - \frac{\sin 2\Omega T}{2\Omega T}\right) \tag{11.8-6}$$

The quantity $[p_o^2(T)/\sigma_o^2]_{\max}$ in Eq. (11.8-6) attains its largest value when Ω is selected so that $2\Omega T = 3\pi/2$. For this value of Ω we find

$$\left[\frac{p_o^2(T)}{\sigma_o^2}\right]_{\max} = 2.42\frac{A^2T}{\eta} = 4.84\frac{(A^2/2)T}{\eta} \tag{11.8-7}$$

The probability of error, calculated using Eq. (11.5-6) with $[p_o^2(T)/\sigma_o^2]_{\max}$ as given in Eq. (11.8-7), is found to be

$$P_e = \frac{1}{2}\,\text{erfc}\left\{\frac{1}{8}\left[\frac{p_o^2(T)}{\sigma_o^2}\right]_{\max}\right\}^{1/2} \simeq \frac{1}{2}\,\text{erfc}\left(0.6\,\frac{E_s}{\eta}\right)^{1/2} \tag{11.8-8}$$

where the signal energy is $E_s = A^2T/2$.

Comparing the probability of error obtained for FSK [Eq. (11.8-8)] with the probability of error obtained for PSK [Eq. (11.7-3)], we see that equal probability of error in each system can be achieved if the signal energy in the PSK signal is 0.6 times as large as the signal energy in FSK. As a result, a 2-dB increase in the transmitted signal power is required for FSK. Why is FSK inferior to PSK? The answer is that in PSK $s_1(t) = -s_2(t)$, while in FSK this condition is not satisfied. Thus, although an optimum filter is used in each case, PSK results in considerable improvement compared with FSK.

When one of two *orthogonal* frequencies are transmitted, $2\Omega T = m\pi$ (m an integer) and

$$P_e = \frac{1}{2}\,\text{erfc}\left(\frac{E_s}{2\eta}\right)^{1/2} \tag{11.8-9}$$

11.9 NONCOHERENT DETECTION OF FSK

Since FSK can be thought of as the transmission of the output of either of two signal sources, the first at the frequency $\omega_1 = \omega_0 + \Omega$, and the second at the frequency $\omega_2 = \omega_0 - \Omega$, we should think that a reasonable detection system would consist of two bandpass filters, with center frequencies at ω_1 and ω_2. The bandwidth of each filter would be adjusted to yield a maximum output when the appropriate signal is received. Thus, when filter H_1 with center frequency ω_1 has a larger output than filter H_2 with center frequency ω_2, we would decide that $s_1(t)$ was transmitted. Similarly we would decide that $s_2(t)$ was transmitted when the output of H_2 was greater than the output of H_1.

When using a filter receiver, we make no use of the *phase* of the incoming signal. Such reception is, therefore, called *noncoherent* detection. The *coherent* matched-filter detector, on the other hand, uses synchronization techniques to determine the phase of the incoming signal. Since some valuable information concerning the signal is not used, the probability of detecting the signal is reduced. The probability of error of noncoherent FSK is found to be[2]

$$P_e = \tfrac{1}{2}e^{-E_s/2\eta} \tag{11.9-1}$$

This probability of error is compared in Sec. 11.20 with the other systems discussed.

11.10 DIFFERENTIAL PSK

The operation of differential phase-shift keying (DPSK) was explained in Sec. 6.3. We shall now show the *suboptimum* nature of DPSK by considering the system in terms of the phasor diagram shown in Fig. 11.10.1. In Fig. 11.10-1*a* we see that

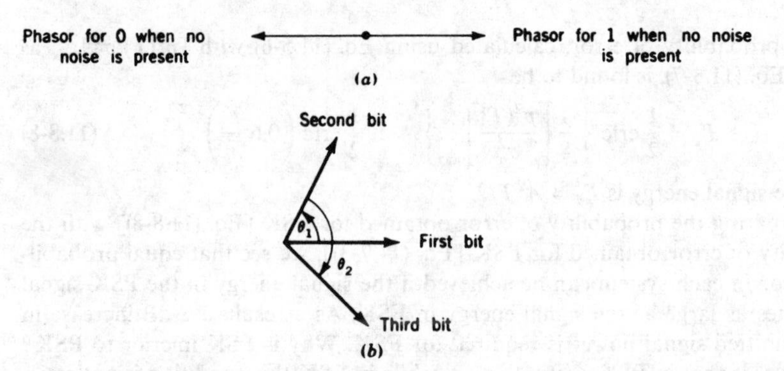

Figure 11.10-1 Reception of DPSK. (*a*) Phasors when no noise is present. (*b*) Phasors when noise is present.

when no noise is present the received phase is either at angle 0 or π. From this we draw a decision boundary at angle $\pi/2$ and decide that a 1 was sent if the phase difference between two consecutive bits differs by less than $\pi/2$, or we decide that a 0 was sent if the phase difference between two consecutive bits differs by more than $\pi/2$.

Figure 11.10-1*b* shows three consecutive received bits. Each bit was transmitted as a 1, but because of noise each is perturbed from the horizontal axis as shown. The DPSK receiver compares bit 2 with bit 1, reads an angle θ_1 which is less than 90°, and decides that bit 2 is a 1. The DPSK receiver then compares bit 3 with bit 2, reads an angle θ_2 which is greater than 90°, and decides that a 0 was transmitted.

The error was due to the fact that the DPSK receiver uses only the previous bit as a reference. This method of operation is analogous to employing *poor* synchronization. If all the previous positive bits were somehow averaged by employing good synchronization, and this average were used as a reference, then the DPSK receiver would have a *stable* reference, and the error described above would not have occurred. As a matter of fact we would then have a PSK system, not a DPSK system. Thus, DPSK is *suboptimum* and results in a higher probability of error than in PSK where we have a *stable* reference phase (when perfect synchronization is assumed).

The calculation of probability of error is complicated and will not be given here.[2] The result is

$$P_e = \frac{1}{2}e^{-E_s/\eta} \tag{11.10-1}$$

The probability of error of DPSK and PSK are compared in Sec. 11.20.

11.11 FOUR-PHASE PSK (QPSK)

In previous sections of this chapter we dealt exclusively with binary communication systems, that is, systems in which, in any interval $0 \leq t \leq T$, one of two possible messages was transmitted. However, data transmission systems allowing many possible messages in the interval T (so-called M-ary systems, the number of messages being M) are also possible and are widely used. As an example of such a M-ary system we consider 4-phase PSK (QPSK) which, because of its relative simplicity, is very popular (see Sec. 6.5).

In QPSK one of four possible waveforms is transmitted during each interval T. These waveforms are

$$s_i(t) = A \cos\left(\omega_0 t + [2m - 1]\frac{\pi}{4}\right) \qquad m = 1, 2, 3, 4 \qquad 0 \leq t \leq T_s = 2T \tag{11.11-1}$$

These four waveforms are represented in the phasor diagram of Fig. 11.11-1. The receiver system is shown in Fig. 11.11-2. Observe that two correlators are required and that the local reference waveforms, as indicated also in Fig. 11.11-1, are $A \cos \omega_0 t$ and $A \sin \omega_0 t$.

Suppose, now, that, in the absence of the noise, signal $s_1(t)$ is received. Let us use the symbol V_o to represent the corresponding output of correlator 1, i.e., $V_o \equiv v_{o1}(T_s)$ when $s_1(t)$ is received. Then, as is readily verified, the output of the two

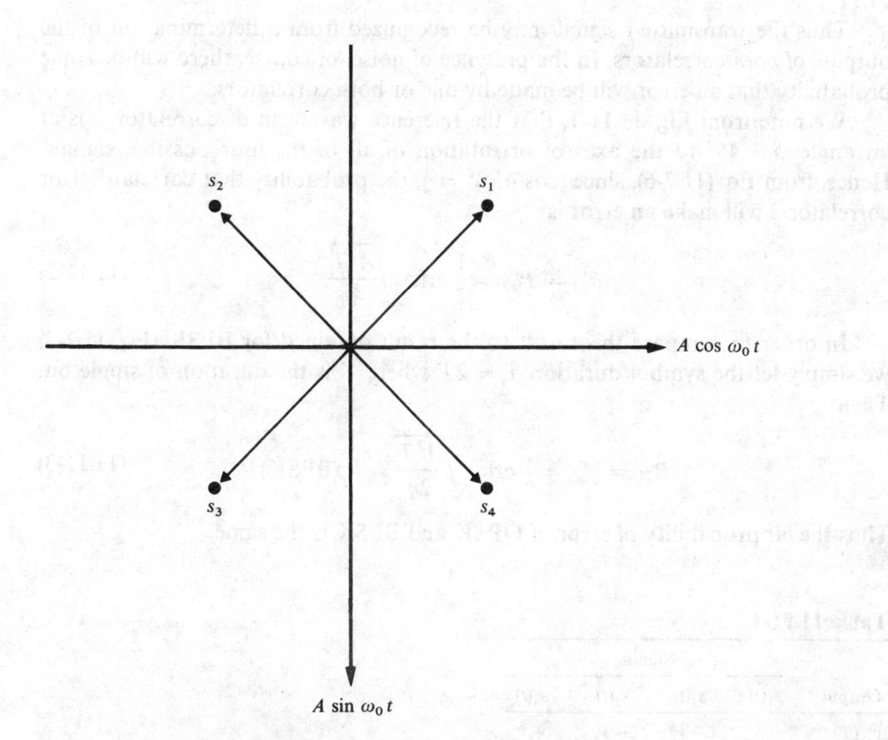

Figure 11.11-1 A phasor diagram representation of the signals in QPSK.

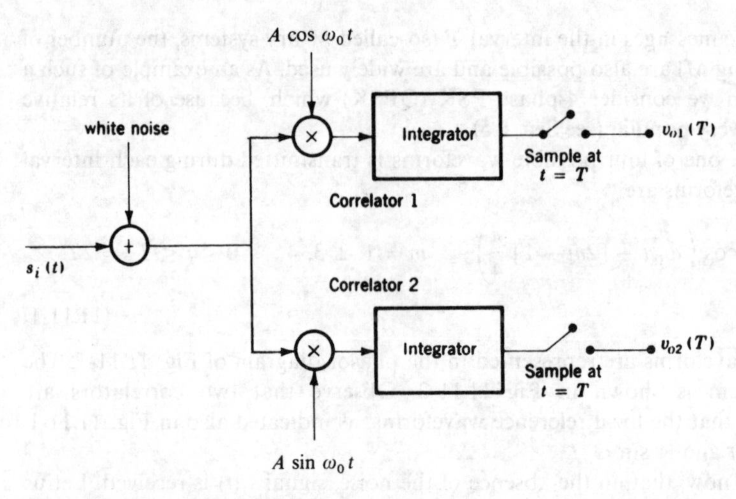

Figure 11.11-2 A correlation receiver for QPSK.

correlators corresponding to each of the four possible signals is as given in Table 11.11-1.

Thus the transmitted signal may be recognized from a determination of the outputs of *both* correlators. In the presence of noise, of course, there will be some probability that an error will be made by one or both correlators.

We note from Fig. 11.11-1, that the reference waveform of correlator 1 is at an angle $\phi = 45°$ to the axes of orientation of all of the four possible signals. Hence, from Eq. (11.7-6), since $(\cos 45°)^2 = \frac{1}{2}$, the probability that correlator 1 or correlator 2 will make an error is

$$P'_{e1} = P'_{e2} = \frac{1}{2} \operatorname{erfc} \sqrt{\frac{A^2 T_s}{4\eta}} \tag{11.11-2}$$

In order to compare this result to the result obtained for BPSK (Eq. 11.7-4) we simply let the symbol duration $T_s = 2T$ where T is the duration of single bit. Then

$$P'_{e1} = P'_{e2} = \tfrac{1}{2} \operatorname{erfc} \sqrt{\frac{A^2 T}{2\eta}} = P_e(\text{BPSK}) \tag{11.11-3}$$

Thus the *bit* probability of error of QPSK and BPSK is the same.

Table 11.11-1

	Signal			
Output	$s_1(t)$	$s_2(t)$	$s_3(t)$	$s_4(t)$
$v_{o1}(T)$	$+V_o$	$-V_o$	$-V_o$	$+V_o$
$v_{o2}(T)$	$-V_o$	$-V_o$	$+V_o$	$+V_o$

The probability P'_{e2} that correlator 2 will make an error is similarly given by the expression in Eq. (11.11-2). The probability P_c that the QPSK receiver will correctly identify the transmitted signal is equal to the product of the probabilities that both correlator 1 and correlator 2 have yielded correct results. Thus, using $P'_e = P'_{e1} = P'_{e2}$

$$P_c = (1 - P'_e)(1 - P'_e) = 1 - 2P'_e + P'^2_e \tag{11.11-4}$$

If, as is normally the case, $P'_e \ll 1$, the last term in Eq. (11.11-4) may be neglected. Finally, then, the probability of error of the system is

$$P_e(\text{QPSK}) = 1 - P_c \simeq 2P'_e = \text{erfc}\sqrt{\frac{A^2 T}{4\eta}} \tag{11.11-5}$$

11.12 USE OF SIGNAL SPACE TO CALCULATE P_e

In Sec. 2.22 we considered a situation in which a number of distinct messages m_k were represented by distinct waveforms s_k each of which has a duration T_b. At the receiver, the signal is processed in such a manner that, in the absence of noise, the received signal s_k generates a response r_k, r_k being a single real number. We saw there how best to determine the message from the generated response in order, with noise present, to minimize the probability of error. We found that if a response r was generated, and if all messages are equally likely, we are to determine the message to be that m_k whose corresponding response r_k yields a minimum difference $|r_k - r|$.

Since message signals can be represented as a linear superposition of orthonormal functions, it is useful to represent the signals as vectors in a signal space. For example, consider the case of BPSK shown in Fig. 11.12-1. In this simple case, the "signal space" is one dimensional. The signals are s_1 and s_2, given by

$$\left.\begin{matrix} s_1 \\ s_2 \end{matrix}\right\} = b(t)\sqrt{2P_s}\cos\omega_0 t \qquad 0 < t \leq T_b \tag{11.12-1}$$

in which, say $b(t) = +1$ for s_1 and $b(t) = -1$ for s_2. P_s is the signal power. If we introduce the unit (normalized) vector $u(t) = \sqrt{2/T_b}\cos\omega_0 t$, then

$$\left.\begin{matrix} s_1 \\ s_2 \end{matrix}\right\} = b(t)\sqrt{P_s T_b}\sqrt{2/T_b}\cos\omega_0 t = b(t)\sqrt{P_s T_b}\, u(t) \qquad 0 \leq t \leq T_b \tag{11.12-2}$$

In Fig. 11.12-1*a* are shown the signal vectors each of length $\sqrt{P_s T_b}$ measured in terms of the unit vector $u(t) = \sqrt{2/T_b}\cos\omega_0 t$. Processing at the correlator receiver shown in Fig. 11.12-1*b* will generate a response r_1 or r_2 for s_1 and s_2 respectively when no noise is present. Now suppose that in some interval, because of noise, a response r ($\neq r_1$ nor r_2) is generated. If we find $|r - r_1| < |r - r_2|$ then we determine that s_1 was transmitted.

We shall now extend this result, without formal proof, to two dimensional signal space which we have used to represent BFSK, QPSK, and 16 QAM. Spe-

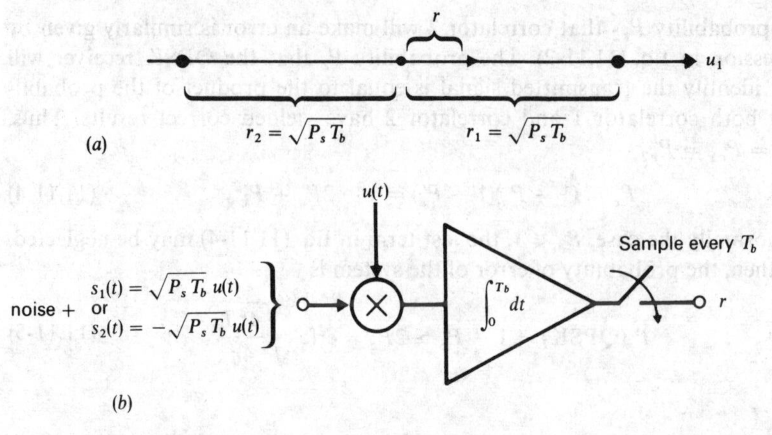

Figure 11.12-1 (*a*) BPSK representation in signal space showing r_1 and r_2. Received r is drawn for the case where $|r - r_1| < |r - r_2|$ and signal s_1 is determined to have been transmitted. (*b*) Correlator receiver for BPSK showing that $r = r_1 + n_o$ or $r_2 + n_o$.

cifically a noisy signal response r having been generated, we shall determine, in signal space, the response r_k corresponding to the signal s_k to which r is closest, and we shall then determine that m_k corresponding to s_k was the message that was intended.

11.13 CALCULATION OF ERROR PROBABILITY FOR BPSK AND BFSK

BPSK Using the considerations of the previous section we shall calculate the error probabilities for synchronous detection in the cases of BPSK. The synchronous detector for BPSK is shown in Fig. 11.12-1*b*. Since the BPSK signal is one dimensional, we have from the discussion of Secs. 7.14 and 7.15 the result that the only relevant noise in the present case is

$$n(t) = n_o u(t) = n_o \sqrt{2/T_b} \cos \omega_0 t \tag{11.13-1}$$

in which n_o is a Gaussian random variable of variance $\sigma_o^2 = \eta/2$. Now let us suppose that s_2 was transmitted. The error probability, i.e., the probability that the signal is mistakenly judged to be s_1 is the probability that $n_o > \sqrt{P_s T_b}$. Thus the error probability P_e is

$$P_e = \frac{1}{\sqrt{2\pi\sigma^2}} \int_{\sqrt{P_s T_b}}^{\infty} e^{-n_o^2/2\sigma_o^2}\, dn_o = \frac{1}{\sqrt{\pi\eta}} \int_{\sqrt{P_s T_b}}^{\infty} e^{-n_o^2/\eta}\, dn_o \tag{11.13-2}$$

We let $y^2 \equiv n_o^2/2\sigma_o^2$ and find that

$$P_e = \frac{1}{\sqrt{\pi}} \int_{\sqrt{P_s T_b/\eta}}^{\infty} e^{-y^2}\, dy = \tfrac{1}{2} \cdot \frac{2}{\sqrt{\pi}} \int_{\sqrt{P_s T_b/\eta}}^{\infty} e^{-y^2}\, dy = \tfrac{1}{2}\, \text{erfc} \sqrt{\frac{P_s T_b}{\eta}} \tag{11.13-3}$$

The signal energy is $E_b = P_s T_b$ and the distance between end points of the signal vectors in Fig. 11.12-1 is $d = 2\sqrt{P_s T_b}$. Accordingly we find that

$$P_e = \frac{1}{2} \text{ erfc } \sqrt{\frac{E_b}{\eta}} = \frac{1}{2} \text{ erfc } \sqrt{\frac{d^2}{4\eta}} \tag{11.13-4}$$

The error probability is thus seen to fall off monotonically with an increase in distance between signals.

BFSK The case of synchronous detection of orthogonal binary FSK is represented in Fig. 11.13-1. The signal space is shown in (*a*). The unit vectors are

$$u_1(t) = \sqrt{\frac{2}{T_b}} \cos \omega_1 t \tag{11.13-5a}$$

and

$$u_2(t) = \sqrt{\frac{2}{T_b}} \cos \omega_2 t \tag{11.13-5b}$$

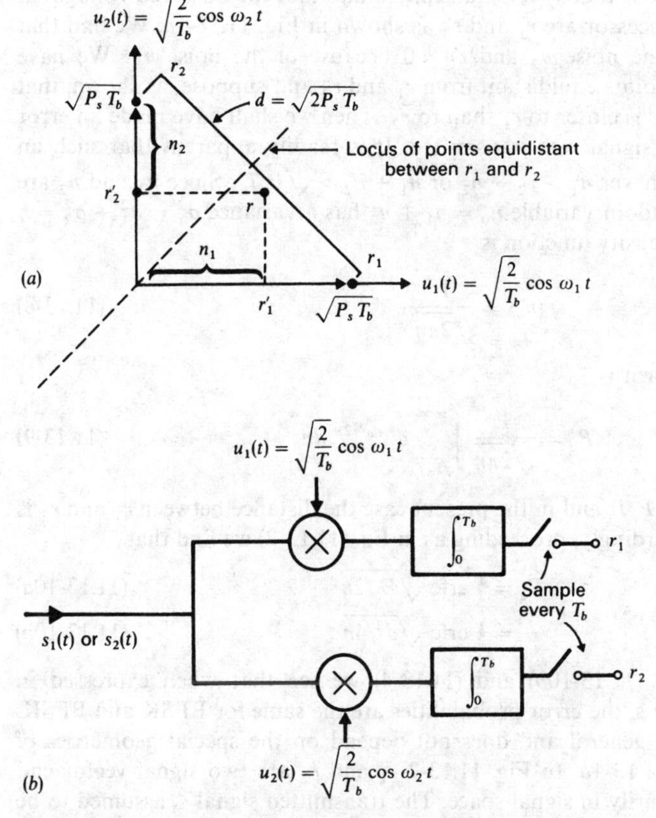

Figure 11.13-1 (*a*) Signal space representation of BFSK. (*b*) Correlator receiver for BFSK.

orthogonality over the interval T_b having been insured by the selection of ω_1 and ω_2 (see Prob. 11.13-1). The transmitted signals s_1 and s_2 are of power P_s and are

$$s_1(t) = \sqrt{2P_s} \cos \omega_1 t = \sqrt{P_s T_b}\sqrt{2/T_b} \cos \omega_1 t = \sqrt{P_s T_b}\, u_1(t) \quad (11.13\text{-}6a)$$

and

$$s_2(t) = \sqrt{2P_s} \cos \omega_2 t = \sqrt{P_s T_b}\sqrt{2/T_b} \cos \omega_2 t = \sqrt{P_s T_b}\, u_2(t) \quad (11.13\text{-}6b)$$

Detection is accomplished in the manner shown in Fig. 11.13-1*b*. The outputs are r_1 and r_2. In the absence of noise when $s_1(t)$ is received, $r_2 = 0$ and $r_1 = \sqrt{P_s T_b}$. For $s_2(t)$, $r_1 = 0$ and $r_2 = \sqrt{P_s T_b}$. Hence (as in PSK) the vectors representing r_1 and r_2 are of length $\sqrt{P_s T_b}$ as shown in Fig. 11.13-1*a*.

Again from the discussion of Sec. 7.14 we have that since the signal is two dimensional the relevant noise in the present case is

$$n(t) = n_1 u_1(t) + n_2 u_2(t) \quad (11.13\text{-}7)$$

in which n_1 and n_2 are Gaussian random variables each of variance $\sigma_1^2 = \sigma_2^2 = \eta/2$. Now let us suppose that $s_2(t)$ is transmitted and that the observed voltages at the output of the processor are r_1' and r_2' as shown in Fig. 11.13-1*a*. We find that $r_2' \neq r_2$ because of the noise n_2 and $r_1' \neq 0$ because of the noise n_1. We have drawn the locus of points equidistant from r_1 and r_2 and suppose, as shown, that the received voltage r is closer to r_1 than to r_2. Then we shall have made an error in estimating which signal was transmitted. It is readily apparent that such an error will occur whenever $n_1 > r_2 - n_2$ or $n_1 + n_2 > \sqrt{P_s T_b}$. Since n_1 and n_2 are uncorrelated, the random variable $n_o = n_1 + n_2$ has a variance $\sigma_o^2 = \sigma_1^2 + \sigma_2^2 = \eta$ and its probability density function is

$$f(n_o) = \frac{1}{\sqrt{2\pi\eta}} e^{-n_o^2/2\eta} \quad (11.13\text{-}8)$$

The probability of error is

$$P_e = \frac{1}{\sqrt{2\pi\eta}} \int_{\sqrt{P_s T_b}}^{\infty} e^{-n_o^2/2\eta}\, dn_o \quad (11.13\text{-}9)$$

Again we have $E_b = P_s T_b$ and in the present case the distance between r_1 and r_2 is $d = \sqrt{2}\sqrt{P_s T}$. Accordingly, proceeding as in Eq. (11.13-2) we find that

$$P_e = \tfrac{1}{2} \operatorname{erfc} \sqrt{E_b/2\eta} \quad (11.13\text{-}10a)$$

$$= \tfrac{1}{2} \operatorname{erfc} \sqrt{d^2/4\eta} \quad (11.13\text{-}10b)$$

Comparing Eqs. (11.13-10*b*) and (11.13-4) we see that when expressed in terms of the distance d, the error probabilities are the same for BPSK and BFSK. This result is rather general and does not depend on the special geometries of Figs. 11.12-1*a* and 11.13-1*a*. In Fig. 11.13-2 r_1 and r_2 are two signal vector end points located arbitrarily in signal space. The transmitted signal is assumed to be r_1. Noise components n_1 and n_2 displace r_1 and if the displacement is large

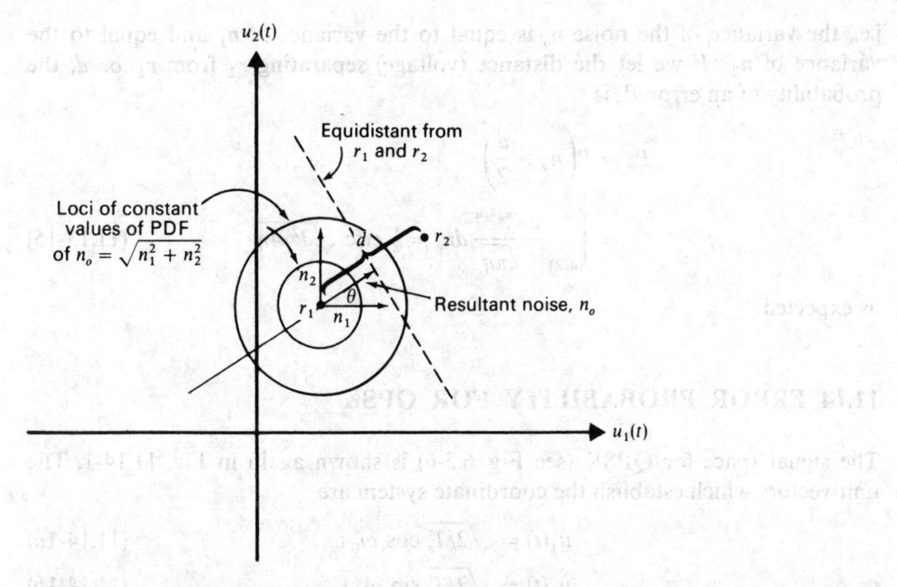

Figure 11.13-2 Signal space representation of two signals r_1 and r_2. The loci of constant values of the probability density function (PDF) of n_o are circles with center at r_1.

enough to carry the resultant (signal r_1 plus noise) across the (dashed) decision boundary we shall mistakenly estimate that r_2 was transmitted. We note that n_1 and n_2, the random noise amplitudes of the $u_1(t)$ and $u_2(t)$ components, are added to the initial position of r_1 wherever it may be. The displacement of the received voltage caused by the noise components n_1 and n_2 in the direction of r_2 is

$$n_o = n_1 \cos \theta + n_2 \sin \theta \tag{11.13-11}$$

where n_1 and n_2 are independent, gaussian, random variables of mean zero and of equal variance, $\eta/2$. Thus n_o is also a gaussian random variable with mean zero and variance σ_0^2, where

$$\begin{aligned}\sigma_0^2 = E(n_o^2) &= E[(n_1 \cos \theta + n_2 \sin \theta)^2] \\ &= E(n_1^2 \cos^2 \theta) + E(n_2^2 \sin^2 \theta) + 2E(n_1 n_2 \sin \theta \cos \theta)\end{aligned} \tag{11.13-12}$$

Noting that n_1 and n_2 are statistically independent, we have

$$E(n_1 n_2) = 0 \tag{11.13-13a}$$

Further, n_1 and n_2 are statistically independent of θ. Thus,

$$E(n_1^2) = E(n_2^2) = \eta/2 \tag{11.13-13b}$$

and since $E(\cos^2 \theta) = E(\sin^2 \theta) = \frac{1}{2}$, we have

$$E(n_o^2) = \eta/2 \tag{11.13-14}$$

i.e., the variance of the noise n_o is equal to the variance of n_1 and equal to the variance of n_2. If we let the distance (voltage) separating r_2 from r_1 be d, the probability of an error P_e is

$$P_e = P\left(n_o > \frac{d}{2}\right)$$

$$= \int_{(d/2)}^{\infty} \frac{e^{-n_o^2/\eta}}{\sqrt{\pi\eta}}\, dn_o = \tfrac{1}{2} \operatorname{erfc} \sqrt{d^2/4\eta} \tag{11.13-15}$$

as expected.

11.14 ERROR PROBABILITY FOR QPSK

The signal space for QPSK (see Fig. 6.5-6) is shown again in Fig. 11.14-1. The unit vectors which establish the coordinate system are

$$u_1(t) = \sqrt{2/T_s} \cos \omega_0 t \tag{11.14-1a}$$

$$u_2(t) = \sqrt{2/T_s} \sin \omega_0 t \tag{11.14-1b}$$

in which $T_s = 2T_b$, T_b being the bit time. The relevant noise is

$$n(t) = n_1 u_1(t) + n_2 u_2(t) \tag{11.14-2}$$

where again n_1 and n_2 are independent, gaussian random variables of variance $\eta/2$. In the present case, for a reason that will appear shortly, it is more convenient to calculate not the error probability P_e but rather the probability P_c that the decision is *correct*. Thereafter, we can calculate P_e since $P_e = 1 - P_c$. If the

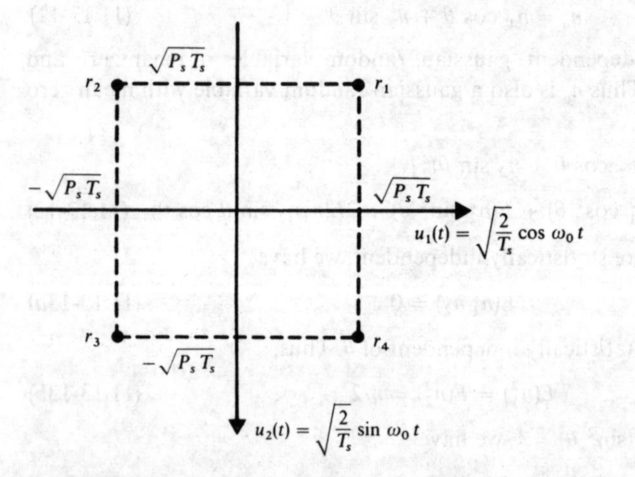

Figure 11.14-1 Signal space for QPSK.

transmitted signal is s_1 then a determination will be correct provided that the noise does not move r_1 out of the first quadrant. If such is to be the case we require that *both* n_1 and n_2 be in the range from $-\sqrt{P_s T_s}$ to infinity. Accordingly, since $d/2 = \sqrt{P_s T_s}$, we have that the probability of a correct decision, given that s_1 was transmitted, is

$$P(c \mid s_1) = P\left(n_1 > -\frac{d}{2}, n_2 > -\frac{d}{2}\right) = \int_{-d/2}^{\infty} \frac{e^{-n_1^2/\eta}}{\sqrt{\pi\eta}}\, dn_1 \int_{-d/2}^{\infty} \frac{e^{-n_2^2/\eta}}{\sqrt{\pi\eta}}\, dn_2 \qquad (11.14\text{-}3a)$$

$$= \left[\frac{1}{\sqrt{\pi\eta}} \int_{-d/2}^{\infty} e^{-n_1^2/\eta}\, dn_1\right]^2 \qquad (11.14\text{-}3b)$$

as can be readily verified (see Prob. 11.14-1), $P(c \mid s_1)$ in Eq. (11.14-3) evaluates to

$$P(c \mid s_1) = [1 - \tfrac{1}{2} \operatorname{erfc} \sqrt{d^2/4\eta}]^2 \qquad (11.14\text{-}4)$$

If each signal s_1, s_2, s_3, s_4 are equally likely to be transmitted

$$P_e = 1 - [1 - \tfrac{1}{2} \operatorname{erfc} \sqrt{d^2/4\eta}]^2 \qquad (11.14\text{-}5)$$

11.15 THE UNION BOUND APPROXIMATION

The signal space for 16-QASK is shown in Fig. 11.15-1. We have also marked off the boundaries to which the displacement due to noise must be restricted if an error in message determination is to be avoided. We observe that in the present case we shall have to calculate three different error probabilities because there are three different configurations to the boundaries. Let us use the symbol $P_e(r_1, r_4, r_{13}, r_{16})$ to represent the error probabilities for the four signals in the corners of Fig. 11.15-1 and similar symbols for the other groupings of signals. Then if all signals are equally likely, the error probability overall would be

$$P_e = \tfrac{1}{16}[4P_e(r_1, r_4, r_{13}, r_{16})$$
$$+ 8P_e(r_2, r_3, r_5, r_8, r_9, r_{12}, r_{14}, r_{15}) + 4P_e(r_6, r_7, r_{10}, r_{11})] \qquad (11.15\text{-}1)$$

It is apparent that Eq. (11.15-1), when evaluated, would lead to a rather formidable and awkward expression whose individual terms would have a significance not easily interpretable. Such indeed would be the result as well in a more general case. Further, it is not hard at all to conceive of a case in which an analytic result is not possible and a numerical calculation must be made. Both of these circumstances, complicated expressions or numerical results, have the disadvantage that they limit insight into the merits and limitations of a system and prevent, as well, an easy comparison of one system with another. Accordingly, we consider now a manner of calculation which leads to an approximate result but does have the feature that it can often circumvent the difficulties we have just described. This calculation leads to the *union-bound approximation*.

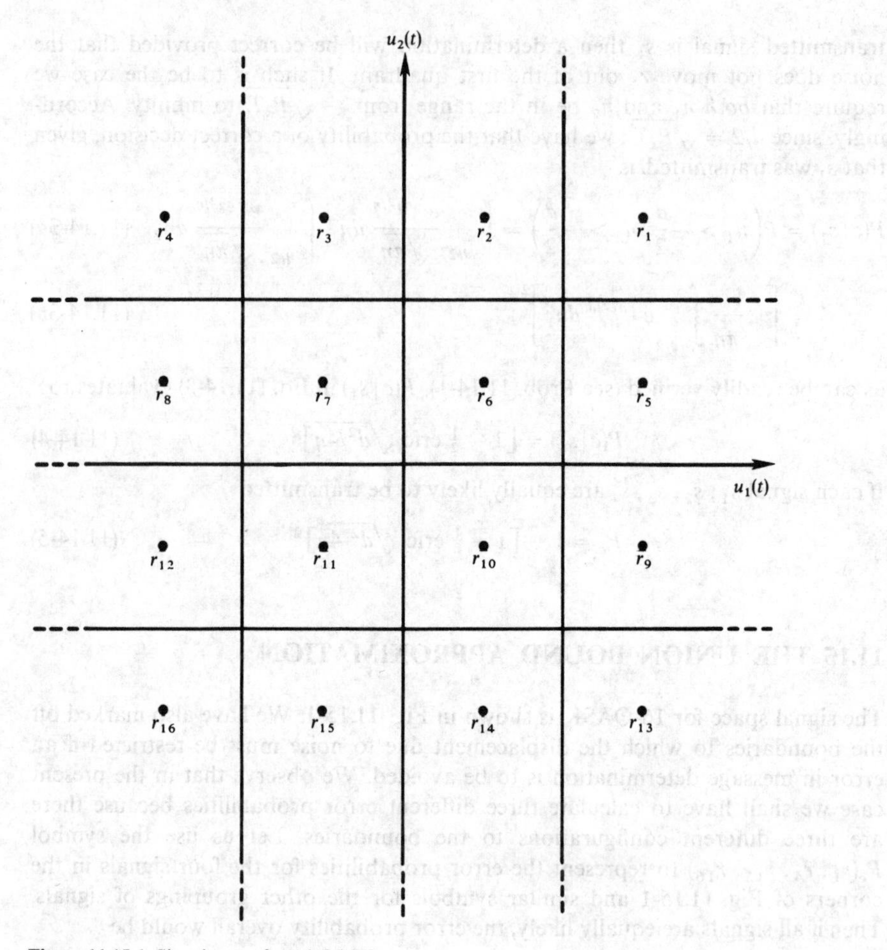

Figure 11.15-1 Signal space for 16-QASK.

To see the nature of the approximation involved we consider the case of QPSK shown again in Fig. 11.15-2. Assuming r_1 was transmitted let us calculate the probability of an error, the probability of error being that the noise, when added to r_1 yields a vector r that falls outside the upper right quadrant. We calculate first the probability $P(r_2, r_1)$ which we define as the probability that the noise added to r_1 carries the resultant closer to r_2 than to r_1. Next we calculate $P(r_3, r_1)$ and $P(r_4, r_1)$. Then we ask whether the probability of an error given that s_1 was transmitted, $P(e \mid s_1)$ is given by

$$P(e \mid s_1) = P(r_2, r_1) + P(r_3, r_1) + P(r_4, r_1) \tag{11.15-2}$$

That is we inquire whether $P(e \mid s_1)$ is given by the *union* (sum) of these three probabilities. We can see from Fig. 11.15-2 that such a union involves an approximation. The probability $P(r_2, r_1)$ is the probability that r_1 has been displaced by noise into the region (second and third quadrant) shaded by shading lines which are parallel to the $u_1(t)$ axis. The corresponding shading lines for $P(r_4, r_1)$ are parallel to the $u_2(t)$ axis and the diagonal shading corresponds to $P(r_3, r_1)$. We note that some areas are shaded more than once and hence the sum given in Eq. (11.15-2) is larger than the correct result and therefore constitutes an *upper* bound on $P(e \mid s_1)$.

For the case where M signals are used, the union bound approximation yields that the probability of error, given s_1 is transmitted is, using Eq. (11.13-10*b*)

$$P(e \mid s_1) \leq \sum_{k=2}^{M} \tfrac{1}{2} \operatorname{erfc} \sqrt{d_{k1}^2/4\eta} \tag{11.15-3}$$

If there is a symmetry in the M signals so that $P(e \mid s_k)$ is the same for all k, then we need go no further than Eq. (11.15-3). Otherwise the error probability without regard to which signal is transmitted will have to be calculated as

$$P_e < \frac{1}{M} \sum_{\substack{j=1 \\ j \neq k}}^{M} \sum_{k=1}^{M} \tfrac{1}{2} \operatorname{erfc} \sqrt{d_{kj}^2/4\eta} \tag{11.15-4}$$

Even when the use of Eq. (11.15-3) is appropriate, the result often turns out to be too complicated to furnish any useful insight. In such cases, it is common to

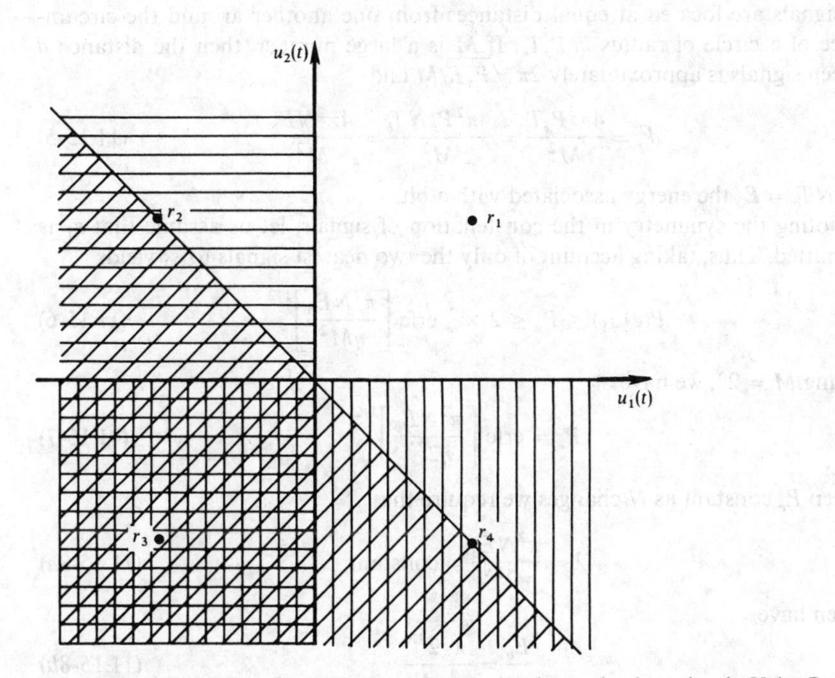

Figure 11.15-2 Signal space for QPSK showing approximation made when using the Union Bound.

make a further approximation. The added approximation consists of including in the summation for $P(e \mid s_1)$ only those terms corresponding to distances between r_1 and its *nearest* neighbors. The justification for such a procedure is to be seen in the following considerations: As a matter of practice, a communications system is useful and of interest only when operated under circumstances where the error probability is very low. Indeed the error probability is usually much less than 10^{-3}. Thus, each term in Eq. (11.15-3) must then be comparable or smaller than 10^{-3}. Now the complementary error function *erfc* x has the property that when erfc x is small it falls off in value extremely rapidly with increasing x. For example (see Sec. 2.28), if $x = 2.2$, erfc $x \cong 2 \times 10^{-3}$ while if x increases by only 50 percent to $x = 3.3$ we find that erfc $x = 3.2 \times 10^{-6}$. This rapid decrease becomes even more pronounced as x increases. A further 50 percent increase to $x = 5$ reduces the function to erfc $x = 1.5 \times 10^{-12}$. Thus, altogether, if the contributions to $P(e \mid s_1)$ from the closest signal to s_1 is small, the contributions from more distant signals is generally negligible. We now consider some useful results which can be obtained by using the approximation we have discussed.

M-ary PSK

The signal space for M-ary PSK is shown in Fig. 6.6-1. Each signal is represented by an N-bit sequence with $2^N = M$. Each signal is located at a distance $\sqrt{P_s T_s}$ from the origin where $T_s = NT_b$. The signals have an angular separation $2\pi/M$. The signals are located at equal distances from one another around the circumference of a circle of radius $\sqrt{P_s T_s}$. If M is a large number, then the distance d between signals is approximately $2\pi\sqrt{P_s T_s}/M$ and

$$d^2 = \frac{4\pi^2 P_s T_s}{M^2} = \frac{4\pi^2 P_s N T_b}{M^2} = \frac{4\pi^2 N E_b}{M^2} \tag{11.15-5}$$

since $NT_b = E_b$ the energy associated with a bit.

Noting the symmetry in the constellation of signals, let us assume that s_1 is transmitted. Thus, taking account of only the two nearest signals to s_1 yields

$$P(e \mid s_1) \leq P_e \leq 2 \times \tfrac{1}{2} \operatorname{erfc} \left[\frac{\pi^2 N E_b}{\eta M^2} \right]^{1/2} \tag{11.15-6}$$

or, using $M = 2^N$, we have

$$P_e = \operatorname{erfc} \left[\frac{\pi^2 N E_b}{\eta 2^{2N}} \right]^{1/2} \tag{11.15-7}$$

To keep P_e constant as N changes we require that

$$\lambda = \frac{\pi^2 N E_b}{\eta 2^{2N}} = \text{constant} \tag{11.15-8a}$$

we then have

$$\frac{E_b}{\eta} = \frac{\lambda}{\pi^2} \frac{2^{2N}}{N} \tag{11.15-8b}$$

Thus the signal energy-to-noise ratio increases nominally in an exponential manner with N for constant P_e.

When M is not so large as to permit the approximation of Eq. (11.15-5) one can readily show that the distance between adjacent signals can always be obtained from the equation

$$d^2 = 4NE_b \sin^2 \pi/M \tag{11.15-9}$$

which reduces to Eq. (11.15-5) for large values of M.

16-QAM (16-QASK)

The signal space for a 16-QASK system is shown in Fig. 11.15-1. As we already noted, because of the lack of symmetry, Eq. (11.15-4) must be used to calculate P_e. Still, as an application of Eq. (11.15-3) let us calculate the error probability for a signal such as s_6 in Fig. 11.15-1 located at (a, a) in Fig. 6.7-1. This signal (in common with s_7, s_{10}, and s_{11}) has the largest error probability because it has four neighboring signals, each at a minimum distance,

$$d = \sqrt{0.4E_s} = \sqrt{1.6E_b} \tag{11.15-10}$$

Accordingly, we find

$$P_e \le 4 \times \tfrac{1}{2} \text{ erfc} \left[\frac{1.6E_b}{4\eta}\right]^{1/2} = 2 \text{ erfc} \left[0.4 \frac{E_b}{\eta}\right]^{1/2} \tag{11.15-11}$$

More exact calculations for the present case, using the union bound approximation and also involving no approximation are left as problems (see Probs. 11.15-3 and 11.15-4).

Orthogonal M-ary FSK

The signal space for orthogonal M-ary FSK for $M = 3$ is shown in Fig. 6.10-3. For any number of orthogonal signals M, the distance between signals is

$$d = \sqrt{2E_s} = \sqrt{2NE_b} \tag{11.15-12}$$

We are not able to draw or visualize a space of more than three dimensions, however, we can conceive of such a space, called a "hyperspace" where the M signals lie. There is complete symmetry among signals; and if we single out any signal, say s_1, for transmission we have

$$P_e = P(e \mid s_1) < \frac{(M-1)}{2} \text{ erfc } \sqrt{2E_s/4\eta} = \frac{2^N - 1}{2} \text{ erfc } \sqrt{NE_b/2\eta} \tag{11.15-13}$$

Observe, in Eq. (11.15-13) that the coefficient $(2^N - 1)/2$ causes P_e to increase with increasing N while the factor N in the argument of erfc causes P_e to decrease with increasing N. It is of interest to note that the erfc can overcome the coefficient at large N and that as N increases P_e can be made to become as small as desired. Using the identity

$$2^N = e^{N \ln 2} \tag{11.15-14}$$

and the approximation valid for large x that

$$\text{erfc } x = \frac{e^{-x^2}}{\sqrt{\pi}\, x} \tag{11.15-15}$$

it can be verified (see Prob. 11.15-6) that P_e may be approximated for large N by

$$P_e < \frac{\exp -N\left[\dfrac{E_b}{2\eta} - \ln 2\right]}{2\sqrt{\pi}\sqrt{NE_b/2\eta}} \tag{11.15-16}$$

Thus, as N increases, P_e decreases so long as

$$\frac{E_b}{\eta} > 2 \ln 2 \cong 1.4 \tag{11.15-17}$$

As a matter of fact, it can be shown by using a tighter bound than the union bound, that P_e decreases with increasing N so long as

$$\frac{E_b}{\eta} > \ln 2 \cong 0.7 \tag{11.15-18}$$

11.16 BIT-BY-BIT ENCODING VERSUS SYMBOL-BY-SYMBOL ENCODING

In BPSK and in BFSK we use only *two* waveforms, each persisting for a time T_b. The two waveforms represent the two possible values of a *single* bit. If our communications system needs to accommodate $M = 2^N$ symbols we need to transmit N bits to convey this symbol. In M-ary PSK or M-ary FSK, where M possible symbols are to be accommodated, we transmit one of M distinct waveforms, each time T_s. If the two systems are to be able to transmit messages at the same rate then we require that $T_s = NT_b$. The first method is described as bit-by-bit encoding, and the second is described as symbol-by-symbol encoding. It is clearly of interest to inquire into which method provides the higher probability that the *entire message* is correctly received.

If the message is transmitted through bit-by-bit encoding the probability that the message is correctly received is the probability that *all* of the N bits are correctly received. If P_{eb} is the probability that a bit is correctly received then the probability that the entire message is correctly received is

$$P_{c,\,\text{message}} = (1 - P_{eb})^N \tag{11.16-1}$$

For BPSK, for example, we have from Eqs. (11.16-1) and (11.13-4)

$$P_{c,\ \text{message}} = \left(1 - \tfrac{1}{2}\operatorname{erfc}\sqrt{E_b/\eta}\right)^N \tag{11.16-2}$$

It is clear that in bit-by-bit encoding, for fixed P_{eb}, the probability of correctly receiving the message *decreases* with increasing N, and therefore the probability that there is some error in the detected message increases with N.

If, on the other hand we employ M-ary FSK and encode each of the M messages into its own distinct and individual waveform, the error probability would be given by Eq. (11.15-13). Then, as we have shown, provided that E_b/η is adequately large, the error probability decreases as N (and hence M) increases. For the case of M-ary PSK, Eq. (11.15-7) applies and we find that the message error probability increases with increasing N.

11.17 RELATIONSHIP BETWEEN BIT ERROR RATE AND SYMBOL ERROR RATE

We have derived a number of expressions which give the *symbol* error probability P_e. If the symbol consists of a single bit, as in BPSK, then, of course, the bit error probability P_{eb} is the same as the symbol error probability P_e. In the more general case, a system encodes N bits into M symbols with $M = 2^N$. In such cases it is of interest to see what may be determined about P_{eb} from calculations of P_e.

When an N-bit symbol is received with error, it may be that 1 bit or 2 bits or that all N bits are in error. Let us assume that the probability P_e of receiving any of these erroneous symbols is the same. With such an assumption we can without difficulty, relate P_{eb} to P_e. As an example of such a calculation let us consider a case with $M = 4$, $N = 2$ as in QPSK. With such a two-bit transmitted symbol, there are three possible received signals which are erroneous. One such received signal will, after detection, have an error in the first bit, a second signal will have an error in the second bit and the third signal will have errors in both bits. The probability of a bit error P_{eb} is then the weighted average:

$$P_{eb} = \frac{\frac{1}{2}P(\text{1st bit error}) + \frac{1}{2}P(\text{2nd bit error}) + \frac{2}{2}P(\text{two-bit error})}{3} \tag{11.17-1}$$

Since we assume that the three symbol error probabilities are the same, P_e, Eq. (11.17-1) becomes:

$$P_{eb} = \tfrac{2}{3}P_e \tag{11.17-2}$$

In the general case, it can be shown (Prob. 11.17-1) that

$$P_{eb} = \frac{M/2}{M-1}P_e \tag{11.17-3}$$

Actually, errors in received symbols which involve many bit errors and which contribute heavily to P_{eb} are less likely than received symbols with fewer bit errors. Hence P_{eb} given by Eq. (11.17-3) is an overestimate, i.e., an upper bound.

Next let us assume that when one of M possible symbols is received and is erroneously decoded, it will be misinterpreted as one or another of only a limited number of the other $M - 1$ symbols. We have already seen in Sec. 11.15 that such an approximation is reasonable. There we noted that, in a realistic situation, the probability is overwhelming that a missed symbol will be read as a symbol which is closest in signal space to the symbol which was transmitted. Finally, let us arrange that the symbols are coded in such manner that closest neighboring symbols differ by only a single bit. Such a coding, using a reflected Grey code, for an eight symbol PSK system is shown in Fig. 11.17-1. Here, if s_0 is misread, it will be almost inevitable that s_1 or s_7 was transmitted. Under the circumstances just described we have the result that a lower bound on P_{eb} is $P_{eb} = P_e/N$. Altogether, using *Grey coding* and assuming that every signal has an *equal likelihood* of transmission, P_{eb} is bounded as follows:

$$\frac{P_e}{N} \leq P_{eb} \leq \frac{M/2}{M-1} P_e \tag{11.17-4}$$

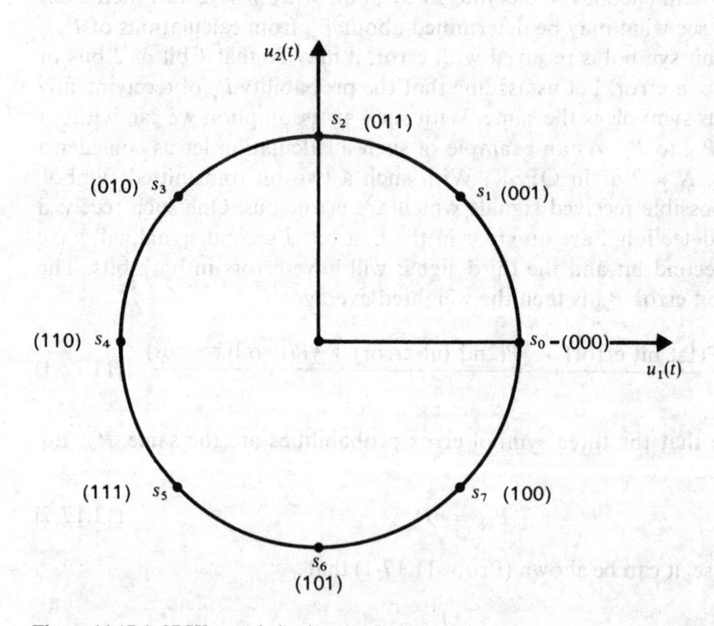

Figure 11.17-1 8PSK encoded using the Grey code.

11.18 PROBABILITY OF ERROR IN A QUADRATURE PARTIAL RESPONSE (QPR) SYSTEM

Figure 11.18-1 shows a radio communication system employing two quadrature carriers. We have already seen the merit of the use of quadrature carriers in our discussion of QPSK (Sec. 6.5) and other systems, i.e., as compared to a single carrier system, the quadrature carrier system provides two independent channels without any increase in required bandwidth. As a result we can use the independent channels to transmit alternate data bits and thereby make available for each bit a time interval twice as long as is available with only one carrier.

The input data bits $d(t)$ of duration T_b are presented alternately to the $\cos \omega_0 t$ and the $\sin \omega_0 t$ channels and a bit so presented to a channel is not changed for a time $2T_b$. This demultiplexing is represented symbolically by the input switch and the details of a scheme by which the bits can be alternated and held are shown in Fig. 6.5-2. The data is encoded into an even and an odd baseband signal and these baseband signals amplitude modulate (suppressed carrier) the two carrier waveforms. At the receiver the baseband signal is recovered by synchronous detection and the data is recovered by the usual procedure of sampling each channel at the proper time in each $2T_b$ interval and then taking the magnitude of the sample and inverting it. The even and odd outputs $d_e(t)$ and $d_o(t)$ are multiplexed (interleaved) to reconstruct the original bit stream $d(t)$.

In Fig. 11.18-1 the duobinary filter with transfer function $H_C(f)$, given by Eq. (6.14-2), has been split into two parts, a part $H_E(f)$ associated with the encoding and a part $H_D(f)$ associated with the decoding. We arrange that

$$H_C(f) = H_E(f) H_D(f) \tag{11.18-1}$$

Accordingly, if we ignore the noise that is added in transmission, then, except for the fact that the baseband signal has been modulated and then demodulated, we still have exactly the same system as in Fig. 6.14-1. In the present case, however, because white noise $n_w(t)$ has been added in transmission we have split the filter so that we can arrange that the decoding filter be the *matched filter* for the waveform generated when an impulse is transmitted through the encoding filter.

In the present case, the transfer function $H_C(f)$ of Eq. (6.14-2) has to be modified by replacing T_b by $2T_b$ (and f_b by $f_b/2$) to take account that each bit is held for a time $2T_b$. Accordingly, we have

$$H_C(f) = 2 \cos 2\pi f T_b \qquad |f| \le f_b/4 \tag{11.18-2}$$

It turns out that if we select

$$H_E(f) = H_D(f) = \sqrt{H_C(f)} \tag{11.18-3}$$

then, as is evident, $H_C(f) = H_E(f) H_D(f)$ and as can be verified (see Prob. 11.18-1) the decoding filter is indeed the matched filter for an impulse transmitted through the encoding filter.

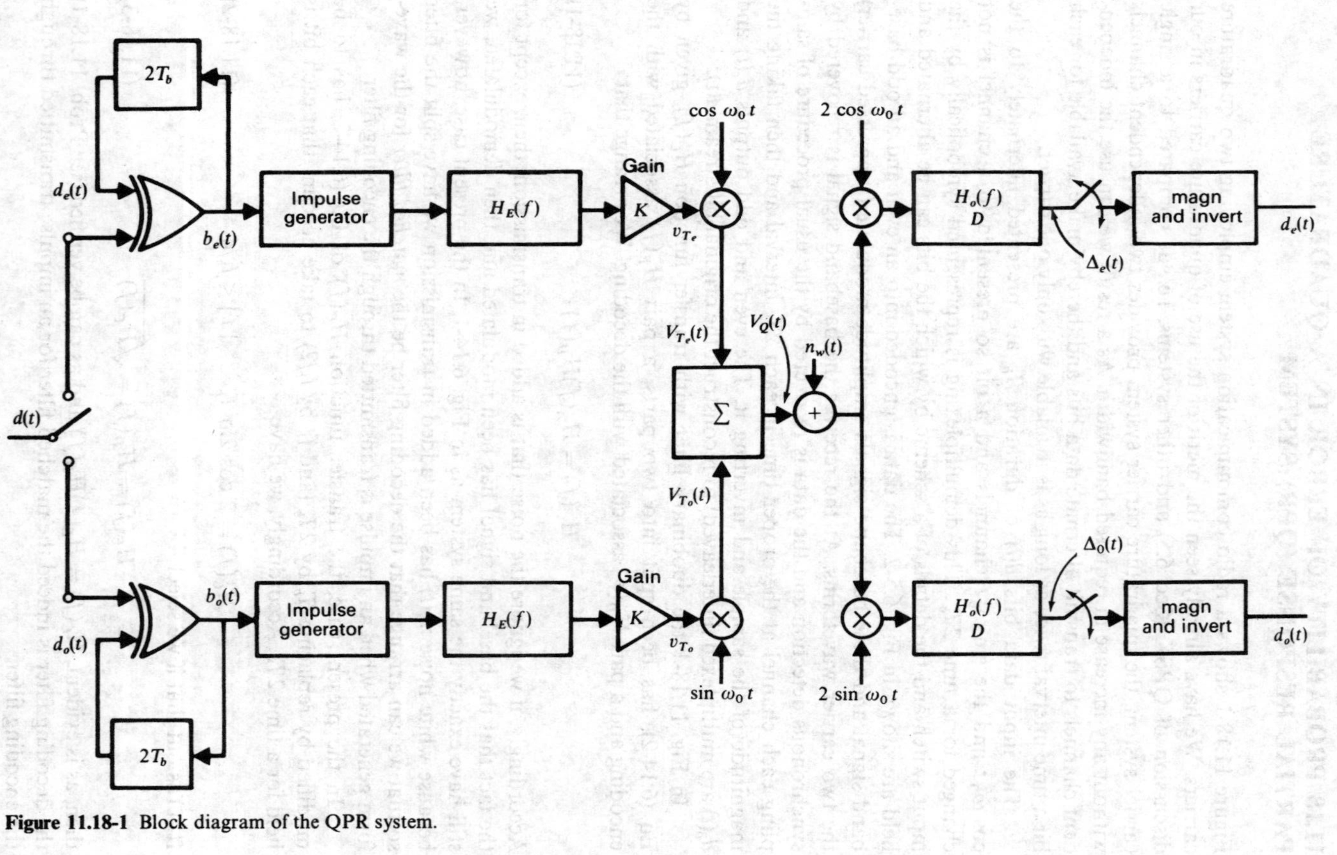

Figure 11.18-1 Block diagram of the QPR system.

We shall now adjust the gain K so that the modulating baseband signals v_{Te} and v_{To} each have a normalized power $P_s = \overline{v_{Te}^2} = \overline{v_{To}^2}$. Assuming impulses of unit strength, the power spectral density of the impulse train applied to the encoding filters is $G(f) = 1/2T_b$. The power spectral density of v_{Te} or of v_{To} is

$$K^2 G(f) |H_E(f)|^2 = \frac{K^2}{2T_b} \cdot 2 \cos 2\pi f T_b = \frac{K^2}{T_b} \cos 2\pi f T_b \tag{11.18-4}$$

The power in v_{Te} or in v_{To} is calculated by integrating the power spectral density in Eq. (11.18-4) over the frequency range from $-f_b/4$ to $+f_b/4$. Accordingly K is determined from the condition that

$$P_s = \int_{-f_b/4}^{+f_b/4} \frac{K^2}{T_b} \cos 2\pi f T_b \, df \tag{11.18-5}$$

from which we find

$$K^2 = \pi T_b^2 P_s \tag{11.18-6}$$

With K^2 selected as in Eq. (11.18-6), v_{Te} and v_{To} are each of power P_s and the transmitted signal $v_Q(t)$ is

$$v_Q(t) = v_{Te} \cos (\omega_0 t + \theta) + v_{To} \sin (\omega_0 t + \theta) \tag{11.18-7}$$

We see, also, that v_Q has a power P_s since

$$\overline{v_Q^2(t)} = \overline{v_{Te}^2}\, \overline{\cos^2 (\omega_0 t + \theta)} + \overline{v_{To}^2}\, \overline{\sin^2 (\omega_0 t + \theta)}$$

and since $\overline{v_{Te}^2} = \overline{v_{To}^2}$, and $\overline{\cos^2 (\omega_0 t + \theta)} = \overline{\sin^2 (\omega_0 t + \theta)} = \frac{1}{2}$,

$$\overline{v_Q^2} = \overline{v_{Te}^2} = \overline{v_{To}^2} = P_s \tag{11.18-8}$$

Equation (11.18-8) is valid because v_{Te} and v_{To} are uncorrelated.

As can be verified, because of the carrier amplitudes selected at the modulator and demodulator as indicated in Fig. 11.18-1, the demodulator multiplier outputs, neglecting the noise, are precisely v_{Te} and v_{To}. That is, as a matter of computational convenience, we have arranged that the baseband signal applied to the modulating multiplier is recovered unaltered at the output of the demodulating multiplier. We have assumed impulses of unit strength. Such impulses have a Fourier transform of magnitude unity over all frequencies. Hence, altogether, the Fourier transform of the signals $\Delta_e(t)$ and $\Delta_o(t)$ which are to be sampled are

$$\Delta_e(f) = \Delta_o(f) = 1 \cdot K \cdot H_E(f) \cdot H_D(f) \tag{11.18-9a}$$

$$= K H_C(f) = \sqrt{\pi T_b^2 P_s} \cdot 2 \cos 2\pi f T_b \tag{11.18-9b}$$

$$= \sqrt{4\pi T_b^2 P_s} \cos 2\pi f T_b \qquad |f| \leq f_b/4 \tag{11.18-9c}$$

Taking the inverse transform (see Prob. 11.18-3) we find

$$\Delta_e(t) = \Delta_o(t) = \sqrt{\frac{4P_s}{\pi}} \frac{\cos \pi f_b t/2}{1 - (f_b t)^2} \tag{11.18-10}$$

Examining Fig. 6.14-1 we note that when the bit has a duration T_b the sample is to be taken at time $t = T_b/2$. Since, in the present case a bit interval has the duration $2T_b$, the sampling is to be done at $t = T_b$. We then find that

$$\Delta_e(t = T_b) = \Delta_o(t = T_b) = \sqrt{\frac{\pi P_s}{4}} \tag{11.18-11}$$

We recall that at each sample time the total output is due to two adjacent input pulses. The total sample value will then be double the value given in Eq. (11.18-11) if two successive inputs are of the same polarity or will be zero if two successive inputs are of opposite polarity. Altogether then, the outputs are

$$d_e(\text{or } d_o) = \pm\sqrt{\pi P_s} \text{ or } 0 \tag{11.18-12}$$

The white noise has a power spectral density $\eta/2$. After passing through a demodulator multiplier whose input carrier is $2 \cos \omega_0 t$ or $2 \sin \omega_0 t$ the noise will have a density η. After filtering by the decoding filter with transfer function $H_D(f)$ the mean square value of the noise, i.e., its variance squared will be

$$\sigma_N^2 = \int_{-f_b/4}^{+f_b/4} \eta\,|H_D(f)|^2\, df = 4\eta \int_0^{f_b/4} \cos 2\pi f T_b\, df \tag{11.18-13a}$$

$$= \frac{2\eta}{\pi T_b} \tag{11.18-13b}$$

The noise adds to the signal, and the sampling process samples the superposition of the two. An error will be made whenever the noise component of the sample is larger than one half the signal component. Both the white noise and the filtered noise are Gaussian, hence the probability of error P_e is the probability that the noise exceeds $\sqrt{\pi P_s}/2$

$$P_e = \int_{\sqrt{\pi P_s}/2}^{\infty} \frac{e^{-n^2/2\sigma_N^2}}{\sqrt{2\pi\sigma_N^2}}\, dn \tag{11.18-14}$$

Defining $x \equiv n/\sqrt{2}\sigma_N$ and using Eq. (11.18-13b), Eq. (11.18-14) becomes

$$P_e = \frac{1}{2}\frac{2}{\sqrt{\pi}} \int_{\sqrt{(\pi^2 P_s T_b/16\eta)} = \sqrt{(\pi^2 E_b/16\eta)}}^{\infty} e^{-x^2}\, dx = \tfrac{1}{2}\,\text{erfc}\left[\frac{\pi^2 E_b}{16\eta}\right]^{1/2} \tag{11.18-15}$$

We note that the channel bandwidth required to transmit a QPR signal is $BW = f_b/2$ which is the same bandwidth required to transmit 16 QAM or 16 phase PSK. If we compare the probability of symbol error for these systems we find that QPR has a significantly lower error rate. For example, compared with 16 QAM, QPR requires approximately two-thirds the signal-to-noise ratio to achieve the same error rate.

11.19 PROBABILITY OF ERROR OF MINIMUM SHIFT KEYING (MSK)

Referring to Fig. 6.11-4 we see that MSK employs synchronous detection to separate the even and odd bit streams $b_e(t)$ and $b_o(t)$. Furthermore, the correlation interval is $2T_b$ as in QPSK. Thus, to all intents and purposes, the synchronous detector (correlator) could be processing a QPSK signal rather than an MSK signal. Hence, we would expect that the probability of error of MSK be the same as the probability of error of QPSK:

$$P_{eb}(\text{MSK}) = P_{eb}(\text{QPSK}) = \frac{1}{2} \text{erfc} \sqrt{\frac{E_b}{\eta}} \tag{11.19-1}$$

To show that such is indeed the case, let us refer to the signal space representation of MSK shown in Fig. 6.11-2. There we see that $d^2 = 4E_b$ and therefore the symbol error rate of MSK is

$$P_e(\text{MSK}) = 2 \cdot \frac{1}{2} \text{erfc} \sqrt{\frac{d^2}{4\eta}} = \text{erfc} \sqrt{\frac{E_b}{\eta}} \tag{11.19-2a}$$

Hence the bit error rate of MSK is

$$P_{eb}(\text{MSK}) = \frac{1}{2} P_e(\text{MSK}) = \frac{1}{2} \text{erfc} \sqrt{\frac{E_b}{\eta}} \tag{11.19-2b}$$

as expected.

11.20 COMPARISON OF MODULATION SYSTEMS

Table 11.20-1 presents a summary of the error probability, P_e, and the half-bandwidth of the digital modulation techniques discussed in this chapter for the case of white noise interference. The half-bandwidth, W is the bandwidth of the

Table 11.20-1 Summary of P_e and W for Digital Modulation Techniques (M = number of symbols, N = number of bits/symbol, $2^N = M$)

Modulation Technique	P_e	W
BPSK, QPSK, MSK	$\frac{1}{2} \text{erfc} [E_b/\eta]^{1/2}$	f_b/N
MPSK	$\text{erfc} \left[\frac{NE_b}{\eta} \sin^2 \pi/M\right]^{1/2}$	f_b/N
16 QAM	$2 \text{erfc} [0.4E_b/\eta]^{1/2}$	$f_b/4$
Orthogonal MFSK	$\frac{M-1}{2} \text{erfc} [NE_b/2\eta]^{1/2}$	Mf_b/N
QPR	$\frac{1}{2} \text{erfc} \left[\frac{\pi^2}{16} \frac{E_b}{\eta}\right]^{1/2}$	$f_b/4$

modulating baseband signal. For SSB modulation W is the bandwidth of both the modulating baseband signal and the modulated carrier. However, all of the other cases are doubled sideband systems and hence:

$$W = \frac{B}{2} \tag{11.20-1}$$

B being the (two-sided) bandwidth of the modulated carrier. For the sake of having a uniform basis of comparison, we have in every case tabulated the error probability P_e.

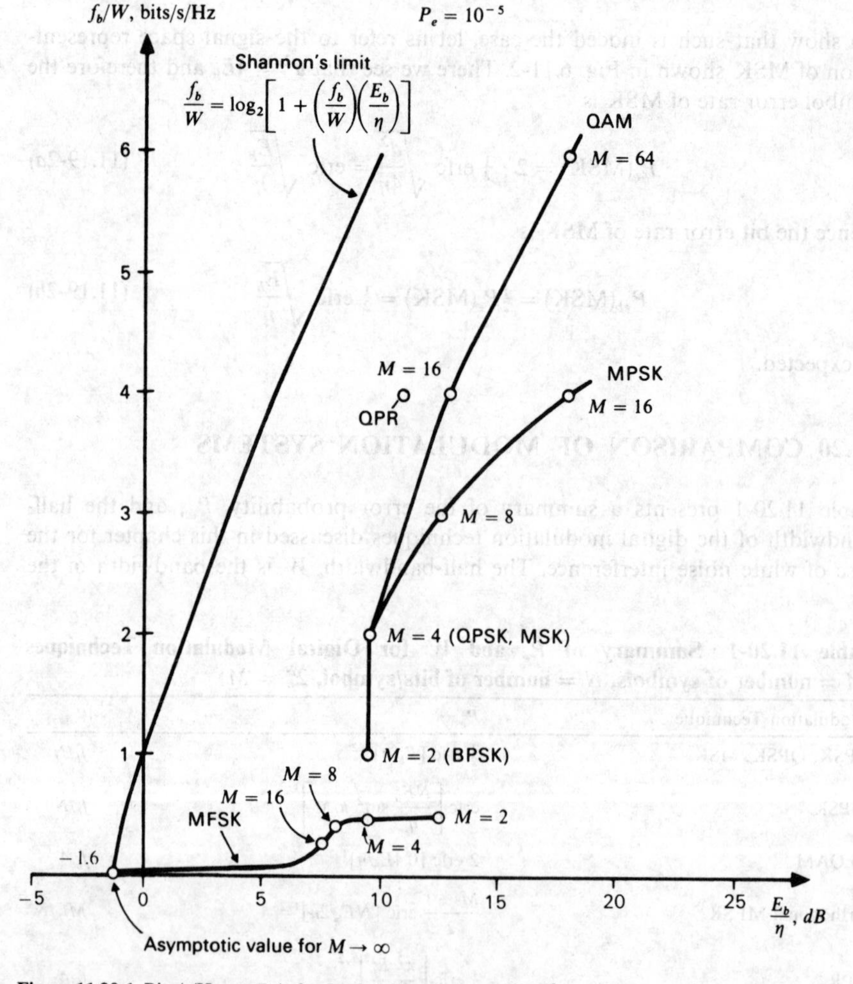

Figure 11.20-1 Bits/s/Hz vs. E_b/η for Probability of error $= 10^{-5}$.

In connection with a communication system which is to be designed, it is common place that, in addition to such practical constraints as cost, volume, etc., there be a specification on allowable error rate. In such circumstances, plots as shown in Fig. 11.20-1 are very useful. Here P_e is fixed at $P_e = 10^{-5}$. As we shall see in Sec. 13.7, there is an ultimate limit to the performance of a communications system. This limit, deduced by Shannon, is given by the inequality:

$$\frac{f_b}{W} \leq \log_2 \left[1 + \left(\frac{f_b}{W}\right)\left(\frac{E_b}{\eta}\right) \right] \tag{11.20-2}$$

and is independant of the error probability. We have included a plot of Eq. (11.20-2) in Fig. 11.20-1. In Fig. 11.20-1, as well as for any given error probability, P_e, we shall always find that all modulation schemes yield plots which lie to the *right* of Shannon's limiting plot.

Figure 11.20-1 shows that QAM more closely approaches Shannon's curve than does MPSK, indicating that QAM is a more efficient system for operation in a white gaussian noise environment. Note that when we compare MPSK with MFSK (see Table 11.20-1 and Fig. 11.20-1) we find that as M increases the bandwidth of MPSK decreases while the SNR required to obtain $P_e = 10^{-5}$ increases. However, with MFSK the bandwidth increases while the SNR decreases. In all cases however we note that the *slope* of each curve is positive, so that an increase in f_b/W requires an increase in E_b/η if a constant bit error rate is to be maintained, i.e., for a constant error rate

$$\frac{f_b}{W} \sim \frac{E_b}{\eta} \tag{11.20-3}$$

If the error rate were, for example, $P_e = 10^{-6}$ each curve would shift approximately 1 dB to the *right*, except of course, for Shannon's curve which is independent of P_e.

REFERENCES

1. Stein, S., and J. Jones: "Modern Communication Principles," McGraw-Hill Book Company, New York, 1967.
2. Schwartz, M., W. R. Bennett, and S. Stein: "Communication Systems and Techniques," McGraw-Hill Book Company, New York, 1966.
3. Wozencraft, J., and I. Jacobs: "Communication Engineering," John Wiley and Sons, New York, 1966.

PROBLEMS

11.1-1. (*a*) Find the power spectral density $G_n(f)$ of noise $n(t)$ which has an autocorrelation function

$$R_n(\tau) = \sigma^2 e^{-|\tau/\tau_0|}$$

(*b*) The noise in (*a*) is applied to an integrator at $t = 0$. Find the mean square value $\overline{[n_o(T)]^2}$ of the noise output of the integrator at $t = T$.

(c) The noise in (a) accompanies a signal which consists of either the voltage $+V$ or the voltage $-V$ sustained for a time T. At time $t = T$ find the ratio of the integrator output due to the signal to the rms noise voltage.

11.1-2. A received signal $s(t) = \pm V$ is held for an interval T. The signal is accompanied by white gaussian noise of power spectral density $\eta/2$. The received signal is to be processed as in Fig. 11.1-2. However, as an approximation to the required integrator we use a low-pass RC circuit of 3-dB bandwidth f_c. Calculate the value of f_c for which the signal-to-noise voltage, at the sampling time, will be a maximum. For this value of f_c calculate the signal-to-rms noise ratio and compare with Eq. (11.1-6) which applies when an integrator is used. Show that for the RC network the signal-to-noise ratio is about 1 dB smaller than for the integrator.

11.2-1. A signal which can assume one of the voltage $+V$ or $-V$ is transmitted. Consider that the probability of transmitting $+V$ is $\frac{3}{4}$ while the probability of transmitting $-V$ is $\frac{1}{4}$. The signal is accompanied by white gaussian noise.

(a) Assume that the threshold voltage for decision between the two possible signals is V_t rather than zero volts. Write an expression for the probability that an error in decision will be made. An integrate and dump receiver is used as in Fig. 11.1-2.

(b) Find V_t such that the probability of error is a minimum and calculate the corresponding probability of error.

11.2-2. A signal, which can take on the voltages $+V$, 0, $-V$ with equal likelihood, is transmitted. When received, it is embedded in white gaussian noise. The receiver integrates the signal and noise for a time T s.

Write an expression for the threshold voltages $\pm V_t$ so that the probability of error is independent of which signal is transmitted.

11.2-3. A received signal is either $+2V$ or $-2V$ held for a time T. The signal is corrupted by white gaussian noise of power spectral density 10^{-4} volt^2/Hz. If the signal is processed by an integrate and dump receiver, what is the minimum time T during which a signal must be sustained if the probability of error is not to exceed 10^{-4}?

11.3-1. A transmitter transmits the signals $\pm V$ with equal probability; the channel noise has the power spectral density $G_n(f) = G_0/[1 + (f/f_1)^2]$.

(a) Find the transfer function $H(f)$ of the *matched filter*, and comment on its realizability.

(b) Find the average probability of error when using the matched filter.

(c) An integrator is employed rather than the optimum filter. Find its P_e and compare with (b).

11.5-1. A signal is either $s_1(t) = A \cos 2\pi f_0 t$ or $s_2(t) = 0$ for an interval $T = n/f_0$ with n an integer. The signal is corrupted by white noise with $G_n(f) = \eta/2$. Find the transfer function of the matched filter for this signal. Write an expression for the probability of error P_e.

11.5-2. Repeat Prob. 11.5-1 if the signal is $s(t) = \pm A(1 - \cos 2\pi f_0 t)$.

11.6-1. Compare the outputs of the MF and the correlator, when the input signal is either $\pm V$, as a function of time t for $0 \leq t \leq T$. Assume white gaussian noise. Are the outputs the same for all t, or just when $t = T$?

11.6-2. A signal is $s(t) = \pm 2(t/T)$ for $0 \leq t \leq T$. The signal is corrupted by white gaussian noise of power spectral density 10^{-6} volt^2/Hz.

(a) Draw the signal waveform at the output of a matched filter receiver.

(b) If the probability of error P_e is to be no larger than 10^{-4}, find the minimum allowable interval T.

11.7-1. Verify Eq. (11.7-5).

11.8-1. Plot Eq. (11.8-8) versus E_s/η.

11.8-2. If the frequency offset Ω in FSK satisfies $\Omega T = n\pi$, $s_1(t)$ and $s_2(t)$ are orthogonal.

(a) Prove this statement.

(b) Calculate P_e.

(c) Plot P_e versus E_s/η and compare with the results given in Eq. (11.8-8).

11.8-3. Plot the P_e in binary FSK as a function of ΩT. Select $E_s/\eta = 15$.

11.9-1. Plot Eq. (11.9-1) versus E_s/η.

11.11-1. M-ary PSK involves choosing signals of the form $(\cos \omega_0 t + \theta_i)$ for M values of i.

(a) Show how to choose the θ_i so that the probability of error of each θ_i is the same.

(b) Find the correlator detector.

(c) Obtain an expression for the probability of error.

11.11-2 Verify the entries in Table 11.11-1.

11.12-1. Verify Eq. (11.12-2).

11.13-1. Refer to Eq. (11.13-5a and b). Determine $\omega_2 - \omega_1$ such that the unit vectors $u_1(t)$ and $u_2(t)$ are orthonormal.

11.13-2. (a) If $u_1(t) = \sqrt{2/T_b}\cos(\omega_1 t + \Theta)$ and $u_2(t) = \sqrt{2/T_b}\cos(\omega_2 t + \phi)$ where Θ and ϕ are independent random variables uniformly distributed between $-\pi \leq \Theta, \phi < \pi$, find $\omega_2 - \omega_1$ so that u_1 and u_2 are orthonormal.

(b) Compare your answer to the answer obtained for Prob. 11.13-1. Why do they differ?

11.13-3. Plot a graph having an ordinate which is f_b/W and an abscisssa which is E_b/η for BPSK and BFSK for $P_e = 10^{-4}, 10^{-5}$ and 10^{-6}. Use Eqs. (11.13-4) and (11.13-10a) and the results,

$$B(\text{BPSK}) = 2f_b \quad \text{and} \quad B(\text{BFSK}) = 4f_b.$$

11.14-1. Verify that Eq. (11.14-3) reduces to Eq. (11.14-4).

11.15-1. Equations (11.15-5) and (11.15-7) assume that the *arc* of a circle is of the same length as the *chord*. (a) Derive an expression for P_e to replace Eq. (11.15-7) if the above approximation is *not* made. (b) If the approximation is to be used but the resulting P_e is to be correct to within 50 percent determine the minimum value of M.

11.15-2. If the probability of BPSK, BFSK and MPSK for $M = 4$, 8, 16, 32, and 64 are to be the same, what is the ratio of their E_b/η compared to $(E_b/\eta)_{\text{BPSK}}$?

11.15-3. Calculate P_e for 16 QAM using the Union Bound. Compare your answer with that of Eq. (11.15-11).

11.15-4. Calculate P_e for 16 QAM without using the Union Bound approximation.

11.15-5. (a) Using the Union Bound approximation, calculate P_e for 64 and 256 QAM.

(b) To obtain the same P_e for 16, 64 and 256 QAM determine the ratio of each E_b/η as compared to $(E_b/\eta)_{\text{BPSK}}$.

11.15-6. Verify Eq. (11.15-16).

11.15-7. A 9.6 kb/s NRZ data stream is to be transmitted over a 2.4 kHz bandwidth channel. What modulation system would you choose if an error rate of 10^{-4} is to be achieved with a minimum signal-to-noise ratio (minimum E_b/η)?

11.15-8. A 9.6 kb/s NRZ data stream is to be transmitted over a 2.4 kHz bandwidth channel. An error rate of 10^{-4} is desired. Due to channel nonlinearities and its phase characteristics, intersymbol interference (ISI) is produced which results in an "eye pattern" which closes by 3 dB compared to the response obtained when the transmitter is connected directly to the receiver (by-passing the channel) when E_b/η is 12 dB.

Can a modulation system be found to achieve this error rate of 10^{-4}? What is the value of E_b/η that is needed?

11.16-1. The probability of a bit being in error is 10^{-3}. (a) If a message consists of 10 bits, calculate the probability of the message being in error. (b) Repeat (a), if the message length is 100, 1000, 10,000. Note how quickly the probability of a message being in error increases toward unity.

11.16-2. The probability of a bit being in error is 10^{-3}. (a) What is the probability of a 6-bit word containing an error? (b) The bits are grouped so that instead of transmitting 6 bits, bit-by-bit, a 64 phase, 64 PSK signal is sent. (The symbol rate is one sixth of the bit rate.) Calculate the probability of a 6-bit symbol being in error. (c) Repeat (b) using 64 QAM. (d) Repeat using 64 FSK. (e) Discuss your results.

11.17-1. Verify Eq. (11.17-2).

11.17-2. Consider that bits are grouped so that there are 4 bits/symbol and then modulated using 16 QAM. Following Fig. 11.17-1 show how to arrange the Grey code so that an error in a symbol will with high probability correspond to an error in only 1 of the 4 bits.

11.18-1. Prove that $H_D(f)$ as defined by Eqs. (11.18-2) and (11.18-3) is the matched filter for an impulse transmitted through $H_E(f)$.

11.18-2. Verify Eq. (11.18-6).

11.18-3. Verify Eq. (11.18-9) and Eq. (11.18-10).

11.18-4. Verify Eq. (11.18-11).

11.18-5. $\Delta_0(kT_s)$ and $\Delta_e(kT_s)$ can each assume one of three values. Thus, a signal space sketch would indicate nine signal points. Determine the nine signal points in signal space to yield Eq. (11.18-15).

11.18-6. A probability of error of 10^{-5} is desired and a channel bandwidth of 20 kHz is available. If the bit rate is 80 kb/s, 16 PSK, 16 QAM, or QPR can be used. Calculate the value of E_b/η required for each of these systems.

11.19-1. MSK can be viewed as an FM system with a peak frequency deviation of $f_b/4$. If coherent FM detection is employed as in MFSK (see Sec. 11.8) calculate the probability of error.

11.19-2. MSK can be detected in a noncoherent manner using an FM discriminator. Referring to Fig. 10.2-1, the 1F filter bandwidth is set to $B = 1.5\,f_b$ (since the spectral density of MSK is zero at $f - f_c = 0.75\,f_b$, very little required signal power lies outside this region and hence ISI can be neglected). The baseband filter is often replaced by an integrate-and-dump filter, which integrates over each bit, i.e. for a time T_b.

(*a*) Using Eqs. (9.2-16), (9.2-21), and (10.6-1) where$|\delta f| = f_b/4$, calculate the probability of error. (*b*) Why is $f = f_b/4$? (*c*) How much degradation is produced when the FM discriminator is employed?

11.20-1. Using Table 11.20-1 and Shannon's limit, plot Fig. 11.20-1 for $P_e = 10^{-6}$.

CHAPTER

TWELVE

NOISE IN PULSE-CODE AND DELTA-MODULATION SYSTEMS

12.1 PCM TRANSMISSION

A binary PCM transmission system is shown in Fig. 12.1-1. The baseband signal $m(t)$ is quantized, giving rise to the quantized signal $m_q(t)$, where

$$m_q(t) = m(t) + e(t) \tag{12.1-1}$$

The term $e(t)$ is the error signal which results from the process of quantization. The quantized signal is next sampled. Sampling takes place at the Nyquist rate. The sampling interval is $T_s = 1/2f_M$, where f_M is the frequency to which the signal $m(t)$ is bandlimited.

Sampling is accomplished by multiplying the signal $m_q(t)$ by a waveform which consists of a periodic train of pulses, the pulses being separated by the sampling interval T_s. We shall assume that the sampling pulses are narrow enough so that the sampling may be considered as instantaneous. It will be recalled (Sec. 5.1) that with such instantaneous sampling, the sampled signal may be reconstructed *exactly* by passing the sequence of samples through a low-pass filter with cut-off frequency at f_M. Now, as a matter of mathematical convenience, we shall represent each sampling pulse as an impulse. Such an impulse is infinitesimally narrow yet is characterized by having a finite area. The area of an impulse is called its *strength*, and an impulse of strength I is written $I\delta(t)$. The sampling-impulse train is therefore $S(t)$, given by

$$S(t) = I \sum_{k=-\infty}^{\infty} \delta(t - kT_s) \qquad T_s = \frac{1}{2f_M} \tag{12.1-2}$$

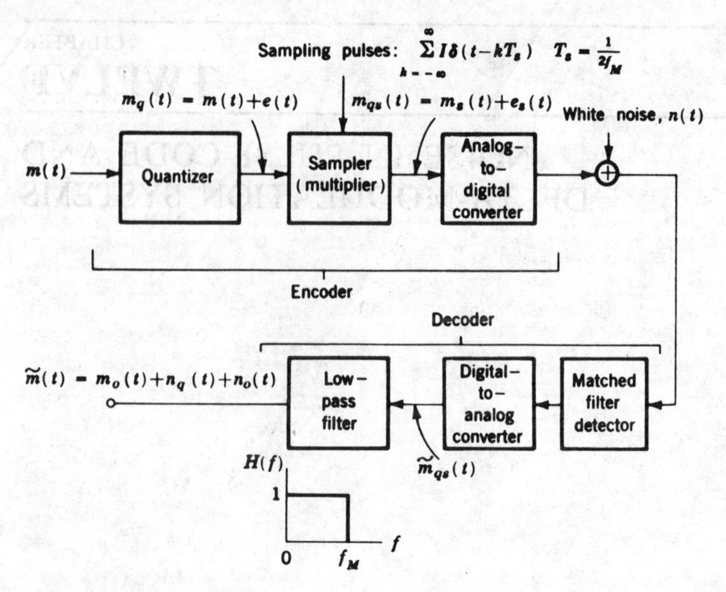

Figure 12.1-1 A binary PCM encoder-decoder.

From Eqs. (12.1-1) and (12.1-2) the quantized signal $m_q(t)$ after sampling becomes $m_{qs}(t)$, written as

$$m_{qs}(t) = m(t)I \sum_{k=-\infty}^{\infty} \delta(t - kT_s) + e(t)I \sum_{k=-\infty}^{\infty} \delta(t - kT_s) \tag{12.1-3a}$$

$$= m_s(t) + e_s(t) \tag{12.1-3b}$$

The quantized, sampled signal $m_{qs}(t)$ which consists of the sum of two strength-modulated impulse trains is applied to an analog-to-digital converter. (In a physical system, the input to the A/D converter is a quantized amplitude-modulated *pulse* train.) The binary output of the A/D converter is transmitted over a communication channel and arrives at the receiver contaminated as a result of the addition of white thermal noise $n(t)$. Transmission may be direct, as indicated in Fig. 12.1-1, or the binary output signal may be used to modulate a carrier as in PSK or FSK. In any event, the received signal is detected by a matched filter to minimize errors in determining each binary bit and thereafter passed on to a D/A converter. The output of the D/A converter is called $\tilde{m}_{qs}(t)$. [In the absence of thermal noise, and assuming unity gain from the input to the A/D converter to the output of the D/A converter, we would have $\tilde{m}_{qs}(t) = m_{qs}(t)$.] Finally, the signal $\tilde{m}_{qs}(t)$ is passed through the low-pass baseband filter. At the output of this filter we find a signal $m_o(t)$ which, aside from a possible difference in amplitude, has exactly the waveform of the original baseband signal $m(t)$. This output signal, however, is accompanied by a noise waveform $n_q(t)$, which is due to the quantization, and an additional noise waveform $n_{th}(t)$, due to the thermal noise.

12.2 CALCULATION OF QUANTIZATION NOISE

We shall now temporarily ignore the effect of the thermal noise and shall calculate the output power due to the quantization noise in the PCM system of Fig. 12.1-1.

The sampled quantization error waveform, as given by Eq. (12.1-3), is

$$e_s(t) = e(t)I \sum_{k=-\infty}^{\infty} \delta(t - kT_s) \tag{12.2-1}$$

It is to be noted that if the sampling rate is selected to be the Nyquist rate for the baseband signal $m(t)$, the sampling rate will be inadequate to allow reconstruction of the error signal $e(t)$ from its samples $e_s(t)$. That such is the case is readily apparent from Fig. 12.2-1. In Fig. 12.2-1*a* is shown the relationship between $m_q(t)$ and $m(t)$, while in Fig. 12.2-1*b* is shown the error waveform $e(t)$ as a function of $m(t)$. The quantization levels are separated by amount S. We observe in Fig. 12.2-1*b* that $e(t)$ executes a complete cycle and exhibits an abrupt discontinuity every time $m(t)$ makes an excursion of amount S. Hence the spectral range of $e(t)$ extends far beyond the bandlimit f_M of $m(t)$.

To find the quantization noise output power N_q, we require the power spectral density of the sampled quantization error $e_s(t)$, given in Eq. (12.2-1). Since $\delta(t - kT_s) = 0$ except when $t = kT_s$, $e_s(t)$ may be written

$$e_s(t) = I \sum_{k=-\infty}^{\infty} e(kT_s)\delta(t - kT_s) \tag{12.2-2}$$

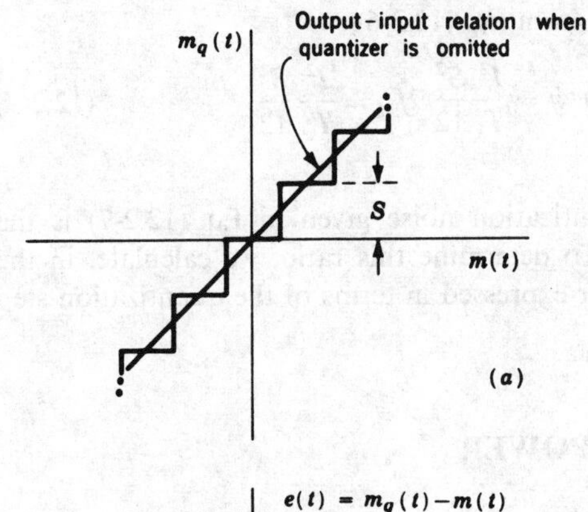

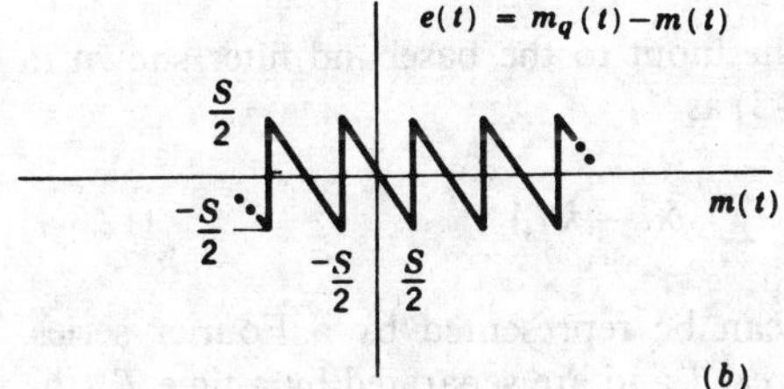

Figure 12.2-1 (*a*) Plot of $m_q(t)$ as a function of $m(t)$. (*b*) Plot of $e(t)$ as a function of $m(t)$.

The waveform of Eq. (12.2-2) consists of a sequence of impulses of area (strength) $A = e(kT_s)I$ occurring at intervals T_s. The quantity $e(kT_s)$ is the quantization error at the sampling time and is a random variable. In Sec. 2.25 we calculated the power spectral density of just such a waveform and arrived at the result given in Eq. (2.25-7). Applying Eq. (2.25-7) to the present case, we find that the power spectral density $G_{e_s}(f)$ of the sampled quantization error is

$$G_{e_s}(f) = \frac{I^2}{T_s} \overline{e^2(kT_s)} \tag{12.2-3}$$

In Sec. 6.3 [see Eq. (6.3-3)] we found that if the quantization levels are separated by amount S, then the quantization error is given by

$$\overline{e^2(t)} = \frac{S^2}{12} \tag{12.2-4}$$

Equation (12.2-3) involves $\overline{e^2(kT_s)}$ rather than $\overline{e^2(t)}$. However, since the probability density of $e(t)$ does not depend on time, the variance of $e(t)$ is equal to the variance of $e(t = kT_s)$. Thus

$$\overline{e^2(t)} = \overline{e^2(kT_s)} = \frac{S^2}{12} \tag{12.2-5}$$

From Eqs. (12.2-3), (12.2-4), and (12.2-5) we have

$$G_{e_s}(f) = \frac{I^2 S^2}{T_s 12} \tag{12.2-6}$$

Finally, the quantization noise N_q is, from Eq. (12.2-6),

$$N_q = \int_{-f_M}^{f_M} G_{e_s}(f)\, df = \frac{I^2}{T_s} \frac{S^2}{12} 2f_M = \frac{I^2}{T_s^2} \frac{S^2}{12} \tag{12.2-7}$$

since $2f_M = 1/T_s$.

Of more interest than the quantization noise given in Eq. (12.2-7) is the signal-to-quantization noise ratio. To determine this ratio, we calculate, in the next section, the output-signal power expressed in terms of the quantization step size S.

12.3 THE OUTPUT-SIGNAL POWER

The sampled signal which appears at the input to the baseband filter shown in Fig. 12.1-1 is given by $m_s(t)$ in Eq. (12.1-3) as

$$m_s(t) = m(t)I \sum_{k=-\infty}^{\infty} \delta(t - kT_s) \tag{12.3-1}$$

Since the impulse train is periodic it can be represented by a Fourier series. Because the impulses have a strength (area) I and are separated by a time T_s, the first term in the Fourier series is the dc component which is $1/T_s$. Hence the signal $m_o(t)$ at the output of the baseband filter is

$$m_o(t) = \frac{I}{T_s} m(t) \tag{12.3-2}$$

Since $T_s = 1/2f_M$, other terms in the series of Eq. (12.3-1) lie outside the passband of the filter. The normalized signal output power is, from Eq. (12.3-2),

$$\overline{m_o^2} = \frac{I^2}{T_s^2} \overline{m^2(t)} \tag{12.3-3}$$

We shall now express $\overline{m^2(t)}$ in terms of the number M of quantization levels and the step size S. To do this we assume that the signal can vary from $-MS/2$ to $+MS/2$, that is, we assume that the instantaneous value of $m(t)$ may fall anywhere in its allowable range of MS volts with equal likelihood. Then the probability density of the instantaneous value of m is $f(m)$ given by

$$f(m) = \frac{1}{MS} \tag{12.3-4}$$

The variance of $m(t)$, that is, $\overline{m^2(t)}$, is

$$\overline{m^2(t)} = \int_{-MS/2}^{MS/2} m^2 f(m)\, dm = \int_{-MS/2}^{MS/2} \frac{m^2 dm}{MS} = \frac{M^2 S^2}{12} \tag{12.3-5}$$

Hence, from Eq. (12.3-3), the output-signal power is

$$S_o = \overline{m_o^2(t)} = \frac{I^2}{T_s^2} \frac{M^2 S^2}{12} \tag{12.3-6}$$

From Eqs. (12.2-7) and (12.3-6) we find that the signal-to-quantization-noise ratio is

$$\frac{S_o}{N_q} = M^2 = (2^N)^2 = 2^{2N} \tag{12.3-7}$$

where N is the number of binary digits needed to assign individual binary-code designations to the M quantization levels.

12.4 THE EFFECT OF THERMAL NOISE

The effect of additive thermal noise is to cause the matched filter detector of Fig. 12.1-1 to make an occasional error in determining whether a binary 1 or a binary 0 was transmitted. As we saw in Chap. 11, if the thermal noise is white and gaussian, the probability of such an error depends on the ratio E_b/η, where E_b is the signal energy transmitted during a bit and $\eta/2$ is the two-sided power spectral density of the noise. The error probability depends also on the type of modulation employed, i.e., direct transmission, PSK, FSK, etc.

Rather typically, PCM systems operate with error probabilities which are small enough so that we may ignore the likelihood that more than a single bit error will occur within a single word. By way of example, if the error probability $P_e = 10^{-5}$ and a word has 8 bits ($N = 8$), we would expect, on the average, that 1 word would be in error for every 12,500 words transmitted. Indeed, the probability of two bits being in error in the same 8 bit word when $P_e = 10^{-5}$ is $P_2 = \binom{8}{2} P_e^2(1 - P_e)^6 \simeq 28 \times 10^{-10}$.

Let us assume that a code word used to identify a quantization level has N binary digits. We assume further that the assignment of code words to levels is in the order of the numerical significance of the word. Thus, we assign 00 ... 00 to the most negative level, 00 ... 01 to the next-higher level until the most positive level is assigned the code word 11 ... 11.

An error which occurs in the least significant bit of the code word corresponds to an incorrect determination by amount S in the quantized value $m_s(t)$ of the sampled signal. An error in the next-higher significant bit corresponds to an error $2S$; in the next-higher, $4S$, etc. Let us call the error Δm_s. Then assuming that an error may occur with equal likelihood in any bit of the word, the variance of the error is

$$\overline{(\Delta m_s)^2} = \frac{1}{N}[S^2 + (2S)^2 + (4S)^2 + (8S)^2 + \cdots + (2^{N-1}S)^2] \tag{12.4-1}$$

The sum of the geometric progression in Eq. (12.4-1) is

$$\overline{(\Delta m_s)^2} = \frac{2^{2N} - 1}{3N} S^2 \cong \frac{2^{2N}}{3N} S^2 \tag{12.4-2}$$

for $N \geq 2$.

The preceding discussion indicates that the effect of thermal-noise errors may be taken into account by adding, at the input to the A/D converter in Fig. 12.1-1, an error voltage Δm_s, and by deleting the white-noise source and the matched filter. We have assumed unity gain from the input to the A/D converter to the output of the D/A converter. Thus, the same error voltage appears at the input to the low-pass baseband filter. The result of a succession of errors is a train of impulses, each of strength $I\, \Delta m_s$. These impulses are of random amplitude and of random time of occurrence.

A thermal-noise error impulse occurs on each occasion when a word is in error. With P_e the probability of a bit error, the mean separation between bits which are in error is $1/P_e$ bits. With N bits per word, the mean separation between words which are in error is $1/NP_e$ words. Words are separated in time by the sampling interval T_s. Hence the mean time between words which are in error is T, given by

$$T = \frac{T_s}{NP_e} \tag{12.4-3}$$

Again using Eq. (2.25-7), we find that the power spectral density of the thermal-noise error impulse train is, using Eqs. (12.4-2) and (12.4-3),

$$G_{th}(f) = \frac{I^2\overline{(\Delta m_s)^2}}{T} = \frac{NP_e I^2\overline{(\Delta m_s)^2}}{T_s} \tag{12.4-4}$$

Using Eq. (12.4-2), we have

$$G_{th}(f) = \frac{2^{2N}S^2P_eI^2}{3T_s} \tag{12.4-5}$$

Finally, the output power due to the thermal error noise is

$$N_{th} = \int_{-f_M}^{f_M} G_{th}(f)\,df = \frac{2^{2N}S^2P_eI^2}{3T_s^2} \tag{12.4-6}$$

since $T_s = 1/2f_M$.

12.5 OUTPUT SIGNAL-TO-NOISE RATIO IN PCM

The output signal-to-noise ratio, including both quantization and thermal noise, is found by combining Eqs. (12.2-7), (12.3-6), and (12.4-6). The result is

$$\frac{S_o}{N_o} = \frac{S_o}{N_q + N_{th}} = \frac{(I^2/T_s^2)(M^2S^2/12)}{(I^2/T_s^2)(S^2/12) + (I^2/T_s^2)(P_e 2^{2N}S^2/3)}$$

$$= \frac{2^{2N}}{1 + 4P_e 2^{2N}} \tag{12.5-1}$$

in which we have used the fact that $M = 2^N$.

In PSK (or for direct transmission) we have, from Eq. (11.7-3), that

$$(P_e)_{\text{PSK}} = \tfrac{1}{2}\,\text{erfc}\sqrt{\frac{E_b}{\eta}} \tag{12.5-2}$$

where E_b is the signal energy of a bit and $\eta/2$ is the two-sided thermal-noise power spectral density. Also, for coherent reception of FSK we have, from Eq. (11.8-8), that

$$(P_e)_{\text{FSK}} = \tfrac{1}{2}\,\text{erfc}\sqrt{0.6\,\frac{E_b}{\eta}} \tag{12.5-3}$$

To calculate E_b, we note that if a sample is taken at intervals of T_s, and the code word of N bits occupies the entire interval between samples, then a bit has a duration T_s/N. If the received signal power is S_i, the energy E_b associated with a single bit is

$$E_b = S_i\frac{T_s}{N} = S_i\frac{1}{2f_M N} \tag{12.5-4}$$

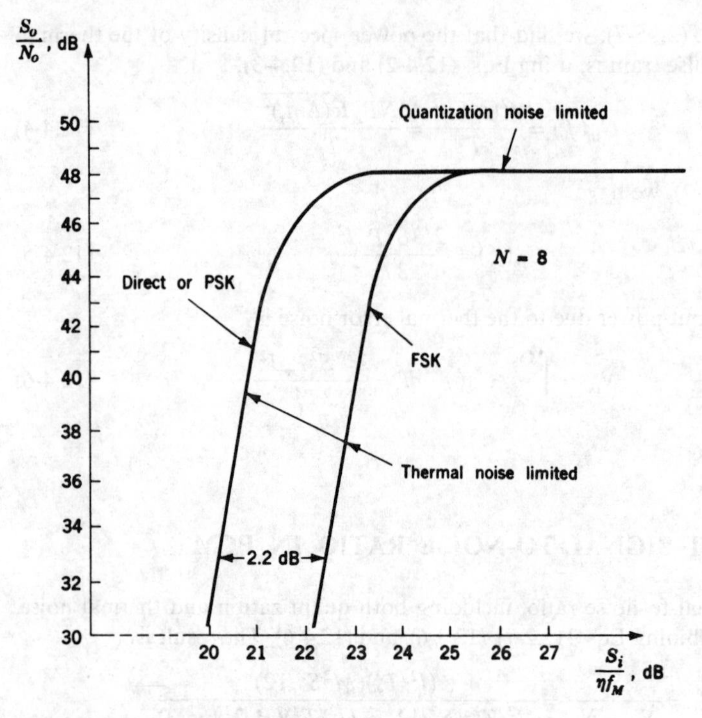

Figure 12.5-1 Comparison of PCM transmission systems.

Combining Eqs. (12.5-1), (12.5-2), and (12.5-4), we find

$$\left(\frac{S_o}{N_o}\right)_{\text{PSK}} = \frac{2^{2N}}{1 + 2^{2N+1} \text{ erfc } \sqrt{(1/2N)(S_i/\eta f_M)}} \tag{12.5-5}$$

Using Eq. (12.5-3) in place of (12.5-2), we have

$$\left(\frac{S_o}{N_o}\right)_{\text{FSK}} = \frac{2^{2N}}{1 + 2^{2N+1} \text{ erfc } \sqrt{(0.3/N)(S_i/\eta f_M)}} \tag{12.5-6}$$

Equations (12.5-5) and (12.5-6) are plotted in Fig. 12.5-1 for $N = 8$. Note that for $S_i/\eta f_M \gg 1$ and $N = 8$

$$\left(\frac{S_o}{N_o}\right)_{\text{PSK,FSK}} = 10 \log (2^{16}) = 48 \text{ dB}$$

Observe that both PCM systems exhibit a threshold, the FSK threshold occurring at a $S_i/\eta f_M$ which is 2.2 dB greater than for PSK. (The threshold point is arbitrarily defined as the $S_i/\eta f_M$ value at which S_o/N_o has fallen 1 dB from the value corresponding to a large $S_i/\eta f_M$.) Experimentally, the onset of threshold in

PCM is marked by an abrupt increase in a *crackling* noise analogous to the clicking noise heard below threshold in analog FM systems.

A comparison of PCM and FM is presented in Sec. 13.28.

12.6 DELTA MODULATION (DM)

The operation of a delta-modulation communication system was discussed in Sec. 6.13. At that point the discussion was entirely qualitative, and we considered neither quantization nor thermal noise. In the following sections we shall calculate the output signal-to-noise ratio of a DM system taking quantization noise and thermal noise into account.

A delta-modulation system, including a thermal-noise source, is shown in Fig. 12.6-1. The impulse generator applies to the modulator a continuous sequence of impulses $p_i(t)$ of time separation τ. The modulator output is a sequence of pulses $p_o(t)$ whose polarity depends on the polarity of the difference signal $\Delta(t) = m(t) - \tilde{m}(t)$, where $\tilde{m}(t)$ is the integrator output. We assume that the integrator has been adjusted so that its response to an input impulse of strength I is a step of size S; that is, $\tilde{m}(t) = (S/I) \int p_o(t)\, dt$.

A typical impulse train $p_o(t)$ is shown in Fig. 12.6-2*a* [actually, of course, the waveform $p_o(t)$ is a train of narrow pulses, each having very small energy]. Before transmission, the impulse waveform will be converted to the two-level waveform of Fig. 12.6-2*b* since this latter waveform has much greater power than a train of narrow pulses. This conversion is accomplished by the block in Fig. 12.6-1 marked "transmitter." The transmitter, in principle, need be nothing more complicated than a bistable multivibrator (flip-flop). We may readily arrange that the positive impulses set the flip-flop into one of its stable states, while the negative impulses reset the flip-flop to its other stable state. The binary waveform of Fig. 12.6-2*b* will be transmitted directly or used to modulate a carrier as in PSK or FSK. After detection by the matched filter shown in Fig. 12.6-1, the binary

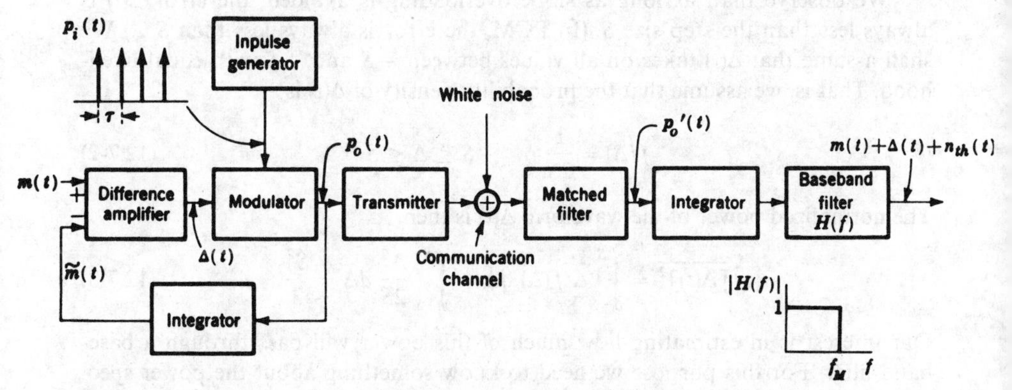

Figure 12.6-1 A delta-modulation system.

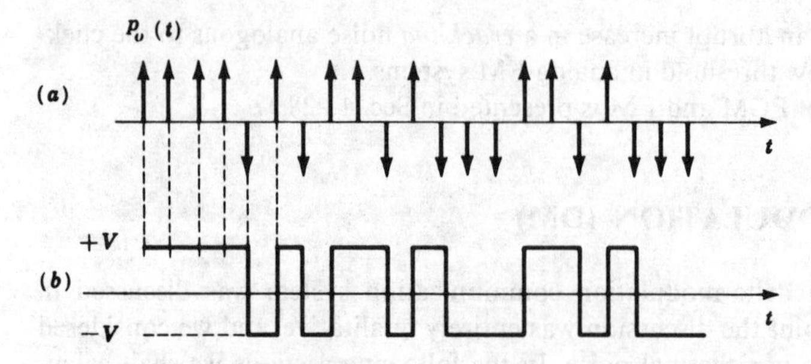

Figure 12.6-2 (*a*) A typical impulse train $p_o(t)$ appearing at the modulator output in Fig. 12.6-1. (*b*) The two-level signal transmitted over the communication channel.

waveform will be reconverted to a sequence of impulses $p_o'(t)$. In the absence of thermal noise $p_o'(t) = p_o(t)$, and the signal $\tilde{m}(t)$ is recovered at the receiver by passing $p_o'(t)$ through an integrator. We assume that transmitter and receiver integrators are identical and that the input to each consists of a train of impulses of strength $\pm I$. Hence in the absence of thermal noise, the outputs of both integrators are identical.

12.7 QUANTIZATION NOISE IN DELTA MODULATION

To arrive at an estimate of the quantization noise power in delta modulation, we consider the situation represented in Fig. 12.7-1. Here, in Fig. 12.7-1*a* is shown a sinusoidal input signal $m(t)$ and the waveform $\tilde{m}(t)$ which is the delta modulator approximation to $m(t)$. In Fig. 12.7-1*b* is shown the error waveform $\Delta(t)$ given by

$$\Delta(t) \equiv m(t) - \tilde{m}(t) \tag{12.7-1}$$

This error waveform is the source of the quantization noise.

We observe that, as long as slope overloading is avoided, the error $\Delta(t)$ is always less than the step size S. (In PCM, the error is always less than $S/2$.) We shall assume that $\Delta(t)$ takes on all values between $-S$ and $+S$ with equal likelihood. That is, we assume that the probability density of $\Delta(t)$ is

$$f(\Delta) = \frac{1}{2S} \qquad -S \le \Delta \le S \tag{12.7-2}$$

The normalized power of the waveform $\Delta(t)$ is then

$$\overline{[\Delta(t)]^2} = \int_{-S}^{S} \Delta^2 f(\Delta)\, d\Delta = \int_{-S}^{S} \frac{\Delta^2}{2S}\, d\Delta = \frac{S^2}{3} \tag{12.7-3}$$

Our interest is in estimating how much of this power will pass through a baseband filter. For this purpose we need to know something about the power spectral density of $\Delta(t)$.

In Fig. 12.7-1 the period T of the sinusoidal waveform $m(t)$ has been selected so that T is an integral multiple of the step duration τ. (Note that the bit rate $f_b = 1/\tau$.) We then observe that $\Delta(t)$ is periodic, with fundamental period T, and is, of course, rich in harmonics. Suppose, however, that the period T is changed very slightly by amount δT. Then the fundamental period of $\Delta(t)$ will not be T but will be instead $T \times \tau/\delta T$ (Prob. 12.7-3) corresponding to a fundamental frequency near zero as $\delta T \to 0$. And again, of course, $\Delta(t)$ will be rich in harmonics. Hence, in the general case, especially with $m(t)$ a random signal, it is reasonable to assume that $\Delta(t)$ has a spectrum which extends continuously over a frequency range which begins near zero.

To get some idea of the upper-frequency range of the spectrum of the waveform $\Delta(t)$, let us contemplate passing $\Delta(t)$ through a low-pass filter of adjustable cutoff frequency. Suppose that initially the cutoff frequency is high enough so that $\Delta(t)$ may pass with nominally no distortion. As we lower the cutoff frequency, the first type of distortion we would note is that the abrupt discontinuities in the waveform would exhibit finite rise and fall times. Such is the case since it is these abrupt changes which contribute to the high-frequency power content of the signal. To keep the distortion within reasonable limits, let us

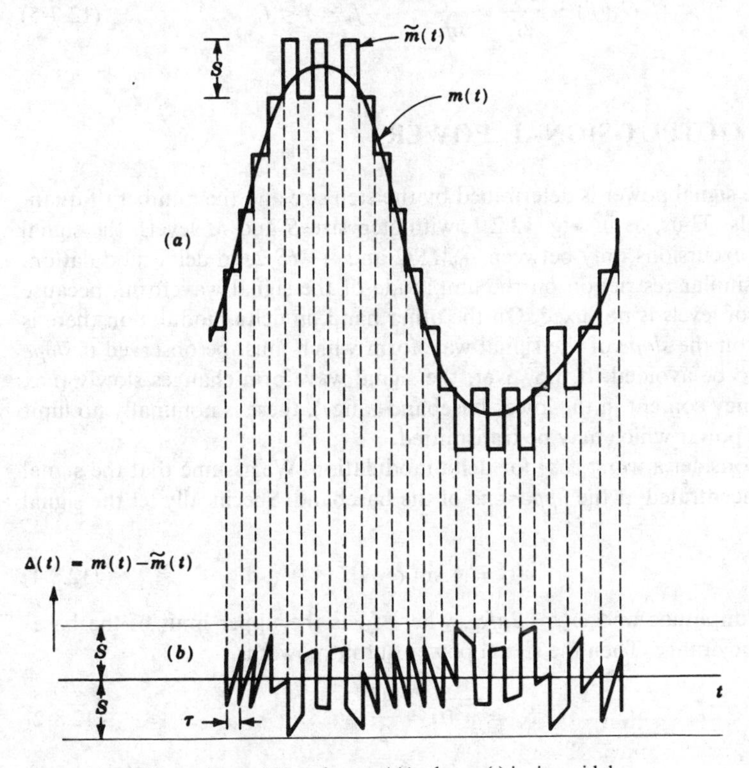

Figure 12.7-1 The estimate $\tilde{m}(t)$ and error $\Delta(t)$ when $m(t)$ is sinusoidal.

arrange that the rise time be rather smaller than the interval τ. As was discussed in Sec. 1.15, to satisfy this condition, we require that the filter cutoff frequency f_c be of the order of $f_c = 1/\tau$. Since the transmitted bit rate $f_b = 1/\tau$, $f_c = f_b$ as expected.

We now have made it appear reasonable, by a rather heuristic argument, that the spectrum of $\Delta(t)$ extends rather continuously from nominally zero up to $f_c = f_b$. We shall assume further that over this range the spectrum is white. It has indeed been established experimentally that the spectrum of $\Delta(t)$ is approximately white over the frequency range indicated.

We may now finally calculate the quantization noise that will appear at the output of a baseband filter of cutoff frequency f_M. Since the quantization noise power in a frequency range f_b is $S^2/3$ as given by Eq. (12.7-3), the output-noise power in the baseband-frequency range f_M is

$$N_q = \frac{S^2}{3}\frac{f_M}{f_b} = \frac{S^2 f_M}{3f_b} \tag{12.7-4}$$

We may note also, in passing, that the two-sided power spectral density of $\Delta(t)$ is

$$G_\Delta(f) \simeq \frac{S^2/3}{2f_b} = \frac{S^2}{6f_b} \qquad -f_b \le f \le f_b \tag{12.7-5}$$

12.8 THE OUTPUT-SIGNAL POWER

In PCM, the signal power is determined by the step size and the number of quantization levels. Thus, as in Fig. 12.2-1, with step size S and M levels, the signal could make excursions only between $-MS/2$ and $+MS/2$. In delta modulation, there is no similar restriction on the amplitude of the signal waveform, because the number of levels is not fixed. On the other hand, in delta modulation there is a limitation on the *slope* of the signal waveform which must be observed if *slope overload* is to be avoided. If, however, the signal waveform changes slowly (i.e., has a frequency content in the lower-baseband range), there is nominally no limit to the signal power which may be transmitted.

Let us consider a *worst case* for delta modulation. We assume that the signal power is concentrated at the *upper end* of the baseband. Specifically let the signal be

$$m(t) = A \sin \omega_M t \tag{12.8-1}$$

with A the amplitude and $\omega_M = 2\pi f_M$, where f_M is the upper limit of the baseband frequency range. Then the signal power output power is

$$S_o = \overline{m^2(t)} = \frac{A^2}{2} \tag{12.8-2}$$

The maximum slope of $m(t)$ is $\omega_M A$. The maximum average slope of the delta modulator approximation $\tilde{m}(t)$ is $S/\tau = Sf_b$, where S is the step size and f_b the bit rate. The limiting value of A just before the onset of slope overload is, therefore, given by the condition

$$\omega_M A = Sf_b \tag{12.8-3}$$

From Eqs. (12.8-2) and (12.8-3) we have that the maximum power which may be transmitted is

$$S_o = \frac{S^2 f_b^2}{2\omega_M^2} \tag{12.8-4}$$

The condition specified in Eq. (12.8-3) is unduly severe. A design procedure, more often employed, is to select the Sf_b product to be equal to the rms value of the *slope* of $m(t)$. In this case the output-signal power can be increased above the value given in Eq. (12.8-4).

12.9 DELTA-MODULATION OUTPUT-SIGNAL-TO-QUANTIZATION-NOISE RATIO

The output-signal-to-quantization-noise ratio for delta modulation is found by dividing Eq. (12.8-4) by Eq. (12.7-4). The result is

$$\frac{S_o}{N_q} = \frac{5}{8\pi^2}\left(\frac{f_b}{f_M}\right)^3 \cong \frac{3}{80}\left(\frac{f_b}{f_M}\right)^3 \tag{12.9-1}$$

It is of interest to note that when our heuristic analysis is replaced by a rigorous analysis,[1] it is found that Eq. (12.9-1) continues to apply, except with the factor 3/80 replaced by 3/64, corresponding to a difference of less than 1 dB.

The dependence of S_o/N_q on the product f_b/f_M should be anticipated. For suppose that the signal amplitude were adjusted to the point of slope overload. If now, say, f_M were increased by some factor, then f_b would have to be increased by this same factor in order to continue to avoid overload [see Eq. (12.8-4)].

Let us now make a comparison of the performance of PCM and DM in the matter of the ratio S_o/N_q. We observe that the transmitted signals in DM (Fig. 12.6-2) and in PCM (Fig. 6.6-1) are of the same waveform, a binary pulse train. In PCM a voltage level corresponding to a single bit persists for the time duration allocated to one bit of a code word. With sampling at the Nyquist rate, every $1/2f_M$ s, and with N bits per code word, the PCM bit rate is $f'_b = 2f_M N$. In DM, a voltage level, corresponding to a single bit, is held for a duration τ which is the interval between samples. Thus the DM system operates at a bit rate $f_b = 1/\tau$.

If the communication channel is of limited bandwidth, then there is the possibility of interference (Sec. 6.8) in either DM or PCM. Whether such intersymbol interference occurs in DM depends on the ratio of f_b to the bandwidth of the channel and similarly, in PCM, on the ratio of f'_b to the channel bandwidth. For a fixed channel bandwidth, if inter-symbol interference is to be equal in the two cases, DM or PCM, we require that both systems operate at the same bit rate or

$$f_b = f'_b = 2f_M N \tag{12.9-2}$$

Combining Eq. (12.3-7) for PCM with Eq. (12.9-2) yields for PCM

$$\frac{S_o}{N_q} = 2^{2N} = 2^{f_b/f_M} \tag{12.9-3}$$

Combining Eq. (12.9-1) with Eq. (12.9-2) yields for delta modulation

$$\frac{S_o}{N_q} = \frac{3}{\pi^2} N^3 \tag{12.9-4}$$

Comparing Eq. (12.9-3) with Eq. (12.9-4), we observe that for a fixed channel bandwidth the performance of DM is always poorer than PCM. By way of example, if a channel is adequate to accommodate code words in PCM with $N = 8$, Eq. (12.9-3) gives $(S_o/N_q) = 48$ dB. The same channel used for DM would, from Eq. (12.9-4), yield $(S_o/N_q) \cong 22$ dB.

Comparison of DM and PCM for Voice

When the signal to be transmitted is the waveform generated by voice, the comparison between DM and PCM is overly pessimistic against DM. For, as appears in the discussion leading to Eq. (12.8-3), in our concern to avoid slope overload under any possible circumstance, we have allowed for the very worst possible case. We have provided for the possibility that all the signal power might be concentrated at the angular frequency ω_M which is the upper edge of the signal bandwidth. Such is certainly not the case for voice. Actually, for speech, a bandwidth $f_M = 3200$ Hz is adequate and the voice spectrum has a pronounced peak at 800 Hz $= f_M/4$. If we replace ω_M by $\omega_M/4$ in Eq. (12.8-3) we have

$$\frac{\omega_M}{4} A = S f_b \tag{12.9-5}$$

The amplitude A will now be four times larger than before and the allowable signal power before slope overload will be increased by a factor of 16 (12 dB). Correspondingly, Eq. (12.9-1) now becomes

$$\frac{S_o}{N_q} = \frac{6}{\pi^2}\left(\frac{f_b}{f_M}\right)^3 \cong 0.6\left(\frac{f_b}{f_M}\right)^3 \simeq 5N^3 \tag{12.9-6}$$

It may be readily verified that for $(f_b/f_M) \leq 8$ the signal-to-noise ratio for delta modulation, SNR(Δ), given by Eq. (12.9-6) is larger than SNR(PCM) given by Eq. (12.9-3). At about $f_b/f_M = 4$ the ratio SNR(Δ)/SNR(PCM) has a maximum value 2.4 corresponding to a 3.8 dB advantage. Thus if we allow $f_M = 4$ kHz for voice, then to avail ourselves of this maximum advantage afforded by delta modulation we would take $f_b = 16$ kHz. (When using PCM, f_M is usually taken to be 4 kHz even though 3.2 kHz would be adequate for voice.)

In our derivation of the signal-to-noise ratio in PCM we assumed that at all times the signal is strong enough to range widely through its allowable excursion. As a matter of fact, we specifically assumed that the distribution function $f(m)$ for the instantaneous signal value $m(t)$ was uniform throughout the allowable signal range. As a matter of practice, such would hardly be the case. Voice levels wax and wane and it is precisely for this reason that companding is employed as we have discussed in Sec. 6.4. The fact is, that commercial PCM systems using companding, are designed so that the signal-to-noise ratio remains at about 30 dB over a 40 dB range of signal power. In short, while Eq. (12.9-3) predicts a continuous increase in SNR(PCM) with increasing f_b/f_M, this result is for uncompanded PCM, and, in practice SNR(PCM) is approximately constant at 30 dB. The linear DM discussed above has a dynamic range of about 15 dB. In order to widen this dynamic range to 40 dB, as in PCM, one employs adaptive DM (ADM) discussed below, which yields advantages similar to the companding of PCM. When adaptive DM is employed, the SNR of ADM is comparable to the SNR of companded PCM. Today, many systems, such as the Satellite Business System (SBS), employ ADM operating at 32 kb/s rather than companded PCM which operates at 64 kb/s, thereby providing twice as many voice channels in a given frequency band.

12.10 DELTA PULSE-CODE MODULATION (DPCM)

In delta modulation, the approximation $\tilde{m}(t)$ is compared with the signal $m(t)$. A correction of fixed magnitude (of step size S) is then made in $\tilde{m}(t)$. The direction of this correction depends on whether $\tilde{m}(t)$ is greater or smaller than $m(t)$. An improved system results if the correction is not of fixed magnitude but rather increases as the error $\Delta(t) = \tilde{m}(t) - m(t)$ increases. Delta pulse-code modulation is just such a system in which, however, the correction is quantized. Delta pulse-code modulation is implemented by replacing the modulator of Fig. 12.6-1 by an M-level quantizer and sampler. The output of this quantizer-sampler is an impulse whose strength is not fixed, but is proportional to the quantized error. This quantized error sample is also applied to the transmitter. Corresponding to each quantized error sample, the transmitter impresses on the communication channel a binary waveform of N bits ($2^N = M$) which is the binary code representation of the quantized error sample.

12.11 THE EFFECT OF THERMAL NOISE IN DELTA MODULATION

When thermal noise is present, the matched filter in the receiver of Fig. 12.6-1 will occasionally make an error in determining the polarity of the transmitted waveform. Whenever such an error occurs, the received impulse stream $p_o'(t)$ will exhibit an impulse of incorrect polarity. The received impulse stream is then

$$p_o'(t) = p_o(t) + p_{th}(t) \tag{12.11-1}$$

in which $p_{th}(t)$ is the error impulse stream due to thermal noise. If the strength (area) of the individual impulses is I, then each impulse in p_{th} is of strength $2I$ and occurs only at each error. The factor of two results from the fact that an error reverses the polarity of the impulse.

The thermal-error noise appears as a stream of impulses of random time of occurrence and of strength $\pm 2I$. The average time of separation between these impulses is τ/P_e, where P_e is the bit error probability, and τ is the time duration of a bit. If the results of Sec. 2.25 [see Eq. (2.25-7)] are used, the power spectral density of the thermal-noise impulses is

$$G_{pth}(f) = \frac{P_e}{\tau}(2I)^2 = \frac{4I^2 P_e}{\tau} \tag{12.11-2}$$

Now we have already characterized the integrators (assumed identical in both the DM transmitter and receiver) as having the property that when the integrator input is an impulse of strength I, the output is a step of amplitude S. The Fourier transform of the impulse is I, and the Fourier transform of a step of amplitude S is [using $u(t) \equiv$ unit step]

$$\begin{aligned} \mathscr{F}\{Su(t)\} &= \frac{S}{j\omega} \qquad \omega \neq 0 \\ &= S\pi\delta(\omega) \qquad \omega = 0 \end{aligned} \tag{12.11-3}$$

We may ignore the dc component in the transform since such dc components will not be transmitted through the baseband filter. Hence we may take the transfer function of the integrator to be $H_I(f)$ given by

$$H_I(f) = \frac{S}{I}\frac{1}{j\omega} \qquad \omega \neq 0 \tag{12.11-4a}$$

and

$$|H_I(f)|^2 = \left(\frac{S}{I}\right)^2 \frac{1}{\omega^2} \qquad \omega \neq 0 \tag{12.11-4b}$$

From Eqs. (12.11-2) and (12.11-4b) we find that the power spectral density of the thermal noise at the input to the baseband filter is $G_{th}(f)$ given by

$$G_{th}(f) = |H_I(f)|^2 G_{pth}(f) = \frac{4S^2 P_e}{\tau\omega^2} \qquad \omega \neq 0 \tag{12.11-5}$$

It would now appear that to find the thermal-noise output, we need but to integrate $G_{th}(f)$ over the passband of the baseband filter. We have performed similar integrations on many occasions. And, in so doing, we have extended the range of integration from $-f_M$ through $f = 0$ to $+f_M$, even though we recognized that the baseband filter does not pass *dc* and actually has a low-frequency cutoff f_1. However, in these other cases the power spectral density of the noise near $f = 0$ is not inordinately large in comparison with the density throughout the baseband range generally. Hence, if as is normally the case, $f_1 \ll f_M$, the procedure is certainly justified as a good approximation. We observe however, that in the present case [Eq. (12.11-5)], $G_{th}(f) \to \infty$ at $\omega \to 0$, and more importantly that the integral of $G_{th}(f)$, over a range which includes $\omega = 0$, is infinite.

Let us then explicitly take account of the low-frequency cutoff f_1 of the baseband filter. The thermal-noise output is, using Eq. (12.11-5) with $\omega = 2\pi f$, and since $f_b = 1/\tau$,

$$N_{th} = \frac{S^2 P_e}{\pi^2 \tau}\left(\int_{-f_M}^{-f_1} \frac{df}{f^2} + \int_{f_1}^{f_M} \frac{df}{f^2}\right) \tag{12.11-6a}$$

$$= \frac{2S^2 P_e}{\pi^2 \tau}\left(\frac{1}{f_1} - \frac{1}{f_M}\right) \tag{12.11-6b}$$

$$\cong \frac{2S^2 P_e}{\pi^2 f_1 \tau} = \frac{2S^2 P_e f_b}{\pi^2 f_1} \tag{12.11-6c}$$

if $f_1 \ll f_M$. Observe that, unlike the situation encountered in all other earlier cases, the thermal-noise output in delta modulation depends on the low-frequency cutoff rather than the high-frequency limit of the baseband range. In many applications such as voice encoding where the voice signal is typically band-limited from 300 to 3200 Hz, the use of a band pass output filter is commonplace. In this case $f_1 = 300$ Hz.

12.12 OUTPUT SIGNAL-TO-NOISE RATIO IN DELTA MODULATION

The output SNR is obtained by combining Eqs. (12.7-4), (12.9-6), and (12.11-6*c*). The result is

$$\frac{S_o}{N_o} = \frac{S_o}{N_q + N_{th}} = \frac{(2S^2/\pi^2)(f_b/f_M)^2}{(S^2 f_M/3f_b) + (2S^2 P_e f_b/\pi^2 f_1)} \tag{12.12-1}$$

which may be rewritten

$$\frac{S_o}{N_o} = \frac{S_o}{N_q + N_{th}} \cong \frac{0.6(f_b/f_M)^3}{1 + 24P_e(f_b^2/f_M f_1)/4\pi^2} = \frac{0.6(f_b/f_M)^3}{1 + 0.6P_e(f_b^2/f_M f_1)} \tag{12.12-2}$$

If transmission is direct or by means of PSK,

$$P_e = \tfrac{1}{2}\,\mathrm{erfc}\sqrt{\frac{E_s}{\eta}} \tag{12.12-3}$$

where E_s, the signal energy in a bit, is related to the received signal power S_i by

$$E_s = S_i T_b = S_i/f_b \tag{12.12-4}$$

Combining Eqs. (12.12-2), (12.12-3), and (12.12-4), we have

$$\frac{S_o}{N_o} = \frac{0.6(f_b/f_M)^3}{1 + [0.3(f_b^2/f_M f_1)]\,\mathrm{erfc}\sqrt{S_i/\eta f_b}} \tag{12.12-5}$$

12.13 COMPARISON OF PCM AND DM[2]

We can now compare the output signal-to-noise ratios in PCM and DM by comparing Eqs. (12.5-5) and (12.12-5). To ensure that the communications channel bandwidth required is the same in the two cases, we use the condition, given in Eq. (12.9-2), that $2N = f_b/f_M$. Then Eq. (12.5-5) may be written

$$\frac{S_o}{N_o} = \frac{2^{f_b/f_M}}{1 + 2(2^{f_b/f_M}) \operatorname{erfc} \sqrt{(S_i/\eta f_b)}} \tag{12.13-1}$$

Equations (12.13-1) and (12.12-5) are compared in Fig. 12.13-1 for $N = 8$ (f_b(DM) = 48 kb/s). To obtain the threshold performance of the delta-modulation system, we assume voice transmission where $f_M = 3000$ Hz and $f_1 = 300$ Hz. Thus

$$\frac{f_b}{f_M} = 16 \tag{12.13-2a}$$

and

$$\frac{f_M}{f_1} = 10 \tag{12.13-2b}$$

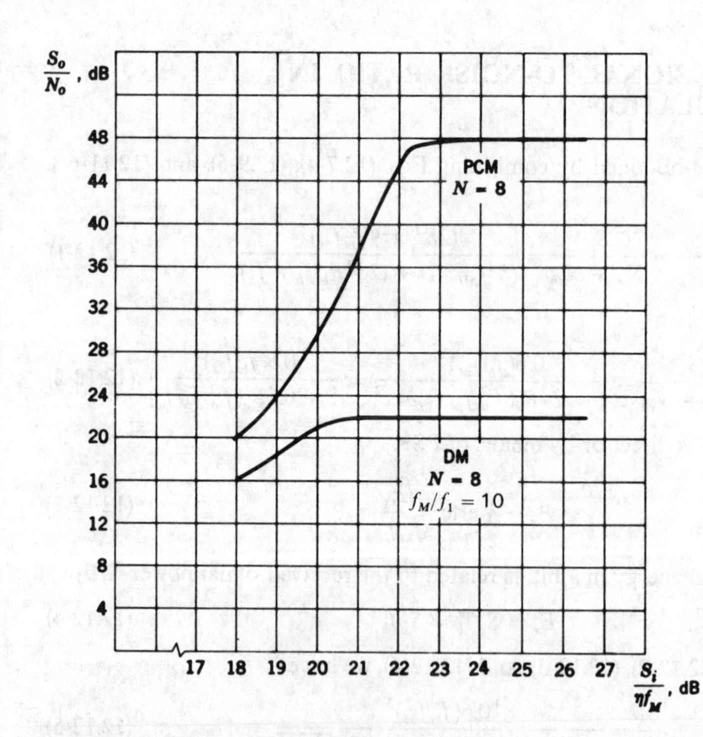

Figure 12.13-1 A comparison of PCM and DM.

Let us compare the ratios S_o/N_o for PCM and DM for the case of voice transmission. We assume $f_M = 3000$ Hz, $f_1 = 300$ Hz and $N = 8$. We then have $f_b = 2Nf_m = 48 \times 10^3$ Hz. Using these numbers and recalling that the probability of an error in a bit is $P_{eb} = \frac{1}{2}\,\text{erfc}\,\sqrt{S_i/\eta f_b}$ we have from Eqs. (12.12-5) and (12.13-1) the result for DM is,

$$\left(\frac{S_o}{N_o}\right)_{\text{DM}} = \frac{2457.6}{1 + 768\,\text{erfc}\,\sqrt{S_i/\eta f_b}} = \frac{2457.6}{1 + 1536P_e} \tag{12.13-3a}$$

and for PCM is

$$\left(\frac{S_o}{N_o}\right)_{\text{PCM}} = \frac{65{,}536}{1 + 131{,}072\,\text{erfc}\,\sqrt{S_i/\eta f_b}} = \frac{65{,}536}{1 + 262{,}144P_e} \tag{12.13-3b}$$

When the probability of a bit error is very small, the PCM system is seen to have a higher output SNR than the DM system. Indeed, the output SNR for the PCM system is 48 dB and only about 33 dB for the DM system. However, an output SNR of 30 dB is all that is required in a communication system. Indeed, if companded PCM is employed, the output SNR will decrease by about 12 dB to 36 dB for the PCM system. Thus, while Eq. 12.13-3*b* indicates that the output SNR is higher for PCM, the output SNR, in practice, can be considered as being *comparable*.

With regard to the threshold, we see that when $P_e \sim 10^{-6}$ the PCM system has reached threshold while the DM system reaches threshold when $P_e \approx 10^{-4}$. In practice, we find that our *ear* does not detect threshold until P_e is about 10^{-4} for PCM and 10^{-2} for DM and ADM. Some ADM systems can actually produce understandable speech at error rates as high as 0.1.

Figure 12.13-1 shows a comparison of PCM and DM for $N = 8$ and $f_M/f_1 = 10$.

12.14 THE SPACE SHUTTLE ADM

The Space Shuttle flown since 1982 employs an adaptive delta modulation system for voice communications. Block diagrams of the transmitter and receiver are shown in Fig. 12.14-1. The adaptive feature of the system is incorporated in the block labeled "impulse-size algorithm." The output of this block, at each bit time, is an impulse whose polarity and amplitude depend on the result of the present comparison of $m(t)$ and $\tilde{m}(t)$ and also on the past history of the results of these comparisons. The algorithm block incorporates the memory and logic required to carry out this adaptive determination. The intention of the algorithm is to arrange that the slope overload be reduced in comparison to the overload encountered in linear delta modulation (LDM), thereby increasing the dynamic range of the ADM.

The adaptive feature operates as follows: Consider first the case in which the baseband signal $m(t)$ is a constant, that is unvarying. Then in the steady state the

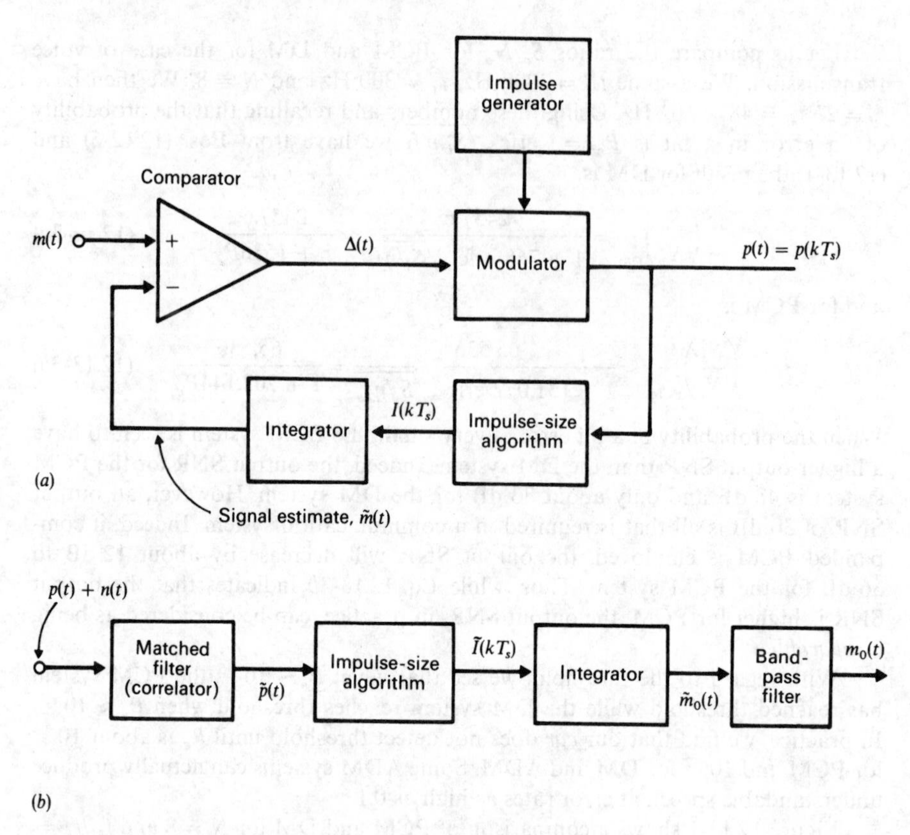

Figure 12.14-1 (*a*) An adaptive delta modulation encoder. (*b*) Decoder.

estimate $\tilde{m}(t)$ simply hunts back and forth across $m(t)$, that is, $\tilde{m}(t)$ alternately overshoots and undershoots $m(t)$. An example of such behavior is seen in Fig. 6.13-2 which applies to LDM. In the present ADM case, whenever there is a *reversal* in the sign of $m(t) - \tilde{m}(t)$, the algorithm block generates an impulse of minimum size $I_o(kT_s)$ which is adequate when integrated to generate a step S_o of correspondingly minimum size. The step is positive or negative according to sgn $[m(t) - \tilde{m}(t)]$. Such a minimum size step is shown at time ② in Fig. 12.14-2*a*. The step is of minimum size S_o, because, as we note the sign of $m(t) - \tilde{m}(t)$ has reversed in comparison with time ①. At time ③ we still find that $m(t) > \tilde{m}(t)$. The algorithm correctly interprets this result to indicate that slope overload is being generated, i.e., $m(t)$ is rising too rapidly for $\tilde{m}(t)$ to keep pace with steps of amplitude S_o. Accordingly the step size is increased to $2S_o$. At time ④ it is found that still $m(t) > \tilde{m}(t)$ so the step size is increased to $3S_o$. In short, starting at a reversal with step S_o, at each successive sample, if $m(t)$ remains greater than $\tilde{m}(t)$ an additional S_o is added to the step size. At time ⑤ there is a reversal in sign of $m(t) - \tilde{m}(t)$ so the step reverts to S_o and, of course reverses polarity. The polarities of the impulses output from the modulation in Fig. 12.14-2a are shown in Fig. 12.14-2b. The algorithm can be expressed precisely by the equations

$$S(kT_s) = \begin{cases} S_o \operatorname{sgn} p(kT_s) & \text{if } p(kT_s) \neq p[(k-1)T_s] \\ \{S[(k-1)T_s] + S_o\} \operatorname{sgn} (kT_s) & \text{if } p(kT_s) = p[(k-1)T_s] \end{cases} \tag{12.14-1}$$

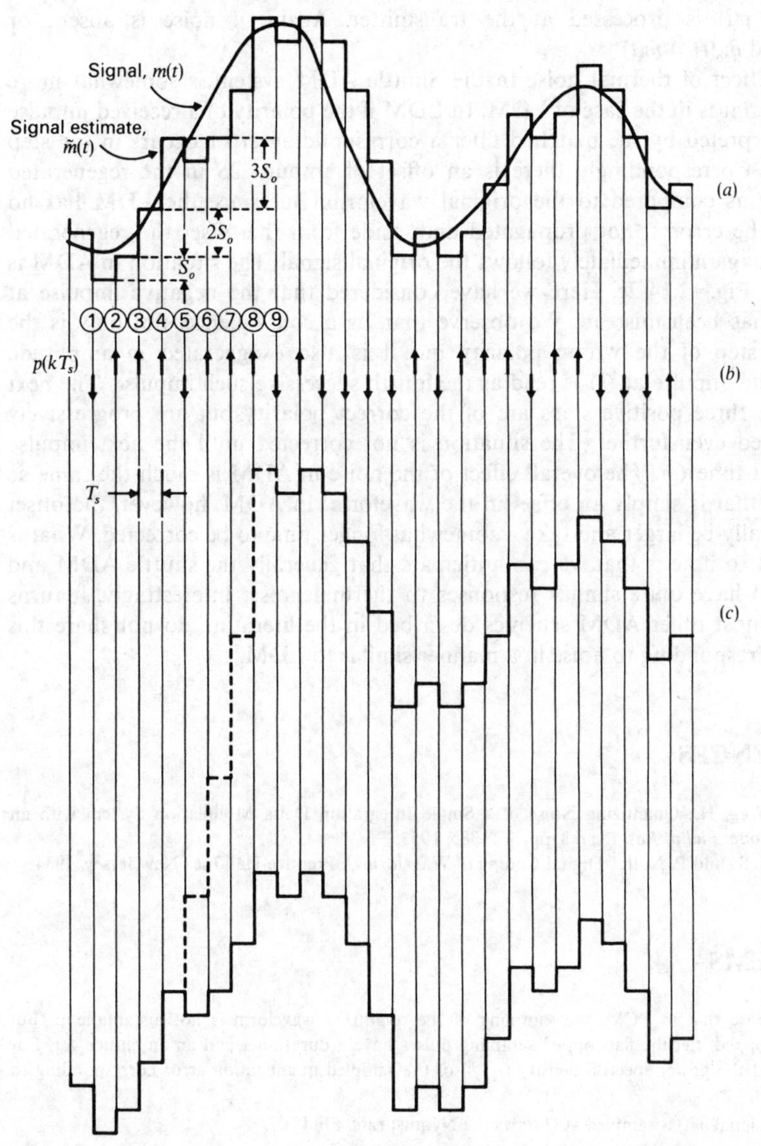

Figure 12.14-2 (a) Signal and estimate. (b) Transmitted bit stream. (c) Decoded signal (with and without error).

Decoder

The ADM receiver (decoder) is shown in Fig. 12.14-1*b*. The matched filter regenerates the impulse stream $\tilde{p}(t)$, hopefully without error, in spite of the presence of the thermal noise $n(t)$ at its input. If the effect of the noise is suppressed completely then $\tilde{p}(t) = p(t)$. The algorithm block and the integrator processes $\tilde{p}(t)$ precisely as $p(t)$ is processed at the transmitter. Again, if noise is absent or suppressed $\tilde{m}_o(t) = \hat{m}(t)$.

The effect of thermal noise in the Shuttle ADM system is somewhat more severe than it is in the case of LDM. In LDM if the polarity of a received impulse is misinterpreted by the matched filter a corresponding error occurs in the step direction. Correspondingly there is an offset of amount $2S$ in the regenerated waveform as compared to the original waveform. But since the LDM has no memory the error is not propagated and, aside from the offset the regenerated waveform again immediately follows the original signal. The situation in ADM is shown in Fig. 12.14-2*c*. Here we have considered that the negative impulse at time ⑤ has been misread. We observe that, as a consequence, not only is the resultant step of the wrong polarity but it is also exaggerated in amplitude because the impulse at ⑤ is read as the fourth successive such impulse. The next successive three positive steps are of the correct polarity but are progressively exaggerated even further. The situation is not corrected until the next impulse reversal at time ⑨. The overall effect of the noise in ADM is much the same as in LDM, that is simply an offset in the waveform. In ADM, however, the offset will generally be larger and take a somewhat longer time to be corrected. What is important to note is that we can anticipate that generally the shuttle ADM and LDM will have quite similar responses to thermal noise. Interestingly, it turns out that most other ADM schemes described in the literature do not share this feature of responding to noise in a manner similar to LDM.

REFERENCES

1. Van de Weg, H.: Quantizing Noise of a Single Integration Delta Modulation System with an *N*-digit Code, *Philips Res. Rept.* 8, pp. 367–385, 1953.
2. Jayant, N. S., and P. Noll: "Digital Coding of Waveforms," Prentice-Hall Inc., New Jersey, 1984.

PROBLEMS

12.2-1. Suppose that in PCM, the sampling of the quantized waveform is not instantaneous but rather flat-topped. Let the flat-topped sampling pulses have a duration τ and an amplitude I/τ. For this case find the power spectral density $G_{e_s}(f)$ of the sampled quantization error corresponding to Eq. (12.2-3).

12.2-2. The signal $m(t)$ is sampled at Q times the Nyquist rate. Find N_q.

12.3-1. Prove that Eq. (12.3-6) is valid when M is an even integer.

12.3-2. Plot Eq. (12.3-7) as a function of N.

12.3-3. A signal $m(t)$ is not strictly bandlimited. We bandlimit $m(t)$ and then sample it. Due to the bandlimiting, distortion results even without quantization.

(*a*) Show that the noise caused by the distortion is $N_D = 2 \int_{f_M}^{\infty} G_m(f)\, df$, where $G_m(f)$ is the power spectral density of $m(t)$, and f_M is the cutoff frequency of the bandlimiting filter.

(*b*) If $G_m(f) = G_o e^{-|f/f_1|}$ find N_D.

(*c*) If the signal $m(t)$ is sampled at the Nyquist rate and quantized, find the total output SNR = $S_o/(N_D + N_q)$.

12.4-1. The thermal-noise error pulse train is *not* an impulse train. Each pulse has the duration $1/2f_M$ if sampling is at the Nyquist rate. The pulse amplitude is proportional to Δm_s.

(*a*) Find $G_{th}(f)$.

(*b*) Calculate N_{th}.

12.5-1. Plot S_o/N_o in Eq. (12.5-5) as a function of $S_i/\eta f_M$ for $N = 4, 8, 100$.

12.5-2. Show that threshold is defined by the equation $P_e \approx 1/[(16)2^{2N}]$. Plot P_e versus N.

12.5-3. Find the output SNR if the binary signal is transmitted using DPSK. Plot your result.

12.5-4. The signal $m(t)$ is *not* bandlimited. Its power spectral density is $G_m(f) = G_o e^{-|f/f_1|}$.

Using the results of Prob. 12.3-3*c*, obtain an expression for the output signal-to-noise ratio $S_o/N_o = S_o/(N_D + N_Q + N_{th})$.

12.5-5. A signal $m(t)$, bandlimited to 4 kHz, is sampled at *twice* the Nyquist rate and the samples transmitted by PCM. An output SNR of 47 dB is required. Find N and the minimum value of S_i/η if operation is to be above threshold.

12.5-6. Two signals $m_1(t)$ and $m_2(t)$, each bandlimited to 4 kHz, are sampled at the Nyquist rate, PCM encoded, and then time-division multiplexed. The output SNR, including thermal-noise effects, of each demultiplexed signal is to be at least 30 dB.

(*a*) Sketch the entire transmitter and receiver structure.

(*b*) Find N and the minimum S_i/η that is required if operation is to be above threshold.

12.7-1. Let $m(t)$ be a constant M_0.

(*a*) Sketch the steady-state error $\Delta(t)$ for the DM shown in Fig. 12.6-1.

(*b*) Calculate the output quantization noise power after filtering. Assume a step size S and that the time between samples is τ. Let the filter bandwidth be f_M.

12.7-2. Let $m(t) = Kt$ where $K = S/\tau$.

(*a*) Sketch $\Delta(t)$ for the DM shown in Fig. 12.6-1.

(*b*) Sketch the filtered output quantization noise power. Let the baseband filter bandwidth be f_M.

12.7-3. Assume initially that $m(t)$ in Fig. 12.6-1 is periodic with a period T which is an integral multiple of the sampling interval τ. Now let T change, very slightly, by an amount δT. Show that the fundamental period of the error waveform $\Delta(t)$ becomes $(T\tau)/\delta T$.

12.8-1. The baseband input signal $m(t)$ in Fig. 12.6-1 is a random waveform with power spectral density which is uniform at $\eta_m/2$ up to the bandlimit f_M. To keep slope overload within tolerable limits, the design criterion to be used is that S/τ is to be equal to the rms value of the slope of $m(t)$. For this condition calculate the maximum power S_o and compare with Eq. (12.8-4).

12.9-1. A signal $m(t)$ is to be encoded using either DM or PCM. The signal-to-quantization-noise ratio $S_o/N_q \geq 30$ dB. Find the ratio of the PCM to DM bandwidths required.

12.9-2. Plot $\left(\dfrac{S_o}{N_q}\right)_{\text{PCM}}$ versus $\left(\dfrac{S_o}{N_q}\right)_{\text{DM}}$ for equal bandwidths.

12.10-1. A DPCM system is shown in Fig. P12.10-1.

(*a*) If $m(t) = Kt$, find $K_{\max}$ to avoid slope overloading.

(*b*) If a two-level quantizer is used, $S/\tau = K_{\max}$. With this value of $K_{\max}$, find τ'/τ, τ' being the time interval allowed for a bit in the transmitted PCM waveform.

(*c*) Sketch $\tilde{m}(t)$ and $\Delta(t)$, if $S/\tau = 2K_{\max}$.

(*d*) Comment on the quantization noise of the DPCM system as compared with the DM system.

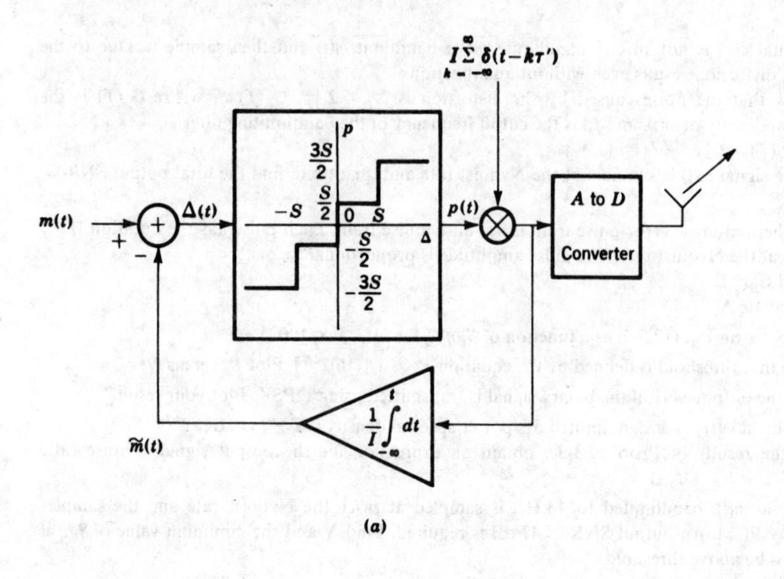

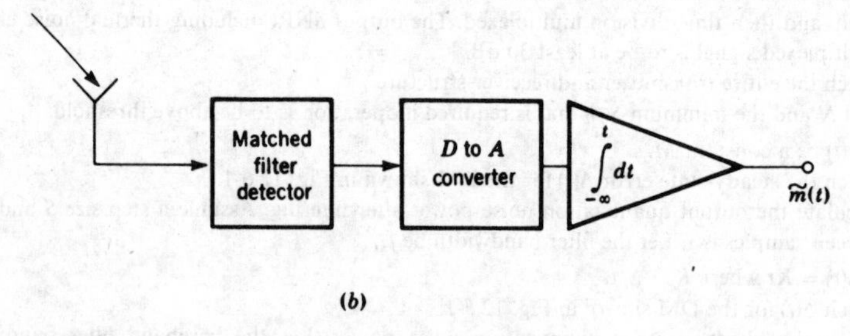

Figure P12.10-1 A DPCM system. (*a*) Encoder. (*b*) Decoder.

12.11-1. Verify Eq. (12.11-6*c*).

12.12-1. (*a*) Show that the threshold in DM occurs when $100\ P_e \approx (\omega_M \tau)^2 (f_1/f_M)$.

(*b*) Plot P_e versus $\omega_M \tau$ for $f_1/f_M = 0.01, 0.04, 0.1$.

12.12-2. Plot Eq. (12.12-5) for $\omega_M \tau = 10$ and $\omega_1 \tau = 0.004$.

12.12-3. (*a*) Derive an expression for the output SNR when the binary signal is transmitted using DPSK.

(*b*) If $\omega_m \tau = 10$ and $\omega_1 \tau = 0.04$, compare the thresholds obtained using PSK and DPSK.

12.13-1. A PCM and a DM system are both designed to yield an output SNR of 30 dB. Let $f_m =$ 4 kHz and assume PCM sampling at 5 times the Nyquist rate.

(*a*) Compare the bandwidths required for each system.

(*b*) If $f_1/f_M = 0.04$, compare the threshold of each system.

CHAPTER

THIRTEEN

INFORMATION THEORY AND CODING

We have seen that communication systems are limited in their performance by the available signal power, the inevitable background noise, and the need to limit bandwidth. The performance of each of the systems studied is not equal; some are better than others. We are naturally led to ask whether there is some system, which we have not yet encountered, which is superior to any so far considered. More specifically, we inquire what would be the characteristics of an *ideal* system, that is, a system which is not limited by our engineering ingenuity and inventiveness but limited rather only by the *fundamental nature of the physical universe*. In the course of this inquiry we shall be led to discuss the following matters:

1. What are the performance characteristics of such an ideal system?
2. How, at least in principle, is such a system to be realized?
3. How do the performances of existing systems compare with that of the ideal system?

In response to the first of these questions we shall have occasion to introduce some elemental concepts of Information Theory,[1,2] a theory which, as employed here, is due principally to Shannon.[3] The second question will lead us to an elementary discussion of *coding*. The third question is discussed beginning with Sec. 13.19.

13.1 DISCRETE MESSAGES

In considering communication systems broadly, we may assume, without loss of generality, that the function of any communication system is to convey, from transmitter to receiver, a sequence of messages which are selected from a finite number of predetermined messages. Thus, within some specified time interval one of these messages is transmitted. During the next time interval another of these messages (or possibly the same message) is transmitted. It should be noted that while the messages are predetermined and hence known by the receiver, the message selected for transmission during a particular interval is not known, a priori, by the receiver. Thus the receiver does not have the burden of extracting an arbitrary signal from a background of noise, but need only perform the operation of identifying which of a number of allowable messages was transmitted. The job of the receiver is not to answer the question "What was the message?" but rather the question "Which one?" As may well be anticipated the receiver will answer this question by determining the correlation of the received noisy signal with all of the possible predetermined signals individually. The receiver will then decide that the transmitted signal is the predetermined signal with which the noisy received signal has the greatest correlation. Rather generally, the probability that a particular message has been selected for transmission will not be the same for all messages. In this case, we assume that the probability of occurrence of each possible message is known at the receiver.

We recognize that this transmission of known messages is actually nothing more than an extension of the concept of quantization discussed in Sec. 6.2. When a signal is quantized, the receiver need determine only which of the quantized levels is encountered at each sampling time. The discrete predetermined messages then consist of those quantization levels which may be transmitted directly or encoded into any one of a number of forms. It is to be noted, of course, that such quantization imposes no restriction on the precision with which an arbitrary signal may be transmitted, since, in principle, the number of quantization levels may be increased without limit.

We may be rather free in our interpretation of the term "message." Thus, suppose that, during some time interval, there is generated at the transmitter one of a number of predetermined waveforms. If the receiver correctly identifies the waveform, then the receiver has received the transmitted message. For example, suppose that a quantization level is encoded into a binary waveform as in binary PCM. Then we may view the quantization level as a message, but we may also view each binary digit as a message. Thus a number of messages conveyed by a succession of binary digits may convey the single message contained in one quantization sample.

13.2 THE CONCEPT OF AMOUNT OF INFORMATION

Let us consider a communication system in which the allowable messages are m_1, m_2, ..., with probabilities of occurrence $p_1, p_2, \ldots$. Of course, $p_1 + p_2 + \cdots = 1$. Let the transmitter select message m_k, of probability p_k; let us further assume that the receiver has correctly identified the message. Then we shall say, by way of *definition* of the term *information*, that the system has conveyed an *amount of information* I_k given by

$$I_k \equiv \log_2 \frac{1}{p_k} \tag{13.2-1}$$

The concept of *amount of information* is so essential to our present interest that it behooves us to examine with some care the implications of Eq. (13.2-1). We note first that while I_k is an entirely dimensionless number, by convention, the "unit" it is assigned is the *bit*. Thus, by way of example, if $p_k = \frac{1}{4}$, $I_k = \log_2 4 = 2$ bits. The unit bit is employed principally as a reminder that in Eq. (13.2-1) the base of the logarithm is 2. (When the natural logarithmic base is used, the unit is the *nat*, and when the base is 10, the unit is the *Hartley* or *decit*. The use of such units in the present case is analogous to the unit *radian* used in angle measure and *decibel* used in connection with power ratios.) The use of the base 2 is especially convenient when binary PCM is employed. For, if the two possible binary digits (bits) may occur with equal likelihood, each with a probability $\frac{1}{2}$, then the correct identification of the binary digit conveys an amount of information $I = \log_2 2 = 1$ bit.

In the past we have used the term bit as an abbreviation for the phrase *binary digit*. When there is an uncertainty whether the word bit is intended as an abbreviation for binary digit or as a unit of information measure, it is customary to refer to a binary digit as a *binit*. Note that if the probabilities of two possible binits are not equally likely, one binit conveys *more* and one conveys *less* than 1 *bit* of information. For example, if the binits 0 and 1 occur with probabilities of $\frac{1}{4}$ and $\frac{3}{4}$, respectively, then binit 0 conveys information in amount $\log_2 4 = 2$ bits, while binit 1 conveys information in amount $\log_2 \frac{4}{3} = 0.42$ bit.

Suppose that there are M equally likely and independent messages and that $M = 2^N$, with N an integer. In this case the information in each message is

$$I = \log_2 M = \log_2 2^N = N \text{ bits} \tag{13.2-2a}$$

Suppose, further, that we choose to identify each message by binary PCM code words. The number of binary digits required for each of the 2^N messages is also N. Hence in this case the information in each message, as measured in bits, is numerically the same as the number of binits needed to encode the messages.

When $p_k = 1$, we have a trivial case since only one possible message is allowed. In this instance, since the receiver knows the message, there is really no need for transmission. We find that $I = \log_2 1 = 0$. As p_k decreases from 1 to 0, I_k increases monotonically, going from 0 to infinity. Thus, a greater amount of information has been conveyed when the receiver correctly identifies a less likely message.

When two independent messages m_k and m_l are correctly identified, we can readily prove that the amount of information conveyed is the sum of the information associated with each of the messages individually. Thus, we note that the individual information amounts are

$$I_k = \log_2 \frac{1}{p_k} \tag{13.2-2b}$$

$$I_l = \log_2 \frac{1}{p_l} \tag{13.2-2c}$$

Since the messages are independent, the probability of the composite message is $p_k p_l$ with corresponding information content of messages m_k and m_l is

$$I_{k,l} = \log_2 \frac{1}{p_k p_l} = \log_2 \frac{1}{p_k} + \log_2 \frac{1}{p_l} = I_k + I_l \tag{13.2-3}$$

It is of interest to note that the term information applied to the symbol I_k in Eq. (13.2-1) is rather aptly chosen, since there is some correspondence between the properties of I_k and the meaning of the word information as used in everyday speech. For example, suppose that an airplane dispatcher calls the weather bureau of a distant city during daylight hours to inquire about the present weather. If he receives, in response, the message, "There is daylight here," he will surely judge that he has received no information since he was certain beforehand that such was the case. On the other hand, if he hears "It is not raining here," he will consider that he has received information since he might anticipate such a situation with less than perfect certainty. Further, suppose the dispatcher receives some weather information on two different days. Then he might indeed consider that the total information received was the sum of the information received in the individual weather reports.

Furthermore, consider that the weather bureau of a city located in the desert, where it has not rained for 25 years, is called before each flight. The call is required although everyone "knows" that the weather will be clear. However, the day the call is made and the reply is that a very heavy rainstorm is in progress and that the flight must be canceled. The information received is then very great.

13.3 AVERAGE INFORMATION, ENTROPY

Suppose we have M different and independent messages $m_1, m_2, \ldots,$ with probabilities of occurrence $p_1, p_2, \ldots$. Suppose further that during a long period of transmission a sequence of L messages has been generated. Then, if L is very large, we may expect that in the L message sequence we transmitted $p_1 L$ messages of m_1, $p_2 L$ messages of m_2, etc. The total information in such a sequence will be

$$I_{\text{total}} = p_1 L \log_2 \frac{1}{p_1} + p_2 L \log_2 \frac{1}{p_2} + \cdots \tag{13.3-1}$$

The average information per message interval, represented by the symbol H, will then be

$$H \equiv \frac{I_{\text{total}}}{L} = p_1 \log_2 \frac{1}{p_1} + p_2 \log_2 \frac{1}{p_2} + \cdots = \sum_{k=1}^{M} p_k \log_2 \frac{1}{p_k} \tag{13.3-2}$$

This average information is also referred to by the term *entropy*.

We have seen that when there is only a single possible message ($p_k = 1$), the receipt of that message conveys no information. At the other extreme, as $p_k \to 0$, $I_k \to \infty$. However, since

$$\lim_{p\to 0} p \log \frac{1}{p} = 0 \tag{13.3-3}$$

the *average* information associated with an extremely unlikely message, as well as an extremely likely message, is zero.

As an example of the dependence of H on the probabilities of messages, let us consider the case of just two messages with probabilities p and $(1-p)$. The average information per message is

$$H = p \log_2 \frac{1}{p} + (1-p) \log_2 \frac{1}{1-p} \tag{13.3-4}$$

A plot of H as a function of p is shown in Fig. 13.3-1. Note that $H = 0$ at $p = 0$ and at $p = 1$. The maximum value of H may be located by setting to zero dH/dp as calculated from Eq. (13.3-4). It is then found that, as indicated in the figure, the maximum occurs at $p = 1/2$, that is, when the two messages are equally likely. The corresponding H is

$$H_{\max} = 1/2 \log_2 2 + 1/2 \log_2 2 = \log_2 2 = 1 \text{ bit/message} \tag{13.3-5}$$

When there are M messages, it may likewise be proved that H becomes a maximum when all the messages are equally likely. (The student is guided through the details of this proof in Prob. 13.3-3.) In this case each message has a probability $p = 1/M$ and

$$H_{\max} = \sum \frac{1}{M} \log_2 M = \log_2 M \tag{13.3-6}$$

since there are M terms in the summation.

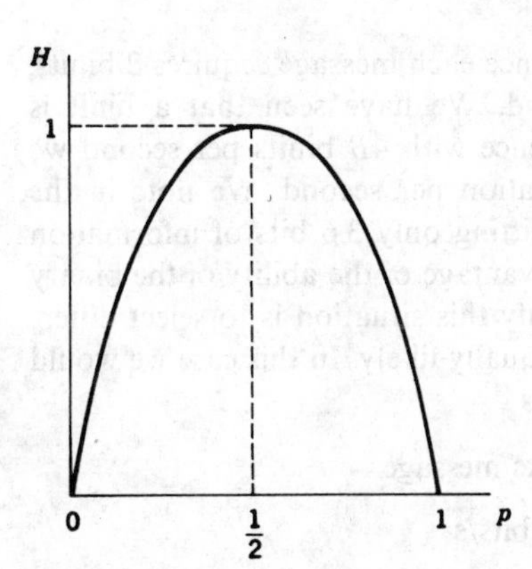

Figure 13.3-1 Average information H, for case of two messages, plotted as a function of the probability p of one of the messages.

13.4 INFORMATION RATE

If the *source* of the messages generates messages at the rate r messages per second, then the *information rate* is defined to be

$$R \equiv rH = \text{average number of bits of information/second} \qquad (13.4\text{-}1)$$

Example 13.4-1 An analog signal is bandlimited to B Hz, sampled at the Nyquist rate, and the samples are quantized into 4 levels. The quantization levels Q_1, Q_2, Q_3, and Q_4 (messages) are assumed independent and occur with probabilities $p_1 = p_4 = \frac{1}{8}$ and $p_2 = p_3 = \frac{3}{8}$. Find the information rate of the source.

SOLUTION The average information H is

$$\begin{aligned} H &= p_1 \log_2 \frac{1}{p_1} + p_2 \log_2 \frac{1}{p_2} + p_3 \log_2 \frac{1}{p_3} + p_4 \log_2 \frac{1}{p_4} \\ &= \tfrac{1}{8} \log_2 8 + \tfrac{3}{8} \log_2 \tfrac{8}{3} + \tfrac{3}{8} \log_2 \tfrac{8}{3} + \tfrac{1}{8} \log_2 8 \\ &= 1.8 \text{ bits/message} \end{aligned} \qquad (13.4\text{-}2)$$

The information rate R is

$$R = rH = 2B(1.8) = 3.6B \text{ bits/s}$$

We might choose to transmit the messages, referred to in the above example, by binary PCM. Thus we might identify each message by a binary code as indicated in the following table.

Message	Probability	Binary code
Q_1	$\frac{1}{8}$	0 0
Q_2	$\frac{3}{8}$	0 1
Q_3	$\frac{3}{8}$	1 0
Q_4	$\frac{1}{8}$	1 1

If we transmit $2B$ messages per second, then, since each message requires 2 binits, we shall be transmitting $4B$ binits per second. We have seen that a binit is capable of conveying 1 bit of information. Hence with $4B$ binits per second we should be able to transmit $4B$ bits of information per second. We note in the above illustration that we are actually transmitting only 3.6 bits of information per second. We are therefore not taking full advantage of the ability of the binary PCM to convey information. One way to rectify this situation is to select different quantization levels such that each level is equally likely. In this case we would find that the average information per message is

$$H = 4(\tfrac{1}{4} \log_2 4) = 2 \text{ bits/message}$$

and

$$R = rH = 2B(2) = 4B \text{ bits/s}$$

Suppose, however, that it was not convenient or appropriate to change the messages. In this case we might seek an alternative coding scheme (rather than binary encoding) in which, on the average, the number of bits per message was fewer than 2 (ideally 1.8). Such a coding scheme is described in Sec. 13.5.

13.5 CODING TO INCREASE AVERAGE INFORMATION PER BIT

Suppose we have $M = 2^N$ messages, each message coded into N bits. We have seen that if the messages are equally likely the average information per message interval is, from Eq. (13.3-6), $H = N$. There being N bits in the message, the average information carried by an individual bit is $H/N = 1$ bit. If, however, the messages are not equally likely then H is less than N and each bit carries less than 1 bit of information. The situation can be improved by using a code in which not all messages are encoded into the same number of bits. The more likely a message is, the fewer the number of bits that should be used in its code word. The merit of such a scheme is to some extent intuitively obvious. Witness the fact that in the *Morse* code of telegraphy the most common letter *e* is represented by a single "dot" while the less frequently occuring letters are represented by longer sequences of "dots" and "dashes".

One basic technique by which we may generate such a more "efficient" code is the Shannon–Fano algorithm. An example of the application of this algorithm is given in Fig. 13.5-1. Here we consider that there are 8 possible messages m_1 through m_8 with probabilities as given. We have ordered the messages in order of decreasing probability. In column I we divide the message into two partitions such that the sum of the probabilities of each group are the same. Thus m_1 in one partition has the probability $\frac{1}{2}$ and the sum of the probabilities of all the other messages in the second partition is also $\frac{1}{2}$. We assign the bit 0 to the message in

Message	Probability	I	II	III	IV	V	No of bits/ message
m_1	1/2	0					1
m_2	1/8	1	0	0			3
m_3	1/8	1	0	1			3
m_4	1/16	1	1	0	0		4
m_5	1/16	1	1	0	1		4
m_6	1/16	1	1	1	0		4
m_7	1/32	1	1	1	1	0	5
m_8	1/32	1	1	1	1	1	5

Figure 13.5-1 An example of the application of the Shannon–Fano algorithm.

one partition and we assign the bit 1 to all the messages in the other partition. This process of dividing groups in the same partition into two partitions each with equal sums of probabilities is continued, until each message finds itself alone in a partition. At each partitioning, one group has a zero assigned while a 1 is assigned to the other. In the example of Fig. 13.5-1 five partitionings are required. We find that m_1 is represented by the single bit 0, m_2 is represented by the 3 bit code 100, etc., up to m_8 whose code is the 5 bit word 11111.

We now find that with this code, the average information per message interval is

$$H = \sum_{i=1}^{\infty} p_i \log_2 \frac{1}{p_i} = \tfrac{1}{2} \log_2 2 + 2 \times \tfrac{1}{8} \log_2 2^3 + 3 \times \tfrac{1}{16} \log_2 2^4 + 2 \times \tfrac{1}{32} \log_2 2^5$$

$$= 2\tfrac{5}{16} \tag{13.5-1}$$

We find further that the average number of bits per message is

$$\text{aver. bits/message} = 1(\tfrac{1}{2}) + 3(\tfrac{1}{8}) + 4(\tfrac{1}{16}) + 4(\tfrac{1}{16}) + 5(\tfrac{1}{32}) + 5(\tfrac{1}{32})$$

$$= 2\tfrac{5}{16} \tag{13.5-2}$$

Hence we have the result from Eqs. (13.5-1) and (13.5-2) that each bit on the average carries 1 bit of information which is the maximum information that can be conveyed by a bit. If we had not used this more efficient code, we would have required 3 bits per message which is about 30 percent larger than the number $2\frac{5}{16}$ calculated above.

The saving in average number of bits per message can be put to good advantage in a number of ways. On one hand we may decide to allow more time per bit thereby reducing bandwidth and improving noise immunity. Or, if we decide to keep the bit rate fixed, we may use bits not used for information transmission to provide error detection, error correction, or both.

The code generation example of Fig. 13.5-1 is special in that we selected probabilities in such a manner that at each partitioning it was possible to arrange that the sum of the probabilities in each group is exactly equal. When such is not the case we partition to satisfy this equal probability condition as nearly as possible. In this more general case the code efficiency will be reduced.

13.6 SHANNON'S THEOREM, CHANNEL CAPACITY

The importance of the concept of information rate is that it enters into a theorem due to Shannon[3] which is fundamental to the theory of communications. This theorem is concerned with the rate of transmission of information over a communication channel. While we have used the term communication channel on many occasions, it is well to emphasize at this point, that the term, which is something of an abstraction, is intended to encompass all the features and component parts of the transmission system which introduce noise or limit the bandwidth.

Shannon's theorem says that it is possible, in principle, to devise a means whereby a communications system will transmit information with an arbitrarily small probability of error provided that the information rate R is less than or equal to a rate C called the *channel capacity*. The technique used to approach this end is called *coding* and is discussed beginning with Sec. 13.9. To put the matter more formally, we have the following:

Theorem Given a source of M equally likely messages, with $M \gg 1$, which is generating information at a rate R. Given a channel with channel capacity C. Then, if

$$R \leq C$$

there exists a *coding* technique such that the output of the source may be transmitted over the channel with a probability of error in the received message which may be made arbitrarily small.

The important feature of the theorem is that it indicates that for $R \leq C$ transmission may be accomplished without error in the presence of noise. This result is surprising. For in our consideration of noise, say, gaussian noise, we have seen that the probability density of the noise extends to *infinity*. We should then imagine that there will be some times, however infrequent, when the noise *must* override the signal thereby resulting in errors. However, Shannon's theorem says that this need not cause a message to be in error.

There is a negative statement associated with Shannon's theorem. It states the following:

Theorem Given a source of M equally likely messages, with $M \gg 1$, which is generating information at a rate R; then if

$$R > C$$

the probability of error is close to unity for every possible set of M transmitter signals.

This *negative* theorem states that if the information rate R exceeds a specified value C, the error probability will increase toward unity as M increases, and that also, generally, in this case where $R > C$, increasing the complexity of the coding results in an increase in the probability of error.

13.7 CAPACITY OF A GAUSSIAN CHANNEL

A theorem which is complementary to Shannon's theorem and applies to a channel in which the noise is gaussian is known as the Shannon–Hartley theorem.

Theorem The channel capacity of a white, bandlimited gaussian channel is

$$C = B \log_2 \left(1 + \frac{S}{N}\right) \text{ bits/s} \tag{13.7-1}$$

where B is the channel bandwidth, S the signal power, and N is the total noise within the channel bandwidth, that is, $N = \eta B$, with $\eta/2$ the (two-sided) power spectral density.

This theorem, although restricted to the gaussian channel, is of fundamental importance. First, we find that channels encountered in physical systems generally are, at least approximately, gaussian. Second, it turns out that the results obtained for a gaussian channel often provide a *lower bound* on the performance of a system operating over a nongaussian channel. Thus, if a particular encoder-decoder is used with a gaussian channel and an error probability P_e results, then with a nongaussian channel another encoder-decoder can be designed so that the P_e will be smaller. We may note that channel capacity equations corresponding to Eq. (13.7-1) have been derived for a number of nongaussian channels.

The derivation of Eq. (13.7-1) for the capacity of a gaussian channel is rather formidable and will not be undertaken. However, the result may be made to appear reasonable by the following considerations. Suppose that, for the purpose of transmission over the channel, the messages are represented by fixed voltage levels. Then, as the source generates one message after another in sequence, the transmitted signal $s(t)$ takes on a waveform similar to that shown in Fig. 13.7-1.

The received signal is accompanied by noise whose root-mean-square voltage is σ. The levels have been separated by an interval $\lambda\sigma$, where λ is a number presumed large enough to allow recognition of individual levels with an acceptable probability of error. Assuming an even number of levels, the levels are located at voltages $\pm\lambda\sigma/2$, $\pm 3\lambda\sigma/2$, etc. If there are to be M possible messages, then there

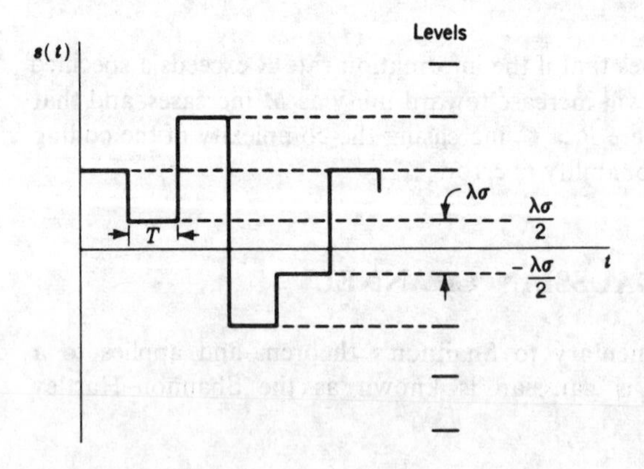

Figure 13.7-1 A sequence of messages is represented by a waveform $s(t)$ which assumes voltage levels corresponding to the messages.

must be M levels. We assume that the messages and hence the levels occur with equal likelihood. Then the average signal power is

$$S = \frac{2}{M}\left\{\left(\frac{\lambda\sigma}{2}\right)^2 + \left(\frac{3\lambda\sigma}{2}\right)^2 + \cdots + \left[\frac{(M-1)\lambda\sigma}{2}\right]^2\right\} \tag{13.7-2a}$$

$$= \frac{M^2 - 1}{12}(\lambda\sigma)^2 \tag{13.7-2b}$$

The number of levels for a given average signal power is, from Eq. (13.7-2*b*),

$$M = \left(1 + \frac{12S}{\lambda^2\sigma^2}\right)^{1/2} = \left(1 + \frac{12}{\lambda^2}\frac{S}{N}\right)^{1/2} \tag{13.7-3}$$

where $N = \sigma^2$ is the noise power. Each message is equally likely and therefore conveys an average amount of information

$$H = \log_2 M = \log_2\left(1 + \frac{12}{\lambda^2}\frac{S}{N}\right)^{1/2}$$

$$= \tfrac{1}{2}\log_2\left(1 + \frac{12}{\lambda^2}\frac{S}{N}\right) \text{ bits/message} \tag{13.7-4}$$

To find the information rate of the signal waveform $s(t)$ of Fig. 13.7-1, we need to estimate how many messages per unit time may be carried by this signal. That is, we need to estimate the interval T which should be assigned to each message to allow the transmitted levels to be recognized individually at the receiver, even though the bandwidth B of the channel is limited. Now the principal effect of limited bandwidth on the signal waveform $s(t)$ will be a *rounding* of the initially abrupt transitions from one level to another. We saw in Sec. 1.15 that when an abrupt step is applied to an ideal low-pass filter of bandwidth B, the response has a 10 to 90 percent rise time τ given by $\tau = 0.44/B$. We may then reasonably estimate that if we set $T = \tau$, we shall be able to distinguish levels *reliably*. Since our discussion is heuristic, let us, as a matter of convenience, take $T = \tau = 0.5/B$. The message rate is then

$$r = \frac{1}{T} = 2B \text{ messages/s} \tag{13.7-5}$$

which is equal, as we might expect, to the Nyquist sampling rate. Since the transmission of any of the M messages is equally likely,

$$H = \log_2 M$$

Thus our channel is transferring information at a rate $R = rH$. Since we have presumably taken what precautions are necessary to ensure that the channel is just able to allow the transmission with acceptable probability of error, $R \approx C$. The channel capacity is, therefore, found by combining Eqs. (13.7-4) and (13.7-5):

$$C \approx R = rH = B \log_2 \left(1 + \frac{12}{\lambda^2} \frac{S}{N}\right) \tag{13.7-6}$$

Comparing Eq. (13.7-6) with Eq. (13.7-1) of the Shannon–Hartley theorem, we observe that the results would be identical if we set $12/\lambda^2 = 1$, that is, $\lambda = 3.5$.

This heuristic discussion leading to Eq. (13.7-6) was undertaken to make the Shannon–Hartley theorem [Eq. (13.7-1)] rather intuitively acceptable. It needs to be emphasized, however, that Eq. (13.7-6) specifies the rate at which information may be transmitted with *small* error, while the Shannon–Hartley theorem contemplates that, with a sufficiently sophisticated transmission technique, transmission at channel capacity is possible with arbitrarily small error.

13.8 BANDWIDTH—S/N TRADEOFF

The Shannon–Hartley theorem, Eq. (13.7-1); indicates that a noiseless gaussian channel ($S/N = \infty$) has an infinite capacity. On the other hand, while the channel capacity does increase, it does not become infinite as the bandwidth becomes infinite because, with an increase in bandwidth, the noise power also increases. Thus for a fixed signal power and in the presence of white gaussian noise the channel capacity approaches an upper limit with increasing bandwidth. We now calculate that limit. Using $N = \eta B$ in Eq. (13.7-1), we have

$$C = B \log_2 \left(1 + \frac{S}{\eta B}\right) = \frac{S}{\eta} \frac{\eta B}{S} \log_2 \left(1 + \frac{S}{\eta B}\right) \tag{13.8-1a}$$

$$= \frac{S}{\eta} \log_2 \left(1 + \frac{S}{\eta B}\right)^{\eta B/S} \tag{13.8-1b}$$

We recall that $\lim_{x \to 0} (1 + x)^{1/x} = e$ (the naperian base), and identifying x as $x = S/\eta B$, we find that Eq. (13.8-1b) becomes

$$C_\infty = \lim_{B \to \infty} C = \frac{S}{\eta} \log_2 e = 1.44 \frac{S}{\eta} \tag{13.8-2}$$

The Shannon–Hartley principle also indicates that we may trade off bandwidth for signal-to-noise ratio and vice versa. For example, if $S/N = 7$ and $B = 4$ kHz, we find $C = 12 \times 10^3$ bits/s. If the SNR is increased to $S/N = 15$ and B decreased to 3 kHz, the channel capacity remains the same. With a 3-kHz bandwidth the noise power will be $\frac{3}{4}$ as large as with 4-kHz. Thus the signal power will have to be increased by the factor $(\frac{3}{4} \times \frac{15}{7}) = 1.6$. Therefore, the 25 percent reduction in bandwidth requires a 60 percent increase in signal power.

It is of great interest, and even a little surprising, to recognize that the trade off between bandwidth and signal-to-noise ratio is not limited by a lower limit on bandwidth. Specifically, suppose that we want to transmit a signal whose spectral range extends up to a frequency f_M. Let us quantize the signal so that we may determine an information rate for the signal, and let us assume that the channel has a channel capacity greater than the signal information rate. Now, however, suppose that it turns out that the bandwidth B of the channel is less than f_M. Say, as an extreme example, that the bandwidth is 1 Hz while f_M is 1000 Hz. Then is it really possible, in principle, to receive the signal with arbitrarily small probability of error? The answer is yes, provided that the channel capacity is greater than the transmitted data bit rate.

To make the idea somewhat more acceptable that such errorless signal reception is indeed possible, consider the extreme case where there is no noise at all. Suppose that the signal with $f_M = 1000$ Hz is transmitted through a channel which can be represented as a low-pass RC circuit with cut off at 1 Hz. Then the received signal will be greatly attenuated and severely distorted. If, however, there is no noise, then we are entirely free to make up for the attenuation by the use of an amplifier and to correct the frequency distortion by the use of an equalizer. The signal is then recoverable precisely as it was transmitted.

13.9 USE OF ORTHOGONAL SIGNALS TO ATTAIN SHANNON'S LIMIT[4]

In the preceding section we saw that in the (trivial) case of no noise the channel capacity was infinite, and that no matter how restricted the bandwidth was, it was always possible to receive a signal without error, as predicted by the Shannon limit. In the present section we shall discuss one possible method of attaining performance predicted by Shannon's theorem in the presence of gaussian noise. We shall see that, to obtain errorless transmission for $R \leq C$ requires an extremely complicated communication system. In physically realizable systems, therefore, we must accept a performance less than optimum.

Orthogonal Signals

The system we are to describe involves the use of a set $s_1(t)$, $s_2(t)$, ... of orthogonal signals. Such signals $s_i(t)$, defined as being orthogonal over the interval 0 to T, have the property that

$$\int_0^T s_1(t)s_2(t)\, dt = 0 \qquad \text{for } i \neq j \tag{13.9-1}$$

A familar example of a set of orthogonal signals is M-ary FSK, whose probability of error we have calculated in Sec. 11.15. Because of the importance of the result and the fact that the probability of error obtained using the Union Bound is "too loose" we now proceed with an alternate derivation of P_e.

We shall assume that the amplitudes of the signals have been adjusted so that in every case

$$\int_0^T s_i^2(t)\, dt = E_s \tag{13.9-2}$$

that is, each signal has the same energy E_s in the interval T and also the same power $P_s = E_s/T$.

Matched-Filter Reception

Let us assume that our message source generates M messages, each with equal likelihood. Let each message be represented by one of the orthogonal sets of signals $s_1(t), s_2(t), \ldots, s_M(t)$. The message interval is T. The signals are transmitted over a communications channel where they are corrupted by additive white gaussian noise. As shown in Fig. 13.9-1, at the receiver a determination of which message has been transmitted is made through the use of M matched filters, that is, correlators. Each correlator consists of a multiplier followed by an integrator. The local inputs to the multipliers are, as shown, the signals $s_i(t)$. Suppose then, in the absence of noise, that the signal $s_i(t)$ is transmitted, and the output of each integrator is sampled at the end of a message interval. Then, because of the orthogonality condition of Eq. (13.9-1), all integrators will have zero output, except that the ith integrator output will be E_s, as given by Eq. (13.9-2).

In the presence of an additive noise waveform $n(t)$, the output of the lth correlator $(l \neq i)$ will be

$$e_l = \int_0^T n(t)s_l(t)\,dt \equiv n_l \tag{13.9-3}$$

This quantity n_l is a random variable. In Prob. 13.9-2 the student is guided through a proof that (a) the random variable n_l is gaussian, (b) that it has a mean value zero, (c) that it has a mean-square value, i.e., a variance σ^2 given by

$$\sigma^2 = \frac{\eta E_s}{2} \tag{13.9-4}$$

and (d) that $E(e_l e_m) = 0$; that is, the outputs of the matched filter are independent. The gaussian character and zero mean value of n_l do not depend on the form of the deterministic signal $s_1(t)$. The variance σ^2 depends only on the signal

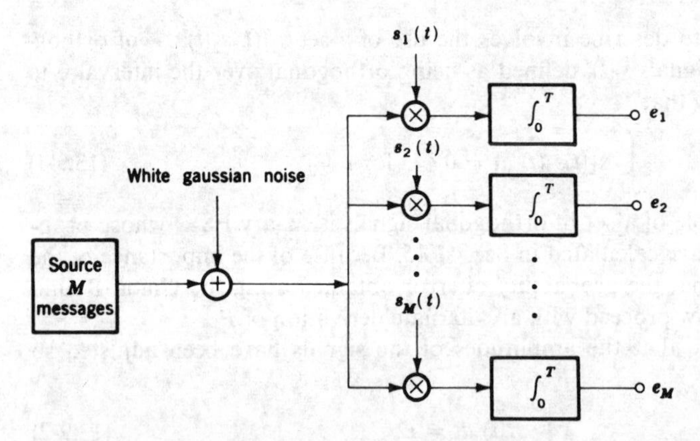

Figure 13.9-1 The messages of a source are represented by orthogonal signals $s_1(t), s_2(t), \ldots, s_M(t)$. Reception is accomplished by the use of correlation detectors (matched filters).

energy. Since we have selected the orthogonal signals $s_1(t)$, $s_2(t)$, ..., to have the same energy [Eq. (13.9-2)], the noise sample value at the output of each correlator will have the same statistical properties.

The correlator corresponding to the transmitted message $s_i(t)$ will have an output

$$e_i = \int_0^T (s_i(t) + n(t))s_i(t)\, dt \tag{13.9-5a}$$

$$= \int_0^T s_i^2(t)\, dt + \int_0^T n(t)s_i(t)\, dt \tag{13.9-5b}$$

$$= E_s + n_i \tag{13.9-5c}$$

from Eqs. (13.9-2) and (13.9-3). Again n_i is a random variable with statistical properties identical with those specified for n_l.

Calculation of Error Probability

To determine which message has been transmitted, we shall compare the matched-filter outputs $e_1, e_2, \ldots, e_M$. We shall decide that $s_i(t)$ has been transmitted if the corresponding output e_i is larger than the output of any other filter. We shall now calculate the probability that such a determination will lead to an error.

The probability that some arbitrarily selected output e_l is less than the output e_i is

$$P(e_l < e_i) = \frac{1}{\sqrt{2\pi\sigma^2}} \int_{-\infty}^{e_i} e^{-e_l^2/2\sigma^2}\, de_l \tag{13.9-6}$$

in which σ^2 is given by Eq. (13.9-4). Observe, in Eq. (13.9-6), that $P(e_l < e_i)$ depends only on e_i and does not depend on which output e_l $(l \neq i)$ has been selected for comparison with e_i.

The probability that, say, e_1 and e_2 are *both* smaller than e_i is

$$P(e_1 < e_i \text{ and } e_2 < e_i) = P(e_1 < e_i)P(e_2 < e_i) \tag{13.9-7a}$$

$$= [P(e_1 < e_i)]^2 = [P(e_2 < e_i)]^2 \tag{13.9-7b}$$

since e_1 and e_2 are independent (Prob. 13.9-2), and as we have noted, $P(e_l < e_i)$ does not depend on e_l. Hence, the probability P_L that e_i is the largest of the outputs is

$$P_L \equiv P(e_i > e_1, e_2, \ldots, e_{i-1}, e_{i+1}, \ldots, e_M)$$

$$= \left(\frac{1}{\sqrt{2\pi\sigma^2}} \int_{-\infty}^{e_i} e^{-e_l^2/2\sigma^2}\, de_l\right)^{M-1} \tag{13.9-8a}$$

$$= \left(\frac{1}{\sqrt{2\pi\sigma^2}} \int_{-\infty}^{E_s+n_i} e^{-e_l^2/2\sigma^2}\, de_l\right)^{M-1} \tag{13.9-8b}$$

from Eq. (13.9-5c). If we let $x \equiv e_l/(\sqrt{2}\,\sigma)$, Eq. (13.9-8b) becomes

$$P_L = \left(\frac{1}{\sqrt{\pi}} \int_{-\infty}^{\sqrt{E_s/\eta} + n_i/\sqrt{2}\sigma} e^{-x^2}\, dx \right)^{M-1} \tag{13.9-9}$$

where we have used Eq. (13.9-4) to show that $E_s/(\sqrt{2}\,\sigma) = \sqrt{E_s/\eta}$. Observe that P_L depends on the two deterministic parameters E_s/η and M and on the single random variable, $n_i/\sqrt{2}\,\sigma$; that is,

$$P_L = P_L\left(\frac{E_s}{\eta}, M, \frac{n_i}{\sqrt{2}\,\sigma}\right)$$

To find the probability that e_i is the largest output without reference to the noise output n_i of the ith correlator, we need but average P_L over all possible values of n_i. This average is the probability that we shall be correct in deciding that the transmitted signal corresponds to the correlator which yields the largest output. Let us call this probability P_c. The probability of an error is then $P_e = 1 - P_c$.

We have seen that n_i is a gaussian random variable with zero mean and variance σ^2. Hence the average value of P_L, considering all possible values of n_i, is

$$P_c = 1 - P_e = \frac{1}{\sqrt{2\pi\sigma^2}} \int_{-\infty}^{\infty} P_L(E_s/\eta, M, n_i/\sqrt{2}\,\sigma) e^{-n_i^2/2\sigma^2}\, dn_i \tag{13.9-10}$$

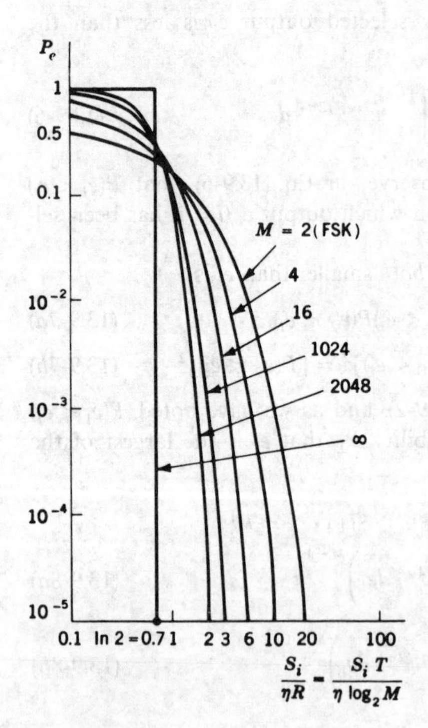

Figure 13.9-2 Probability of error with orthogonal signals. (*Courtesy of A. J. Viterbi, "Principles of Coherent Communications," McGraw-Hill Book Co.,* 1966.)

If we let $y \equiv n_i/(\sqrt{2}\,\sigma)$, Eq. (13.9-10) becomes, using Eq. (13.9-9),

$$1 - P_e \equiv P_c = \left(\frac{1}{\sqrt{\pi}}\right)^M \int_{-\infty}^{\infty} e^{-y^2} \left(\int_{-\infty}^{\sqrt{E_s/\eta}+y} e^{-x^2}\, dx\right)^{M-1} dy \quad (13.9\text{-}11)$$

The integral in Eq. (13.9-11) is rather formidable. It has been approximated using the Union Bound and the result presented in Eq. (11.15-16). It has also been evaluated[4] by numerical integration by computer, and the results for several values of M are shown in Fig. 13.9-2. Here the ordinate is the error probability P_e. We may note that it appears from Eq. (13.9-11) that P_e is a function of E_s/η and M. The abscissa of the plot of Fig. 13.9-2 has been marked off in units of the ratio $E_s/\eta \log_2 M$). Note that $(\log_2 M)/T = R$, the rate of information transfer (since each of the M signals has equal probability of occurrence), so that the quantity $E_s/(\eta \log_2 M) = S_i/\eta R$, where $S_i = E_s/T$ is the signal power.

13.10 EFFICIENCY OF ORTHOGONAL SIGNAL TRANSMISSION

We observe that for all M, P_e decreases as $S_i/\eta R$ increases. Also, we note that as $S_i/\eta R \rightarrow \infty$, $P_e \rightarrow 0$. The axis $P_e = 0$ does not appear on the plot because of the logarithmic scale of the ordinate. In the plot for $M \rightarrow \infty$, $P_e = 0$, provided that

$$\frac{S_i}{\eta R} \geq \ln 2 \quad (13.10\text{-}1)$$

and $P_e = 1$ otherwise. It is interesting to note that using the Union Bound, we found that $P_e \rightarrow 0$, provided that [see Eq. (11.15-17)]

$$\frac{E_s}{\eta \log_2 M} = \frac{S_i}{\eta R} > 2 \ln 2 \quad (13.10\text{-}2)$$

Thus the Union Bound was too loose by 3 dB. In general, as M increases, S_i/η must also increase to keep P_e small, otherwise P_e increases.

Some of the features in Fig. 13.9-2 are certainly to have been anticipated readily enough on a qualitative basis. Thus we note that for fixed M and R, P_e decreases as the signal power S_i goes up or as the background noise density η goes down. Similarly, we note that for fixed S_i, η, and M, the P_e decreases as we allow more time T (that is, decreasing R) for the transmission of a single message. This result is to have been anticipated as an easy generalization of our earlier discussion (Sec. 11.1) concerning baseband binary PCM transmission [see particularly Eq. (11.1-6) and the ensuing comments].

However, the real usefulness to us of the results given in Fig. 13.9-2 is that, in the first place, they allow us the satisfaction of verifying, in this one special case, the validity of Shannon's principle and of the Shannon–Hartley result for channel capacity. In the second place, they allow us a basis on which to make an engineering judgement concerning how much complexity is warranted in terms of the reduction in error which results therefrom. We consider now the first point.

Shannon Limit

Let us assume that our signal source generates four equally likely messages $m_1(t)$, $m_2(t)$, $m_3(t)$, $m_4(t)$, and a time $T = 1$ s is to be allowed as the transmission interval for a message. We shall refer to these messages as elemental messages in the sense that they extend over the smallest time interval in which some information may be conveyed. The transmission rate is $R = (\log_2 4)/1 = 2$ bits/s. Let us assume that $S_i/\eta = 10\ \text{s}^{-1}$. Then $S_i/\eta R = 5$, and, as indicated on the plot of Fig. 13.9-2, we may expect an error probability $P_e \approx 10^{-2}$. At the receiver we require four matched filters. If we assume for simplicity that 4-ary FSK is being used the channel bandwidth is approximately 8 Hz.

Now let us join two successive intervals into an extended interval of length $2T = 2$ s. Since four messages are possible in each interval T, there are $4 \times 4 = 16$ message combinations, i.e., new orthogonal messages possible in the extended interval. We now consider $2T$ as the message interval, and the transmission rate R is, of course, unaltered, for we now have $R = (\log_2 16)/2 = 2$ bits/s as before. If we transmit using these sixteen new orthogonal messages, then we shall complicate the receiver very seriously, since we now require sixteen matched filters in place of the original four. The advantage which accrues to us in return for this added complexity is to be seen by referring again to Fig. 13.9-2. For now we find, with $S_i/\eta R = 5$ again, that the error probability has been reduced to $P_e \cong 5 \times 10^{-4}$. However, now the required bandwidth is approximately 16 Hz.

If we extend our message intervals to $3T$, $4T$, etc., the number of composite messages increases correspondingly to 64, 256, etc., reducing the error probability further, but correspondingly increasing the number of matched filters required, as well as the bandwidth needed. In the limit as the number of message intervals increases without bound, $M \to \infty$, and the error probability P_e will approach zero provided that Eq. (13.10-1) is satisfied. The required bandwidth is now infinite. From this condition we may calculate the maximum allowable errorless transmission rate R_{max}, which, by definition, is the channel capacity C. We find

$$R_{\text{max}} = C = \frac{S_i}{\eta} \frac{1}{\ln 2} = \frac{S_i}{\eta} \log_2 e = 1.44 S_i/\eta \tag{13.10-3}$$

Equation (13.10-3) is in agreement with Eq. (13.8-2) which we deduced from the Shannon–Hartley theorem for a channel of unlimited bandwidth.

When we transmit elemental messages, then, of course, some information is received in each elemental interval T. Suppose, however, we form composite messages each of which extends over many time intervals. Then no information will be received except at the termination of the extended interval. At this termination, all the information in all the intervals will be received at once. In the limit, as we increase the number of intervals without bound, and hence let $M \to \infty$, to reduce the error probability to zero, we find that we must wait a time $T \to \infty$ before we receive any information.

In summary, we have examined one possible way of achieving errorless transmission. We have shown that such transmission is possible through the use of orthogonal signals and matched filters over a channel of unlimited bandwidth. We have also calculated the channel capacity and arrived at a result which is consistent with the Shannon–Hartley theorem.

Tradeoffs

The plots of Fig. 13.9-2 may be used not only to illustrate some general principles as we have already done, but also to determine what advantage results in one direction as a result of a sacrifice in another direction. By way of example, suppose we are required to transmit information at some fixed information rate R, in the presence of a fixed background-noise power spectral density, with an error probability $P_e = 10^{-5}$. Suppose we use $M = 2$. Then, from Fig. 13.9-2 we find that $S_i/\eta R \approx 20$. Now consider what happens if M is increased to $M = 4$. Then to keep R fixed, T must be increased by the factor $(\log_2 4/\log_2 2) = 2$. Again referring to Fig. 13.9-2, we see that $S_i/\eta R$ can now be reduced to 10 and still result in $P_e = 10^{-5}$. Since η and R are fixed, S_i can be halved (reduced by 3 dB). Thus, to increase the number of messages while maintaining the same information rate R and P_e required that the time T be increased; however, the required signal power can be decreased.

13.11 CODING: INTRODUCTION

In Secs. 13.9 and 13.10 we studied one method of improving the efficiency of operation of a communications channel by transmitting a coded signal which consisted of 1 of M orthogonal signals. That is, given a channel of capacity C, we examined one procedure for transmitting information at a rate as nearly equal to C as possible with a minimum likelihood of error. However, this technique required infinite bandwidth. We consider now alternative, more efficient, methods of coding signals, to permit us to detect errors and to correct errors caused by noise.

Coding accomplishes its purpose through the deliberate introduction of redundancy into messages. For example, *orthogonal* signaling required the transmission of 1 of M orthogonal messages. If each of these orthogonal messages were transmitted using binary digits, it can be shown that M binary digits/message are required. If, however, the original M quantization levels were binary-encoded, then only $N = \log_2 M$ binary digits/message are required. The excess number of binary digits required for orthogonal encoded signals is $M - N$. Hence the coding will produce an effective redundancy of $M - N$. As an additional example consider that we are transmitting information by means of binary PCM. Then we transmit a stream of binary digits, 0 or 1, and our concern is not to confuse a 0 for a 1, or a 1 for a 0. Suppose that when a 0 is to be transmitted, we transmit instead a sequence of three 0's, that is, 000, and transmit 111 to represent the digit 1. These triplets of 0's or 1's are certainly redundant, for two of the 0's in 000 add no information to the message. Suppose, however, that the signal-to-noise ratio on the channel is such that we can be nearly certain that not more than one error will be made in a triplet. Then, if we received 001, 010, or 100, we would be rather certain that the transmitted message was actually 000. Similarly, if we received 011, 101, or 110, we would be rather certain that the message was actually 111. Thus the redundancy, *deliberately introduced*, has enabled us to detect the occurrence of an error and even to correct the error.

The introduction of redundancy, however, can not *guarantee* that an error will be either detectable or correctable, for errors are caused by the *unpredictable* random process called noise. Hence, while the noise level, as we assumed, may be low enough so that more than a single error is *unlikely*, there is always a finite possibility that two errors will occur. In this case we would know that an error had occurred but we would be inclined to read a 0 as a 1 and a 1 as a 0. Even more, there is always the possibility, however small, that three errors had been made. In this case, not only would we misread the digit, but we would not even suspect that an error had been made. We are thus led to the conclusion that while coding may allow a great deal of detection and correction, it ordinarily cannot detect or correct all errors.

An essential feature which results from the introduction of redundancy is that not all sequences of symbols constitute a bona fide message. For example, with a triplet of binary digits, eight combinations are possible. However, only two of these combinations, that is, 000 and 111, are recognized as words which convey a message. The remaining combinations are not in our "dictionary." It is this fact that certain words are not in our "dictionary" that allows us to detect errors. Corrections are made on the basis of a similarity between unacceptable and acceptable words.

There is a correspondence between the redundancy deliberately introduced in coding messages prior to transmission over a channel and the redundancy which is part of language. For example, suppose that on a page of printed text we encounter the word *ekpensive.* We would immediately recognize that the printer had made an error since we know that there is no such word. If we were inclined to judge it unlikely that the printer had made an error in more than one letter, we would easily recognize the word to be *expensive.* On the other hand there is always the possibility that more than one error was made and that the intended word was, say, *offensive.* And in this latter case, of course, we would not make the proper correction.

In language, redundancy extends beyond combinations of letters in a word. It extends to entire words, and beyond to phrases and even sentences. For example, if we had come across the sentence "the army launched an ekpensive," we would not have much difficulty in recognizing that the proper correction of the last word is *offensive*, in spite of the fact that corrections in three letters are called for. In this case we would be taking advantage of the redundancy in the sentence.

Now let us return briefly to the binary PCM transmission scheme we described above, in which three bits 000 or 111 were transmitted as a code to represent the bits 0 and 1, respectively. If the redundant message is to be transmitted at the same rate as the original binary signal, we shall have to transmit three bits in the time T otherwise allocated to a single bit. As we have seen on many occasions when the time allocated to a bit transmission decreases, the error rate increases. Hence, the required increased bit rate will undo some of the advantage that will accrue from redundancy coding. We shall however, of course, find that coding may yield a very worthwhile net advantage.

Error Probability with Repetition in the Binary Symmetric Channel

In Fig. 13.11-1 we represent the transmission of a bit over a *binary symmetric channel* (BSC). The probability $P(0)$ that a 0 was transmitted is the same as the probability $P(1)$ that a 1 was transmitted, i.e., $P(0) = P(1) = \frac{1}{2}$. There is a probability p of an error in transmission, i.e., that a 0 was transmitted and a 1 received or vice versa. The probability of correct transmission is $1 - p$. Suppose, as we have described, when a 0 is to be transmitted we transmit instead a sequence of M0's and similarly a 1 is replaced by a sequence of M1's. At the receiver we shall interpret the received sequence to represent a 0 or a 1 if more than half the bits in the sequence are 0's or 1's respectively. To avoid ambiguity we shall select M to be an odd number.

Consider, as an example, the case $M = 3$. An error will result if all three bits of the sequence are in error or if any two bits of the sequence are in error. The probability that all three are wrong is p^3. Now suppose next that a 0 was transmitted and was incorrectly read at the receiver because the received sequence was 110. The probability of such an occurrence is the probability that two bits in the sequence were incorrectly received while one bit was correctly received. This probability is $p^2(1 - p)$. There are two additional similar sequences, making a total of three, that will yield an error, i.e., 101 and 011. Hence altogether the error probability that the transmitted bit will be received in error is

$$P_e(M = 3) = p^3 + 3p^2(1 - p) \tag{13.11-1}$$

We find that if $p = 0.01$, $P_e(M = 3) = 3 \times 10^{-4}$. If $M = 5$ we find by a similar calculation that

$$P_e(M = 5) = p^5 + 5p^4(1 - p) + \frac{5.4}{2!}p^3(1 - p)^2 \tag{13.11-2}$$

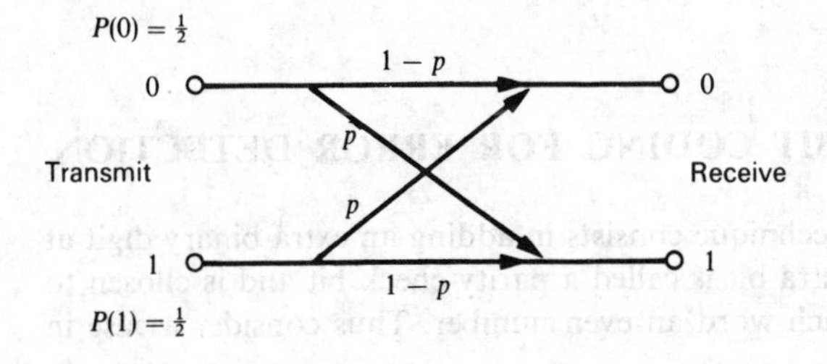

Figure 13.11-1 The binary symmetric channel.

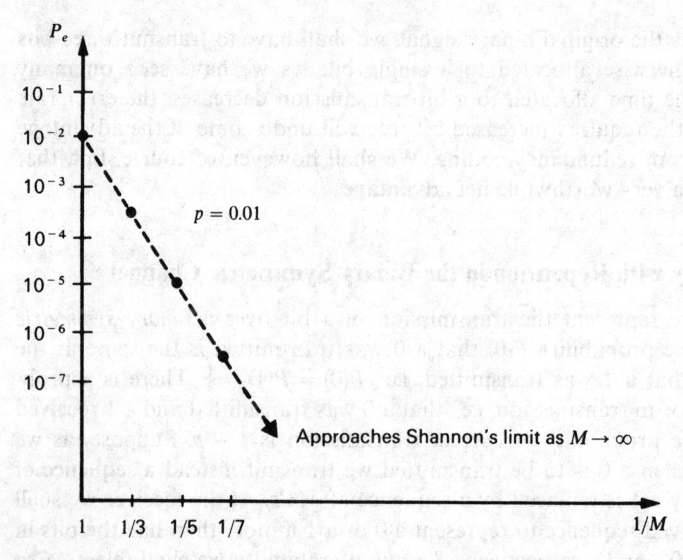

Figure 13.11-2 Plot of P_e of BSC as a function of number of repetitions M.

In Eq. (13.11-2) the first term is the probability that all five bits in the sequence are wrong, while the second and third terms are, respectively, the probabilities that four and that three bits in the sequence are wrong. In the general case, as can be verified (see Prob. 13.11-2)

$$P_e(M) = \sum_{i=(M+1)/2}^{M} \binom{M}{i} p^i (1-p)^{M-i} \tag{13.11-3}$$

where, as usual the symbol $\binom{M}{i}$ represents the number of combinations of M things taken i at a time. In Fig. 13.11-2 we have plotted $P_e(M)$ as a function of $1/M$ for the case $p = 0.01$. As is suggested by the plot and as can be verified, with increasing M, P_e decreases without limit as is predicted by Shannon's theorem.

When a bit of duration τ is replaced by a sequence of M bits each of the same duration τ, then, of course, the rate at which information is transmitted is reduced by the factor $1/M$. For this reason such a code is called a "rate $1/M$" code. If we propose to keep the information rate constant then the bit duration must be reduced to τ/M. Such a reduction will require that the bandwidth be increased by the factor M.

13.12 PARITY CHECK BIT CODING FOR ERROR DETECTION

The simplest error-detecting technique consists in adding an extra binary digit at the end of each word. This extra bit is called a parity-check bit and is chosen to make the number of 1's in each word an even number. Thus consider a case in which there are only sixteen ($=2^4$) possible messages to be transmitted. These sixteen messages could be encoded into four bits. In adding an extra bit, the fifth bit, we now have $2^5 = 32$ possible words, only sixteen of which are to be recognized as messages conveying information. Thus the parity-check bit has introduced redundancy. If an odd number of the first four bits were 1's, we would add a 1. Thus 1011 would become 10111. If an even number of the first four bits were 1's, we would add a 0. Thus 1010 would become 10100.

The parity-check bit is rather universally used in digital computers for error detection. This check bit will be effective if the probability of error in a bit is low enough so that we may ignore the likelihood of more than a single bit error in a word. If such a single error occurs, it will change a 0 to a 1 or a 1 to a 0 and the word, when received, would have an odd number of 1's. Hence it will be known that an error has occurred. We shall not, however, be able to determine which bit is in error. An error in two bits in a word is not detectable, since such a double error will again yield an even number of 1's.

13.13 CODING FOR ERROR DETECTION AND CORRECTION

Coding has the usefulness that it allows us to increase the rate at which information may be transmitted over a channel while maintaining a fixed error rate. Alternatively, coding allows us to reduce the information bit error rate while maintaining a fixed transmission rate. More generally, coding allows us, in principle (up to the Shannon limit) to design a communication system in which both information bit rate and error rate are independently and arbitrarily specified but subject to a constraint on bandwidth. The price we pay, for seeking to reach closer to the Shannon limit, is increased hardware complexity both at the transmitter where the encoding is done and at the receiver where the decoding is effected. In principle, with ingenious enough coding and unlimited complexity we would be able to reach the Shannon limit. That is, we would be able to transmit at channel capacity and with an error rate which may be made as small as desired. One measure of the efficiency of a code is precisely the extent to which it allows us to approach the Shannon limit.

Many different types of codes have been developed and are in use. The design of codes is a rather sophisticated topic which we shall not discuss. We shall, however, describe some of the more commonly employed codes.

13.14 BLOCK CODES

Consider that a message source can generate M equally likely messages. Then initially we represent each message by k binary digits with $2^k = M$. These k bits are the information bearing bits. We next add, to each k bit message, r redundant bits (parity check bits). Thus, each message has been expanded into a codeword of length n bits with

$$n = k + r \tag{13.14-1}$$

the total number of possible n bit codewords is 2^n while the total number of possible messages is 2^k. There are, therefore, $2^n - 2^k$ possible n bit words which do not represent possible messages.

Codes formed by taking a *block* of k information bits and "adding" r $(= n - k)$ redundant bits to form a codeword are called *block* codes and designated (n, k) codes. In this section we shall discuss *systematic* codes only. When an n bit codeword consists of k information bits and r redundant bits the code is said to be *systematic*. A *nonsystematic* code has n bits in the codeword and k information bits which are *not* explicitly presented in the codeword. Such codes are discussed in Sec. 13.17.

A (7, 4) systematic code has four information bits which distinguish from one another the sixteen $(= 2^4)$ possible messages. There are seven bits in each codeword so that each codeword has three redundant bits. If the codeword with n bits is to be transmitted in no more time than is required for the transmission of the k information bits and if T_b and T_c are the bit durations in the uncoded and coded word, then it is required that

$$nT_c = kT_b \tag{13.14-2}$$

We define the *rate of the code* to be

$$R_c \equiv k/n$$

Accordingly, with $f_b = 1/T_b$ and $f_c = 1/T_c$ we have

$$\frac{f_c}{f_b} = \frac{T_b}{T_c} = \frac{n}{k} = \frac{1}{R_c} \tag{13.14-3}$$

The Hamming Distance, $d_{\min}$

Consider that C_i and C_j are any two codewords in a particular block code. Then these codewords will differ in some bit positions and we represent by the symbol d_{ij} the number of such positions with differences. Thus suppose

$$C_i = 1 \quad 0 \quad 0 \quad 0 \quad 1 \quad 1 \quad 1$$

$$C_j = 0 \quad 0 \quad 0 \quad 1 \quad 0 \quad 1 \quad 1$$

Then we observe that these codewords differ in the leftmost bit position and in the bit positions fourth and fifth from the left. Accordingly $d_{ij} = 3$. Assume that we have determined d_{ij} for each pair of codewords. Then the minimum value of the d_{ij}'s is called the *Hamming distance*, $d_{\min}$. It is rather intuitively clear that when we try to determine, which of two or more codewords, a received "codeword" represents when some of the received codeword's bits may be misread because of noise, the likelihood of a successful determination will be greater for words which have a larger number of bit differences. The greatest likelihood of confusion between words will be encountered for a codeword pair for which d_{ij} is a minimum. Hence the Hamming distance $d_{\min}$ establishes an upper limit to the effectiveness of a code. We shall prove below two important properties of the parameter $d_{\min}$:

1. Suppose that there are D errors in a received codeword. Then provided that

$$D \leq d_{\min} - 1 \tag{13.14-4}$$

   we shall be able to *detect*, with certainty, that the received codeword is not valid, i.e., not a word in our vocabulary.
2. If there are t errors in the received word, then provided

$$2t + 1 \leq d_{\min} \leq 2t + 2 \tag{13.14-5}$$

   we shall not only be able to establish that the received word is not valid but also we shall be able to *correct* the errors, i.e., we shall be able to regenerate the original correct codeword.

Thus with $d_{\min} = 7$ we can detect a non valid word even if as many as six errors have been made and, if no more than three errors have been made we can determine the correct codeword. Finally, it is to be noted that Eqs. (3.14-4) and (3.14-5) apply in worst-case conditions. In many cases even more errors can be detected or corrected than indicated by those equations.

13.15 UPPER BOUNDS OF THE PROBABILITY OF ERROR WITH CODING

In Fig. 13.15-1 we have codewords transmitted over a communications channel. We should like to direct our attention to the matters pertaining to the use of codes without the distraction of having to take into account the characteristics of modulation–demodulation scheme. For this reason we assume that transmission takes place at baseband. The waveform $c_i(t)$ makes excursions between $+1$ and -1 corresponding to logic 1 and 0 respectively. The waveform $\sqrt{P_s}\,c_i(t)$ therefore

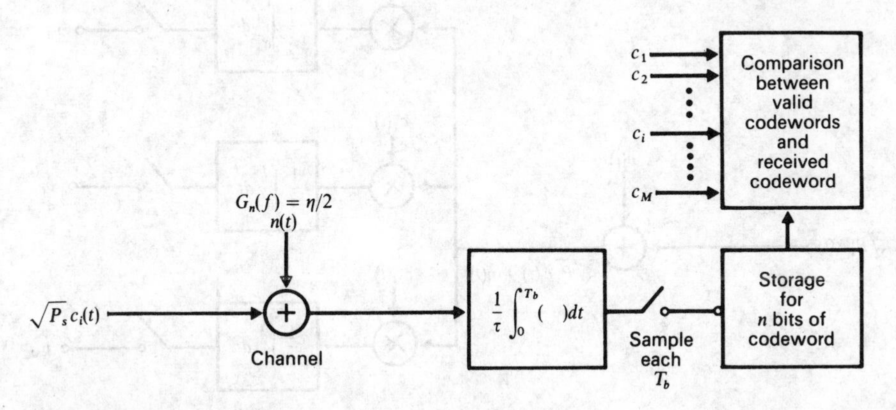

Figure 13.15-1 A system incorporating hard-decision decoding.

has a signal power P_s. A complete codeword is comprised of a sequence of n bits each allocated the bit time T_c. White noise of power spectral density $\eta/2$ is added to the signal waveform $\sqrt{P_s}\,c_i(t)$. At each sampling time we shall have made a *hard* decision, i.e., an irrevocable decision about *each individual bit*. We then store bit after bit until we have assembled an entire codeword. Finally we pass the codeword to appropriate hardware which compares the received codeword with all the valid codewords. The codeword chosen is the one which differs from the received codeword in the fewest number of bits. In this comparison we shall correct errors if they do not exceed the limits of Eq. (13.14-5).

We note that two decisions are made when using hard decision decoding: the first is a decision concerning each bit and the second is a decision to determine which codeword compares most favorably to our reconstructed word. It is well known that to minimize the probability of an error one should correlate the received signal and noise with all possible signals and then choose the signal yielding the highest correlation. Such a technique does not suggest making a decision in each bit but only *one* final decision. The problem with such a single correlation technique, which is called *soft-decision* decoding is that the resulting hardware is extremely complex. However, it does provide approximately 2 to 3 dB of improvement and is therefore used when needed.

The scheme for soft-decision decoding is shown in Fig. 13.15-2. Here the demodulated baseband signal plus noise $\sqrt{P_s}\,c_i(t) + n(t)$ is correlated with each of the possible waveforms $c_1, c_2, \ldots, c_i, \ldots, c_M$. Decoding is carried out by seeking maximum correlation between the incoming signal and the locally generated codewords. The received codeword $\sqrt{P_s}\,c_i(t)$ (plus noise) is applied to a bank of correlators consisting of a multiplier followed by an integrator. The integration interval is $nT_c = T_n$ and the integrator output is sampled each T_n. In the correlator the incoming signal in noise $\sqrt{P_s}\,c_i(t) + n(t)$ is first multiplied by each of the

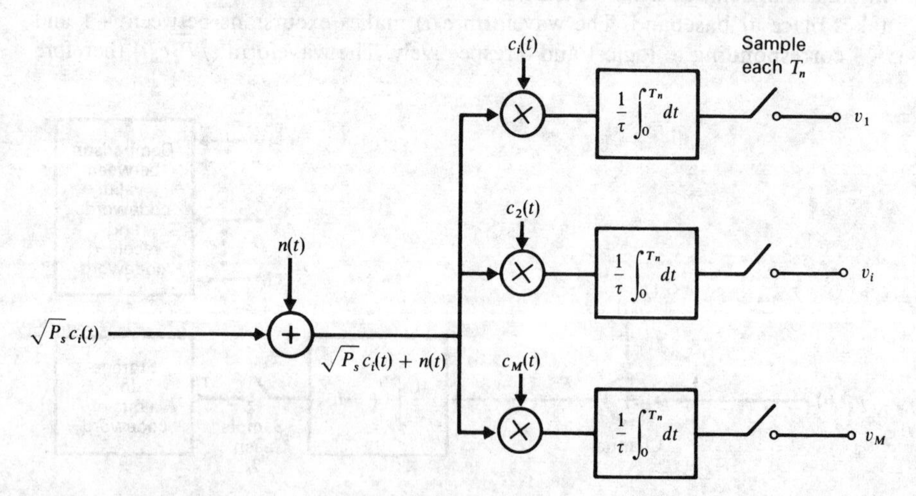

Figure 13.15-2 A system incorporating soft-decision decoding.

locally generated waveforms $c_i, c_2, \ldots, c_i, \ldots, c_M$. We assume that $c_i(t)$ takes on the voltages ± 1. In general if the transmitted codeword is c_i, the sampled outputs are

$$v_j = \begin{cases} \dfrac{\sqrt{P_s T_n}}{\tau} + n_i & i = j \quad (13.15\text{-}1a) \\[2ex] \dfrac{\sqrt{P_s T_b}}{\tau}(n - 2d_{ij}) + n_j & i \neq j \quad (13.15\text{-}1b) \\[2ex] = \dfrac{\sqrt{P_s T_n}}{\tau}\left(1 - \dfrac{2d_{ij}}{n}\right) + n_j & \quad (13.15\text{-}1c) \end{cases}$$

where d_{ij} is the distance (number of bit differences) between c_i and c_j and

$$n_i = \frac{1}{\tau}\int_0^{T_n} n(t)c_i(t)\,dt \tag{13.15-2a}$$

$$n_j = \frac{1}{\tau}\int_0^{T_n} n(t)c_j(t)\,dt \tag{13.15-2b}$$

are the values of the noise output at the sampling times.

Equation (13.15-1) confirms that, so far as the deterministic portions of v_i and v_j are concerned, the larger is d_{ij} the greater is the difference between v_i and v_j. If we should arrange that for all codewords the distance $d_{ij} = n/2$ then we find from Eq. (13.15-1b) that the deterministic portion of $v_j = 0$. In such coding the number n is an even number and any two codewords differ by $n/2$ bits. In this case the integrated product of any two different codewords yields a zero result. Hence such a code is referred to as an *orthogonal* code.

We have from Eq. (13.15-2)

$$n_i^2 = \frac{1}{\tau}\int_0^{T_n} n(t)c_i(t)\,dt \cdot \frac{1}{\tau}\int_0^{T_n} n(t)c_i(t)\,dt \tag{13.15-3}$$

Since the name of the variable of integration does not affect the value of the definite integral we have

$$n_i^2 = \frac{1}{\tau}\int_0^{T_n} n(t)c_i(t)\,dt \cdot \frac{1}{\tau}\int_0^{T_n} n(\lambda)c_i(\lambda)\,d\lambda \tag{13.15-4a}$$

$$= \frac{1}{\tau^2}\int_0^{T_n} dt \int_0^{T_n} n(t)n(\lambda)c_i(t)c_i(\lambda)\,d\lambda \tag{13.15-4b}$$

and the expected value $E(n_i^2)$ is

$$E(n_i^2) = E\left[\frac{1}{\tau^2}\int_0^{T_n} dt \int_0^{T_n} n(t)n(\lambda)c_i(t)c_i(\lambda)\,d\lambda\right] \tag{13.15-5}$$

We now take account of the facts: (1) since $c_i(t)c_i(\lambda)$ are deterministic waveforms

$E[c_i(t)c_i(\lambda)] = c_i(t)c_i(\lambda)$, (2) that $c_i(t)$ is not correlated with the noise $n(t)$, so that after interchanging expectation with integration we find

$$E(n_i^2) = \frac{1}{\tau^2}\int_0^{T_n} dt \int_0^{T_n} E[n(t)n(\lambda)][c_i(t)c_i(\lambda)]\, d\lambda \tag{13.15-6}$$

For white gaussian noise of power spectral density $\eta/2$ we have

$$E[n(t)n(\lambda)] = (\eta/2)\,\delta(t-\lambda) \tag{13.15-7}$$

Hence Eq. (13.15-6) becomes

$$E(n_i^2) = \frac{1}{\tau^2}\int_0^{T_n} dt \int_0^{T_n} (\eta/2)c_i(t)c_i(\lambda)\,\delta(t-\lambda)\, d\lambda \tag{13.15-8a}$$

$$= \frac{1}{\tau^2}\int_0^{T_n} (\eta/2)c_i^2(t)\, dt \tag{13.15-8b}$$

Correspondingly we find that

$$E(n_j^2) = \frac{1}{\tau^2}\int_0^{T_n} (\eta/2)c_j^2(t)\, dt \tag{13.15-9}$$

and since $c_i^2(t) = c_j^2(t) = 1$ we have

$$E(n_i^2) = E(n_j^2) = \frac{\eta T_n}{2\tau^2} \tag{13.15-10}$$

If we now undertake to calculate $E(n_i n_j)$ we find that all of the steps beginning with Eq. (13.15-3) and leading to Eq. (13.15-10) are duplicated with the exception that one of the c_i's in each equation is replaced by a c_j. Thus Eq. (13.15-8*b*) becomes

$$E(n_i n_j) = \frac{1}{\tau^2}\int_0^{T_n} (\eta/2)c_i c_j\, dt = \frac{\eta}{2\tau^2}\int_0^{T_n} c_i c_j\, dt \tag{13.15-11}$$

Applying now the same consideration which led to Eq. (13.15-1*b*) we have the result that

$$E(n_i n_j) = \frac{\eta T_b}{2\tau^2}(n - 2d_{ij}) = \frac{\eta T_n}{2\tau^2}\left(1 - \frac{2d_{ij}}{n}\right) \tag{13.15-12}$$

In the system of Fig. 13.15-2, if there were no noise and if c_i were the transmitted codeword then v_i would be larger than any other output. In the presence of noise we shall then make the judgement that, if v_i is the largest of the outputs at the sampling time, then the transmitted codeword was c_i. An upper bound to the probability P_e, that an error is made in decoding a codeword, can be obtained using the *union bound*, which was discussed in Sec. 11.15. The Union Bound is the *sum* of the probabilities that $v_1 > v_i$, that $v_2 > v_i$, etc. Thus

$$P_e \le \sum_{j=1}^{M} P(v_j > v_i) \qquad i \ne j \tag{13.15-13}$$

The probability that $v_j > v_i$ is

$$P(v_j > v_i) = P\left[\frac{\sqrt{P_s T_n}}{\tau}\left(1 - \frac{2d_{ij}}{n}\right) + n_j > \frac{\sqrt{P_s T_n}}{\tau} + n_i\right] \tag{13.15-14}$$

Letting $n_0 = n_j - n_i$, Eq. (13.15-14) becomes

$$P(v_j > v_i) = P\left(n_0 > \frac{2\sqrt{P_s T_n}}{\tau n}\, d_{ij}\right) \tag{13.15-15}$$

The variance σ_{n0}^2 of the gaussian random variable n_0 is

$$\sigma_{n0}^2 = E(n_0^2) = E[(n_j - n_i)^2] = E(n_j^2) + E(n_i^2) - 2E(n_j n_i) \tag{13.15-16}$$

Using the results given in Eqs. (13.15-10) and (13.15-12) we find that

$$\sigma_{n0}^2 = \frac{2\eta T_n}{\tau^2}\frac{d_{ij}}{n} \tag{13.15-17}$$

Therefore

$$P(v_j > v_i) = \int_{2\sqrt{P_s T_n} d_{ij}/\tau n}^{\infty} \frac{\exp\left(-n_0^2/2\sigma_{n0}^2\right)}{\sqrt{2\pi\sigma_{n0}^2}}\, dn_0 \tag{13.15-18}$$

Proceeding as we have on a number of other occasions we make the substitution $x^2 = n_0^2/2\sigma_0^2$ and we find that Eq. (13.15-18) becomes

$$P(v_j > v_i) = \frac{1}{2}\frac{2}{\sqrt{\pi}}\int_{\sqrt{T_n d_{ij} P_s/\eta n}}^{\infty} e^{-x^2}\, dx \tag{13.15-19a}$$

$$= \tfrac{1}{2}\,\text{erfc}\left[\frac{P_s T_n d_{ij}}{\eta n}\right]^{1/2} \tag{13.15-19b}$$

Finally, from Eqs. (13.15-13) and (13.15-19*b*) we have

$$P_e \leq \sum_{j=1,\ i\neq j}^{M} \tfrac{1}{2}\,\text{erfc}\left[\frac{P_s T_n d_{ij}}{\eta n}\right]^{1/2} \tag{13.15-20}$$

Since $d_{ij} \geq d_{\min}$ we also have

$$P_e \leq \frac{(M-1)}{2}\,\text{erfc}\left[\frac{P_s T_n d_{\min}}{\eta n}\right]^{1/2} \tag{13.15-21}$$

To put this result in a more useful form we note that for large M, $M - 1 \approx M$ and that $M = 2^k$. Also we recall that $T_n = nT_c$ and from Eq. (13.14-2) $nT_c = kT_b$, T_c being the bit duration in the coded word and T_b the bit duration in the uncoded word. Further, P_s is the normalized signal power and $P_s T_b$ is the normalized bit energy E_b in a bit of the uncoded word. Altogether we find that

$$P_e \leq \frac{2^k}{2}\,\text{erfc}\left[\frac{E_b}{\eta}\left(\frac{k}{n}\right) d_{\min}\right]^{1/2} \tag{13.15-22}$$

Hard Decision Decoding

As we have noted, in hard-decision decoding an irrevocable decision is made about each individual bit, the entire codeword is then assembled and this "codeword" is then compared with each valid codeword. From Sec. 11.13 we have the result that with coherent detection, the probability that an individual bit is erroneously decoded is

$$p = \tfrac{1}{2}\,\mathrm{erfc}\,\sqrt{\frac{P_s T_c}{\eta}} = \tfrac{1}{2}\,\mathrm{erfc}\,\sqrt{\frac{E_b}{\eta}\cdot\left(\frac{k}{n}\right)} \tag{13.15-23}$$

If a coding scheme can correct t errors, then the probability that an n-bit codeword is in error is the probability that more than t errors have occurred in n bits. This probability is

$$P_e = \sum_{i=t+1}^{n} \binom{n}{i} p^i (1-p)^{n-i} \tag{13.15-24}$$

where p^i is the probability that i bits are in error, $(1-p)^{n-i}$ is the probability that the remaining $n-i$ bits are correct and $\binom{n}{i}$ is the number of combinations of n things taken i at a time. This number is

$$\binom{n}{i} = \frac{n!}{(n-i)!\,(i)!} \tag{13.15-25}$$

and is the number of ways in which i bits can be incorrect in a block of n-bits.

For small P_e (in which case p must also be small) P_e can be approximated (see Prob. 11.15-4) by the first term of the sum, so that

$$P_e \cong \binom{n}{t+1} p^{t+1} (1-p)^{n-t-1} \tag{13.15-26}$$

In Prob. 13.15-5 the student is guided through a calculation to verify that Eq. (13.15-26) can be approximated by

$$P_e \approx \frac{2^k}{2}\exp - \left\{(t+1)\left(\frac{E_b}{\eta}\right)\left(\frac{k}{n}\right)\right\} \approx \frac{2^k}{2}\exp - \left\{\frac{E_b}{\eta}\left(\frac{k}{n}\right)\frac{d_{\min}}{2}\right\} \tag{13.15-27}$$

since $d_{\min} \approx 2t$.

For large values of E_b/η, the probability of error for soft and hard decision decoding are seen from Eqs. (13.15-22) and (13.15-27), to behave asymptotically (i.e., for large values of E_b/η) as

$$P_{e_{\text{soft}}} < \frac{2^k}{2} e^{(-k d_{\min}/n)(E_b/\eta)} \tag{13.15-28}$$

$$P_{e_{\text{hard}}} < \frac{2^k}{2} e^{(-k d_{\min}/n)(E_b/2\eta)} \tag{13.15-29}$$

Thus soft-decision decoding can provide a 3 dB coding gain compared to hard-decision decoding.

The difficulty encountered in performing the n correlations, each for a time $T_n = kT_b$ which is required for soft-decision decoding often inclines code designers to employ hard-decision decoding.

13.16 BLOCK CODES—CODING AND DECODING

A block code (like any code) is an *invention*. Its merit depends on the ingenuity and insight of its inventor. There are few rules for creating good and effective codes. The mathematical framework in terms of which we shall describe the block codes is useful for systematization, organization, and comparison of codes and also for the purpose of clarifying the mechanics of encoding and decoding.

In a block code, as we have noted, a codeword has the form

$$a_1\, a_2\, a_3 \ldots a_k\, c_1\, c_2 \ldots c_r \tag{13.16-1}$$

There are k information bits, r parity bits and the codeword has $n = k + r$ bits. There are $M = 2^k$ valid codewords. The uncoded word is

$$\bar{A} = [a_1\, a_2 \ldots a_k] \tag{13.16-2}$$

The generation of a block code starts with a selection of the number r of parity bits to be added and thereafter with the specification of an $\bar{H}$ matrix

$$\bar{H} = \left|\begin{array}{cccc|cccc} h_{11} & h_{12} & \ldots & h_{1k} & 1 & 0 & 0 \;\ldots & 0 \\ h_{21} & h_{22} & \ldots & h_{2k} & 0 & 1 & 0 \;\ldots & 0 \\ \multicolumn{4}{c}{\ldots\ldots\ldots\ldots} & \multicolumn{4}{c}{\ldots\ldots\ldots\ldots} \\ h_{r1} & h_{r2} & \ldots & h_{rk} & 0 & 0 & 1 \;\ldots & 0 \end{array}\right| \tag{13.16-3}$$

$\underbrace{\qquad\qquad}_{r \times k}$ $\bar{h}$ submatrix — $\underbrace{\qquad\qquad}_{r \times r}$ identity submatrix, $\bar{I}$

We observe that $\bar{H}$ consists of an $r \times k$ submatrix $\bar{h}$, and an $r \times r$ identity submatrix. In the expression (13.16-1) and Eq. (13.16-2) the a's and c's are logical variables with possible values 0 or 1. Correspondingly the h's in Eq. (13.16-3) are *constant* logical variables with values 0 or 1. The merit of the block code depends on the ingenuity with which the h's have been selected. If the entries in the matrix of Eq. (13.16-3) and in other matrices we shall encounter were numbers, then the process of finding an element in a product matrix would consist in forming the arithmetic sum of arithmetic products. In the present case, since we deal with logical variables it turns out that the arithmetic product is replaced by the logical product, i.e., the AND operation and the arithmetic sum is replaced by the

EXCLUSIVE-OR operation. We recall that these logical operations are defined as follows

$$\begin{array}{ll} \text{AND} & \text{EXCLUSIVE-OR} \\ 0 \cdot 0 = 0 & 0 \oplus 0 = 0 \\ 0 \cdot 1 = 0 & 0 \oplus 1 = 1 \\ 1 \cdot 0 = 0 & 1 \oplus 0 = 1 \\ 1 \cdot 1 = 1 & 1 \oplus 1 = 0 \end{array} \tag{13.16-4}$$

The transpose of the $\bar{H}$ matrix, that is the matrix derived from $\bar{H}$ by interchanging rows and columns is

$$\bar{H}^T = \left| \begin{array}{cccc} h_{11} & h_{21} & \dots & h_{r1} \\ h_{12} & h_{22} & \dots & h_{r2} \\ \dots & \dots & \dots & \dots \\ h_{1k} & h_{2k} & \dots & h_{rk} \\ 1 & 0 & \dots & 0 \\ 0 & 1 & \dots & 0 \\ 0 & 0 & \dots & 0 \\ 0 & 0 & \dots & 1 \end{array} \right| \begin{array}{l} \left.\vphantom{\begin{array}{c}1\\1\\1\\1\end{array}}\right\} k \times r \;\; \bar{h}^T \text{ submatrix} \\ \left.\vphantom{\begin{array}{c}1\\1\\1\\1\end{array}}\right\} r \times r \;\; \text{identity submatrix} \end{array} \tag{13.16-5}$$

This transposed matrix $\bar{H}^T$ consists of an $r \times r$ identity submatrix and a submatrix $\bar{h}^T$ which is the transpose of $\bar{h}$.

To generate a codeword $\bar{T}$ from the uncoded word $\bar{A}$ given by Eq. (13.16-2) we form a generator matrix $\bar{G}$ where

$$\bar{G} = \left| \begin{array}{ccccc ccccc} 1 & 0 & 0 & \dots & 0 & h_{11} & h_{21} & \dots & h_{r1} \\ 0 & 1 & 0 & \dots & 0 & h_{12} & h_{22} & \dots & h_{r2} \\ \dots & & & & & \dots & & & \\ 0 & 0 & 0 & \dots & 1 & h_{1k} & h_{2k} & \dots & h_{rk} \end{array} \right| \tag{13.16-6}$$

$$\underbrace{\qquad\qquad}_{k \times k} \qquad \underbrace{\qquad\qquad}_{k \times r}$$

Observe that $\bar{G}$ consists of an identity submatrix, this time of order $k \times k$ and a second submatrix which is the submatrix $\bar{h}^T$ encountered in Eq. (13.16-5). We can readily verify that

$$\bar{G}\bar{H}^T = 0 \tag{13.16-7}$$

Thus, if for example, we form $(\bar{G}\bar{H}^T)_{11}$ we get

$$(\bar{G}\bar{H}^T)_{11} = h_{11} \cdot 1 \oplus 1 \cdot h_{11} = 0 \tag{13.16-8}$$

Similarly, all other elements of $\bar{G}\bar{H}^T$ are zero.

The generator matrix $\bar{G}$ having been formed, the codeword $\bar{T}$ corresponding to each uncoded word $\bar{A}$ is

$$\bar{T} = \bar{A}\bar{G} \tag{13.16-9}$$

Most importantly, it turns out that the coded words so formed have the property that [see Eq. (13.16-7)]

$$\bar{T}\bar{H}^T = \bar{A}\bar{G}\bar{H}^T = 0 \tag{13.16-10a}$$

or (see Prob. 13.16-8)

$$\bar{H}\bar{T}^T = 0 \tag{13.16-10b}$$

which, written out at length, means

$$\begin{array}{l} h_{11}a_1 \oplus h_{12}a_2 \oplus \ldots h_{1k}a_k \oplus 1 \cdot c_1 \oplus 0 \cdot c_2 \oplus \ldots \oplus 0 \cdot c_r = 0 \\ h_{21}a_1 \oplus h_{22}a_2 \oplus \ldots h_{2k}a_k \oplus 0 \cdot c_1 \oplus 1 \cdot c_2 \oplus \ldots \oplus 0 \cdot c_2 = 0 \\ \cdots\cdots\cdots\cdots\cdots\cdots\cdots\cdots\cdots\cdots\cdots\cdots \\ h_{r1}a_1 \oplus h_{r2}a_2 \oplus \ldots h_{rk}a_k \oplus 0 \cdot c_1 \oplus 0 \cdot c_2 \oplus \ldots \oplus 1 \cdot c_r = 0 \end{array} \tag{13.16-11a}$$

For example, the $\bar{H}$ matrix of a (7.4) Hamming code is

$$\bar{H} = \begin{bmatrix} 1 & 1 & 1 & 0 & 1 & 0 & 0 \\ 1 & 1 & 0 & 1 & 0 & 1 & 0 \\ 1 & 0 & 1 & 1 & 0 & 0 & 1 \end{bmatrix} \tag{13.16-11b}$$

$$\underbrace{\qquad\qquad}_{\bar{h}} \qquad \underbrace{\qquad\qquad}_{r \times r = 3 \times 3\ \bar{I}\ \text{matrix}}$$

Here the identity matrix is $r \times r = 3 \times 3$ because there are $7 - 4 = 3$ parity bits. The total matrix has seven columns leaving a 3×4 $\bar{h}$ submatrix. The generator matrix $\bar{G}$ then consists of a $k \times k = 4 \times 4$ identity matrix together with a submatrix which is the transpose $\bar{h}^T$ of $\bar{h}$ so that

$$\bar{G} = \begin{bmatrix} 1 & 0 & 0 & 0 & 1 & 1 & 1 \\ 0 & 1 & 0 & 0 & 1 & 1 & 0 \\ 0 & 0 & 1 & 0 & 1 & 0 & 1 \\ 0 & 0 & 0 & 1 & 0 & 1 & 1 \end{bmatrix} \tag{13.16-12}$$

$$\underbrace{\qquad\qquad}_{4 \times 4\ \bar{I}\ \text{matrix}} \qquad \underbrace{\qquad\qquad}_{\bar{h}^T}$$

It is readily verified that $\bar{G}\bar{H}^T = 0$. The coded words are now determined as

$$\bar{T} = \bar{A}\bar{G} = \overbrace{[a_1 \quad a_2 \quad a_3 \quad a_4]}^{k\ \text{bits}} \left[\begin{array}{cccc|ccc} 1 & 0 & 0 & 0 & 1 & 1 & 1 \\ 0 & 1 & 0 & 0 & 1 & 1 & 0 \\ 0 & 0 & 1 & 0 & 1 & 0 & 1 \\ 0 & 0 & 0 & 1 & 0 & 1 & 1 \end{array}\right] \tag{13.16-13}$$

Thus, the uncoded word $\bar{A} = [a_1\, a_2\, a_3\, a_4]$ having been specified, the corresponding coded word is

$$\bar{T} = [a_1 \quad a_2 \quad a_3 \quad a_4 \quad c_1 \quad c_2 \quad c_3] \tag{13.16-14}$$

where

$$\begin{aligned} c_1 &= a_1 \oplus a_2 \oplus a_3 \\ c_2 &= a_1 \oplus a_2 \oplus a_4 \\ c_3 &= a_1 \oplus a_3 \oplus a_4 \end{aligned} \tag{13.16-15}$$

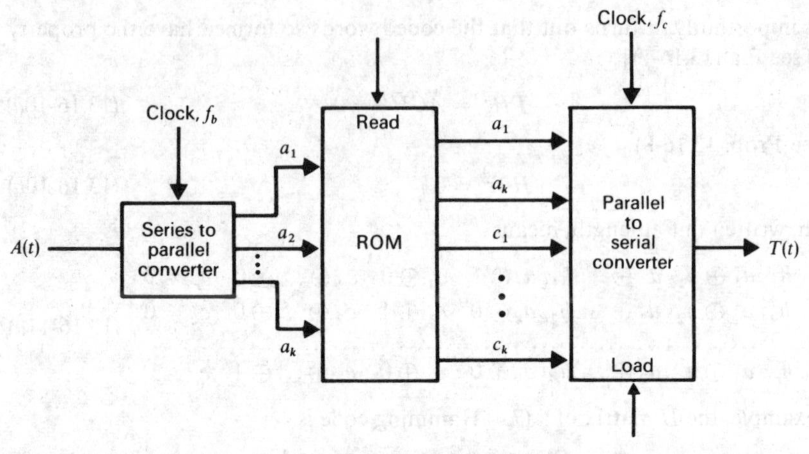

Figure 13.16-1 Showing that a read-only memory (ROM) can perform the encoding function.

It is left as a problem (Prob. 13.16-6) to verify that for each $\bar{T}$ so generated, the product $\bar{H}\bar{T}^T = 0$ as indicated by Eq. (13.16-10).

When the codeword corresponding to each uncoded word has been determined, the results can be stored in a memory. Such a memory, in this case a read-only memory (ROM), is shown in Fig. 13.16-1. The bits $a_1 a_2 \ldots a_k$ of the uncoded word are presented as an address to the memory. At the memory location specified by that address we have stored the n $(= k + r)$ bits corresponding to the coded word. The uncoded word bit *serial* bit stream $A(t)$ is converted to a *parallel* bit array (i.e., all bits available at the same time) by a serial-to-parallel converter (i.e., a shift register). The converter is driven by a clock at the uncoded bit rate f_b. When the bits $a_1 a_2 \ldots a_k$ are all available, a READ command is given to the ROM and a LOAD command to the parallel-to-series converter (i.e., another shift register with a parallel-load facility). The coded word $a_1 a_2 \ldots a_k c_1 c_2 \ldots c_r$ is thus loaded into the converter. The converter then presents the bits serially at its output. The clock rate for this second converter is f_c, the bit rate for the coded bits. As we have noted before $f_c/f_b = n/k$.

Decoding the Received Codeword

Now let the received message be $\bar{R}$ which may or may not be the transmitted codeword. Suppose that at the receiver we have an appropriate apparatus for forming the product $\bar{H}\bar{R}^T$. If $\bar{H}\bar{R}^T \neq 0$ we know that $\bar{R}$ is not a possible message and one or more bits are in error. Under these circumstances we shall be interested in inquiring into whether we can determine the transmitted codeword and what is the probability that our determination is correct. Even, however, if we find that $\bar{H}\bar{R}^T = 0$ we cannot be absolutely certain that $\bar{R}$ was transmitted, since there is always the possibility that the number of bit errors was so large that the transmitted codeword was transformed into another possible codeword.

We look now into the matter of decoding the received codewords. One possible technique for decoding is to evaluate the correlation of the received codeword with all the possible codewords. This was illustrated earlier in Figs. 13.15-1 and 13.15-2. We then assume that the transmitted word was the codeword with which the received codeword exhibits the closest correlation. We have already considered such a procedure in Sec. 13.15. However the procedure gets out of hand when the number of bits in a word is large. If the uncoded word has k bits then, excluding the word in which all bits are zero, there are $2^k - 1$ codewords with which comparisons (correlations) have to be made. In some schemes of computer communications, messages (called packets) may contain as many as 1000 bits in which case $2^{1000} - 1$ correlations would be required. One of the primary advantages of *algebraic* codes, whose generation we have just described, is that they allow alternative techniques which reduce very considerably the complexity of decoding.

If $\bar{T}$ is transmitted and $\bar{R} \neq \bar{T}$ is received then

$$\bar{R} = \bar{T} \oplus \bar{E} \tag{13.16-16}$$

in which $\bar{E}$ is the *error* word. For example if

$$\bar{T}^T = \begin{vmatrix} 1 \\ 0 \\ 0 \\ 0 \\ 1 \\ 1 \\ 1 \end{vmatrix} \tag{13.16-17}$$

and an error is made in the fifth place so that

$$\bar{E}^T = \begin{vmatrix} 0 \\ 0 \\ 0 \\ 0 \\ 1 \\ 0 \\ 0 \end{vmatrix} \tag{13.16-18}$$

then the received codeword will be

$$\bar{R}^T = \begin{vmatrix} 1 \\ 0 \\ 0 \\ 0 \\ 1 \\ 1 \\ 1 \end{vmatrix} \oplus \begin{vmatrix} 0 \\ 0 \\ 0 \\ 0 \\ 1 \\ 0 \\ 0 \end{vmatrix} = \begin{vmatrix} 1 \\ 0 \\ 0 \\ 0 \\ 0 \\ 1 \\ 1 \end{vmatrix} \tag{13.16-19}$$

Observe from this example that our algebra is such that a 1 in the error word $\bar{E}$ indicates an error in the corresponding bit position and that a 0 indicates that no error has been made.

A first step in the process of decoding is to evaluate the ***syndrome*** $\bar{S}$ of the received codeword $\bar{R}$. The syndrome is defined as

$$\bar{S} = \bar{H}\bar{R}^T \tag{13.16-20}$$

If the received word $\bar{R}$ is $\bar{T}$ as transmitted then $\bar{S} = \bar{H}\bar{R}^T = \bar{H}\bar{T}^T = 0$. But if there are errors we find

$$\bar{S} = \bar{H}\bar{R}^T = \bar{H}(\bar{T}^T + \bar{E}^T) \tag{13.16-21a}$$

$$= \bar{H}\bar{T}^T + \bar{H}\bar{E}^T \tag{13.16-21b}$$

$$= \bar{H}\bar{E}^T \tag{13.16-21c}$$

since $\bar{H}\bar{T}^T = 0$. Accordingly, if the syndrome $\bar{S}$ *is not* zero then we have an indication that there is one or more errors. If the syndrome *is* zero then either there are no errors or the errors are so many that a transmitted codeword has been changed to a different codeword. In this latter unlikely, but possible, case, we have from Eq. (13.16-21*c*) that $\bar{H}\bar{E}^T = 0$ so that the error word itself is the same as one of the valid codewords.

When $\bar{S}$ is not zero, the syndrome provides information about the bit positions in which errors may have been made. For example, let $\bar{H}$ be the matrix given in Eq. (13.16-11). Let us assume that a received word is $\bar{R} = [1000011]$. Forming the syndrome we find

$$\bar{S} = \bar{H}\bar{R}^T = \begin{vmatrix} 1 & 1 & 1 & 0 & 1 & 0 & 0 \\ 1 & 1 & 0 & 1 & 0 & 1 & 0 \\ 1 & 0 & 1 & 1 & 0 & 0 & 1 \end{vmatrix} \begin{vmatrix} 1 \\ 0 \\ 0 \\ 0 \\ 0 \\ 1 \\ 1 \end{vmatrix} = \begin{vmatrix} 1 \\ 0 \\ 0 \end{vmatrix} \tag{13.16-22}$$

Since $\bar{S} = \bar{H}\bar{E}^T$ we now have

$$\begin{vmatrix} 1 \\ 0 \\ 0 \end{vmatrix} = \begin{vmatrix} 1 & 1 & 1 & 0 & 1 & 0 & 1 \\ 1 & 1 & 0 & 1 & 0 & 1 & 0 \\ 1 & 0 & 1 & 1 & 0 & 0 & 1 \end{vmatrix} \begin{vmatrix} E_1 \\ E_2 \\ E_3 \\ E_4 \\ E_5 \\ E_6 \\ E_7 \end{vmatrix} \tag{13.16-23}$$

From Eq. (13.16-23) we have the three simultaneous equations

$$\begin{aligned} 1 &= E_1 \oplus E_2 \oplus E_3 \oplus E_5 \\ 0 &= E_1 \oplus E_2 \oplus E_4 \oplus E_6 \\ 0 &= E_1 \oplus E_3 \oplus E_4 \oplus E_7 \end{aligned} \tag{13.16-24}$$

Table 13.16-1 Five patterns of E_1, E_2 ... E_7 which satisfy Eq. (13.16-24)

E_1	E_2	E_3	E_4	E_5	E_6	E_7
0	0	0	0	1	0	0
1	0	0	1	0	0	0
1	1	1	0	0	0	0
1	0	1	0	1	1	0
0	1	1	0	1	1	1

Since there are seven variables E_1 through E_7 and only three equations, a unique solution for the E's in Eq. (13.16-24) is therefore not possible. As a matter of fact, it can be shown that there are 2^k different sets of values for the E's, that is, 2^k different patterns of errors which will satisfy Eq. (13.16-24). It will be recalled that k is the number of bits in the uncoded word. In the present case $k = 4$ so that there are $2^4 = 16$ different possible error patterns. Five of these patterns are given in Table (13.16-1). It is left as a student exercise (Prob. 13.16-7) to verify that the solutions given in Table 13.16-1 are valid and to find the remaining 11 solutions.

If the probability that a bit in a word is received with an error is p (say $p = 10^{-3}$) then the probability that two bits are in error is $p^2(10^{-6})$ and so on. Accordingly, when we undertake to decide which error pattern is valid it is eminently most reasonable to decide on that pattern which involves the fewest errors. In the present case where Eq. (13.16-24) applies it can be verified that there is only a single error pattern that involves only one error. That pattern is the one given in the first row of Table 13.16-1. All of the other error patterns involves two or more errors. Accordingly, we accept that the error pattern which is most likely is $\bar{E} = \bar{E}_m = [0\ 0\ 0\ 0\ 1\ 0\ 0]$ i.e., the error is in the fifth bit position. Thus the best estimate $\hat{T}$ about $\bar{T}$ is the sum of $\bar{R}$ and the most likely error pattern $\bar{E}_m$, so that

$$\hat{T} = \bar{R} + \bar{E}_m \qquad (13.16\text{-}25)$$

In the present case, this yields

$$\begin{aligned}\hat{T} &= [1\quad 0\quad 0\quad 0\quad 0\quad 1\quad 1] \oplus [0\quad 0\quad 0\quad 0\quad 1\quad 0\quad 0]\\ &= [1\quad 0\quad 0\quad 0\quad 1\quad 1\quad 1]\end{aligned} \qquad (13.16\text{-}26)$$

In the general case, the number, 2^k, of possible error patterns can turn out to be astronomical. For example in the (63,45) BCH code, where $k = 45$ the number of patterns is $2^{45} = 35 \times 10^{12}$. Fortunately, we do not have to find them all to make certain that we have not missed the most likely one. First we note that there is a maximum number t of errors that a code can correct. For this reason let us simply ignore the possibility of errors larger in number than t, since we can do nothing about the matter. For errors fewer or equal to t let us simply explore the possible error patterns by trial and error. For example in the (7,4) code which we have just used as an example, with $\bar{H}$ given in Eq. (13.16-11), $t = 1$. Hence rather than to undertake to find all error pattern solutions to Eq. (13.16-24) let us simply try, in turn, all the seven possible error patterns involving only a single error, i.e.,

[1 0 0 0 0 0 0], [0 1 0 0 0 0 0], ..., [0 0 0 0 0 0 1].

We would then find, as we already have, that [0 0 0 0 1 0 0] is the most likely pattern. In the (63, 45) BCH code it turns out that $t = 3$. Hence we would first try the 63 error patterns of one error. If none of these was found valid we would then try all possible error patterns of two errors. The number of these is $\binom{63}{2}$ i.e., the number of combinations of 63 things taken two at a time. If again we met with no success we would look at the patterns with three errors. In total the number of trials would be

$$\sum_{i=1}^{3} \binom{63}{i} = 63 + \frac{63 \cdot 62}{2} + \frac{63 \cdot 62 \cdot 61}{6} = 41{,}727 \qquad (13.16\text{-}27)$$

which is quite an improvement over finding all 35×10^{12} error patterns. There are algorithms which reduces still further the computational complexity.

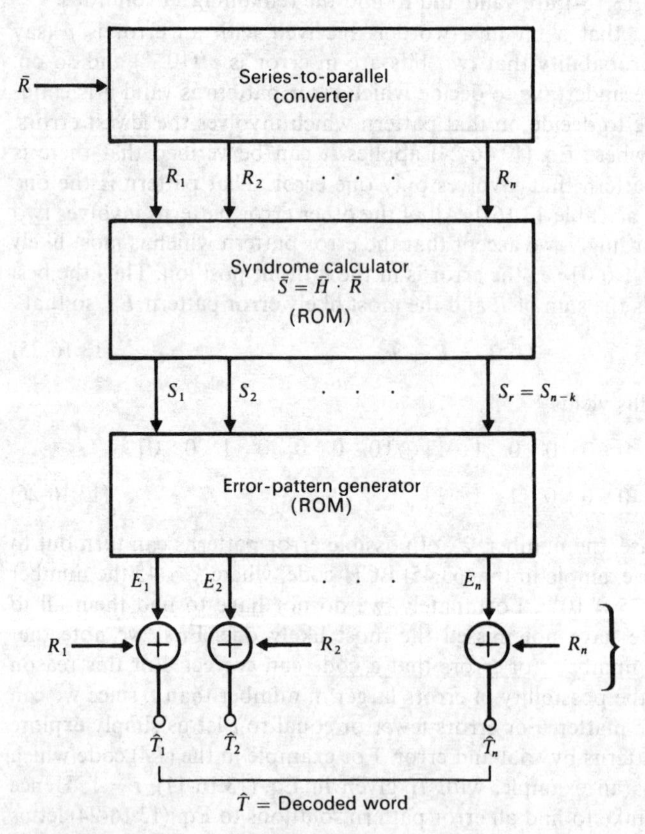

Figure 13.16-2 Generic circuit to decode a block code.

A block diagram of a method for decoding a received signal is shown in Fig. 13.16-2. The received serial bit stream is first converted to parallel form so that all bits of the codeword $R_1 R_2 \ldots R_m$ are available simultaneously. Since $\bar{H}$ is fixed and known, the syndrome corresponding to any $\bar{R}$ can be stored in a ROM, the syndrome calculator. This syndrome is stored at a memory location whose address is arranged to be $\bar{R}$ itself. Hence when $\bar{R}$ appears at the address inputs of the ROM the syndrome is read out. The syndrome has r bits with $r = n - k$. Corresponding to each syndrome we shall have to determine, as described above, the most likely error pattern. The error pattern, so determined for each syndrome, will be stored in the error-pattern generator ROM at an address which is arranged to be the same as the syndrome. Each output bit, $E_1 E_2 \ldots E_n$ is applied to one input of an EXCLUSIVE-OR logic gate, the other input of the gate being the corresponding bit of $\bar{R}$. If there is an error in a bit, say R_i, the corresponding error bit E_i will be $E_i = 1$ and the output $\hat{T}$ will be $\hat{T}_i = R_i^*$ (where * denotes the *complement*) as required.

It is apparent from the discussion and from the diagram of Fig. 13.16-2 that the complexity of a decoder grows with increasing codeword size n. However, from our discussion of Shannon's theorem we saw that the error rate can be reduced efficiently by increasing the message size, that is, by making n large. For this reason the key to efficient communication is linked to the development of electronics which incorporates very large numbers of components and operates at high speed such as VLSI and VHSIC. Also required are efficient algorithms.

13.17 EXAMPLES OF ALGEBRAIC CODES

Single Parity-Check Bit Code

The single parity-check bit code is an example of a block code. In this case the parity-check is selected to satisfy the equation

$$a_1 \oplus a_2 \oplus \cdots \oplus a_k \oplus c_1 = 0 \qquad (13.17\text{-}1)$$

Equation (13.17-1) is the first of the set of Eqs. (13.16-11) with

$$h_{11} = h_{12} = \cdots = h_{1k} = 1$$

All other equations of the set are identically zero. In this case, then, $r = 1$ and $n = k + 1$. Keeping in mind the definition of EXCLUSIVE-OR addition given in Eq. (13.16-4), we may see that Eq. (13.17-1) requires that the number of 1's in a word be even, for only the sum of an even number of 1's will add up to zero. If there were an odd number of 1's in the sum of Eq. (13.17-1), the result would be 1 rather than 0. This code is a single error detecting code and will not correct errors.

Repeated Codes

In a repeated code a binary 0 is encoded as a sequence of $(2t + 1)$ zeros, and a binary 1 as a similar number of 1's. Thus $k = 1$, $r = 2t$, and $n = 2t + 1$. We considered such encoding in Sec. 13.11. For the case $t = 1$, so that $r = 2t = 2$ and $n = 2t + 1 = 3$. A repeated code is a block code since the redundant bits are determined by Eqs. (13.16-11). We have $2t$ equations with a_1 equal to either 0 or 1, and

$$h_{11} = h_{21} = \cdots = h_{r1} = 1$$

while all other h's are set to zero. The redundant bits are therefore given by

$$\begin{aligned} 0 &= a_1 \oplus c_1 \\ 0 &= a_1 \qquad \oplus c_2 \\ &\cdots\cdots\cdots\cdots \\ 0 &= a_1 \qquad\qquad \oplus c_{2t} \end{aligned} \tag{13.17-2}$$

Using again the coding arithmetic of Eq. (13.16-4), we find that if $a_1 = 0$, all the c's are 0, and if $a_1 = 1$, all the c's are 1. The matrix $\bar{H}$ for the repeated code with $n = 2t + 1 = 3$ is given by [see Eq. (13.16-3)]:

$$\bar{H} = \begin{vmatrix} h_{11} & 1 & 0 \\ h_{21} & 0 & 1 \end{vmatrix} = \begin{vmatrix} 1 & 1 & 0 \\ 1 & 0 & 1 \end{vmatrix} \tag{13.17-3}$$

The ($n = 2t + 1$, $k = 1$) repeated code is capable of correcting t errors. However it requires the use of significant bandwidth as it is a rate $1/(2t + 1)$ code, and therefore such codes are inefficient.

Hadamard Code

The codeword in a Hadamard code are the rows of a Hadamard matrix. The Hadamard matrix is a square ($n \times n$) matrix in which $n = 2^k$ where, as usual, k is the number of bits in the uncoded word. One codeword consists of all zeros and all the other codewords have $n/2$, 0's and $n/2$, 1's. Further each codeword differs from every other codeword in $n/2$ places and for this reason the codewords are orthogonal to one another.

The Hadamard matrix which provides two codewords is

$$M_2 = \begin{vmatrix} 0 & 0 \\ 0 & 1 \end{vmatrix} \tag{13.17-4}$$

and the codewords are 00 and 01. The matrix M_4 which provides four codewords is

$$M_4 = \begin{vmatrix} M_2 & M_2 \\ M_2 & M_2^* \end{vmatrix} = \begin{vmatrix} 0 & 0 & 0 & 0 \\ 0 & 1 & 0 & 1 \\ 0 & 0 & 1 & 1 \\ 0 & 1 & 1 & 0 \end{vmatrix} \tag{13.17-5}$$

in which, as appears, M_2^* is the M_2 matrix with each element replaced by its complement. In general

$$M_{2n} = \begin{vmatrix} M_n & M_n \\ M_n & M_n^* \end{vmatrix} \tag{13.17-6}$$

Observe that an $n \times n$ Hadamard matrix provides n codewords each of n bits. To provide for the n different codewords, each with k information bits, each codeword contains $n = 2^k$ bits. Hence in the n bit codeword there are $r = n - k = 2^k - k$ parity bits. Accordingly as k increases, the number of parity bits becomes overwhelmingly large in comparison with the number k of information bits. Correspondingly the rate of the code becomes quite small since

$$R_c = \frac{k}{n} = \frac{k}{2^k} = 2^{k-1} \tag{13.17-7}$$

If, with large k, we propose to transmit coded words at the same rate as uncoded words, the coded transmission will require a bit rate larger than the uncoded bit rate by the factor $1/R_c$. Hence there would be a corresponding increase in the bandwidth required. This result is consistent with the conclusions we drew in Sec. 13.9 where we took note of the large bandwidth required in orthogonal signalling. Because Hadamard coding makes such demands on bandwidth it is generally used only where there are no bandwidth restrictions such as in deep space probes.

As we have noted earlier, the Hamming distance of an orthogonal code is

$$d_{\min} = \frac{n}{2} = 2^{k-1} \tag{13.17-8}$$

Thus the number of errors that can be corrected with a Hadamard code is from Eq. (13.14-5)

$$t = \frac{n}{4} - 1 = 2^{k-2} - 1 \tag{13.17-9}$$

Hence to provide for error correction we require that $k > 2$. However, for large k, t increases as 2^k so that significant error correction is possible.

Hamming Code

The Hamming code is a code in which $d_{\min} = 3$ so that $t = 1$, i.e., a single error can be corrected. The number n of bits in the codeword, the number k of bits in the uncoded word and the number r of parity bits are related by

$$n = 2^r - 1 \tag{13.17-10}$$

and

$$k = 2^r - 1 - r \tag{13.17-11}$$

For $r = 3$ we have a (7,4) code and for $r = 4$ we have a (15,11) code.

The parity check matrix $\bar{H}$ has r rows and n columns. No column consists of all zeros, each column is unique and each column has r elements: 1's or 0's. The matrix for the (7,4) code is

$$\bar{H} = \left| \underbrace{\begin{matrix} 1 & 1 & 1 & 0 \\ 1 & 1 & 0 & 1 \\ 1 & 0 & 1 & 1 \end{matrix}}_{\bar{h}} \quad \underbrace{\begin{matrix} 1 & 0 & 0 \\ 0 & 1 & 0 \\ 0 & 0 & 1 \end{matrix}}_{\bar{I}} \right| \tag{13.17-12}$$

Observe, that if we read the 0's and 1's in Eq. (13.17-12) as binary numbers then there is a single column corresponding to each of the decimal numbers from 1 through 7. The order of the columns is not of consequence. Of course, different orderings of the rows of $\bar{H}$ will yield different sets of codewords, but each such set is equally valid. The ordering established in Eq. (13.17-12) has been selected to separate the $\bar{h}$ submatrix from the identify submatrix $\bar{I}$ and hence give $\bar{H}$ the form shown in Eq. (13.16-3). This form is described as the *systematic* form and the corresponding code is called a *systematic code.*

The systematic $\bar{H}$ matrix for the (15,11) code is

$$\bar{H} = \left|\begin{matrix} \underbrace{\begin{matrix} 1 & 1 & 1 & 1 & 1 & 1 & 1 & 0 & 0 & 0 & 0 \\ 1 & 1 & 1 & 1 & 0 & 0 & 0 & 1 & 0 & 1 & 0 \\ 1 & 1 & 0 & 0 & 1 & 1 & 0 & 1 & 1 & 0 & 1 \\ 1 & 0 & 1 & 0 & 1 & 0 & 1 & 1 & 1 & 1 & 1 \end{matrix}}_{\bar{h}} & \underbrace{\begin{matrix} 1 & 0 & 0 & 0 \\ 0 & 1 & 0 & 0 \\ 0 & 0 & 1 & 0 \\ 0 & 0 & 0 & 1 \end{matrix}}_{\bar{I}} \end{matrix}\right| \tag{13.17-13}$$

Observe again that all 4-bit column patterns except the all-zero pattern each appear just once. It is left as a student exercise (Prob. 13.17-3) to determine the generating matrices for the $\bar{H}$ given above.

Extended Codes

All codes can be extended, that is, starting with a parity check matrix $\bar{H}$, a new, extended, matrix $\bar{H}_e$ can be formed as follows

$$\bar{H}_e = \left|\begin{array}{ccc:c} & & & 0 \\ & \bar{H} & & 0 \\ & & & \vdots \\ & & & 0 \\ \hdashline 1 & 1 & \cdots & 1 & 1 \end{array}\right| \tag{13.17-14}$$

Observe that the extended matrix $\bar{H}_e$ consists of the $\bar{H}$ matrix with an added row consisting of all 1's and an added right-hand column consisting of all 0's except for the bottommost element which remains a 1. The extended (7,4) Hamming code matrix is

$$\bar{H}_e = \left|\begin{matrix} 1 & 1 & 1 & 0 & 1 & 0 & 0 & 0 \\ 1 & 1 & 0 & 1 & 0 & 1 & 0 & 0 \\ 1 & 0 & 1 & 1 & 0 & 0 & 1 & 0 \\ 1 & 1 & 1 & 1 & 1 & 1 & 1 & 1 \end{matrix}\right| \tag{13.17-15}$$

The matrix $\bar{H}_e$ defines a $(n + 1, k)$ code. It can be verified (Prob. 13.17-4) that the extension of a code increases the minimum distance by 1, so that

$$d_{e,\,\min} = d_{\min} + 1 \tag{13.17-16}$$

For the extended Hamming code, $d_{e,\,\min} = 4$. The increase in this case does not improve the error-correcting ability of the code, since with $d_{e,\,\min} = 4$ we shall still have $t = 1$. However it does allow triple errors to be detected, while only double errors could be detected with $d_{\min} = 3$.

Cyclic Codes

Coding theory is a very sophisticated discipline. We shall not, accordingly, pursue the matter in detail. However, in order to convey an idea of a very large class of codes which are widely used we shall describe rather qualitatively some of the features of *cyclic codes*, stating a number of results and procedures without proof or justification.

A cyclic code has the property that a cyclic shift of one codeword of the code forms another codeword. The meaning of the term "cyclic shift" is to be seen in Fig. 13.17-1. Here a 7-bit word, instead of being written out horizontally, is written out around a circle. Starting at an arbitrary point, say at A, the 7-bit word encountered by a clockwise rotation is 1 1 0 1 0 0 1. Starting at some other arbitrary point, say B, we would read 0 1 1 1 0 1 0. The two words are related in that one is derived from the other by a cyclic shift. There are seven possible starting places in Fig. 13.17-1. Hence seven words can be read, each related to the other by a cyclic shift. The order in which the words are generated depends on the direction, clockwise or counterclockwise, of the shift, but the end result of the resultant collection of words is not affected by the shift direction.

Cyclic codes are important because they have algebraic properties which allow them to be easily encoded and decoded. The Hamming code turns out to be an example of such a code type. In the (7,4) Hamming code there are $2^k = 2^4 = 16$ codewords. It can be verified (Prob. 13.17-5) that seven of these words are precisely the cyclic-shift related words read from Fig. 13.17-1. Seven other words in the code are similarly cyclic-shift related. The fifteenth and sixteenth words are 0 0 0 0 0 0 0 and 1 1 1 1 1 1 1 in which a cyclic shift leaves the words unchanged.

A procedure for generating an (n, k) cyclic code is the following: The bits of the uncoded word $\bar{A} = [A_0\, A_1 \cdots A_{k-1}]$ are written as the coefficients of the polynomial:

$$A(x) = A_0 \oplus A_1\, x \oplus A_2\, x^2 \oplus \cdots \oplus A_{k-1}\, x^{k-1} \qquad (13.17\text{-}17)$$

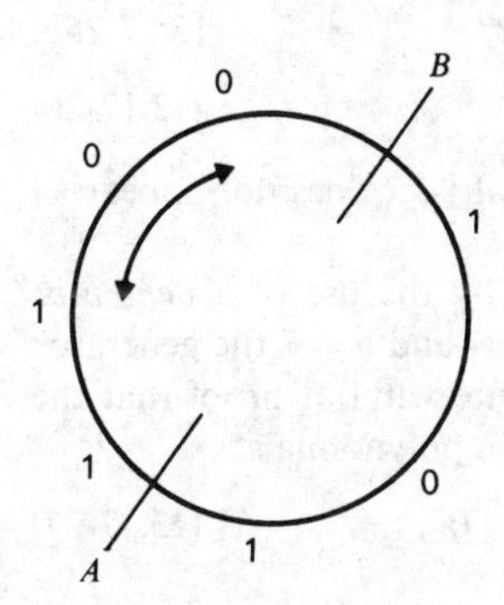

Figure 13.17-1 Showing that for a (7,4) Hamming code, a cycle shift in either direction produces a new codeword.

The bits of the coded word $\bar{T} = [T_0\, T_1 \cdots T_{n-1}]$ are written as the coefficients of the polynomial:

$$T(x) = T_0 \oplus T_1\, x \oplus T_2\, x^2 \oplus \cdots \oplus T_{n-1}\, x^{n-1} \tag{13.17-18}$$

We next form the "generating" polynomial $g(x)$ of degree $r = n - k$:

$$g(x) = 1 \oplus g_1\, x \oplus g_2\, x^2 \oplus \cdots \oplus g_{r-1}\, x^{r-1} \oplus x^r \tag{13.17-19}$$

and we determine the values of the coefficients $g_1, g_2 \cdots g_{r-1}$ from the condition that $g(x)$ be a factor of the polynomial

$$f(x) = x^n \oplus 1 \tag{13.17-20}$$

where n is the number of bits in the codeword. Finally, when $g(x)$ is determined, $T(x)$ is found from the equation

$$T(x) = g(x)A(x) \tag{13.17-21}$$

As an example of the application of this procedure, let us generate a (7,4) code. Since $n = 7$

$$f(x) = x^7 \oplus 1 \tag{13.17-22}$$

It can be verified (Prob. 13.17-6) that the factors of $f(x)$ are

$$f(x) = \lambda_1(x) \cdot \lambda_2(x) \cdot \lambda_3(x) \tag{13.17-23a}$$

$$= (1 \oplus x)(1 \oplus x \oplus x^3)(1 \oplus x^2 \oplus x^3) \tag{13.17-23b}$$

To generate a code with $n = 7$ bits, $T(x)$ in Eq. (13.17-18) must be a polynomial of degree $n - 1 = 6$. Since $\bar{A}$ is a $k = 4$ bit word, $A(x)$ is of degree $k - 1 = 3$. Hence we must select a factor from Eq. (13.17-23) which is of degree 3 so that $T(x)$ be of degree $3 + 3 = 6$. There are two such factors, either is suitable, and we arbitrarily select the factor $\lambda_2(x) = (1 \oplus x \oplus x^3)$. Now suppose the uncoded word is $\bar{A} = [1001]$ then

$$A(x) = 1 \oplus 0x \oplus 0x^2 + 1 \cdot x^3 = 1 \oplus x^3 \tag{13.17-24}$$

and the corresponding coded word is

$$T(x) = g(x)A(x) = (1 \oplus x \oplus x^3)(1 \oplus x^3) \tag{13.17-25a}$$

$$= 1 \oplus x \oplus x^4 \oplus x^6 \tag{13.17-25b}$$

so that

$$\bar{T} = [1 \quad 1 \quad 0 \quad 0 \quad 1 \quad 0 \quad 1] \tag{13.17-26}$$

Note that the generated codeword is *not systematic*, that is, it does not appear as the original uncoded word with parity bits added on.

Alternatively the coded words may be generated by the use of a *generator matrix* as in Eq. (13.16-9). In the present case, since $k = 4$ and $n = 7$ the generator matrix will have four rows and seven columns. We state without proof that the elements of the matrix are given by the coefficients of the polynomials

$$x^i \lambda(x); \quad i = k - 1, k - 2, \cdots, 0 \tag{13.17-27}$$

$\lambda(x)$ being $\lambda_2(x)$ or $\lambda_3(x)$ of Eq. (13.17-23*b*). Each value of i yields the elements of one row. Let us again arbitrarily select $\lambda_2(x)$. Then we find, setting $i = k - 1 = 3$

$$\begin{aligned} x^i\lambda_2(x) &= x^3(1 \oplus x \oplus x^3) = x^3 \oplus x^4 \oplus x^6 \\ &= 0 \cdot x^0 \oplus 0 \cdot x^1 \oplus 0 \cdot x^2 \oplus 1 \cdot x^3 \oplus 1 \cdot x^4 \oplus 0 \cdot x^5 \oplus 1 \cdot x^6 \end{aligned} \tag{13.17-28}$$

Thus the elements of the first row are 0 0 0 1 1 0 1. The elements of the second row are found by setting $i = 2$, etc. Altogether we find that the generator matrix $\bar{G}_2$, corresponding to $\lambda_2(x)$, is

$$\bar{G}_2 = \begin{vmatrix} 0 & 0 & 0 & 1 & 1 & 0 & 1 \\ 0 & 0 & 1 & 1 & 0 & 1 & 0 \\ 0 & 1 & 1 & 0 & 1 & 0 & 1 \\ 1 & 1 & 0 & 1 & 0 & 0 & 0 \end{vmatrix} \tag{13.17-29}$$

It is now readily verified that forming $\bar{T} = \bar{A}\bar{G}_2$ does indeed yield the codeword given above in Eq. (13.17-26).

The Golay Code

Another example of a cyclic code is the *Golay code*. The Golay code is a (23,12) cyclic code whose generating function is

$$g(x) = x^{11} \oplus x^9 \oplus x^7 \oplus x^6 \oplus x^5 \oplus x \oplus 1 \tag{13.17-30}$$

The extended form of the code is a (24,12) code. For the unextended Golay code $d_{\min} = 7$ and for the extended code $d_{\min} = 8$. The distinctive feature of this code is that it is the only known code of codeword length 23 which is able to correct $t = 3$ errors.

BCH Codes

The *Bose*, *Chaudhuri*, and *Hocquenghem* (BCH) codes form a large class of error correcting codes. The Hamming code, discussed earlier, and the Reed–Solomon code to be discussed subsequently are two special cases of this powerful error correction coding technique.

The BCH code employs k information bits, r parity check bits and therefore the number of bits in a codeword is $n = k + r$. Furthermore, the number of errors t which can be corrected in an n-bit codeword is

$$t = r/m \tag{13.17-31}$$

where m is an integer related to the number of bits n in the codeword by the formula

$$n = 2^m - 1 \tag{13.17-32}$$

The minimum distance between BCH codes is related to t by the inequality

$$2t + 1 \le d_{\min} \le 2t + 2 \tag{13.17-33}$$

The Hamming code, where $n = 2^r - 1$, is seen to be a BCH code where $m = r$ so that $t = 1$.

To see the capability of the BCH code relative to the (23,12) Golay code let $t = 3$. Then, from Eq. (13.17-3), $m = r/3$. If $n = 31$ (no integer value of m, when substituted into Eq. (13.17-22) will make n equal to 23) then $m = 5$ and $r = 15$ thus $k = 16$ and the code is a (31,16) BCH code. Note that the code rate is $R_c = \frac{16}{31} \approx \frac{1}{2}$ but if we define the code efficiency as $E = t/n$ then $E = \frac{3}{31} \simeq 10\%$; for the (23,12) Golay code, $R_c = \frac{12}{23} \simeq \frac{1}{2}$ but $E = \frac{3}{23} \simeq 13\%$ and is therefore 30% more efficient.

Note that a BCH code can correct $t = 3$ errors and employ $k = 45$ information bits in a 63-bit codeword. The code rate is now $R_c = \frac{45}{63} \simeq 0.7$ however, $E = \frac{3}{63} \simeq 5\%$. Thus we see the tradeoff between the code rate and its capability to correct errors.

13.18 BURST ERROR CORRECTION

The parity bits added in the block codes discussed above will correct a limited number of bit errors in each codeword. However, there are occasions when the *average* bit error rate is small, yet the error correcting codes discussed in Sec. 13.17 are not effective in correcting the errors because the errors are *clustered*. That is, examination of a received bit stream will show that in one region a large percentage of bits are in error, thereby precluding error correction using the techniques of Sec. 13.17, while in another region there are few and perhaps no errors.

Since the bit errors are closely clustered we say that the errors occur in *bursts*. Examples of the sources of such bursts are the following: (*a*) In magnetic tape recording and playback there may occur, for mechanical reasons, intervals when the spacing between head and tape is incorrect or it may be that some small section of the tape is simply defective. In either case a burst of errors will occur. (*b*) Static (i.e., non-thermal noise), such as results in radio transmission due to lightning, causes bursts of errors as long as it lasts. (*c*) There are channels in which the level of the received signal power waxes and wanes with time. In such *fading* channels bursts of errors are likely to occur when the received power is low.

Block Interleaving

A primary technique which is effective in overcoming error bursts, is *interleaving*. The principle of interleaving is illustrated in Fig. 13.18-1. As shown in Fig. 13.18-1*a*, before the data stream is applied to the channel, the data goes through a process of interleaving and error correction coding. At the receiving end the data is decoded i.e., the data bits are evaluated in a manner to take advantage of the error correcting and detecting features which result from the coding and the process of interleaving is undone. As represented in Fig. 13.18-1*b*

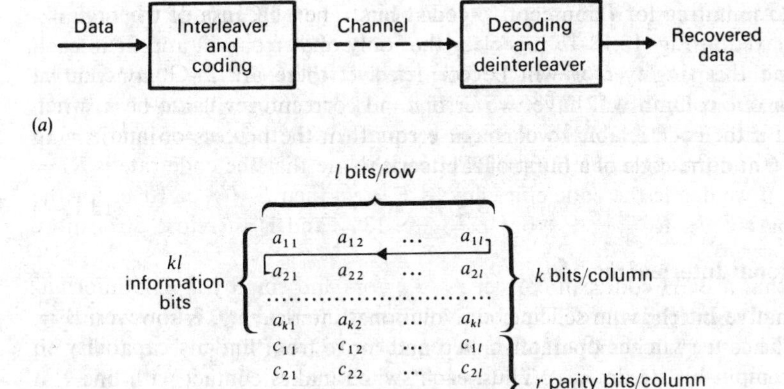

Figure 13.18-1 (*a*) Interleaving used to contend with burst errors. (b) Organization of information and parity bits in interleaving.

a group of kl data bits is loaded into a shift register which is organized into k rows with l bits per row. The data stream is entered into the storage element at a_{11}. At each shift each bit moves one position to the right while the bit in the rightmost storage element moves to the leftmost stage of the next row. Thus, for example, as indicated by the arrow, the content of a_{1l} moves to a_{21}. When kl data bits have been entered, the register is full, the first bit being in a_{kl} and the last bit in a_{11}. At this point the data stream is diverted to a second similar shift register and a process of coding is applied to the data held stored in the first register. In this coding process, the information bits in a column (e.g., a_{11}, a_{21}, . . . , a_{k1}) are viewed as the bits of an uncoded word to which parity bits are to be added. Thus the codeword $a_{11}\, a_{21} \cdots a_{k1}\, c_{11}\, c_{21} \cdots c_{r1}$ is formed, thereby generating a codeword with k information bits and r parity bits. Observe that the information bits in this codeword were l bits apart in the original bits stream.

When the coding is completed, the entire content of the $k \times l$ information register as well as the $r \times l$ parity bits are transmitted over the channel. Generally the bit-by-bit serial transmission is carried out row by row, that is in the order

$$c_{rl} \cdots c_{r1} \cdots c_{1l} \cdots c_{11}\, a_{kl} \cdots a_{k1} \cdots a_{2l} \cdots a_{21}\, a_{1l} \cdots a_{12}\, a_{11}.$$

Note that the data is transmitted in exactly the same order it entered the register, however, now parity bits are also transmitted. The received data is again stored in the same order as in the transmitter and error correction decoding is performed. The parity bits are then discarded and the data bits shifted out of the register.

To see how interleaving affects bursts of errors, consider that the code incorporated into a codeword (a column in Fig. 13.18-1*b*) is adequate to correct a single error. Next suppose that in the transmitted data stream there occurs a burst of noise lasting for l consecutive coded bits. Then because of the organization displayed in Fig. 13.18-1*b* it is clear that only one error will appear in each column and this single error will be corrected. If there are $l + 1$ consecutive errors then one column will have two errors and correction will not be assured. In general, if the code is able to correct t errors then the process of interleaving will permit the correction of a burst of B bits with

$$B \leq tl \tag{13.18-1}$$

Convolutional Interleaving

An alternative interleaving scheme, convolutional interleaving, is shown in Fig. 13.18-2. The four switches operate in step and move from line-to-line at the bit rate of the input bit stream $d(k)$. Thus, each switch makes contact with line 1 at the same time, then moves to line 2 together, etc., returning to line 1 after line l. The cascade of storage elements in the lines are shift registers. Starting with line 1, on the transmitter side, which has no storage elements, the number of elements increases by s as we progress from line to line. The last line l has $(l - 1)s$ storage elements. The total number of storage elements in each line (transmitter plus receiver side) is, in every case, the same, Hence, in each line there are a total of $(l - 1)s$ storage elements.

To describe the timing of operations in the interleaver, let us consider a single line l_i on the transmitter side. Suppose that during the particular bit interval of bit $d(k)$ there is switch contact, at input and output sides of line l_i. At the end of the bit interval, a clock signal causes the shift register of line l_i (and line l_i only), to enter into the leftmost storage element the bit on the input side of the line and to start moving the contents of each of its storage elements one bit to the right. That process having started, a synchronous clock advances the switches to the next line l_{i+1}. When the shift register response is completed, there will be a new bit at the output end of the line, l_i. However, because of the propagation delay through the storage elements, the switch at the output end of the line will have already lost contact with line l_i, before the new bit has appeared at the line output. In summary, during the interval of input $d(k)$, during which the switches were connected to line l_i, there is one-bit shift of the shift register on line l_i which accepts bit $d(k)$ into the register. However, the fact that such a shift took place is not noticed on the output switch until the *next* time the switch makes contact with line l_i. We observe also that while the clock that drives the switches has a clock rate f_b, which is the bit rate, the clocks that drives the shift registers have a rate f_b/l. The shift registers are not driven in unison, but rather in sequence, each register being activated as the switches are about to break contact with its line.

Now let us consider that, initially, all the shift registers in the transmitter and the receiver are short circuited. If bit $d(k)$ occurs when all the switches are on line l_i, then the corresponding received bit $\hat{d}(k)$ will appear immediately. The next input bit $d(k + 1)$ will be the next received bit except that it will be transmitted over line l_{i+1}, and so on. In short, the received sequence $\hat{d}(k)$ will be the same as the transmitted sequence. With the shift registers in place in the transmitter and

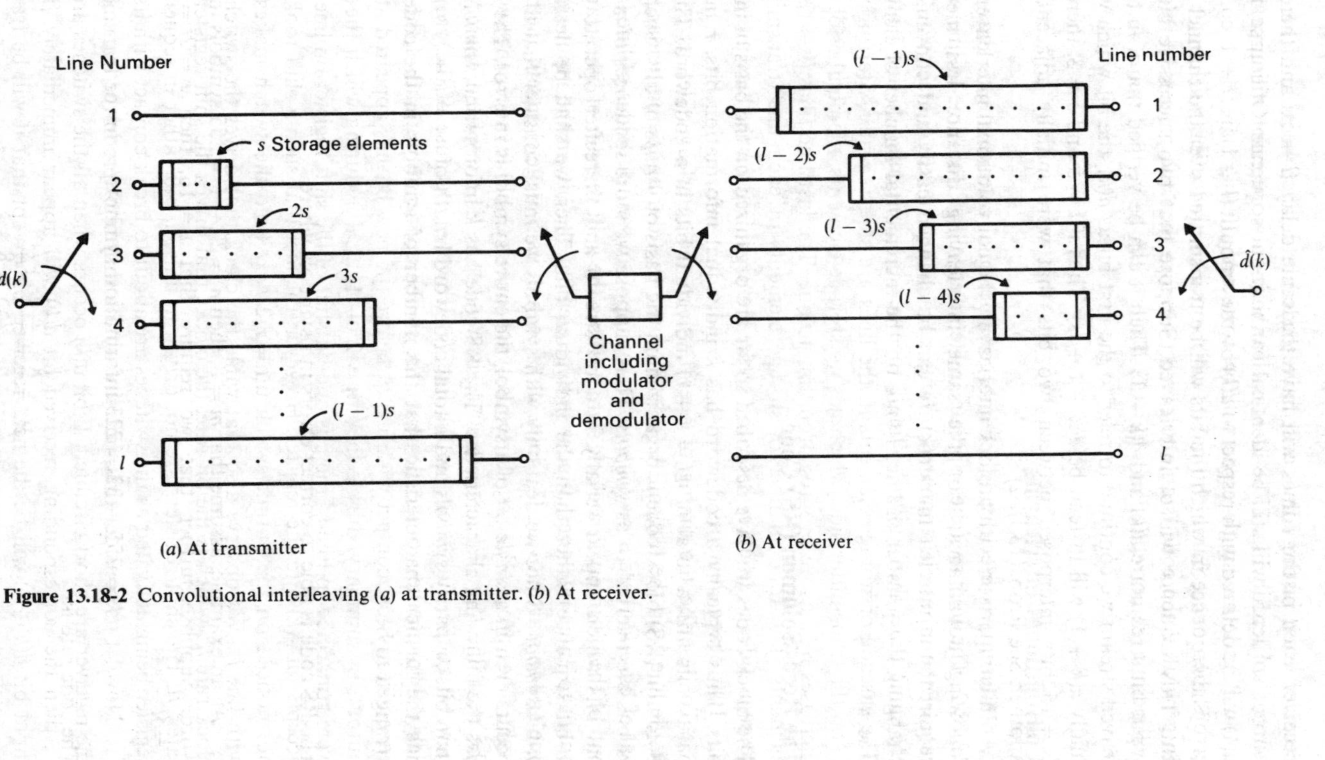

Figure 13.18-2 Convolutional interleaving (*a*) at transmitter. (*b*) At receiver.

receiver, each of the l lines will have the same delay $(l - 1)s$ and therefore the output sequence will still be identical to the input sequence. Of course, however, $\hat{d}(k)$ will be delayed with respect to $d(k)$ by the amount $(l - 1)s$.

The sequence in which the bits will be transmitted over the channel is different. This sequence will be interleaved. Suppose that two successive bits in the input bit stream are $d(k)$ and $d(k + 1)$. Then it can be verified that if in the interleaved stream we continue to refer to the first bit as $d(k)$, the bit which was originally $d(k + 1)$ will instead be $d(k + 1 + ls)$. Thus if $l = 5$ and $s = 3$, there will be $ls = 15$ bits interposed between two bits that were initially adjacent to one another. See Prob. (13.18-2).

In comparison with a block interleaver, the convolutional interleaver has the following advantages: For the same interleaving distance less memory is required; the interleaving structure can be changed easily and conveniently by changing the number of lines l and, or, the incremental number of elements per line, s.

The Reed–Solomon (RS) Code

The block codes we have described so far are organized on the basis of individual bits. Thus typically a codeword has k individual information bits, r individual parity bits and a total of $n\,(= k + r)$ individual bits in a codeword. The Reed–Solomon (RS) block code is organized on the basis of *groups* of bits. Such groups of bits are referred to as *symbols.* Thus suppose we store sequences of m individual bits which appear serially in a bit stream and thereafter operate with the m-bit sequence rather than the individual bits. Then we shall be dealing with m-bit *symbols.* Since we deal only with symbols we must consider that if an error occurs even in a *single* bit of a symbol, the entire symbol is in error. The RS code has the following characteristics: The RS code has k information *symbols* (rather than bits), r parity *symbols* and a total codeword length of $n\,(= k + r)$ *symbols.* It has the further characteristic that the number of symbols in the codeword is arranged to be

$$n = 2^m - 1 \tag{13.18-2}$$

The RS code is able to correct errors in t symbols where

$$t = r/2 \tag{13.18-3}$$

As an example, assume that $m = 8$, then $n = 2^8 - 1 = 255$ symbols in a codeword. Suppose further that we require that $t = 16$, then $r = 2t = 32$ and $k = n - r$ is

$$k = 255 - 32 = 223 \text{ information symbols/codeword} \tag{13.18-4}$$

The code rate is

$$R_c = \frac{k}{n} = \frac{223}{255} \cong \frac{7}{8} \tag{13.18-5}$$

The total number of bits in the codeword is $255 \times 8 = 2040$ bits/codeword.

Since the RS code of our example can correct sixteen symbols it can correct a burst of $16 \times 8 = 128$ consecutive bit errors. If we use the RS code with the interleaving as described above, then using Eq. (13.18-1) the number of correctable *symbols* is lt and the number of correctable bits is

$$B = mlt \tag{13.18-6}$$

with $m = 8$, $t = 16$ and $l = 10$, $B = 1280$ bits.

The organization of an RS block code with $m = 8$, $k = 233$, $r = 32$ utilizing interleaving to a depth of $l = 4$ is shown in Fig. 13.18-3.

It is interesting to note that while the (255,223) RS code can correct 128 consecutive bit errors it must then have an *error-free* region of $255 - 16 = 239$ symbols (1912 bits). Further, if the errors are random, and there is, at most, one error per symbol, then the RS code can correct only sixteen bit errors in 2040 bits. Clearly the RS code is *not* an efficient code for correcting random errors.

Concatenated Codes

The use of nonbinary codes (i.e., codes based on symbols rather than on individual bits such as the RS code) is an effective way of dealing with error bursts. As we have noted, the RS code turns out to be an especially effective and useful code when m (the symbol length) is much larger than unity. The RS code is important because there is available an efficient hard-decision decoding algorithm which makes it possible to employ long codes. A second, effective, way of dealing with bursts is interleaving. However, nonbinary codes and interleaving are not particularly effective in dealing with random errors which affect only a single bit or only a small number of consecutive bits. When we must contend with both burst and random errors it is useful to cascade coding which is effective with bursts and coding useful with random errors. Such cascading of codes is called *concatenation.*

To illustrate the technique of concatenation, let us start with the coding represented in Fig. 13.18-3. Here we have used an RS code with $k = 223$, $r = 32$, $m = 8$ and we have used interleaving with $l = 4$. Now let us add further coding to

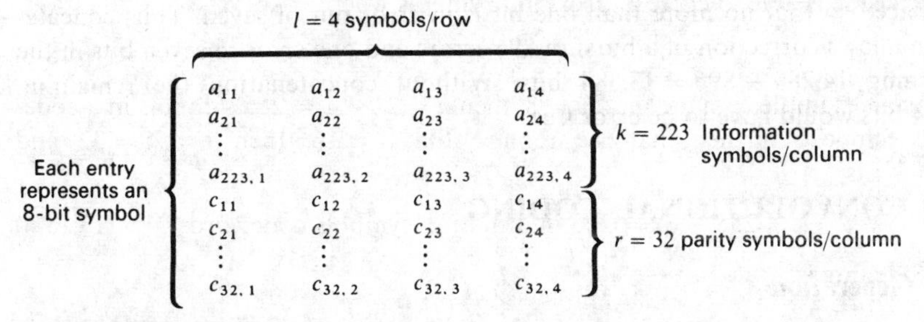

Figure 13.18-3 The organization of an RS code with $m = 8$, $k = 233$, and $r = 32$. Here a_{11}, etc. is an 8-bit information symbol and c_{11} etc. is an 8-bit parity symbol. The interleaving depth is $l = 4$. There are $233 \times 8 \times 4 = 7136$ information bits and $32 \times 8 \times 4 = 1024$ parity bits.

each of the 255 rows of Fig. 13.18-3 using the (7,4) Hamming code. To effect this additional coding we add three parity symbols to each row so that now the pattern of Fig. 13.18-3 is expanded to 4 + 3 = 7 columns. The number of information bits remains

$$k = 223 \times 8 \times 4 = 7136 \text{ information bits}$$

The number of parity bits due to the RS coding is, as before

$$r(\text{RS}) = 32 \times 8 \times 4 = 1024 \text{ RS parity bits}$$

The number of bits added by the Hamming coding is

$$r(\text{H}) = 255 \times 8 \times 3 = 6120 \text{ Hamming parity bits}$$

The code rate of the concatenated coding is

$$R_c = \frac{k}{k + r(\text{RS}) + r(\text{H})} \cong \frac{1}{2}$$

Without concatenation, the correctable burst error is, using Eq. (13.18-6)

$$B = 8 \times 4 \times 16 = 512 \text{ bits}$$

with concatenation we have

$$B = 8 \times 7 \times 16 = 896 \text{ bits}$$

However, in the first case we have a block length

$$n = 8 \times 4 \times 255 = 8160 \text{ bits}$$

In the second case we have

$$n = 8 \times 7 \times 255 = 14{,}280 \text{ bits}$$

The effectiveness of the coding against bursts is reasonably measured by the ratio B/n and we find that this ratio is the same with or without concatenation, that is, 512/8160 = 896/14,280. Since, for the sake of minimizing hardware we might well be inclined to use the shorter block length, it appears that, so far as burst error connection is concerned, the concatenation is actually somewhat disadvantageous. However, the concatenated code can also correct random errors if they are spaced so that no more than one bit is in error out of seven. Thus concatenation allows correction of a burst of 896 errors and one error in seven bits in the remaining 14,280 − 896 = 13,384 bits. Without concatenation the remaining 13,384 bits would have to be error free.

13.19 CONVOLUTIONAL CODING

Code Generation

A convolutional code is generated by combining the outputs of a K-stage shift register through the employment of v EXCLUSIVE-OR logic summers. Such a coder is illustrated in Fig. 13.19-1 for the case $K = 4$ and $v = 3$. Here M_1 through M_4

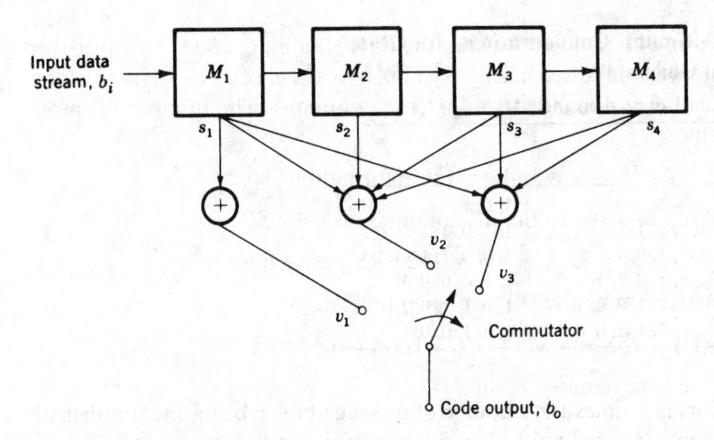

Figure 13.19-1 An example of a convolutional coder.

are 1-bit storage (memory) devices such as flip-flops. The outputs v_1, v_2, and v_3 of the adders in Fig. 13.19-1 are

$$v_1 = s_1 \tag{13.19-1a}$$

$$v_2 = s_1 \oplus s_2 \oplus s_3 \oplus s_4 \tag{13.19-1b}$$

$$v_3 = s_1 \oplus s_3 \oplus s_4 \tag{13.19-1c}$$

The operation of the encoder proceeds as follows: We assume that initially the shift register is clear. The first bit of the input data stream is entered into M_1. During this message bit interval the commutator samples, in turn, the adder outputs v_1, v_2, and v_3. Thus a single bit yields, in the present case, three coded output bits. The encoder is therefore of rate 1/3. The next message bit then enters M_1, while the bit initially in M_1 transfers to M_2, and the commutator again samples all the v adder outputs. This process continues until eventually the last bit of the message has been entered into M_1. Thereafter, in order that every message bit may proceed entirely through the shift register, and hence be involved in the complete coding process, enough 0's are added to the message to transfer the last message bit through M_4, and, hence, out of the shift register. The shift register then finds itself in its initial "clear" condition.

It may be verified, by way of example, that, for the encoder of Fig. 13.19-1, if the input bit stream to the encoder is given by the 5-bit sequence

$$m = 1 \quad 0 \quad 1 \quad 1 \quad 0 \tag{13.19-2}$$

then the coded output bit stream is

$$c = 111 \quad 010 \quad 100 \quad 110 \quad 001 \quad 000 \quad 011 \quad 000 \quad 000 \tag{13.19-3}$$

If the number of bits in the message stream is L, the number of bits in the output code is $v(L + K)$. As a matter of practice however, L is ordinarily a very large number while K is a relatively small number. Hence $v(L + K) \simeq vL$. Thus the

Table 13.19-1 Optimum Configurations for Rate 1/2 Convolutional Coders

	Code Generator Connections ($M_1, M_2, \ldots, M_K$)	
K (Number of Register Stages)	v_1	v_2
3	1, 1, 1	1, 0, 1
4	1, 1, 1, 1	1, 1, 0, 1
5	1, 1, 1, 0, 1	1, 0, 0, 1, 1
6	1, 1, 1, 0, 1, 1	1, 1, 0, 0, 0, 1
7	1, 1, 1, 1, 0, 0, 1	1, 0, 1, 1, 0, 1, 1
8	1, 1, 1, 1, 1, 0, 0, 1	1, 0, 1, 0, 0, 1, 1, 1

number of code bits is v times the number of message bits, v being the number of commutator segments. Accordingly, also, the rate of the code is $1/v$.

Note the continuity of operation of the convolutional encoder. Even if the input message bit stream were to consist of millions of bits, the stream would be run continuously through the encoder. Each bit remains in the shift register for as many message bit intervals as there are stages in the shift register. Hence each input bit influences K groups of v bits.

The coded output depends on the number K of shift-register stages, on the number of EXCLUSIVE-OR adders used and on the connections of the register stages to the adders. With a view toward searching out optimal encoders, exhaustive computer analyses have been undertaken. Some results, for rate 1/2 coders (two adders) are given in Table 13.19-1. A 1 in the table indicates a connection to an adder while a 0 indicates that no connection is made.

Thus, by way of example, for the case $K = 5$, stages 1, 2, 3, and 5 should be connected to the adder which generates v_1 while stages 1, 4, and 5 should be connected to the adder for v_2. More extensive tables are available in the literature.

13.20 DECODING A CONVOLUTIONAL CODE

The Code Tree

With a view toward exploring procedures for decoding a convolutional code, we consider the code tree of Fig. 13.20-1. This code tree applies to the convolutional encoder of Fig. 13.19-1, for which $K = 4$ and $v = 3$, and which is constructed for the case $L = 5$ corresponding to a 5-bit message sequence. We now discuss the interpretation of the code tree.

The starting point on the code tree is at the left and corresponds to the situation before the occurrence of the first message bit. The first message bit may be either a 1 or a 0. We adopt the convention that, when an input bit is a 0, we shall diverge upward from a node of the tree, and when the input bit is a 1 we shall diverge downward. Suppose then that the first input bit is 1. Then entering the tree at node A, we move downward to the lower branch and to node B. Now

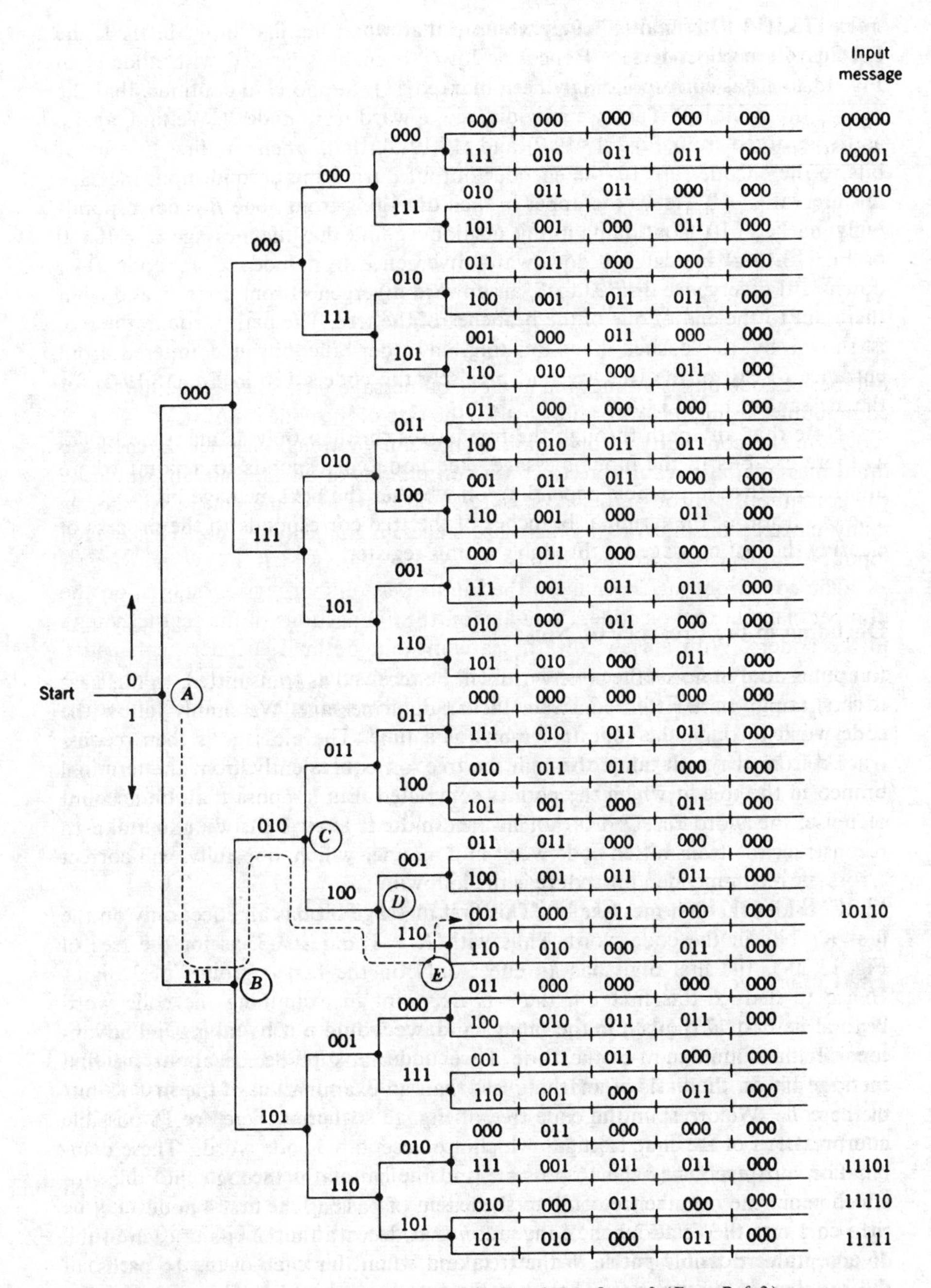

Figure 13.20-1 Code tree for encoder of Fig. 13.19-1: $K = 4$, $L = 5$, $v = 3$. (*From Ref. 2.*)

from Eqs. (13.19-2) and (13.19-3) we note that when the first input bit is 1, the output of the coder is 111. Hence the lower branch associated with node A in Fig. 13.20-1 has correspondingly been marked 111. Suppose, to continue, that the second input bit is 0. Then we now diverge upward from node B. We find, again, as to be noted from Eqs. (13.19-2) and (13.19-3), that, when the first two input bits to the encoder are 10, the encoder output during this second input message bit interval is 010. Hence the upper branch diverging from node B is correspondingly marked 010. Continuing in this fashion we find that the message $m = 10110$ of Eq. (13.19-2) indicates a downward divergence from node C to node D, a downward divergence from D to E, an upward divergence from node E, and from there out to the end of one of the branches of the tree. The path through the tree is shown by the dashed line. Reading, in order, the bits encountered from entrance to exit of the tree, we find precisely the code given in Eq. (13.19-3) for the message of Eq. (13.19-2).

Note that any path through the tree passes through only as many nodes (L) as there are bits in the input message. The node corresponds to a point where alternate paths are possible depending on whether the next message bit is 1 or 0. The extension of the terminal branches of the tree corresponds to the process of clearing the last message bit through the shift register.

Decoding in the Presence of Noise

In the absence of noise, the code word will be received as transmitted. In this case it is a simple matter to reconstruct the original message. We simply follow the code word through the code tree v bits at a time. The message is then reconstructed from the path taken through the tree, or, equivalently, from the terminal branch of the tree at which the path is completed. But suppose that, on account of noise, the word received is not as transmitted. How shall we undertake to reconstruct the transmitted code word in a manner which, hopefully, will correct errors? A recommended procedure is the following:

Consider the first message bit. This first message bit has an effect only on the first Kv bits in the code word. Thus with $K = 4$ and $v = 3$, as for the tree of Fig. 13.20-1, the first digit has an effect only on the first 4 groups of 3 digits. Hence to deduce this first bit, there is no point in examining the code word beyond its first 12 digits. On the other hand, we would not be taking full advantage of the redundancy of the code if we undertook to decide about the first message bit on the basis of anything less than an examination of the first 12 bits of the code. We see from the code tree of Fig. 13.20-1 that there are 16 possible combinations of the first 12 digits which are acceptable code words. These combinations correspond to the 16 possible (incomplete) paths through the code tree which penetrate into the tree only to the extent of passing the first 4 nodes. Let us then compare the first 12 bits of the received code with the 12 bits of each of the 16 acceptable possible paths. We next take a count, for each of the 16 paths, of the number of discrepancies between bits of the received code word and the acceptable code word corresponding to each path. If now the path that yields the minimum number of discrepancies is a path which diverges upward from the first node (node A in Fig. 13.20-1), we make the decision that the first message bit is 0. And, of course, if this path diverges downward from node A, we decide that the first bit is 1. Thus we see that a decision about the first message bit is not made until after a complete exploration of all possible paths which penetrate into the tree to the extent of including K nodes or equivalently Kv digits. It will be recognized that the operation of convolutional decoding described here is a digital operation which corresponds to the analog correlation detection process described in Sec. 11.6.

The second message bit is decided upon in the same manner. Thus, referring to Fig. 13.20-1, suppose it turns out that, using the above procedure, we decide that the first message bit is 1. Then we start at node B and again examine the 16 paths starting at node B and penetrating into the code tree to the extent of $Kv = 12$ code bits. We compare the 12 code bits corresponding to each of these 16 paths with the 12 received code bits after *discarding the first* 3 *received code bits.* A decision about the second message bit is now made on the basis of the same criterion used to decide about the first bit. Successive message bits are decoded in the same manner.

In summary, the decoding procedure is the following: The ith message bit m_i is decoded on the basis of previous decisions about the message bits $m_1, m_2, \ldots, m_{i-1}$ and on the basis of an examination of the Kv digit span of the received code word that is influenced by m_i. The previous decisions determine the starting node, the ith starting node, for the K node section of the code tree to be examined. Thereafter, m_i itself is decoded by determining which one of the 2^K branch sections of the code tree diverging from the ith starting node exhibits the fewest discrepancies when compared with the Kv digit span of the received code word that is influenced by m_i. If this section of the code tree diverges upward from the ith starting node, the ith message digit is 0; otherwise it is 1.

It turns out that, when a message is decoded in the manner we have just described, the probability that a bit in the decoded message is in error decreases exponentially with K. Hence there is an interest in making K as large as possible. On the other hand, as noted, the decoding of each bit requires an examination of the 2^K branch sections of the code tree. Hence for large K the above, general, decoding procedure may involve such a lengthy procedure that it is simply not feasible. We therefore consider next a *sequential* decoding scheme that remains manageable even when K is large.

Sequential Decoding

The principal advantage of sequential decoding is that such decoding generally allows the decoder to avoid the lengthy process of examining *every* branch of the 2^K possible branches of the code tree in the course of decoding a single message bit. In sequential decoding, at the arrival of the first v code bits, the encoder compares these bits with the two branches which diverge from the starting node. If one of the branches agrees exactly with these v code bits, then the encoder follows this branch. If, because of noise, there are errors in the received bits, the encoder follows the branch, in comparison with which the received bits exhibit the fewer discrepancies. At the second node a similar comparison is made between the diverging branches and the second set of v code bits, and so on, at succeeding nodes.

Now suppose that in the transmission of any v bits, corresponding to some tree, branch errors have found their way into more than half the bits. Then at the node from which this branch diverges the decoder will make a mistake. In such a case, the entire continuation of the path taken by the encoder must be in error. Consider, then, that the decoder keeps a record, as it progresses, of the total number of discrepancies between the received code bits and the corresponding bits encountered along its path. Then, after having taken a wrong turn at some node, the likelihood is very great that this total number of discrepancies will grow much more rapidly than would be the case if the decoder were following the correct path. The decoder may be programmed to respond to such a situation by retracing its path to the node at which an apparent error has been made and then taking the alternate branch out of that node. In this way the decoder will eventually find a path through K nodes. When such a path is found, the decoder decides about the first message digit on the basis of the direction of divergence of this path from the first node. Similarly, as before, the second message is determined on the basis of a path searched out by the decoder, again K branches long, but starting at the second starting node and on the basis of the received code bit sequence with the first v bits discarded.

We consider now how the decoder may judge that it has made an error and taken a wrong turn. Let the probability be p that a received code bit is in error, and let the encoder already have traced out a path through l nodes. We assume $l < K$, so that the decoder has not yet made a decision about the message bit in question. Since every branch is associated with v bits, then, on the average over a long path of many branches, we would expect the total number of bit differences between the decoder path and the corresponding received bit sequence to be $d(l) = plv$ even when the correct path is being followed. In Fig. 13.20-2 we have

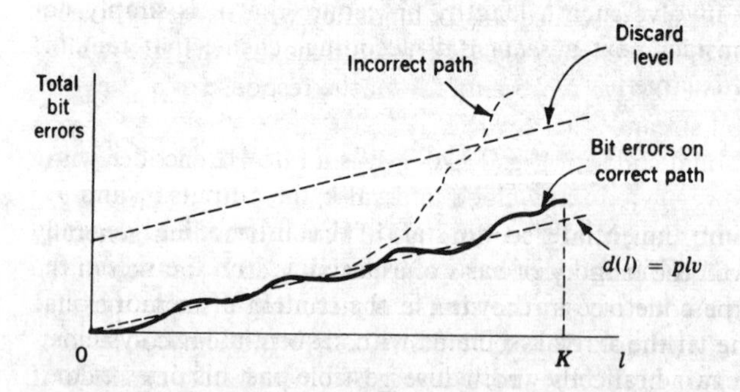

Figure 13.20-2 Setting the threshold in sequential decoding.

plotted $d(l)$ in a coordinate system in which the abscissa is l, the number of nodes traversed, and the ordinate is the total bit error that has accumulated. We would expect that, if the encoder were following the correct path, the total bit errors accumulated would oscillate about $d(l)$. On the other hand, shortly after a wrong decoder decision has been made, we expect the accumulated bit error to diverge sharply as indicated. A *discard level* has also been indicated in the figure. When the plot of accumulated errors crosses the discard level, the decoder judges that a decoding error has been made. The decoder then returns to the nearest *unexplored* branch and starts moving forward again until, possibly, it is again reversed because the discard level is crossed. In this way, after some trial and error, an entire K-node section of the code tree is navigated. And at this point a decision is made about the message bit associated with this K-node section of the tree. In Fig. 13.20-2 the discard level line does not start at the origin in order to allow for the possibility that the initial bits of the received code sequence may be accompanied by a burst of noise.

The great advantage of sequential decoding of a convolutional code is that such decoding makes it unnecessary, generally, to explore every one of the 2^K paths in the code tree section. Thus, suppose it should happen that the decoder takes the correct turn at each node. Then, in this case, the decoder will be able to make a decision about the message bit in question on the basis of a *single* path. Let us assume that at some node the decoder errs and must return to this node to take an alternative branch. Then even the information that an error has been made is useful because thereafter the decoder may exclude from its searchings all paths which diverge in the original direction from this node. The end result is that sequential decoding may generally be accomplished with very much less computation than the direct decoding discussed earlier.

State and Trellis Diagrams

A node, in the tree diagram of Fig. 13.20-1, is a point from which a branch diverges in one direction or the other depending on whether the next message bit is a 1 or a 0. Thus the number of new nodes at each stage increases exponentially with the number of input bits. However, as we shall see, there is a great deal of redundancy in the tree diagram. To reduce this redundancy we consider a structure referred to as a *trellis*. The trellis avoids the redundancy of the tree while allowing, as does the tree, an effective representation of the response of a convolutional coder to an input bit stream.

Consider the convolutional coder of Fig. 13.20-3. It is a rate 1/2 encoder with $v_1 = s_1 \oplus s_3$ and $v_2 = s_1 \oplus s_2 \oplus s_3$. In any clock interval k, the outputs v_1 and v_2 depend on the bit moved into the encoder at the start of that interval and depend also on the past history which the encoder has experienced, i.e., on the sequence of earlier input bits. This past history is recorded in the content of memory bits M_1 and M_2. Following the terminology associated with sequential logic systems, we use the term *states* to characterize the individual possible past histories of the system which are relevant to the future response of the system. Accordingly, we

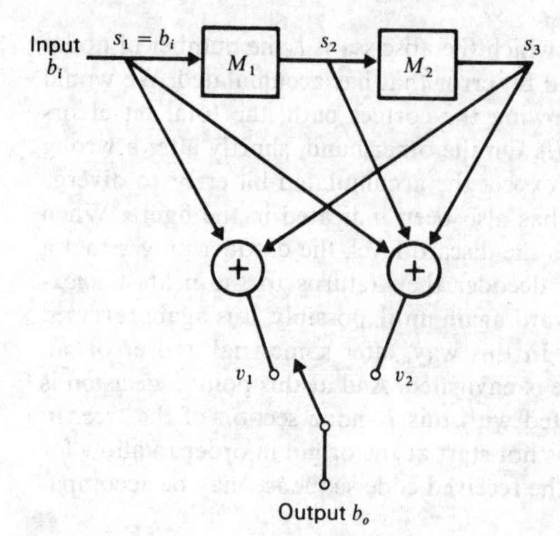

Figure 13.20-3 A convolutional coder.

find that the encoder of Fig. 13.20-3 has four states a, b, c, and d corresponding to $M_1M_2 = 00$, 01, 10, and 11, respectively.

One way in which we can represent the response of the encoder to an arbitrary input sequence is through the *state diagram* of Fig. 13.20-4. Here the four states are shown, and transitions from state-to-state are represented by arrows. The dashed arrows represent state-to-state transitions which result from 0 inputs and the solid arrows from 1 inputs. Each arrow is also marked with the encoder output v_1v_2 determined by the state and the next input. Thus, for example, suppose that during clock interval k the encoder is in state a ($M_1\,M_2 = 00$) and the next input bit is a 1 transferred into M_1 at the beginning of clock interval $k + 1$. Then this input will cause the encoder to make a transition to state c

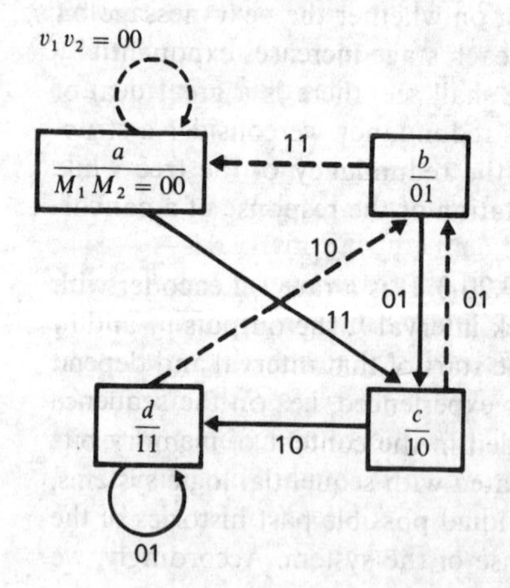

Figure 13.20-4 The state diagram for the coder of Fig. 13.20-3.

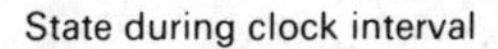

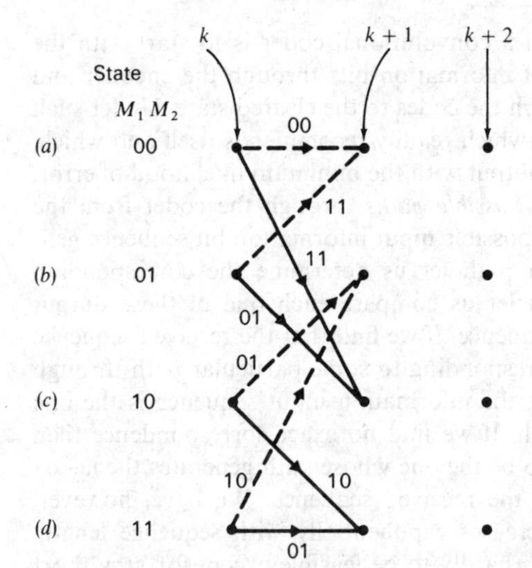

Figure 13.20-5 A trellis diagram for the coder of Fig. 13.20-3.

$(M_1 M_2 = 10)$ in which state it will remain for the duration of clock interval $k + 1$ and further, during this interval the output will be $v_1 v_2 = 11$.

An alternative representation is shown in Fig. 13.20-5. In this *trellis structure* the states, each of which appears just once in the state diagram of Fig. 13.20-4, are duplicated for each new clock cycle. The possible state transitions from clock cycle k to clock cycle $k + 1$ are shown as are the corresponding outputs. The symbolism follows the convention used in Fig. 13.20-4. The transitions and outputs from $k + 1$ to $k + 2$ (not shown) are of course duplicates of the transitions and output from k to $k + 1$. Finally, we note that, if the encoder has n memory elements, the number of states is 2^n. There is a one-to-one correspondence between the states of Figs. 13.20-4 and 13.20-5 and the nodes of the tree diagram of Fig. 13.20-1. Since, in the tree diagram, the nodes proliferate without limit, it is clear that, as the tree expands, the same set of nodes reappear time and again.

Suppose, that we start at an arbitrary state, say state b during interval k in Fig. 13.20-5. The next input bit will carry the encoder over one of 2 branches either to state a or to state c where it will remain during interval $k + 1$. The second bit will carry the encoder from $k + 1$ to $k + 2$ over one of four possible branches, two from state a and two from state c. During interval $k + 2$ any of the four states is possible. Since the number of branches leaving each state is two, the number of branches from $k + 2$ to $k + 3$ is $4 \times 2 = 8$. Thereafter, the number of branches available is always eight. Altogether then the total number of paths through the trellis is $2 \times 4 \times 8 \times 8 \cdots$. If then we go through the trellis from k to $k + l$ the total number of possible paths is $8^{l-1} = 2^{3(l-1)}$ (except for $l = 1$). We note the rapid (exponential) increase of paths with increasing l.

The Viterbi Algorithm

As noted, a common procedure with a convolutional coder is to start with the coder cleared, then to run a block of information bits through the encoder and finally to add 0 bits as needed to return the codes to the cleared state. Under such circumstances, a decoding procedure which readily recommends itself and which, as a matter of fact, yields a decoded output with the minimum likelihood of error, is the following: Let us consider *all possible paths* through the coder from the starting point to the end point. Each possible input information bit sequence generates its own path. For each such path let us determine the corresponding sequence of coder output bits. Next let us compare each one of these output sequences with the actual received sequence. If we find that the received sequence is exactly identical to a sequence corresponding to some particular path through the coder, then we shall assume that the information input sequence is the one corresponding to that particular path. If we find no exact correspondence then we shall assume the input sequence to be the one whose path generates the *fewest bit discrepancies* when compared to the received sequence. We have, however, noted that the number of paths increases exponentially with sequence length. Hence the suggested procedure seems hardly to be feasible except for very short sequences. However, it turns out, through the application of an algorithm due to Viterbi, that many paths may be discarded thereby rendering the procedure much more useful.

To illustrate and explain the Viterbi algorithm, let us use the encoder of Fig. 13.20-3. We assume that the encoder is initially cleared i.e., the encoder is in state *a* (Fig. 13.20-5) with $M_1 M_2 = 00$. Now let there be presented at the encoder a sequence of five information bits, and let it be that the corresponding *received* v_{1R}, v_{2R} bits (not the transmitted encoder output bits) are

$$v_{1R}, v_{2R} = 10\ 00\ 10\ 00\ 00 \tag{13.20-1}$$

It is readily apparent at the outset that at least one and possibly more errors must have been introduced in transmission. For, as appears in Fig. 13.20-5 starting from state *a*, if the first information bit were a 0 the first two received bits should have been $v_{1R}, v_{2R} = 00$, while if the first information bit had been a 1 the first two received bits should have been $v_{1R}, v_{2R} = 11$. In either case $v_{1R}, v_{2R} \neq 10$.

First, without reference to the received sequence, let us trace the possible paths through the encoder states as shown in the trellis of Fig. 13.20-6. Starting at state *a*, in clock interval $k = 1$, a 0 (dashed line) will cause an output 00, as indicated and will carry the encoder again to state *a*. A 1 (solid line) will generate an output 11 and will carry the encoder to state *d*. Using the information provided in Fig. 13.20-5, the remainder of the transitions shown in Fig. 13.20-6 can be verified. Next, taking account of the received sequence, we have indicated in parenthesis in Fig. 13.20-6 the number of bit discrepancies in each clock cycle between the bits associated with the paths in the trellis diagram and the actual received bits. Thus starting at state *a* with $k = 1$, a 0 input to the encoder should

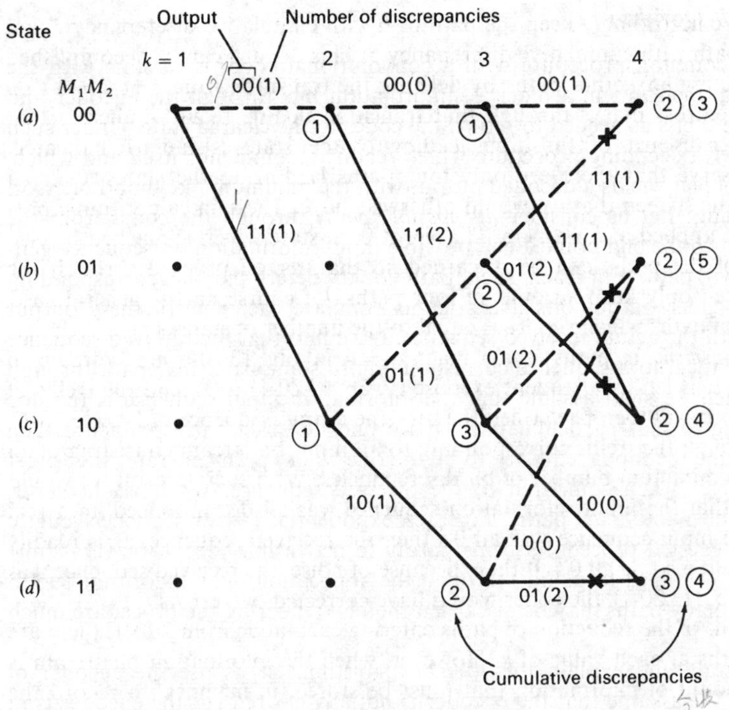

Figure 13.20-6 Showing the possible paths through the trellis of Fig. 13.20-5.

generate an output 00. Since the actual output is 10 the number of bit discrepancies is one. In the next interval the input 0 should again yield an output 00 and since the corresponding set of received bits is indeed 00, there are no discrepancies. Finally, we have also indicated in Fig. 13.20-6, in circles, the *cumulative* number of bit discrepencies encountered in reaching any state. Thus, to reach state c at $k = 3$ we go first from a at $k = 1$ to a at $k = 2$ encountering one discrepancy and then to c encountering two more discrepancies for a cumulative sum of three discrepancies. The states at $k = 4$ can be reached by *two* paths and we have noted the cumulative discrepancies for *both* paths.

The important point that needs now to be noted is the following: Suppose we consider any complete path through the entire trellis from state a at $k = 1$ back to state a at $k = L$. Suppose, for example that the path passes through state b at $k = 4$. Let us assume that the path from state b at $k = 4$ to the end of the trellis is known, but, that to go from state a at $k = 0$ to state b at $k = 4$, we are free to select either of the two available paths. Our interest lies in the *complete* path which has the smallest cumulative discrepancy. The number of discrepancies accumulated beyond state b is fixed by the fixed path selected. Hence the *complete* path of interest is the path whose cumulative discrepancy up to state b at $k = 4$ is the smaller. On this basis, one of the two paths leading to state b can be discarded. We accordingly keep the path to b with cumulative discrepancy 2 and *discard* the path with cumulative discrepancy 5. This discard can be accomplished without affecting any other paths by deleting the transition from c at $k = 3$ to b at $k = 4$. The cross marks through that transition in Fig. 13.20-6 indicates that discard. Other discards of transitions leading to other states is similarly indicated. We now observe that there are only four paths leading to the states at $k = 4$ rather than the sixteen that we would otherwise have. The same algorithmic processes can be applied at $k = 5$, etc. That is, at each state which is reached by two paths, one of the paths can be discarded so that, as we proceed through the trellis, we need only keep account of four paths. Of course, in the general case, the number of paths which *survive* is equal to the number of states.

The four *surviving* paths which reach $k = 4$ in Fig. 13.20-6 are redrawn in Fig. 13.20-7. It is left as a student exercise (Prob. 13.20-1) to extend the trellis of Figs. 13.20-7. It will then be verified that, for the output sequence of Eq. (13.20-1), the path through the trellis corresponding to an input bit stream consisting of all 0's yields the minimum number of bit discrepancies. With such a result we would then decide that the input information sequence was all 0's. If, indeed, in a real situation, the input sequence was all 0's then the received sequence, as is readily verified, should also be all 0's. If then, because of noise, the received sequence was as given in Eq. (13.20-1), the coder would have corrected two errors.

As a result of the reduction of paths entering each node from 2 to 1, there are only four paths at each value of k. However, when the information bit stream is long, the amount of information that must be stored, in memory, to record the

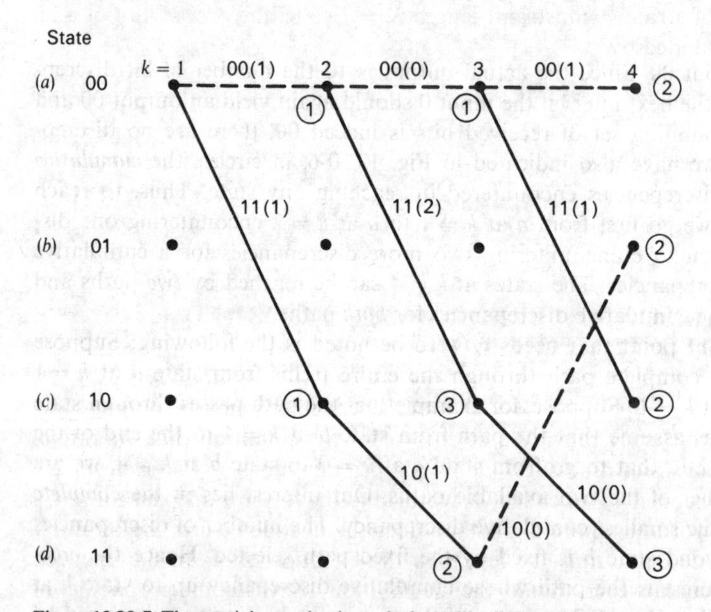

Figure 13.20-7 The surviving paths through the trellis of Fig. 13.20-6.

four possible bit streams, and the error associated with each, becomes enormous. For example, if data is sent at the rate of 10 Mb/s for one second, 10 M bits of data is transmitted. The memory must therefore store more than 40 M bits of data.

To reduce the data handling capability required of the decoder, the Viterbi algorithm presents a sub-optimum procedure which truncates the trellis after L nodes. Typically $L = 5K$ where K is the *constraint length*, i.e., the number of memory elements. Thus if $K = 7$, $L = 35$ bits. When truncating the trellis we look at the trellis after L bits are received. We then choose that path (remember there are four paths) with the fewest errors and delete the remaining three paths. In this manner the memory is reduced to somewhat more than L bits as after each decision the $L/2$ data information bits are outputted from the system.

The reason that the Viterbi truncation technique is sub-optimum is that we are choosing the minimum number of errors from among the paths terminating at nodes a, b, c, and d when $K = L$. Such a comparison is not optimum since each node yields a distinct path and is then not comparable. An optimum system would make a decision only after receiving the entire message.

13.21 PROBABILITY OF ERROR OF CONVOLUTIONAL CODES

The calculation of the probability of error of a convolutional code is beyond the scope of this text and the interested reader is referred to the references [5]. For the special case of a rate $\frac{1}{2}$, constraint length $K = 7$ code, the probability of a bit error is upper bounded by

$$P_b < 36D^{10} + 211D^{12} + 1404D^{14} + 11{,}633D^{16} + \cdots \tag{13.21-1}$$

where

$$D = 2\sqrt{p(1-p)} \tag{13.21-2a}$$

and

$$p = \frac{1}{2}\operatorname{erfc}\sqrt{\frac{E_b}{\eta}} \tag{13.21-2b}$$

13.22 COMPARISON OF ERROR RATES IN CODED AND UNCODED TRANSMISSION

In this section we compare block-coded and uncoded systems to obtain some idea of their relative error rates. We do this under the condition that the rate of information transmission (messages or words per unit time) is the same in the two cases. Then, coded or uncoded, a word must have the same duration T_w. Since the coded word has more digits than does the uncoded word, the bit duration T_c in the coded case must be less than the bit duration T_b in the uncoded case. We have seen, however, that as the bit duration decreases, the probability of error in a bit increases. We have then to inquire whether coding yields a net advantage. For the sake of being specific, we shall compare k-bit uncoded transmission with transmission using an (n, k) block code. We shall, of course, assume that the signal power P_s and the thermal-noise power spectral density η are the same in both cases.

We shall use the symbols p and p_c to represent, respectively, the probability of error in a bit in the uncoded and coded cases, reserving the symbols P_e and $P_e^{(c)}$ for the probability of error of an uncoded or coded word. The energy in a bit in the uncoded case is $E_s = P_s T_w/k$ and $E_s^{(c)} = P_s T_w/n$ in the coded case. Hence, from Eq. (11.5-8), assuming matched-filter reception,

$$p = \frac{1}{2} \operatorname{erfc} \sqrt{\frac{P_s T_w}{k\eta}} \tag{13.22-1}$$

and

$$p_c = \frac{1}{2} \operatorname{erfc} \sqrt{\frac{P_s T_w}{n\eta}} \tag{13.22-2}$$

In the uncoded case a word will be in error if one or more digits are in error. The probability that a digit is not in error is $1 - p$. The probability that all k digits are not in error is $(1 - p)^k$. The probability of at least one bit error, and hence a word error, is

$$P_e = 1 - (1 - p)^k \cong kp \tag{13.22-3}$$

since typically $kp \approx 10^{-3}$ or smaller.

Consider using a (7,4) single error correcting Hamming code. Then a word error occurs only if two or more errors occur. It may be verified (Prob. 13.22-1) that if $p_c \ll 1$ the likelihood of more than two errors is entirely negligible in comparison with the likelihood that one or two errors will occur. For $n = 7$ (7 digits in the coded word), the probability of just two errors, and hence the probability of a word error, is

$$P_e^{(c)} = \binom{7}{2} (1 - p_c)^5 p_c^2 \tag{13.22-4}$$

where $\binom{7}{2}$ is the number of combinations of seven things taken two at a time. Since $\binom{7}{2} = (7 \cdot 6)/2 = 21$ and assuming that $p_c \ll 1$, we have

$$P_e^{(c)} \cong 21 p_c^2 \tag{13.22-5}$$

From Eqs. (13.22-1) and (13.22-3) with $k = 4$ we have

$$P_e = 2 \operatorname{erfc} \sqrt{\frac{P_s T_w}{4\eta}} \tag{13.22-6}$$

while from Eqs. (13.22-2) and (13.22-5) we find

$$P_e^{(c)} = 5.25 \left(\operatorname{erfc} \sqrt{\frac{P_s T_w}{7\eta}} \right)^2 \tag{13.22-7}$$

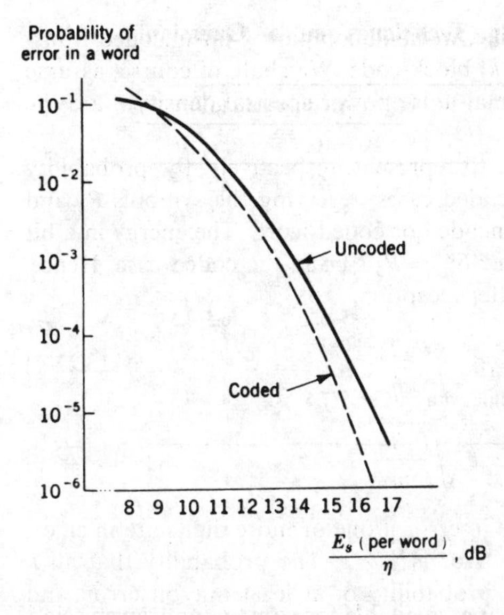

Figure 13.22-1 Probability of error for an uncoded (4 bit) and a coded (7 bit) system.

For the purpose of comparing the performance of coded and uncoded transmission, Eqs. (13.22-6) and (13.22-7) are plotted in Fig. 13.22-1. Note that when $P_s T_w/\eta = E_s/\eta$ is greater than 9 dB, coded transmission results in lower probability of error than uncoded transmission. The difference, however, is not significant until E_s/η exceeds 16 dB. When E_s/η is less than 9 dB, the coded system has a higher probability of error than does the uncoded system.

Note that coding is not a cure-all. The complicated and expensive coding and decoding equipment needed does not yield significant improvement in this example until E_s/η is large. Here we see that at a probability of error in a word of 10^{-6}, the E_s/η difference between systems is only about 1 dB. Before deciding to employ coding, we should therefore carefully analyze the system response, with and without coding, to determine if the improvement obtained is worth the complication and expense.

The reader may well inquire at this point about why one studies coding if our improvement is only 1 dB. The answer is that the improvement may be considerably greater than 1 dB depending on the coding technique employed and the length of the code.

Table 13.22-1 compares several coding techniques when using BPSK or QPSK modulation on a gaussian channel. The *coding gain* is used as the basis of comparison. The coding gain refers to the number of dB that the signal-to-noise ratio, E_b/η, can be reduced from the value required when there is no coding and still provide the same bit error rate, P_b. Then to achieve the *same* P_b, the coding gain A is

$$A = \frac{(E_b/\eta)_{\text{uncoded}}}{(E_b/\eta)_{\text{coded}}} \tag{13.22-8}$$

Table 13.22-1 Comparison of Coding Techniques on a Gaussian Channel

Coding technique	Coding gain (dB) $P_b = 10^{-5}$	Coding gain (dB) $P_b = 10^{-8}$
Block codes (hard decision)	3–4	4.5–5.5
Convolutional coding with sequential decoding (hard decision)	4–5	6–7
Convolutional coding with Viterbi decoding (hard decision)	4–5.5	5–6.5
Block codes (soft decision)	5–6	6.5–7.5
Convolution encoding with sequential decoding (soft decisions)	6–7	8–9
Concatenated codes (RS and convolutional coding with Viterbi decoding) (hard decisions)	6.5–7.5	8.5–9.5

13.23 AUTOMATIC-REPEAT-REQUEST

There are two basically different techniques available for controlling transmission errors. The first technique involving the coding schemes, which have been the subject of our interest up to the present, is referred to as *forward-error correction* (FEC). In an FEC scheme we depend on the coding to allow error *correction.* The FEC method has the limitation that if the errors are too numerous, the code will not be effective. Further, to achieve low error rates, it is necessary to add a relatively large number of redundant bits. As a result the efficiency of the code (i.e., the rate $= k/n$) is low. It also turns out that good codes invariably involve long code words whose processing requires complex and expensive hardware. Hence it is not surprising that, in spite of the very great merit and extensive application of coding, an alternative scheme is also extensively employed.

The alternative scheme is referred to as *automatic-repeat-request* (ARQ) and is used primarily when extremely low error rates are required. In this system, the receiver is not called upon to *correct* but only to *detect* errors. When an error is detected in a word, the receiver signals back to the transmitter and the word is transmitted again. In an ARQ system, then, a *feedback* channel must be provided. Since coding allows more error detection than error correction, ARQ makes more effective use of the coding.

There are three basic ARQ systems. They are: (*a*) *stop-and-wait ARQ*, (*b*) *go-back N ARQ*, and (*c*) *selective-repeat ARQ*.

The *stop-and-wait* ARQ system is the simplest to implement and is represented in Fig. 13.23-1*a*. The transmitter sends a codeword to the receiver during the time T_w. The receiver receives and processes the received word and if the receiver detects no error, it then sends back to the transmitter a positive-acknowledgement (ACK) signal. Upon receipt of the ACK signal, the transmitter sends the next word. If the receiver does detect an error, it returns to the transmitter a *negative-acknowledgement* (NAK) signal. In this case the transmitter transmits the same message and then waits again for an ACK or NAK response before undertaking further transmission. The elapsed time between the end of transmission of one word and the start of transmission of the next word is T_I. Clearly the limitation of such a system is that it must stand by idly without transmission while waiting for an ACK or NAK. Nonetheless this system is effectively used in many data systems including IBM's Binary Synchronous Communication (BISYNC) protocol (see Sec. 16.6).

The *go-back N* ARQ scheme is represented in Fig. 13.23-1*b*. The transmitter sends messages, one after another, without delay and does not wait for an ACK

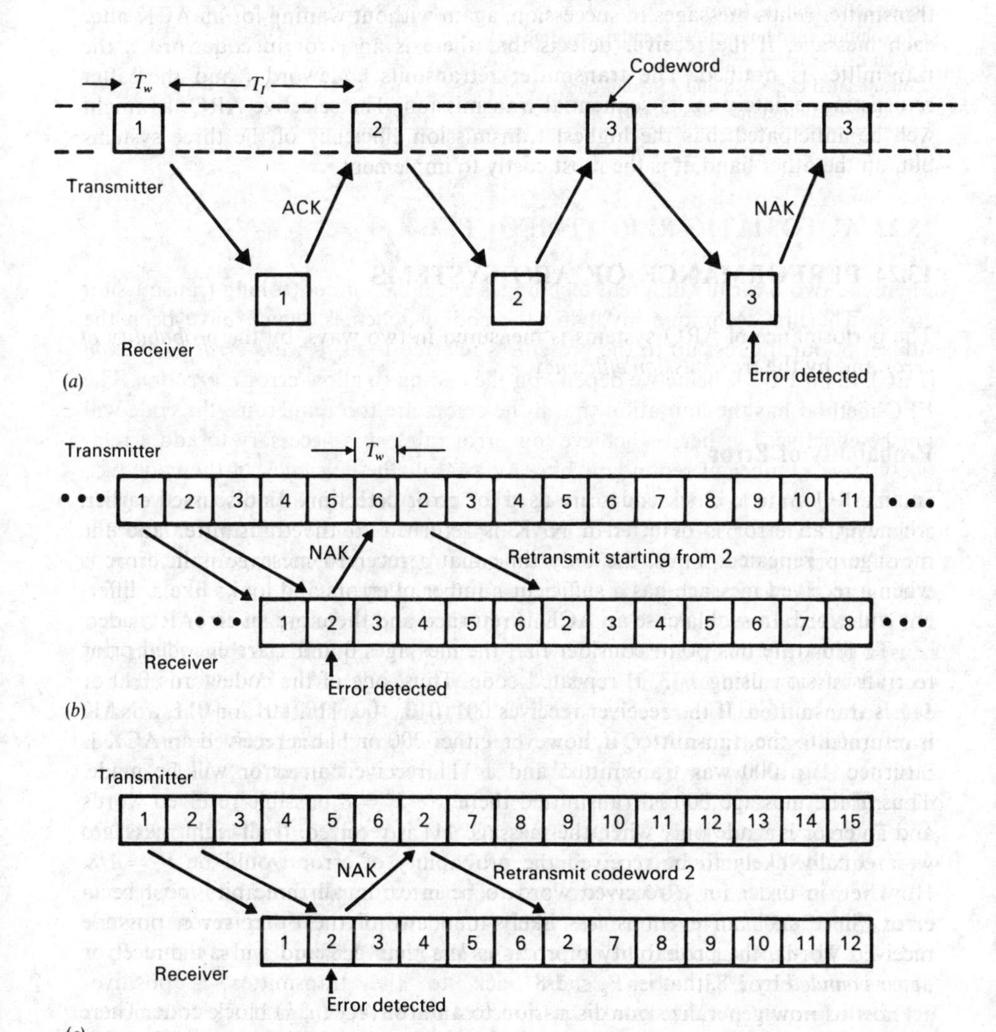

Figure 13.23-1 (*a*) Stop-and-wait ARQ. (*b*) Go-back *N* ARQ. (*c*) Selective-repeat ARQ.

signal. When, however, the receiver detects an error in a message, say message i, a NAK signal is returned to the transmitter. In response to the NAK the transmitter returns to codeword i and starts all over again at that word. In Fig. 13.23-1*b* we have assumed that the propagation delay and the processing at the receiver occupies an interval of such length that when an error is detected in word i the number N or words that are sent over again is $N = 5$. The go-back-N system is readily implemented, and, as we shall see, it is a significant improvement over the stop-and-wait system. The go-back-N scheme is used in many data systems.

The *selective-repeat* ARQ system is represented in Fig. 13.23-1*c*. Here the transmitter sends messages in succession, again without waiting for an ACK after each message. If the receiver detects that there is an error in codeword i, the transmitter is notified. The transmitter retransmits codeword i and thereafter returns immediately to its sequential transmission. The selective ARQ, as might well be anticipated, has the highest transmission efficiency of the three systems but, on the other hand, it is the most costly to implement.

13.24 PERFORMANCE OF ARQ SYSTEMS

The performance of ARQ systems is measured in two ways, by the *probability of error* and by the *transmission efficiency*.

Probability of Error

In an ARQ system, block codes are used for error detection. As discussed earlier, whenever an error is detected a NAK is returned to the transmitter and the message is repeated. Thus, the only time that a received message is in error is when a received message has a sufficient number of errors and looks like a different codeword. In such a case an ACK is returned and therefore an error is made.

To illustrate this point consider that the messages 0 and 1 are encoded prior to transmission using a (3, 1) repeated code. Thus, one of the codewords 000 or 111 is transmitted. If the receiver receives 001, 010, 100, 110, 101, or 011, a NAK is returned to the transmitter. If, however, either 000 or 111 is received an ACK is returned. If a 000 was transmitted and a 111 received an error will be made. Thus, if the message 000 is transmitted there are $2^3 = 8$ possible received words and an error is made only when the message 111 is received. If all eight messages were equally likely to be received the probability of error would be $P_e = 1/8$. However, in order for a received word to be in error, all three bits must be in error. Since such an event is *less* likely than any of the other seven possible received words, the probability of error is *less* than 1/8 and we say that P_e is *upper bounded* by 1/8, that is, $P_e \leq 1/8$.

Let us now generalize our discussion to an arbitrary (n, k) block code. There are now 2^n possible received words and of these there are 2^k codewords. Thus, if a codeword is transmitted, an error will occur if one of the other $2^k - 1$ codewords is received. To upper bound P_e we again assume that all 2^n possible received words are equally likely. Then P_e is

$$P_e \leq \frac{2^k - 1}{2^n} \simeq 2^{-(n-k)} \tag{13.24-1}$$

Example If a (1023,973) BCH code is used for error detection, the probability of error P_e is bounded by

$$P_e \leq 2^{-50} \cong 10^{-15}$$

Throughput

The *throughput efficiency* is defined as the ratio of the average number of information bits accepted at the receiver per unit of time to the number of information bits that would be accepted per unit of time if ARQ were not used. While all the ARQ systems yield the same error rate, the throughput efficiencies are different.

Throughput of the Stop-and-Wait ARQ

Let P_A be the probability that the receiver accepts the message on any particular transmission. Then the probability that only a single transmission is all that is needed for acceptance is P_A. The probability that two transmissions will be required is $(1 - P_A)P_A$ that is, the product of the probability $(1 - P_A)$ that the first transmission was rejected and P_A the probability that it was accepted on the second try. The average number of transmissions required for acceptance of a single word is the sum of the products of the number of transmissons j and the probability of requiring j transmissions, $P_A(1 - P_A)^{j-1}$. Thus (see Prob. 13.24-5)

$$\bar{N}_{sw} = 1 \cdot P_A + 2 \cdot P_A(1 - P_A) + 3P_A(1 - P_A)^2 + \cdots \tag{13.24-2a}$$

$$= \frac{1}{P_A} \tag{13.24-2b}$$

We note from Fig. 13.23-1 that the total time devoted to a single attempt to get the receiver to accept a word is $T_w + T_I$. Hence, on average, the time required to transmit one word is

$$\bar{T}_{sw} = \frac{T_w + T_I}{P_A} \tag{13.24-3}$$

If ARQ is not used and no coding bits were added to the k information bits, the time needed to transmit the k bits would be

$$T_k = \frac{k}{n} T_w \tag{13.24-4}$$

The throughput efficiency of the stop-and-wait ARQ system is

$$\eta_{\text{S\&W}} = \frac{T_k}{\bar{T}_{sw}} = \frac{k}{n} \frac{P_A}{1 + \dfrac{T_I}{T_w}} \tag{13.24-5}$$

Throughput of Go-Back-*N* ARQ

In this system, when the transmitter is informed that an error has been detected in a particular word, retransmission is required of that word and of the $N - 1$ words that followed. Hence the retransmission involves N words. Thus, if a word is received in error, N words are retransmitted. Thus, the total number of words transmitted is $N + 1$. If the same word is again in error, the N words are repeated once again, etc. Following the analysis used in the stop-and-wait ARQ we find that the average number of word transmissions required for the acceptance of a single word is

$$N_{\text{GBN}} = 1 \cdot P_A + (N + 1)P_A(1 - P_A) + (2N + 1)P_A(1 - P_A)^2 + \cdots \tag{13.24-6a}$$

$$= 1 + \frac{N(1 - P_A)}{P_A} \tag{13.24-6b}$$

Correspondingly the average time to transmit one word is

$$\bar{T}_{\text{GBN}} = T_w\left(1 + \frac{N(1 - P_A)}{P_A}\right) \tag{13.24-7}$$

and
$$\eta_{\text{GBW}} = \frac{T_k}{\bar{T}_{\text{GBN}}} = \frac{k}{n} \frac{1}{1 + \dfrac{N(1 - P_A)}{P_A}} \tag{13.24-8}$$

Selective Repeat ARQ

The mean time for transmission of a word $\bar{T}_{\text{SR}}$ in this selective-repeat case is calculated exactly as in the stop-and-wait case except that T_I is set to zero. Hence we find

$$\bar{T}_{\text{SR}} = T_w/P_A \tag{13.24-9}$$

and
$$\eta_{\text{SR}} = \frac{T_k}{\bar{T}_{\text{SR}}} = \frac{k}{n} P_A \tag{13.24-10}$$

Example

As an example, to compare throughput efficiencies, let us assume that $T_w = 10\ \mu s$, that a BCH (1023, 973) code is used, that $P_A = 0.99$ and that $T_I = 40\ \mu s$. Let us further assume that $N = 4$ so that the retransmission time in the go-back-N system is the same as the idle time in the stop-and-wait system. We then find,

$$\eta_{\text{S\&W}} = \frac{k}{n} \cdot \frac{P_A}{1 + \dfrac{T_I}{T_W}} = \frac{973}{1023} \cdot \frac{0.99}{1+4} = 0.188 \tag{13.24-11}$$

$$\eta_{\text{GBN}} = \frac{k}{n} \cdot \frac{1}{1 + \dfrac{N(1-P_A)}{P_A}} = \frac{973}{1023} \cdot \frac{1}{1 + \dfrac{4(0.01)}{0.99}} = 0.915 \tag{13.24-12}$$

and
$$\eta_{\text{SR}} = \frac{k}{n} P_A = \frac{973}{1023}(0.99) = 0.942 \tag{13.24-13}$$

Thus there is a significant improvement obtained by using the go-back-N algorithm rather than the stop-and-wait algorithm. However, the improvement made in using the selective repeat algorithm is often not deemed worth the additional complexity.

13.25 AN APPLICATION OF INFORMATION THEORY: AN OPTIMUM MODULATION SYSTEM

A generalized communication system is shown in Fig. 13.25-1. The baseband signal $m(t)$ of bandwidth f_M is encoded or modulated, or both, onto a carrier and is then transmitted over a channel of bandwidth B. Noise is added on the channel, and when the modulated/encoded signal arrives at the receiver it has a signal-to-noise ratio S_i/N_i at the receiver input. Here N_i is the noise power, at the receiver input, in the bandwidth B, that is $N_i = \eta B$. The output signal obtained after decoding or demodulation, or both, is $\hat{m}(t) = m(t) + n_o(t)$, where $n_o(t)$ is the noise that accompanies the *output* signal. The output signal-to-noise ratio is S_o/N_o, and the output waveform $m(t)$ is bandlimited to f_M.

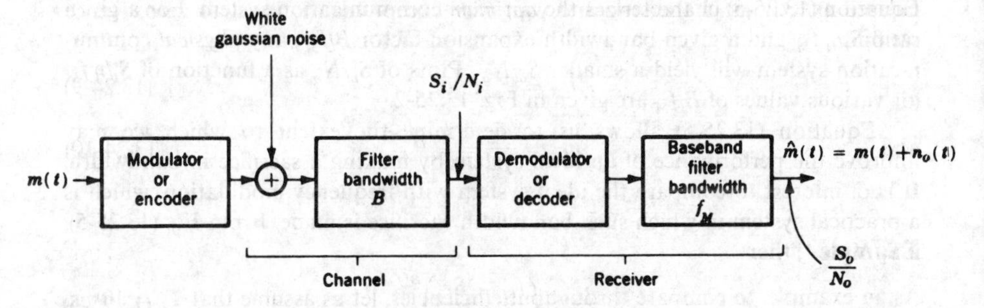

Figure 13.25-1 A generalized communication system.

From the Shannon–Hartley theorem, Eq. (13.6-1), the maximum rate at which information may be arriving at the receiver is

$$C_i = B \log_2 \left(1 + \frac{S_i}{N_i}\right) \tag{13.25-1}$$

while the maximum rate at which information may be issuing from the receiver is

$$C_o = f_M \log_2 \left(1 + \frac{S_o}{N_o}\right) \tag{13.25-2}$$

We now introduce the concept of an *optimum* or *ideal* modulation or encoding system as one in which $C_o = C_i$. Such a system is defined as one in which the rate of information output of the receiver is equal to the rate of information input to the receiver. That is, no information is lost as the information passes through the receiver, nor is any information accumulated in the receiver. We assume that this feature of the receiver persists at all rates of information flow up to the maximum as determined by the Shannon–Hartley theorem.

Setting $C_i = C_o$ and solving for S_o/N_o, we find

$$\frac{S_o}{N_o} = \left(1 + \frac{S_i}{N_i}\right)^{B/f_M} - 1 \tag{13.25-3}$$

An essential feature of the communication system represented in Fig. 13.25-1 is that while the baseband signal is bandlimited to f_M, we deliberately arranged that the encoded signal should occupy a larger bandwidth B. The *bandwidth expansion factor* B/f_M is a most important parameter of the system, and it is useful to rewrite Eq. (13.25-3) in a manner to emphasize this point. The input noise $N_i = \eta B$, and if we call the input-signal power S_i, we may write

$$\frac{S_i}{N_i} = \frac{S_i}{\eta B} = \frac{f_M}{B} \frac{S_i}{\eta f_M} \tag{13.25-4}$$

and from Eq. (13.25-3)

$$\frac{S_o}{N_o} = \left(1 + \frac{f_M}{B} \frac{S_i}{\eta f_M}\right)^{B/f_M} - 1 \tag{13.25-5}$$

Equation (13.25-5) characterizes the *optimum* communication system. For a given ratio $S_i/\eta f_M$ and a given bandwidth expansion factor B/f_M, any *physical* communication system will yield a smaller S_o/N_o. Plots of S_o/N_o as a function of $S_i/\eta f_M$ for various values of B/f_M are given in Fig. 13.25-2.

Equation (13.25-5) allows us to determine the extent to which we may improve the performance of an ideal system by making a sacrifice in bandwidth. It is of interest to compare the ideal system with frequency modulation, which is a practical system in which such bandwidth sacrifice is made. From Eq. (13.25-5), if $S_i/N_i \gg 1$, then

$$\frac{S_o}{N_o} \approx \left(\frac{f_M}{B} \cdot \frac{S_i}{\eta f_M}\right)^{B/f_M} \tag{13.25-6}$$

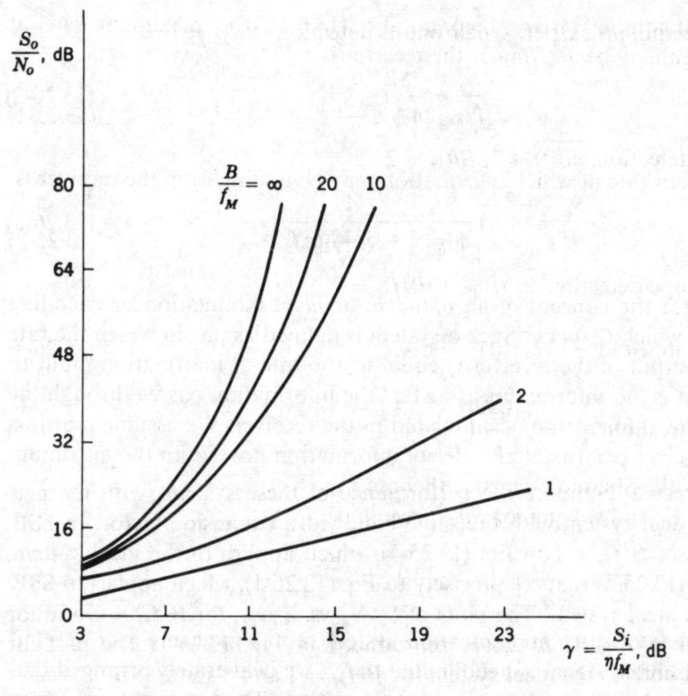

Figure 13.25-2 Signal-to-noise ratio characteristic of an optimum communication system.

On the other hand, it may be verified (Prob. 13.25-2) that in a wideband FM system

$$\frac{S_o}{N_o} = \frac{3}{4}\left(\frac{B}{f_M}\right)^2 \frac{S_i}{\eta f_M} \tag{13.25-7}$$

Thus, while the performance of the ideal system increases exponentially with bandwidth expansion, the performance of an FM system increases only with the square of the bandwidth expansion factor. Thus, for $B/f_M > 1$ the performance of the ideal system increases much more rapidly with increase in $S_i/\eta f_M$ than does the FM system.

13.26 A COMPARISON OF AMPLITUDE-MODULATION SYSTEMS WITH THE OPTIMUM SYSTEM

In Chap. 8 we arrived at the following results for the amplitude-modulation systems studied there.

Single sideband, $B/f_M = 1$

$$\frac{S_o}{N_o} = \frac{S_i}{\eta f_M} \tag{13.26-1}$$

Double sideband, no carrier, synchronous detection, $B/f_M = 2$

$$\frac{S_o}{N_o} = \frac{S_i}{\eta f_M} \tag{13.26-2}$$

Square-law detection, $\overline{m^2(t)} \ll 1$, $B/f_M = 2$

$$\frac{S_o}{N_o} = \overline{m^2(t)} \frac{S_i}{\eta f_M} \frac{1}{1 + 3/4S_i/\eta f_M} \tag{13.26-3}$$

Linear envelope detection, $\overline{m^2(t)} \ll 1$, $B/f_M = 2$

$$\text{Above threshold } \frac{S_o}{N_o} = \overline{m^2(t)} \frac{S_i}{\eta f_M} \tag{13.26-4}$$

$$\text{Below threshold } \frac{S_o}{N_o} = \frac{\overline{m^2(t)}}{1.1} \left(\frac{S_i}{\eta f_M}\right)^2 \tag{13.26-5}$$

It is of interest to compare the performance of these systems with the performance of an ideal system with the same bandwidth expansion factor. In SSB, $B/f_M = 1$. If we set $B/f_M = 1$ in Eq. (13.25-5), which applies to the ideal system, we find that Eq. (13.25-5) reduces precisely to Eq. (13.26-1), which applies to SSB. Hence SSB is an *ideal* system. The plots of S_o/N_o vs. S_i/N_i, for $B/f_M = 1$ and for $B/f_M = 2$, shown in Fig. 13.25-2, are reproduced in Fig. 13.26-1. The plot for $B/f_M = 1$ applies both to the ideal system for $B/f_M = 1$ and also to SSB.

Equation (13.26-2) for the DSB system is identical with Eq. (13.26-1) for SSB. Hence again the plot in Fig. 13.26-1 for $B/f_M = 1$ applies to DSB. However, in DSB, $B/f_M = 2$. Hence if DSB were an optimum system, the plot corresponding to $B/f_M = 2$ would apply. DSB is therefore not an optimum system. By way of

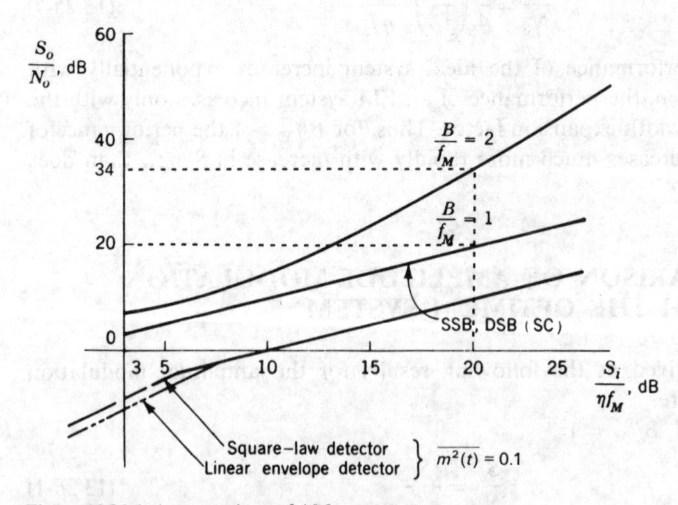

Figure 13.26-1 A comparison of AM systems.

example, if $S_i/\eta f_M$ is chosen to be 20 dB, then the output S_o/N_o is also 20 dB. If, however, DSB were optimum, the output signal-to-noise ratio would would be larger by 14 dB.

The performance curves for the asynchronous square-law and linear envelope detector are also given in Fig. 13.26-1 for $\overline{m^2(t)} = 0.1$ ($= -10$ dB). The performance of these systems is seen to be poor in comparison with the optimum system for $B/f_M = 2$.

13.27 A COMPARISON OF FM SYSTEMS

The output SNR of each of the FM systems considered in Chaps. 9 and 10 are the same above threshold. The result, for sinusoidal modulation, is [see Eq. (9.2-21)]

$$\frac{S_o}{N_o} = \frac{3}{2}\beta^2 \frac{S_i}{\eta f_M} \tag{13.27-1}$$

where $\beta \equiv \Delta f/f_M$, and Δf is the frequency deviation. To relate β and the bandwidth expansion factor, we note that $B = 2(\beta + 1) f_M$. Thus

$$\frac{B}{f_M} = 2(\beta + 1) \tag{13.27-2}$$

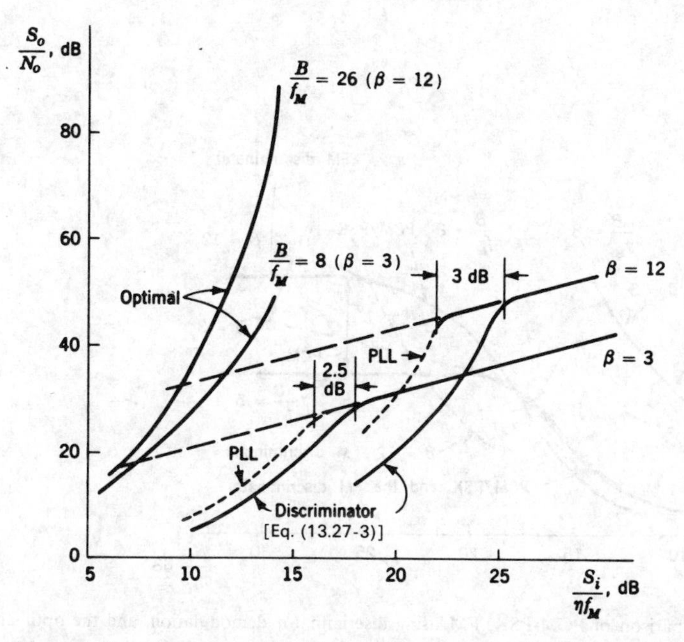

Figure 13.27-1 Comparison of FM demodulators including the effect of sinusoidal modulation.

The expression for the output signal-to-noise ratio of the discriminator, including the effect of sinusoidal modulation and valid both above and below threshold, is found by combining Eqs. (10.6-6) and (13.27-2). We find

$$\frac{S_o}{N_o} = \frac{(3/2)\beta^2(S_i/\eta f_M)}{1 + (12\beta/\pi)(S_i/\eta f_M)\exp[-(f_M/B)(S_i/\eta f_M)]} \tag{13.27-3}$$

This equation is plotted in Fig. 13.27-1 for $\beta = 3$ and 12. The output SNR characteristics of the optimal demodulator and the second-order phase-locked loop are sketched on the same set of axes. The threshold extension presented for the PLL was obtained experimentally and theoretically verified.[6] We note that the PLL results in a 3-dB threshold extension when $\beta = 12$, and a 2-dB threshold extension when $\beta = 3$. We note that the performance of FM systems falls substantially short of the performance of ideal systems.

13.28 COMPARISON OF PCM AND FM COMMUNICATION SYSTEMS

The transmission of analog signals by modulating a binary PCM signal onto a PSK or FSK carrier was studied in Sec. 12.5. In Fig. 12.5-1 we saw that the threshold of FSK was 2.2 dB greater than the threshold using PSK. In Fig. 13.28-1 we compare PCM-PSK, the FM discriminator, and the optimal demodulator. The PCM characteristics are obtained from Eq. (12.5-1).

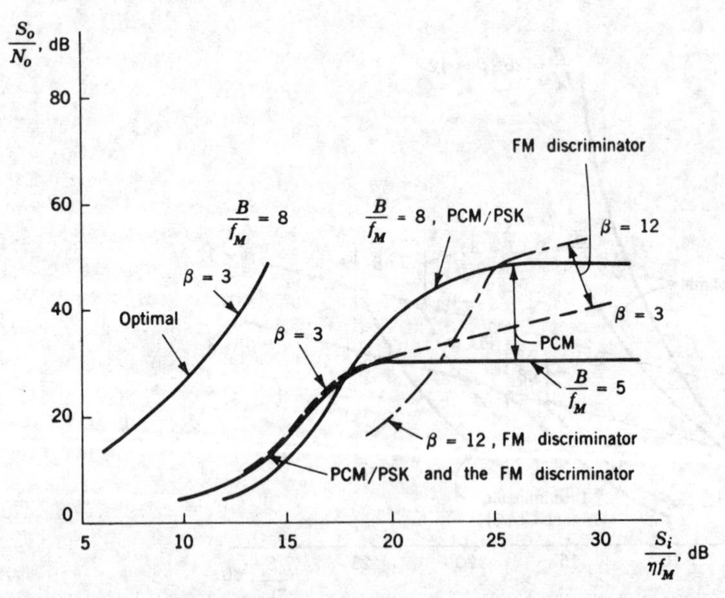

Figure 13.28-1 Comparison of PCM-PSK, FM using discriminator demodulation, and the optimal system.

Consider the two FM discriminator characteristics first. When $\beta = 12$, the threshold occurs at 25 dB, and when $\beta = 3$, threshold occurs at 18 dB. To compare these results with PCM, we consider both threshold and the bandwidths necessary for transmission, i.e., the bandwidth expansion factor. We note that a bandwidth expansion factor $B/f_M = 8$ in PCM (which corresponds to $\beta = 3$) results in an output SNR which is approximately equal to that of the discriminator operating with $\beta = 12$, $B/f_M = 2(12 + 1) = 26$. Thus, to obtain an output SNR of 48 dB requires a $B/f_M = 26$ when using a discriminator, and a $B/f_M = 8$ when employing PCM-PSK. Hence 3.25 times more bandwidth is required of the discriminator.

If an output SNR of 28 dB is all that is required, the FM discriminator with $\beta = 3$, $B/f_M = 8$, can be employed. However, the results can be obtained using PCM-PSK with $B/f_M = 5$. In this case the FM discriminator requires only 1.6 times more bandwidth. Thus, we see that the improvement of PCM over the discriminator increases with increased bandwidth expansion.

Comparing the PCM system with the optimal system for $B/f_M = 8$ indicates that the PCM system operating at threshold requires a signal-to-noise ratio which is 11 dB greater than required by the optimal demodulator threshold.

13.29 FEEDBACK COMMUNICATION

Many communication systems such as the ARQ system studied in Sec. 13.24 provide a two-way communication link. That is, not only is station A able to transmit and station B to receive, but station A is able also to receive and station B to transmit. In such a two-way link it is possible to incorporate *feedback* and thereby improve the performance of the communication system. We shall consider a type of feedback communication system, referred to as an *information-feedback* system, which is especially effective when transmission in one direction is much more reliable than transmission in the other direction. Such a situation might well result if one station were a fixed ground-based station while the other station was located in an airplane or a satellite. The ground-based station, having no essential limitations on weight, might be able to transmit hundreds of kilowatts of power while the station in space might be limited to tens of watts. We discuss now how feedback may be used to reduce the average energy required to transmit a bit of information from the low-powered station to the high-powered station, while maintaining a low probability of error.

System Description

An information-feedback communication system is shown in Fig. 13.29-1. The two transceivers (transmitter-receiver combinations) are coupled through two communication channels. Over one channel, transceiver 1 (with low-power transmitter) transmits information to transceiver 2. We have included a noise source in this channel. Over the second channel (the feedback channel) trans-

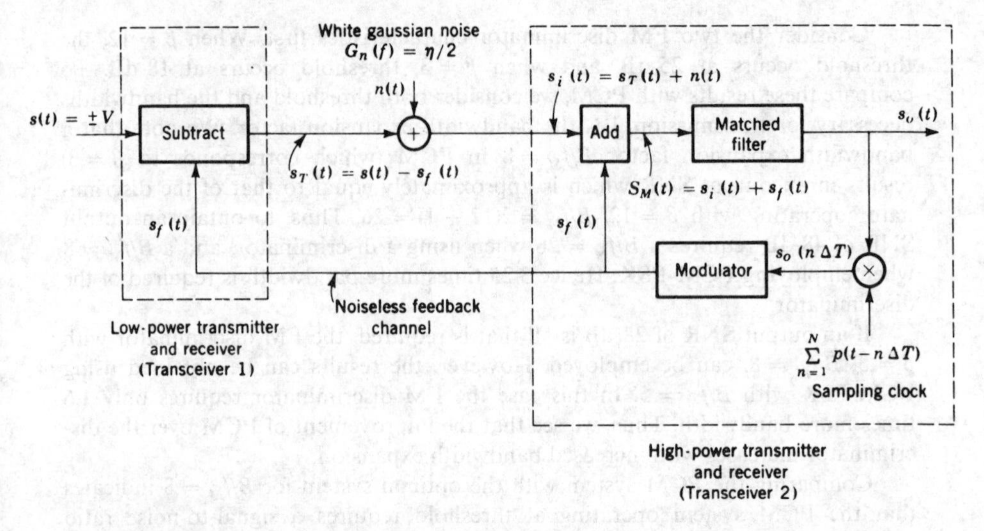

Figure 13.29-1 An information-feedback communication system.

ceiver 2 transmits back to transceiver 1. We have assumed that the transmitter in transceiver 2 has so much transmitter power available that we may ignore the effect of noise on the feedback channel. Thus we allow the possibility that receiver 2 may make an error in determining the transmission of transmitter 1. However, we assume that receiver 1 will receive the transmission of transmitter 2 with perfect reliability, that is, with negligible probability of error.

We assume transmission by binary pulse-code modulation. The baseband signal $s(t)$ is a sequence of voltage levels $+V$ or $-V$ held for intervals T and representing binary 1's and 0's. For simplicity we assume transmission at baseband, although in a physical system the baseband signal may be used to modulate a carrier, as in PSK or FSK. We now describe the operation of the system.

Since transmission is at baseband, the matched filter in the receiver is the *integrate and dump* filter described in Sec. 11.1. At the beginning of the interval T allocated to a bit, the output of the matched filter is zero. The interval T is divided into N subintervals, each of duration $\Delta T = T/N$, by the sampling clock pulses which occur at intervals of ΔT. The modulation provides a feedback voltage $s_f(t)$ which, like the baseband signal $s(t)$, is $s_f(t) = +V$ or $s_f(t) = -V$. The feedback signal is $s_f(t) = +V$ when the modulator input $s_o(n\,\Delta T)$ is positive and $s_f(t) = -V$ when $s_o(n\,\Delta T)$ is *negative*.

Now suppose that, during some bit interval T, the signal $s(t) = +V$ is transmitted. At the beginning of this interval, i.e., at $t = 0$, $s_o(t) = 0$. This output is sampled and the sample $s_o(t = 0) = 0$. The modulator output $s_f(t) = 0$. Thus, during the first subinterval, from $t = 0$ to $t = \Delta T$, there is no feedback signal. During this subinterval the input to the matched filter is (refer to Fig. 13.29-1)

$$s_M(t) = s_i(t) = s_T(t) + n(t) = s(t) + n(t) \tag{13.29-1}$$

At the end of his subinterval, the output of the matched filter is again sampled. Let us suppose that it turns out that $s_o(\Delta T)$ is positive. That is, let us suppose that at $t = \Delta T$ the matched filter *correctly* decides that the transmission is $s(t) = +V$. Then $s_f(t) = +V$. The transmitted signal in the second subinterval, from $t = \Delta T$ to $t = 2\Delta T$, would then be

$$s_T(t) = s(t) - s_f(t) = V - V = 0 \tag{13.29-2}$$

Accordingly, we see that during this second subinterval the transmitter will be *turned off*. On the other hand, the input to the matched filter will be

$$s_M(t) = s_i(t) + s_f(t) = s_T(t) + n(t) + s_f(t) \tag{13.29-3a}$$

$$= s(t) - s_f(t) + n(t) + s_f(t) \tag{13.29-3b}$$

$$= s(t) + n(t) \tag{13.29-3c}$$

precisely as in Eq. (13.29-1). Thus the matched filter will not "know" that the transmitter has been turned off. The signal plus noise present at the input to the filter will always be independent of the feedback signal. This result is apparent from Eq. (13.29-3) as well as from Fig. 13.29-1. For we see that the feedback signal is both added and subtracted from the filter input. Thus the matched filter will continue to integrate the signal plus noise $s(t) + n(t)$ in a manner which is entirely independent of the feedback signal.

Next, let us suppose that it turns out that $s_o(\Delta T)$ is negative. That is, let us suppose that at $t = \Delta T$ the matched filter *incorrectly* decides that the transmission is $s(t) = -V$. In this case we find $s_f(t) = -V$, and the transmitted signal during the subinterval $t = \Delta T$ to $t = 2\Delta T$ will be

$$s_T(t) = s(t) - s_f(t) = V - (-V) = 2V \tag{13.29-4}$$

In this case the transmitted signal voltage is doubled, and the transmitted signal power is increased by a factor of four. But again, as in the previous case, the input to the matched filter will remain as $s_M(t) = s(t) + n(t)$, independent of the feedback.

The process we have described over the first two subintervals continues over the remaining intervals up to the time T. Interval by interval the transmitter is turned off or turned on to a high power, depending on the preliminary and tentative estimates made by the matched filter at intervals of ΔT. However, a final determination of the transmitted bit is made only by examining the matched filter at the end of the bit interval, at $t = T$. As noted, the input to the matched filter is independent of the feedback. Hence the probability of an error in the determination of a bit is *exactly the same* as if no feedback were employed. This error probability is given by [see Eq. (11.2-3)]

$$P_e = \frac{1}{2} \operatorname{erfc} \sqrt{\frac{V^2 T}{\eta}} \tag{13.29-5}$$

The merit of the information feedback system, then, lies not in that it reduces the probability of error, but rather that, for a fixed probability of error, the feedback system may use less average energy per bit than a system without feedback. For if the preliminary estimates of the matched filter, made at intervals ΔT, are rather more frequently right than wrong, the transmitter may be turned off often enough to produce a saving in energy.

Calculation of Average Transmitted Signal Energy Per Bit

We now calculate the average signal energy per bit, that is, the energy transmitted in the interval T. We assume arbitrarily that $s(t) = +V$. Since $+V$ and $-V$ are transmitted with equal probability, this assumption does not affect the generality of the result.

During the first interval ΔT, the signal $+V$ is transmitted and the corresponding normalized energy is

$$E_1 = V^2 \, \Delta T \tag{13.29-6}$$

During the second interval ΔT, $s_T(t) = 0$ if the decision made by the matched filter at $t = \Delta T$ was correct. In this case no energy is transmitted. If however, the decision was incorrect, $s_T(t) = 2V$ and the energy $(2V)^2 \, \Delta T$ is transmitted. Since the probability of an incorrect decision after processing the signal and noise for time ΔT is q_1, where

$$q_1 = \frac{1}{2} \operatorname{erfc} \sqrt{\frac{V^2 \, \Delta T}{\eta}} \tag{13.29-7}$$

the average energy transmitted during the second ΔT second interval is

$$E_2 = (2V)^2 \, \Delta T q_1 \tag{13.29-8}$$

Similarly, the average energy transmitted during the nth ΔT second interval is

$$E_n = (2V)^2 \, \Delta T q_{n-1} \tag{13.29-9}$$

where

$$q_{n-1} = \frac{1}{2} \operatorname{erfc} \sqrt{\frac{V^2(n-1) \, \Delta T}{\eta}} \tag{13.29-10}$$

The average energy transmitted in the interval T is E_T, given by

$$E_T = E_1 + E_2 + \cdots + E_N \tag{13.29-11a}$$

$$= V^2 \, \Delta T + (2V)^2 \, \Delta T q_1 + \cdots + (2V)^2 \, \Delta T q_{N-1} \tag{13.29-11b}$$

$$= V^2 \, \Delta T + 2V^2 \, \Delta T \sum_{n=2}^{N} \operatorname{erfc} \sqrt{\frac{V^2(n-1) \, \Delta T}{\eta}} \tag{13.29-11c}$$

For any given N, the total energy per bit E_T may be calculated from Eq. (13.29-11c). We find, for example (Prob. 13.29-1), that for $N = 2$, and if $V^2T/2\eta \gg 1$, that the feedback provides a 3-dB improvement. That is, the energy E_T is only one-half the energy that would be required, without feedback, for the same error probability.

It is of interest to explore the improvement possible in the limiting case as N becomes very large. For this purpose we introduce the variable

$$x_n \equiv \frac{V^2(n-1)\,\Delta T}{\eta} \tag{13.29-12}$$

so that

$$\Delta x \equiv x_{n+1} - x_n = \frac{V^2\,\Delta T}{\eta} \tag{13.29-13}$$

We note also that

$$\operatorname{erfc}\sqrt{x_n} \equiv \frac{2}{\sqrt{\pi}}\int_{\sqrt{x_n}}^{\infty} e^{-u^2}\,du \tag{13.29-14}$$

Substituting Eqs. (13.29-13) and (13.29-14) into Eq. (13.29-11*c*), we find

$$E_T = \eta\left(\Delta x + \sum_{n=2}^{N}\frac{4}{\sqrt{\pi}}\,\Delta x\int_{\sqrt{x_n}}^{\infty} e^{-u^2}\,du\right) \tag{13.29-15}$$

In the limit as $N \to \infty$, Δx and ΔT may be replaced by the differentials dx and dt and the summation in Eq. (13.29-15) replaced by an integral. We then find (Prob. 13.29-2) that

$$E_T = \frac{4\eta}{\sqrt{\pi}}\int_0^{V^2T/\eta} dx\int_{\sqrt{x}}^{\infty} e^{-u^2}\,du \tag{13.29-16}$$

This integral may be evaluated (Prob. 13.29-3) with the result

$$E_T = \eta\left(\frac{-2}{\sqrt{\pi}}\sqrt{\frac{V^2T}{\eta}}\,e^{-V^2T/\eta} + \operatorname{erf}\sqrt{\frac{V^2T}{\eta}} + \frac{2V^2T}{\eta}\operatorname{erfc}\sqrt{\frac{V^2T}{\eta}}\right) \tag{13.29-17}$$

For $V^2T/\eta \gg 1$ (or, more specifically, for $V^2T/\eta > 3$) the first and third terms in Eq. (13.29-17) may be neglected in comparison with the second term. Furthermore, $\operatorname{erf}\sqrt{V^2T/\eta} \simeq 1$. Hence we then find

$$E_T \simeq \eta \tag{13.29-18}$$

From Eqs. (13.29-5) and (13.29-18) we may now deduce the following interesting result. The error probability may be made arbitrarily small by increasing V. However, provided $V^2T/\eta \gg 1$, the required average energy per bit is constant at $E_T = \eta$ and does not depend on the error probability P_e. Thus the characteristic of the transmitter which determines the extent to which the error probability may be reduced is not the power it can transmit but rather its *peak power*, i.e., the peak value V^2 which the transmitter can attain.

Comparison of Information Rate with Channel Capacity

It is of interest to compare the performance of an information-feedback communication system with the performance of an ideal system. The system of Fig. 13.29-1 is a binary PCM system. For such a system, the rate of information transmission R as given by Eq. (13.4-1) with $M = 2$ is

$$R = \frac{1}{T} \log_2 M = \frac{1}{T} \tag{13.29-19}$$

the channel capacity for a gaussian channel is [Eq. (13.8-2)]

$$C = 1.44 \frac{P_s}{\eta} \tag{13.29-20}$$

in which P_s is the input power at the receiver. From Eqs. (13.23-19) and (13.29-20), setting $R = C$, we find that in a system with ideal coding an arbitrarily small error probability is attainable with a power P_s that satisfies the condition

$$P_s T = \frac{1}{1.44} \eta = 0.69\eta \tag{13.29-21}$$

The quantity $P_s T$ is the energy associated with a bit of duration T. We have found that the information-feedback system is capable of attaining arbitrarily small error with an energy per bit given by $E_T = \eta$. Hence we find that the feedback system requires an energy per bit which is only 1.44 times greater ($\simeq$ 1.5 dB) than required by an optimal system.

13.30 TRELLIS-DECODED MODULATION[7,8]

In this section we shall describe a communication scheme which involves trellis decoding (Sec. 13.20) together with a judicious selection of the signal waveforms used to represent messages. We shall consider a simple example of the method which, when applied in a more sophisticated manner, can yield the same noise immunity as would result from increasing the signal power of uncoded data by more than 6 dB without increasing the channel bandwidth.

To illustrate the communication technique, we employ, as an example, the transmission system shown in Fig. 13.30-1*a*. The data stream $d(k)$, as well as the data stream *delayed* by one bit time, $d_D(k) = d(k-1)$ are applied to the signal encoder shown in Fig. 13.30-1*b*. It consists of a code converter and a pair of multipliers (mixers). The code converter is characterized by the truth table given in Fig. 13.30-1*c*. We shall subsequently discuss the reason for using this particular truth table. As usual, corresponding to the logic levels 1 and 0, we take the voltage levels of $\alpha(k)$ and $\beta(k)$ to be $+1$ and -1 volt. The carrier inputs to the signal encoder are $\sqrt{P_s} \cos \omega_0 t$ and $\sqrt{P_s} \sin \omega_0 t$, P_s being the carrier power and ω_0 being the carrier angular frequency. The signals which can be generated by the encoder are shown in the signal space diagram of Fig. 13.30-1*d*. The values of $d(k)$ and $d_D(k)$ corresponding to each signal are also specified. A signal persists for one bit interval and has an energy $E_s = P_s T_b$, T_b being the duration of the bit interval.

In Fig. 13.30-2 are shown a typical bit stream $d(k)$, the corresponding delayed stream $d_D(k)$, and the generated signals $s_0, s_1, s_2, s_3, s_1, \ldots$. As can be observed

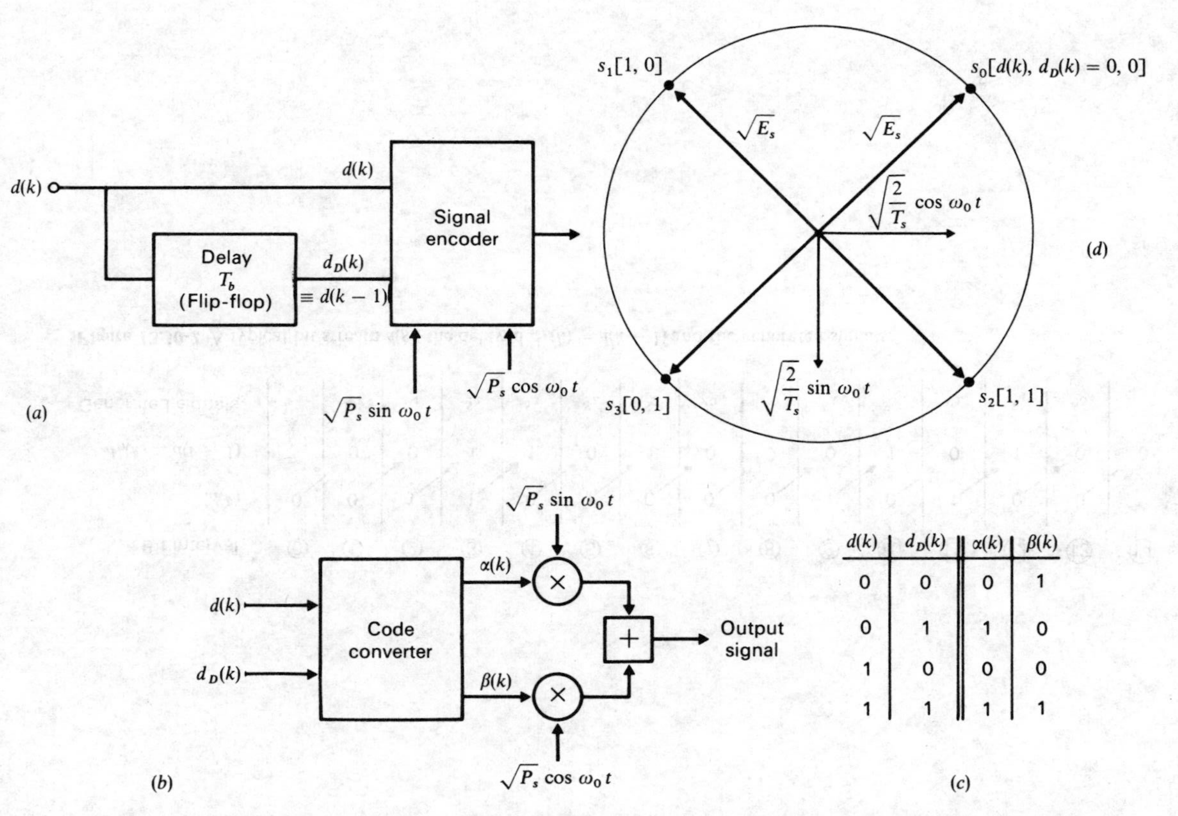

$d(k)$	$d_D(k)$	$\alpha(k)$	$\beta(k)$
0	0	0	1
0	1	1	0
1	0	0	0
1	1	1	1

Figure 13.30-1 (*a*) Transmitter of Trellis-demodulation communication system. (*b*) The signal encoder. (*c*) Truth table of code converter of (*b*). (*d*) Signal space representation of generated signals.

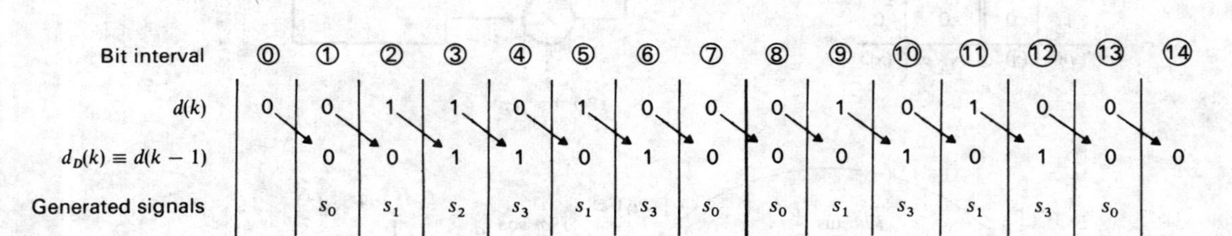

Figure 13.30-2 A typical bit stream $d(k)$, the delayed $d_D(k) = d(k-1)$ and the generated signals.

from Figs. 13.30-2 and 13.30-1*d*, s_0 corresponds to $d(k)$, $d_D(k) = 00$, s_1 to 10, etc. Note that each signal s_i corresponds to a pair of bits and not just to the data bit itself.

The correlator portion of the receiver is shown in Fig. 13.30-3. It consists of four correlators which determine the correlation of the received signal with each of the signals s_0 through s_3. As usual we assume that, in transmission, white noise $n(t)$ of power spectral density $G(f) = \eta/2$ has been added to the signal. In the absence of noise, at each sampling time we would find that one correlator yielded an output voltage corresponding to complete correlation while the other three generated an output of zero volts corresponding to zero correlation. With noise, we would still anticipate that generally the output of the correlator corresponding to the transmitted signal would yield an output higher than the output of each of the other three correlators. In any event, at each sampling time we observe and record in a memory all the correlator *discrepancies*. We use the term correlator "discrepancy" to refer to the amount by which the correlator output voltage is different from what it would be if the correlation were perfect. (The hardware, memory, etc., needed to take note of and to store these discrepancies are not indicated on the figure.)

The transmitter shares with convolutional encoders the feature of a storage element (i.e., the delay element, generally a flip-flop). The output signal depends

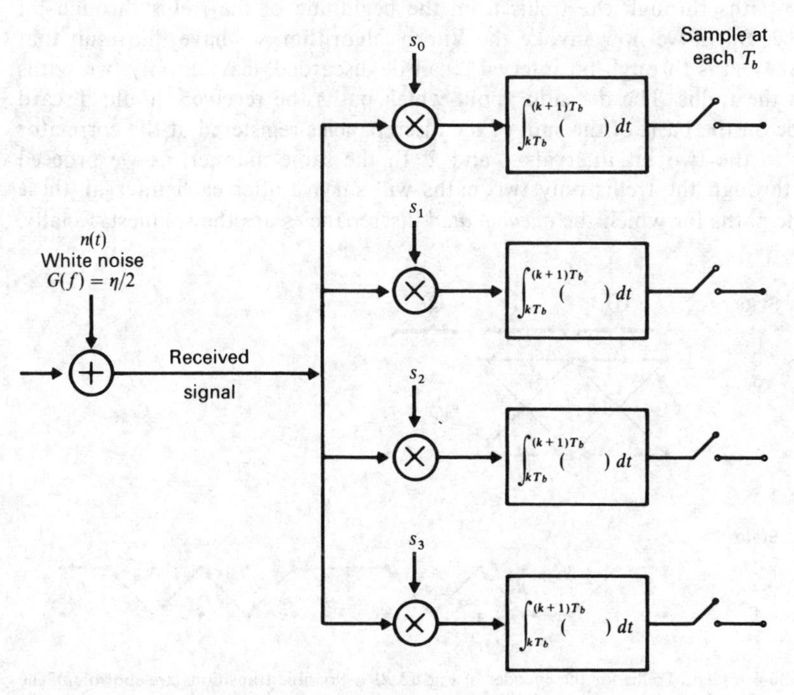

Figure 13.30-3 The correlator portion of the Trellis demodulation receiver.

on both the bits at the input and at the output of the storage element. Accordingly, as in the case of convolutional encoder–decoder schemes, the sequence of events from one bit interval to the next bit interval can be represented on a trellis diagram. The trellis diagram, shown in Fig. 13.30-4*a* has two states, states 0 and 1, corresponding to the state of the flip-flop, shown in Fig. 13.30-1*a*.

Starting at state 0 at the beginning of interval 1, suppose that the data bit $d(1)$ is $d(1) = 0$. Then, as appears in Fig. 13.30-2, the signal generated during interval 1 will be s_0 and at the end of the interval the system will again be in state 0. The starting points and transitions corresponding to the other signals s_1, s_2, and s_3 are also shown in Fig. 13.30-1*a*. Observe that not every sequence of signals is possible. For, when the system is in state 0, the next signal can be only s_0 or s_1; and when the system is in state 1, the next signal can be only s_2 or s_3. The path through the trellis, corresponding to the signal sequence shown in Fig. 13.30-2, is shown in Fig. 13.30-4*b*.

Now let us consider the situation represented in Fig. 13.30-5. Here, before transmitting a message, we start at the transmitter with a bit sequence of all 0's so that the message signal sequence is all s_0's. The first *message* signal will be either s_0 or s_1 so that in the receiver, at the end of message bit interval 1, we need only inquire about the outputs of the s_0 and the s_1 correlators. At the end of interval 1 the receiver may find the system to be in state 0 or state 1 so that in the second interval the message signal may be s_0, s_1, s_2, or s_3. It appears that there are four paths through the trellis from the beginning of the trellis through bit interval 2. But if we now invoke the Viterbi algorithm, we have the result that two of the paths through bit interval 2 can be discarded, leaving only two paths through the trellis. The decision about which paths the receiver should discard are made on the basis of the sum of the discrepancies registered at the correlator outputs in the two bit intervals 1 and 2. In the same manner, as we proceed further through the trellis only two paths will survive after each interval, these being the paths for which the *accumulated* discrepancies are the smallest. Finally,

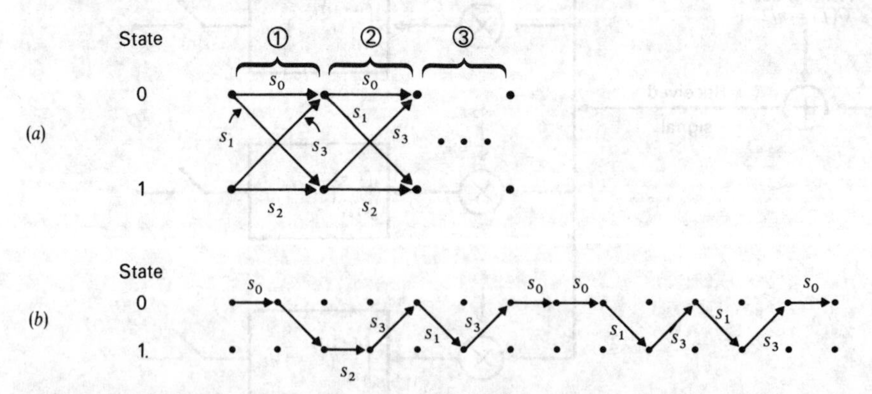

Figure 13.30-4 (*a*) The Trellis for the encoder of Fig. 13.30-1. Possible transitions are shown. (*b*) The path through the Trellis corresponding to the signal sequence of Fig. 13.30-2.

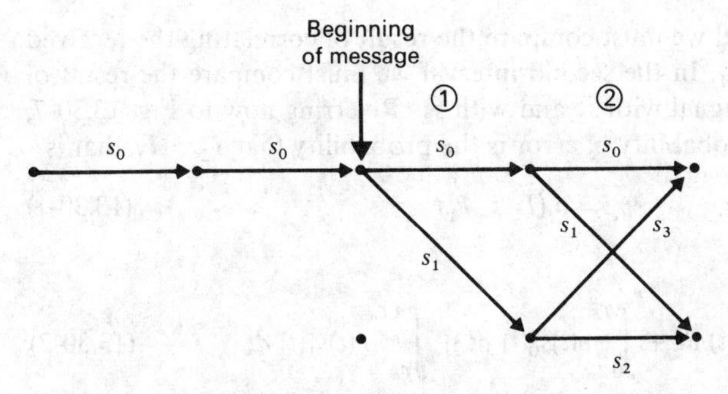

Figure 13.30-5 Possible paths through the Trellis for a message starting at the point indicated after a sequence of all zeros.

at message end, the system will return permanently to state 0 and we shall then select as the valid path, the path with minimum discrepancy.

The path we finally select as valid, is of course, not guaranteed to be completely correct. Rather it is, based on our observations, the most likely set of signals that could be transmitted. We shall now consider the types of errors that are possible, their probabilities of occurrence and their effect on the average probability that a transmitted bit is received in error.

Figure 13.30-6 shows a possible received signal sequence, The solid line indicates the transmitted sequence while the dashed lines indicate possible receiver determinations. In Fig. 13.30-6*a*, because of noise, the receiver has incorrectly decided that the path $s_1 - s_2$ was transmitted rather than $s_0 - s_1$. In Fig. 13.30-6*b* the receiver has incorrectly decided that path $s_1 - s_3 - s_1$ was transmitted. Note that an incorrect path involving only one bit interval is not possible. We would expect that errors in a "good" communication system would occur rather infrequently. Hence path errors involving many successive signals will be progressively less probable as the number of successive signals increases. Hence the most likely error is as shown in Fig. 13.30-6*a*. In this case, to calculate the error probability we consider that in two successive intervals the transmitted signals are s_0

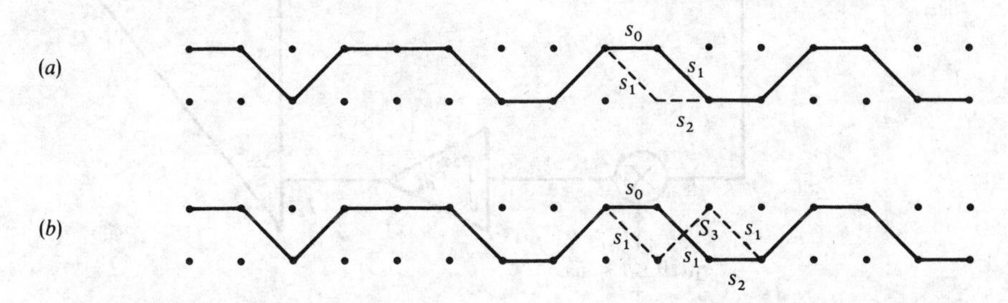

Figure 13.30-6 The solid line is the transmitted sequence. The dashed line departure represents errors in the receiver determination. (*a*) Errors in two successive signal determinations. (*b*) Three successive errors.

and s_1. In the first interval we must compare the result of correlating the received signal with s_0 and with s_1. In the second interval we must compare the result of correlating the received signal with s_1 and with s_2. Referring now to Fig. 13.30-7, it is to be seen that the probability of error is the probability that $I_2 > I_1$, that is

$$P_e = p\,(I_2 > I_1) \tag{13.30-1}$$

We first evaluate I_1:

$$I_1 = \int_0^{T_b} s_0^2(t)\,dt + \int_{T_b}^{2T_b} s_1^2(t)\,dt + \int_0^{T_b} n(t)s_0(t)\,dt + \int_{T_b}^{2T_b} n(t)s_1(t)\,dt \tag{13.30-2}$$

Since each signal s_0, s_1, s_2, and s_3 have equal power P_s,

$$I_1 = 2E_b + N_0^{(1)} + N_1^{(2)} \tag{13.30-3}$$

where
$$N_0^{(1)} = \int_0^{T_b} n(t)s_0(t)\,dt \tag{13.30-4}$$

and where the notation $N_0^{(1)}$ refers to the correlation between $n(t)$ and $s_0(t)$ over the *first* interval and $N_1^{(2)}$ to the correlation between $n(t)$ and $s_1(t)$ over the *second* interval.

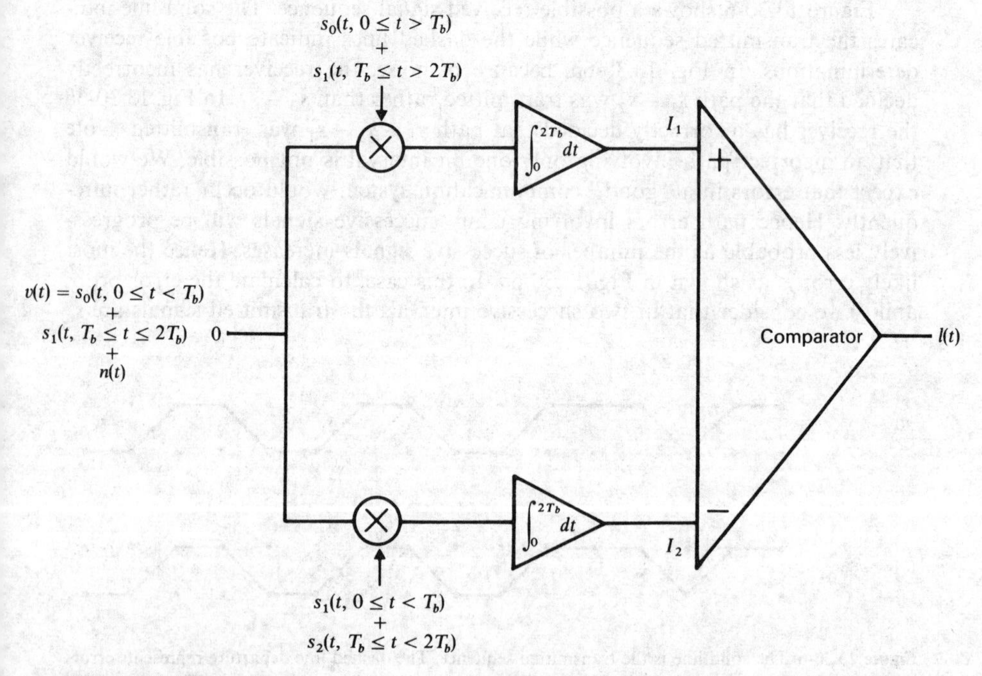

Figure 13.30-7 Indicating the correlation and comparison which is required to determine the most likely path in Fig. 13.30-6*a*.

In the *second* interval the correlation between $n(t)$ and $s_1(t)$ is

$$N_1^{(2)} = \int_{T_b}^{2T_b} n(t)s_1(t)\,dt \tag{13.30-5}$$

The noise terms $N_0^{(1)}$ and $N_1^{(2)}$ are uncorrelated, gaussian, random variables, each with mean zero and variance σ^2 where

$$\sigma^2 = E[N_0^{(1)}]^2 = E[N_1^{(2)}]^2 = E\left\{\int_0^{T_b} dt \int_0^{T_b} d\lambda s_0(t)s_0(\lambda)n(t)n(\lambda)\right\}$$

$$= \int_0^{T_b} dt \int_0^{T_b} d\lambda s_0(t)s_0(\lambda)E(n(t)n(\lambda)) \tag{13.30-6}$$

We have from Eq. (7.9-27) that

$$E(n(t)n(\lambda)) = (\eta/2)\,\delta(t - \lambda) \tag{13.30-7}$$

Substitution of Eq. (13.30-7) into Eq. (13.30-6) yields

$$\sigma^2 = \eta E_b/2 \tag{13.30-8}$$

Similarly, taking into account that $s_0(t)$ and $s_1(t)$ are orthogonal and that $s_2(t) = -s_1(t)$ we find that

$$I_2 = \int_0^{T_b} s_0(t)s_1(t)\,dt + \int_{T_b}^{2T_b} s_1(t)s_2(t)\,dt$$

$$+ \int_0^{T_b} n(t)s_1(t)\,dt + \int_{T_b}^{2T_b} n(t)s_2(t)\,dt = 0 - E_b + N_1^{(1)} + N_2^{(2)} \tag{13.30-9}$$

where

$$N_1^{(1)} = \int_0^{T_b} n(t)s_1(t)\,dt \tag{13.30-10}$$

and

$$N_2^{(2)} = \int_{T_b}^{2T_b} n(t)s_2(t)\,dt \tag{13.30-11}$$

we note that $N_1^{(1)}$ and $N_2^{(2)}$ are uncorrelated gaussian random variables, each with zero mean and variance

$$\sigma^2 = \eta E_b/2 \tag{13.30-12}$$

It is important to note here that while $N_0^{(1)}$ and $N_1^{(2)}$, $N_1^{(1)}$ and $N_2^{(2)}$, $N_0^{(1)}$ and $N_2^{(2)}$, $N_1^{(2)}$ and $N_2^{(1)}$, and, $N_0^{(1)}$ and $N_2^{(1)}$ are each uncorrelated pairs, $N_1^{(2)}$ and $N_2^{(2)}$ are correlated. The correlation between them is

$$E(N_1^{(2)}N_2^{(2)}) = \int_{T_b}^{2T_b} dt \int_{T_b}^{2T_b} d\lambda s_1(t)s_2(t)E(n(t)n(\lambda)) \tag{13.30-13}$$

Using Eq. (13.30-5*b*) and $s_2(t) = -s_1(t)$, we have

$$E[N_1^{(2)}N_2^{(2)}] = -(\eta/2)\int_{T_b}^{2T_b} s_1^2(t)\,dt = -\eta E_b/2 \tag{13.30-14}$$

The probability of error is the probability that $I_2 > I_1$. Using Eqs. (13.30-9) and (13.30-3) yields

$$P(I_2 > I_1) = P(-E_b + N_1^{(1)} + N_2^{(2)} > 2E_b + N_0^{(1)} + N_1^{(2)})$$

$$= P(N_1^{(1)} + N_2^{(2)} - N_0^{(1)} - N_1^{(2)} > 3E_b) \tag{13.30-15}$$

The superposition of the noise components

$$N_T = N_1^{(1)} + N_2^{(2)} - N_0^{(1)} - N_1^{(2)} \tag{13.30-16}$$

that appears in Eq. (13.30-15) is a gaussian random variable with zero mean and variance $\sigma_T^2 = E(N_T)^2$. This variance is readily evaluated by taking account of the fact that only $N_1^{(2)}$ and $N_2^{(2)}$ are correlated and that the variance of each of the four noise components is given by Eq. (13.30-12). We then find that

$$\sigma_T^2 = E(N_T^2) = 4\sigma^2 - 2E(N_1^{(2)}N_2^{(2)})$$

$$= 4\sigma^2 - 2(-\sigma^2)$$

$$= 6\sigma^2 = 3\eta E_b \tag{13.30-17}$$

Hence P_e is

$$P_e = P(N_T > 3E_b)$$

$$= \int_{3E_b}^{\infty} \frac{e^{-N_T^2/2\sigma_T^2}}{\sqrt{2\pi\sigma_T^2}}\, dN_T \tag{13.30-18}$$

Letting $x^2 = N_T^2/2\sigma_T^2$ yields

$$P_e = \frac{1}{2}\frac{2}{\sqrt{\pi}}\int_{\sqrt{(9E_b^2/6\eta E_b)}}^{\infty} e^{-x^2}\, dx = \frac{1}{2}\,\text{erfc}\left[1.5\,\frac{E_b}{\eta}\right]^{1/2} \tag{13.30-19}$$

If data were being transmitted at the same rate using QPSK we would have the result

$$P_e = \frac{1}{2}\,\text{erfc}\left[\frac{E_b}{\eta}\right]^{1/2} \tag{13.30-20}$$

Hence for the same P_e, QPSK would require signal power 1.5 (1.7 dB) times larger than trellis-decoded modulation.

We note that in general, the relationship between P_e and minimum distance, d_{min}, is

$$P_e = \frac{1}{2}\,\text{erfc}\sqrt{d_{\text{min}}^2/4\eta} \tag{13.30-21}$$

Comparing Eq. (13.30-21) with Eq. (13.30-19) yields

$$d_{\text{min}}^2 = 6E_b \tag{13.30-22}$$

This value of d_{min}^2 is not simply related to the distances between signals shown in Fig. 13.30-1*d*. The reason for this situation is that P_e in Eq. (13.30-19) results from a correlation calculation which extends over *two* bit intervals (see Fig. 13.30-6*a*). If d^2_{min} were computed using the equation:

$$d^2_{min} = (\text{distance between } s_0 \text{ and } s_1)^2 + (\text{distance between } s_1 \text{ and } s_2)^2 \tag{13.30-23}$$

then we would find from Fig. 13.30-1*d*, with $E_s = E_b$, that

$$d^2(s_0 \text{ and } s_1) = 2E_b$$

and

$$d^2(s_1 \text{ and } s_2) = 4E_b$$

so that

$$d^2_{min} = 6E_b$$

From these considerations we generalize and state, without proof[7,8], the valid result that where multiple bit intervals are involved in trellis demodulation, the quantity d^2_{min} is the sum of the square of the distance between signals in each bit interval. (d_{min} is thus the "Euclidian" distance between signals.)

Finally we note that the result given in Eq. (13.30-19) depends on the placement of the signals in signal space (see Fig. 13.30-2*d*). For example if the code converter were modified so that s_2 were to correspond to $d(k)$, $d_D(k) = 0, 1$ and s_3 were to correspond to $d(k)$, $d_D(k) = 1, 1$ then P_e would increase significantly (see Prob. 13.30-1). In a more sophisticated case, the placement of signals in signal space is determined by a computer search.

REFERENCES

1. Sakrison, D.: "Communication Theory," John Wiley & Sons, Inc., New York, 1968.
2. Wozencraft, J., and I. Jacobs: "Principles of Communication Engineering," John Wiley & Sons, Inc., New York, 1965.
3. Shannon, C. E.: A Mathematical Theory of Communications, *BSTJ*, vol. 27, pp. 379–623, 1948. Shannon, C. E.: Communication in the Presence of Noise, *Proc. IRE*, vol. 37, p. 10, 1949.
4. Viterbi, A.: "Principles of Coherent Communications," McGraw-Hill Book Company, New York, 1966.
5. Viterbi, A. T., and J. K. Omura: "Principles of Digital Communications," McGraw-Hill Book Company, New York, 1979, Chap. 5.
6. Osborne, P., and D. L. Schilling: Threshold Response of a Phase Locked Loop, *Proc. Intl. Conf. Commun.*, 1968.
7. Ungerboeck, G.: "Channel Coding with Multilevel/Phase Signals," *IEEE Trans. on Information Theory*, January 1982, pp. 55–67.
8. Calderbank, R., and J. E. Mazo: "A New Description of Trellis Codes," *IEEE Trans. on Information Theory*, November 1984, pp. 784–791.

PROBLEMS

13.3-1. One of four possible messages Q_1, Q_2, Q_3, and Q_4, having probabilities 1/8, 3/8, 3/8, and 1/8, respectively, is transmitted. Calculate the average information per message.

13.3-2. One of five possible messages Q_1 to Q_5 having probabilities 1/2, 1/4, 1/8, 1/16, 1/16, respectively, is transmitted. Calculate the average information.

13.3-3. Messages $Q_1, \ldots, Q_M$ have probabilities $p_1, \ldots, p_M$ of occurring.

(*a*) Write an expression for H.

(*b*) If $M = 3$, write H in terms of p_1 and p_2, by using the result that $p_1 + p_2 + p_3 = 1$.

(*c*) Find p_1 and p_2 for $H = H_{max}$ by setting $\partial H/\partial p_1 = 0$ and $\partial H/\partial p_2 = 0$.

(*d*) Extend the result of (*c*) to the case of M messages.

13.3-4. A code is composed of dots and dashes. Assume that the dash is 3 times as long as the dot, and has one-third the probability of occurrence.

(*a*) Calculate the information in a dot and that in a dash.

(*b*) Calculate the average information in the dot-dash code.

(*c*) Assume that a dot lasts for 10 ms and that this same time interval is allowed between symbols. Calculate the average rate of information transmission.

13.4-1. In Example 13.4-1 we saw that the four messages have different probabilities and that as a result we transmit $4B$ binary digits per second and convey only $3.6B$ bits of information. Consider transmitting Q_1, Q_2, Q_3, and Q_4 by the symbols 0, 10, 110, 111.

(*a*) Is this *code* uniquely decipherable? That is, for every possible sequence is there only one way of interpreting the message?

(*b*) Calculate the average number of code bits per message. How does it compare with $H = 1.8$ bits per message?

13.4-2. Consider the four messages of Example 13.4-1. Let Q_1, Q_2, Q_3, Q_4 have probabilities 1/2, 1/4, 1/8, 1/8.

(*a*) Calculate H.

(*b*) Find R if $r = 1$ message per second.

(*c*) What is the rate at which binary digits are transmitted if the signal is sent after encoding $Q_1, \ldots, Q_4$ as 00, 01, 10, 11?

(*d*) What is the rate, if the code employed is 0, 10, 110, 111?

13.5-1. Consider five messages given by the probabilities 1/2, 1/4, 1/8, 1/16, 1/16.

(*a*) Calculate H.

(*b*) Use the Shannon–Fano algorithm to develop an efficient code and, for that code, calculate the average number of bits/message. Compare with H.

13.7-1. A gaussian channel has a 1-MHz bandwidth. If the signal-power-to-noise power spectral density $S/\eta = 10^5$ Hz, calculate the channel capacity C and the maximum information transfer rate R.

13.7-2. Suppose 100 voltage levels are employed to transmit 100 equally likely messages. Assume $\lambda = 3.5$ and the system bandwidth $B = 10^4$ Hz.

(*a*) Calculate S/η, using Eq. (13.6-3).

(*b*) If an integrate-and-dump filter is employed to determine which level is sent, calculate the probability of an error when sending the kth level. Assume that the only errors possible are in choosing the $k - 1$ or the $k + 1$ levels.

13.8-1. (*a*) Plot channel capacity C versus B, with $S/\eta =$ constant, for the gaussian channel.

(*b*) If the channel bandwidth $B = 5$ kHz and a message is being transmitted with $R = 10^6$ bits per sec, find S/η for $R \leq C$.

13.9-1. (*a*) Why is n_l, in Eq. (13.9-3), gaussian?

(*b*) Find $E(n_l)$ by interchanging integration and ensemble averaging. Why is this permitted?

(*c*) Show that $E(n_l^2) = \int_0^T dt \int_0^T d\lambda\, E[n(t)n(\lambda)]s_l(t)s_l(\lambda)$.

(*d*) Using the result that $E[n(t)n(\lambda)] = R_n(t - \lambda) = (\eta/2)\delta(t - \lambda)$, show that $E(n_l^2) = \eta E_s/2$, and that $E(n_l n_k) = 0$ if $l \neq k$.

13.9-2. Show the following by changing variables:

(*a*) That as in Eq. (13.9-9)

$$P_L = \left(\frac{1}{\sqrt{\pi}} \int_{-\infty}^{\sqrt{E_s/\eta} + n_l/\sqrt{2}\sigma} e^{-x^2}\, dx\right)^{M-1}$$

(*b*) That as in Eq. (13.8-11)

$$P_c = \left(\frac{1}{\sqrt{\pi}}\right)^M \int_{-\infty}^{\infty} dy\, e^{-y^2} \left(\int_{-\infty}^{\sqrt{E_s/\eta}+y} e^{-x^2}\, dx\right)^{M-1}$$

(*c*) $\dfrac{E_s}{\sqrt{2}\sigma} = \dfrac{E_s}{\eta} = \dfrac{S_i T}{\eta} = \dfrac{S_i}{\eta R} \log_2 M$

13.10-1. If S_i/η = constant, plot T versus M for $P_c = 10^{-5}$ from Fig. 13.9-2.

13.10-2. An analog signal has a 4-kHz bandwidth. The signal is sampled at 3 times the Nyquist rate and quantized using 256 quantization levels. The S_i/η ratio at the receiver is $S_i/\eta = 1$ MHz.

(*a*) Calculate the time T between samples.

(*b*) The quantized and sampled signal is encoded into a binary PCM waveform. Calculate the number of bits N and find the P_e.

(*c*) The quantized and sampled signal is encoded into 1 of 256 orthogonal signals. Find P_e.

13.10-3. Show that $T/\log_2 M$ = duration of a bit of a binary encoded signal.

13.11-1. We transmit either a 1 or a 0, and add redundancy by repeating the bit.

(*a*) Show that if we transmit 11111 or 00000, then 2 errors can be corrected.

(*b*) Show that in general if we transmit the same bit $2t + 1$ times we can correct up to t errors.

13.11-2. Verify Eq. (13.11-3).

13.12-1. Show that a parity-check bit can be inserted which makes the number of 1s *odd* and that this detects errors.

13.12-2. An 8-bit binary code has a parity-check bit added. How many code words are now available that are not used.

13.15-1. Verify Eq. (13.15-10).

13.15-2. Verify Eq. (13.15-12).

13.15-3. Consider that an orthogonal code is used with soft decision decoding. (*a*) If the probability of error in a codeword is to be 10^{-7}, find the number of information bits/codeword k as a function of E_b/η. (*b*) Will coding always improve performance? *Hint*: Use Eq. (13.15-22).

13.15-4. For small values of p, Eq. (13.15-24) can be approximated by Eq. (13.15-26). To verify this, evaluate the ratio of the second term of the sum in Eq. (13.15-24) to the first term. When this ratio is much less than unity, Eq. (13.15-26) can be employed. Show that if $p \ll t/n$, Eq. (13.15-26) is valid.

13.15-5. (*a*) Show that we can approximate p in Eq. (13.15-26) by

$$p \simeq \frac{1}{2} e^{-(E_b/\eta)(k/n)}$$

for large values of E_b/η (i.e., small values of p), and let

$$(1 - p)^{n-t-1} \simeq 1 - (n - t - 1)p$$

(*b*) Now show that the number of ways that t errors can occur $\binom{n}{t+1}$ can be bounded by $2^k - 1$.

(*c*) Verify Eq. (13.15-27).

13.15-6. A (24,12) code capable of correcting $t = 3$ errors using hard decisions, has a $d_{min} = 7$. (*a*) Plot P_e as a function of E_b/η using Eqs. (13.15-22) and (13.15-26). (*b*) When $P_e = 10^{-5}$ what is the difference, in dB, between E_b/η required for soft and hard decisions?

13.16-1. The numbers 0 to 7 are binary-encoded.

(*a*) Write the 3 binary digits for each decimal number.

(*b*) Add a single parity-check bit to each code word.

(*c*) Each 4-bit code word forms a $\bar{T}$ matrix. Show that if $\bar{H} = [1111]$, $\bar{H}\bar{T} = 0$ for each $\bar{T}$. Also show that if a single error occurs, $\bar{H}\bar{T} = 1$.

13.16-2. The repeated code symbols 111 and 000 are the two values of $\bar{T}$. Find H using Eq. (13.16-3). Show that $\bar{H}\bar{T} = 0$.

13.16-3. The repeated code symbols are 11111 and 00000. Find the $\bar{H}$ matrix.

13.16-4. Using $\bar{H}$ as given by Eq. (13.16-11) and $\bar{G}$ given by Eq. (13.16-12), show that $\bar{G}\bar{H}^T = 0$.

13.16-5. Generate all 16 codewords for the Hamming code described by the $\bar{G}$ matrix given by Eq. (13.16-12).

13.16-6. Show that $\bar{H}\bar{T}^T = 0$ where $\bar{H}$ is given by Eq. (13.16-11) and $\bar{T}$ is given in Prob. 13.16.5.

13.16-7. Find all 16 entries to Table 13.16-1.

13.16-8 (*a*) Verify that the transpose of the transpose of a matrix is the original matrix, i.e., $(\bar{X}^T)^T = \bar{X}$.
(*b*) Show that if the product $\bar{X}\bar{Y} = 0$ then also $\bar{Y}^T\bar{X}^T = 0$.

13.17-1. The code word transmitted is either $T_1 = 111$ or $T_2 = 000$. The $\bar{H}$ matrix is given by Eq. (13.17-3). The signal 101 is received. Show that $\bar{H}\bar{R} = \bar{S}$ shows where the error is located.

13.17-2. (*a*) Construct all eight, 8-bit codewords for the Hadamard code. (*b*) How many errors can be detected? (*c*) How many errors can be corrected? (*d*) Show that the four, 4-bit Hadamard codewords given in Eq. (13.17-5) can detect one error but can correct no errors.

13.17-3. (*a*) Find the generating matrix of the $\bar{H}$ matrix given in Eq. (13.17-13). (*b*) If the information word is $\bar{A} = 00101100111$ find the codeword $\bar{T}$.

13.17-4. Verify that $\bar{H}_e$ as given in Eq. (13.17-15) has a minimum distance of 4.

13.17-5. Verify that cyclic shifts of the Hamming codeword 1110100 are also Hamming codewords.

13.17-6. Show that the factors of $f(x)$ of Eq. (13.17-22) are as shown in Eq. (13.17-23).

13.17-7. If an uncoded word is 111100101110 (*a*) find the 23 bit Golay codeword using Eq. (13.17-30). (*b*) How would you implement codeword generation using a ROM?

13.17-8. It is desired to determine a BCH code having *approximately* 1000 bits in a codeword with the capability of correcting three errors. (*a*) Find $d_{\min}$. (*b*) Find *m*. (*c*) Find *r*. (*d*) Find *k*. (*e*) What is the rate of the code?

13.17-9. The ability of a code to correct errors can be increased by the use of *erasure* information. For example, consider that a received codeword is $R = 1 \times 00111$ where the symbol x denotes that the receiver "was not sure" if the second bit was a 1 or a 0. If the code employed is a Hamming code, then the codeword $\bar{T} = \bar{A}\bar{G}$ is given by Eq. (13.16-13). Equation (13.16-13) provides seven equations, one for each bit in T. However, we only want to determine, $\bar{A} = a_1 a_2 a_3 a_4$, which consists of only four bits. Hence, we can discard any three equations. In this case we assume $\bar{R} = \bar{T}$ but the second bit is unknown.

(*a*) Show that using equations 1, 3, 4, and 5 yields the correct decoded word.

(*b*) In general, for a (*n,k*) code capable of correcting *t* errors, how many erasures can occur in a received codeword and still result in correctly decoding the word if there are no bits detected incorrectly?

(*c*) Show that, in general, a decoder can correct a codeword having E erasures and ε errors if $r \geq n - k = E + 2\varepsilon$ where $\varepsilon \leq t$.

(*d*) Usually, when erasures are employed, thresholds are set so that the probability of a bit error is much less than the error rate required. For example, if the error rate of a bit in the codeword is $P_{ec} = 10^{-11}$ and the probability of an erasure is $q = 10^{-3}$, calculate the probability of a received codeword being incorrectly decoded. Assume that the code can correct $t = 5$ errors.

13.18-1. The (31,15) RS code is very popular. (*a*) How many bits are there in a symbol? (*b*) How many symbols can be corrected in a codeword?

13.18-2. (*a*) Sketch the convolutional interleaver and deinterleaver if $l = 5$ and $s = 3$.

(*b*) Show that there are 15 bits now separating two bits that were initially adjacent to one another.

13.19-1. Find the output sequence of the shift register connected as in Eq. (13.19-1) for the input given in Eq. (13.19-2). Thereby verify Eq. (13.19-3).

13.19-2. Verify that the tree shown in Fig. 13.20-1 yields the same output as the encoder shown in Fig. 13.19-1.

13.19-3. The signal 001 101 001 000 100 is received by the tree shown in Fig. 13.20-1. How many discrepancies are found by the time the output is reached? How many discrepancies if the correct path is always taken?

13.19-4. In connection with the tree of Fig. 13.20-1, we note that a double error coming first leads us along an incorrect path. What is the effect of a double error coming last? Consider, for example, that the signal 111 101 001 000 100 is transmitted and the signal 111 101 001 000 111 is received.

13.20-1. Extend the trellis of Fig. 13.20-6. For the output sequence of Eq. (13.20-1), find the path through the trellis having the minimum number of bit discrepancies.

13.21-1. (*a*) Calculate the upper bound on the bit error rate for a rate $\frac{1}{2}$, constraint length 7 convolutional code using Eq. (13.21-1) when $p = 10^{-2}$.

(*b*) Plot P_b vs. p for $p = 10^{-2}, 10^{-3}, 10^{-4}$.

(*c*) Compare (*b*) with the result obtained for the (24,12), $t = 3$, Golay code.

13.22-1. Show that for hard decision decoding of a (7,4) Hamming code the probability of 2 or more errors is at least a factor of 10 less than the probability of a single error if $p \leq P_0$. Find P_0.

13.22-2. The binary digits 1 and 0 are transmitted by repeating the digits three times: 111 and 000.

(*a*) Calculate the probability of error in receiving the uncoded 1, in terms of ST/η, where T is the duration of the word.

(*b*) Calculate the probability of error in receiving the coded 1(111), in terms of ST/η.

(*c*) Plot the probability of error of (*a*) and (*b*) as a function of ST/η, and compare.

13.22-3. Data having a bit rate 20 kb/s is to be transmitted. The value of $E_b/\eta = 8$ dB. Compare the probability of a *bit* being in error if (*a*) no coding is used, but QPSK is employed; (*b*) rate $-\frac{1}{2}$ Golay code is used and 16 PSK is then employed (interleaving is used so that bit errors are independent). Note that the bandwidths required of (*a*) and (*b*) are the same.

13.24-1. A 1 or 0 is to be sent using a repeated code 111 or 000 respectively. The probability of a bit being in error is p. Find the probability of an undetected error. *Hint:*

$$P_e = P(\text{undetected error} \mid 111 \text{ is sent})P(111 \text{ being sent}) + P(\text{undetected error} \mid 000)P(000 \text{ being sent})$$

Next find

$$P(\text{undetected error} \mid 111) = P(\text{undetected error} \mid 000)$$

Show that

$$P_e = p^3 \text{ and not } 2^{n-k} = 1/4$$

Explain why the bound given by Eq. (13.24-1) is so loose.

13.24-2. A (7,4) Hamming code is used for error detection. The probability of a bit error is p. (*a*) Calculate the probability of an undetected error. (*b*) How does your result compare to the bound given by Eq. (13.24-1)? Explain.

13.24-3. A 16 cycle redundancy check (CRC) is an IC which supplies 16 parity check bits to each k information bit word to form an $n = k + 16$ bit codeword for error detection: (*a*) What is the probability of an undetected error? (*b*) What is the probability of detecting an error?

13.24-4. Assume that a BCH (1023,973) code is used for error detection and that $P_A = 0.99$ and $T_I = 50\ \mu s$. If $T_W = 50$ ms, compare the thruput efficiencies of the stop-and-wait, go-back N and selective-repeat systems. Assume that the time to transmit or to receive is 25 μs.

13.24-5. Verify Eq. (13.24-2*b*). *Hint:* From the series given in Eq. (13.24-2*a*), factor out P_A, then let $(1 - P_A) = y$ and show that the resulting series in y is given by

$$\frac{d}{dy}(y/(1 - y)).$$

13.25-1. (*a*) Using Eq. (13.25-5), find $\lim_{B/f_M \to \infty} S_o/N_o$.

13.25-2. Verify Eq. (13.25-7). *Hint:* Use Eq. (9.2-17) and recognize that $k^2\overline{m^2(t)} = 4\pi^2\, \Delta f_{\text{rms}}^2$. Assume that $B = 2\, \Delta f_{\text{rms}}$.

13.25-3. Plot S_o/N_o versus S_i/N_i for the ideal and FM systems. Assume that $B/f_M = 2\beta$. Choose $\beta = 10$ and 100.

13.27-1. (*a*) Plot S_o/N_o and B/f_M versus $S_i/\eta f_M$ at the threshold of an FM discriminator. The abscissa of the graph should be $S_i/\eta f_M$ at threshold, and the two ordinate axes should be S_o/N_o and B/f_M.

(*b*) For each value of S_o/N_o (and its corresponding value of B/f_M), calculate and plot the value of $S_i/\eta f_M$ required by the ideal demodulator.

(*c*) Compare the results obtained in (*a*) and (*b*).

13.27-2. A signal $m(t)$ is gaussian and bandlimited to 100 kHz. The sensitivity of the FM modulator $k = 10^{-6}$ rad(sec)(volt). An output SNR of 40 db is required.

(*a*) Calculate the rms frequency deviation, if a minimum $S_i/\eta f_M$ is to be employed.

(*b*) Calculate $S_i/\eta f_M$.

(*c*) Calculate B, the IF bandwidth.

(*d*) Calculate the $S_i/\eta f_M$ required by the ideal demodulator to give $S_o/N_o = 40$ dB for the same value of B. Compare results.

13.28-1. (*a*) Repeat Prob. 13.27-2 assuming that $m(t)$ is PCM-encoded and then transmitted using PSK.

(*b*) Compare results for Probs. 13.27-2 and 13.28-1.

13.28-2. (*a*) Plot S_o/N_o and N versus $S_i/\eta f_M$ at threshold for PCM.

(*b*) Compare your results with those obtained in Prob. 13.27-1.

(*c*) When would you use PCM rather than FM?

13.29-1. Show that, for $N = 2$ and when $V^2T/\eta \gg 1$, the information-feedback communication system allows a 3-dB reduction in energy per bit for a fixed error probability.

13.29-2. Show that in the limit as $N \to \infty$, Eq. (13.29-15) may be replaced by Eq. (13.29-16).

13.29-3. Evaluate Eq. (13.29-16):

$$E_T = \frac{4\eta}{\sqrt{\pi}} \int_0^{V^2T/\eta} dx \int_{\sqrt{x}}^{\infty} du\, e^{-u^2}$$

This evaluation is most easily accomplished by interchanging the order of integration. Refer to Fig. P13.29-3*a*. Here we see that the shaded area represents the region of integration of Eq. (13.29-16), i.e., we integrate over x for $0 \le x \le V^2T/\eta$, and over u for $\sqrt{x} \le u < \infty$. The solid line at $x = x_1$, of thickness dx, illustrates that in Eq. (13.29-16) we integrate first over u and then over x. Now refer to Fig. P13.29-3*b*. Here the shaded area is the same as in Fig. P13.29-3a. However, we are now integrating first over x, from $0 \le x \le u^2 \le V^2T/\eta$ and then over u, from 0 to infinity.

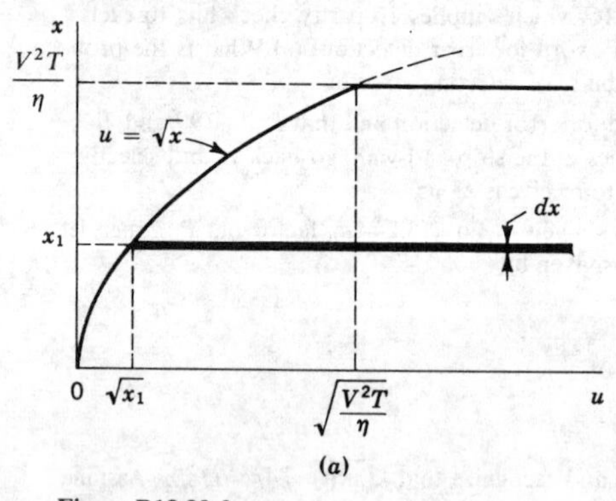

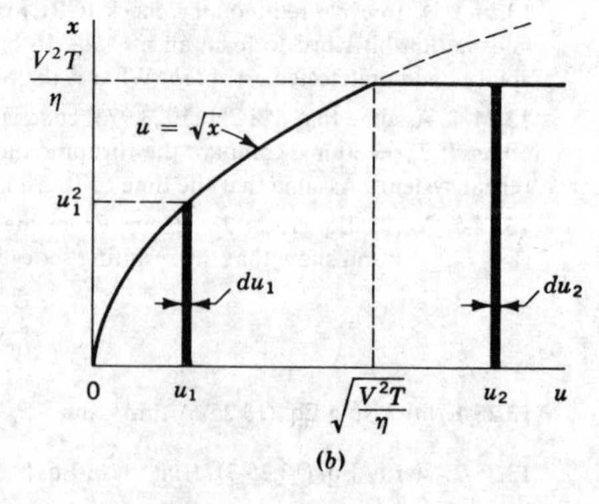

Figure P13.29-3

Show that using this new order of integration yields

$$E_T = \frac{4\eta}{\sqrt{\pi}}\left(\int_0^{\sqrt{V^2T/\eta}} du_1 e^{-u_1{}^2} \int_0^{u_1{}^2} dx + \int_{\sqrt{V^2T/\eta}}^{\infty} du_2\, e^{-u_2{}^2} \int_0^{\sqrt{V^2T/\eta}} dx\right)$$

$$= \frac{4\eta}{\sqrt{\pi}}\left(\int_0^{\sqrt{V^2T/\eta}} du_1 u_1^2 e^{-u_1{}^2} + \frac{V^2T}{\eta}\int_{\sqrt{V^2T/\eta}}^{\infty} du_2\, e^{-u_2{}^2}\right)$$

Complete your evaluation of E_T by integrating both integrals. To do this change variables, so that

$$x = u_1 \qquad \text{and} \qquad dy = 2u_1\, e^{-u_1{}^2}\, du_1$$

Show that

$$E_T = \eta\left(\frac{2}{\sqrt{\pi}}\sqrt{\frac{V^2T}{\eta}}\, e^{-V^2T/\eta} + \operatorname{erf}\sqrt{\frac{V^2T}{\eta}} + \frac{2V^2T}{\eta}\operatorname{erfc}\sqrt{\frac{V^2T}{\eta}}\right)$$

13.30-1. Fig. 13.30-6*a* shows only one path containing double errors. (*a*) How many paths exist that contain double errors? (*b*) Calculate the minimum distance of each path? (*c*) Is any distance found in (*a*) less than $d_{\min} = \sqrt{6E_b/\eta}$.

13.30-2. Consider the signal, $A(t)$, generated by transmitting a voltage $A(k)$ during the time interval $kT_b < t < (k+1)T_b$, where $k = 0, 1, 2, \ldots$, and $A(k) = d(k-1) - 3\, d(k)\, d(k-1)\, d(k-2)$. Here $d(k)$ is the present data bit, $d(k-1)$ is the preceding bit and $d(k-2)$ precedes $d(k)$ by two bits. Let $d(k)$, $d(k-1)$ and $d(k-2)$ have the values ± 1 and occur with equal likelihood.

(*a*) Show that $A(k)$ can have one of four voltages, ± 2 V and ± 4 V.

(*b*) Find the average power, P_s, in $A(t)$.

(*c*) Show that $A(k)$ can be generated using *two* delay units, two EXCLUSIVE-OR gates and an adder.

(*d*) Since there are two delay (storage) elements, the trellis code will have four states: 00, 01, 10, 11. Draw the trellis.

(*e*) Show that the minimum error paths have a duration $3T_b$.

(*f*) Show that $d_{\min}^2 = 8E_b$, where $E_b = P_s T_b$. Note that by using this trellis code we gain 3 dB in signal power as contrasted with using BPSK.

CHAPTER

FOURTEEN

COMMUNICATION SYSTEM AND NOISE CALCULATIONS

A received signal, whether it arrives over a wire communication channel or is received by an antenna, is accompanied by noise. As the signal is processed by the various stages of the receiver, each stage superimposes additional noise on the signal. We shall discuss in this chapter the parameters (noise temperature, noise figure, etc.) which are used to describe the extent to which the various component stages of a receiving system degrade the signal-to-noise ratio. We shall discuss, as well, how these parameters, relating to individual component stages, are combined to determine the overall signal-to-noise performance of a communication system. We shall then be able to determine, among other things, the minimum power level of the signal at the point of transmission which will ensure a final receiver output signal with an acceptable signal-to-noise ratio.

14.1 RESISTOR NOISE

As we have already noted in Sec. 7.1, resistors are a source of noise. The conductivity of a resistor results from the availability, within the resistor, of electrons which are free to move, and the resistor noise is due precisely to the random motion of these electrons. This random agitation at the atomic level is a universal characteristic of matter. It accounts for the ability of matter to store the energy which is supplied through the flow of heat into the matter. This energy, stored in the random agitation, is made manifest generally by an increase in temperature. Thus resistor noise as well as other noise of similar origin is called *thermal* noise.

It has been determined experimentally that the noise voltage $v_n(t)$ that appears across the terminals of a resistor R is gaussian and has a mean-square voltage, in a narrow frequency band df, equal to

$$\overline{v_n^2} = 4kTR\,df \tag{14.1-1}$$

where T is the temperature in degrees Kelvin, and k is the Boltzmann constant $k = 1.37 \times 10^{23}$ J/°K. At room temperature, taken as $T_0 = 290°K$, we have $kT_0 \approx 4 \times 10^{-21}$ W/s. Experiments indicate further that $\overline{v_n^2}$ is independent of the center frequency f_0 of the filter having the bandwidth df, for values of f_0 between 0 Hz and approximately 10 GHz [see Sec. 14.5 for a discussion of the limitations to Eq. (14.1-1)]. Thus, we conclude that for the communication systems employed today (excluding optical communication systems), the normalized power supplied by $v_n(t)$ is also independent of f_0, and hence the power spectral density of the noise source is approximately white and equal to

$$G_v(f) = \frac{\overline{v_n^2}}{2\,df} = 2kTR \tag{14.1-2}$$

The *noisy* resistor R can be represented by the equivalent circuit shown in Fig. 14.1-1. Here the physical resistor shown in Fig. 14.1-1*a* has been replaced in Fig. 14.1-1*b* by a Thévenin circuit representation consisting of a *noiseless* resistor in series with a noise voltage source $v_n(t)$ having the mean-square voltage $\overline{v_n^2} = 4kTR\,df$. If the terminals in Fig. 14.1-1*b* were short-circuited, a noise current $i_n(t)$ would flow having a mean-square value

$$\overline{i_n^2} = \frac{\overline{v_n^2}}{R^2} = \frac{4kTR\,df}{R^2} = 4kTG\,df \tag{14.1-3}$$

where $G = 1/R$ is the conductance of the resistor. The Thévenin equivalent in Fig. 14.1-1*b* therefore corresponds to the Norton equivalent representation shown in Fig. 14.1-1*c*.

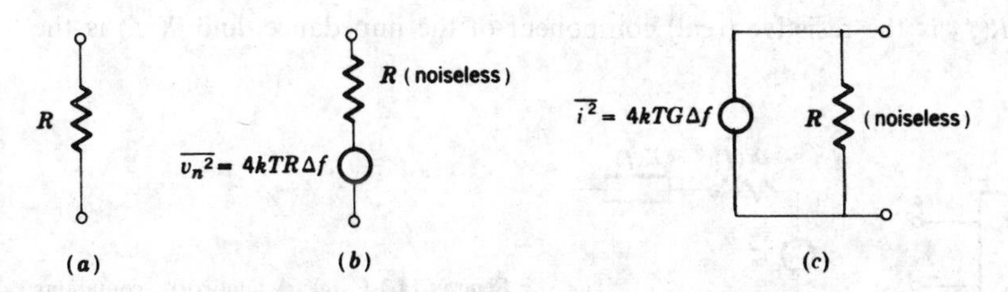

Figure 14.1-1 A physical resistor in (*a*) of resistance R and at a temperature T is replaced in (*b*) by a Thévenin equivalent representation consisting of the noiseless resistor in series with a noise source. In (*c*) a Norton equivalent representation is shown.

14.2 MULTIPLE-RESISTOR NOISE SOURCES

We saw in Sec. 2.17 that the sum of two (or more) independent gaussian random processes is itself a gaussian random process and that, further, the variance of the sum is equal to the sum of the variances of the individual processes. This result may be applied to the calculation of the noise power generated by combinations of resistors. Consider, for example, the case of two resistors, of resistances R_1 and R_2 both at the temperature T, connected in series. The mean-square values of the noise generated by each of the resistors and measured in the frequency band df are $\overline{v_{n1}^2} = 4kTR_1\,df$ and $\overline{v_{n2}^2} = 4kTR_2\,df$. Then, the mean-square voltage measured across the series combination of R_1 and R_2 is $\overline{v_n^2} = \overline{v_{n1}^2} + \overline{v_{n2}^2}$, where

$$\overline{v_n^2} = \overline{v_{n1}^2} + \overline{v_{n2}^2} = 4kTR_1\,df + 4kTR_2\,df$$
$$= 4kT(R_1 + R_2)\,df \qquad (14.2\text{-}1)$$

This result is to have been anticipated since it indicates that two noisy resistors R_1 and R_2 can be replaced by a single noisy resistor $R = R_1 + R_2$. The mean-square noise voltage due to R is given by Eq. (14.2-1).

Rather obviously, this discussion can be extended to an arbitrary network of resistors, all at the same temperature T. If the resistance seen looking into some set of terminals is $R = 1/G$, then the open-circuit mean-square noise voltage at those terminals is $\overline{v_n^2} = 4kTR\,df$, and the mean-square short-circuit current is $\overline{i_n^2} = 4kTG\,df$.

14.3 NETWORKS WITH REACTIVE ELEMENTS

Consider a network composed of resistors, inductors, and capacitors as indicated in Fig. 14.3-1*a*. We arbitrarily select a set of terminals *a*-*b* and inquire now about an appropriate equivalent circuit to represent these terminals as a noise source.

The impedance $Z(f)$ seen looking back into these terminals *a*-*b* is generally a function of frequency and has a real and an imaginary part. That is,

$$Z(f) = R(f) + jX(f) \qquad (14.3\text{-}1)$$

where $R(f)$ is the resistive (real) component of the impedance, and $X(f)$ is the

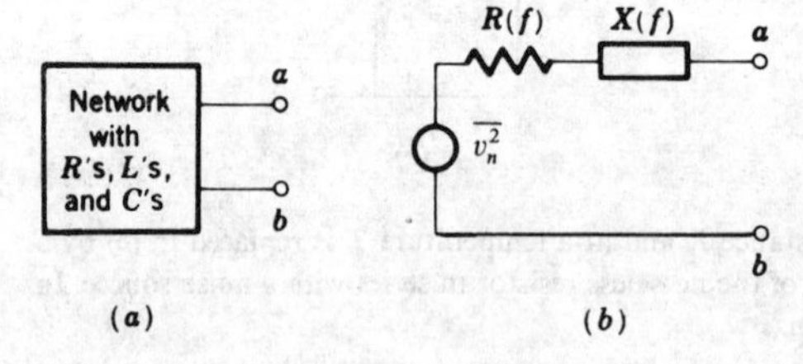

Figure 14.3-1 (*a*) A network containing resistors, capacitors, and inductors with terminals *a* and *b*. (*b*) An equivalent circuit representing the terminals as a noise source.

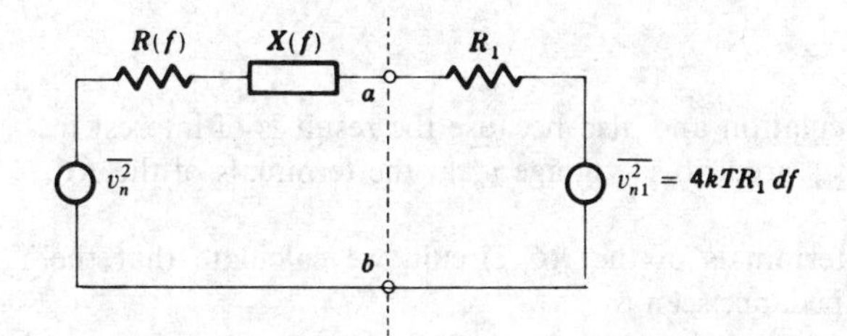

Figure 14.3-2 A load resistor R_1 is bridged across the terminals $a - b$ of Fig. 14.3-1.

imaginary (reactive) component. We therefore replace the network by the equivalent circuit shown in Fig. 14.3-1*b* in which the resistance $R(f)$ and the reactance $X(f)$ are noiseless and the voltage source has a mean-squared value $\overline{v_n^2}$ which is to be determined.

In Fig. 14.3-2 we have bridged a load resistor R_1 across the terminals *a-b* of the network of Fig. 14.3-1. This load resistor is represented by a noiseless resistor R_1 in series with a noise generator of mean-squared voltage $\overline{v_{n1}^2}$. We assume that all the parts of the circuit of Fig. 14.3-2 are at the same temperature in thermal equilibrium. In this case the power delivered by $Z(f)$ to R_1 must equal the power delivered by R_1 to $Z(f)$. To show that this is indeed the case, consider what occurs if there is a net power flow from $Z(f)$ to R_1. Then the temperature of R_1 would increase, thereby exceeding the temperature of $Z(f)$. However, we have assumed that $Z(f)$ and R_1 are the same temperature in thermal equilibrium. Hence the power flow to R_1 from $Z(f)$ must equal the power flow to $Z(f)$ from R_1 to maintain this equilibrium condition.

The mean-squared current due to the source $\overline{v_{n1}^2}$ is

$$\overline{i_{n1}^2} = \frac{\overline{v_{n1}^2}}{[R_1 + R(f)]^2 + [X(f)]^2} \tag{14.3-2}$$

and that due to the source $\overline{v_n^2}$ is

$$\overline{i_n^2} = \frac{\overline{v_n^2}}{[R_1 + R(f)]^2 + [X(f)]^2} \tag{14.3-3}$$

From the condition that there be no net transfer of power we have

$$\overline{i_{n1}^2}\, R(f) = \overline{i_n^2}\, R_1 \tag{14.3-4}$$

From Eqs. (14.3-2), (14.3-3), and (14.3-4), and using $\overline{v_{n1}^2} = 4kTR_1\, df$, we find

$$\overline{v_n^2} = 4kTR(f)\, df \tag{14.3-5}$$

Correspondingly, the two-sided power spectral density of the open-circuit voltage in Fig. 14.3-1 is

$$G_v(f) = 2kTR(f) \tag{14.3-6}$$

14.4 AN EXAMPLE

As an example of a noise calculation and also because the result is of interest in itself, we calculate the mean-squared noise voltage $\overline{v_n^2}$ at the terminals of the *RC* circuit shown in Fig. 14.4-1*a*.

Looking back into the terminals of the *RC* circuit, we calculate that the resistive component of the impedance seen is

$$R(f) = \frac{R}{1 + 4\pi^2 f^2 R^2 C^2} \tag{14.4-1}$$

Applying Eq. (14.3-6) to the entire frequency spectrum, we find

$$\overline{v_n^2} = 2kT \int_{-\infty}^{\infty} \frac{R\, df}{1 + 4\pi^2 f^2 R^2 C^2} = \frac{kT}{C} \tag{14.4-2}$$

The calculation of $\overline{v_n^2}$ may be performed in an alternate manner. In Fig. 14.4-1*b* we have replaced the noisy resistor by a noise generator of power spectral density $G_v(f) = 2kTR$ and a noiseless resistor. The *RC* combination has a transfer function $H(f)$ from noise generator to output terminals given by

$$H(f) = \frac{1}{1 + j2\pi f RC} \tag{14.4-3}$$

The power spectral density of the output noise is $G_v(f)\,|H(f)|^2$, and $\overline{v_n^2}$ is given, as before, by

$$\overline{v_n^2} = 2kTR \int_{-\infty}^{\infty} \frac{df}{1 + 4\pi^2 f^2 R^2 C^2} = \frac{kT}{C} \tag{14.4-4}$$

Equation (14.4-4) may be written

$$\tfrac{1}{2} C \overline{v_n^2} = \tfrac{1}{2} kT \tag{14.4-5}$$

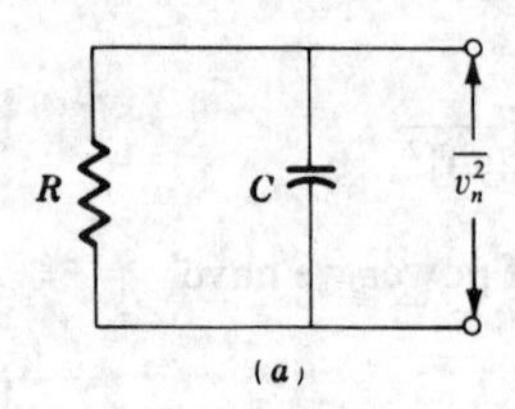

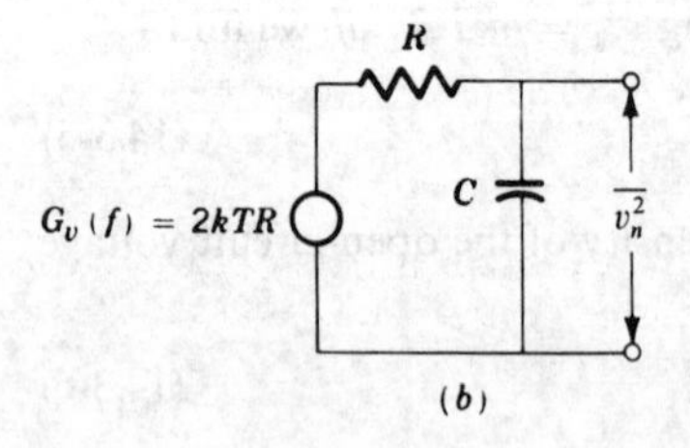

Figure 14.4-1 (*a*) An *RC* circuit. (*b*) The equivalent representation in which the resistor is replaced by a noise generator and a noiseless resistor.

in which the left-hand member is the average energy stored on the capacitor. This result is an example of the famous equipartition theorem of classical statistical mechanics. The equipartition theorem states that a system in equilibrium with its surroundings, all at a temperature T, shares in the general molecular agitation and has an average energy which is $\frac{1}{2}kT$ for each degree of freedom of the system. Thus, an atom of a gas, which is free to move in three directions, has three degrees of freedom and correspondingly has an average kinetic energy which is $3 \times \frac{1}{2}kT = \frac{3}{2}kT$. At the other extreme, a macroscopic system such as a speck of dust suspended in a gas similarly flits about erratically and has an average energy associated with this random motion of $\frac{3}{2}kT$. Since the dust speck is much more massive than an atom, the average velocity of the dust speck will be correspondingly much smaller. As another example, consider a wall galvanometer, which, being free only to rotate, has a single degree of freedom. The kinetic energy associated with such rotation is $\frac{1}{2}I\dot{\theta}^2$ where I is the moment of inertia and $\dot{\theta}$ is the angular velocity. Such a galvanometer shares in the thermal agitation of the air in which it is suspended, and $\frac{1}{2}I\overline{\dot{\theta}^2} = \frac{1}{2}kT$. If the beam of light reflected from the galvanometer mirror is brought to focus on a scale sufficiently far removed, the slight random rotation of the galvanometer may be observed with the naked eye. Altogether, it is interesting to note that the noise generated by a resistor is not a phenomenon restricted to electrical systems alone, but is a manifestation of, and obeys, the same physical laws that characterize the general thermal agitation of the entire universe.

Returning now to the RC circuit of Fig. 14.4-1, we observe that it has one degree of freedom, i.e., the circuit has one mesh, and a single current is adequate to describe the behavior of the system. On this basis, then, Eq. (14.4-5) is seen to be an example of the equipartition theorem.

14.5 AVAILABLE POWER

The *available power* of a source is defined as the maximum power which may be drawn from the source. If, as in Fig. 14.5-1, the source consists of a generator v_s in series with a source impedance $Z_s = R + jX$, then maximum power is drawn when the load is $Z_L = R - jX$, that is, $Z_L = Z_s^*$, the complex conjugate of Z_s. The available power is, therefore,

$$P_a = \frac{\overline{v_s^2}}{4R_s} \tag{14.5-1}$$

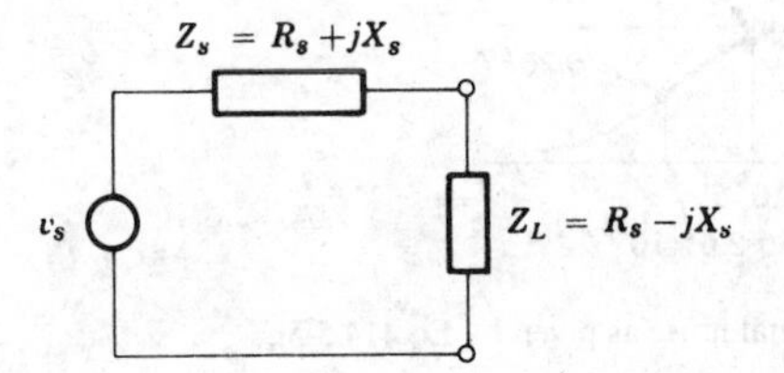

Figure 14.5-1 A source of impedance Z_s is loaded by a complex conjugate impedance $Z_L = Z_s^*$ in order to draw maximum power.

Note that the available power depends only on the resistive component of the source impedance.

Using Eq. (14.5-1), we have that the available thermal-noise power (actual power, not normalized power) of a resistor R in the frequency range df is

$$P_a = \frac{4kTR\,df}{4R} = kT\,df \tag{14.5-2}$$

The two-sided available thermal-noise power spectral density is

$$G_a = \frac{P_a}{2\,df} = \frac{kT}{2} \tag{14.5-3}$$

Observe that G_a does not depend on the resistance of the resistor but only on the physical constant k and on the temperature. If the source consists of a combination of resistors (all at temperature T) together with inductors and capacitors, then in Eq. (14.5-2) the R in the numerator and the R in the denominator are both replaced by $R(f)$, where $R(f)$ is the (usually frequency-dependent) resistive component of the impedance seen looking back into the network. These $R(f)$'s will cancel, as do the R's. Hence, whether the network is a single resistor or a complicated RLC network, the available noise-power spectral density is $G_a = kT/2$ quite independently of its component values and circuit configuration.

Equation (14.5-3) expresses the available noise-power spectral density as predicted by the principles of classical physics, which also predict that this value of G_a applies at all frequencies; i.e., the noise is white. This result is manifestly untenable, since it predicts that the total available power

$$P_a = \int_{-\infty}^{\infty} G_a(f)\,df \tag{14.5-4}$$

is infinite. This prediction was one of a series of similar inconsistencies which were, in part, responsible for the development of the branch of physics called quantum mechanics. The quantum mechanical expression for $G_a(f)$ is

$$G_a(f) = \frac{hf/2}{e^{hf/kT} - 1} \tag{14.5-5}$$

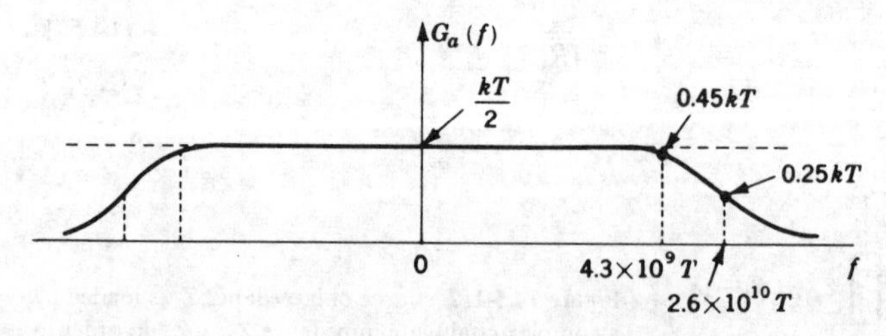

Figure 14.5-2 Available power spectral density of thermal noise as given by Eq. (14.5-5).

in which $h = 6.62 \times 10^{-34}$ J/s is Planck's constant. Equation (14.5-5) yields a finite value for P_a and reduces to Eq. (14.5-3) when $hf \ll kT$.

The power spectral density of Eq. (14.5-5) is plotted in Fig. 14.5-2. Note that the density is lower than $kT/2$ by 1 dB or more only when $f \geq 4.3 \times 10^9 T$, which at room temperature, $T_0 \cong 290°\text{K}$, corresponds to $f \geq 1.3 \times 10^{12} = 1.3 \times 10^3$ GHz. Hence we may certainly use $G_a = kT/2$ at radio and even microwave frequencies (≈ 10 GHz). Note that a microwave receiver may employ a maser amplifier operating at a temperature as low as 4°K in order to minimize the noise due to the amplifier. Even at these low temperatures, it is still appropriate to assume that the noise is white. At optical frequencies this assumption is no longer valid, and Eq. (14.5-5) must be employed.

14.6 NOISE TEMPERATURE

Solving Eq. (14.5-2) for T, we have

$$T = \frac{P_a}{k\, df} \tag{14.6-1}$$

When we apply Eq. (14.6-1) to a passive *RLC* circuit in which the noise is due entirely to the resistors, then T is the actual common temperature of the resistors. Consider, however, the noise which may appear across a set of terminals connected to a more general type of circuit, including possibly active devices. Suppose that we measure the available power at the terminals and find that the noise is white, i.e., the available power P_a increases in proportion to the bandwidth, so that P_a/df is the same at all frequencies. We may then take Eq. (14.6-1) to be the definition of the noise temperature of the network. The noise temperature of the network need not be the temperature of any part of the network.

Consider, for example, the simple idealized situation represented in Fig. 14.6-1. Here a resistor R, which is a thermal-noise source at a temperature T, is connected to the input terminals of an amplifier of gain A. We assume that the input impedance of the amplifier is infinite and assume further, for simplicity, that the amplifier output resistance is a *noiseless* resistor R_o. Then the noise power, in a frequency range df, available at the amplifier output terminals is

$$P_a = \frac{\overline{v_o^2}}{R_o} = \frac{kTRA^2\, df}{R_o} \tag{14.6-2}$$

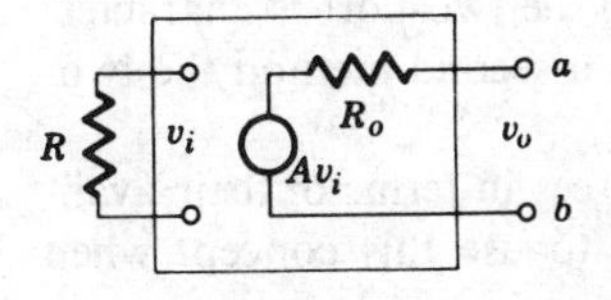

Figure 14.6-1 Illustrating that the noise temperature seen looking back into a set of terminals $a - b$ may assume any value.

The noise temperature seen looking back into these terminals is, using Eq. (14.6-1),

$$T_n = \frac{P_a}{k\,df} = A^2\left(\frac{R}{R_o}\right)T \tag{14.6-3}$$

Thus, depending on the gain A and the ratio R/R_o, this noise temperature may assume any value including even $T_n = 0$.

As a further example to illustrate the concept of noise temperature, consider an antenna which may consist of nothing more than a loop of wire. If we assume that the wire has zero resistance, then the antenna by itself will generate no noise. Noise may, however, be induced in the antenna from a number of sources, some atmospheric and some man-made, including lightning, automobile ignition systems, fluorescent lights, etc. The spectral density of such noise falls off above about 50 MHz. Thus, while an AM radio may be affected by such noise, commercial FM, operating at higher carrier frequencies, is not appreciably affected. A second source of noise is the thermal radiation of any physical body which is at a temperature other than 0°K. Thus the earth, the atmosphere, the sun, the stars, and other cosmic bodies are all sources of noise. If it were possible to shield an antenna completely from all noise sources, then the antenna noise temperature would be zero. (It should also be noted that in this case the antenna would receive no signals either.) Otherwise, if, in the frequency range of interest, the available power spectral density of the antenna noise is constant, then Eq. (14.6-1) may be used to determine the antenna noise temperature. That is, if in a frequency band B the available noise power is P_a, then the antenna noise temperature is T (antenna) $= P_a/kB$.

14.7 TWO-PORTS

Received signals may be processed in a variety of ways. For example, the signal may need to be amplified through a number of amplifier stages or the signal may need to be subjected to frequency conversion to an intermediate frequency, as in a superheterodyne receiver. Each of these processing stages has input and output terminals, i.e., each is a two-port network. Each such two-port may, in general, contain resistors and active devices which are sources of noise. The signal at the input to a two-port will be accompanied by noise and will be characterized by some signal-to-noise ratio. Because of the noise sources within the two-port, the signal-to-noise ratio at the output will be lower than at the input. It is a great convenience to have available a means of describing the extent to which a signal is degraded in passing through a two-port. There are two methods which are commonly employed for this purpose. In one method, the two-port is characterized in terms of an *effective input noise temperature*; in a second method the two-port is characterized by a *noise figure*.

In preceding sections we have described noise sources in terms of their available noise power. In order to continue conveniently to use this concept when two-ports are involved, it is useful to introduce, in connection with two-ports, the idea of the *available gain* of a two-port. The available gain $g_a(f)$ is generally frequency-dependent and is defined as

$$g_a(f) = \frac{\text{available power spectral density at the two-port output}}{\text{available power spectral density at the source output}} \tag{14.7-1}$$

Thus $g_a(f)$ is the ratio of available output power in a small frequency range df to the available source power in this same range.

A point which is immediately apparent from the definition of Eq. (14.7-1) is that the available gain is not a characteristic of a two-port alone, but depends on the driving source as well. The available gain does not depend on the two-port load. To consider the matter further, consider the situation represented in Fig. 14.7-1. Here 2 two-port networks, say two amplifiers, are in cascade, the first being driven by a source of impedance R. The available gain of the first two-port depends on R as well as on the two-port itself. The available gain of the second depends on the output impedance R_{o1} of the first two-port, which may in turn depend on R. In the general case of a cascade of N two-ports, the available gain of the last two-port depends, in principle, on all the preceding $N - 1$ two-ports as well as on the source. Of course, in a two-port cascade encountered in practice, the situation may not be as complicated as suggested here. It may turn out, to a good approximation, that the output impedance of a stage is influenced only slightly by its driving source. In spite of the possible complexity associated with calculating the available gain, particularly of a later stage in a cascade, fortunately the overall available gain of a cascade is related in a very simple manner to the available gains of the individual stages. It may be verified (Prob. 14.7-1) that if the available gains of an N-stage cascade are $g_{a1}, g_{a2}, \ldots, g_{aN}$, then the overall available gain is the product

$$g_a = g_{a1}, g_{a2}, \ldots, g_{aN} \tag{14.7-2}$$

The concept of available gain is especially useful because it permits us to write in very simple form generalized results for the noise characteristics of two-ports.

For example, we note that if a source of available-power spectral density $G_a(f)$ is connected to the input of a two-port of available gain $g_a(f)$, the available-power spectral density at the output is $G_a(f)g_a(f)$, and the total available output power is

$$P_{ao} = \int_{-\infty}^{\infty} G_a(f)g_a(f)\, df \tag{14.7-3}$$

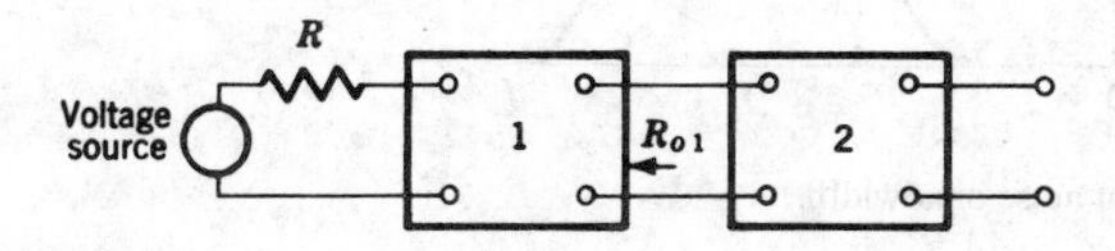

Figure 14.7-1 A cascade of two-ports driven by a source of impedance R.

If the source has a noise temperature T, then $G_a(f) = kT/2$, and

$$P_{ao} = \frac{kT}{2} \int_{-\infty}^{\infty} g_a(f)\, df \tag{14.7-4}$$

14.8 NOISE BANDWIDTH

Bandpass amplifiers, as well as other two-ports of restricted bandpass, will typically have available gains with a frequency dependence such as illustrated by the solid-line plot of Fig. 14.8-1. We have indicated a passband characteristic which is symmetrical about some frequency f_o, since this is the type of characteristic most frequently encountered. However, this symmetry is not essential to the present discussion.

If this two-port is driven by a thermal-noise source of temperature T, then the two-port output available power will be

$$P_{ao} = \frac{kT}{2} \int_{-\infty}^{\infty} g_a(f)\, df \tag{14.8-1}$$

It is frequently convenient to replace the actual available-gain characteristic by a rectangular characteristic, as shown by the dashed plot, which is equivalent for the purpose of computing available output-noise power. Such a rectangular characteristic would have to have a bandwidth B_N determined by the condition of equal available noise output for the two cases, that is, B_N would be determined by

$$P_{ao} = g_{ao}\, kTB_N = \frac{kT}{2} \int_{-\infty}^{\infty} g_a(f)\, df \tag{14.8-2}$$

where g_{ao} is the constant value of $g(f)$ over the passband for the rectangular characteristic. The bandwidth B_N is called the *noise bandwidth* and, from Eq. (14.8-2), is given by

$$B_N = \frac{1}{2g_{ao}} \int_{-\infty}^{\infty} g_a(f)\, df \tag{14.8-3}$$

In Fig. 14.8-1 we have selected g_{ao} to equal the value of g_a at $f = f_0$. Hence the

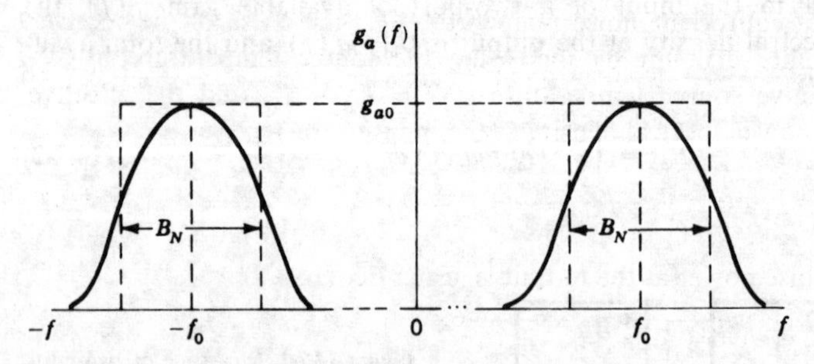

Figure 14.8-1 Illustrating the concept of noise bandwidth.

noise bandwidth which results is the noise bandwidth with respect to the frequency f_0. Customarily, as indicated, f_0 is selected to be the frequency at which $g_a(f)$ is a maximum.

14.9 EFFECTIVE INPUT-NOISE TEMPERATURE

Consider that a two-port is driven by a noise source that has a noise temperature T. Then the source behaves like a resistor at temperature T, and from Eq. (14.5-3) the available noise-power spectral density of the source will be $kT/2$. If the available gain of the two-port is g_a, and if the two-port itself is entirely noise-free, then the available two-sided power spectral density at the two-port output would be

$$G'_{ao} = g_a(f)\frac{kT}{2} \tag{14.9-1}$$

However, because the two-port itself will contribute noise, the available output noise power will be G_{ao}, which is larger than G'_{ao}. We may choose to make it appear that the two-port itself is noise-free, and to account for the increased noise by assigning to the source a new noise temperature higher than T by an amount T_e. We would then have

$$G_{ao} = g_a(f)\frac{k}{2}(T + T_e) \tag{14.9-2}$$

The temperature T_e is called the *effective input-noise temperature* of the two-port. It is to be kept in mind, however, that T_e, like the available gain, depends on the source as well as on the two-port itself.

Example 14.9-1 An antenna has a noise temperature $T_{\text{ant}} = 10°\text{K}$. It is connected to a receiver which has an equivalent noise temperature $T_e = 140°\text{K}$. The midband available gain of the receiver is $g_{ao} = 10^{10}$ and, with respect to its midband frequency, the noise bandwidth is $B_N = 1.5 \times 10^5$ Hz. Find the available output noise power.

SOLUTION The available power spectral density of the antenna is $kT_{\text{ant}}/2$. By the definition of equivalent noise temperature T_e, the noise of the receiver may be taken into account by increasing the source temperature by T_e. Hence the effective source temperature is $T = T_{\text{ant}} + T_e$, and the effective available noise-power spectral density at the input to the receiver is

$$G_a(f) = \frac{k}{2}(T_{\text{ant}} + T_e) \tag{14.9-3}$$

The available noise power at the output is, using Eq. (14.8-2),

$$\begin{aligned} P_{ao} &= g_{ao}\,k(T_{\text{ant}} + T_e)B_N \\ &= 10^{10} \times 1.38 \times 10^{-23} \times 150 \times 1.5 \times 10^5 = 3.1\ \mu\text{W} \end{aligned} \tag{14.9-4}$$

14.10 NOISE FIGURE

Let us assume that the noise present at the input to a two-port may be represented as being due to a resistor at the two-port input, the resistor being at room temperature T_0 (usually taken to be $T_0 = 290°K$). If the two-port itself were entirely noiseless, the output available noise-power spectral density would be $G'_{ao} = g_a(f)(kT_0/2)$. However, the actual output noise-power spectral density is G_{ao}, which is greater than G'_{ao}. The ratio $G_{ao}/G'_{ao} \equiv F$ is the *noise figure* of the two-port, that is,

$$F(f) \equiv \frac{G_{ao}}{G'_{ao}} = \frac{G_{ao}}{g_a(f)(kT_0/2)} \tag{14.10-1}$$

If the two-port were noiseless, we would have $F = 1$ (0 dB). Otherwise $F > 1$. Using Eq. (14.9-2) with $T = T_0$, and Eq. (14.10-1), we find that the noise figure F and the effective temperature T_e are related by

$$T_e = T_0(F - 1) \tag{14.10-2}$$

or

$$F = 1 + \frac{T_e}{T_0} = \frac{T_e + T_0}{T_0} \tag{14.10-3}$$

The noise figure as defined by Eq. (14.10-1) is referred to as the *spot noise figure*, since it refers to the noise figure at a particular "spot" in the frequency spectrum. If we should be interested in the *average noise figure* over a frequency range from f_1 to f_2, then, as may be verified (Prob. 14.10-3), this average noise figure F is related to $F(f)$ by

$$\bar{F} = \frac{\int_{f_1}^{f_2} g_a(f)F(f)\,df}{\int_{f_1}^{f_2} g_a(f)\,df} \tag{14.10-4}$$

Two-ports are most commonly characterized in terms of noise figure when the driving noise source is at or near T_0, while the concept of effective noise temperature T_e is generally more convenient when the noise temperature is not near T_0.

When following a signal through a two-port, we are not so much interested in the noise level as in the signal-to-noise ratio. Consider, then, the situation indicated in Fig. 14.10-1. Here the noise at the two-port input is represented as being

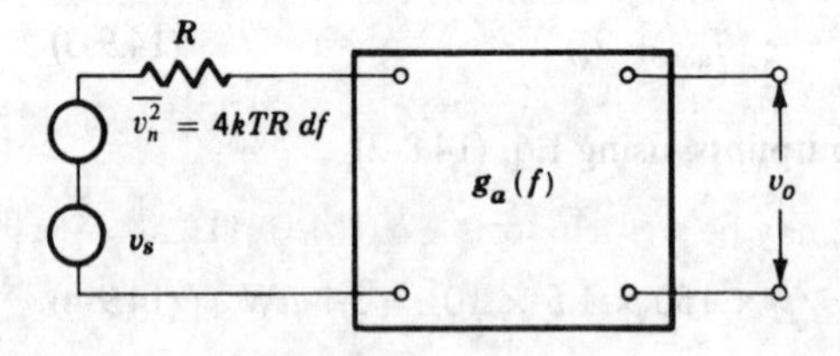

Figure 14.10-1 A signal v_s and a noise source are superimposed and applied at the input of a two-port of available gain $g_a(f)$.

due to a resistor R so that the available input-noise-power spectral density is $G_{ai}^{(n)} = kT/2$. A signal is also present at the input with available power spectral density $G_{ai}^{(s)}$. The available output-signal-power spectral density is

$$G_{ao}^{(s)} = g_a G_{ai}^{(s)} \tag{14.10-5}$$

However, because of the noise added by the two-port itself, the available output-noise spectral density is

$$G_{ao}^{(n)} = g_a F G_{ai}^{(n)} \tag{14.10-6}$$

Combining Eqs. (14.10-5) and (14.10-6), we have an alternative interpretation of the spot noise figure, that is,

$$F(f) = \frac{G_{ai}^{(s)}/G_{ai}^{(n)}}{G_{ao}^{(s)}/G_{ao}^{(n)}} \tag{14.10-7}$$

Thus F is a ratio of ratios. The numerator in Eq. (14.10-7) is the input-signal-to-noise power spectral density ratio, while the denominator is the output-signal-to-noise power spectral density ratio.

Let us assume that in a frequency range from f_1 to f_2 the power spectral densities of signal and noise are uniform. In this case it may be verified (Prob. 14.10-5) that the average noise figure $\bar{F}$ defined by Eq. (14.10-4) has the significance

$$\bar{F} = \frac{S_i/N_i}{S_o/N_o} \tag{14.10-8}$$

where S_i and N_i are, respectively, the total input available signal and noise powers in the frequency range f_1 to f_2, and similarly S_o and N_o are the total output available signal and noise powers.

The noise figure F (or $\bar{F}$) may be expressed in a number of alternative forms which are of interest. If the available gain g_a is constant over the frequency range of interest, so that $F = \bar{F}$, then $S_o = g_a S_i$. In this case Eq. (14.10-8) may be written

$$F = \frac{1}{g_a} \frac{N_o}{N_i} \tag{14.10-9}$$

Further, the output noise N_o is

$$N_o = g_a N_i + N_{tp} \tag{14.10-10}$$

where $g_a N_i$ is the output noise due to the noise present at the input, and N_{tp} is the additional noise due to the two-port itself. Combining Eqs. (14.10-9) and (14.10-10), we have

$$F = 1 + \frac{N_{tp}}{g_a N_i} \tag{14.10-11}$$

or, the noise due to the two-port itself may be written, from Eq. (14.10-11), as

$$N_{tp} = g_a(F - 1)N_i \tag{14.10-12}$$

14.11 NOISE FIGURE AND EQUIVALENT NOISE TEMPERATURE OF A CASCADE

In Fig. 14.11-1 is shown a cascade of 2 two-ports with a noise source at the input of noise temperature T_0. The individual two-ports have available gains g_{a1} and g_{a2} and noise figures F_1 and F_2. If the input-source noise power is N_i, the output noise due to this source is $g_{a1} g_{a2} N_i$. The noise output of the first stage due to the noise generated within this first two-port is $g_{a1}(F_1 - 1)N_i$ from Eq. (14.10-12). The corresponding noise at the output of the second stage is $g_{a1} g_{a2}(F_1 - 1)N_i$. Again, using Eq. (14.10-12), we find that the noise output due to the noise generated within the second two-port is $g_{a2}(F_2 - 1)N_i$. The total noise output is therefore

$$N_o = g_{a1} g_{a2} N_i + g_{a1} g_{a2}(F_1 - 1)N_i + g_{a2}(F_2 - 1)N_i \tag{14.11-1}$$

If we use Eq. (14.10-9), the overall noise figure of the cascade is

$$F = \frac{1}{g_a} \frac{N_o}{N_i} = \frac{1}{g_{a1} g_{a2}} \frac{N_o}{N_i} \tag{14.11-2}$$

$$= F_1 + \frac{F_2 - 1}{g_{a1}} \tag{14.11-3}$$

from Eq. (14.11-1). If the calculation leading to Eq. (14.11-3) is extended to a cascade of k stages, the result is

$$F = F_1 + \frac{F_2 - 1}{g_{a1}} + \frac{F_3 - 1}{g_{a1} g_{a2}} + \cdots + \frac{F_k - 1}{g_{a1} g_{a2} \cdots g_{a(k-1)}} \tag{14.11-4}$$

If the two-ports are characterized by equivalent temperatures rather than noise figures, then the equivalent temperature of the cascade, T_e, is related to the equivalent temperatures and available gains of the individual stages by

$$T_e = T_{e1} + \frac{T_{e2}}{g_{a1}} + \frac{T_{e3}}{g_{a1} g_{a2}} + \cdots + \frac{T_{ek}}{g_{a1} g_{a2} \cdots g_{a(k-1)}} \tag{14.11-5}$$

Equation (14.11-5) may be established by combining Eq. (14.11-4) with Eq. (14.10-3) (see Prob. 14.11-2).

Suppose that the individual two-ports have comparable noise figures or equivalent temperatures. Then, especially if the gains are large, the contribution to the net output noise of succeeding stages in the cascade becomes progressively smaller. A very effective practice, for the purpose of securing a low-noise receiving

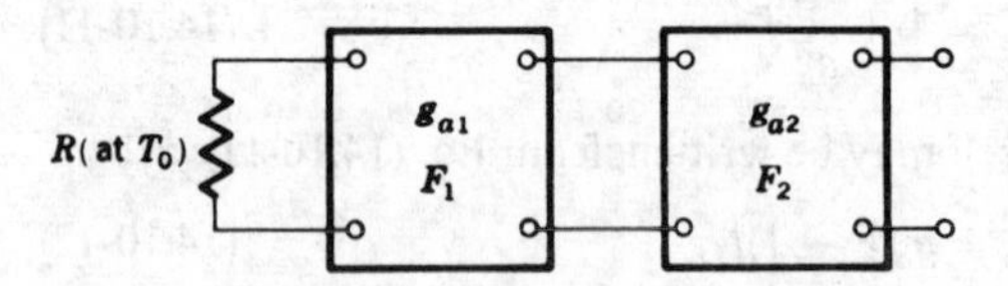

Figure 14.11-1 A noise source at temperature T_0 drives a cascade of two-ports.

system, is to design the first stage of the cascade with a low equivalent temperature and a high gain. A gain of 30 dB is not uncommon. Similarly, equivalent temperatures as low as $T_s = 4°\text{K}$ are obtained by cooling the amplifier with liquid nitrogen.

14.12 AN EXAMPLE OF A RECEIVING SYSTEM

The receiver shown in block diagram form in Fig. 14.12-1 is rather typical of microwave receivers such as are used for satellite communication. In such cases, it is certainly justifiable to take considerable pains to keep the noise figure of the receiver as low as possible. For variety, and also to be consistent with practice, we have characterized the noisiness of the first amplifier in terms of a noise temperature, while the other stages have been characterized by a noise figure. We calculate now the overall noise figure of the receiver. The antenna does not enter the calculation, since it is considered the driving source and not part of the receiver. Using Eqs. (14.10-3) and (14.11-4), we have

$$F(\text{receiver}) = \left(1 + \frac{4}{290}\right) + \frac{4-1}{1000} + \frac{16-1}{100{,}000} = 1.017\ (= 0.05\ \text{dB}) \tag{14.12-1}$$

From Eq. (14.10-2) the equivalent temperature of the receiver is

$$\begin{aligned} T_e(\text{receiver}) &= T_0[F(\text{receiver}) - 1] \\ &= 290(0.017) = 4.93°\text{K} \end{aligned} \tag{14.12-2}$$

Note that, because of the high gain of the first amplifier stage, the travelling wave tube amplifier, the mixer, and the IF amplifier increase the effective receiver temperature only 0.93°K above the temperature of the maser amplifier.

The available noise power present at the demodulator input in the bandwidth B is

$$P_a = \frac{k}{2}[T_{\text{ant}} + T_e(\text{receiver})](2B)g_a(\text{receiver}) \tag{14.12-3}$$

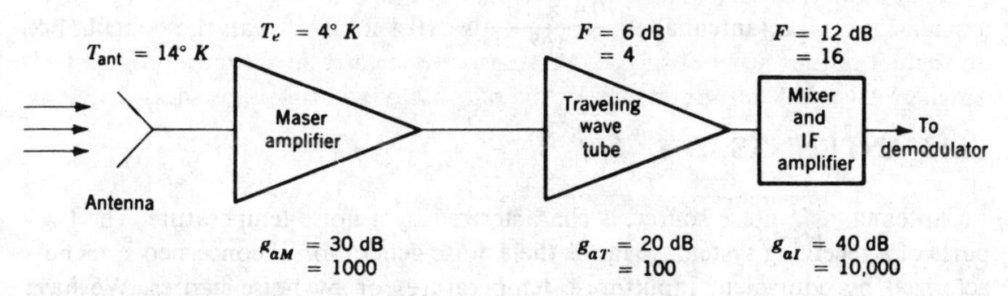

Figure 14.12-1 A typical microwave receiver.

The available gain of the receiver is 30 + 20 + 40 = 90 dB = 10^9; hence

$$P_a = 1.38 \times 10^{-23}(14 + 4.93) \times 10^9 B$$
$$\approx 2.6 \times 10^{-13} B \text{ watts} \tag{14.12-4}$$

Because of the high gain which precedes the demodulator, the noise which may be introduced by the demodulator or any succeeding processing will not further degrade the signal-to-noise ratio of the signal.

In the receiver of Fig. 14.12-1 each of the stages provides some bandlimiting, and the IF amplifier may well consist of a number of stages, each one of which is bandlimited. Conceptually, however, it is very convenient to consider that all the stages of the system are of unlimited bandwidth and that bandlimiting is done in a single IF filter at the output end of the IF amplifier. Under these circumstances the noise input to this IF filter would have a white (uniform) spectral density. We have, as a matter of fact, throughout this text assumed that such was the case when we assumed that our communication channel was characterized by a two-sided noise-power spectral density $G_n(f) \equiv \eta/2$ [see Eq. (7.9-1)]. It is now of interest, in connection with the receiver of Fig. 14.12-1 to inquire into the magnitude of $\eta/2$. From the result given in Eq. (14.12-4) we have

$$G_n(f) \equiv \frac{\eta}{2} = \frac{P_a}{2B} = \frac{2.6 \times 10^{13} B}{2B} = 1.3 \times 10^{-13} \text{ watt/Hz} \tag{14.12-5}$$

As a further example of the performance of the system of Fig. 14.12-1 consider that the receiver is being used to receive a frequency-modulated signal with a baseband frequency range $f_M = 4$ MHz. The received signal power at the demodulator is S_i. Let us assume that, in order to keep the discriminator operating above threshold, we require $S_i/\eta f_M = 20$ dB. What then must be the value of available signal power at the output of the antenna? Since

$$\frac{S_i}{\eta f_M} = \frac{S_i}{2.6 \times 10^{-13} \times 4 \times 10^6} \geq 100\ (= 20 \text{ dB}) \tag{14.12-6}$$

we find

$$S_i \geq 10.4 \times 10^{-5} \text{ watt} \tag{14.12-7}$$

Since the receiver gain is 90 dB (= 10^9), the required minimum available signal power at the antenna must be

$$S_i(\text{antenna}) = \frac{10.4 \times 10^{-5}}{10^9} = 10.4 \times 10^{-14} \text{ watt} \tag{14.12-8}$$

14.13 ANTENNAS

An antenna, as a noise source, is characterized by a noise temperature. The two-ports of a receiving system, so far as their noise generation is concerned, are characterized by equivalent input-noise temperatures or by noise figures. We have seen how, in terms of these characterizations, we may determine the signal power required to be available from a receiving antenna to ensure an acceptable signal-to-noise ratio. We refer briefly now to the manner in which the transmission performance of an antenna system is characterized, so that, given the required available power at the receiving antenna, we may determine the power required to be radiated by the transmitting antenna.

Consider a transmitting antenna which radiates a power P_T, and assume that the power is radiated uniformly in all directions, that is, *isotropically*. The power incident on an area A oriented perpendicularly to the direction of power flow, and at a distance d from the transmitting antenna, is

$$P_R = \frac{P_T}{4\pi d^2} A \tag{14.13-1}$$

Equation (14.13-1) may be used to define an effective area A_e of an antenna. Thus, if the available power from a receiving antenna is P_R when the antenna is a distance d from an isotropic antenna transmitting a power P_T, then the effective area of the receiving antenna is

$$A_e \equiv 4\pi d^2 \frac{P_R}{P_T} \tag{14.13-2}$$

The effective area of an antenna is related principally to the physical shape and dimensions of the antenna. Thus, for example, for a parabolic disk antenna the effective area is generally in the range 0.5 to 0.6 of the physical area of the disk.

Real antennas do not radiate isotropically but are rather directional. This directivity is of advantage when we are interested in transmitting from one antenna to a *particular* receiving antenna. In such a case we would be interested in making the transmitting antenna as directional as possible, and we would orient the transmitting antenna to radiate with maximum intensity toward the receiving antenna. A typical antenna-radiation pattern of a directional antenna is shown in Fig. 14.13-1. If we draw a line, in an arbitrary direction, from the

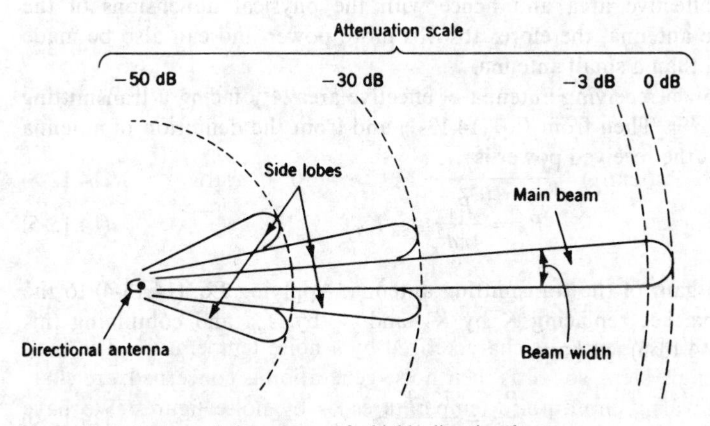

Figure 14.13-1 Transmission pattern of a highly directional antenna.

antenna to the antenna pattern, the length of the line is proportional to the radiant power density in that direction. As indicated, the radiated power is principally confined to a *main beam*, while some power is also radiated in the direction of the *side lobes.* The *beam width* of the antenna is defined as the angle, at the antenna, between directions in which the radiated-power density is down 3 dB from the maximum. The direction of maximum radiation is referred to as the 0 dB direction. In a highly directional antenna, beam widths of 1°, with side lobes down 30 dB to 50 dB as indicated, are feasible.

There is a reciprocity relationship between an antenna used for radiation and the same antenna used for reception. An antenna has the same directivity pattern in the two cases. Thus a highly directional antenna radiates principally in one direction, and when used for reception it similarly absorbs most of the radiant energy from this same direction. Hence, for communication between two particular antennas, and to mimize interference and spurious signals, it is advantageous that both antennas be directional and, of course, with main lobes oriented toward one another.

The extent to which the principal direction of an antenna is favored is measured by the *antenna gain.* Thus suppose that an isotropic radiator would radiate a power per unit solid angle of p_i when furnished with a power P (that is, $p_i = P/4\pi$). If a directional antenna radiates a power per unit solid angle p_m in the direction of most intense radiation, then the antenna gain K is defined by

$$K \equiv \frac{p_m}{p_i} \tag{14.13-3}$$

It turns out that the gain of an antenna and its effective area are related by

$$K = \frac{4\pi A_e}{\lambda^2} \tag{14.13-4}$$

where λ is the wavelength of the radiation. Note that, for fixed λ, the gain increases with effective area and hence with the physical dimensions of the antenna. A large antenna, therefore, absorbs more power and can also be made more directional than a small antenna.

Consider now a receiving antenna of effective area A_{eR} facing a transmitting antenna of gain K_T. Then from Eq. (14.13-1) and from the definition of antenna gain we find that the received power is

$$P_R = \frac{P_T}{4\pi d^2} A_{eR} K_T \tag{14.13-5}$$

where K_T is the gain of the transmitting antenna. Applying Eq. (14.13-4) to the receiving antenna, i.e., replacing K by K_R and A_e by A_{eR} and combining this result with Eq. (14.13-5), we have

$$\frac{P_R}{P_T} = \frac{K_T K_R}{(4\pi d/\lambda)^2} \tag{14.13-6}$$

Thus we find that the received-to-transmitted power ratio P_R/P_T depends on the ratio d/λ which is called the *effective distance* and on the gains K_T and K_R of the two antennas.

Example 14.13-1 The available power required at a receiving antenna is 10^{-6} watt (that is, -60 dB with respect to 1 watt). Transmitting and receiving antennas have gains of 40 dB each. The carrier frequency used is 4 GHz, and the distance between antennas is 30 miles. Find the required transmitter power.

SOLUTION Using Eq. (14.13-6), we have (1.6×10^3 m = 1 mile)

$$10 \log P_R = 10 \log P_T + 10 \log K_T + 10 \log K_R - 20 \log \left(4\pi \frac{d}{\lambda}\right) \tag{14.13-7}$$

$$-60 = 10 \log P_T + 40 + 40 - 20 \log \left[4\pi \frac{30 \times 1.6 \times 10^3}{(3 \times 10^8)/(4 \times 10^9)}\right]$$

$$-60 = 10 \log P_T + 40 + 40 - 138$$

so that

$$10 \log P_T = -2$$

or

$$P_T = -2 \text{ dB} \tag{14.13-8}$$

That is, P_T is 2 dB below 1 watt or $P_T = 0.64$ watt.

14.14 SYSTEM CALCULATION

For the sake of tying together a number of the ideas developed in this chapter as well as several of the concepts encountered in connection with frequency modulation, let us undertake some calculations on a proposed satellite-to-earth communication system as represented in Fig. 14.14-1. We propose a system with the following specifications:

1. Frequency modulation is to be used to transmit from the satellite a TV signal with $f_M = 4$ MHz on a carrier of frequency $f_c = 3$ GHz.
2. The satellite antenna gain is to be $K_T = 20$ dB, while the receiving antenna on the ground is to have $K_R = 50$ dB. (Note that the earth-bound antenna may be much larger than the satellite antenna and hence may have a larger gain.) The satellite is assumed to be at a distance of 32×10^6 m (= 20,000 miles).
3. The noise temperature of the receiving antenna is to be $T_{\text{ant}} = 14°$K.
4. The receiver (a cooled maser amplifier is employed) is to have a total noise figure $F = 0.2$ dB (1.047), and the overall available gain of the receiver up to the demodulator is to be $g_a = 70$ dB.
5. The demodulated signal-to-noise ratio is to be $S_o/N_o = 40$ dB.

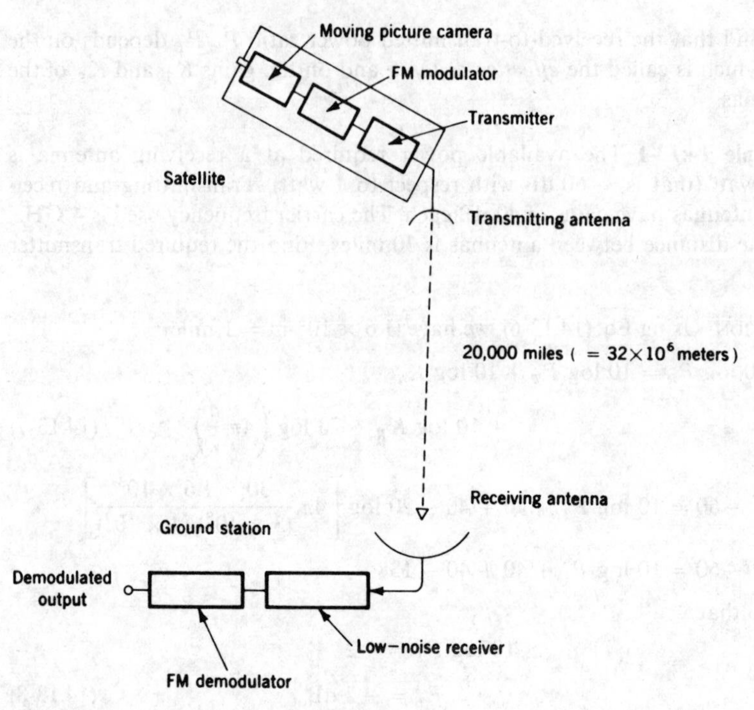

Figure 14.14-1 A satellite communication system.

We are to find:

1. The rms frequency deviation (Δf_{rms})
2. The IF bandwidth
3. The minimum required transmitter power

For simplicity we shall assume that demodulation is performed by an FM discriminator. To minimize the required transmitted power, we assume operation at or just above threshold. We shall further assume that the TV signal can be approximated by a gaussian process. Thus, the received FM signal can be written as

$$v(t) = A \cos \left[\omega_c t + k \int_{-\infty}^{t} m(\lambda)\, d\lambda \right] + n(t) \tag{14.14-1}$$

where $m(t)$ is the TV signal. Hence the rms frequency deviation produced is

$$(\Delta f)_{\text{rms}} = \frac{\sqrt{k^2 \overline{m^2(t)}}}{2\pi} \tag{14.14-2}$$

From Eq. (10.6-16) we have

$$\frac{S_o}{N_o} = \frac{3(\Delta f_{\text{rms}}/f_M)^2(S_i/\eta f_M)}{1 + 6\sqrt{2/\pi}\,(\Delta f_{\text{rms}}/f_M)(S_i/\eta f_M)e^{-(f_M/B)(S_i/\eta f_M)}} \tag{14.14-3}$$

Equation (14.14-3) is a function of the input SNR $S_i/\eta f_M$ and the rms modulation index $\Delta f_{\text{rms}}/f_M$ [note that using Eq. (4.13-5) $B/f_M = 4.6\Delta f_{\text{rms}}/f_M$]. Thus, for a given output SNR, an infinite number of combinations of these ratios are possible. However, we have a constraint in our problem; that is, we are to operate at or above threshold. Let us solve Eq. (14.14-3) subject to the constraint that we are operating *at* threshold. The 1-dB dropoff associated with threshold, as noted in Fig. 10.1-1, occurs when the denominator in Eq. (14.14-3) has the value 1.26. Hence, at threshold, since an output SNR of 40 dB ($= 10^4$) is required, we have, from Eq. (14.14-3), the following two results:

$$0.26 = 6\sqrt{\frac{2}{\pi}}\,\frac{\Delta f_{\text{rms}}}{f_M}\,\frac{S_i}{\eta f_M}\,e^{-(f_M/4.6\Delta f_{\text{rms}})(S_i/\eta f_M)} \tag{14.14-4}$$

$$3\left(\frac{\Delta f_{\text{rms}}}{f_M}\right)^2 \frac{S_i}{\eta f_M} = 1.26 \times 10^4 \tag{14.14-5}$$

Equations (14.14-4) and (14.14-5) can be solved by solving Eq. (14.14-5) for $S_i/\eta f_M$ and substituting this result in Eq. (14.14-4), thereby eliminating $S_i/\eta f_M$ (Prob. 14.14-2). The resulting equation is

$$13 \times 10^{-6}\,\frac{\Delta f_{\text{rms}}}{f_M} \simeq e^{-913(f_M/\Delta f_{\text{rms}})^3} \tag{14.14-6}$$

It may be shown (Prob. 14.14-3) that this equation has the solution

$$\frac{\Delta f_{\text{rms}}}{f_M} \approx 4.5 \tag{14.14-7}$$

The corresponding input SNR is

$$\frac{S_i}{\eta f_M} = 23 \text{ dB} \tag{14.14-8}$$

Good design requires that a margin of safety be employed to ensure operation at or above but not below threshold. We therefore decrease the ratio $\Delta f_{\text{rms}}/f_M$, making it

$$\frac{\Delta f_{\text{rms}}}{f_M} = 4 \tag{14.14-9}$$

The input SNR required to obtain an $S_0/N_0 = 40$ dB is still

$$\frac{S_i}{\eta f_M} = 23 \text{ dB} \tag{14.14-10}$$

since we are now operating above threshold.

The IF bandwidth can now be calculated by letting

$$B = 4.6\Delta f_{\text{rms}} = 18.4 f_M = 73.6 \text{ MHz} \tag{14.14-11}$$

We need now determine only the required transmitter power, to complete our design. We begin this calculation with Eq. (14.14-10). Thus

$$\frac{S_i}{\eta} = 23 \text{ dB} + 10 \log f_M = 89 \text{ dB} \tag{14.14-12}$$

The noise-power spectral density $\eta/2$ at the input to the IF filter was shown in Sec. 14.7 to be a function of the noise figure (equivalent temperature) of the receiver and the receiver gain. Using Eqs. (14.9-3) and (14.12-5), we have

$$\eta = k(T_{\text{ant}} + T_e)g_a(f) \tag{14.14-13}$$

where, from the problem specifications,

$$T_{\text{ant}} = 14°\text{K}$$

$$T_e = T_0(F - 1) = 290(1.047 - 1) \approx 13.6°\text{K}$$

$$g_a(f) = 70 \text{ dB}$$

Thus

$$\eta = 1.38 \times 10^{-23}(27.6)g_a(f) = 38 \times 10^{-23} g_a(f) \tag{14.14-14}$$

We can now compute the signal power measured at the IF filter. We find [Eq. (14.14-12)]

$$S_i = 89 \text{ dB} + 10 \log \eta \approx (89 - 224) \text{ dB} + 10 \log g_a(f) \tag{14.14-15}$$

The input-signal power measured at the *antenna* is $S_i/g_a(f)$. This result, in decibels, is

$$(S_i)_{\text{antenna}} = (S_i)_{\text{IF filter}} - 10 \log g_a(f) = -135 \text{ dB} \tag{14.14-16}$$

The transmitter power required to deliver a signal power of -135 dB and meeting the problem specifications is found from Eq. (14.13-6).

$$P_T = P_R \frac{(4\pi\, d/\lambda)^2}{K_T K_R} \tag{14.14-17}$$

Performing the calculations in decibels yields

$$10 \log P_T = 10 \log P_R + 20 \log \left(\frac{4\pi\, d}{\lambda}\right) - 10 \log K_T - 10 \log K_R$$

$$= -135 + 20 \log \left[\frac{4\pi \times 32 \times 10^6}{(3 \times 10^8)/(3 \times 10^9)}\right] - 20 - 50$$

Thus

$$P_T = -13 \text{ dB} \tag{14.14-18}$$

Hence $P_T \geq 50$ mW must be transmitted.

PROBLEMS

14.2-1. The three resistors are at a temperature T. A bandlimited rms voltmeter is placed across ab, bc, and ac in succession.

(a) Is $V_{ac}(\text{rms}) = V_{ab}(\text{rms}) + V_{bc}(\text{rms})$? Why, or why not?

(b) Is $\overline{V_{ac}^2} = \overline{V_{ab}^2} + \overline{V_{bc}^2}$? Why or why not?

(c) Calculate $V_{ac}(\text{rms})$ read by a meter having a bandwidth B, if R_a and R_c are at temperature T, and R_b is at temperature T_b. Give your answer in terms of symbols.

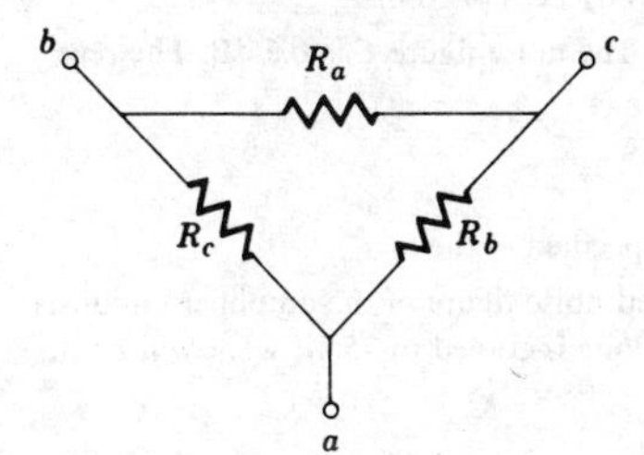

Figure P14.2-1

14.3-1. A parallel RLC circuit centered at 3 GHz has a bandwidth of 10 MHz. If the resistance R is 10 kilohms, calculate $R(f)$ and the power spectral density $G_v(f)$ of the noisy circuit.

14.3-2. (a) Develop an expression for the power spectral density of the noise voltage e_n.

(b) The noise voltage e_n is passed through a low-pass filter with cutoff frequency at ω_c and then through an amplifier of gain $A = 9$. Develop an expression for the total noise output power of the amplifier.

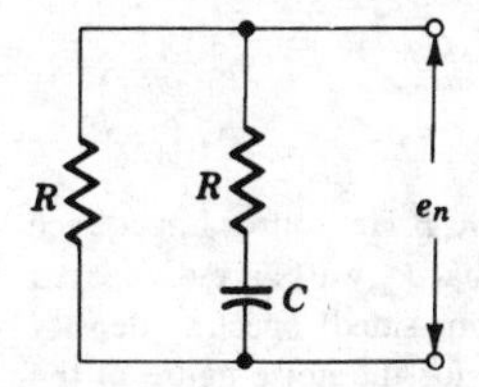

Figure P14.3-2

14.4-1. Refer to Fig. 14.4-1a. Assume that the resistor has an inductance L in series with it. Find $R(f)$, $G_v(f)$, and $\overline{v_n^2}$ when the integral is over all frequencies.

14.5-1. Comment on the difference between $G_a(f) = kT/2$ in Eq. (14.5-3) and $G_v(f) = 2kTR(f)$ in Eq. (14.3-6).

14.5-2. If $T = 4°\text{K}$, find $G_a(f)$ from Eq. (14.5-5) when the wavelength $\lambda \approx 1$ mm, 1 μm. Is the noise white in these regions?

14.7-1. Verify Eq. (14.7-2).

14.8-1. Calculate the noise bandwidth of a parallel RLC filter having a 3-dB bandwidth B.

14.8-2. Calculate the noise bandwidth of an RC low-pass filter having a 3-dB bandwidth f_c.

14.8-3. A gaussian filter has the characteristic

$$H(f) = e^{-k^2 f^2} \qquad -\infty \leq f \leq \infty$$

(a) Calculate the 3-dB bandwidth.

(b) Calculate the noise bandwidth.

14.9-1. An antenna is connected to a receiver having an equivalent noise temperature $T_e = 100°\text{K}$. The available gain of the receiver is $g_a(f) = 10^8$ and the noise bandwidth is $B_N = 10$ MHz. If the available output-noise power is 10 μW, find the antenna temperature.

14.9-2. An antenna has a noise temperature $T_{\text{ant}} = 4°\text{K}$. It is connected to a receiver which has an equivalent noise temperature $T_e = 100°\text{K}$; the midband available gain of the receiver is $g_{a0} = 10^{10}$, and can be represented by a parallel *RLC* filter having a 3-dB bandwidth of 10 MHz. Find:

(*a*) B_N.

(*b*) The available output-noise power.

14.10-1. (*a*) Explain why *F* cannot be less than 1.

(*b*) Verify Eqs. (14.10-2) and (14.10-3).

14.10-2. The noise figure of an amplifier is 0.2 dB. Find the equivalent temperature T_e.

14.10-3. Show that the average value of the noise figure is given by Eq. (14.10-4).

14.10-4. The noise present at the input to a two-port is 1 μW. The noise figure *F* is 0.5 dB. The receiver gain $g_a = 10^{10}$. Calculate:

(*a*) The available noise power contributed by the two-port.

(*b*) The output available noise power.

14.10-5. Show that Eq. (14.10-8) applies under the conditions specified in the text.

14.10-6. With the entire setup operating at 290°K the measured noise figure of the amplifier circuit is $F = 3$ and the voltmeter reads 12 volts. When the resistor R_s alone is cooled to 25°K, what will be the reading of the voltmeter?

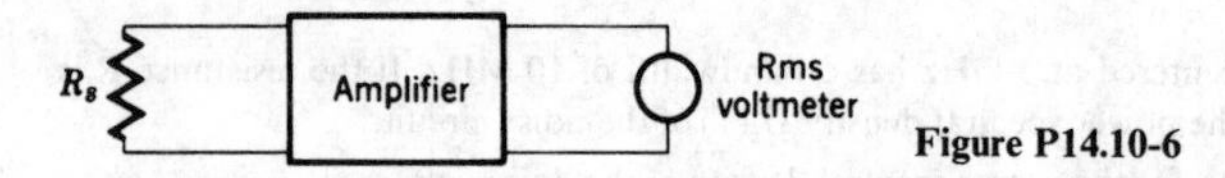

Figure P14.10-6

14.10-7. The entire setup is at $T = 290°\text{K}$. Boltzmann's constant $k = 1.38 \times 10^{-23}$. When $e = 0$ V, the voltmeter reads 3 volts. When $e = 10\ \mu$V rms, the voltmeter reads 5 volts. Find the noise figure *F*.

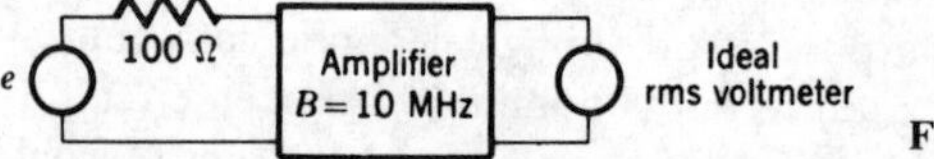

Figure P14.10-7

14.10-8. A noisy amplifier has flat gain from 0 to *B* Hz and zero gain above *B* Hz, source impedance R_s, and load impedance R_L. When the input voltage is zero, it is found that P_o watts are dissipated in the load impedance. When the input voltage is white noise with (one-sided) spectral density η (volts)²/Hz, the power dissipated in the load is $2P_o$. Find an expression for the noise figure of the amplifier.

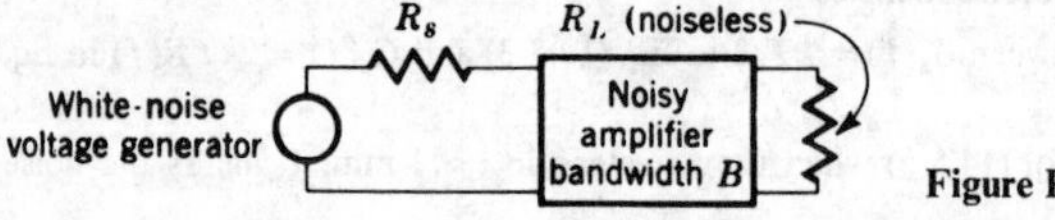

Figure P14.10-8

14.11-1. Derive Eqs. (14.11-3) and (14.11-4).

14.11-2. Derive Eq. (14.11-5).

14.12-1. Refer to Fig. 14.12-1. Let $T_{\text{ant}} = 10°\text{K}$, $T_e(\text{maser}) = 4°\text{K}$ with a gain of 20 dB, the Travelling Wave Tube (TWT) has a noise figure $F = 3$ dB and a gain $g_a = 40$ dB, and a mixer and IF amplifier with $F = 10$ dB and a gain $g_{aT} = 60$ dB. Calculate the available noise power at the receiver output.

14.12-2. Repeat Prob. 14.12-1. Now, however, consider that the available noise power at the receiver output, measured in a noise bandwidth of 10 MHz, is 1 μW. Find the value of T_{ant}.

14.14-1. Verify Eqs. (14.14-4) and (14.14-5).

14.14-2. Verify Eq. (14.14-6).

14.14-3. Verify Eq. (14.14-7).

CHAPTER

FIFTEEN

TELEPHONE SWITCHING[1-10]

15.1 ELEMENTAL PHONE SYSTEM

If we require only two telephone stations, then the system in Fig. 15.1-1 is, at least in principle, adequate for the purpose of providing telephone communications. The inductor L offers no dc resistance but is nominally an open circuit at voice frequencies. The transmitter consists of a box containing a powder of small carbon granules. One side of the enclosure is flexible and is mechanically attached to a diaphragm on which sound waves impinge. The sound vibrates the diaphragm, the diaphragm causes the carbon granules to compress or allow them to expand, and there is a consequent decrease or increase in the resistance of the carbon granules in the box. The carbon granule box with diaphragm is, of course, a microphone. It is not a microphone intended for high fidelity sound reproduction, but when used with an external battery, the resistance changes cause corresponding changes in current and the signal so generated is very much stronger than is available from more sophisticated high-fidelity microphones.

The receiver is an electromagnet with an accompanying magnetic diaphragm. The received changes in current, cause changes in the force of the electromagnet which produce corresponding displacements of the receiver diaphragm. In the circuit of Fig. 15.1-1 there is a quiescent current flowing even in the absence of sound. This quiescent current is necessary for faithful sound reproduction (see Prob. 15.1-3). The receiver diaphragm must always be displaced in one direction from its unstressed position. In more sophisticated and more practical systems than indicated in Fig. 15.1-1, where there is no quiescent current in the receiver, this biasing "tug" on the diaphragm is provided by supplementing the electromagnet with a permanent magnet.

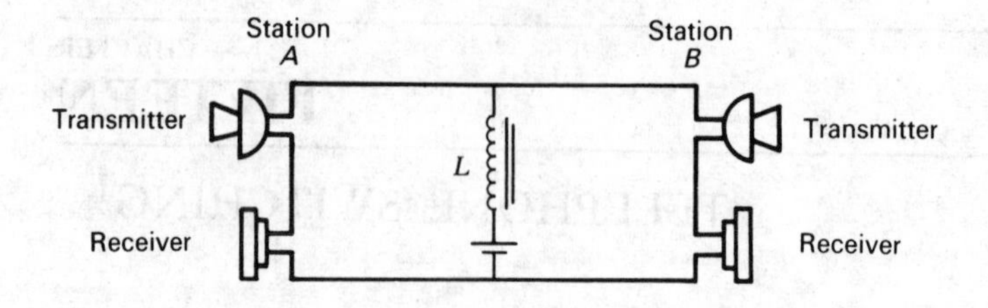

Figure 15.1-1 A telephone communication system between two stations.

In the system of Fig. 15.1-1 the sound incident on the transmitter at station A will be heard not only at the receiver at station B but at the receiver at A as well. This sound, heard at the sound generating station, is called a *sidetone*. Some sidetone is useful. In the absence of sidetone, when the speaker does not hear himself in the earpiece, he unconsciously raises his voice to an unnecessarily high level. Too much sidetone, correspondingly, prompts the talker to reduce his voice level excessively.

A *station set* is the hardware provided to a telephone subscriber and includes, among other things, the receiver and carbon granule transmitter. Some of the features incorporated into a modern station set are shown in Fig. 15.1-2. It provides facility for reducing the sidetone and for taking account of the fact that the resistance between the battery (located at the telephone company's local or "end" office) and station set varies with the physical distance between the two.

The minimum current required for proper operation of a modern carbon microphone is about 23 mA. The battery used with such systems provides about 50 V. Hence the maximum resistance allowable in the battery-station set loop is $50/23 \times 10^{-3} \simeq 2200\ \Omega$. A resistance of about 400 Ω is maintained at the battery to protect against short circuits in the wire between the local office and the subscriber, which is called the "loop," and the station-set resistance may be as much as 200 Ω. Hence the copper wire loop connecting the battery to the station set

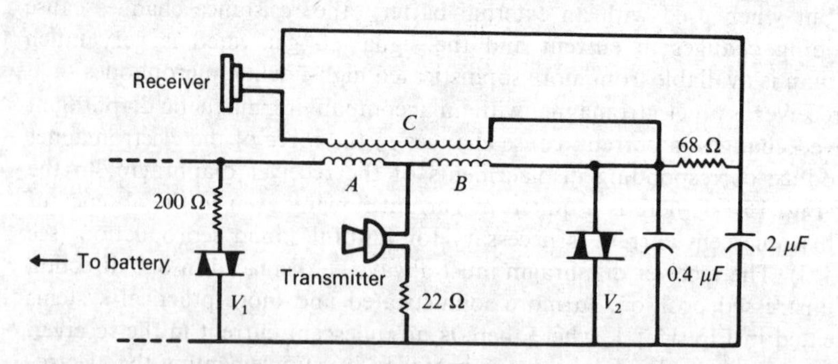

Figure 15.1-2 Illustrating some features of a modern station set. Provision is included to reduce side tone and to take account of variation in resistance of the connection between station set and battery.

can be about 1600 Ω at most. Telephone connections are provided with wire whose size ranges from 19 to 26 gauge. (The higher the gauge number, the smaller the wire diameter.) With 26 gauge wire, a loop distance of only about four miles is possible. With 19 gauge wire the loop distance might be extended to as much as about 18 miles. In any event it is clear that the transmitter current and hence the signal strength generated and received by a station set would be a marked function of the distance between the battery and the station set. To keep this variability within acceptable bounds the *varistor* V_1 has been incorporated into the set. A varistor is a semiconductor device whose resistance is a function of the current flowing through it. The resistance decreases with increasing current. It is now to be seen in Fig. 15.1-2 that, as the resistance between battery and station set decreases, the current shunted through V_1 will increase thereby shunting current away from the microphone.

The other essential feature of the circuit of Fig. 15.1-2 is that it provides a mechanism for reducing the sidetone. When a signal is generated by the microphone in the transmitter the resultant signal current flows through the transformer windings A and B in opposite directions. The varistor V_2 and the associated circuitry to the right simulates approximately the impedance of varistor V_1 and the wire connection to the battery. Hence the currents in A and B are not only oppositely directed but are also nominally of equal magnitude. Accordingly the voltages induced in winding C due to the current in A and B oppose one another and the sidetone is thereby kept within acceptable limits. The signal from the battery side, i.e., from the transmitter of another station set, flows through A and B in the same direction and the resulting signal is therefore transmitted from the primary winding AB to the secondary winding C and hence to the receiving ear piece without attenuation.

15.2 CENTRAL SWITCHING

We do not intend to make light of the marvel furnished by the simple system of Fig. 15.1-1 which allows telephony, literally, sound transmission over a distance. Yet even more marvelous and certainly limitlessly more complicated is the system of switching which allows any of the hundreds of millions of phones in the world to be connected to any other. The remainder of this chapter is devoted to a discussion of some features of such switching systems.

In Fig. 15.2-1 there is shown a system with four telephone stations. The scheme shown allows any station to call any other station to undertake a conversation while leaving the remaining two stations to communicate at the same time if they choose. Each station is furnished with two inputs, one by which it is *signaled* that a call is being made to it and one over which it can talk. The *signal* and *talk* lines and switches are shown only in connection with station A, but it is to be understood that each station is similarly equipped. While only a single wire interconnection between stations is shown, at least two wires are needed, and more may be used if multiple connections between stations are intended. The

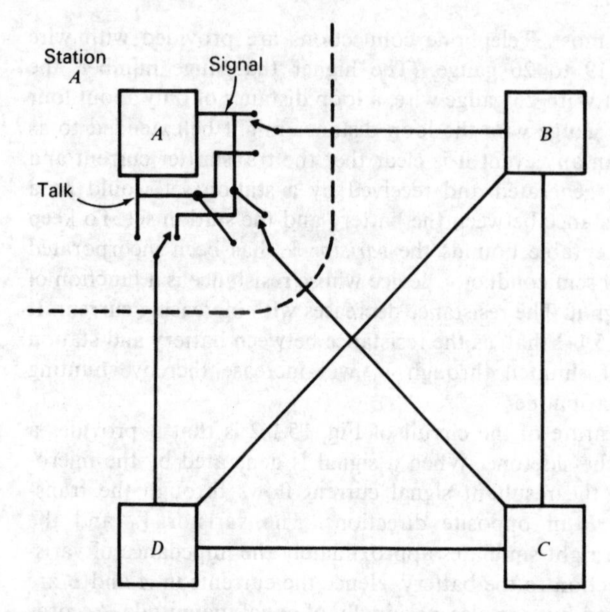

Figure 15.2-1 A switching system in which each station provides its own switching.

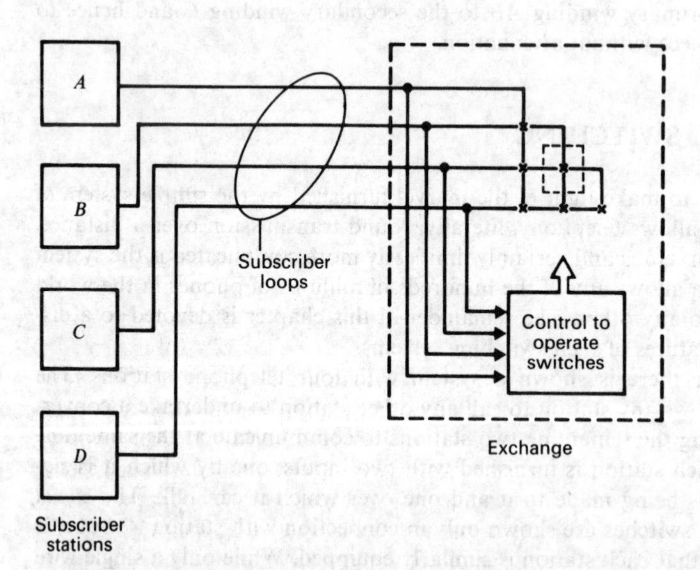

Figure 15.2-2 A central exchange system. Closing the contacts in the dashed square connects subscribers *B* and *C*.

switches will need to be correspondingly modified to show these extra connections. When the system is not carrying a conversation, all switches are in the signalling position so that any party can call any other party. Signals between the calling and called party having been exchanged, both parties will switch to the talk position. Presumably they will disable their call indicator (light, bell, etc.) to avoid interuption. The internal workings of the station are not of present interest to us. With some ingenuity we could devise a number of feasible schemes to allow both signalling and talking over the same wires, to provide "station busy" indications, etc. As a matter of fact, the system can be made effective and, at the present time, is used for intraoffice communication in small businesses. However, our present interest is simply to note that in this system *each station provides its own switching*. If there are N stations then the number of switches is $N(N-1)$ and the number of interconnecting wires between stations is $N(N-1)/2$. Note that since it is communications across a *distance* that we are dealing with, the path length of such interconnections will be *long*.

An improvement with respect to the total length of the interconnecting wires and number of switches results when individual station switching is replaced by *central switching*. A central switching arrangement is shown in Fig. 15.2-2. All of the switches are now located at one central location called a switching *exchange* and the system is referred to as *central* switching. The crosses at the exchange indicate normally open switches (for two or more wires) which can be closed to provide a completed communication path for two subscribers. For the purpose of signalling, to indicate that a connection is called for, each subscriber is connected to a *control* facility which will close switches as required. This control can be provided by a human operator or by automatic machinery. The connections from the subscribers to the exchange are called subscriber loops. The system of switches which effect the interconnections between subscribers is referred to as the switch *matrix*. Thus the essential components of an exchange are the *control* and the *switch matrix*. We observe that with central switching, as compared with station switching, the number of long interconnecting wires is reduced from $N(N-1)/2$ to N and the number of switches is reduced from $N(N-1)$ to $N(N-1)/2$.

15.3 A SIMPLE (HUMAN) EXCHANGE

A very simple exchange is illustrated in Fig. 15.3-1. Here the control is provided by a human operator and the elements of the switch assemblies are plugs and jacks. Each subscriber has a transmitter-receiver which is bridged across the subscriber's loop and a capacitatively-coupled bell similarly bridged. The line lamp relay coil serves as the inductor L shown in Fig. 15.1-1. All of the remaining hardware is located at the exchange, and, with the exception of the ringing generator, the operator's head set and the battery, all of the exchange equipment is duplicated for each subscriber. We have identified the subscriber whose equipment appears in the diagram as subscriber A and a second subscriber as subcriber B.

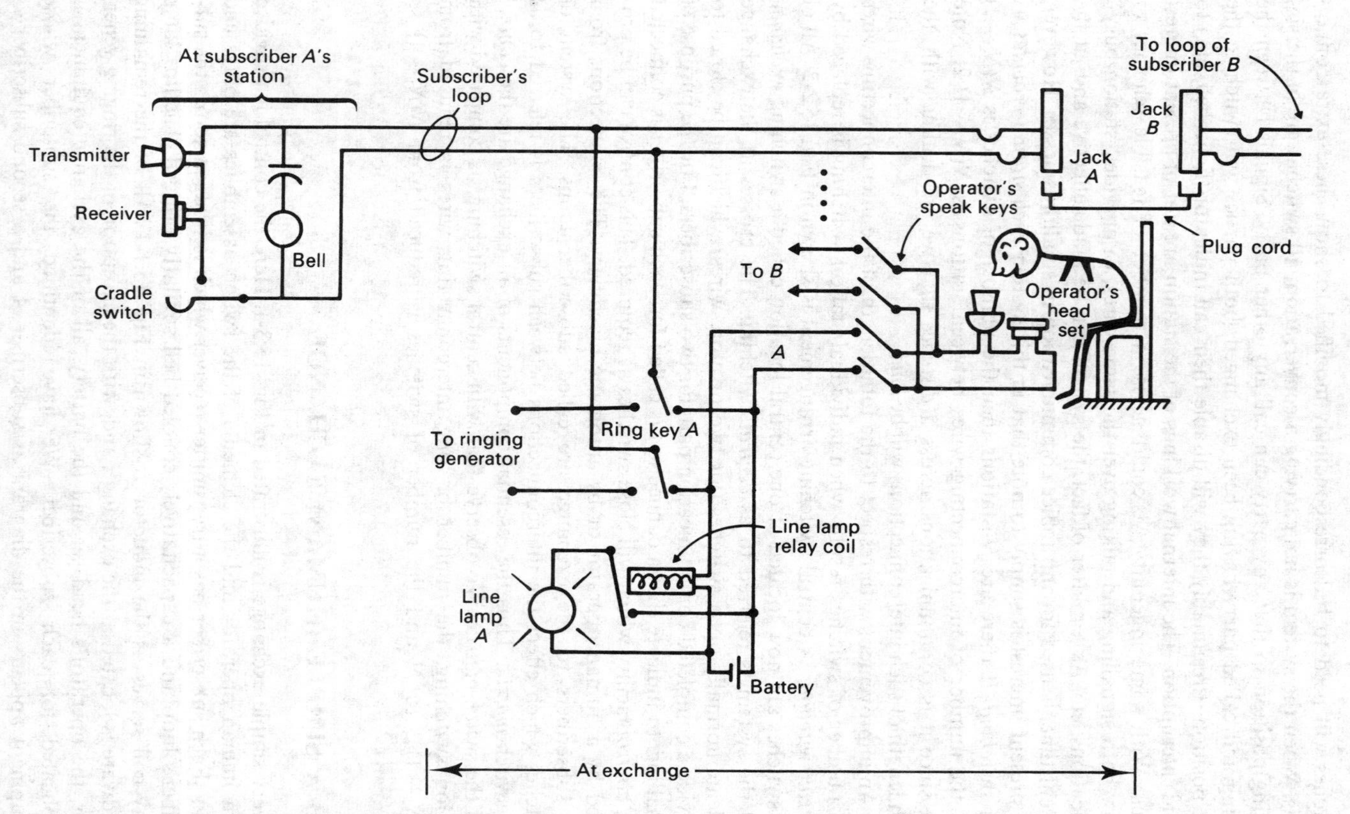

Figure 15.3-1 A simple exchange operated by a human operator.

To initiate a call, subscriber A lifts the transmitter-receiver from its *cradle*. The cradle switch, held open by the weight of the transmitter-receiver, then closes, bridging the transmitter-receiver across the subscriber's loop. The *ring* key is normally connected to the plug side so that the closing of the cradle switch causes a dc current provided by the battery to flow through the line lamp relay coil. (The capacitor in series with the bell assures that no dc bell current will flow except when the cradle switch closes.) The excitation of the line lamp relay causes the line lamp to illuminate. The lamps are labelled to identify the caller. When the operator sees subscriber A's lamp light, she closes the "speak" key connected to subscriber A and asks "number please." A says he wants to talk to B. The operator then throws ring key B (not shown) to the ringing generator side. (This key is spring loaded so that it remains in the ringing position only while being held.) The ringing generator provides an ac current which passes through the bell capacitor at B's subscriber station. If B does not pick up the phone after a reasonable time, the operator reports to A that the call cannot be completed. (The speak keys are also spring loaded and provide connection to the subscribers only while being held.)

If B lifts his phone from its cradle, the operator will be so informed because B's line lamp (not shown) will go on. In this case, the operator will connect jack A to jack B, and then say to A "go ahead please." So long as A and B are in communication, the line lamps of both will be lit. When both hang up, the lamps will go out, and the operator will disconnect the jacks. If only one party hangs up, the operator can ring the other party to remind him to do the same.

15.4 THE STROWGER AUTOMATIC DIALING SYSTEM

Early in the history of telephony it became clear that the demand for telephone service was increasing at such a pace that it was necessary to find some machine replacement for the telephone operator. Otherwise there would simply not be enough people available to operate switch boards. There were accordingly invented a wide range of electromechanical switching devices which could be used to facilitate and execute operations to connect a caller to a called party without human intervention.

One of these switching devices is the Strowger switch invented in 1899 and named after its inventer. Of all the electromechanical switching devices that have become available over the years, the Strowger switch was the most popular and widely used. While new switching centers being installed at the present time no longer employ Strowger switching, the fact is that so widespread were Strowger installations that even in the present day much of the switching is still done in Strowger equipped central offices.

A Strowger switch, including the essential switching mechanism together with the relays and other electromechanical features necessary to make it operate, is a very formidable piece of equipment. It may weigh in excess of 40 pounds and its physical dimensions may be measured in feet. Because of its complexity, we shall content ourselves with a qualitative description of what it does.

The essential switching feature of the Strowger switch is shown in Fig. 15.4-1. It consists of an array of 100 terminals arranged in ten rows stacked vertically, each row having 10 contacts. The terminals, as shown, are placed in circular fashion at constant distance from a shaft on which is mounted a wiper arm. The shaft, and hence the wiper, can be moved vertically as well as circularly. There is a "home" position for the wiper arm below and to one side of the lowest row of terminals. It can be arranged that the wiper be moved to make contact with any one of the 100 terminals. The procedure is to move the shaft vertically one row at a time to the desired row and thereafter to rotate the shaft, one terminal at a time, to the terminal to be selected. Not shown in the diagram are the relays and other mechanisms used to produce this displacement and rotation in response to signals from external sources.

There are three modes in which a Strowger switch is arranged to operate:

(*a*) In response to an external signal it can search out a specific terminal distinguished by some voltage differential. A Strowger switch that operates so is referred to as being "self-driven" in search of a "mark." In this mode the shaft will automatically search out the right row and then rotate to land the wiper on the mark.
(*b*) It can respond to an external signal by an advance of just a single row in the vertical direction. Thus it advances *step-by-step* and an advance to row k would require k individual signals. Thereafter the circular motion is *self-driven* to a mark.
(*c*) It can advance step-by-step, circularly and vertically, in response to individual circular or vertical driving signals from an external source.

The wiper once having arrived at its intended terminal requires no power to keep it there. It will return to its home position in response to a "knock-down" signal. The shaft is rotated back by spring tension and then simply falls back down due to gravity.

The Strowger switch is intended to be used in connection with a dialing mechanism at the calling station set. The process by which the switching system connects a *calling* party to a *called* party is illustrated in Fig. 15.4-2. We shall assume that the telephone number of the called party is 5123. Observe that four switches are being used, one described as a "line finder" two as "selectors" and one as a "connector." In the line finder, the terminal bank faces the caller while with all of the other switches the terminal bank faces the called party. When the caller lifts the phone from its cradle (the phone is then described as being "off hook") a contact is made and current flows in the subscriber's loop. This current signals the line-finder control mechanism to set the line finder switch on a self-driven search to find the terminal on its bank to which the caller is connected. The switch moves upward until it senses the proper row and then rotates until it finds its "mark." (As a matter of practice line finders are often equipped with two parallel, 100 terminal, banks. The wiper arm is then equipped with two contacts and an additional relay is needed to distinguish one bank from the other.) It is, of

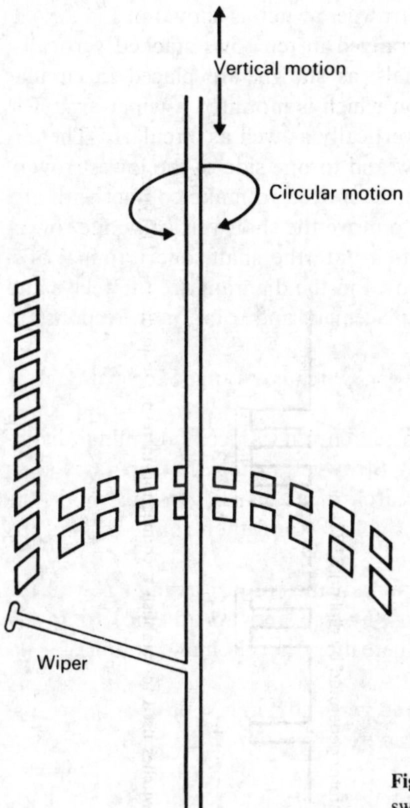

Figure 15.4-1 The essential mechanism of the Strowger switch.

course, necessary that an incoming subscriber's line be connected to a multitude of line finders since, at any particular time, an arbitrary line finder may be busy with some other call. The line finder now connects the calling party to a multitude of selector switches (with associated controls). Only a selector in the "home" position, i.e., not already busy will respond to this connection. If there is available such an idle selector, there will be returned to the caller a dial tone which tells the caller that he may begin dialing.

The caller puts his finger in the *rotary* station-set dial at position 5, rotates the dial to the stop and then releases the dial. Spring tension returns the dial to its rest position. As the dial returns, a cam mechanism opens the line 5 times, i.e., a number of times equal to the number dialed. The waveform seen by the switching system is shown in Fig. 15.4-2*b*. The first change in voltage level corresponds to taking the phone "of hook." The sequence of 5 negative pulses represents the effect of dialing the number 5. The response of the first selector to these 5 pulses is to advance vertically one step for each pulse so that it arrives at the 5th row of the Strowger. All the terminals on this row can connect eventually through other

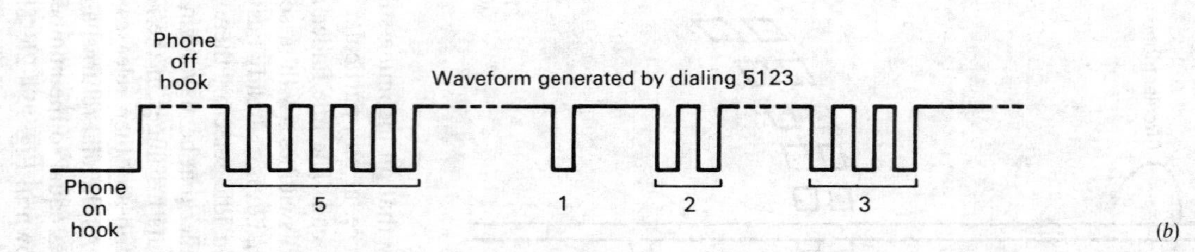

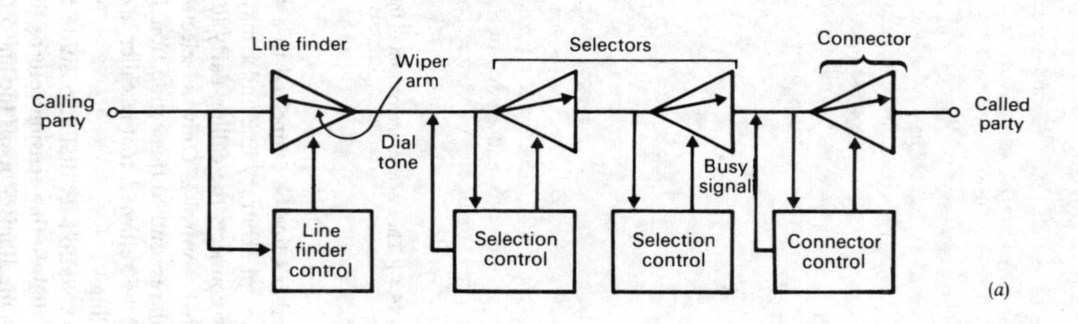

Figure 15.4-2 (*a*) Arrangement of Strowger switches used to establish a connection in response to dialing. (*b*) Typical waveform generated by dialing.

Strowger switches only to phones the first digit of whose number is 5. Now this first selector must connect the incoming line to a second connector which can respond to the second digit. Some second level selectors may well be busy, hence the first connector (on row 5) rotates, testing terminal after terminal on that 5th row until it finds one connected to an idle switch. This second stage selector responds to the second dialed digit. The last switch (the connector) takes care of the last two digits. It can do so because only 100 numbers are possible with two decimal digits and there are 100 terminals on the switch. Accordingly, the last two digits being 23, the dialing of 2 advances the switch to row 2 and then the dialing of 3 rotates the switch to the 3rd column. If the called phone is idle the call is then completed; otherwise a busy signal is returned to the caller. It is clear that if the amount of switching equipment is limited, it may well happen that a call connection may not be completed simply because there is no available idle switching gear. In such a case the call is described as being *blocked*.

15.5 TRAFFIC LOAD AND SERVICE GRADE

In a community of N subscribers it is extremely unlikely that all N will use the phone system at the same time. Also, while the usage of a telephone system is not deterministic, it is still predictable, e.g., on *Mother's Day* and on *New Year's Eve* the call intensity is always much greater than at 3 AM on a weekday. It is, accordingly, not economically sound to provide sufficient switching facility to allow for such an improbable occurrence. Instead phone companies provide a lesser amount of switching gear and correspondingly a facility is rated by the *grade of service* that it can provide i.e., by the probability that it can establish a called-for interconnection. Thus, if the probability is 0.01 that a subscriber's call will not be able to be serviced, that is, the call will be blocked, then the grade of service is characterized as P.01.

The elements of the hardware that make up a switching center and its associated facilities (circuits, toll lines etc.) are collectively referred to as *servers*. The *load*, that is the traffic, which such a switching facility must bear during some time interval from t_1 to t_2 is defined as the average number of servers which are in use during that interval. If during the interval no new calls are initiated and no previously begun calls are terminated the load is fixed and is equal to the number of servers that are in use. The "unit" of load is called the *Erlang* so named to honor a Danish engineer A. K. Erlang. (Actually the *load* is a dimensionless number and hence falls in a class with other, nonetheless, useful dimensionless "dimensions" such as the degree, the radian, the decibel, etc.) Equivalently the load ρ (in Erlangs) can be measured as the product of the number of calls λ initiated during a specified time interval (i.e., the calling *rate*) and the average duration of a call h, i.e.,

$$\rho = \lambda h \tag{15.5-1}$$

Thus, suppose that during a 25 minute interval it is observed the 50 subscribers initiate calls and that the total combined duration of these calls is 5400 s. Then the load is

$$\rho = \frac{50 \text{ calls}}{25 \text{ min} \times 60 \text{ s/min}} \times \frac{5400 \text{ s}}{50} = 3.6 \text{ Erlangs} \qquad (15.5\text{-}2)$$

The average load imposed by each caller is 3.6/50 = 0.072 Erlangs. The unit of time used to specify λ and h is irrelevant, of course, provided only that the same unit be used for both parameters. Some engineers prefer, however, to specify the calls per unit time by the number of *calls per hour* and to measure the average duration in units of *100 s*. In this case we have

$$1 \text{ Erlang} \equiv 1 \frac{\text{call}}{\text{s}} \cdot 1 \text{ s per average call} = 1 \frac{\text{call}}{1/3600 \text{ hr}} \cdot \frac{100 \text{ s}}{100}$$

$$= 36 \text{ hundred call s/hr}$$

$$= 36 \text{ HCS or } 36 \text{ CCS} \qquad (15.5\text{-}3)$$

If we assume that calls are independently initiated at an average rate of λ calls/s, and the time that a call lasts, on the average, is h (this is called the *service time*), then $\rho = \lambda h$ (see Eq. (15.5-1)). If the number of *servers* is s, then the probability that all s servers are busy, i.e., that the call is *blocked* can be shown to be,

$$B(s, \rho) = \frac{\rho^s/s!}{\sum_{j=0}^{s} \rho^j/j!} \qquad (15.5\text{-}4)$$

The probability $B(s, \rho)$ is called the *Erlang B Congestion Formula*. It can be verified (see Prob. 15.5-5) that a recursive relation for the Erlang B probability is

$$B(s, \rho) = \frac{\rho B(s-1, \rho)}{s + \rho B(s-1, \rho)} \qquad (15.5\text{-}5)$$

Note that, as expected,

$$B(0, \rho) = 1 \qquad (15.5\text{-}6)$$

Let us measure the load ρ imposed on a switching facility by determining the parameters λ and h which appear in Eq. (15.5-1). This quantity ρ will measure the *average* load imposed on the facility during the time interval selected, which may be a half hour or so. As we have noted, the load ρ is the average number of servers which the facility must provide in order to accommodate all calls. Let us repeat the load measurement at various times during the day, on various days, etc., and suppose we find that the maximum value of ρ we encounter is, say, $\rho = 14$ Erlangs. Then, if the switching facility has 14 servers will it always be able to accommodate all calls, that is will blocking be prevented? In general, the answer is *no*. For at various times during an extended interval, the load instantaneously may be smaller or, more importantly, larger than the average load. It can be established, using Eq. (15.5-4), that for $\rho = 14$ and for a grade of service

Table 15.5-1

Average Load, ρ	10 Erlangs		19 Erlangs	
Grade of service $B(s, \rho)$	P.01	P.001	P.01	P.001
Number of servers needed, s	18	21	89	100
Average number of servers idle	8	11	14	25
Percentage of servers idle	44%	52%	16%	25%

$B = \text{P}0.01$, then actually 23 servers are needed, while 27 are needed for $B = \text{P}0.001$ service.

It is, of course, rather apparent, as we have just noted, that for better service more equipment is needed. What is a bit less obvious and is brought out by Table 15.5-1 and Eq (15.5-4), is that a *larger* facility carrying a *larger* load will operate *more efficiently* than a *smaller* facility.

15.6 HIERARCHY OF SWITCHING OFFICES

There are in excess of 150 million phones in the United States alone. While, in principle, it is possible to arrange that a single switching exchange accommodate all these phones, we need hardly belabor the point that it is not practical to do so. For, suppose that there were a single exchange at some central location, say at Chicago. Then with $N \cong 150 \times 10^6$ subscribers there would be N connections to the exchange and the number of switch assemblies would be $N(N-1)/2 \cong 10^{16}$ switches. Two subscribers A and B located on the same street in Los Angeles would communicate over a 4000 mile line extending to Chicago and back again. Over such a long path the voice signal levels would frequently have to be passed through repeaters (amplifiers). The need for repeaters would in turn require one communication path to carry A's voice to B and a second path from B to A. Hence, at a minimum, four wires would be required. On the other hand, we have noted, as in Fig. 15.3-1, that when no amplification is required, two wires are adequate to accommodate a two-way communication.

To circumvent the difficulties just described and to accommodate to changes in available equipment and subscriber requirements, operating switching systems employ a hierarchy of switching offices. The first levels of the hierarchy is to be seen in Fig. 15.6-1. The solid circles represent *end offices* to which individual subscribers are directly connected. The connection from subscriber to end office is referred to as a *subscriber loop*. If two subscribers are connected to the same end office then that office alone provides the necessary switching service for these two. In Fig. 15.6-1 we contemplate a number of end offices clustered into three relatively distantly separated locations. Where a number of end offices are located in the same general area, the need for multiple offices will have arisen either because the subscribers are not close enough to one another or are too numerous to be serviced by a single end office or for both of these reasons. The end offices in the same area will be interconnected by trunks (shown dashed) so that subscribers

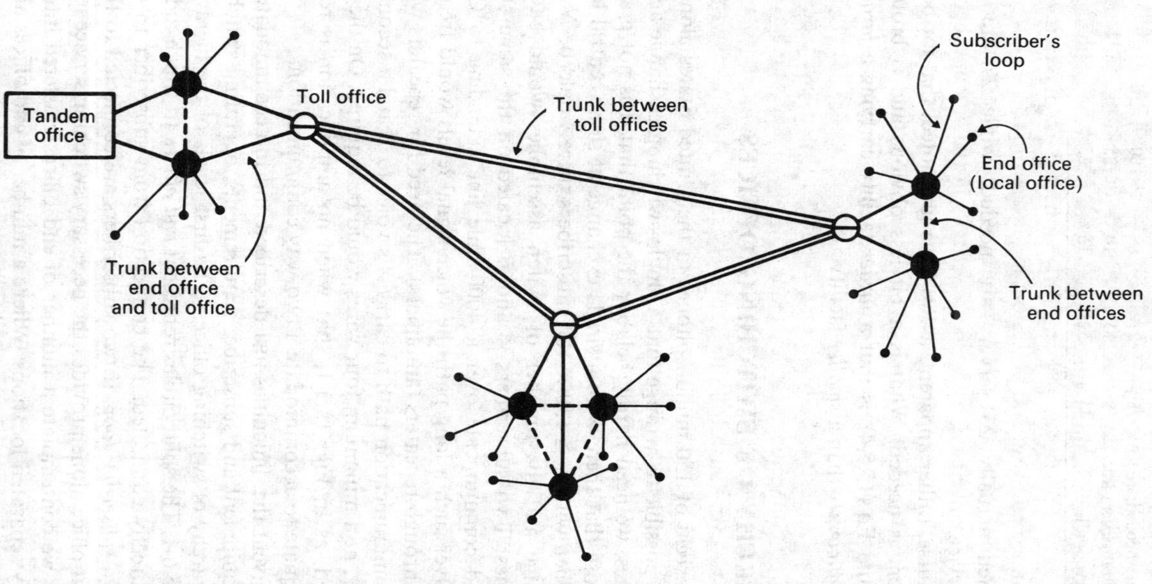

Figure 15.6-1 Illustrating interconnections of subscribers to end offices, end offices to toll offices, and trunks between toll offices.

serviced by nearby end offices may be interconnected. For the purpose of allowing communication between distant clusters of end offices, toll offices have been provided. The end offices in one area are connected to a nearby toll office and the toll offices are, in turn, interconnected through toll-connecting trunks. While, for the sake of simplicity, we have not so indicated, depending on geography and customer usage, it may happen that an end station will be connected to more than one toll station and also end stations connected to different toll stations may also be connected by an end office to end office trunk.

A very populous area, such as a large city, will have many end offices. Much of the traffic will involve calls between subscribers in the city and hence between end offices. To facilitate this end office to end office traffic a *tandem* switching office may be provided as shown in Fig. 15.6-1 for one group of end offices. Calls between subscribers connected to different end offices in the city will go by way of a tandem office. Intercity calls will use a toll office.

In the United States there are about 10,000 Bell system end offices, each one of which must be connected to at least one toll office. The number of such toll offices is consequently so large that it would still be prohibitive to provide interconnections from every such toll office to every other. Accordingly the hierarchy is continued so that altogether there are five *classes* of switching facilities. These are identified by name and class number in Fig. 15.6-2. As noted, there are about 10,000 end offices in the Bell system in the United States and progressively fewer centers as we go up in the hierarchy, there being only ten regional centers. The tandem offices, which may interconnect end offices, are not shown.

Suppose that subscriber *A* in Fig. 15.6-2 wants to call subscriber *B* in a distant location and there is no direct connection between their end offices. End

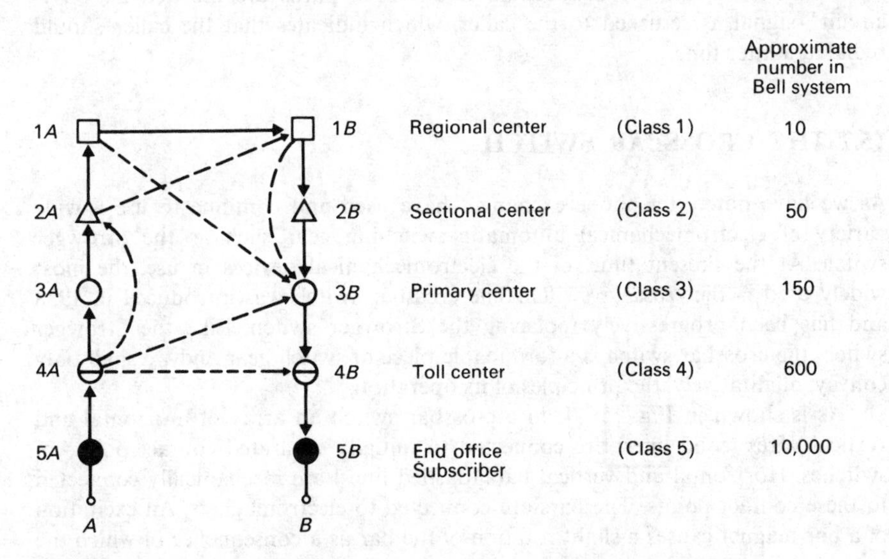

Figure 15.6-2 The hierarchy of switching centers showing possible interconnections.

office *A* is certainly connected to some toll center (4*A*) which is certainly connected to some primary center (3*A*), etc, and finally all regional centers are accessible to one another. Hence if there is no alternative, the call can be routed through all of the centers shown: from 5*A* to 1*A* through all the intermediate centers then to 1*B* and finally to 5*B* again through all the *B* intermediate centers. This path, being a route of last resort, adequate facility is incorporated to assure that the possibility of blocking is very small.

On the other hand, depending on the locations of the various centers, it may be that toll centers 4*A* and 4*B* are connected. In such a case the call will be routed over the shorter path $5A \rightarrow 4A \rightarrow 4B \rightarrow 5B$ as shown by one of the dashed paths in Fig. 15.6-2. If there is no 4*A* to 4*B* connection there may be a connection from toll center 4*A* to primary center 3*B* giving rise to an alternative route which again circumvents the very long route up and down the entire hierarchy. As we have noted, every toll center will have a direct route to a primary center. But, again depending on location, a toll center may also have a connection to a sectional center such as from 4*A* to 2*A*. And it may be that center 2*A* is connected not only to 1*A* but to 1*B* as well. We have assumed that there is no direct connection between end offices 5*A* and 5*B*, but otherwise many other interconnections may be available, some of which are shown by the dashed lines in Fig. 15.6-2, any one or combination of which may shorten the path from *A* to *B*. The dashed connections will generally be heavily utilized and hence attempted calls through them may frequently be blocked. As noted, if all attempts to find a shortened path meet with failure, the long path up and down the hierarchy will be employed. Typically, each caller may be routed through as many as seven possible paths, starting with the shortest and proceeding to the longest, in an attempt to complete the connection. If all seven paths are blocked a "busy-circuit" signal, is returned to the caller, which indicates that the caller should redial at a later time.

15.7 THE CROSSBAR SWITCH

As we have noted, telephone exchanges have used and continue to use a wide variety of electromechanical automatic switching gear such as the Strowger switch. At the present time, of the electromechanical devices in use, the most widely used is the *crossbar* switch. The crossbar switch was introduced in 1938 and has been progressively replacing the Strowger switch. Like the Strowger switch, the crossbar switch is a formidable piece of switch gear and we shall only convey, qualitatively, the principles of its operation.

As is shown in Fig. 15.7-1, in a crossbar switch an array of horizontal and vertical wires (solid lines) are connected to initially separated contact points of switches. Horizontal and vertical bars (dashed lines) are mechanically connected to these contact points. The bars are connected to electromagnets. An excitation of a bar magnet causes a slight rotation of the bar as a consequence of which the contact points connected to the bar move closer to its facing contact points. The

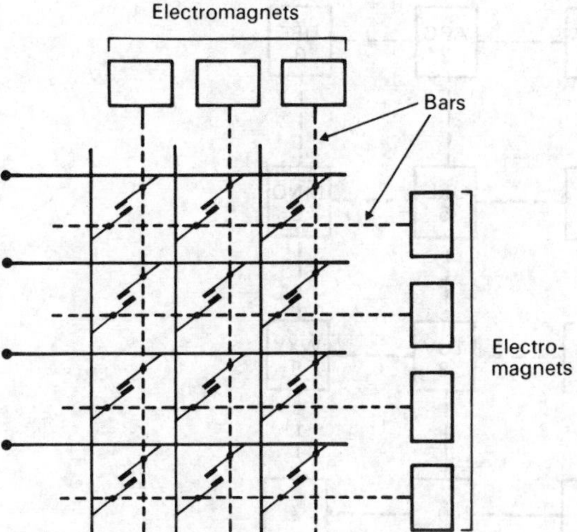

Figure 15.7-1 Illustrating the principle of the crossbar switch.

motion is not enough to cause contact. If however, an electromagnet connected to a horizontal bar and an electromagnet connected to a vertical bar are *both* activated, the contact of the intersection of the two bars will close because both of the contact points will move, each toward the other. There is mechanical latching incorporated into crossbar switches so that when a horizontal and vertical bar are activated, a contact is made and continues to hold even if the horizontal bar activation is released. The contact will not open until the vertical activation is also released.

15.8 COMMON CONTROL

The Strowger switching system which was described in Sec. 15.5 is an example of a step-by-step system. It has a number of limitations. In the first place, for the entire call duration, the entire set of switches including contacts and all the relay mechanisms are " busy " and are not available to service other calls. Secondly, a subscriber's physical location, i.e., the particular loop to which he is connected, imposes constraints on the telphone number which can be assigned to him. If the subscriber moves, even to a relatively nearby location, it may be necessary to change the number. Finally, the mechanism by which the equipment at the exchange finds the correct path for the call is rather slow acting and, since the system responds to one number at a time, there is limited facility for analyzing the information contained in the entire number. Many step-by-step exchanges cannot readily distinguish a toll call from a local call and the dialer must alert the exchange that a toll call is intended by dialing a " 1 " before dialing the called number itself.

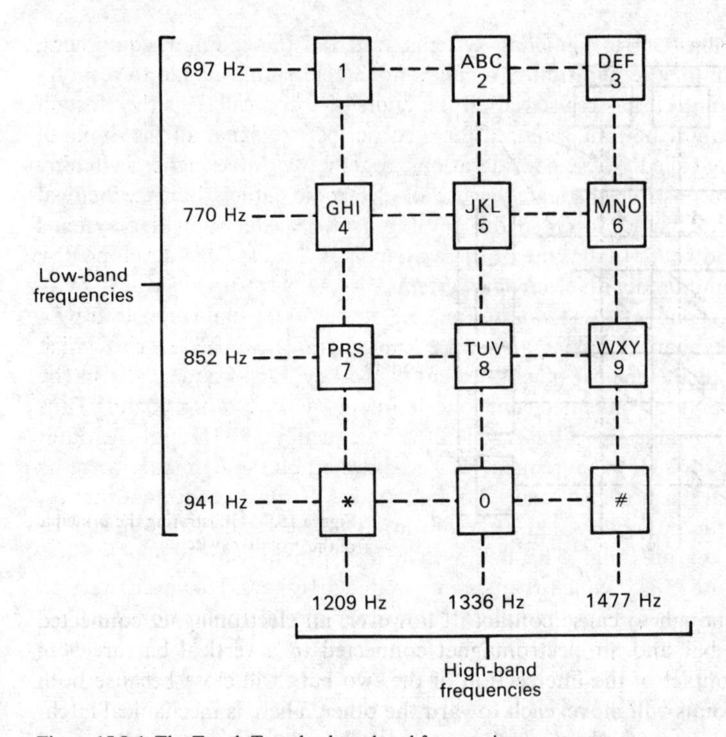

Figure 15.8-1 The Touch Tone keyboard and frequencies.

The *common control* method of switching overcomes the disadvantages of step-by-step switching. With common control the switching system makes no connections until it has received the entire number. The number is stored in a memory. The information conveyed by the number is then translated by logic circuitry into instructions to switch contact mechanisms which then establish the appropriate connections. The connections having been made, the logic circuitry becomes available to service other calls so that only the contact mechanisms are tied up for the duration of the call. It is possible to use common control in conjunction with Strowger switches but such control is used much more effectively with crossbar switches and this feature of crossbar switches constitutes one of its important merits.

Pushbutton Dialing

The pulses produced by rotary dials have a frequency which depends on the speed with which the dial spins back to its rest position and is of the order of 10 per second. Typically it takes about 12 seconds to dial a 7 digit telephone number. If the dialing pulses were generated at a higher frequency, the higher speed might well make it difficult for the relatively massive Strowger switches to follow. In common-control signalling systems, much of the common equipment, while not tied up for the duration of a call is nonetheless unavailable to respond to a second call until it has registered all the digits of a first call. It is therefore of advantage to reduce the time required for a subscriber to signal all the digits of the number being called. These considerations, together with the higher switching speeds which are possible through the use of electronic rather than mechanical switching elements, have prompted the development of faster signalling systems than the standard rotary dial. One of these systems is *Touch Tone* developed by the Bell system and used with electronic systems.

On a Touch Tone set, as shown in Fig. 15.8-1, the rotary dial is replaced by a twelve button keyboard. Pressing any single key generates two frequencies which are transmitted to the central office to identify the key. One frequency is in the low-frequency band, as indicated, and one frequency is in the high band. Thus pressing key 9 transmits two signals, one at the frequency 852 Hz and the other at the frequency 1477 Hz. The frequencies were selected on the basis that there be no likelihood that harmonics of one set of frequencies be mistaken for another set and also very little possibility that an audio input to the handset transmitter be mistaken for a key pressing. (The Bell system also provides for a fourth high-band frequency, 1633 Hz, which then accommodates 16 keys. However, such 16 button keyboards are used only in military and other special applications.)

15.9 SWITCHING MATRICES

A special merit of the crossbar switch is that it lends itself readily to the implementation of a basic array of switches called a *switching matrix* which is found to be very effective in interconnecting many subscribers. Such a maxtrix is shown in Fig. 15.9-1. Subscribers *A*, *B*, *C* and *D* are connected to the horizontal wires and subscribers *a*, *b* and *c* to the vertical wires. The crosses represent the (normally open) contacts. Each horizontal and vertical line in Fig. 15.9-1 represents at least

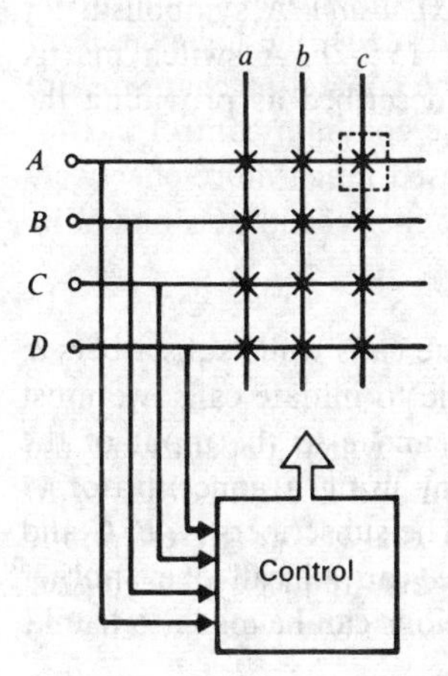

Figure 15.9-1 A one-way matrix. Station *A* reaches station *C* when the contacts enclosed in the dotted box are closed.

two (and possibly many more) wires and each cross represents a corresponding number of contact points. We have provided that subscribers *A*, *B*, *C* and *D* exercise control over the switch closures. Hence these subscribers can initiate a call while *a*, *b*, *c* can only receive. Of course, once a connection has been established, conversation can proceed symmetrically in both directions. The matrix of Fig. 15.9-1 is called a *one-way* matrix since it allows that the direction of the call be in one direction only, from a "horizontal" caller to a "vertical" called party. The potential callers are called matrix inputs or *inlets*, the called parties are connected to matrix outputs or *outlets*.

On the matrix of Fig. 15.9-1 it is intended that in any row or in any column no more than one switch be closed at any given time. Hence, the inlet subscribers cannot be connected to one another and neither can the outlet subscribers be connected. Even with this limitation the matrix is entirely effective in a *transit* facility. Thus, in a toll station, subscribers in city *A* need to be connected to subscribers in city *B* but the toll station need provide no facility for interconnecting subscribers in the same city. We note finally that if the number of inlets is N and the number of outlets is M then the number of switches is NM. If the matrix is square, i.e., $M = N$ then the number of switches is N^2.

Concentration and Expansion

Even though it is necessary as a matter of practicality, to allow blocking, nonetheless it is still necessary that a telephone switching system allow *full access* of each potential called party from each potential caller. In Fig. 15.9-2*a* a system using matrices as in Fig. 15.9-1 interconnects six subscribers in one city to six subscribers in a second city. The arrangement provides full access, since any subscriber located in one city can talk to any subscriber in the other city. However, blocking may occur because we have provided only three links between cities. The arrangement on the left in Fig. 15.9-2*a* where a larger number of subscribers has access to a fewer number of trunks is referred to as a *concentration* and correspondingly, on the right there is a complementary *expansion*. A symbolism for such concentration and expansion is shown in Fig. 15.9-2*b*. A switch matrix which has the same number of inlets and outlets is described as providing the function of *distribution*.

Two-Way Call Initiation

In Fig. 15.9-1 the subscribers *A*, *B*, *C* and *D* can initiate calls while subscribers *a*, *b* and *c* cannot do so. If we want *a*, *b* and *c* to be able to initiate calls, we must expand the matrix and provide connections for *a*, *b* and *c* to the input of the matrix. In Fig. 15.9-3 is shown such an arrangement using a concentrator (4 inlets, 2 outlets) and an expander (2 inlets, 8 outlets). The subscribers *A*, *B*, *C* and *D* on the inlets, all serviced by the same local office, can all call one another because of the "loop-back" connections. Also connections can be made to trunks

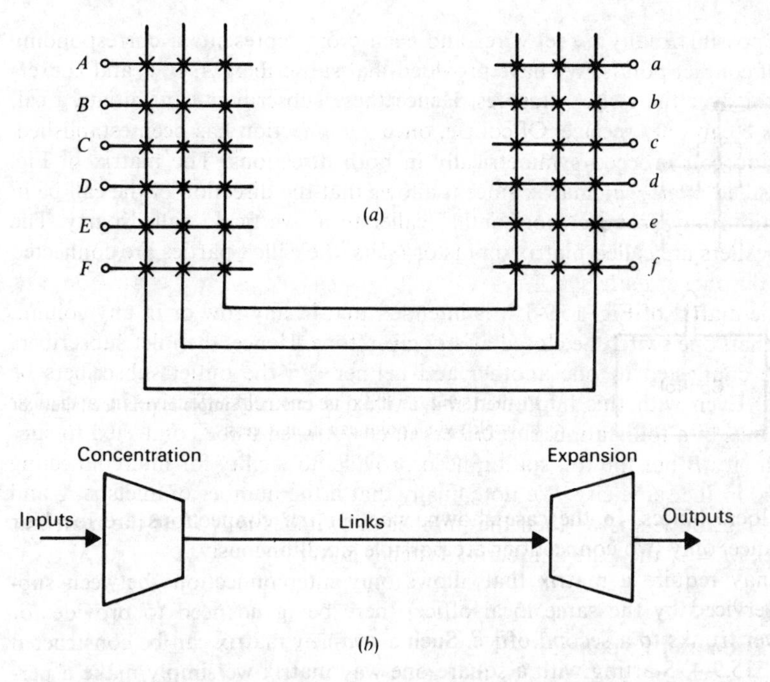

Figure 15.9-2 (*a*) Rectangular matrices used to provide *concentration* and *expansion.* (*b*) Symbols for concentration and expansion.

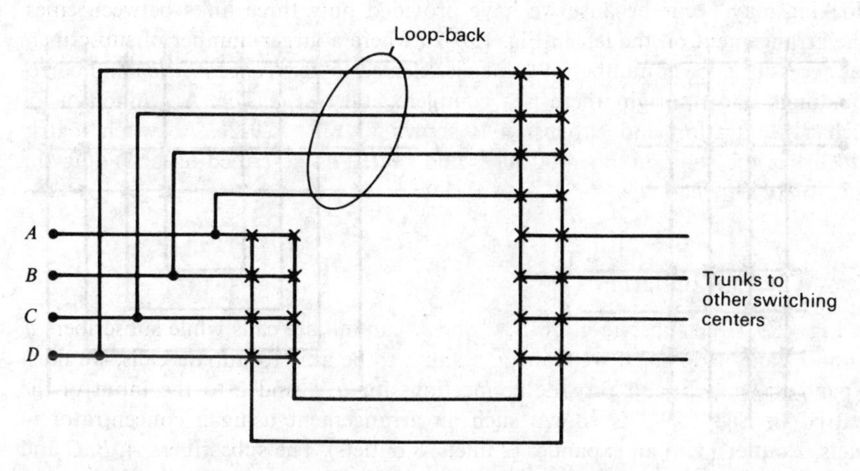

Figure 15.9-3 A concentrator and expander used to allow communication both between subscribers served by the same local office and between subscribers served by different offices.

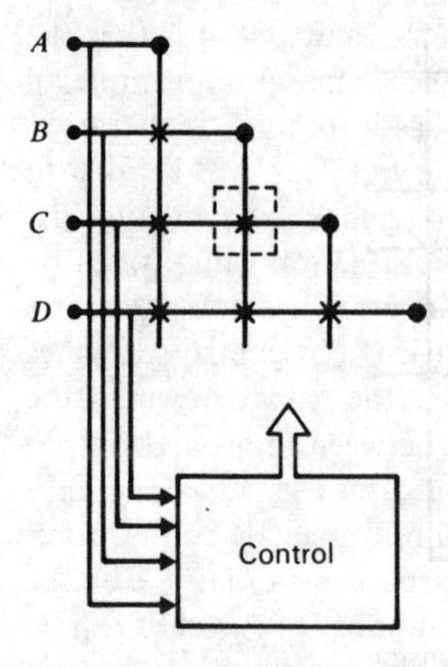

Figure 15.9-4 To illustrate that in a simple matrix the failure of a simple switch point can limit access.

to other local offices. In the case shown, whether the connections are local or lòng distance, only two connections are possible simultaneously.

We may require a matrix that allows only interconnections between subscribers serviced by the same local office, there being no need to provide for transit over trunks to a second office. Such a *two-way* matrix can be constructed as in Fig. 15.9-4. Starting with a square one-way matrix we simply make a permanent connection between a horizontal and a corresponding vertical wire. This permanent connection is represented by each of the four heavy dots at the intersection of wires. We observe that while a one-way rectangular matrix requires N^2 switches, the two-way matrix (also called a triangular matrix because of its geometrical appearance) requires $N(N-1)/2$ switches.

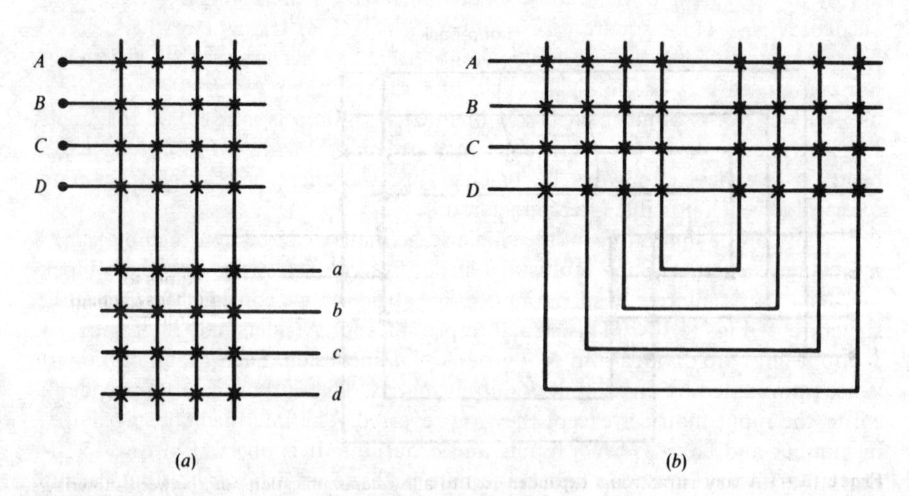

Figure 15.9-5 Two stage matrices used to provide multiple paths.

A difficulty associated with the matrices of Fig. 15.9-1 and 15.9-4 is that the failure of a single switch will make some subscribers inaccessible. For example, in Fig. 15.9-4, if the switch enclosed in the dashed box should become inoperative, then stations B and C will not be able to communicate but each will be able to communicate with other stations. Similarly a failure of the enclosed switch in Fig. 15.9-1 will prevent station A from calling station C. The situation is improved by using two-stage matrices as shown in Fig. 15.9-5*a* and *b* for a rectangular and loop-back matrix, respectively. In such two-stage matrices the control provides for the operation of one switch in each assembly. For the cases shown in the figure it can be verified that there are now four paths between input and output in Fig. 15.9-5*a* and four paths between any of the stations in Fig. 15.9-5*b*. Hence the random failure of a limited number of switches will not preclude connections. Of course the required number of switches has increased substantially. For the rectangular matrices with N inputs and N outputs the number of switches is now $2N^2$ compared to N^2 for the single stage case. For the loop-back, two-stage matrices the increase is proportionately much larger, increasing from $N(N-1)/2$ to $2N^2$.

15.10 MULTIPLE STAGE SWITCHING

The switching matrices of Fig. 15.9-1 and 15.9-4 establish a connection between parties by closing a single crosspoint. For this reason, such switching structures are called *single-stage* structures. It is an advantage that only a single crosspoint closing is required but it is a disadvantage that a specific crosspoint is needed to establish a connection between specific subscribers, for if that crosspoint should fail there is no way to make the connection. Further, the number of crosspoints needed, $N(N-1)$ for square arrays and $N(N-1)/2$ for triangular arrays, is prohibitively large for large N. Finally, we note that the hardware of the matrices is not being used very economically. For, in a square matrix, even if all lines are in use, only one crosspoint in each row or in each column is in use. Still the single-stage structures have the merit that they are non-blocking. Provided a called party is idle he can always be reached because there is a single crosspoint reserved for each possible interconnection.

Many of the limitations of the single stage matrix can be remedied by using a multistage structure. Such multistage structures can, however, introduce blocking. As a representative of such a multistage structure we consider the three stage structure of Fig. 15.10-1. This structure provides for N inlets and N outlets. The N input lines are divided into N/n groups of n lines each. Each group of n inputs is accommodated by an n-input, k output matrix. The output matrices are identical to the input matrices except they are reversed. The intermediate stages are k in number and have $\alpha = N/n$ inputs and α outputs. It is oberved in Fig. 15.10-1 that an arbitrary input can find k alternate paths to an arbitrary output because the connection can be established through any one of the k intermediate stages.

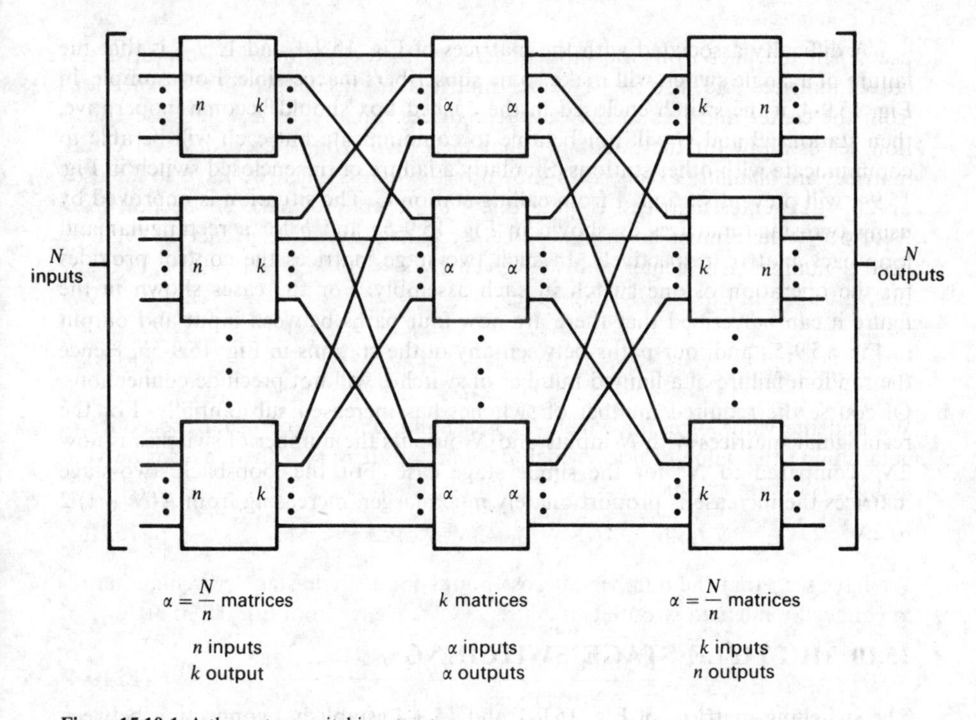

Figure 15.10-1 A three-stage switching structure.

To determine the number N_x of cross points in the three-stage matrix we note first that each of the first stage and third stage matrices has $n \times k$ cross points. The total of these input and output stages combined is $2N/n$, yielding a number of cross points $(2N/n)(n \times k) = 2Nk$. The total number of cross points in the center stages is $k(N/n)^2$ so that altogether the total number of cross points is

$$N_x = 2Nk + k\left(\frac{N}{n}\right)^2 = k\left[2N + \left(\frac{N}{n}\right)^2\right] \tag{15.10-1}$$

It is now of interest to determine how large N_x in Eq. (15.10-1) must be, if, like the single stage matrix, the three stage matrix is to be nonblocking. We note, first of all, that to allow an outlet for each input of the first stage matrices, even if all the inputs are being used simultaneously, it is necessary but not sufficient that k be as large as n. To see that $k = n$ is not sufficient, let each of the input matrices be called X_i $(i = 1, 2, \ldots, \alpha)$ and each of the output matrices be called Y_i as shown in Fig. 15.10-2. Suppose that we want to make a connection from the input X_{11} of the first stage matrix X_1 to the output Y_{11} of the third stage matrix Y_1. We keep in mind that from each input matrix there is only *one connection* to each center stage and there is also *one connection* from each center stage to *each* output matrix. Now let it be that all $n - 1$ inputs, other than X_{11} of matrix X_1 are in use and also that all $n - 1$ outputs, other than Y_{11} of matrix Y_1 are also in use. Then the lines from the input matrix to the center matrices use up $n-1$ center matrices and also $n-1$ center matrices are used up by the connections from the center matrices to the output matrices. Finally, as a worst case condition, let us allow for the possibility that these two groups, each of $n-1$ center matrices are *mutually exclusive* as shown in the figure. Then we have so far used up $2(n-1)$ center matrices. If we are still to be able to make connection from X_{11} to Y_{11} there must be one more center matrix available. Hence in total we need a number k of center matrices given by $k = 2(n-1) + 1 = 2n - 1$. Using this value of k in Eq. (15.10-1) we have

$$N_x = 2N(2n-1) + (2n-1)\left(\frac{N}{n}\right)^2 \tag{15.10-2}$$

We find the value of n, for which N_x is a minimum, by the usual procedure of determining the value of n for which $dN_x/dn = 0$. If $N \gg 1$, then, as can be verified, N_x is minimum when $n \simeq \sqrt{N/2}$ and N_x becomes

$$N_x(\text{min}) \simeq 4N\sqrt{N/2} + 2N^2/\sqrt{N/2} \simeq 4\sqrt{2}\,N^{3/2} \tag{15.10-3}$$

We have seen that the number of cross points for a single-stage switching matrix to connect N inlets to N outlets is $N_x(\text{s.s.}) = N^2$. Hence from Eq. (15.10-3)

$$\lambda \equiv \frac{N_x(\text{min, 3 stage})}{N_x(\text{s.s.})} = \frac{4\sqrt{2}\,N^{3/2}}{N^2} \simeq \frac{4\sqrt{2}}{\sqrt{N}} \tag{15.10-4}$$

for large N. For $N = 128$, $\lambda = \frac{1}{2}$. For $N = 2048$, $\lambda = \frac{1}{8}$. The saving in cross points becoming more pronounced with increasing N. Still, even with such savings, the

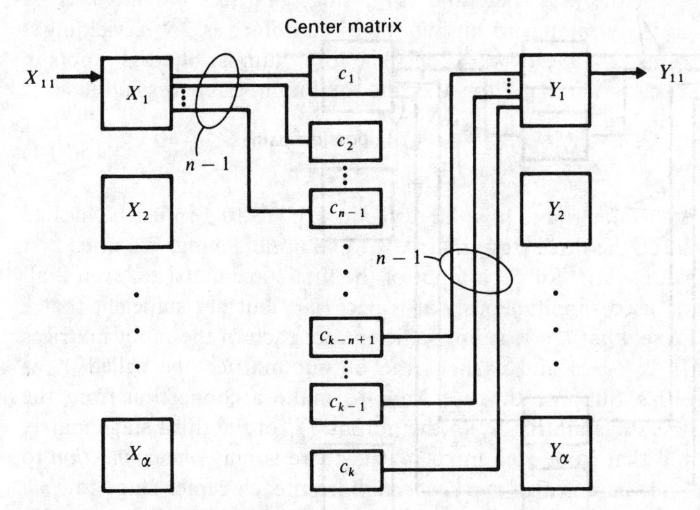

Figure 15.10-2 Three-stage switching matrix showing that $k - n + 1 \geq n$ to insure a connection between X_{11} and Y_{11}.

number of cross points can become prohibitively large for large N. Thus, for $N = (256)^2 = 65{,}536$, Eq. (15.10-3) yields $N_x(\text{min}) \simeq 95 \times 10^6$. Hence, when a very large number of lines must be accommodated, switching structures are used which have more stages, even up to eight stages.

We have noted earlier that as a practical matter it is both reasonable and necessary to accept occasional blocking of a call. Allowing such blocking will allow us to make further reductions in the cross point number. We have seen that in the three stage switching matrix we require that $k = 2n - 1$ to guarantee no blocking. Let us inquire into the extent to which k can be reduced without introducing unacceptable blocking and into the extent to which cross points will be saved thereby.

Suppose that we have made observations on the demands on the switching structure over an extended period and we have determined that an arbitrarily selected input terminal is in use p percent of the time. Of course, it is to be understood that the reason the input terminal is in use is that there is a completed connection to some output terminal and that there is presumably a conversation in progress over a link joining an input and an output terminal. We turn our attention to the X_1 matrix shown in Fig. 15.10-3. If only one of the n inputs is in use, then only one of the k outputs must be in use. Since there are k outputs, the probability that a *particular* one of the outputs is in use is p/k. Since there are n inputs, each with *occupancy* probability p, the probability that a particular one of the k paths from input to center stage is in use is $P = np/k$, providing, of course, that $P \leq 1$. If $pn/k > 1$, $P = 1$. The probability that the path is available is $1 - P$. Similarly the probability that a particular path from center stage to output stage

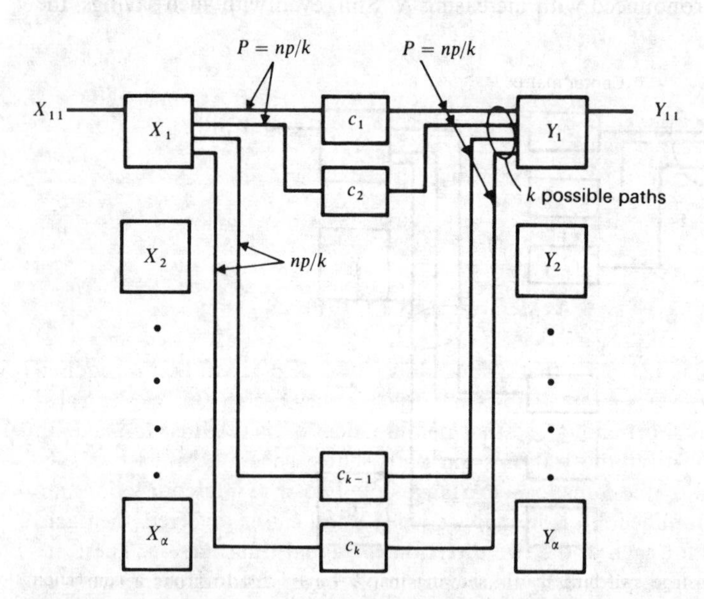

Figure 15.10-3 The k paths between X_i and Y_j (illustration is for $i = j = 1$).

is available is $1 - P$. The probability that both of these particular paths are available, thereby forming a particular idle link from input to output is $(1 - P)^2$. The probability that this particular link is busy is $[1 - (1 - P)^2]$. Now let us select one input X_{11} and one output Y_{11} in Fig. 15.10-3. Mr. Jones, whose phone is connected to X_{11} wants to talk to Mr. Smith whose phone is connected to Y_{11}. The end terminals having been selected, there are only k links between them. The probability that all k links are busy is the probability B that the attempt of Mr. Jones to reach Mr. Smith will be blocked, and is then given by

$$B = [1 - (1 - P)^2]^k \tag{15.10-5}$$

As an example of the use of Eq. (15.10-5) consider the following:

Problem A three-stage switching structure is to accommodate $N = 128$ input and 128 output terminals. There are to be 16 first stage and third stage matrices. (*a*) If the structure is to be nonblocking how many cross points are required. (*b*) It is known that at peak traffic periods the utilization probability $p = 10\%$. Suppose that the number of cross points is reduced by a factor of 3 below the number required to avoid blocking. What is the probability that a call will be blocked?

SOLUTION We have (see Fig. 15.10-1) that

$$n = \frac{N}{16} = \frac{128}{16} = 8$$

To avoid blocking we require $k = 2n - 1 = 15$. Using Eq. (15.10-1) we have

$$N_x = 15(256 + 16^2) = 7680 \text{ cross points}$$

To reduce the number of cross points to $7680/3 = 2560$ we must reduce k from 15 to 5. Using Eq. (15.10-5) we find, with $P = np/k = 8(0.1)/5 = 0.16$,

$$B = [1 - (1 - 0.16)^2]^5 = 0.002 = 0.2\%$$

15.11 TWO AND FOUR-WIRE CONNECTIONS

We have seen as in Fig. 15.3-1 that a single loop (two wires) is all that is required to provide complete service to a subscriber. Such is the case so long as the transmission distance is short enough so that no amplification is required to make up for the attenuation over the transmission path. Thus, generally when two subscribers are served by the same local exchange only two-wire interconnections are required. When amplification is needed, as when a toll call is involved, then gain must be provided for each of the two directions of signal transmission. There are available two terminal repeaters, which can simply be bridged across a two-wire line, which can provide gain in the two directions simultaneously. However, as a matter of practicality such two-way repeaters lack reliability and dependable stability and are generally not used. Instead, as shown in Fig. 15.11-1 separate one-way repeater amplifiers are used, one to amplify the signal generated at station A and to be transmitted to B and a second for the B to A transmission.

We observe that each amplifier's output is connected to the input of the other amplifier. Such an arrangement will result in oscillations, in spite of the *hybrid*, which ideally should isolate the input and output lines. The oscillations produce a loud, sustained, tone referred to frequently as "singing." In a telephone system, the transmitted voice signal from A reaches B and is amplified and returned to the receiver at A. Thus, this system also produces a loud and annoying "echo."

An echo produced by the circuit shown in Fig. 15.11-1 can be suppressed using the circuit shown in Fig. 15.11-2. In this circuit assume that A is transmitting to B. If the received voltage, at point q, exceeds the preset threshold voltage V_q then a loss L_B is inserted in the path from B to A to attenuate the "echo" signal. When A is not talking, the threshold voltage is not exceeded and the loss is reduced so that speaker B can send a signal to A.

One major drawback of this technique is that if A and B speak simultaneously (a situation which occurs more often than you may think), speaker B's voice may be attenuated by the large loss L_B introduced by speaker A, thereby producing *clipping*. "Clipping" may sometimes be heard on long distance calls when the *echo-suppressor* is not operating properly.

Figure 15.11-3 is an echo suppressor designed to permit two speakers to talk "naturally," i.e. both speakers are able to talk simultaneously without noticeable clipping and without noticeable echo.

In this figure, a differential amplifier measures the relative voltage levels between points a and b. If speaker A is talking, the voltage at a exceeds the voltage at b and L_A is reduced while L_B is increased to reduce the echo. If speaker B interrupts so that both speakers are talking simultaneously the voltage at b becomes comparable to the voltage at a, the difference amplifier output becomes

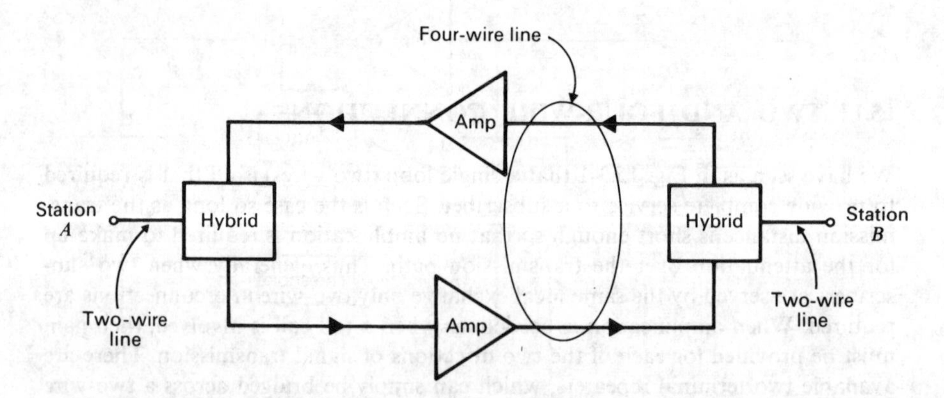

Figure 15.11-1 The four-wire connection which is required when amplification must be provided.

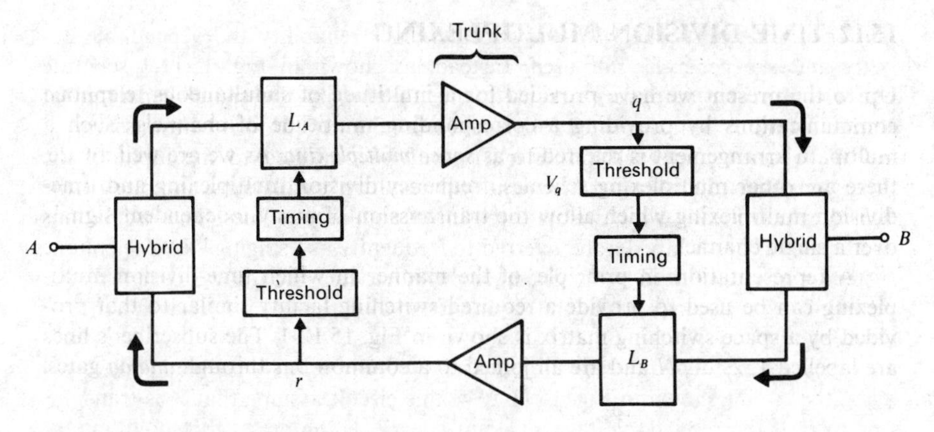

Figure 15.11-2 Basic concept of an echo suppressor.

small and both loss elements L_A and L_B are reduced. The echo now is not heard since the voice signal of speaker B is louder than the echo of speaker A.

It is clearly more expensive and complicated to switch four-wire lines than two-wire lines. Two separate paths through the switching matrices must be established and twice as many contact points are involved. We can however avoid four-wire switching by converting to two-wire lines, performing the required switching and then, if necessary convert back to four-wire lines. Such a procedure is used very sparingly however, because each such conversion saves switching complexity only at the expense of adding echo suppressors. Further, the balance provided by the echo suppressors which is necessary to avoid echo is never perfect so that as a matter of practice it is difficult to suppress an echo if too many conversions are used.

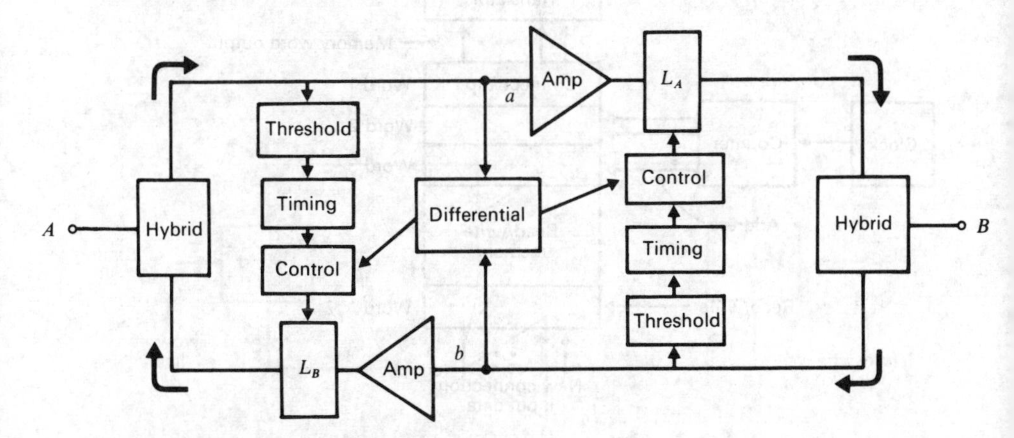

Figure 15.11-3 Balanced echo suppressor.

15.12 TIME-DIVISION MULTIPLEXING

Up to the present we have provided for a multitude of simultaneous telephone communications by providing a corresponding multitude of channels. Such a multipath arrangement is referred to as *space multiplexing.* As we are well aware, there are other multiplexing schemes, frequency-division multiplexing and time-division multiplexing which allow the transmission of many independent signals over a single channel.

A representation, in principle, of the manner in which time-division multiplexing can be used to provide a required switching facility similar to that provided by a space switching matrix is shown in Fig. 15.12-1. The subscriber's lines are labelled 1, 2, ..., N and are all joined to a common bus through *analog* gates.

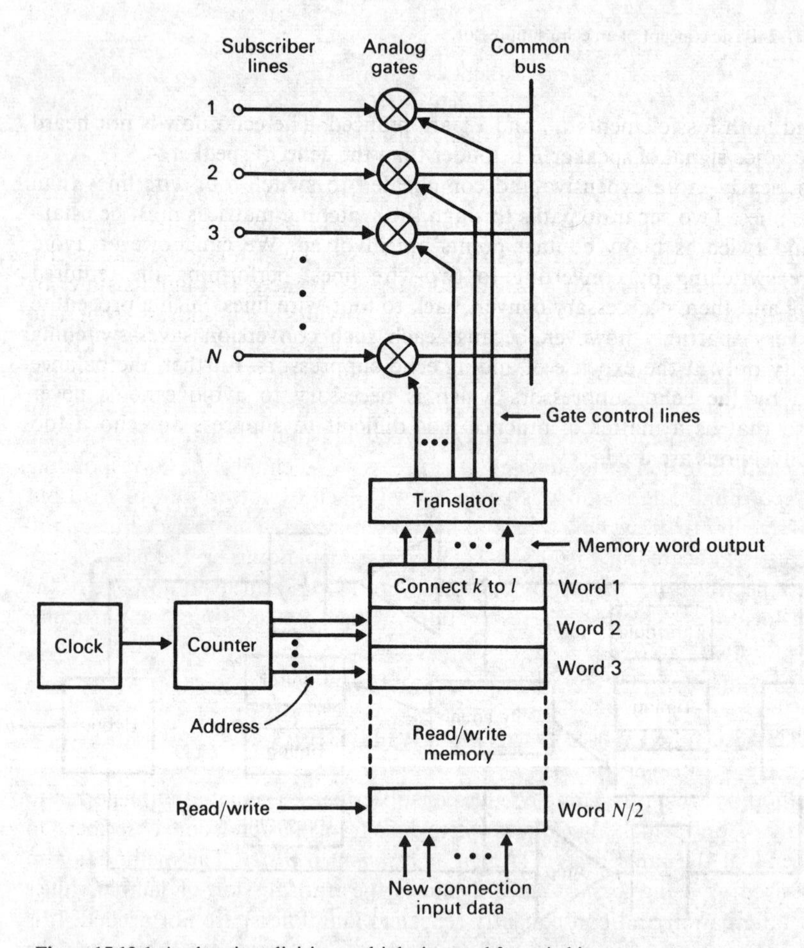

Figure 15.12-1 Analog time-division multiplexing used for switching.

The gates can be *open* to allow for signal transmission or *closed* to prevent transmission. (These gates can be viewed simply as switches. If so, then the gate is *open* when the switch is *closed* and the gate is *closed* when the switch is *open*.) The *gate control lines* serve to control the gates, i.e., to close and open the switches.

If we want to allow subscriber k to transmit a signal to subscriber l then we must allow subscriber l to sample the signal from subscriber k at the Nyquist rate or more frequently. We have already noted that in voice telephony a sampling rate of 8000 Hz (i.e., every 125 μs) is adequate. To permit the sampling we must open for signal transmission the two gates, one in subscriber k's line and one in the line of subscriber l. The length of time τ during which this connection can be allowed to persist is, of course, determined by the number of subscribers who share the bus. If there are N subscribers then there can be at most $N/2$ simultaneous connections. If, say, we require that $\tau = 5\ \mu$s then the number of connections will be $N/2 = 125/5 = 25$ and the number of subscribers who can be served is $N = 50$.

The read/write memory (also called a random-access memory, RAM) in Fig. 15.12-1 has a capacity of $N/2$ words. Each word is a coded instruction indicating which analog gates are to be open for transmission. Thus word 1, as shown, holds the instruction that subscriber k is to be connected to subscriber l. When this word is read out of the memory it is applied to a *translator* which decodes the instruction and sets its output *gate control lines* so that only gates k and l are open for transmission. The address specifying the word which is to be read is provided by a counter of modulo $N/2$. The counter is advanced through its states by a clock which runs at a rate to insure that it provides each address from 1 to $N/2$ every 125 μs. Thus the clock frequency is $8000\ (N/2) = 4000N = 200$ kHz, if $N = 50$, as in the example above. Each word in the memory holds an instruction specifying which gates are to be open. Of course when some subscribers lines are idle there will be memory word locations specifying that no gate is to be open. When a call is initiated or terminated, it is necessary to change the corresponding word. Accordingly the memory is provided with facility not only to be read but also with facility to allow new words to be written into it. This "new connection" data input is indicated in Fig. 15.12-1. Of course, depending on the past history of call connections the instruction "connect k to l" can appear in any one of the memory locations. There is no reservation of memory locations for particular subscriber pairs.

15.13 ANALOG TIME-DIVISION SWITCHING

One method by which the sample values of subscriber k's signal are transferred to a subscriber l, and which also allows the reverse transmission from subscriber l to subscriber k at the same time is known as *resonant transfer*. The method is also characterized as being *lossless* since it allows the transmission of sample values without attenuation and consequently repeaters (amplifiers) are not needed. It is an effective method when the overall transmission distances are short as in the

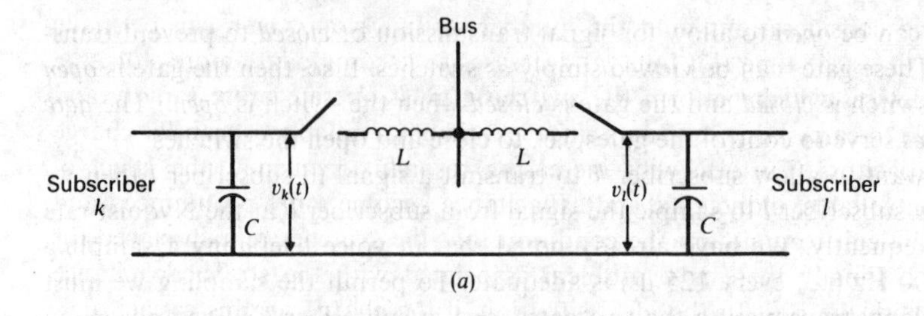

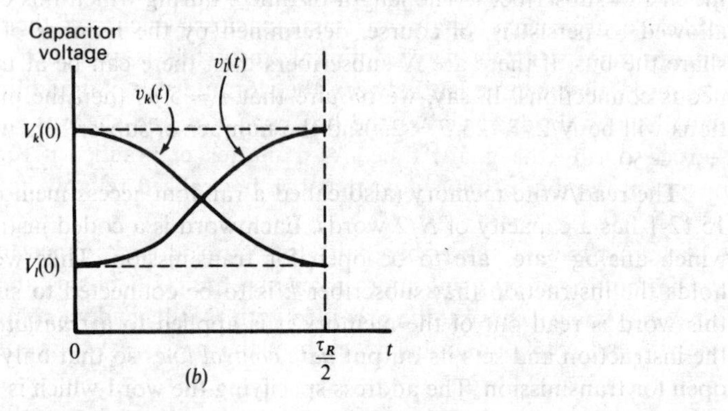

Figure 15.13-1 (*a*) Illustrating the principle of resonant transfer of samples. (*b*) Waveforms of capacitor voltages.

case of a *private branch exchange* (PBX). PBX switching facilities serve phones which are spatially in close proximity such as the phones of a company all of whose phones are in the same building. Resonant transfer is used in the Bell System 101 ESS (electronic switching system) system.

In Fig. 15.13-1*a* we have represented the controlled analog gates by switches which close during the time slot allocated to the connection between k and l. Inductors of inductance L have been added as shown to form a resonant circuit. The capacitors C on each subscriber's side are the output elements of the low-pass filters (cutoff at about 3.4 kHz) used to bandlimit the signal to be transmitted. Immediately before the switches close, each capacitor holds a sample voltage which is to be transferred to the other side.

The resonant frequency of the circuit is $f_R = \sqrt{1/LC}$ with a corresponding resonant period $\tau_R = 1/f_R$. It is left as a student exercise (Prob. 15.13-1) to verify that if the switches remain closed for a time of one-half τ_R and are then opened, then the charges on the two capacitors and the corresponding capacitor charges will have interchanged. Typical waveforms for the capacitor voltages $v_k(t)$ and $v_l(t)$ are shown in Fig. 15.13-1*b*. Sample values have been interchanged, and each subscriber can regenerate, from the received samples, the signal which was transmitted.

Practical considerations limit the number of time slots which can be made available. For as the time slot interval decreases to accommodate more slots, the electronic devices which perform the switching must operate more rapidly and must be able to carry larger currents since less time is available to transfer charge between capacitors. Further, with increasing resonant frequency, the effect of stray, uncontrollable, inductance and capacitance become more pronounced and it is more difficult to maintain the condition that the switches remain closed for exactly $\tau_R/2$. Finally there is a problem associated with the stray capacitance associated with the bus. In the ideal situation, if the charge on this capacitor is zero initially it will be zero after the charge transfer. In the non-ideal case, a small residual charge may remain on this capacitor which will then cause crosstalk between time slots. One way to minimize crosstalk is to reduce τ_R slightly so that a "guard" time can be left between the slot intervals. An additional controllable switch is now bridged across the bus capacitor and the bus capacitance is discharged during the guard time. As a matter of practice it turns out that the maximum number of time slots that can be realized reliably lies in the range 24 to 32.

The number of subscriber pairs that can be interconnected simultaneously can be multiplied by using more than one bus. If B busses are used, then each subscriber line will have to be provided with B independent controllable switches, one to connect the subscriber to each of the busses. Correspondingly, there is a

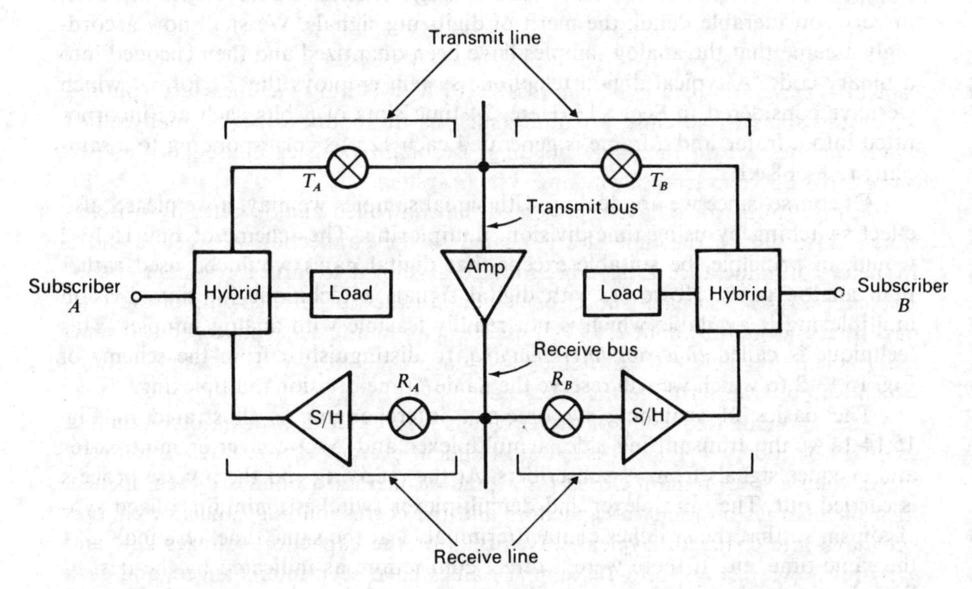

Figure 15.13-2 An analog time-division multiplex switch which uses only one amplifier for two-way transmission.

greater multiplicity of switches to control, and the number of bits in a word in the read-write memory of Fig. 15.12-1 will also have to be increased. With such multiplication of switches and busses, more than one subscriber pair will be able to use a particular time slot simultaneously. Thus one subscriber pair uses time slot 1 while both are connected to bus 1, a second pair uses the same slot while both are connected through different switches to bus 2, etc.

The system we have described is suitable when the physical distance between subscribers is short enough so that attenuation of samples is not a concern. In the more general case, where amplification is required, a four-wire system with its associated echo suppressors must be used. A block diagram of such a system which has the feature that only a single amplifier is used is shown in Fig. 15.13-2. In this arrangement two time slots are needed to accommodate a pair of subscribers. During one time slot, gates T_A and R_B are open for transmission so that A can transmit to B. During a different time slot gates T_B and R_A allow transmission and B transmits to A. The sample and hold circuits (S/H) hold the sample values from sample time to sample time and hence, as usual, provide equivalently a measure of low pass filtering and raise the received signal power level.

15.14 TIME SLOT INTERCHANGING (TSI)

In the switching mode described in the preceding section the signal samples were analog and the controlled switches were analog switches. We have seen, however, in very considerable detail, the merit of digitizing signals. We shall now accordingly assume that the analog samples have been quantized and then encoded into a binary code. A typical digital telephone system employs the $T1$ format which we have considered in Sec. 5.13. Here, 24 time slots of 8 bits each are incorporated into a frame and a frame is generated each 125 μs corresponding to a sampling rate of 8 kHz.

Of course, since we are dealing with signal samples we may, if we please, also effect switching by using time-division multiplexing. The scheme of Fig. 15.13-2 would, in principle, be suitable except that digital gates would be used rather than analog gates. However, with digital signals a technique for time-division multiplexing is available which is not readily feasible with analog samples. This technique is called *time-slot interchanging*, to distinguish it from the scheme of Fig. 15.13-2 to which we will reserve the name "time-division multiplexing."

The basics of switching by time-slot interchanging is illustrated in Fig. 15.14-1. At the transmitting side, a multiplexer and A/D converter multiplexes and encodes signals from N subscribers. At the receiving end the inverse process is carried out. The multiplexer and demultiplexer switches maintain a fixed synchronism so that the switches contact terminals 1 at the same time, terminals 2 at the same time, etc. If there were a direct connection, as indicated by the dashed line, the single channel shown would carry N signals but there would be no provision for switching.

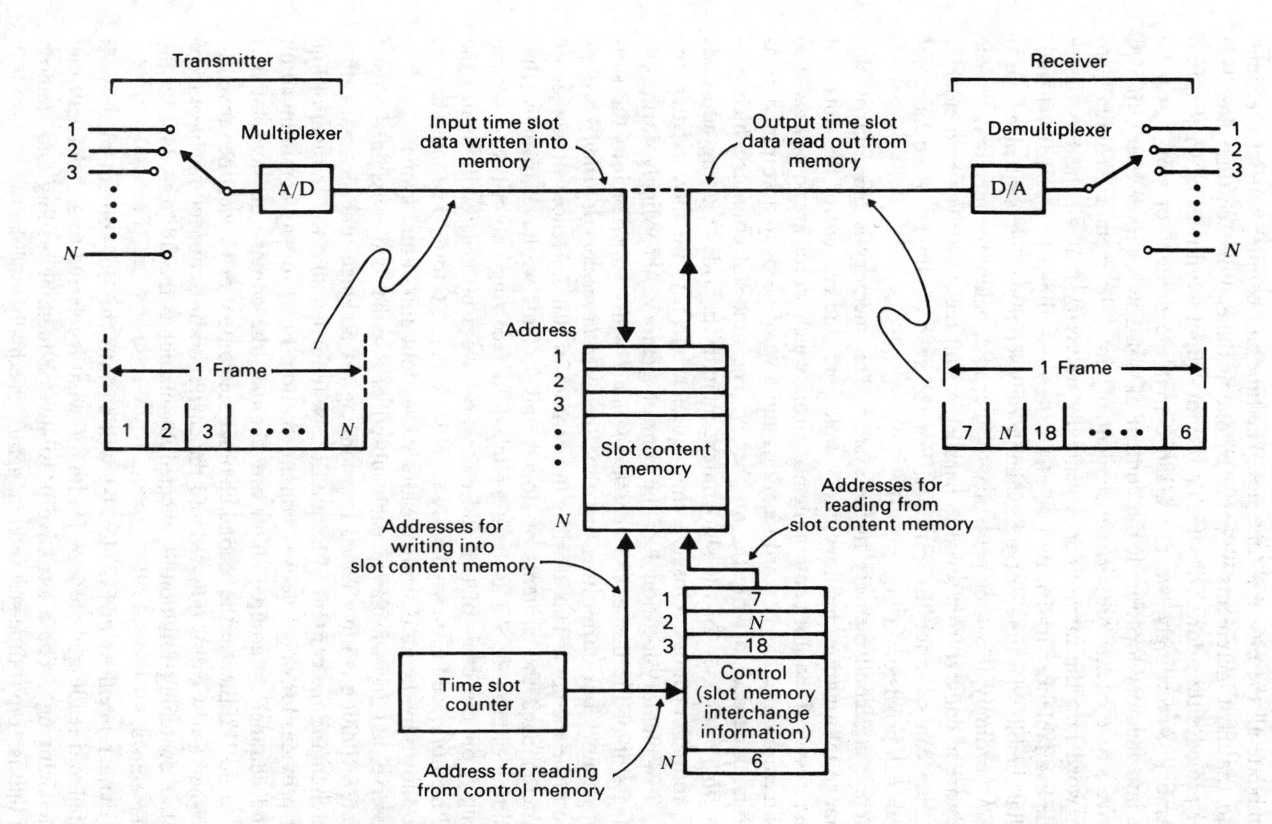

Figure 15.14-1 Illustrating the principle of time-slot interchanging.

To provide for switching, a *slot-content memory* is provided. This memory has N word locations, i.e., as many as there are time slots, and each location can accommodate all the bits in a time slot. The time slot counter runs at the time-slot rate, that is, it increments its count by one at the end of each time slot. This counter provides the address at which a memory input word is to be written into the memory. Accordingly, we can arrange that the content of slot 1 will be *written* into memory location 1, the content of slot 2 in location 2, etc, that is, successive slot contents are *written* sequentially into the memory. (Actually to write a word into the memory it is generally necessary that all the bits of the word be available in parallel, that is, all at the same time.) Figure 15.14-1 indicates that the bits in a time slot are available only serially, i.e., in sequence, one at a time. Accordingly, it will be necessary to interpose, before the memory, a piece of hardware to convert from serial to parallel. A shift register driven by a clock at the bit rate can serve for this purpose. For the sake of simplicity we have not included such provision in Fig. 15.14-1.

There is a second memory in the system. This memory is the control (slot interchange information) memory and it holds, not the content of the slots, but holds addresses of the slot content memory from which words are to be *read*. The control memory also has N word locations and a word needs to accommodate as many bits as needed to write the number N. The time-slot counter which provides *writing* addresses for the slot-content memory provides reading addresses for the control memory. Observe that, quite arbitrarily, we have written the address 7 in memory location 1 of the control memory. Accordingly during the first time slot of the frame, the *content of slot 1 will be written* into the slot-content memory but during this first time the *word in memory location 7 will be read* from the memory. This word is the content of slot 7. Hence, altogether we have interchanged the contents of slots 1 and 7. Thus we have arranged that transmitting subscriber 7 has been switched to receiving subscriber 1. Correspondingly, on the basis of the addresses we have arbitrarily written into the control memory, we have switched N to 2, 18 to 3, ..., 6 to N. The interchanges are also shown by the numbering of slots in the input and output frames.

Observe that, in this scheme, each subscriber at either end is assigned a fixed time slot. Switching is not effected by changing the assigned time slot but rather by interchanging the content of a time slot. Note further that the channel of Fig. 15.14-1 provides for only one-way communication. For two-way communication a second channel is required. If in one channel, the content of, say, slot 3 is moved to slot 12, then in the second channel the content of 12 must be moved to slot 3. When a new call is initiated and the corresponding connections have to be made, this switching function is effected by changing the address data in the control memory.

In the arrangement of Fig. 15.14-1 writing into the slot-content memory is sequential and reading is random, leading to what is referred to as "output time-slot interchanging". The alternative possibility, sequential reading and random writing that is "input time-slot interchanging" is equally feasible.

15.15 COMPARISON OF TSI WITH SPACE SWITCHING

During the course of a frame the slot content of Fig. 15.14-1 must be written-into and read-from a number of times N, equal to the number of slots per frame. The nature of memories and in particular semiconductor random-access, read-write, memories is that a read operation and a write operation cannot be done simultaneously but must be done at separate times. If then there are N slots per frame, then $2N$ read and/or write operations must be performed in the time of a frame. If sampling is done at the rate 8000 samples per second then the frame time is 125 μs and the time available for one read or write operation is 125 $\mu s/2N$. The time required to access a memory location to effect a read or write operation is called the cycle time t_C. (Actually there is a write cycle time t_{WC} and a read cycle time t_{RC} with $t_{WC} > t_{RC}$, but we shall not concern ourselves with this detail.) At the present time, semiconductor memories with cycle times about $t_C = 500$ ns are readily available. Setting $t_C = 500 \text{ ns} = 125\ \mu s/2N$ we find $N = 125 \simeq 128$. The TSI system we have considered is nonblocking. Let us then compare this system with the system described in the illustrative problem of Sec. 15.10. There we found that a three stage nonblocking space switching system with 128 inlets and 128 outlets requires 7,680 cross points. This result suggests, and we state without detailed justification, that to the extent that it is feasible to do so, it is advantageous to use time division switching (TSI) rather than space switching.

15.16 SPACE ARRAY FOR DIGITAL SIGNALS

Even if we choose to effect switching by using space arrays of switches rather than by the method of TSI, the fact that we are dealing with digital signals constitutes an important advantage. A space switch matrix suitable for use with digital signals is shown in Fig. 15.16-1 where we have provided for N inlets and M outlets. Note that bulky expensive mechanical contact points operated by equally bulky, expensive, and power consuming, electromagnets have been replaced by digital gates. In this age of digital integrated electronics the advantages need hardly be belabored. A connection between an inlet and an outlet is made by the simple expedient of enabling the gate connecting the two lines. Such enabling requires no more than maintaining an enabling voltage on the selection terminal s_{ij}, of the appropriate gate. Only one gate in a column is enabled at any time. The electronics of the gates can be designed (and such is assumed in the present case) so that a gate which is not enabled has no effect on the outlet line to which the gate outlet is connected. Again, of course, we must note that because logic gates are unidirectional, two paths through a switching matrix must be established to accommodate a two-way conversation. Like mechanical space arrays, the logic gate array can serve for *concentration*, *expansion* or *distribution* depending on whether N is larger, smaller or equal to M.

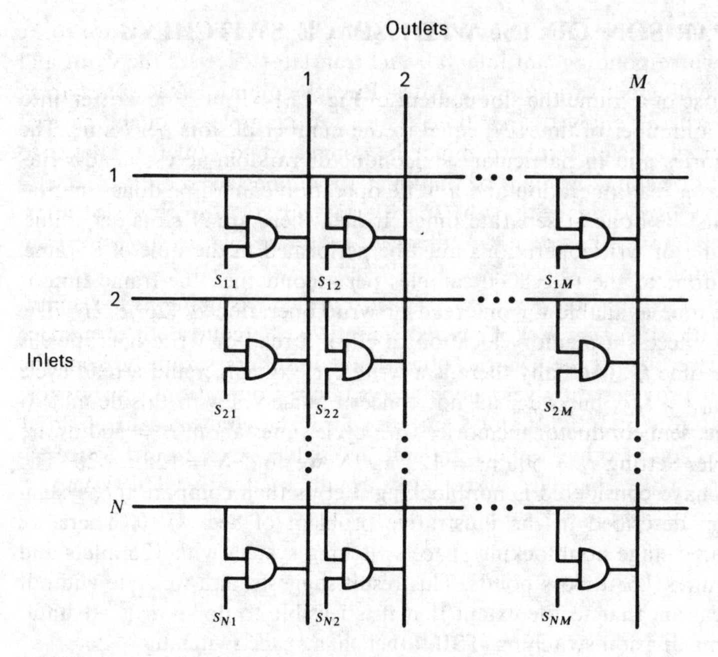

Figure 15.16-1 A space switch array for digital signals.

15.17 COMBINED SPACE AND TIME SWITCHING

If we have a digital channel which accommodates T time slots then we can simultaneously transmit T signals from T sources to T receivers using only a multiplexer (a single pole T position switch) and a demultiplexer (again a single pole T position switch). If two-way transmissions are required then only $T/2$ simultaneous transmissions are possible. Such multiplexing and demultiplexing establishes permanent connections between stations. It makes no provision for switching.

Space (S) Switching

The use of a space array to provide some switching is shown in Fig. 15.17-1. Here we have assumed an equal number N of inlets and outlets. A number N, of T-slot multiplexers, provide the inputs, and the outlets are connected to N, T-slot demultiplexers. There is no slot interchange between multiplexers and demultiplexers. Thus when the multiplexers are connected to slot k the demultiplexers are also connected to slot k. The gate select memory has T locations, that is, as many locations as there are slots. The address for reading the memory is a counter which increments at the slot rate. The content of the word so read out of the memory is the information about which gates of the space array are to be enabled in the corresponding slot interval. The translator decodes the word and provides the enabling activation for the gates. In each column of the space array only one gate will be enabled at a time. Thus, for example, it might be that during the interval of slot 1 inlet 1 might be connected to outlet 7, inlet 2 to outlet 9, etc. During slot 2, it might be that inlet 1 is connected to outlet 4, inlet 2 to outlet 5, etc. Observe again that the space array provides no time-slot interchange. It simply moves the content of time-slot k on one inlet to time-slot k on any (available) outlet.

It is possible for the system of Fig. 15.17-1 to provide as many as NT simultaneous connections. In this very restricted sense it is the equivalent of a space matrix of $(NT)^2$ cross points. On the other hand it suffers seriously from the handicap that it provides only very poor accessibility. Connections are possible from slot k only to slot k. Thus slot 1 on inlet 1 can be connected to N slot 1 outlets while $N(T-1)$ other outlets are not accessible. As a result, Space (S) Switching is not usually employed alone, but is usually used with a TSI system.

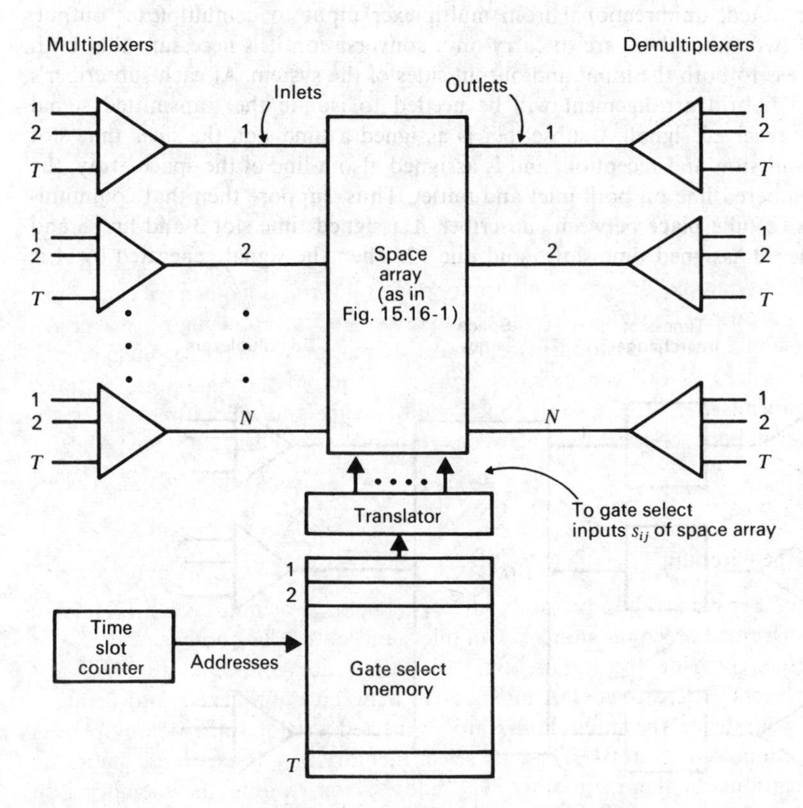

Figure 15.17-1 A space array is used to provide switching of digital signals.

Time (T) Switching

The time-slot interchange (TSI) system of Sec. 15.14-1 is referred to as time switching (T-Switching). It has the merit of providing full availability and is also nonblocking. It has the limitation that an increase in slot numbers to increase the number of allowable simultaneous connections is limited by the switching speed of logic gates and most particularly by the read and write cycle times of memories.

Time-Space (T-S) Switching

One way in which the number of simultaneous connections can be increased is shown in Fig. 15.17-2. Here we have a space array with N inlets and N outlets. For each inlet line we have introduced a time slot interchanger with T slots. Time slot memories have to be provided for each interchanger and a gate select memory needs to be provided for the space array. These memories are not shown in the figure.

The transmission of signals through the system shown in Fig. 15.17-2 is, as we have noted, unidirectional from multiplexer input to demultiplexer output. Hence if two subscribers are to carry on a conversation it is necessary that both have access to both the input and output sides of the system. At each subscriber's station a hybrid arrangement will be needed to isolate the transmitted signal from the received signal. A subscriber is assigned a time slot, the *same* time slot for transmission and reception, and is assigned also a line of the space array, the *same* numbered line on both inlet and outlet. Thus, suppose then that communication is to take place between subscriber A assigned time slot 3 and line 8 and subscriber B assigned time slot 5 and line 12. Then the signal generated by A is

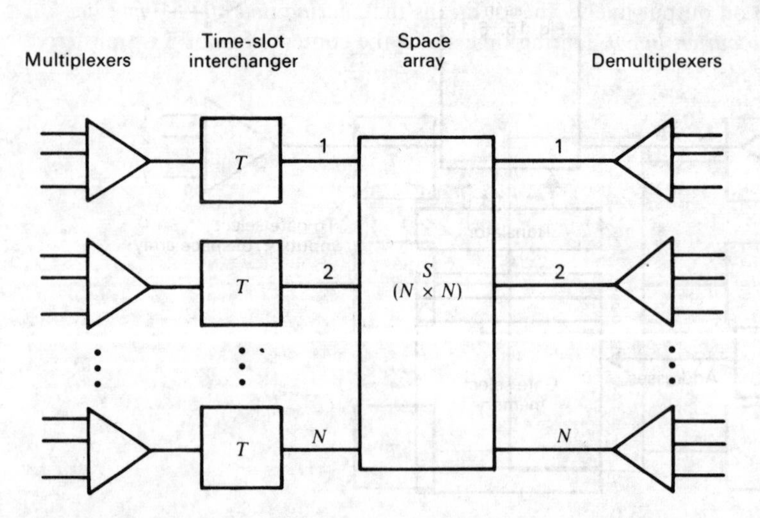

Figure 15.17-2 Time-space (T-S) switching.

moved from time slot 3 to time slot 5 by the time-slot exchanger and is transferred from line 8 to line 12 in the space array. Similarly the signal originated by B is moved from slot 5 to slot 3 and from line 12 to line 8.

The number of simultaneous connections possible with the arrangement of Fig. 15.17-2 is NT. Thus if $T = 128$ and if the space array is a 16×16, i.e., $N = 16$, matrix then $128 \times 16 = 2048$ connections can be supported. The structure is *not* free of blocking. That such is the case can be seen from the following: suppose that two subscribers A and B using different time slots on the *same line* want to connect to two subscribers C and D using the *same time slot* on different lines. We can, of course, move A and B to the same time slot but during that time slot the inlet line can be connected to C's line or D's line but not both.

An alternate arrangement to Fig. 15.17-2 consists of moving the time slot interchangers to the outlet side of the space array. This forms an S-T switch. The two arrangements are equally effective.

Time-Space-Time (T-S-T) Switching

An elaboration of the T-S system is shown in Fig. 15.17-3. Here a space array is sandwiched between sets of time-slot interchangers yielding a T-S-T switching arrangement. This scheme provides more alternative paths and hence, while blocking is still possible, the likelihood of blocking is reduced. (An S-T-S arrangement, in which a time-slot interchanger is preceded and followed by space arrays, has similar properties.)

To see the relative merit of the T-S-T (or S-T-S) scheme over a T-S (or S-T) scheme let us start with the arrangement of Fig. 15.17-3 except with the input T stages omitted so that we have an S-T arrangement. Suppose that in this S-T switch a connection is already established between time slot 5 on input line 7 and time slot 13 on output line 9. Then it means that, *during time slot* 5, *input line* 7 *is connected to output line* 9. During time slot 5 the contents of slot 5 is transferred

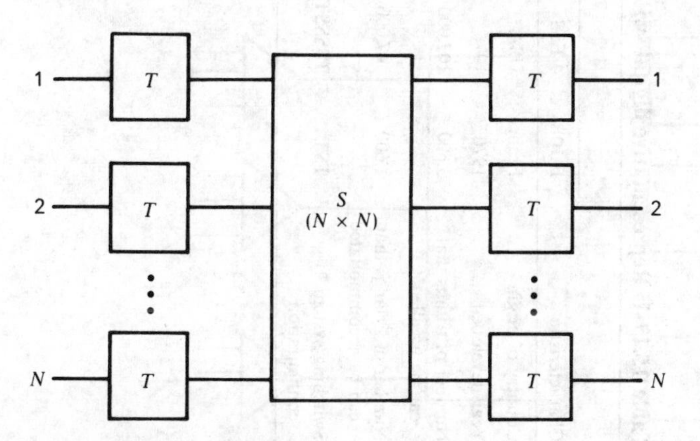

Figure 15.17-3 Time-space-time (T-S-T) switching.

Table 15.17-1 Representative digital switching systems

	Type								
Characteristic	E 10B	4 ESS	ITS 4/5	NEAX 61	AXE 10	DMS 100	EWSD	SYSTEM 12	GTD-5EAX
Country of origin	France	USA	USA	Japan	Sweden	Canada	Germany	USA	USA
Year developed	1970	1976	1976	1979	1978	1979	1980	1978	1982
Number of trunks that can be handled	3,600	107,000	3,000	60,000	65,000	61,000	60,000	60,000	49,000
Number of Erlangs that can be accommodated	1,600	47,000	1,500	22,000	30,000	39,000	30,000	25,000	36,000
Switching system configuration	TST	TSSSST	STS	TSST	TST	TS	TST	TSTSTSTSTSTS	TST

from inlet 7 to outlet 9. The exchanger on outlet 9 will move the content of slot 5 to slot 13. Now suppose further that while this first connection is in effect, there is a need to connect slot 5 on inlet line 8 to slot 14 on outlet line 9. The point is that there is a need to transfer two time slots 5, on different inlets, to different slots on the same outlet. It is clear that a block will occur against the second proposed connection because during slot 5 the outlet line is busy. If, however, we have exchangers on the input side of the space array as in Fig. 15.17-3 then the connection in the space array need not be made in slot 5. Instead the input exchanger can move the content of slot 5 to some other slot, say slot 6 and if the outlet line is not busy during slot 6 the transfer from inlet to outlet can be accomplished in that time slot. The outlet exchanger will now be called upon to move slot 6 to slot 14. Thus the slot interchange from 5 to 14 will take place in two stages, from 5 to 6 and then from 6 to 14.

Examples

Representative systems, employing combinations of time and space switching have characteristics as listed in Table 15.17-1.

15.18 MOBILE TELEPHONE COMMUNICATION—THE CELLULAR CONCEPT[11]

In this section we briefly describe the fundamental principles behind the Bell System's *Advanced Mobile Phone Service* (AMPS). This system was introduced in the 1980's after many years of research and development to meet the following objectives:

(1) provide telephone communications to thousands of mobile users within a greater metropolitan area;
(2) accomplish this goal while using a minimal amount of the frequency spectrum;
(3) be compatible throughout the United States so that as the demand increased, subscribers could use it throughout the country;
(4) permit hand-held portable operation as well as vehicle operation of the telephone;
(5) provide regular telephone services and quality of service.

To accomplish these ends a *cellular radio network concept* was formed. In this type of network each geographical area is divided into small regions called cells. Within a cell all communications is performed using a given bandwidth and center frequency. Thus, referring to the cellular layout shown in Fig. 11.18-1 we arrange that in cell A center frequency f_A is used, in cell B center frequency f_B is used, etc. If two cells are widely separated, so that a receiving antenna in one cell cannot detect the signal transmitted from the other cell, both cells can be given

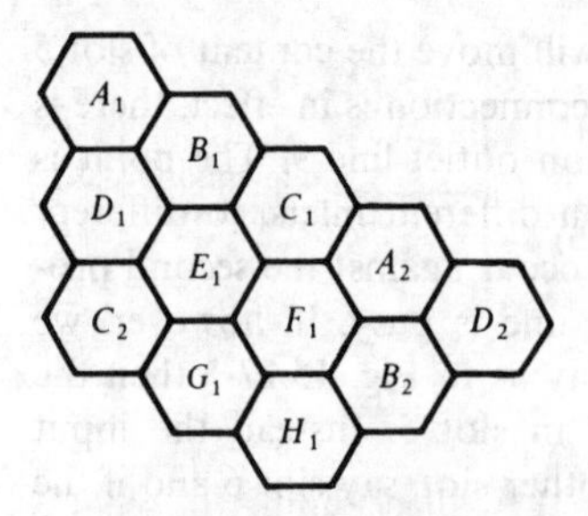

Figure 15.18-1 A cellular layout.

the same center frequency allocation. In Fig. 15.18-1 each pair of cells A_1 and A_2, B_1 and B_2, C_1 and C_2, and D_1 and D_2, use the same center frequency. This technique, which is designed to conserve the frequency spectrum, is called *frequency reuse*.

The bandwidth allocated to each cell is divided into N channels. Thus if the bandwidth of a cell is B_c, the bandwidth allocated to each user is $B_u = B_c/N$. In the AMPS system, communication is by FM and each user's channel is allocated a bandwidth $B_u = 30$ kHz (a peak frequency deviation $\Delta f = 12$ kHz is employed and $f_M = 3$ kHz). The total spectrum allocation per cell is $B_c = 40$ MHz so that $N = 666$ *full-duplex* channels (each user can talk and listen simultaneously just like on a regular telephone) can be provided in a cell. If more than 666 users must be accommodated, the cell is split into several cells each with different center frequency. Thus, the system is robust to the number of possible users.

Mobile-to-Mobile Calls

When a mobile subscriber wants to place a call, the mobile unit first inquires as to the availability of a channel and its center frequency. There are 21 special, *set-up* channels, not intended for communications, but intended instead to determine which center frequency is available. After the appropriate center frequency is determined, the unit then sends a data message containing the user's identification and the number to be called. The bit rate of this message is 10 kb/s. A (63,51) BCH error correcting code is used and the message is repeated 5 times with majority-rule logic employed to minimize errors in signal amplitude fluctuation (called *fading*) which may occur when tall office buildings are present.

The cell site transmits this data message to a *mobile telecommunications switching office* (MTSO) as shown in Fig. 15.18-2. The MTSO analyzes the identi-

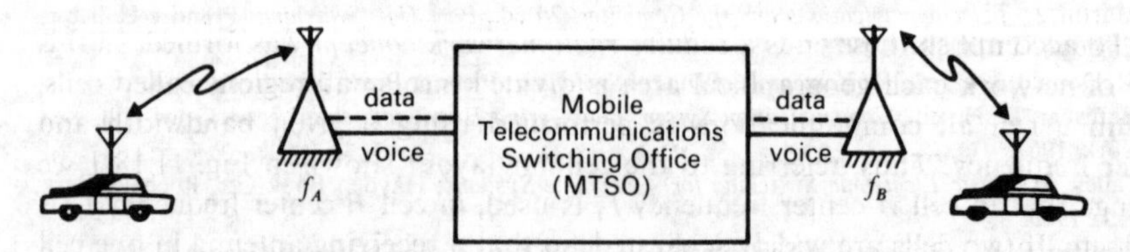

Figure 15.18-2 Mobile-to-mobile call.

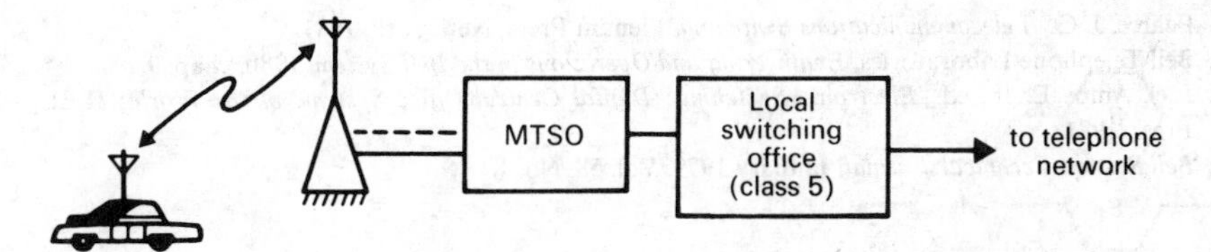

Figure 15.18-3 Mobile-to-fixed call.

fication to insure that it is proper and then analyzes the called number to see if it refers to a mobile or fixed subscriber. If the call is to a mobile unit the MTSO initiates a *paging* process by sending a data message containing the called number to each cell site. Each mobile subscriber's unit scans, once each minute the set-up channels in the cell in which it is located. The unit, sensing the data message indicating it is being called, responds to the cell site in which the unit is located and then the cell site responds to the MTSO. The MTSO then provides the caller's cell and the called unit's cell with voice line connections and the connection is completed and the conversation can begin. When the conversation is terminated and *either* user "hangs-up" the MTSO releases both voice lines.

Mobile-to-Fixed Subscriber Calls

When a mobile subscriber calls a fixed subscriber, the cell enters the MTSO as above. Now, however, the MTSO recognizes the number called as being "fixed." The call is then routed to a class 5 switching office where it is forwarded to its destination. This situation is depicted in Fig. 15.18-3.

Fixed-to-Mobile Subscriber Calls

When a fixed subscriber calls a mobile subscriber the call is routed to the appropriate toll office as determined by the area code of the called number. The number dialled indicates that a mobile subscriber is being called and the toll office routes the call to the local office and MTSO. The MTSO then pages the called subscriber as described above and connection is finally completed.

REFERENCES

1. Bellamy, J., *Digital Telephony*, chap. 5; John Wiley and Sons, New York, 1982.
2. Martin, J., *Telecommunications and the Computer*, 2d ed., *Part III: Switching*; Prentice Hall Inc., Englewood Cliffs, N.J., 1976.
3. McDonald, J. C., ed., *Fundamentals of Digital Switching*; Plenum Press, New York, 1983.
4. Hamsher, D. H., ed., *Communication System Engineering Handbook*, chap. 7; McGraw-Hill, New York, 1967.
5. Talley, D., *Basic Electronic Switching for Telephone Systems*; Hayden Book Co., Rochelle Park, N.J., 1975.
6. Fleming, P. Jr., "Principles of Switching," vol. 10 *Lee's abc of the Telephone*; Geneva Ill, 1979.
7. Flowers, T. H., *Introduction to Exchange Systems*, John Wiley and Sons, New York, 1976.

8. Pearce, J. G., *Telecommunications Switching*, Plenum Press, New York, 1981.
9. Bell Telephone Laboratories, *Engineering and Operations in the Bell System*, 1980, chap. 9.
10. Joel, Amos E. Jr., ed., *Electronic Switching: Digital Central Office Systems of the World*; IEEE Press, 1981.
11. *Bell System Technical Journal*, January 1979, Vol. 58, No. 1.

PROBLEMS

15.1-1. The current i_R in the receiver of station B of Fig. 15.1-1 is

$$i_R = I \frac{R_t}{R_r + R_t}$$

where I is the dc current flowing in the battery, R_t is the variable, voice sensitive resistance of the transmitting microphone and R_r is the receiver impedance.

(*a*) Explain why I is approximately constant.

(*b*) Why should $R_t \ll R_r$?

15.1-2. 23 ma is required for the carbon transmitter. A 50 V battery with a 400 Ω short-circuit protection resistance is employed with 26 gauge wire. Neglecting the resistance of the transmitter, calculate the loop distance. *Hint*: Consult a handbook to determine the resistance/ft of 26 gauge wire.

15.1-3. Explain why it is necessary that, in the absence of voice-generated signal current, the diaphram of a telephone receiver must be displaced from its unstressed position either by a quiescent current or by the presence of a permanent magnet.

15.2-1. Show that if in Fig. 15.2-1 there are N stations, there are (*a*) $N(N-1)$ switches and (*b*) $N(N-1)/2$ interconnections.

15.2-2. Show that with a central exchange system as in Fig. 15.2-2 there are (*a*) $N(N-1)/2$ switches and N interconnections.

15.4-1. The Strowger switch shown in Fig. 15.4-1 contains 10 rows, each with 10 contacts. Show by means of a sketch the switch position for the number 84.

15.5-1. A study of company ABC indicates that from 10 a.m. to 4 p.m. 180 calls are initiated each hour and each call lasts 200 s on the average. (*a*) Calculate the call load in Erlangs. (*b*) If the grade of service required is P.01, how many servers will be needed?

15.5-2. If there are 100 possible users of a telephone system and no more than 10 will be using the system at any one time, calculate the probability of blocking. Assume that each call lasts T s, each initiated call is independent of the others, and the average number of calls initiated per second is τ.

15.5-3. Verify Eq. (15.5-5).

15.9-1. Show that in Fig. 15.9-3 only two connections are possible simultaneously.

15.9-2. Show that in the switch arrangement shown in Fig. 15.9-4 only $N(N-1)/2$ switches are required in general.

15.9-3. Refer to Fig. 15.9-5*a*.

(*a*) How many switches must fail to prevent connection between A and a?

(*b*) In general, how many switches are required to interconnect N callers?

15.10-1. Assume that there are $N = 8$ inputs and 8 outputs in a three-stage switching system. Assume that there are 4 first stage and 4 third stage switches so that $n = 2$.

(*a*) Verify, by means of a sketch, that if $k \geq 3$ there is no blocking.

(*b*) Verify, by means of a sketch, that if $k = 2$ there is blocking.

15.10-2. Prove that setting $dN_x/dn = 0$ in Eq. (15.10-2) yields $n \simeq \sqrt{N/2}$ for large N.

15.10-3. 10,000 input callers are to be able to be connected to 10,000 outputs. Using the 3-stage switching structure shown in Fig. 15.10-1

(*a*) Calculate k and n to minimize the number of cross points. Calculate the minimum value of N_x.

(*b*) If $k = n$ find the minimum value of N_x, k, and n.

(*c*) Compare the results of (*a*) and (*b*).

15.10-4. 10,000 input callers are to be able to be connected to 10,000 outputs using the structure shown in Fig. 15.10-1.

(*a*) If $n = k$ and the probability that an arbitrarily selected input terminal is in use $p = 1$ percent of the time, calculate the probability B that the call is blocked.

(*b*) Repeat (*a*) if $p = 0.1$ percent.

(*c*) Plot B as a function of p.

15.12-1. The system shown in Fig. 15.12-1 is to be employed. There are 24 subscriber lines which are to be capable of connection with a different set of 24 subscribers.

(*a*) How many simultaneous connections can occur?

(*b*) What is the duration τ that each connection can persist?

(*c*) What is the capacity of the R/W memory?

(*d*) What is the clock rate?

15.13-1. Refer to Fig. 15.3-1. Assume that both switches are initially open and that the voltage on each capacitor is $v_k(0) = 5$ volts and $v_l(0) = 7$ volts. The resonant frequency

$$\omega_R = \sqrt{1/LC} = 5k \text{ Hz} \qquad \text{and} \qquad \tau_R = 200\ \mu s$$

Show that if the switches are simultaneously closed for 100 μs, and then opened, that $v_k(100\ \mu s) = 7$ volts and $v_l(100\ \mu s) = 5$ volts.

15.14-1. A 1000 line, 1000 trunk digital switch is to be built using TSI. Using 8-bit PCM specify

(*a*) the size of each memory needed,

(*b*) the speed at which the data must be accessed.

15.17-1. (*a*) Show that T-S switching results in blocking. To do this, assume that in Fig. 15.17-2, subscribers A and B use different time slots on the same line and want to connect to subscribers C and D using the same time slot of different lines. Show that blocking occurs.

(*b*) If $T = 128$ and $N = 16$, and the number of input subscribers is only $0.1NT$ what is the probability of blocking. Assume that all $0.1NT$ input connections are made at the same time.

15.17-2. Can you design a TS network which is non-blocking? If you can, show your answer, if not explain.

15.17-3. The blocking probability of a TST network is

$$P_B = \frac{[(T-1)!]^2}{N!(2T-2-N)!}\rho^N(2-\rho)^{(2T-2-N)}$$

where ρ is the circuit occupancy in Erlangs (see Sec. 15.5).

If the T stages contained 100 time slots and the S stages contained 100 slots calculate the blocking probability if $\rho = 0.5$ Erlangs.

CHAPTER

SIXTEEN

COMPUTER COMMUNICATION SYSTEMS

INTRODUCTION

There are many establishments, such as large corporations, the military establishment, etc., which are served by a substantial number of computers which are situated at different locations. Inevitably, then, there develops the need to ensure that these computers be able to communicate with one another. Such communication allows sharing of data and computational output, and improves overall reliability of the entire computer system by permitting other computers to take over the work of a machine which is temporarily out of service. When an array of computers is interconnected, the hardware and facilities located at each facility is referred to as a *host* while the resources devoted to providing the inter-host communication is referred to as a *communication network*. Certainly, the following are features which are to be desired in such a network.

(1) The network should be able to accept promptly (no excessive delay) an intended transmission from a host,
(2) the network should be *efficient*, that is, no substantial part of the network should be idle for any extended period,
(3) data arriving at a receiving host should be capable of being rapidly processed,
(4) when a transmission calls for a response, the delay of this response should be minimal,
(5) and, of course, the cost of establishing, maintaining and operating the network should be kept to a minimum.

A computer communication network is represented in Fig. 16.1. The square boxes are the computers. The small open circles are terminals, while the larger

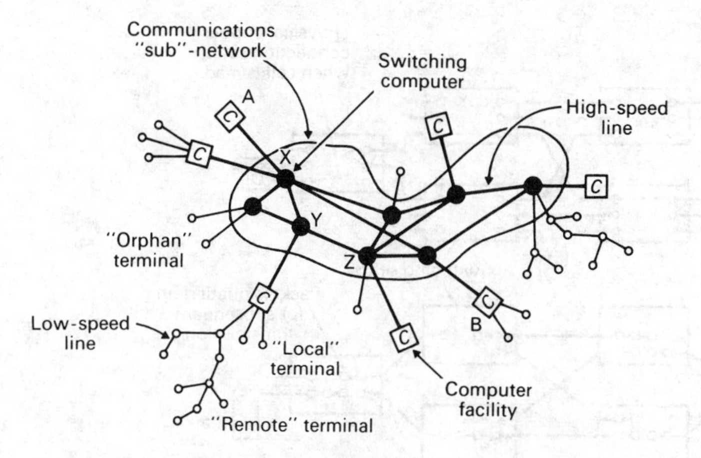

Figure 16.1 The structure of a computer-communication network.

shaded circles are the switching centers which provide the switching facility needed to make the connection as required. These switching centers are themselves sophisticated, albeit specialized computers, and are hence appropriately referred to as *switching computers.* Some computers serve a number of terminals and some of the lines connecting terminals to their local computer serve more than one terminal. Generally lines joining terminals to a local computer need not be required to transmit data at a high rate (i.e., a relatively modest bandwidth is adequate) and these lines are characterized as "low speed" lines. High speed (wide bandwidth) lines are needed to interconnect switching centers. Some terminals, so-called "orphan" terminals have no associated local computer and are instead connected directly to a switching center. Similarly some computers have no directly connected terminals and are associated with the network only through their connection to a switching center. Finally we note that the part of the entire communication system which encompasses only the switching computers is referred to as the "subnetwork".

16.1 TYPES OF NETWORKS

There are three basic types of communication networks in service. These are: *circuit* (*line*) *switching*, *message switching* and *packet switching.*

Circuit Switching

In *circuit switching* the hosts which are to communicate are connected by a wire path which is dedicated to the communicating pair, without interruption, for the entire duration of the transmission. This type of system is entirely analogous to a telephone switching system. The entire wire path remains allocated to the single

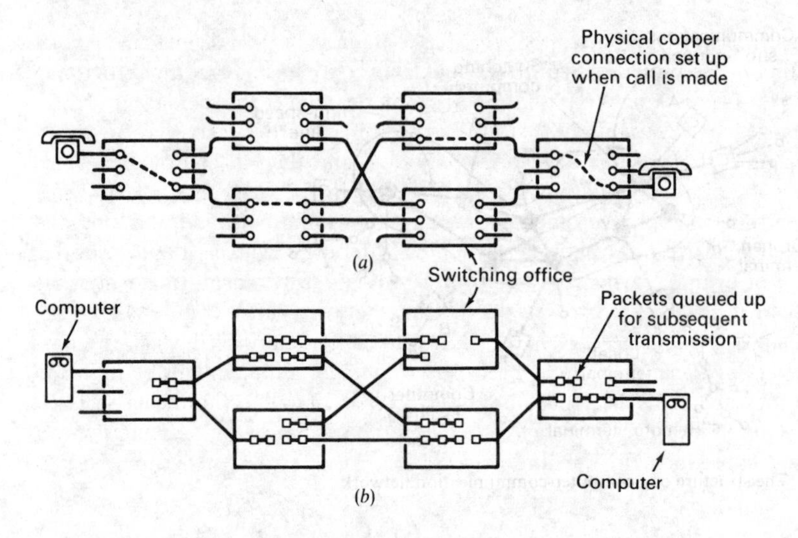

Figure 16.1-1 (*a*) Circuit switching. (*b*) Packet switching.

set of users until released and, during this time, no other potential user can use any part of the wire path, not even during intervals when the path happens to be idle. Figure 16.1-1*a* illustrates the major principle of circuit switching. Here a suitable path is established by physical wire connections and maintained until disconnected by the users.

Message Switching

A distinctive feature of computer communication which makes it quite different from telephone communications, is that, in computer communication, transmission generally occurs in relatively short bursts separated by long intervals of no transmission. Typically a line connecting two computers may be in use only 1 percent of the time allocated. Such a manner of use is clearly uneconomical since, of course, the users must be charged for the entire time of allocation.

Message switching used in conjunction with a network such as is represented in Fig. 16.1 offers the possibility of greatly improved economy. To see how message switching works consider, in Fig. 16.1 that computer *A* needs to transmit data to computer *B*. Consider, further, that the communication will consist of relatively short messages separated by long intervals when there is no transmission. Suppose that the transmission path selected is $A \rightarrow X \rightarrow Y \rightarrow Z \rightarrow B$. Note that the lines from *X* to *Y* and from *Y* to *Z* can be used as part of a connection from many computers to many other computers.

In circuit switching there is no transmission until a *complete* connection is established between users. In message switching, at any particular time, no such complete connection is needed. Instead computer *A* will transmit a message to switching center X where it will be *stored in a buffer*. This stored message will be kept waiting in the buffer at X until the line from X to Y is not required for some other prior transmission. When that line becomes available the message at X will be transmitted to Y. Again, after storage and a possible delay to wait for available open time, the message will go from Y to Z and then to computer C. This system has the disadvantage that there will be a delay between the time of transmission and the time of reception of a message. On the other hand, the long distance, fast, expensive lines between switching computers will be idle for shorter intervals. For example, when the line from X to Y is not transmitting a message from A to B it can be used to relay a message between other computers. In the total transmission of a message from A to B it will generally be that, at any particular time, only one of the lines between switching computers will be involved with that message, the other lines in the overall path of connection being released to handle other messages between other computers. Since the system involves *storing* and subsequent forwarding of messages, a message-switched network is referred to as a *store* and *forward* network.

Packet-Switched Network

Message switching has a number of features which are less than desirable. (1) A message may be very long (many hundreds of thousands of bits) or very short (of the order of thousands of bits). To accommodate the long message, large buffers need to be available. The hardware of these buffers is wasted when short messages are received. (2) Even if a line is available, a switching computer will not begin to relay a message until the entire message has first been received. Accordingly, if the message duration is T_m and it must be relayed R times, then even ignoring waiting time, the delay will be RT_m since at each relay the entire message is stored before being forwarded to the next node. For long messages and many relays this long delay time can become inconvenient. (3) It may be that a link between centers is needed very briefly to accommodate a very short message but the link is busy transmitting a very long message. It would be useful if it were possible to interrupt the long message briefly to allow transmission of the short message. Unfortunately, it is not easily feasible to incorporate such a feature in a message-switched system. Hence it is possible that short messages may be delayed for a time which is long in comparison with the message lengths while the system is "tied-up" processing a long message.

A *packet-switched* network undertakes to circumvent the less desirable features of message switching by subdividing messages into packets. A typical packet may be 1024 bits long. The message is then transmitted packet by packet. Like the messages in message switching, each packet must be stored in buffers at the nodes of the subsystem and then forwarded. Packet switching like message switching, is therefore a *store and forward* system. In some packet switched systems different packets of a single message may arrive at a destination by different routes and with different delays. It is also entirely possible that the packets of a message may arrive out of order. Packet switching is represented in Fig. 16.1-1*b* where the small squares in the switching computers represent the individual packets.

Each unit of transmission, whether the message in message switching, or the packet in packet switching, must include not only the information bits but additional bits referred to as *overhead* information. These overhead bits must identify the *destination* of a unit (message or packet) so that each switching center will know how further to route the unit, the *source* of the unit so that acknowledgement is possible, and the *user identification* so that the user can be charged for services. Further, *synchronization bits*, also part of the overhead, must be included to identify the *beginning* and *end* of a unit. In packet switching, as mentioned above, it may happen that different packets of a message may travel over different routes and therefore arrive out of order. Accordingly packets must be *numbered* so that they may be reassembled in proper sequence.

In a message switched system the overhead information must be appended to each message, while in packet switching each packet must be accompanied by overhead bits. Thus, if there is more than one packet/message the packetized message had more overhead. Accordingly with respect to message switching, packet switching has two disadvantages: (1) To transmit a given amount of information per unit time, packet switching requires that bits be transmitted at a more rapid rate than is required in message switching. (2) The switching hardware needed to packetize, add overhead, depacketize and reassemble is more complicated and must operate more rapidly than the corresponding hardware needed in message switching.

On the other hand, in spite of the fact that packet switching requires more overhead bits, it is entirely possible that packet switching will transmit a given amount of information in less time than message switching. This feature results from the fact that a switching computer cannot begin to retransmit a message (relatively long) or a packet (relatively short) until the *entire* message or packet has been received. Let us assume that a message of B bits is to be relayed from one to another of a series of switching computers using *message switching*. There are to be $K + 1$ computers involved in the overall transmission from the source computer C_0 to the destination computer C_K over K transmission facilities (links). We neglect the time of propagation over the links and assume that the only delay is caused by the bit duration $T_b = 1/f_b$. The number of overhead bits needed is b. The total transmission has $B + b$ bits and will be transmitted from one computer to the next in a time $(B + b)/f_b$. There being K links, the total time required for transmission of the message without packetizing is

$$T_M = K[(B + b)/f_b] \cong K \frac{B}{f_b} \qquad (16.1\text{-}1)$$

since, ordinarily $B \gg b$. Next, let us consider that the message has been divided into P packets each with a number of bits per packet B/P, and let us assume that the number of overhead bits needed is again b as before. The total number of bits in the packet is then $(B/P) + b$. To transmit this packet over the K links will require a time as given by Eq. (16.1-1) except with B replaced by B/P. Hence the time required for the first transmitted packet to arrive at its destination is $K[(B/P) + b]/f_b$. The next packet is processed immediately after the first and will arrive after the first by a time $[(B/P) + b]/f_b$. Since, after the first, there are $p - 1$ further packets, the total time needed to transmit the entire message by packets is

$$T_P = K[(B/P) + b]/f_b + (P - 1)[(B/P) + b]/f_b \qquad (16.1\text{-}2)$$

To find the value of P which minimizes T_P we set to zero the derivation of T_P with respect to P. We find that T_P is minimum when

$$P^2 = (K - 1)\frac{B}{b} \qquad (16.1\text{-}3)$$

and that correspondingly

$$T_{P,\,\min} = \frac{b}{f_b}\left[\sqrt{\frac{B}{b}} + \sqrt{K - 1}\right]^2 \qquad (16.1\text{-}4)$$

If $B/b \gg K$ then

$$T_{P,\,\min} \cong \frac{B}{f_b} \qquad (16.1\text{-}5)$$

and, in this case, referring to Eq. (16.1-1) we note $T_{P,\,\min}/T_M = 1/K$. If there is a finite propagation time τ associated with propagation over a link, then T_P and T_M will be increased by the amount $K\tau$.

16.2 DESIGN FEATURES OF A COMPUTER COMMUNICATION NETWORK

A computer communication network is designed such that:

(*a*) communications between sources and destinations are available to a large number of users at a relatively low cost;

(*b*) the time required to process a transmission from source-to-destination, which is called the *response time* should be less than some specified value. There are several ways of specifying response time such as: (1) that the average response time of the network be less than some given value, (2) that the probability that a response time exceeds a time T_0 is less than, say, 1 percent, etc.;

(*c*) the *throughput* S is greater than some specified value. The throughput is the number of packets that are successfully *received* per unit of time. In any computer communication system, we find that as more users enter the network and the total number of packets transmitted per unit time by these users increases, the more packets are received per unit time until a point is reached where congestion limits the rate at which packets can be received correctly. Clearly, we would like this rate to be as high as possible;

(*d*) the network should continue to operate even if a few nodes or links fail, and that the response time and throughput in such a "stressed" network should not be significantly degraded.

In order to accomplish these design objectives the designer of a computer communication network must consider:

(a) Topological design Topological analysis investigates different ways of interconnecting terminals and computers i.e., different network designs. Such design is very dependent on the geographical location of these *resources* (as the terminals and computers are called).

Figure 16.2-1 shows the four basic topological structures. Figure 16.2-1*a* is a *star* connection. Here, the outputs of terminals located in one location are multiplexed to form a single high speed bit stream. The multiplexer shown is called a *remote* concentrator. The outputs of the remote concentrators are then connected to the computer (central processor).

Figure 16.2-1*b* shows a *tree* configuration. In this system terminals are multiplexed into remote concentrators which are then multiplexed into another remote concentrator, etc., until we finally reach the central processor which routes the message to the appropriate host.

Figure 16.2-1*c* is a *loop* configuration, where the remote concentrators form a loop with the central concentrator. Figure 16.2-1*d* is a *mesh* configuration. Most networks employ the mesh configuration either by itself or in combination with one or more of the other configurations thereby forming a hybrid configuration.

A basic problem with the star and tree configuration is that if a node or link fails, computers on one side of the failure are unable to communicate with computers on the other side of the failure. The mesh configuration has low response time and is robust against node or link failure.

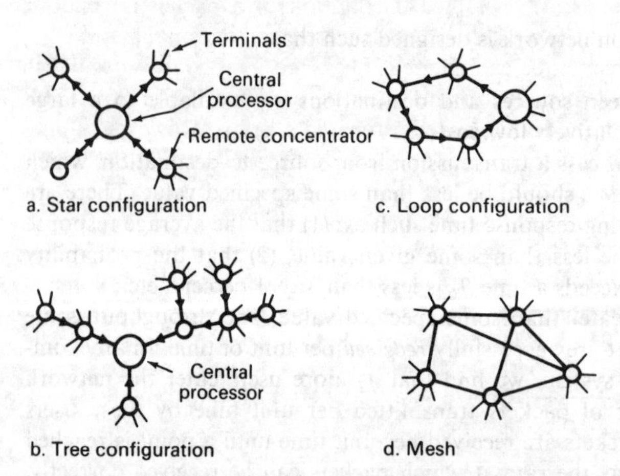

Figure 16.2-1 Different topological configurations.

(b) Line capacity allocation The tree connection and the star connection result in heavy traffic along links that input the central processor. Such links must be capable of carrying the traffic, and wideband cable, filter optic cable or a wideband radio or satellite channel is often used. The network cost depends to some degree on the type of links employed and the bandwidth of these links.

(c) Routing procedures A routing procedure is required to insure that the sender reaches its destination within the specified response time and takes a path which enables the network to carry a maximum number of users. There are many different routing procedures (algorithms), several of which are briefly discussed below, for the typical mesh network shown in Fig. 16.2-2:

1. *Flooding.* This technique is used primarily when a message from one node must be certain to reach one or more specified nodes in a network. Suppose, for example, that in Fig. 16.2-2 a message from node *A* must reach node *L*. Then to *flood* the network, the message is sent on each outgoing link, in this case to *E*, *J*, and *B*. Each of these nodes forward the message on each outgoing link except for the link over which it arrived, i.e., *B* forwards the message to *G* and *C*, etc. As a result of flooding the network, the message has an extremely high probability of being received correctly in a reasonable time at its destination. The technique is robust, i.e., node or link failure will not dramatically affect the response time period. However, flooding results in traffic congestion since a large number of duplicate packets are transmitted.
2. *Random walk.* Using this technique a packet stored at a node is forwarded at random over any available path. For example, a packet at *H* will be forwarded at random to either node *G*, *D*, *L*, or *J*. This technique is robust since node failure does not alter possible transmission to other nodes. In addition, only one packet is transmitted from node-to-node thereby helping to avoid unnecessary congestion. However, since the motion of a packet is random, there is no guarantee that the packet will be delivered in a reasonable time.
3. *Fixed routing.* In this technique each node is furnished a table of routes which has been predetermined, usually on the basis of previous traffic flow. This technique is useful when the traffic is well estimated a priori. However, it does not operate properly when there are large random variations in the traffic pattern.

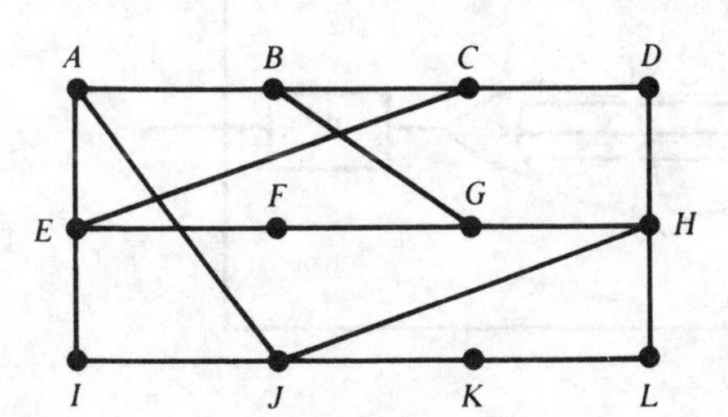

Figure 16.2-2 A typical data communications network.

4. *Centralized adaptive routing.* In this technique each node is furnished a table of routes. This table is continually altered as a result of instructions furnished by a "supervisory" computer which monitors the traffic at each node. This technique is adaptive to network traffic changes and therefore helps to minimize congestion. However, there is a delay in the determination and distribution of table updates.
5. *Isolated adaptive routing.* Here the table of routes at each node is continually upgraded by information the node gathers by itself. One way for this to occur is (see Fig. 16.2-2) for packets from any node, say from nodes B, E, and J, to node A to carry within the packet an entry as to the buffer availability at that node. If, say, node B has not transmitted to node A in a "long" time interval, A assumes that node B is not working and will not route messages to node B. This procedure does not require a supervisory computer, but, on the other hand no node has a global picture of the traffic in the network.

(d) Flow control procedures After the routing technique has been decided upon it is necessary to establish the procedure by which we can assure smooth traffic flow and avoid "bottlenecks" and "deadlocks." Flow control procedures include selecting the multiplexing scheme including determining the number of links that can be combined at a multiplexer, the *access* technique used to input the messages on each link to the multiplexer, as well as the buffer size needed at each multiplexer.

To illustrate two *access* techniques, *polling* and *random access*, refer to Fig. 16.2-3. Here, signal packets $S_1, S_2, \ldots, S_k$ are stored in the input buffers of the concentrator. The access technique determines the strategy by which the stored packets are outputted through the output buffer B_o. In a *polling* technique each of the input buffers are interrogated, in order, to determine if a packet is present. If a packet is present in, say, buffer B_i it is forwarded to B_o and then the next buffer, B_{i+1}, is interrogated, etc. This technique is often used when packets are present on a regular basis. If packets occur only rarely on any of the input links, so that a single packet may be stored in only one of the input buffers, but the probability is very small of more than one packet being stored simultaneously in

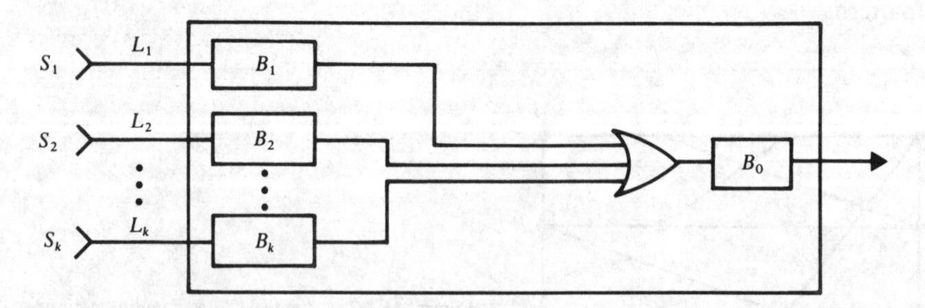

Figure 16.2-3 Concentrator (multiplexer) access.

the concentrator's buffers, then *random access* can be employed to speed throughput through the concentrator. In a random access technique, any packet in any input buffer moves directly to the output buffer. It does *not* wait. One way to avoid a collision between a packet entering B_o from say, B_i and a packet which might be just leaving B_j is to disable each buffer once a packet starts to enter B_o. This type of monitoring and control is often done using a microcomputer which is a part of the concentrator.

16.3 EXAMPLES OF COMPUTER COMMUNICATIONS NETWORKS[1]

In this section we describe some of the features of a few of the more popular computer communication networks.

16.3.1 TYMNET

Tymnet is a packet switched computer communication network which allows users to connect to the Tymnet computers for interactive processing and also provides connection facilities to those users wanting to communicate with other users. The Tymnet company owns the nodes, which are linked together to form the network. The interconnecting channels are leased from *common carriers*, such as AT&T. Such networks are often called *value-added* networks (VAN's) since the (Tymnet) company uses an already existing set of channels and "adds value," e.g., packet switching. In most countries, the government owns and operates a VAN called the *public data network* (PDN).

Tymnet employs a mesh topology and operates throughout the United States and Europe using cable with satellite backup. Such a network is called *distributed*, and insures robust operation in case of a node or link failure. The average time for a transmitted 1024-bit packet to reach its destination is 300 ms.

A user entering Tymnet must first identify himself and his reason for using the network, i.e., for communication to a destination outside the network or, for the use of the Tymnet computer facility. The route is then determined by a centralized supervisory computer and is deterministic. Several "supervisors" are always available in case of failure. Each time the user enters the network he may be given a new path, however, while in the network the route remains unchanged. Thus the user "sees" a *virtual* circuit. On the other hand, many users can share a link between two nodes. Along the designated path a packet is transmitted from node to node. At each node the packet is stored; the node forwards the packets to the next node using the principle of *first-come-first-served.* The supervisory program chooses each transmission path to provide least cost and lowest response time.

As a result of the centralized routing procedure employed, packets always arrive in order. If, due to an error, a packet fails to arrive (e.g. packet j arrives and then packet $j + 2$ arrives) we know that packet $j + 1$ is missing and a *negative acknowledgement* (NAK) is transmitted to the user who then repeats the packet (this technique is described in Sec. 13.23 on ARQ).

16.3-2 ARPANET

The ARPANET is perhaps the best known experimental packet-switched network in the world. It began in 1969 with four nodes and today has more than 100 nodes. The ARPANET derives its name from the fact that its development and operation has been paid for by the Defense Department's Advanced Research Projects Agency (DARPA). The ARPANET is used by experimenters to study the operating characteristics of packet switching, it is used to connect two or more users so that HOST computer capabilities can be shared, and is also used as a means of sending messages, reports, etc, between users (*electronic mail*). The ARPANET services Government Agencies, Departments, Universities and Corporations. While the network is distributed, there are portions of the network that can *not* be accessed except by users having a special identification. If the entire network is considered, as shown in Fig. 16.3-1, we see that the ARPANET extends from Hawaii to Europe. Recently the ARPANET has been divided into two separate networks, one devoted to research and development which retains the name ARPANET and the other network being the Defense Data Network (DDN). Similar technology is used on each.

The ARPANET was designed to transmit data and voice. Data packets are approximately 1000 bits long and voice packets are of variable length, being less than approximately 2000 bits. The ARPANET was designed to provide minimum cost and no more than 200 ms time delay independently of the packet size. To accomplish this end, *adaptive routing* is employed. In this technique any node A decides (without using a centralized supervisory computer) which node j should next be sent a packet being ultimately forwarded to destination node W. This decision is made by maintaining a *table* based on *shortest-time* estimates to the destination node. This, shortest-time, of course, is based on buffer availability as well as path length. Periodic updates of the table are made (at most every 10 s) using information obtained from all nodes. To do this, the packet delay time on every link eminating from node A is recorded at node A. Every 10 s, the processor at node A determines if there is a significant change in throughput time. If there is, each node is so notified by "flooding" the network.

Since each packet may be forwarded from source to destination using different nodes and therefore different paths, the packets may arrive at the HOST (destination) *out of sequence* and must therefore be stored and re-ordered before being processed by the HOST. Such a message is said to be transmitted as a *datagram*. Note that unlike TYMNET there is no complete *virtual* circuit ever set up between source and HOST. As a consequence of the storage needed at the HOST prior to processing, the HOST must tell the source when it can accept the next message. To do this, the HOST sends to the source, after receipt of a

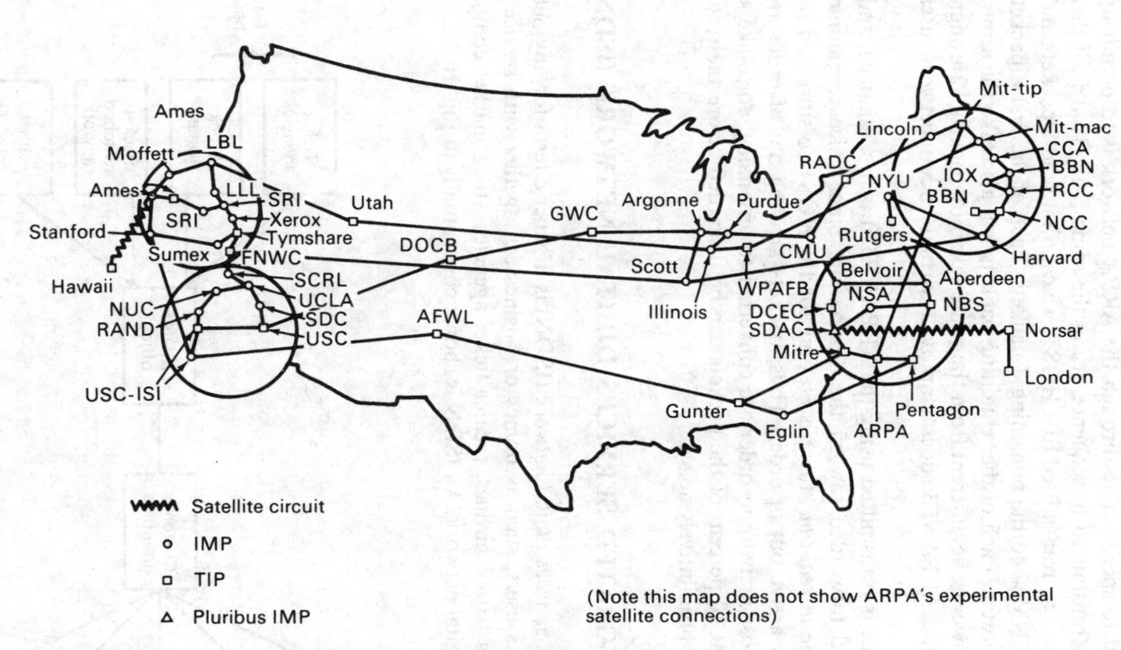

Figure 16.3-1 ARPA network, geographic map, February 1976.

complete message, a short packet which tells the source that the HOST is *Ready-For-The-Next-Message*. This is called the *RFNM*. Such a control message, between source and destination, is called an end-to-end or host-*to-host* protocol.

The end-to-end protocol is a *flow control* algorithm. Another flow-control technique used to speed traffic through the ARPA network is to permit up to 8 packets to be transmitted in sequence, using the same path. That is, if there are say 15 packets to be transmitted to a HOST, we can first send 8 packets and after receiving a RFNM send the remaining 7 packets. Each of the first 8 packets will take the same route. Each of the remaining packets will also take the same route which may, however, be different from the route taken by packets 1 through 8. In such a case the ARPANET does employ a virtual circuit between user and HOST.

When voice is transmitted using the ARPANET it is first digitized, and then packetized and forwarded toward the destination. Since voice communication should be done in *real-time*, there is *no* time for the HOST to store and reorder packets which arrive out of order. Thus, packets arriving out of order are discarded and those arriving in order are converted to an analog, voice-like signal. It is found that 10 percent of the packets can be discarded before the recovered voice becomes prohibitively noisy.

16.3-3 INTEGRATED SERVICES DIGITAL NETWORK (ISDN)

The *integrated services digital network* (ISDN) is a generic term for any network which connects users, such as homes or businesses, together with service companies such as banks, airlines, theater ticket agencies, stock market companies, etc, using a digital network. An ISDN is shown conceptually in Fig. 16.3-2.

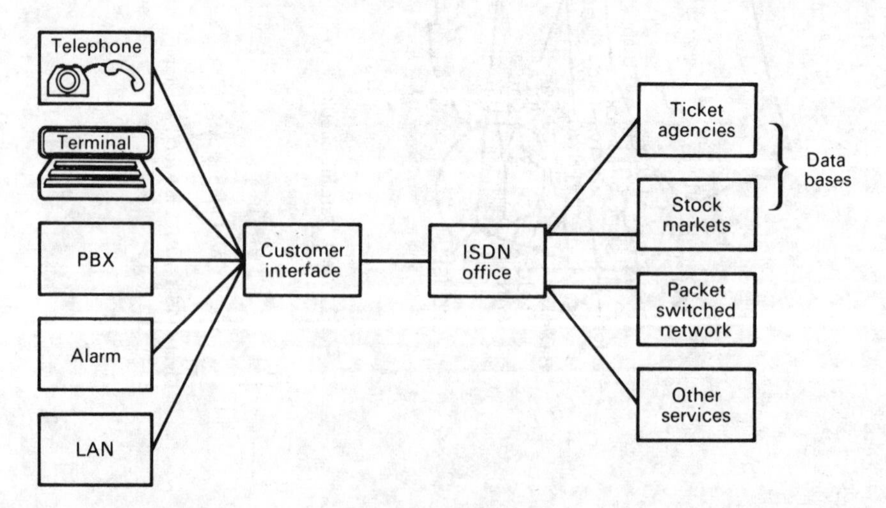

Figure 16.3-2 An integrated services digital network (ISDN).

The "owner" of the network must be able to accommodate a wide variety of computers and terminals thereby making it easy and inexpensive to access the network. Often the network will guarantee payment to the service companies and will issue a single bill to the user. In Canada, the user is billed in his regular telephone bill.

One of the most important aspects of such a network is its ability to connect a wide variety of services and users to the network. This is done using an international standard called the *X*.25 interface protocol. This protocol is described in Sec. 16.6-4.

16.3-4 LOCAL AREA NETWORKS (LAN)

A relatively small and inexpensive digital network used to provide a business access to a digital network is called a *local area network* (LAN). LAN's employ coaxial cable, twisted pair and most recently optical fibers. Transmission bit rates are typically 1 to 20 Mb/s and packet transmission is employed. The LAN is usually 1 to 25 km in length and supports up to several thousand users.

The LAN, such as Xerox's *Ethernet*, is often a single loop network as shown in Fig. 16.3-3. Here four nodes are connected to a loop network which allows transmission only in the clockwise direction. Before transmitting, the user's receiver, shown by the receiver box marked *R*, listens to the channel to see if there are any other transmissions. If no other transmission is sensed the user transmits. Such a technique is called *carrier sense multiple access* CSMA and is explained in Sec. 16.4.

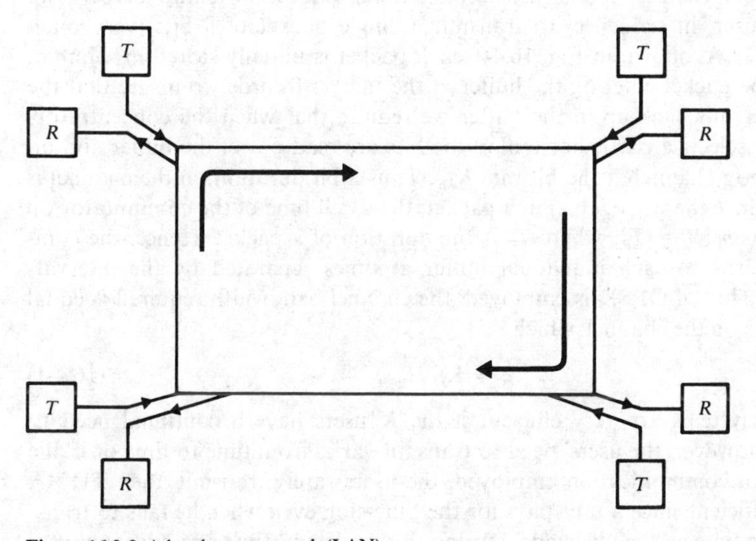

Figure 16.3-3 A local area network (LAN).

Another LAN flow control technique is "token passing." In this technique a token packet is introduced to the network. This packet continues to circulate through the network as long as no user accepts it. When a user wants to transmit he waits for the token packet to reach him and then accepts the token, which acceptance removes the token packet from the network. The user then transmits and when his transmission is complete he restores the token to the network.

LAN's are finding wide application today and their use will continue to grow.

16.4 PACKET RADIO AND SATELLITES

A packet radio network is one in which communication between nodes is accomplished using "radio frequencies." Packet radio can operate from high frequency HF (1-30 MHz) to the extra-high frequency EHF (3-10 GHz) region used by satellites. End-to-end connections may be direct, or in many cases the network topologies described above in Sec. 16.3 may be used. One of the most interesting aspects of packet radio communication is the access and flow-control procedure employed. Several of the more popular techniques are: time division multiple access (TDMA), frequency division multiple access (FDMA), ALOHA and carrier sense multiple access (CSMA). These techniques are described below:

16.4-1 TIME DIVISION MULTIPLE ACCESS (TOMA)

Consider a radio system where each of K users wish to transmit over a channel and each user generates f_b bits per second. Using packet switching TDMA, we allow each user, in sequence, to transmit a single packet of b bits (very often $b \cong 1000$ bits). As shown in Fig. 16.4-1 each packet is initially stored in a buffer, the bits of the packet entering the buffer at the rate f_b. In order to insure that the message does not back-up in the buffer we require that when the concentrator-commutator selects a particular buffer, the bits are read out of the buffer and on to the common channel at the bit rate Kf_b. Thus a bit duration on the channel is $1/Kf_b$, and since there are b bits in a packet, the dwell time of the commutator on each buffer is $b/Kf_b \equiv T_P$, where T_P is the duration of a packet. Hence, the commutator returns to each individual buffer at times separated by the intervals $KT_P = b/f_b$. Thus, if QPSK is employed, the channel bandwidth required is equal to the bit rate on the channel, which is

$$B = Kf_b H_z \tag{16.4-1}$$

TDMA systems are very efficient if the K users have a continual need to transmit. If however, the users' need to transmit varies from time to time or if due to the type of communication employed, the users rarely transmit, then TDMA becomes inefficient since a user pays for the time-slot, even when he fails to transmit, since the time-slot is "wasted." During this "wasted" time an additional user could have been admitted to the network. To improve the system efficiency under

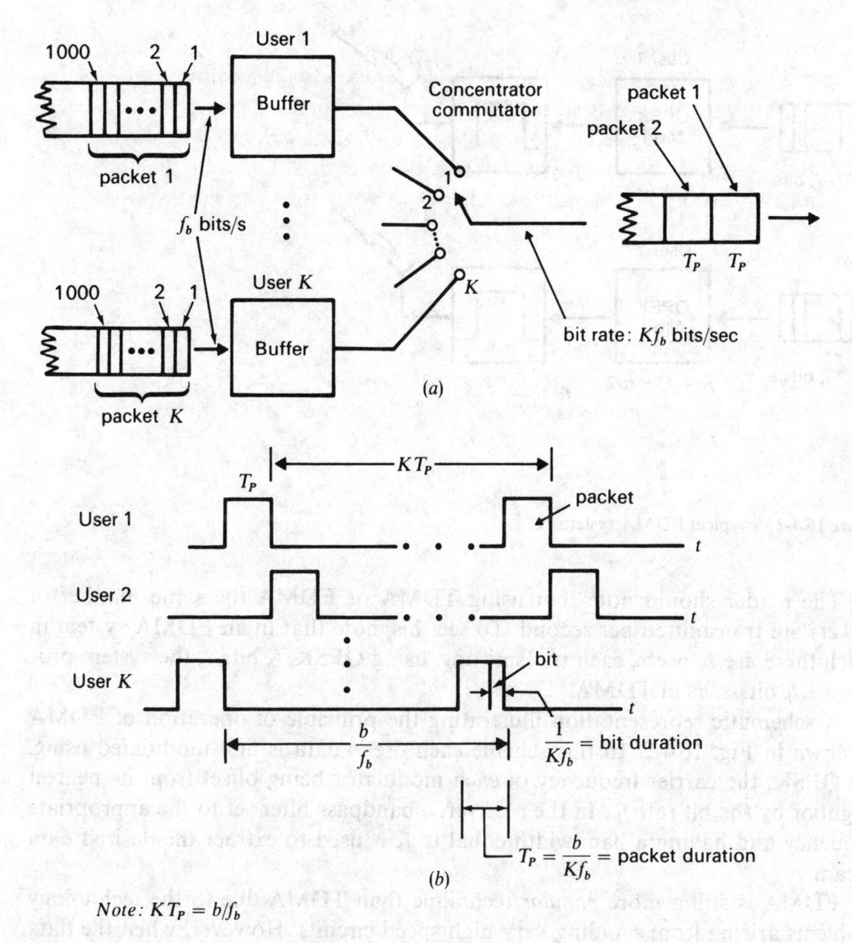

Figure 16.4-1 (*a*) A typical TDMA system. (*b*) Timing patterns. The higher level of each "waveform" marks the dwell time of the commutator on each switch position.

the above circumstances one uses alternate techniques such as ALOHA or CSMA schemes which are described below or a *demand assignment multiple access* (DAMA) scheme (see Prob. 16.4-2).

16.4-2 FREQUENCY DIVISION MULTIPLE ACCESS (FOMA)

Frequency division multiple access (FDMA) is the dual technique of TDMA in which each of the K users can transmit all of the time, but each must use a portion B_u of the total bandwidth, B, such that [see Eq. (16.4-1)]

$$B_u = \frac{B}{K} = f_b \, Hz \tag{16.4-2}$$

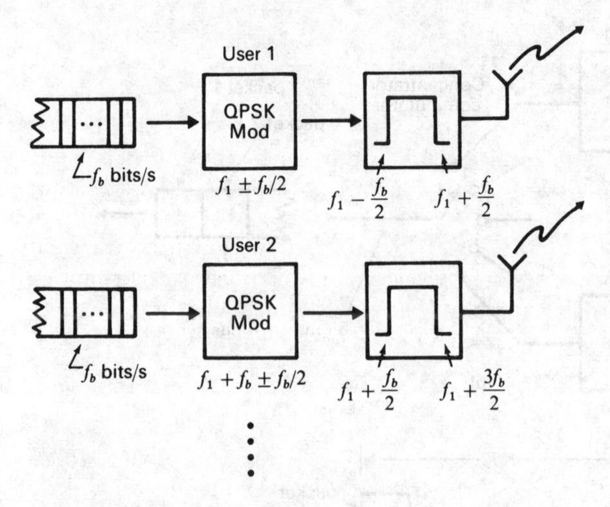

Figure 16.4-2 A typical FDMA system.

The reader should note that using TDMA or FDMA the same number of *packets* are transmitted per second. To see this, note that in an FDMA system in which there are K users, each transmitting, using QPSK, f_b bits/s, the system processes Kf_b bits/s, as in TDMA.

A schematic representation illustrating the principle of operation of FDMA is shown in Fig. 16.4-2. In this scheme each user's data is first modulated using, say QPSK, the carrier frequency of each modulator being offset from its nearest neighbor by the bit rate f_b. In the receiver, a bandpass filter set to the appropriate frequency and having a bandwidth equal to f_b is used to extract the desired data stream.

FDMA is still a more *popular* technique than TDMA due to the technology problems arising from building very high speed circuits. However, when the data is digital, TDMA is more "natural" and as technology improves, more systems are converting to TDMA. As is the case for TDMA, FDMA is efficient as long as each user requires the use of the network a large percentage of the time. If such is not the case, available bandwidth is being wasted since other users could have transmitted on these bandwidths. Using FDMA these extra users were denied access to the network.

16.4-3 ALOHA[2]

To correct the deficiencies found to be inherent in TDMA and FDMA, a group of engineering professors from the University of Hawaii developed a random access protocol which they aptly called ALOHA. In the basic ALOHA system there are K users and each user transmits a packet of duration T_P, every NT_P seconds, where $N \gg K$. Note that in the TDMA or FDMA system the packet rate was one packet each KT_P seconds (see Fig. 16.4-1) which is a much greater rate than for the ALOHA system.

The reader can see that the ALOHA system is one in which the channel is usually not being used. If however, any of the K users wishes to transmit, the user transmits immediately, using the available bandwidth which is now, $B = f_b$. Note that the user does not wait his "turn" as in TDMA nor is he assigned a narrow bandwidth as in FDMA.

The problem faced when using ALOHA is that if two or more users transmit in a time interval such that the receiver receives more than one packet simultaneously, the receiver is unable to separate the packets, since they are now not *orthogonal* in time (TDMA) or in frequency (FDMA). A *collision* is said to occur since being unable to separate the packets the receiver receives none of the packets correctly.

If, the users rarely use the network the probability of a collision is small. To quantify this discussion we shall assume that each user transmits at an average rate equal to $1/NT_P$ so that the average rate at which packets enter the network is

$$\lambda = K/NT_P \text{ packets/s} \tag{16.4-3}$$

We shall also assume that each user transmits independently of each other so that knowledge of the past history of what user u_j transmitted, and when, will yield no information concerning when the next packet will enter the network. That is, we assume that packets enter the network *randomly*. In this case as will be established in Sec. 16.5, the probability that m packets will enter the network in a time τ is given by the Poisson distribution

$$P(m, \tau) = \frac{e^{-\lambda\tau}(\lambda\tau)^m}{m!} \tag{16.4-4}$$

Thus, consider that user 1 transmits a packet at some time T_i. Then, referring to Fig. 16.4-3 we see that if any other user transmits a packet in the time interval

$$T_i - T_P < t < T_i + T_P \tag{16.4-5}$$

which is over an interval of duration $2T_P$, there will be a "collision" and neither of the packets will be received correctly. The probability that no other user is employing the channel over an interval $2T_P$ is found by setting $m = 0$ and $\tau = 2T_P$ in Eq. (16.4-4). The result is, taking account of the fact that $K \simeq K - 1$,

$$P_0 = P(m = 0, \tau = 2T_P) = \frac{e^{-2K/N}(2K/N)^0}{0!} = e^{-2K/N} \tag{16.4-6}$$

If, on the average, K/NT_P packets enter the network per second and the probability of being successfully received is $\exp(-2K/N)$, then the average number of packets which are successfully received per second, is called the *throughput*, S, and is equal to

$$S = \frac{K}{NT_P} P_0 = \frac{K}{NT_P} e^{-2K/N} = \frac{\rho}{T_P} e^{-2\rho} \tag{16.4-7}$$

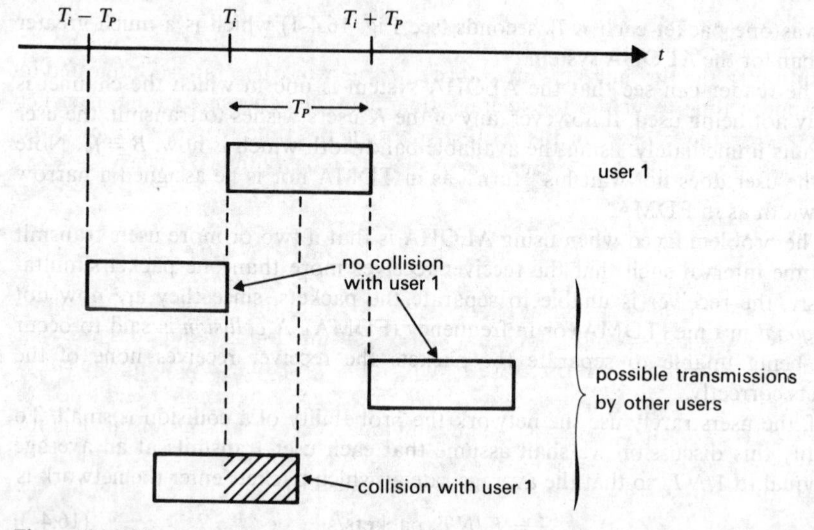

Figure 16.4-3 Illustrating the region over which a collision can occur when a packet is transmitted.

where we have defined $\rho \equiv K/N$. Note that $\rho = K/N$ represents the input density to the network since for a TDMA or FDMA system K can be made equal to N.

In the ALOHA system, the throughput S is a maximum when

$$\rho = \frac{K}{N} = \frac{1}{2} \tag{16.4-8}$$

To prove this result we need only differentiate S in Eq. (16.4-9) with respect to ρ and set the result equal to zero:

$$\frac{dS}{d\rho} = 0 = \frac{e^{-2\rho}}{T_P} - \frac{2\rho}{T_P} e^{-2\rho} \tag{16.4-9}$$

and therefore $\rho = \frac{1}{2}$. Thus when ρ is 50 percent, the maximum throughput is

$$S_{\max} = \frac{1}{2T_P} e^{-1} \tag{16.4-10}$$

In a TDMA or FDMA system $S_{\max} = 1/T_P$ packets/s, or looking at it another way, one packet is transmitted each T_P s. Thus, if we compare the product ST_P for ALOHA to ST_P for TDMA or FDMA, we get

$$\frac{(S_{\max} T_P)_{\text{ALOHA}}}{(S_{\max} T_P)_{\text{TDMA}}} = \frac{1}{2e} \simeq 0.182 \tag{16.4-11}$$

The above discussion shows that as the number of packets entering the network increases, the throughput S increases until $K = N/2$. When K increases above $N/2$ the number of collisions increase so rapidly that the throughput S decreases rapidly.

Response Time (Delay)

In addition to the throughput found above, the network user is also interested in the response time, which is the average time elapsed between sending a packet and receiving the packet. In the ALOHA system the time to receive a message when there is no collision, is merely the propagation delay time T_d, where

$$T_d = \frac{d}{c} \tag{16.4-12}$$

where d is the distance travelled and c is the speed of light. Figure 16.4-4 illustrates the case where a collision occurs and retransmission is required. Note that packet i is transmitted, and is received after a time T_d. After examining the entire message, an error was detected, and a NAK message of duration T_n returned to each user. The system restrains each individual user from retransmitting for random and independent time intervals T_W, so that, hopefully, the users who collided before will not do so again. (Note that this technique is the same as the "stop-and-wait" ARQ technique studied in Sec. 13.24.) Thus, the time between the second reception and the first reception of packet i, is

$$\Delta = T_P + T_n + T_W + 2T_d \tag{16.4-13}$$

The average response time D is then given by

$$D = T_d P_0 + (T_d + \Delta)(1 - P_0)P_0 + (T_d + 2\Delta)(1 - P_0)^2 P_0 + \cdots$$
$$+ (T_d + q\Delta)(1 - P_0)^q P_0 + \cdots \tag{16.4-14}$$

In Eq. (16.4-14), $T_d + q\Delta$ is the elapsed time between the transmission of a packet for the first time and its reception after being transmitted a total of $q + 1$ times; and $(1 - P_0)^q P_0$ is the probability that a packet was received in error q times and

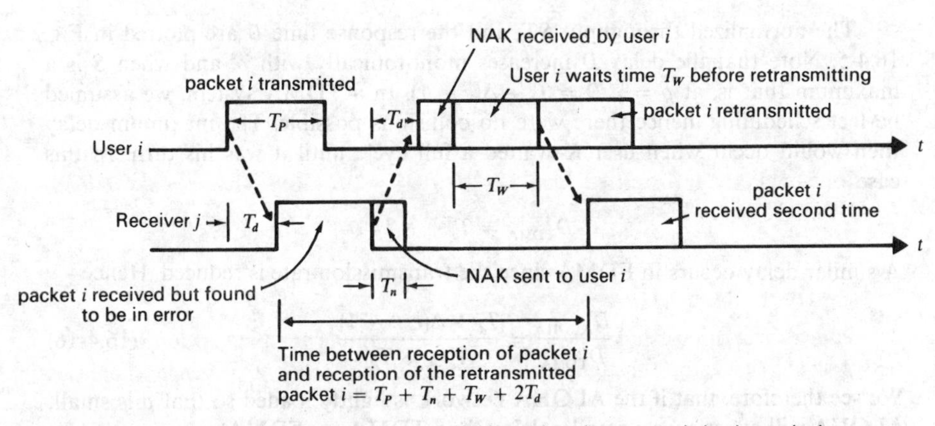

Figure 16.4-4 Showing the elapsed time between packets when the retransmission is required.

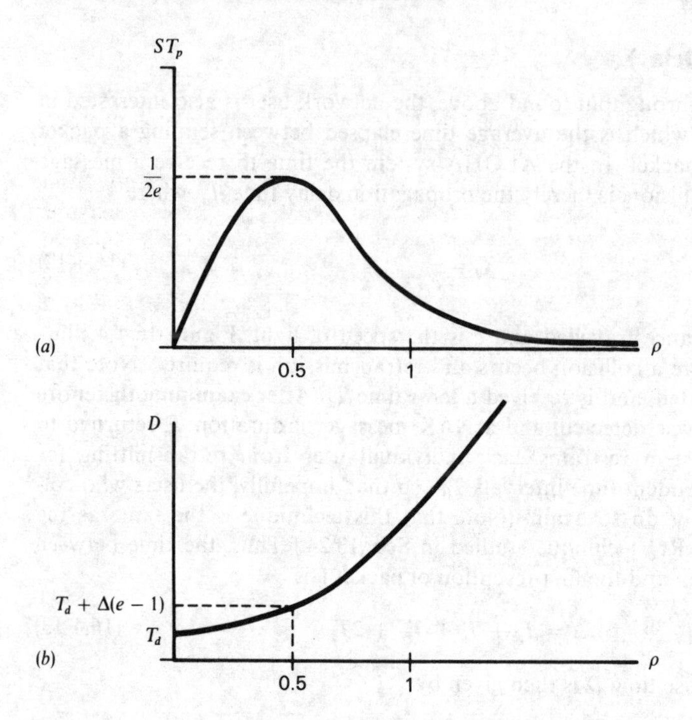

Figure 16.4-5 (*a*) Throughput of the ALOHA system. (*b*) Response time in an ALOHA system.

received correctly on the $q + 1$st time. It is left to the problems to show that Eq. (16.4-14) can be reduced to [see Eq. (16.4-8)]:

$$D = T_d + \frac{\Delta}{P_0}(1 - P_0) = T_d + \Delta(e^{2\rho} - 1) \tag{16.4-15}$$

The normalized throughput ST_ρ and the response time D are plotted in Fig. 16.4-5. Note that the delay D increases monotonically with ρ, and when S is a maximum that is, at $\rho = \frac{1}{2}$, $D = T_d + \Delta(e - 1)$. In a TDMA system, we assumed perfect scheduling, hence there were no collisions possible. The maximum delay then would occur when user K waited a full cycle until it was his turn. In this case

$$D_{\text{TDMA}} = T_d + KT_P$$

A similar delay occurs in FDMA since the transmission rate is reduced. Hence

$$\frac{D_{\text{ALOHA}}}{D_{\text{TDMA}}} = \frac{T_d + \Delta(e^{2\rho} - 1)}{T_d + KT_P} \tag{16.4-16}$$

We see therefore, that if the ALOHA network is lightly loaded so that ρ is small, ALOHA will experience a smaller delay than TDMA (or FDMA).

16.4-4 SLOTTED ALOHA

To increase the throughput S of an ALOHA type system the following protocol has been adopted: Divide the time scale into intervals, each of duration T_P, as shown in Fig. 16.4-6*a*. Then, for example, every user wishing to transmit from, say, $t = T_i$ to $t = T_{i+1}$ must wait and transmit at time T_{i+1}, that is, each user must synchronize his transmission so that it starts at the beginning of an interval. As a result, collisions occur as shown in Fig. 16.4-6*b* only between packets wishing to transmit in a T_P second interval and not in a $2T_P$ second interval as in the "pure" ALOHA system described earlier.

The throughput S of such a time *slotted ALOHA* system is given by Eq. (16.4-7) where P_0, the probability of no collision, is found by setting $\tau = T_P$ rather than $2T_P$ in Eq. (16.4-4). The result is

$$P_0 = P(m = 0, \tau = T_P) = e^{-K/N} = e^{-\rho} \tag{16.4-17}$$

and

$$S = \frac{K}{NT_P} P_0 = \frac{\rho}{T_P} e^{-\rho} \tag{16.4-18}$$

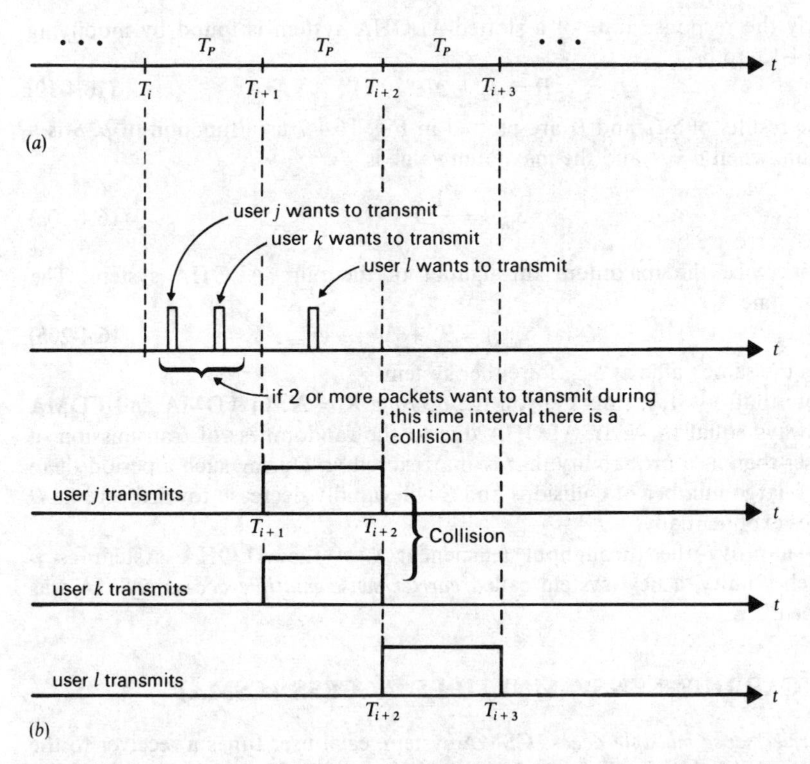

Figure 16.4-6 (*a*) Showing the time slots for a slotted ALOHA system. (*b*) Showing that a collision occurs when more than one packet wants to transmit during a T_P time interval.

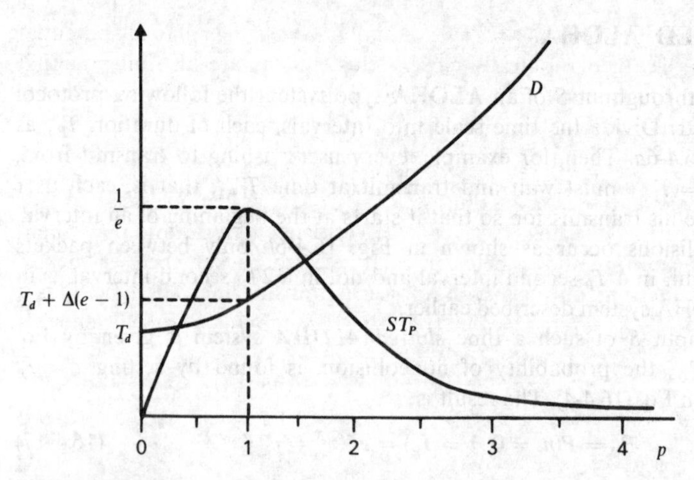

Figure 16.4-7 Throughput and delay as a function of input "density" to the network, $\rho = K/N$ for slotted ALOHA.

Similarly the response time of a slotted-ALOHA system is found by modifying Eq. (16.4-15) to be

$$D = T_d + \Delta(e^{\rho} - 1) \tag{16.4-19}$$

The results of ST_P and D are plotted in Fig. 16.4-7 as a function of ρ. S is a maximum when $\rho = 1$ and the maximum value is

$$S_{\max} = \frac{1}{T_P} e^{-1} \tag{16.4-20a}$$

which is twice the maximum throughput of the pure ALOHA system. The response time is

$$D \text{ (at } S_{\max}) = T_d + \Delta(e - 1) \tag{16.4-20b}$$

and has the same value at $S_{\max}$ for either system.

Note that $\rho = 1$, means that on the *average* $K = N$. In TDMA and FDMA K is always equal to N. In ALOHA due to the randomness of transmission of each user there is a probability that K may exceed N. During such a period there will be a large number of collisions and S will rapidly decrease toward zero as D increases exponentially.

To improve the throughput (efficiency) S of the ALOHA system as ρ approaches unity, a new system called *carrier sense multiple access*, CSMA was developed.

16.4-5 CARRIER SENSE MULTIPLE ACCESS (CSMA)

In a *carrier sense multiple access* CSMA system, each user tunes a receiver to the common carrier frequency f_c employed by all users. Before transmitting, a user will *listen*, to determine if any of the other users is transmitting; if no other user is transmitting he will transmit his packet. Initially we assume that all potential transmitting users are geographically closely clustered as are also all potential receiving users. In this case we may ignore the propagation delay between transmitting users and there will be no collisions.

Throughput To calculate the throughput, S, of a CSMA system, we refer to Fig. 16.4-8. Here we see that the host receiver, receives packets of duration T_P whenever a user is transmitting and in between each received packet, there is "dead" time T_B, which is the time during which no user is transmitting. The time T_B is a random process with an average value $\bar{T}_B$. The throughput, S, or efficiency, is therefore

$$S = \frac{1}{T_P + \bar{T}_B} = \frac{1}{T_P}\left(\frac{1}{1 + \bar{T}_B/T_P}\right) \tag{16.4-21}$$

It is shown in Sec. 16.6-1 that the time between received packets T_B, which is called the *interarrival time*, is given by the probability density

$$P_{\text{interarrival time}} = P_{\text{dead time}} = P(T_B) = \frac{\rho}{T_P}\, e^{-\rho T_B/T_P} \tag{16.4-22}$$

Thus, the ratio of the dead-time-to-packet-duration is a random process with average value

$$\left(\frac{T_B}{T_P}\right)_{\text{average}} = \frac{\bar{T}_B}{T_P} = \int_0^\infty \frac{T_B}{T_P}\frac{\rho}{T_P}\, e^{-\rho T_B/T_P}\, dT_B = \frac{1}{\rho} \tag{16.4-23a}$$

Thus the throughput is found by substituting Eq. (16.4-23*a*) into Eq. (16.4-21) yielding,

$$S = \frac{1}{T_P}\left(\frac{1}{1 + 1/\rho}\right) = \frac{1}{T_P}\left(\frac{\rho}{\rho + 1}\right) \tag{16.4-23b}$$

It is seen from Eq. (16.4-23*b*) and seen in Fig. 16.4-9*a* that the normalized throughput ST_P in a CSMA system can actually approach unity. However, the above calculations ignore the fact that there may be a delay between the transmission of a packet by user i and its being sensed by user j. This matter is discussed below. Before addressing that matter though, we shall first determine the response time, or average delay, associated with the reception of a packet which has been transmitted.

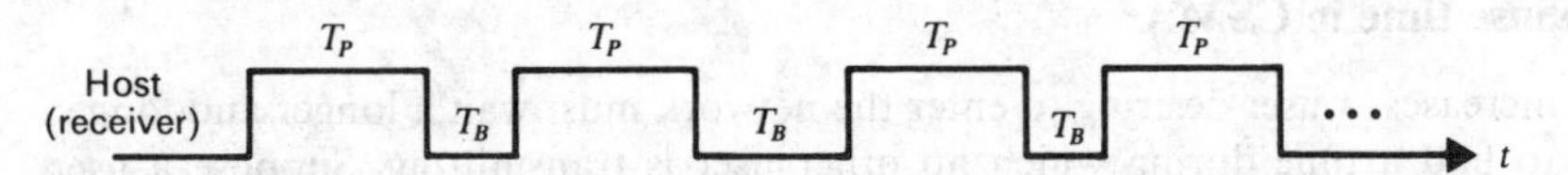

Figure 16.4-8 Host receiver packets T_P and "dead" time, T_B.

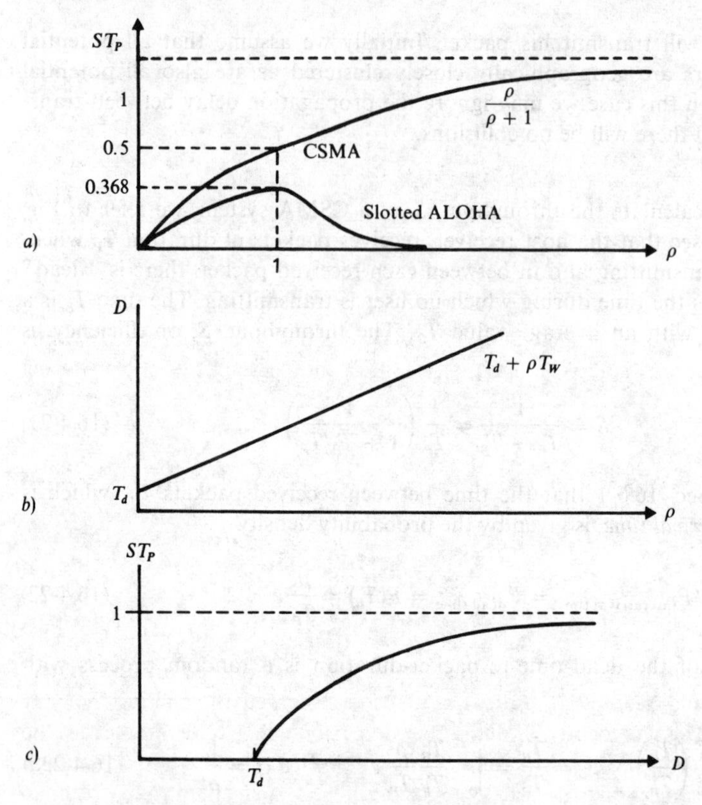

Figure 16.4-9 (*a*) Comparison of CSMA and slotted ALOHA throughput. (*b*) Delay in CSMA. (*c*) Variation of throughput as a function of delay for CSMA.

Waiting time in CSMA

In CSMA, a prospective user does not monitor the system continuously to find a time when the system is available for his use. For, with such continuous monitoring all prospective, waiting, users may try simultaneously to use the system when first it becomes available. Instead, each user, having found the system busy, will wait a random time before trying again. We shall now calculate the average time T_E a prospective user must wait before he can *enter* the system. The total time between the moment a packet is ready for transmission and the moment when it is finally received is then the delay $D = T_d + T_E$ where T_d is the propagation delay between transmitter and receiver.

Response time in CSMA

As ρ increases, a user desiring to enter the network must wait a longer and longer time to find a time during which no other user is transmitting. Suppose a user listens at random times separated by random time intervals T_W having an average value $\bar{T}_W$. The entrance time T_E is then

$$T_E = \bar{T}_W P_{\text{on}} P_{\text{off}} + 2\bar{T}_w P_{\text{on}}^2 P_{\text{off}} + \cdots \tag{16.4-24}$$

where

$$P_{\text{off}} = \frac{\bar{T}_B}{T_P + \bar{T}_B} = \begin{cases}\text{probability of being in a time interval}\\ \text{where no user is transmitting}\end{cases} \tag{16.4-25a}$$

$$P_{\text{on}} = 1 - P_{\text{off}} = \frac{T_P}{T_P + \bar{T}_B} = \begin{cases}\text{probability of being in a time interval}\\ \text{where some user is transmitting}\end{cases} \tag{16.4-25b}$$

The first term $\bar{T}_W P_{\text{on}} P_{\text{off}}$ is the time required if we did not transmit the first time we *sensed* the channel since the channel was busy but transmitted after waiting an average time $\bar{T}_W$. The product $P_{\text{on}} P_{\text{off}}$ is the probability of the channel being occupied the first time and being vacant the second time the channel is sensed, etc.

Equation (16.4-24) is readily simplified (see Prob. 16.4-10) to:

$$T_E = \bar{T}_W/(\bar{T}_B/T_P) = \rho \bar{T}_W \tag{16.4-26a}$$

and hence

$$D = T_d + T_E = T_d + \rho \bar{T}_W \tag{16.4-26b}$$

In a CSMA system the average waiting time $\bar{T}_W$ is established by the system designers. It would seem that users would be best served by making $\bar{T}_W$ very small. In the system presently under discussion in which we have assumed no propagation delay between users, minimizing $\bar{T}_W$ would indeed be to advantage. However, as we shall see in the next section, in a real system there is an optimum value for $\bar{T}_W$ depending on the required throughput.

Figure 16.4-9*b* shows the variation of delay D as a function of ρ and Fig. 16.4-9*c* shows that for ST_d to approach unity requires a very long delay. This rapid increase in delay, limits the useful throughput level to values much less than unity. Note, however, that in the present case the delay D, given by Eq. (16.4-26*b*), increases linearly with ρ while for ALOHA, the delay D, given in Eq. (16.4-15) increases exponentially with ρ.

The Effect of Propagation Delay Between Users: Collisions

In "real-life" a packet travels with the speed of light, c, and therefore when one user transmits a packet, it takes another user (d/c) seconds to receive the packet, where d is the distance between users. Thus, the transmission of one user is not sensed immediately by the other users. As a result of this propagation delay, collisions can occur in a real CSMA system.

To illustrate how collisions occur let us refer to Fig. 16.4-10. Here we see that user i transmits at time t_1. If a propagation delay T_d exists between users i and l, and say user l transmits at time t_2, he will not observe the transmission of user i until time t_3, at which time it is too late since user l began transmitting at time

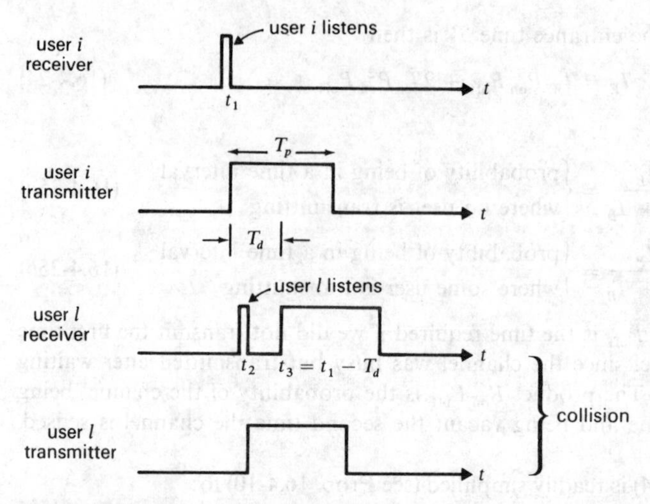

Figure 16.4-10 Showing collision occurs in CSMA.

t_2. Thus, assuming that the distances from the transmitting users i and l to a receiver (HOST) are comparable, the HOST sees a *collision.* The probability that *no* packets enter the network after user i transmits is given by $P(0, \tau_d)$ in Eq. (16.4-4). Thus the probability of a collision P_c is the probability that one or more packets attempted to transmit in the T_d second interval. Thus $P_c = 1 - P(0, \tau_d)$ and using $\lambda = K/NT_P$ and $\rho = K/N$ we find

$$P_c = 1 - e^{-\rho T_d/T_P} \tag{16.4-27a}$$

and the probability of no collision

$$P_{nc} = e^{-\rho T_d/T_P} \tag{16.4-27b}$$

To calculate the throughput of the CSMA system including the effect of collisions we assume that each user senses any other user's transmission after a time delay T_d, Figure 16.4-11 shows that the time T between successfully received packets depends on the frequency of collisions. Let the time x, measured from the beginning of a received packet, be the time at which a collision occurs. Clearly, $x < T_d$ and is a random variable. Thus, the average time $\bar{T}$ between successfully (correctly) received packets is

$$\bar{T} = (T_P + \bar{T}_B)P_{nc} + [2(T_P + \bar{T}_B) + \bar{x}]P_c P_{nc} + [3(T_P + \bar{T}_B) + 2\bar{x}]P_c^2 P_{nc} + \cdots \tag{16.4-28}$$

where the first term states that the time $\bar{T}$ between correctly received packets is $T_P + \bar{T}_B$ if there are no collisions, the second term states that the time $\bar{T}$ between correctly received packets is $2(T_P + \bar{T}_B) + \bar{x}$ if there is no collision during the first packet but there is a collision during the second packet, etc.

The throughput S is equal to

$$S = \frac{1}{\bar{T}} \text{ packets received correctly/s} \tag{16.4-29}$$

To compute $\bar{T}$ we first note that Eq. (16.4-28) can be reduced to

$$\bar{T} = (T_P + \bar{T}_B)\frac{1}{P_{nc}} + \bar{x}P_c/P_{nc} \tag{16.4-30}$$

where $P_{nc} = 1 - P_c$. The average value of T_B was found in Eq. (16.4-23*a*) to be

$$\bar{T}_B = T_P/\rho \tag{16.4-31}$$

and so we must now determine $\bar{x}$ to find the throughput. We note that

$$0 < \bar{x} < T_d \tag{16.4-32}$$

and therefore $\bar{T}$ can be upper bounded by replacing $\bar{x}$ with T_d in Eq. (16.4-30). Hence S is *lower bounded* by such a substitution. Substituting Eqs. (16.4-27*a* and *b*) and Eq. (16.4-31) into Eq. (16.4-30) yields

$$S > \frac{\rho}{T_P}\left(\frac{e^{-\rho T_d/T_P}}{1 + \rho + (\rho T_d/T_P)(1 - e^{-\rho T_d/T_P})}\right) \tag{16.4-33}$$

If we replace $\bar{x}$ by 0 we *upper bound S*. The result is readily shown to be

$$S < \frac{\rho}{T_P}\left(\frac{e^{-\rho T_d/T_P}}{1 + \rho}\right) \tag{16.4-34}$$

Note that when T_d goes to zero, Eqs. (16.4-33 and 34) reduce to Eq. (16.4-23*b*). However, note that as the propagation delay time τ_d increases, the throughput

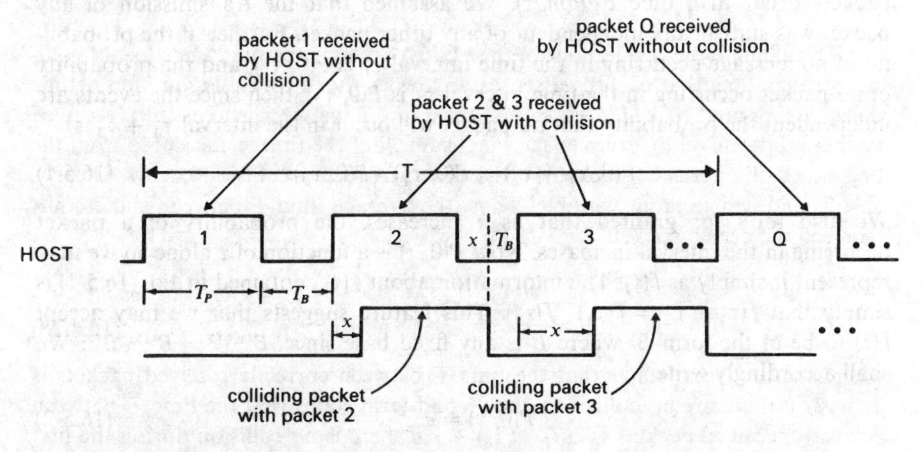

Figure 16.4-11 Illustration that collisions reduce throughput.

decreases rapidly and in some applications ALOHA can yield a higher throughput than CSMA. To illustrate this point consider $T_d = 2T_P$. Then

$$\frac{\rho}{T_P}\left(\frac{e^{-2\rho}}{1+\rho+2\rho(1-e^{-2\rho})}\right) < S_{\text{CSMA}} < \frac{1}{T_P}\left(\frac{\rho}{1+\rho}\right)e^{-2\rho} \tag{16.4-35}$$

while

$$S_{\text{ALOHA}} = \frac{\rho}{T_P}e^{-2\rho} \tag{16.4-36}$$

Here $S_{\text{ALOHA}} \geq S_{\text{CSMA}}$ for any value of ρ. The reader should remember that a propagation delay $T_d = 2T_P$ corresponds to a rather long distance between users. For example, if a packet consists of 1000 bits which are transmitted at a bit rate of 10 Mb/s then $T_P = 100\ \mu$s. The packet travels with the speed of light and in $T_d = 2T_P = 200\ \mu$s travels a distance d of

$$d = cT_d = 3 \times 10^8 \text{ ms} \times 2 \times 10^{-4} \text{ s} = 60 \text{ km} \tag{16.4-37}$$

Conclusions

For values of $T_d < 2T_P$, CSMA has a significantly higher throughput than ALOHA. As a result CSMA has found numerous applications in the computer communications field.

There are many other interesting and useful access techniques which are to be found in the literature.[1,3]

16.5 THE POISSON DISTRIBUTION

In the previous section we had the need to determine the probability that m packets occur in a time τ, $P(m, \tau)$. We assumed that the transmission of any packet was statistically independent of any other packet. Further, if the probability of no message occurring in the time interval τ_1 is $P(0, \tau_1)$, and the probability of no packet occurring in the time interval τ_2 is $P(0, \tau_2)$, then since the events are independent the probability that no packet will occur in the interval $\tau_1 + \tau_2$ is

$$P(0, \tau_1 + \tau_2) = P(0, \tau_1) \cdot P(0, \tau_2) \tag{16.5-1}$$

We also took for granted that as τ increases, the probability of a packet occurring in the interval increases. Now $P(0, \tau)$ is a function of τ alone so we may represent it simply as $f(\tau)$. The information about $f(\tau)$ contained in Eq. (16.5-1) is simply that $f(\tau_1 + \tau_2) = f(\tau_1) \cdot f(\tau_2)$. This feature suggests that we may accept $f(\tau)$ to be of the form B^τ where B is any fixed base since, $B^{\tau_1+\tau_2} = B^{\tau_1} \cdot B^{\tau_2}$. We shall accordingly write

$$P(0, \tau) = e^{-\lambda\tau} \tag{16.5-2}$$

where λ is yet to be determined.

If we let τ decrease to a differentially small value, $\Delta\tau$, then Eq. (16.5-2) reduces to

$$P(0, \Delta\tau) \cong 1 - \lambda\Delta\tau \tag{16.5-3}$$

which approaches unity as $\Delta\tau$ approaches zero. It therefore appears reasonable to assume that in an interval $\Delta\tau$ at most 1 packet can be received.

The probability that m packets are received in a time interval $\tau + \Delta\tau$ can then be written as

$$P(m, \tau + \Delta\tau) = P(m, \tau) \cdot P(0, \Delta\tau) + P(m-1, \tau) \cdot P(1, \Delta\tau) \tag{16.5-4}$$

where the first product term is the probability that all m packets were received in the τ second interval and none in the interval $\Delta\tau$, and the second product term is the probability that $m - 1$ packets are received in the time interval τ and one packet received in $\Delta\tau$. Note that we have neglected the possibility of two or more packets in the interval $\Delta\tau$. Since

$$P(0, \Delta\tau) = 1 - \lambda\Delta\tau \tag{16.5-5a}$$

and

$$P(1, \Delta\tau) = 1 - P(0, \Delta\tau) = \lambda\Delta\tau \tag{16.5-5b}$$

We have, using Eq. (16.5-4),

$$P(m, \tau + \Delta\tau) = P(m, \tau)[1 - \lambda\Delta\tau] + P(m-1, \tau)[\lambda\Delta\tau] \tag{16.5-6}$$

Transposing Eq. (16.5-6) we get

$$\frac{P(m, \tau + \Delta\tau) - P(m, \tau)}{\Delta\tau} = \lambda[P(m-1, \tau) - P(m, \tau)] \tag{16.5-7a}$$

which, in the limit as $\Delta\tau \to d\tau$, becomes:

$$\frac{dP(m, \tau)}{d\tau} = \lambda[P(m-1, \tau) - P(m, \tau)] \tag{16.5-7b}$$

The reader can, by direct substitution, verify that the solution of the differential-difference equation given by Eq. (16.5-7*b*) is the *poisson* distribution

$$P(m, \tau) = \frac{(\lambda\tau)^m e^{-\lambda\tau}}{m!} \tag{16.5-8}$$

The average value of m is shown in the problems (Prob. 16.5-1) to be

$$\bar{m} = E(m) = \lambda\tau \tag{16.5-9}$$

Thus λ is defined as the average number of packets per second.

16.5-1 THE INTERARRIVAL PROCESS

In Fig. 16.4-8 we had to refer to a time T_B during which no packets arrived. To find the probability density of T_B we note that the reason that no packet arrived in precisely the time interval T_B rather than in a longer time interval, was that a

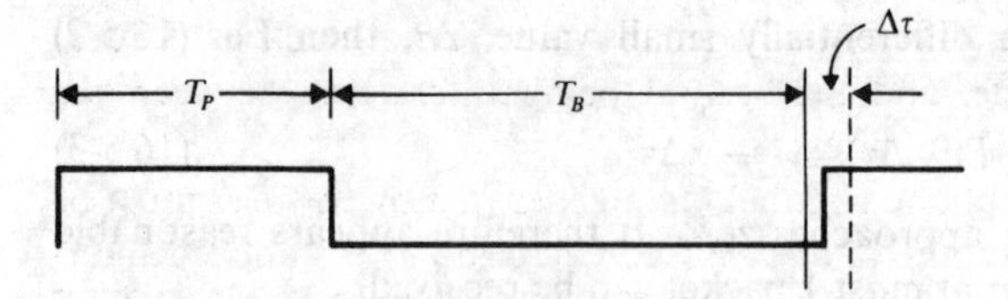

Figure 16.5-1 The interarrival process.

packet did indeed arrive in a time interval $\Delta\tau$ just following the "dead" interval T_B. This point is illustrated in Fig. 16.5-1. The probability of zero packets in a time τ $(= T_B)$ is

$$
\begin{aligned}
P(\tau) &= P(0, \tau) \cdot P(1, \Delta\tau) \\
&= e^{-\lambda\tau}[\lambda\Delta\tau] = \lambda e^{-\lambda\tau}\Delta\tau
\end{aligned}
\tag{16.5-10}
$$

Thus, the probability density of the time during which no packet is received (this is called the interarrival time) is

$$p(\tau) = \frac{P(\tau)}{\Delta\tau} = \lambda e^{-\lambda\tau} \tag{16.5-11}$$

The average value of τ is

$$\bar{\tau} = E(\tau) = \frac{1}{\lambda} \tag{16.5-12}$$

16.6 PROTOCOLS[3]

As we saw earlier, in order for a network to accommodate a number of users efficiently, routing and flow control procedures have to be established and *all* users of the network must follow these procedures. Actually there are many additional rules that must be followed. Indeed, to reduce the system design complexity, these rules, called *protocols*, are divided into approximately seven areas called *layers* or *levels*. This set of layers is called the *network architecture*. A complete, indepth, discussion of these protocols is beyond the intent of this text and we shall accordingly, discuss the matter only briefly.

16.6-1 THE FIRST LAYER: THE PHYSICAL LAYER

The first protocol is called the *physical layer* and is concerned with the actual transmission of data over the physical communication channel. Specifically the *physical layer* considers how to set-up a connection and once a connection is set-up what bit rate to use to insure that the resulting error rate is acceptable.

To see how a connection is set-up we will consider two examples: the first being interfacing the *data terminal equipment* (DTE) provided by the *user* to the *data communications equipment* (DCE) provided by the carrier (such as AT&T) using an analog telephone line, and the second being interfacing between the DTE and DCE when using a digital line.

When using an analog telephone channel, data generated by the user must be interfaced to a DCE which contains a *modem* (see Chap. 6). The standard interface used in the United States is called the RS232C. (The international CCITT equivalent is called the V.24.) The RS232C is a 24 pin integrated circuit (IC). Depending upon which pins are employed the DTE can be restricted to: only transmission, only reception, full duplex, or any of a large number of different transmit/receive strategies.

When using a digital line, the modem is not needed, however, the interface into the DCE is still needed to insure that the correct signals are sent from the DTE at the correct time and in the correct form and received by the DCE at the correct time and in the correct form and *vice versa.* The digital signaling interface standard is called the X.21 and is a 15 pin IC of which 9 pins are used.

To illustrate the use of both interfaces assume that initially the system is idle. When the user (DTE) wishes to place a call it sends a *desire-to-transmit* signal to the DCE. When the DCE is ready to receive the call address, it sends to the DTE an *OK-to-transmit* signal. The DTE now sends the address of the receiver to the DCE and the host's DCE and DTE are notified. If the call is accepted, data is sent from the user's DTE to its DCE, and to the host's DCE and DTE. When the message is completed the DTE sends a *termination of transmission* message to his DCE which is sent to the host's DCE and DTE. When the host's DTE is finished processing the information the connection is terminated. All of these signals are generated, transmitted, received and interpreted by the X.21 interface.

16.6-2 THE SECOND LAYER: THE DATA-LINK LAYER

The first, *physical* layer insured that the proper signaling took place to set up the connection between users and the network. The second, *data-link* layer, is concerned with communications between the IMPs (Interface Message Processor). The transmitting IMP takes the data from the DCE, packetizes it and transmits this packet, to a receiving IMP. The receiving IMP, receives the packet, depacketizes it and passes the resulting data to the end-users DCE (and then to its DTE). The packets, in addition to containing the data, contain the frame synchronization and error control information as well.

The error control procedure used is usually ARQ (see Sec. 13.23) with *stop-and-wait* employed for packets travelling short distances and *go-back-n* for packets transmitted over satellites. To see the need for the go-back-n, when satellite communication is involved, assume transmitting at a rate of 100 kb/s and a round trip propagation time delay of 500 ms. If each packet contains 1000 bits the first frame is completely sent after 10 ms. Hence 50 packets could be completely sent before an acknowledgement is received for packet 1. Thus, the efficiency of a stop-and-wait system is only 2 percent, even if no error is detected.

It is interesting to note that due to noise the ACK of a packet by the receiving IMP may not always be detected. In that case the DTE will assume that the packet was lost and would retransmit. The receiver will then receive two identical packets. In order to know that the second packet is redundant each packet must be numbered.

A third concept included in the data-link layer is that bandwidth can be used more efficiently if ACKs (called a *control* signal) and data are transmitted in the same packet. For example, in full duplex operation between two IMPs *A* and *B*, an acknowledgement of *A*'s transmission to *B* is contained in a packet of data that *B* sends to *A*. When both data and control information are sent in the same packet it is called *piggybacking*.

An Example: The Data-Link Layer of the ARPANET

The basic packet structure used in the ARPANET is shown in Fig. 16.6-1. The first three characters in an ARPANET packet are generated by hardware, not software, and include the frame synchronization bits (SYN), a *Data-link* escape (DLE) character to indicate going from synch to the *start of text* (STX), which is the third character.

The IMP software multiplexes eight independent full-duplex channels onto each physical line (16 channels in the case of a satellite link). It uses a stop-and-wait protocol but without NAKs. If a packet is not received correctly nothing is returned and after waiting 125 ms for an ACK the sender will retransmit. The 16 byte packet header contains the packet number and the number of the last packet received. Other information is contained in the header and will be discussed in Sec. 16.6-3: *The Third Layer*. The 5-byte transport header indicates the onset of the data. After the data ends, six terminating characters are employed for synchronization and error control: The ETX is an *end-of-text* character and the cyclic redundancy checks (CRCs) form a 24-bit checksum for error detection.

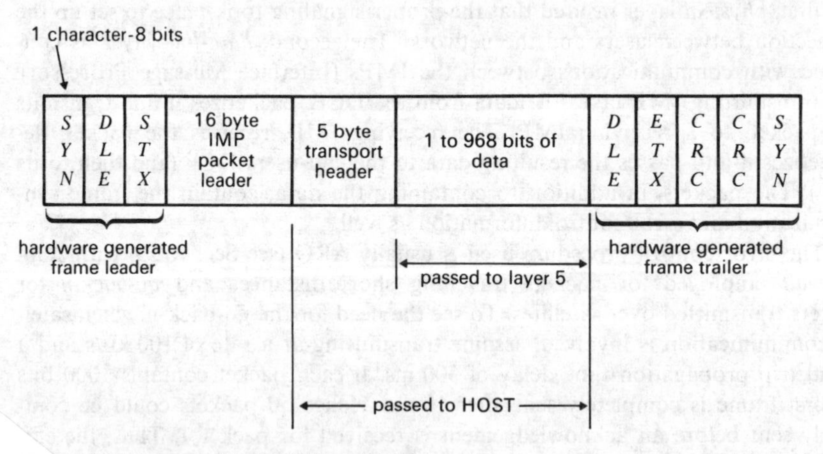

16.6-1 Format of ARPANET packet.

16.6-3 THE THIRD LAYER: NETWORK LAYER

The third or *network* layer's main concern is *routing* to control congestion in the network and to control congestion at the receiving IMPs. Congestion control deals with the problem that develops when more packets arrive at an IMP than there are buffers to store them. One way to control congestion is to monitor the traffic at each IMP and not allow data to be forwarded to an IMP that does not have adequate buffer space.

An Example: The ARPANET

The routing algorithm employed for the ARPANET requires that each IMP maintain an entire representation of the net including the *current* delays on each line. The delay information stored is a 10 s average on each line. These results are transmitted using *flooding*. Using this information the IMP determines the shortest time delay to every other IMP in the net. The source IMP chooses the path with minimum delay to the destination IMP and then transmits the packet to an intermediate IMP on the selected path. The intermediate IMP computes the shortest delay to the destination IMP and transmits the packet to the next appropriate destination IMP, etc. To insure a successful transmission, ACKs are returned from each receiving IMP to its sending IMP, as well as from the destination IMP to the source IMP.

The ACK from the destination IMP to the source IMP is also used for *congestion* control. The source IMP, prior to transmission, asks the destination IMP to reserve the appropriate buffer space. When the buffer space is available an *ALLOCATE* packet is sent to the source. The source now transmits the packet. The destination IMP, after correctly receiving the packet, sends an ACK and also reserves the same number of buffers for 250 ms, in case the source has another packet to send. If the source IMP does not transmit in that time the destination IMP uses the buffers for other users.

Routing for Satellite and Packet Radio

Today, satellite systems primarily employ TDMA or FDMA routing to insure efficient use of the channel. However, when asynchronous communication is required or the channel is lightly loaded, the same algorithms used for packet radio are employed. These consist of the ALOHA and CSMA algorithms which have previously been discussed (see Sec. 16.3).

16.6-4 THE X.25 PROTOCOL

The CCITT has generated an international standard protocol for layers 1, 2, and 3 called X.25. The X.25 interface allows a wide variety of *users services* to access a given network independently of the terminals and computers that each employs.

The X.25 defines the format and meaning of the information exchanged between the DTE-DCE interface for the layer 1, 2, and 3 protocols. Thus the X.21 protocol used in the physical layer is a part of X.25. To illustrate the operation of X.25 consider that DTE *A* wishes to communicate with DTE *B*. Following X.25 DTE *A* first sends a *CALL REQUEST* packet to DCE *A* which is forwarded to DCE *B* and then to DTE *B*. If DTE *B* accepts the call it returns a *CALL ACCEPTED* packet. When this packet is received a *virtual path* has been set-up and communication begins. When the communication ends, the side desiring to terminate the link sends a *CLEAR REQUEST* packet to the other DTE who returns a *CLEAR CONFIRMATION*. The path is now terminated. Note the redundancy used in X.25. This is to insure that a wide variety of terminals and computers can communicate with each other. However, to insure this generally requires an excessive use of bandwidth.

16.6-5 THE FOURTH OR TRANSPORT LAYER

The *transport* layer has the responsibility of providing a reliable and efficient end-to-end transport between users rather than between IMPs. The programs that provide the transport service run on *HOST* computers and not on IMPs.

The transport layer makes sure that packets are delivered in order, without loss or duplication. In addition, the transport layer acts as a multiplexer, sending packets from the HOST to their correct user locations. Further, HOSTS have *resources*, such as files, mailboxes, terminals, etc, that can be addressed using the transport layer. The transport layer also provides buffering, ordering and flow control, if packets are delayed or out of order. Synchronization algorithms are contained in this layer. Finally, the transport layer is used to connect a HOST in one network to a HOST in another network.

16.6-6 THE FIFTH OR SESSION LAYER

The *session* (or *connection*) layer is used to allow users to identify themselves when wanting access to the network. In some networks there is no session layer. In these cases the session layer's functions are performed by the transport layer. Such is the case in the ARPANET. This layer is the USER's interface into the network.

16.6-7 THE SIXTH LAYER: THE PRESENTATION LAYER

The *presentation* layer performs two main functions: *data encryption* and *word compression.*

Data encryption is necessary to insure that any messages transmitted will be understood only by the intended receiver. Indeed, illegal wire tapping is far more prevalent than one can imagine. Furthermore, in a satellite system all of the signals are available to anyone with a satellite antenna. Encryption, can be performed at the presentation layer; or even at the seventh, *application* layer. In either case the encryption process is called *end-to-end* encryption. If it is performed at the data-link level by the IMPs, it is called *data link encryption*. When *end-to-end* encryption is employed the *user* selects the encryption technique, while in data-link encryption the network chooses the encryption procedure.

Another function of the presentation layer is to provide word compression. A user would like to transmit his signal using as few bits as possible so as to save money. For example, rather than using 64 kb/s PCM, the use of 16 kb/s adaptive DM can reduce the users data flow by a factor of 4. In addition, if the time interval when the speaker is not talking is detected and the data-bits generated during the silent intervals not transmitted, an additional factor of 2 can be achieved. In general, any data stream can be compressed using techniques called *run-length* coding in which repetitive patterns are detected and replaced by shorter patterns. (See Sec. 13.5, the Shannon–Fano Code).

16.6-8 THE FINAL, SEVENTH LAYER: THE APPLICATION LAYER

The *application* layer is the user's protocol layer. The application layer presents the set of allowed messages, programs employed and the partitioning of a problem or data in a data base. Industry specific protocols for say the banking and airline industries are typical examples.

Conclusions

To conclude, it is important to note that while the International Standardization Organization (ISO) have developed protocol standards, called the *Open Systems Interconnection* OSI protocols, which were described above, IBM, DEC, and others currently have their own protocol standards which differ slightly with the seven layer ISO–OSI standard.

REFERENCES

1. Stallings, W.: "Data and Computer Communications," Macmillan Publishing Co., New York, 1985.
2. Kleinrock, L.: "Queueing Systems," Vol 2, John Wiley & Sons. Inc., New York, 1976.
3. Tanenbaum, A.: "Computer Networks," Prentice-Hall, Inc., Englewood Cliffs, N.J., 1981.

PROBLEMS

16.1-1. An 8192-bit message is to be transmitted, using message switching over an 8-link path to its destination. After each link there is a switching mode in which a minicomputer stores the message in a buffer and then forwards the message as soon as the message reaches the front of the buffer. The message contains 256 bits used for "overhead."

Assume that there is no queue in any of the 8 buffers so that as soon as a message is received, the overhead bits are processed and the message is forwarded. Neglect this processing time and the propagation time in a link.

If the bit rate is $f_b = 1$ Mb/s calculate the time for the message to be transmitted from the source to the user.

16.1-2. If a link is 25 km long, calculate the time required for an 8192 bit message, transmitted at 1 Mb/s to be completely received at the switching mode following the link. Assume the velocity of propagation is 3×10^8 m/s.

16.1-3. If in Prob. 16.1-1, propagation delay is not neglected, then calculate the time for the message to be transmitted from source to user if the distance separating switching nodes is 25 km and the velocity of propagation is 3×10^8 m/s.

16.1-4. A 7936-bit message is to be transmitted over an 8-link path to its destination. Packet switching is to be used and the overhead/packet is 256 bits.

Neglect the processing time and the propagation time. If the bit rate is $f_b = 1$ Mb/s

(*a*) Calculate the number of packets that the message should be divided into if the time from transmission to reception is to be minimized.

(*b*) Calculate this minimum time.

(*c*) Compare it to the result obtained for message switching. Which moves more quickly? Why?

16.2-1. A computer network is shown in Fig. 16.2-2. Assume the time to travel over any link is T. Flooding is used to transmit a message from A to L. Plot the number of times the same message reaches L as a function of time, for integral values of T. If the same message arrives at any node i from two different nodes j and k simultaneously, a collision occurs and one can assume that the message never reached node i. Assume that the message was transmitted from A only one time.

16.2-2. A shortest route is to be selected to transmit a message from A to L. Find the route.

16.2-3. Random walk is to be used to transmit a message from A to L. For example, the message leaving node A could go to B, E or J, each with probability 1/3. If the system is designed to avoid returning to a past node, e.g., once at J the packet could go to H, I or L with probability 1/3, but not return to node A, calculate the probability of a message sent once from node A, reaching node L, as a function of time for integral values of T.

16.2-4. Signal packets $S_1, S_2, \ldots, S_{10}$ enter a concentrator located at node R where they are multiplexed. Packets arrive independently, each with a Poisson density, at the average rate of 1 packet/s, and the time to transmit a packet is 1 ms. What is the probability of two or more packets being in the queue simultaneously?

16.4-1. 500 users employ TDMA to transmit 1000-bit packets of data. The channel bandwidth is 100 MHz and QPSK is employed.

(*a*) What is the maximum allowable bit rate on the channel?

(*b*) What is the packet rate?

(*c*) How long does it take to transmit one packet from each user? This time is called the frame time.

16.4-2. 500 users employ FDMA to transmit 1000-bit packets of data. The channel bandwidth is 100 MHz and QPSK is used at each of the 500 carrier frequencies employed.

(*a*) What is the maximum bandwidth allocated to each user?

(*b*) What is the bit rate employed by each user?

(*c*) How long does it take to transmit a packet?

16.4-3. (*a*) Plot ST_P as a function of ρ.

(*b*) Verify Eq. (16.4-8).

16.4-4. Verify Eq. (16.4-14).

16.4-5. Verify Eq. (16.4-15).

16.4-6. Ten users employ ALOHA to transmit 1000-bit packets of data. The data rate is 100 Mb/s. The data is to be transmitted a distance $d = 10{,}000$ km. Assume NAK messages contain 100 bits and that in the event of a collision the average waiting time prior to retransmission is 50 μs.

(*a*) Calculate the delay D as a function of ρ.
(*b*) Plot D *vs* ρ.

16.4-7. (*a*) Using Eqs. (16.4-7) and (16.4-15), eliminate ρ and plot S as a function of D.
(*b*) When $\rho > 0.5$ the ALOHA system is said to be "unstable." Why?

16.4-8. (*a*) Using Eqs. (16.4-18) and (16.4-19) eliminate ρ and plot S as a function of D.
(*b*) Compare (*a*) to the result of Prob. 16.4-8*a*.

16.4-9. (*a*) Verify Eq. (16.4-24).
(*b*) Verify Eq. (16.4-26).
(*c*) Plot S as a function of D using Eqs. (16.4-23*b*) and (16.4-26).

16.4-10. Verify Eq. (16.4-28).

16.4-11. Verify Eq. (16.4-33).

16.4-12. (*a*) Verify Eq. (16.4-34).
(*b*) Verify Eq. (16.4-35).

16.4-13. (*a*) Plot the throughput ST_P in Eq. (16.4-35) as a function of T_D/T_P for $\rho = 0.25$.
(*b*) At what value of T_D/T_P would slotted ALOHA yield a higher throughput?

16.5-1. Verify Eq. (16.5-9).

CHAPTER

SEVENTEEN

SPREAD SPECTRUM MODULATION

17.1 INTRODUCTION

Spread spectrum is a technique whereby an already modulated signal is modulated a *second* time in such a way as to produce a waveform which interferes in a barely noticeable way with any other signal operating in the same frequency band. Thus, a receiver tuned to receive a specific AM or FM broadcast would probably not notice the presence of a spread spectrum signal operating over the same frequency band. Similarly, the receiver of the spread spectrum signal would not notice the presence of the AM or FM signal. Thus, we say that interfering signals are *transparent* to spread spectrum signals and spread spectrum signals are transparent to interfering signals.

To provide the "transparency" described above the spread spectrum technique is to modulate an already modulated waveform, either using amplitude modulation or wideband frequency modulation, so as to produce a very wideband signal. For example, an ordinary AM signal utilizes a bandwidth of 10 kHz. Consider that a spread spectrum signal is operating at the same carrier frequency as the AM signal and has the same power P_s as the AM signal but a bandwidth of 1 MHz. Then, in the 10 kHz bandwidth of the AM signal, the power of the second signal is $P_s \times (10^4/10^6) = P_s/100$. Since the AM signal has a power P_s, the interfering spread spectrum signal provides noise which is 20 dB *below* the AM signal.

What are some of the applications of spread spectrum modulation? The widest application at this time is its use in military communications systems where spread spectrum serves two functions. The first is that it allows a transmitter to transmit a message to a receiver without the message being detected by a receiver for which it is not intended i.e., the transmission is transparent to an unfriendly receiver. To achieve this transparency the spread spectrum modulation decreases the transmitted power spectral density so that it lies well below the thermal noise level of any unfriendly receiver. The second major application of spread spectrum is found, when, as a matter of fact, it turns out not to be possible to conceal the transmission. Police radars can employ spread spectrum to avoid detection by radar detectors employed by drivers. In such a case the operator of an unfriendly receiver might attempt to begin transmitting an interfering signal to block communication between transmitter and receiver. Here again spread spectrum acts to reduce the effective power of the interference so that communication can proceed with minimal interference.

In the commercial communications field spread spectrum has many applications, a major application being the transmission of a spread spectrum signal on the same carrier frequency as an already existing microwave signal. By communicating in this manner additional signals can be transmitted over the same band thereby increasing the number of users. In addition, spread spectrum is used in satellite communications and is being considered for use in local area networks.

17.2 DIRECT SEQUENCE (DS) SPREAD SPECTRUM

A Direct Sequence (DS) spread spectrum signal is one in which the amplitude of an already modulated signal is amplitude modulated by a very high rate NRZ binary stream of digits. Thus, if the original signal is $s(t)$, where

$$s(t) = \sqrt{2P_s}\, d(t) \cos \omega_0 t \tag{17.2-1}$$

(a binary PSK signal), the DS spread spectrum signal is

$$v(t) = g(t)s(t) = \sqrt{2P_s}\, g(t)d(t) \cos \omega_0 t \tag{17.2-2}$$

where $g(t)$ is a *pseudo-random* noise (PN) binary sequence having the values ± 1.

The characteristics of $g(t)$ are extremely interesting and are discussed in some detail in Sec. 17.4. Here we merely assume that $g(t)$ is a binary sequence as is the data $d(t)$. The sequence $g(t)$ is generated in a deterministic manner and is repetitive. However, the sequence length before repetition is usually extremely long and to all intents and purposes, and without serious error, we can assume that the sequence is truly random, i.e., there is no correlation at all between the value of a particular bit and the value of any other bits. Furthermore, the bit rate f_c of $g(t)$ is usually much greater than the bit rate f_b of $d(t)$. As a matter of fact the rate of $g(t)$ is usually so much greater than f_b, we say that $g(t)$ "chops the bits of data into *chips*", and we call the rate of $g(t)$ the *chip rate* f_c, retaining the words, *bit rate*, to represent f_b.

To see that multiplying the BPSK sequence $s(t)$ by $g(t)$ spreads the spectrum we refer to Fig. 17.2-1 which shows a data sequence $d(t)$, a pseudo-random (often called a pseudo-noise or PN) sequence $g(t)$ and the product sequence $g(t)$ $d(t)$. Note that (as is standard practice) the edges of $g(t)$ and $d(t)$ are aligned, that is,

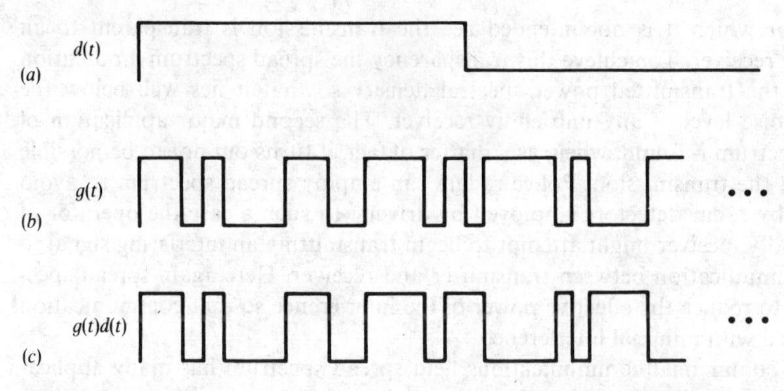

Figure 17.2-1 (*a*) The waveform of the data but stream $d(t)$. (*b*) The chipping waveform $g(t)$. (*c*) The waveform of the product $g(t)d(t)$.

each transition in $d(t)$ coincides with a transition on $g(t)$. The product sequence is seen to be similar to $g(t)$, indeed if $g(t)$ were truly random, the product sequence would be another random sequence $g'(t)$ having the *same* chip rate f_c as $g(t)$. Since the bandwidth of the BPSK signal $s(t)$ is nominally $2f_b$ (see Fig. 6.2-2) the bandwidth of the BPSK spread spectrum signal $v(t)$ is $2f_c$ and the spectrum has been spread by the ratio f_c/f_b. Since the power transmitted by $s(t)$ and $v(t)$ is the same, i.e., P_s, the power spectral density $G_s(f)$ is reduced by the factor f_b/f_c.

To recover the DS spread spectrum signal, the receiver shown in Fig. 17.2-2 first multiplies the incoming signal with the waveform $g(t)$ and then by the carrier $\sqrt{2}\cos\omega_0 t$. The resulting waveform is then integrated for the bit duration and the output of the integrator is sampled, yielding the data $d(kT_b)$. We note that at the receiver it is necessary to regenerate both the sinusoidal carrier of frequency ω_0 and also to regenerate the PN waveform $g(t)$.

Effect of Thermal Noise

Having noted that spread spectrum techniques can suppress the effect of an interfering deterministic signal, we may wonder whether the technique can serve, as well to reduce the disturbance generated by thermal noise. As we shall now see, such is not the case.

In Fig. 17.2-2 is shown a BPSK communication system incorporating a spread-spectrum technique. The data waveform $d(t)$ is an NRZ bit stream which makes excursions between $+1$ and -1 at the rate f_b while the chipping waveform makes excursions between $+1$ and -1 at the rate f_c. Observe that, in the overall transmission, the input signal is twice multiplied by $g(t)$, and, since $g^2(t) = 1$ there is no net effect on the received output signal. The noise $n(t)$ introduced in the channel is chipped at the receiver before reaching the integrator. That is, at nominally random times, the polarity of the noise waveform is reversed. It is intuitively clear that the reversal has no effect on the power spectral density or the probability density function of the gaussian noise. Hence both the signal and the

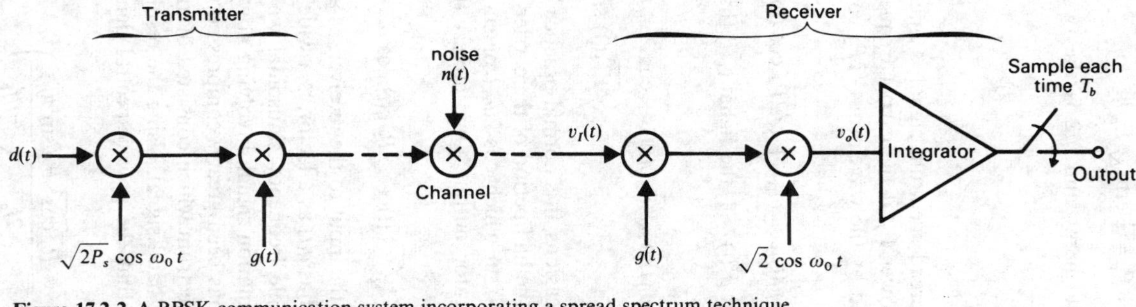

Figure 17.2-2 A BPSK communication system incorporating a spread spectrum technique.

statistical properties of the noise are unaffected by the spread-spectrum technique and the overall performance of the system is not affected. As in BPSK without spread spectrum, the error probability (see Sec. 11.7) is

$$P_e = \tfrac{1}{2} \operatorname{erfc} \sqrt{E_b/\eta} \tag{17.2-3}$$

where as before E_b is the bit energy and $\eta/2$ is the two-sided power-spectral density of the noise.

Single-Tone Interference

Consider next that the DS spread spectrum signal is interfered with by a sinusoidal signal of normalized power P_J at the carrier frequency f_0. That is in Fig. 17.2-2 the noise $n(t)$ is replaced by the waveform $\sqrt{2P_J} \cos(\omega_0 t + \theta)$. The input to the receiver is then

$$v_I(t) = \sqrt{2P_s}\, d(t) g(t) \cos \omega_0 t + \sqrt{2P_J} \cos(\omega_0 t + \theta) \tag{17.2-4}$$

Taking account of the fact that $g^2(t) = 1$, the signal $v_0(t)$ that appears at the input to the integrator is

$$v_0(t) = \sqrt{P_s}\, d(t)(1 + \cos 2\omega_0 t) + \sqrt{P_J}\, g(t)(1 + \cos 2\omega_0 t) \cos \theta - \sqrt{P_J}\, g(t) \sin 2\omega_0 t \sin \theta \tag{17.2-5}$$

As we have noted on other occasions, it is normal practice to arrange that the bit duration be some multiple of the half period of the carrier period $1/f_0$. Hence, as usual, the terms in Eq. (17.2-4), in which $\cos 2\omega_0 t$ or $\sin 2\omega_0 t$ is a factor, will yield no voltage at the integrator output. Our interest then remains with the waveform

$$v_0'(t) = \sqrt{P_s}\, d(t) + \sqrt{P_J}\, g(t) \cos \theta \tag{17.2-6}$$

It is of interest to note that at the input to the receiver, as appears in Eq. (17.2-4), the information signal, which has $g(t)$ as a factor, is a wide spectrum signal while the interfering signal is a single frequency sinusoidal waveform. On the other hand, in Eq. (17.2-6) the information signal, involving now only $d(t)$ and not $g(t)$ is a signal whose bandwidth has been greatly compressed. The interfering signal which in Eq. (17.2-4) had a spectrum which consisted of an impulse at $f = f_0$, in Eq. (17.2-6) appears as a wide band signal. Using the results of Sec. 6.2 (see also Prob. 17.2-4) we readily find that the power spectral density of the interfering signal in Eq. (17.2-6) is

$$G_J(f) = \frac{P_J \overline{\cos^2 \theta}}{2f_c} \left(\frac{\sin \pi f/f_c}{\pi f/f_c} \right)^2 \tag{17.2-7}$$

The integrator has characteristics not unlike a low-pass filter. As a matter of fact, it can be verified (Prob. 17.2-5) that an integrator whose integration period is T_b is approximately equivalent to a low-pass filter with cutoff at frequency

f_b $(=1/T_b)$. Since $f_b \ll f_c$, then, in the frequency range $\pm f_b$, $G_J(f)$ in Eq. (17.2-7) has, approximately, the constant value

$$G_J(f) = \frac{P_J \overline{\cos^2 \theta}}{2f_c}; \quad |f| \le f_b \tag{17.2-8}$$

Now we have noted that the effective information signal which appears at the input to the integrator in Fig. 17.2-2 is the same as it would be if the spread spectrum feature had not been incorporated in the system. If the interfering signal were thermal noise $n(t)$ of power spectral density $\eta/2$, the noise signal at the integrator would also have the power spectral density $\eta/2$, In this case, the probability of error at the output of the integrate and dump filter would be, as we have often noted,

$$P_e = \frac{1}{2} \operatorname{erfc} \sqrt{\frac{E_b}{\eta}} \tag{17.2-9}$$

In the present case, where the interfering signal is sinusoidal, we can continue to apply Eq. (17.2-9). We need simply replace $\eta/2$ by $G_J(f)$, from Eq. (17.2-8), in Eq. (17.2-9). Hence we find

$$\begin{aligned} P_e &= \frac{1}{2} \operatorname{erfc} \sqrt{\frac{E_b}{\eta}} = \frac{1}{2} \operatorname{erfc} \sqrt{\frac{E_b f_c}{P_J \overline{\cos^2 \theta}}} = \frac{1}{2} \operatorname{erfc} \sqrt{\frac{P_s T_b f_c}{P_J \overline{\cos^2 \theta}}} \\ &= \frac{1}{2} \operatorname{erfc} \sqrt{\left(\frac{P_s}{P_J}\right)\left(\frac{f_c}{f_b}\right) \cdot \frac{1}{\overline{\cos^2 \theta}}} \end{aligned} \tag{17.2-10}$$

The angle θ is the phase of the jamming sinusoidal waveform with respect to the phase of the information signal carrier. There is no reason to expect any correlation between the phases of jamming and signal carriers. Hence we consider that θ is a random variable all of whose possible values are equally likely. In this case we have that $\overline{\cos^2 \theta} = \frac{1}{2}$ and we have

$$P_e = \frac{1}{2} \operatorname{erfc} \sqrt{2\left(\frac{P_s}{P_J}\right)\left(\frac{f_c}{f_b}\right)} \tag{17.2-11a}$$

$$= \frac{1}{2} \operatorname{erfc} \sqrt{P_s \Big/ \frac{P_J}{2(f_c/f_b)}} \tag{17.2-11b}$$

The quantity

$$P_{J,\,\text{eff}} = \frac{P_J}{2\left(\dfrac{f_c}{f_b}\right)} \tag{17.2-12}$$

is called the *effective jamming power* since it is this power, in comparison to the signal power P_s that determines the error probability generated by the jamming.

The ratio f_c/f_b measures the extent to which the effect of the (mean) jamming power $P_J/2$ is reduced by the chipping and is called the *processing gain* G_p, i.e.,

$$G_P \equiv f_c/f_b \tag{17.2-13}$$

The results presented above are entirely valid if the chipping waveform $g(t)$ is of frequency very much higher than the bit frequency and if $g(t)$ is a truly random sequence. On the other hand when these conditions are not met it may actually turn out that Eq. (17.2-10) is overly pessimistic. For example, if the length of the pseudo-random sequence before repeating is no longer than the duration of a bit, the error probability is significantly smaller than given in Eq. (17.2-10).

17.3 USE OF SPREAD SPECTRUM WITH CODE DIVISION MULTIPLE ACCESS (CDMA)

In addition to the ALOHA/CSMA techniques discussed in Chapter 16, there is an alternate scheme to provide multiple access communications. This scheme employs spread spectrum and is called *Code Division Multiple Access* (CDMA). The advantage of a CDMA system is that collisions are not destructive, i.e., each of the signals involved in a collision would be received with only a slight increase in error rate.

In CDMA, each user is provided with an individual and distinctive PN code. These codes are almost uncorrelated with one another (see Sec. 17.6) and for the purpose of our explanation we shall assume that the sequences are indeed uncorrelated. To illustrate the principle of operation of CDMA consider that at a given time, each of k users is transmitting data at the same carrier frequency f_0, using DS spread spectrum, and his particular code $g_i(t)$. Then, each receiver is presented with the same input waveform,

$$v(t) = \sum_{i=1}^{k} \sqrt{2P_s}\, g_i(t) d_i(t) \cos(\omega_0 t + \theta_i) \tag{17.3-1}$$

where each signal is assumed to present the same power P_s to the receiver, each pseudo-random sequence $g_i(t)$ has the same chip rate f_c and $d_i(t)$ is the data transmitted by user i. The data rate for each user is the same, f_b. Also θ_i is a random phase, statistically independent of the phase of each of the other users. Thermal noise is omitted to simplify the discussion.

If the receiver is required to receive each of the k users it needs k correlators. At receiver 1, the signal of Eq. (17.3-1) will be multiplied by $g_1(t)$ and also by $\sqrt{2}\cos(\omega_0 t + \theta_1)$ to generate the signal v_{01} which is to be applied to the integrate and dump circuit. As before, we drop from v_{01} those components which will not make their way through the integrator and we shall be left with

$$v'_{01} = \sum_{i=1}^{k} \sqrt{P_s}\, g_1(t) g_i(t) d_i(t) \cos(\theta_i - \theta_1) \tag{17.3-2a}$$

$$= \sqrt{P_s}\, d_1(t) + \sum_{i=2}^{k} \sqrt{P_s}\, g_1(t) g_i(t) d_i(t) \cos(\theta_i - \theta_1) \tag{17.3-2b}$$

We assume that all the g_i's make transitions at the same time. In this case the product $g_1(t)g_i(t)$ has the same chip rate f_c as has any of the g_i's individually. And to the extent that the g_i's individually are random sequences so also is the product $g_1(t)g_i(t) \equiv g_{1i}(t)$. Writing also $\cos(\theta_i - \theta_1) \equiv \cos\theta_{1i}$, Eq. (17.3-2*b*) becomes

$$v'_{01}(t) = \sqrt{P_s}\, d_1(t) + \sum_{i=2}^{k} \sqrt{P_s}\, g_{1i}(t) \cos\theta_{1i} \tag{17.3-3}$$

Comparing Eq. (17.3-3) with Eq. (17.2-6) we see that they are similar, except that Eq. (17.3-3) has $k - 1$ independent interfering signals while Eq. (17.2-6) is for a single interfering signal. Since the power spectral density of one interferer is given by Eq. (17.2-8), letting $P_J = P_s$ and $\cos^2\theta = \frac{1}{2}$, the total power spectral density of $k - 1$ independent interferers is the sum of the power spectral densities. The total power spectral density of the $k - 1$ interferers in Eq. (17.3-3) is therefore

$$G_J(f) \simeq (k-1)\frac{P_s}{4f_c} \qquad |f| \leq f_b \tag{17.3-4}$$

The probability of a bit being in error is found using Eq. (17.2-11*a*), by letting $P_J = (k-1)P_s$:

$$P_e = \frac{1}{2}\,\text{erfc}\,\sqrt{2\left(\frac{1}{k-1}\right)\left(\frac{f_c}{f_b}\right)} \tag{17.3-5}$$

Thus in a slotted CDMA system, to insure a low probability of error, the gain, f_c/f_b, must be adjusted so that

$$\frac{f_c}{f_b} \gg (k-1)/2 \tag{17.3-6}$$

In Eq. (17.3-1) we assumed that each transmitted signal presented the same power P_s to the receiver. There are many practical situations where such is not the case. When transmitting to a satellite, there may be high-power as well as low-power transmitters. Also, in ground communication, one user may be close to the receiver while another user may be far from the receiver.

When an unwanted user's received power is much larger than the received power presented by the desired user, errors can occur. This problem is referred to in the literature as the *near-far problem*, and limits the utility of DS systems to applications where each user's received power is approximately the same.

17.4 RANGING USING DS SPREAD SPECTRUM

Another application of DS spread spectrum is for *ranging*. In this application, illustrated in Fig. 17.4-1, a DS signal

$$s(t) = \sqrt{2P_s}\, g(t) \cos\omega_0 t \tag{17.4-1}$$

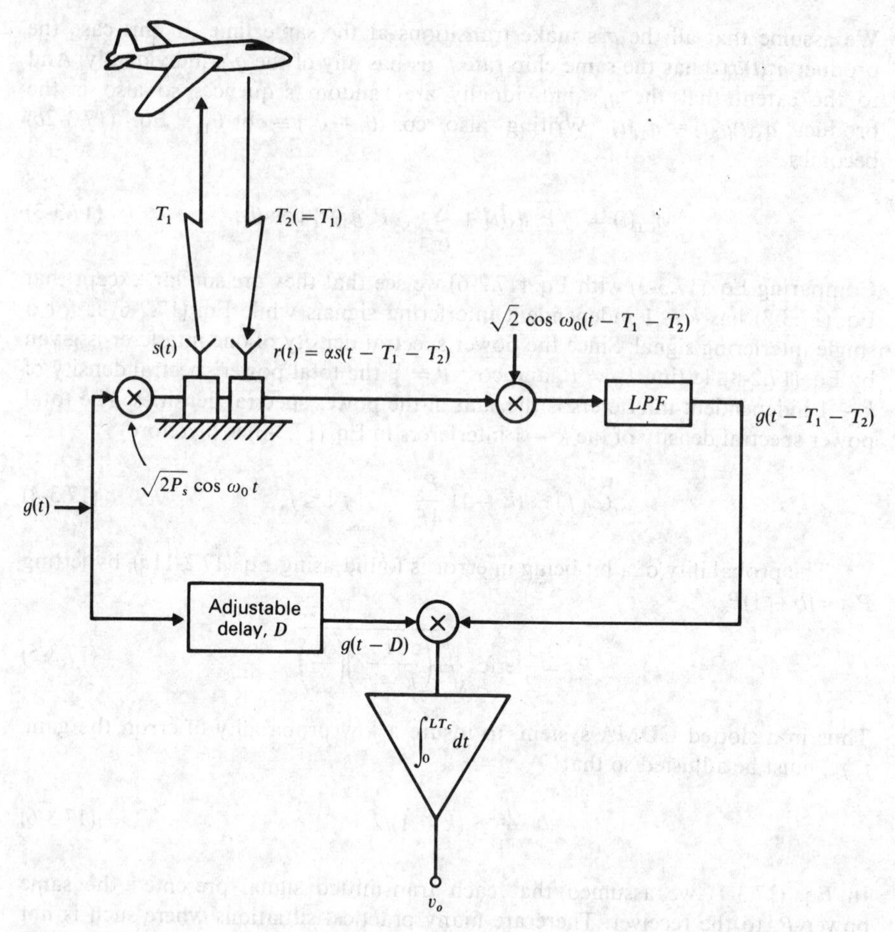

Figure 17.4-1 Ranging using DS spread spectrum.

is transmitted. The signal is reflected from the intended target and received $T_1 + T_2$ seconds later as $r(t)$,

$$r(t) = \alpha s(t - T_1 - T_2) \tag{17.4-2a}$$

$$= \alpha\sqrt{2P_s}\, g(t - T_1 - T_2) \cos(\omega_0 t + \theta) \tag{17.4-2b}$$

where α represents the signal attenuation and θ is a random phase caused by the time delay, i.e., $\theta = -\omega_0 (T_1 + T_2)$. The carrier $\sqrt{2} \cos \omega_0(t - T_1 - T_2)$ is determined using a squaring circuit (not shown in Fig. 17.4-1) or equivalent (see Sec. 10.16) and the PN sequence $g(t - T_1 - T_2)$ is extracted.

The PN waveform $g(t - T_1 - T_2)$ having been recovered, is correlated with the waveform $g(t - D)$ which is the PN waveform delayed by a *known* time interval D. The correlation function $R_g(D)$, which is the output v_o of the integrator in Fig. 17.4-1 is

$$v_o = R_g(D) = \int_0^{LT_c} g(t - T_1 - T_2)g(t - D)\,dt \tag{17.4-3}$$

The integration interval in Eq. (17.4-3) extends over the length of the PN sequence. This length is given by the product of the number of bits in the sequence (i.e., the sequence length L) and the period T_c of the chipping.

The special and important characteristic of the PN waveform in the present application is that the correlation $R_g(D)$ is negligibly small if the difference in delays, i.e., $D - (T_1 + T_2)$ exceeds the chip duration and $R_g(D)$ is a maximum when $D = T_1 + T_2$. Accordingly, in Fig. 17.4-1, the range is determined by adjusting D to maximize v_o. If the velocity of light is $c = 3 \times 10^8$ m/s and $T_1 = T_2$, the range of the target is

$$d = \tfrac{1}{2}cD \tag{17.4-4}$$

Furthermore, the precision of the measurement is the chip duration $\pm T_c$ which corresponds to a distance $cT_c/2$ so that

$$d = \frac{c}{2}(D \pm T_c) \tag{17.4-5}$$

and the measurement accuracy can be improved by decreasing T_c relative to D.

17.5 FREQUENCY HOPPING (FH) SPREAD SPECTRUM

Frequency Hopping (FH) spread spectrum is a FM or FSK technique while DS spread spectrum, described in Sec. 17.2, is an AM (or PSK) technique. The signal to be frequency hopped is usually a BFSK signal although M-ary FSK, MSK or TFM can be employed. Considering that BFSK is used, we have that the original signal, before spread spectrum is applied, is

$$s(t) = \sqrt{2P_s}\cos(\omega_0 t + d(t)\Omega t + \theta) \tag{17.5-1}$$

where $d(t)$ is the data to be transmitted. The FH modulation is then applied by varying the carrier frequency so that the resulting FH spread spectrum is

$$v(t) = \sqrt{2P_s}\cos(\omega_i t + d(t)\Omega t + \theta) \tag{17.5-2}$$

Here the FH signal has a carrier frequency $f_i = \omega_i/2\pi$ which changes at the hopping rate f_H, i.e., the carrier frequency f_i changes each T_H seconds. The frequency chosen each T_H is selected in a pseudo-random manner from a specified set of frequencies. Typically 32–500 different frequencies are used to form this set.

The primary advantage of FH is that it enables the transmitter to change its carrier frequency and thereby avoid an otherwise in-band interfering signal. For example, consider that the FH signal spends an equal time at each of 1000 frequencies $f_1, f_2, \ldots, f_{1000}$. Also assume that BFSK is employed. Then if the bit rate of the data is f_b, the bandwidth used by the signal (see Sec. 6.8), at any carrier frequency f_i, is $B = 4f_b$. Now assume that there is an interfering signal having a bandwidth $B = 4f_b$ and a fixed center frequency f_j. If frequency hopping were not employed and the interference were located at the transmitted signal's carrier frequency, i.e., $j = i$, then if the interfering signal power were sufficiently large the probability of error would be $P_e = 0.5$ since under these circumstances we could do no better than to guess. (In a military system, the interferer determines the signal's carrier frequency and purposely transmits at that frequency to block or *jam* the communications.) By employing FH and using say 1000 frequencies, the probability of the same interferer causing an error is reduced to $P_e = 1/2000 = 5 \times 10^{-4}$ (note that thermal noise is ignored since it is considered to be a second-order effect when "jamming" is present).

The Need for Coding

A probability of error of 5×10^{-4} is often considered to be very large and would render such a system useless for data transmission. As a result FH systems employ error correction coding to reduce the probability of error.

In the above discussion we assumed that when the FH signal "lands" at f_j a single error could occur. Very often, however, the bit rate f_b of the BFSK data is much greater than the hopping rate f_H. For example, it is not uncommon to transmit 10 bits/hop. In such a case, errors, due to jamming, occur in *bursts* and either interleaving and/or Reed-Solomon coding (Sec. 13.18) is employed. To further protect the data from random errors produced by thermal noise, which we have previously omitted from our discussion, one often employs concatenated coding (Sec. 13.18).

The Near-Far Problem

In Sec. 17.2-3 we noted that DS systems were affected by a near-far problem in which the receiver will receive significant noise from a signal transmitted by a nearby interferer and may not receive successfully the signal transmitted by a distant desired transmitter. The cause of the problem is that both signals are located in the *same* frequency band. In a FH system, if the receiver and transmitter are tuned to frequency f_i and the interfering signal at frequency $f_j \neq f_i$ then BFSK demodulator does not even know of the presence of the interferer since its spectrum is not in the signal bandwidth. Thus, there is *no* near-far problem when using FH.

Spectrum of FH Spread Spectrum

To determine the spectrum of a FH/BFSK system consider that the bit rate is equal to the hop rate, i.e., there is 1 bit transmitted/hop. Then, if there were no hopping, the spectrum would be that of BFSK which is shown in Fig. 17.5-1*a*. When FH is employed the BFSK signal is repeated at each of the K hopping fre-

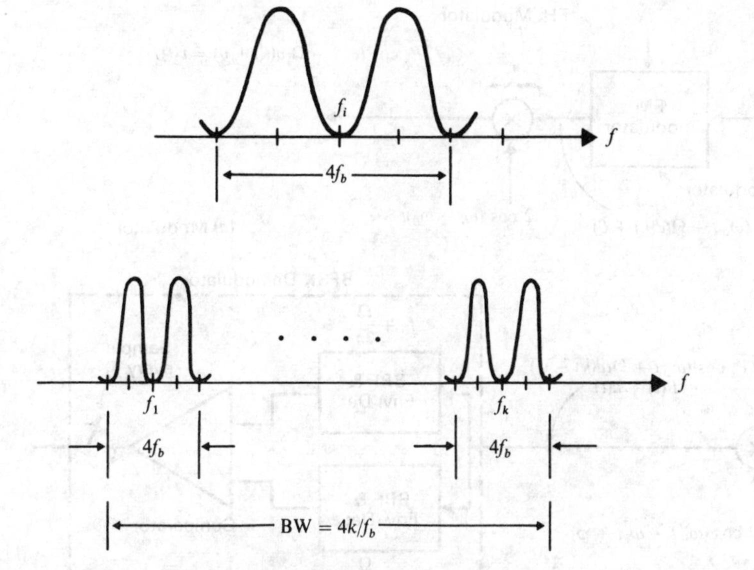

Figure 17.5-1 Spectrum of FH/BFSK.

quencies $f_1, f_2, \ldots, f_k$. If the frequencies are contiguous the spectrum is as shown in Fig. 17.3-1*b* and the bandwidth is

$$\mathrm{BW} = 4kf_b \tag{17.5-3}$$

Detection of a FH/BFSK Signal

A FH/BFSK modulation/demodulation system is shown in Fig. 17.5-2. Note that there are two sections to the modulator, the first being the BFSK modulator and the second the frequency hopped modulator. In practice, the two modulators are combined into a single, FM, modulator using a *frequency synthesizer.*

The received signal $v(t)$ is embedded in both the thermal noise $n(t)$ and an interference $I(t)$, which we assumed, earlier, to be

$$I(t) = \sqrt{2P_J} \cos \omega_j t \tag{17.5-4}$$

The received waveform, shown in Fig. 17.5-2*b*, is first detected and shifted to a constant intermediate frequency (IF), f_0. The resulting waveform is passed through the BFSK detector shown. Note that the interference $I(t)$ can affect the output only if its spectral density is large at frequency f_i.

The major problem area in the use of spread spectrum is synchronization. Note that in Fig. 17.5-2*b* we have assumed that the received carrier frequency and the receiver's locally generated carrier frequency is the same; i.e., ω_i. In addition, we have assumed that whenever the transmitted carrier frequency changes, the receiver's locally generated carrier frequency also changes by an appropriate

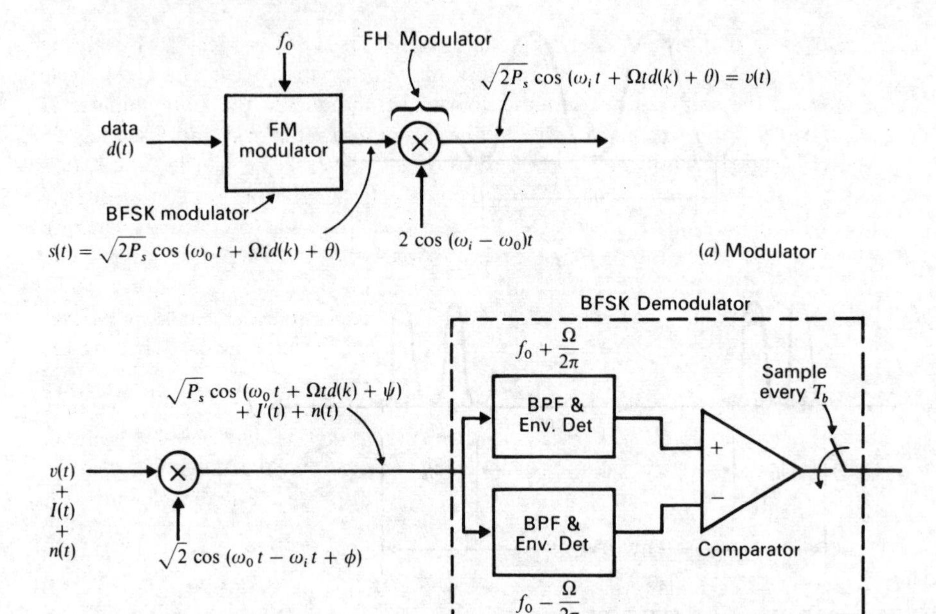

Figure 17.5-2 FH/BFSK modulation/demodulation system.

amount. Should this not be the case, the error rate would increase significantly, rendering the system unusable. The synchronization problem is studied in Secs. 17.7 and 17.8.

17.6 GENERATION AND CHARACTERISTICS OF PN SEQUENCES

A piece of hardware which is widely used to generate PN sequences is shown in Fig. 17.6-1. It consists initially of a shift register. We have selected type-D flip-flops and arranged that each data input except D_o is the Q output of the preceding flip-flop. The input D_o is the output of a parity generator. A parity generator (generally constructed of an array of EXCLUSIVE-OR logic gates, see Prob. 17.6-1) generates an output which is at logic 0 when an *even* number of inputs are at logic 0 and generates an output which is at logic 1 when an *odd* number of inputs are at logic 1. The parity generator inputs are the outputs of the flip-flops. We have shown a portion of each connection to the parity generator input as dashed in order to indicate that not all outputs Q need be connected to the parity generator. As a matter of fact, the character of the PN sequence generated depends on the number N of flip-flops employed and on the selection of which flip-flop outputs are connected to the parity generator. (An alternative PN generator is shown in Prob. 17.6-2).

The *state* of a sequential system such as in Fig. 17.6-1 is specified by stating the logic values of all the Q's of the flip-flops. During the course of a clock cycle the state of the shift register remains fixed, but, in general, the state changes at each advance from one clock cycle to the next. A register of N flip flops has 2^N states from $Q_0 Q_1 Q_2 \cdots Q_{N-1} = 000 \cdots 0$ to $Q_0 Q_1 Q_2 \cdots Q_{N-1} = 111 \cdots 1$. It is clear that the hardware of Fig. 17.6-1 cannot generate a truly random sequence since it is a deterministic structure. It is also clear that whatever the sequence of states through which the sequence generator progresses, the sequence will repeat. That is, it will be periodic. Such is the case because each state is uniquely determined by the immediately preceding state. Thus, each time the generator arrives at some particular state, the subsequent sequences of states will always be the same.

While, as we have noted, a truly random sequence is not possible, we may intuitively anticipate that a sequence with a long enough period will have some of the characteristics of a random sequence. The first step in that direction is taken by using a large number of flip-flops. For, if we can make the sequence generator go through all its states the sequence length will be 2^N. With present-day MOS large scale integration it is reasonable to have a 2000 flip-flop register on a single chip. Actually the maximum sequence length is $2^N - 1$ since the state $000 \cdots 0$ must be excluded. Such is the case because, as is easily verified, if the register should ever arrive at this all zero state, it will remain in that state permanently. We shall now set down some of the characteristics (in some cases, without proof) of the PN sequences produced by the generator of Fig. 17.6-1.

Sequence Length

It is always possible to find a set of connections from flip-flop outputs to parity generator which will yield a maximal length of sequence,

$$L = 2^N - 1 \tag{17.6-1}$$

For the cases $N = 1$ through $N = 15$ one logic design for maximal length sequences is given in Table 17.6-1.

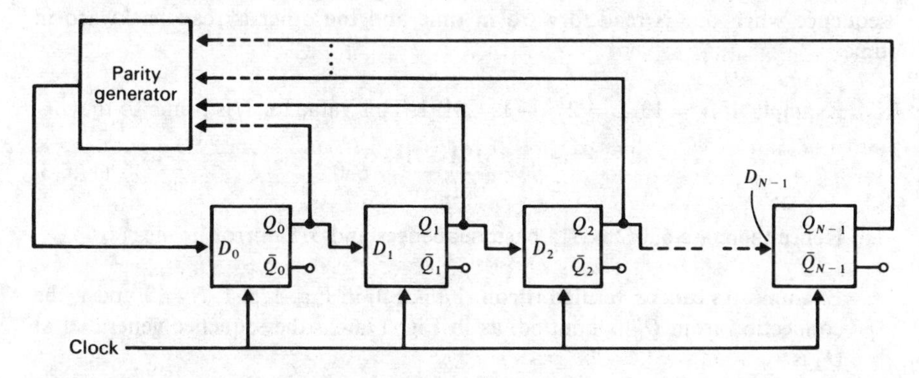

Figure 17.6-1 A pseudo-random sequence generator.

Table 17.6-1

N	D_o for $L = 2^N - 1$
1	Q_0
2	$Q_0 \oplus Q_1$
3	$Q_1 \oplus Q_2$
4	$Q_2 \oplus Q_3$
5	$Q_2 \oplus Q_4$
6	$Q_4 \oplus Q_5$
7	$Q_5 \oplus Q_6$
8	$Q_1 \oplus Q_2 \oplus Q_3 \oplus Q_7$
9	$Q_4 \oplus Q_8$
10	$Q_6 \oplus Q_9$
11	$Q_8 \oplus Q_{10}$
12	$Q_1 \oplus Q_9 \oplus Q_{10} \oplus Q_{11}$
13	$Q_0 \oplus Q_{10} \oplus Q_{11} \oplus Q_{12}$
14	$Q_1 \oplus Q_{11} \oplus Q_{12} \oplus Q_{13}$
15	$Q_{13} \oplus Q_{14}$

Independence of Sequences

For a particular logic design of the parity generator, as for example the designs of Table 17.6-1, sequences are available at each flip-flop output Q and at each flip-flop complementary output $\bar{Q}$. Clearly these sequences are not independent. One may be derived from another by a simple shift in time or by complementing each bit or both. There are, however, other logic designs (i.e., other connections to the parity generator) which do yield sequences which have small correlation to one another. The number of such sequences has an upper bound S given by

$$S \leq \frac{L-1}{N} \tag{17.6-2}$$

The equal sign applies when L is a prime number. These independent sequences can be divided into two equal groups so that each number of one group has a *mirror image* in the other group. Mirror image sequences have the same bit sequence when one is read forward in time and the other is read backward in time.

Example If $N = 13$, $L = 2^{13} - 1 = 8191$. This value for L is prime so that

$$S = \frac{8191 - 1}{13} = 630 \tag{17.6-3}$$

Hence there are $630/2 = 315$ basic sequences and 315 mirror images.

Example As can be verified (Prob. 17.6-3) if, in Fig. 17.6-1, $N = 3$ and if the connection from Q_0 is omitted, as in Fig. 17.6-2*a* the sequence generated at Q_2 is

$$1 \quad 1 \quad 1 \quad 0 \quad 0 \quad 1 \quad 0 \quad 1 \quad 1 \quad 1 \cdots$$

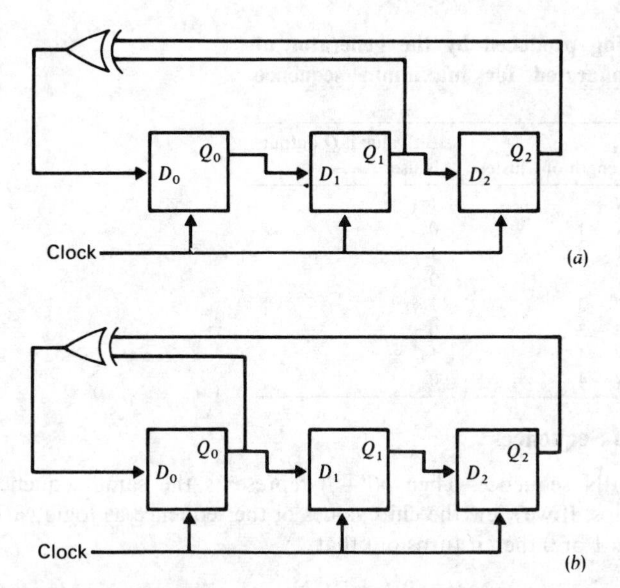

Figure 17.6-2 PN generation which generate mirror image sequences.

If, on the other hand the connection from Q_1 is omitted, as in Fig. 17.6-2*b*, the sequence is

$$1\ 1\ 1\ 0\ 1\ 0\ 0\ 1\ 1\ 1\cdots$$

which is recognized as the mirror image.

Number of 1's and 0's in a maximal sequence

If the PN generator sequenced through all its states, then the number of states would be an even number and the sequence available at any Q or $\bar{Q}$ output would contain the same number of 1's and 0's. Since, however, the all-zero state is excluded, there are an odd number of states. Correspondingly, the sequence will have one more 1's than 0's or one more 0's than 1's depending on when the sequence is taken from a Q or $\bar{Q}$ output.

Clustering in a PN Sequence

One way in which the generator of Fig. 17.6-1 gives evidence of the fact that its sequences are not truly random is that its sequences display *clusters*. A cluster is a run of identical bits that occur in sequence. The clustering exhibited by a generator as in Fig. 17.6-1 using N flip-flops is given in Table 17.6-2. Thus, in the example given above, we encounter the sequence 1 1 1 0 0 1 0 in the case $N = 3$. This sequence has one cluster of $N(=3)$ 1's, one cluster of $N - 1(=2)$ 0's and one cluster of $N - 2(=1)$ 1's and one cluster of $N - 2(=1)$ 0's.

Table 17.6-2 Clustering produced by the generator of Fig. 17.6-1 when connected for maximum sequence length

Number of Clusters	Length of Cluster	Bit value if Q output is used
1	N	1
1	$N-1$	0
1	$N-2$	1
1	$N-2$	0
2	$N-3$	1
2	$N-3$	0
4	$N-4$	1
4	$N-4$	0

Properties of Shifted Sequences

Let $g(i)$ represent a PN sequence. Then $g(i+j)$ represents the same sequence except shifted by j chips. If we view the chip values of the sequence as logic variables with logic values 1 or 0 then it turns out that

$$g(i) \oplus g(i+j) = g(i+k) \tag{17.6-4}$$

That is, the EXCLUSIVE-OR operation, applied chip by chip to a sequence $g(i)$ and the same sequence except shifted by j bits, gives rise to the sequence $g(i)$ shifted by k bits. It turns out further, that in Eq. (17.6-4), we cannot further specify k except to note that $k \neq 0$ and $k \neq j$.

Next let us view $g(i)$ as a waveform which makes excursions between $+1$ volt and -1 volt corresponding to logic 1 and logic 0. Then using Eq. (17.6-4) it can be shown (Prob. 17.6-5) that

$$g(i) \times g(i+j) = -g(i+k) \tag{17.6-5}$$

Here, the left-hand side involves the arithmetic product.

Autocorrelation of a PN Sequence

The autocorrelation function $R_d(\tau)$ of a truly random data sequence $d(t)$ with bit duration T_b is defined as

$$R_d(\tau) = E\{d(t)d(t+\tau)\} \tag{17.6-6}$$

and has the form shown in Fig. 17.6-3 when $d(t) = \pm 1$ volt.

The autocorrelation function $R_{PN}(\tau)$ of a PN sequence is

$$R_{PN}(\tau) = E\{g(t)g(t+\tau)\} \tag{17.6-7}$$

where $g(t)$ assumes the values $g(t) = \pm 1$ volt. As might be expected $R_{PN}(0) = 1$. We now calculate $R_{PN}(\tau)$ for $\tau = nT_c$, where n is an integer and T_c is the chip duration. From Eq. (17.6-5) we have that

$$R_{PN}(\tau = nT_c) = E\{g(t)g(t+nT_c)\} = E\{-g(t+kT_c)\} \tag{17.6-8}$$

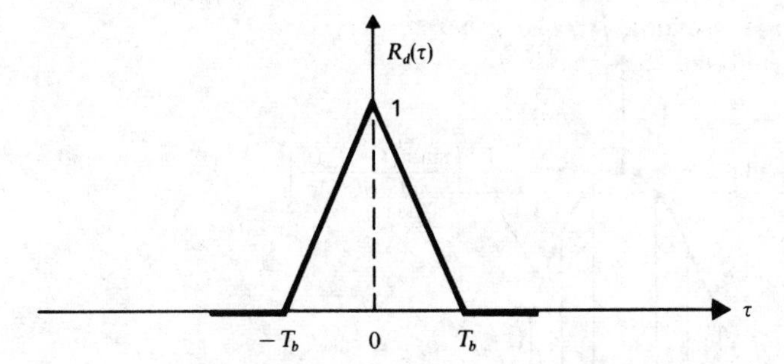

Figure 17.6-3 Autocorrelation function of random data $d(t)$ having the values, $d = +1$.

Now $g(t + kT_c)$ is a PN sequence and, in the course of L chips there is one more 1 than 0. Hence the average value of $-g(t + kT_c)$ is $-1/L$ as shown in Fig. 17.6-4. Finally, since the sequence has a period LT_c so too has $R_{PN}(\tau)$.

Power Spectral Density

We now determine the power spectral density $G_{PN}(f)$ of the pseudo-random noise sequence. The power spectral density $G_{PN}(f)$ is the Fourier transform of $R_{PN}(\tau)$. Since $R_{PN}(\tau)$ is periodic with period LT_c, $G_{PN}(f)$ must consist of impulses at multiples of the frequency $1/LT_c$. $G_{PN}(f)$ will also display an impulse at $f = 0$ since the impulse $G_{PN}(0)$ is the dc power of the PN frequency. The PN sequence $g(t)$ consists of excursions between $+V$ (logic 1) and $-V$ (logic 0). In a sequence of L chips there is one more logic 1 than logic 0, hence the dc voltage of $g(t)$ is V/L and the normalized dc power in V^2/L^2. Hence $G_{PN}(0) = (V^2/L^2)\delta(f)$. Finally, we recall that if $g(t)$ were truly random its power spectral density would have the form $[(\sin \pi f/f_c)/(\pi f/f_c)]^2$. Altogether, then, the spectral density $G_{PN}(f)$ is as shown in Fig. 17.6-5 and can be shown to be

$$G_{PN}(f) \simeq \frac{V^2}{L^2}\,\delta(f) + \frac{V^2}{L}\sum_{i=-\infty}^{\infty}\delta\left(f + i\frac{f_c}{L}\right)\left[\frac{\sin \pi(f + if_c/L)}{\pi(f + if_c/L)}\right]^2 \tag{17.6-9}$$

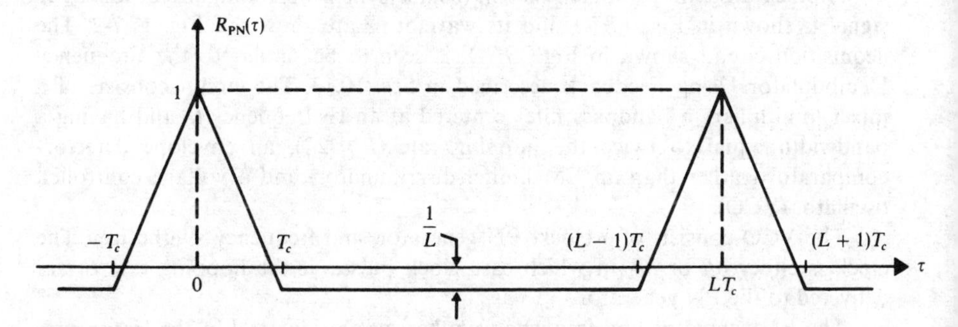

Figure 17.6-4 Autocorrelation function of a PN sequence.

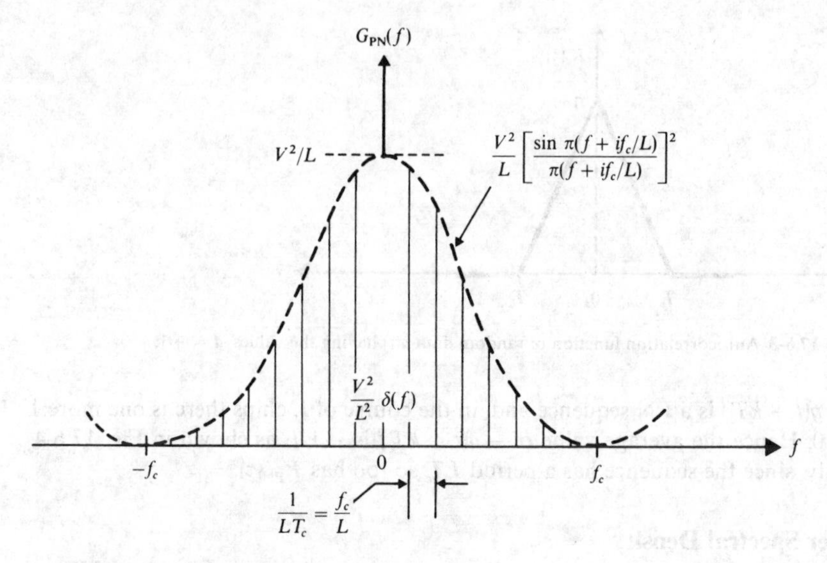

Figure 17.6-5 Power spectral density of a PN sequence.

17.7 ACQUISITION (COARSE SYNCHRONIZATION) OF A FH SIGNAL

As we saw in Chap. 6, noncoherent communication systems such as FSK require *bit synchronizers* to allow signal recovery at the receiver. Coherent systems require, in addition, *carrier and phase synchronization* to permit the modulated received signal to be mixed down to baseband. In spread-spectrum systems, a third synchronizer is required to allow the regeneration at the receiver of a duplicate of the chipping waveform used at the transmitter. We consider, in this and in succeeding sections, synchronization procedures for FH and DS spread-spectrum systems.

A block diagram of the *acquisition* (coarse synchronization) circuit for a FH signal is shown in Fig. 17.7-1 and its waveforms are shown in Fig. 17.7-2. The acquisition circuit shown in Fig. 17.7-1, is seen to be similar to the Frequency Demodulator Using Feedback described in Sec. 10.13. The circuit consists of a mixer (multiplier), a bandpass filter centered at an IF frequency f_0 and having a bandwidth equal to twice the hopping rate ($B = 2f_h$), an envelope detector-comparator (rather than an FM limiter-discriminator) and a voltage controlled oscillator (VCO).

The VCO consists of a clock, PN generator, and frequency synthesizer. The clock is either *off* or *on*, in which case clock pulses at the hopping rate f_H are delivered to the PN generator.

The PN generator and frequency synthesizer are identical in the transmitter and receiver. The frequency synthesizer is merely an oscillator whose frequency

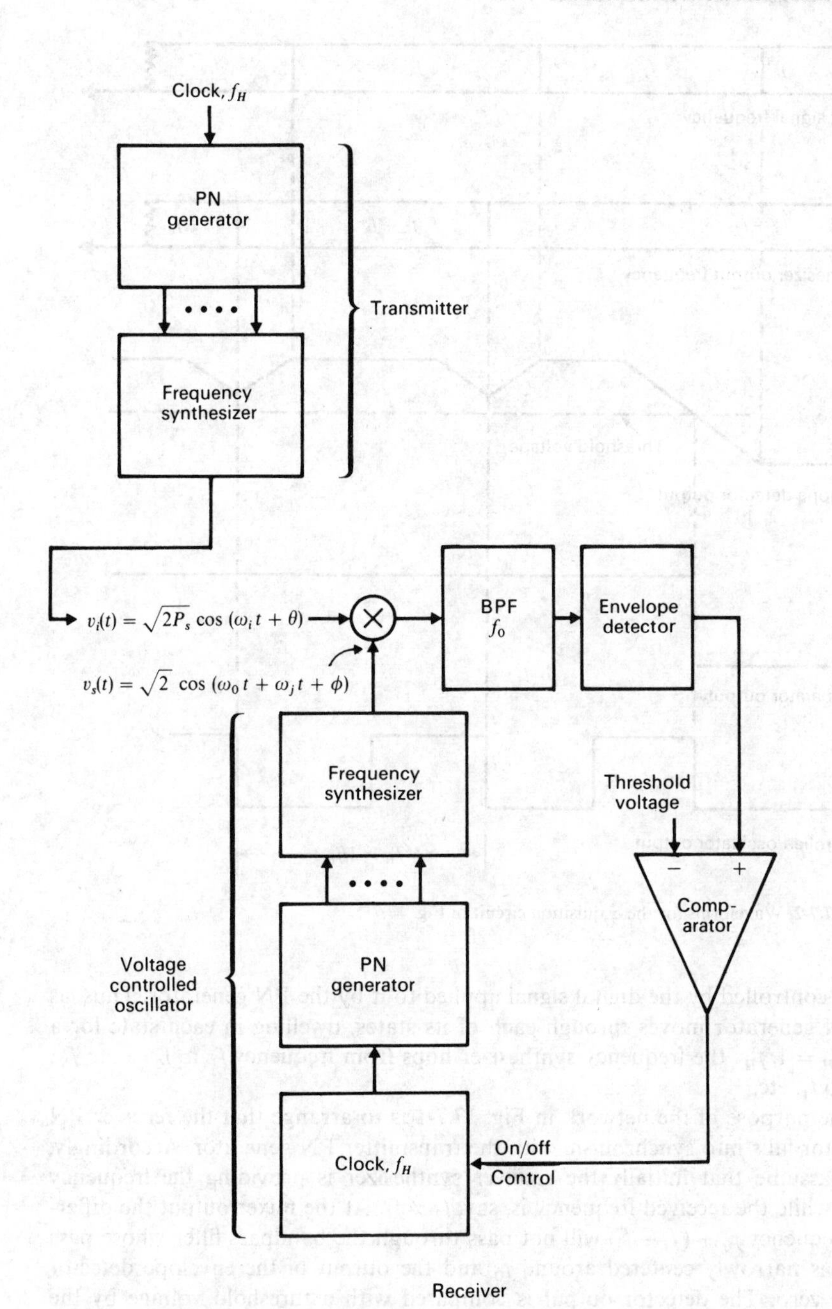

Figure 17.7-1 Acquisition circuit (camp-and-wait) for a FH signal.

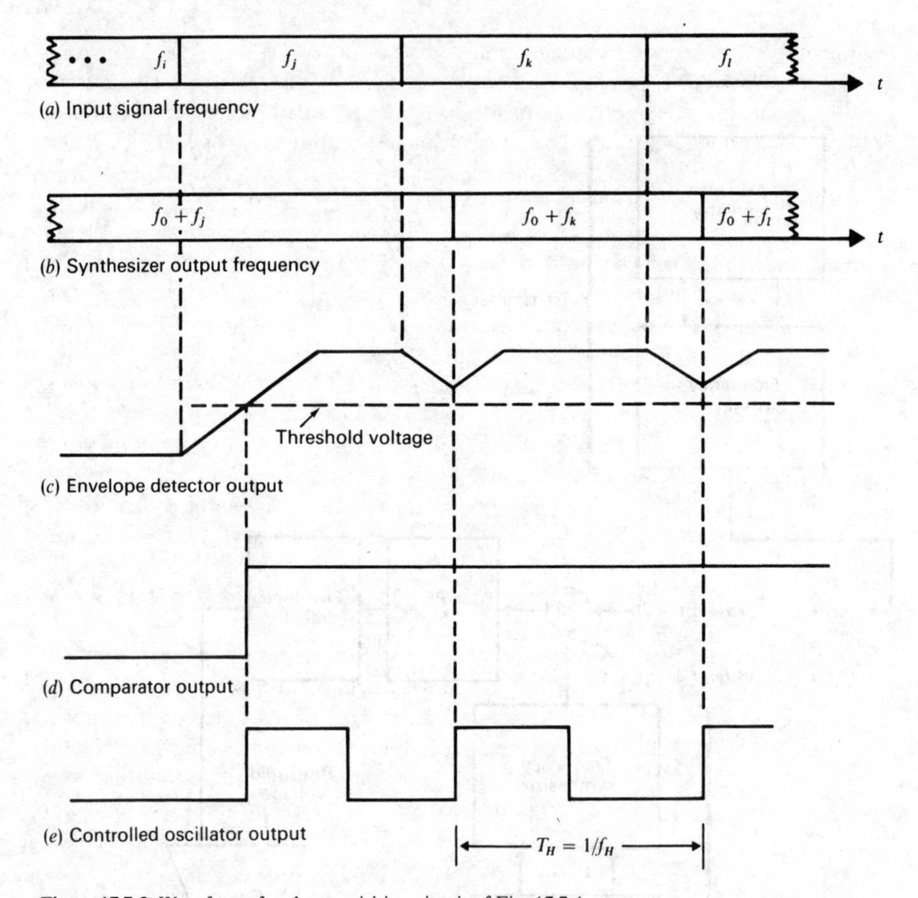

Figure 17.7-2 Waveforms for the acquisition circuit of Fig. 17.7-1.

can be controlled by the digital signal applied to it by the PN generator. Thus, as the PN generator moves through each of its states, dwelling in each state for a time $T_H = 1/f_H$, the frequency synthesizer hops from frequency f_1 to $f_2 \cdots$ to f_N, back to f_1, etc.

The purpose of the network in Fig. 17.7-1 is to arrange that the receiver PN generator falls into synchronism with the transmitter PN generator. Accordingly, let us assume that initially the receiver synthesizer is providing the frequency $f_0 + f_j$ while the received frequency is, say, $f_\lambda \neq f_j$. At the mixer output the difference frequency $f_0 + (f_j - f_\lambda)$ will not pass through the bandpass filter whose pass region is narrowly centered around f_0 and the output of the envelope detector will be zero. The detector output is compared with a threshold voltage by the comparator. A comparator, it will be recalled accept two input voltages and provides an output which assumes one of two voltages, depending on which input is the larger. It is arranged that when the detector output is less than the threshold voltage, the on/off control voltage applied to the controlled oscillator keeps that oscillator *turned off*. Hence, as long as the received frequency is *not* equal to f_j, the PN generator does not cycle through its states and the receiver synthesizer holds at the frequency $f_0 + f_j$. Thus, the synthesizer remains *camped* at $f_0 + f_j$, on account of which the name *Camp-and-Wait* is used to characterize the technique. When finally the received frequency is f_j the difference signal passes through the filter, the detector output rises above the threshold, the comparator output changes, the controlled oscillator is turned on and the receiver PN generator is advanced through its states in synchronism with the transmitter generator. Waveforms for the system are shown in Fig. 17.7-2. In (*a*) is shown the received signal hopping from frequency to frequency while as in (*b*) the receiver synthesizer frequency remains camped at $f_0 + f_j$. When, in (*a*), the awaited frequency f_j appears, then as shown in (*c*) the envelope detector output begins to change. When the threshold voltage is reached the comparator output changes as in (*d*) and, as in (*e*), the voltage controlled clock turns on and provides the clock waveform needed to drive the PN generator. At each positive transition of the clock, except the first transition, the PN generator advances to its next state and the frequency synthesizer advances to its next frequency. (It is necessary to incorporate provisions, not shown, to arrange that the first clock transition should not advance the PN generator.)

We have deliberately provided, as is to be seen in the waveforms, that the change in detector output be relatively slow. Such a feature is not difficult to incorporate because the output hardware component of the detector is a capacitor. Because of this relative sluggishness, the time at which the receiver PN generator changes state lags behind the time of state change of the transmitter generator. On the other hand because of this very sluggishness the detector output will remain above threshold and will keep the controlled oscillator running even during the intervals, which occur one per oscillator cycle, when the two PN generators are not in the same state.

17.8 TRACKING (FINE SYNCHRONIZATION) OF A FH SIGNAL

To effect fine synchronization we need to replace the voltage controlled clock shown in Fig. 17.7-1 (which can only be turned on and off) by a voltage-controlled oscillator (VCO) whose frequency can be adjusted above and below the hopping frequency. We require further that the VCO frequency be controlled by a voltage which measures that phase difference between the two PN generators. The phase discrepancy can then be used to speed up or slow down the VCO as required to reduce the phase error. Such a system is, in short, an automatic feedback control system, or more specifically, a phase-locked loop.

Such a phase-locked loop is shown in Fig. 17.8-1. The frequency synthesizer and PN generator shown here are the same as in Fig. 17.7-1. The bandpass filter bandwidth is now widened to pass the data, i.e., $B = 2\Delta f$. The on-off controlled

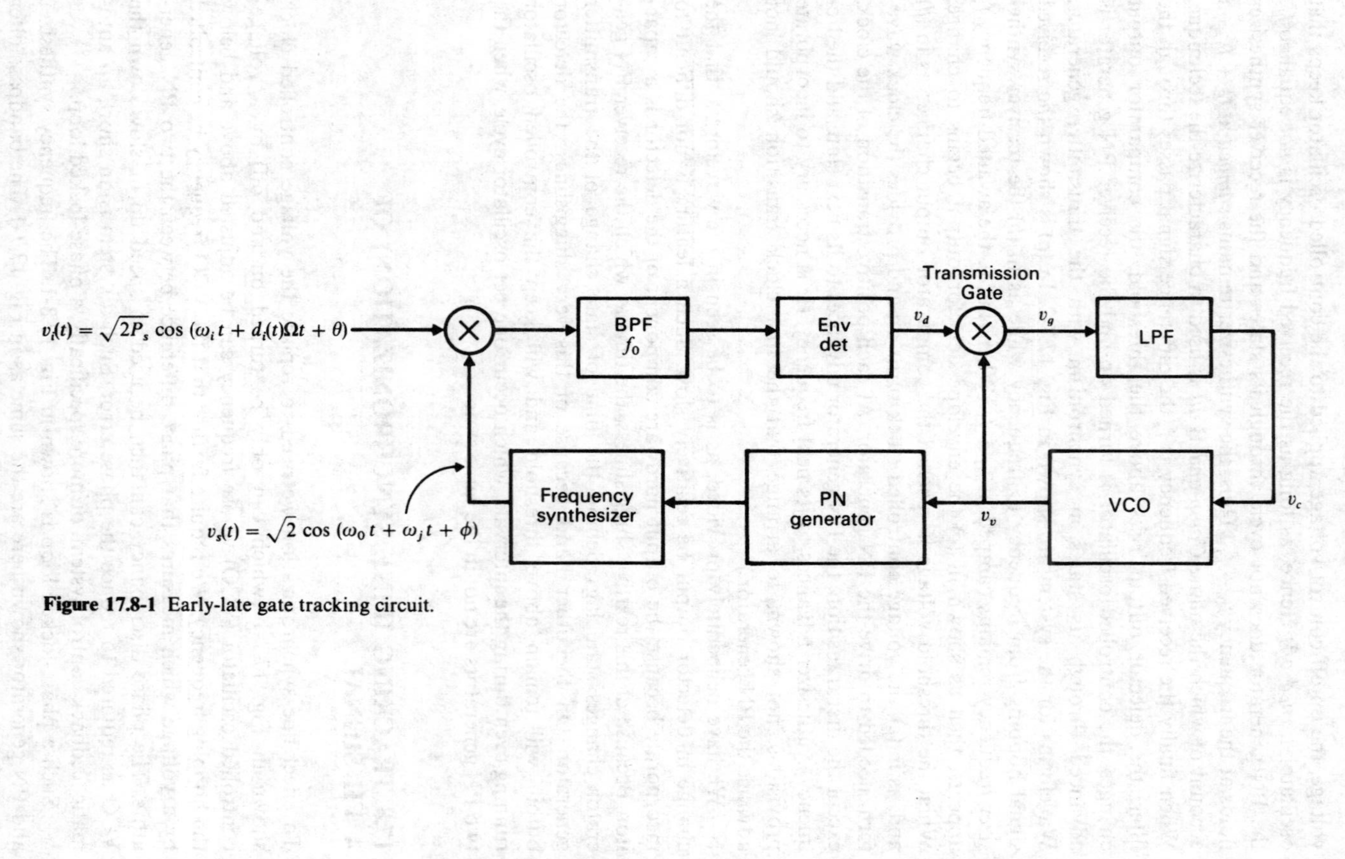

Figure 17.8-1 Early-late gate tracking circuit.

oscillator is replaced by a VCO which operates at a frequency which is nominally f_H but which can be varied continuously about f_H by the control signal furnished by the low-pass filter. As a matter of convenience we shall assume that the clock waveform furnished by the VCO makes excursions between +1 volt and −1 volt. The envelope detector in the present case, unlike the situation which prevailed in the coarse synchronizer, has a fast response. As a matter of fact, for simplicity in the waveforms of Fig. 17.8-2 we have assumed that the envelope detector output responds immediately to the bandpass filter output. However, this feature is not essential for operation of the system. The *gate* between envelope detector and low-pass filter is a *transmission gate*. When there is an output provided by the envelope detector, the VCO clock waveform is transmitted through

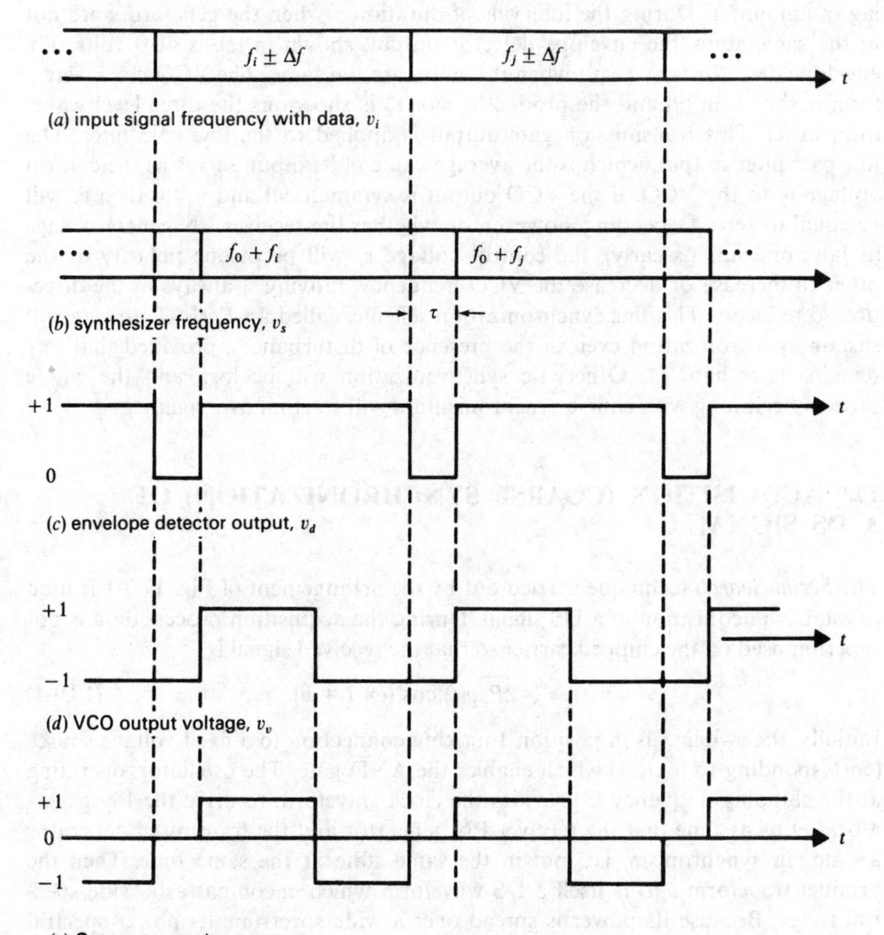

Figure 17.8-2 E-L gate waveforms.

the gate and is not transmitted when the detector output is zero. Altogether we can characterize the operation of this gate as a process of multiplication if we take the VCO clock to make excursions between ± 1 V and the envelope detector output to be 0 when there is no output and 1 when there is an output.

Waveforms describing the circuit operation are shown in Fig. 17.8-2. Here, as in Fig. 17.8-1, we assume the presence of an input data stream which shifts the transmitter frequencies by amount $\Omega/2\pi = \Delta f$ from the frequencies provided by the frequency synthesizer. So long as these shifts keep the received frequencies within the range of the bandpass filter, the operation of the synchronizer will be in no way influenced.

In Fig. 17.8-2*a* and *b* we assume that coarse synchronization has been established between the receiver PN generator and the transmitter generator with a lag of amount τ. During the intervals of duration τ when the generators are not in the same state, the envelope detector output, shown in (*c*), is at 0 volts; the envelope detector is at 1 volt when the states are the same. The VCO clock waveform is shown in (*d*) and the product $v_g = v_v v_d$ is shown as the three level waveform in (*e*). This transmission gate output is applied to the low pass filter. The low pass filter output, which is the average value of its input, serves as the control voltage v_c to the VCO. If the VCO output is symmetrical and $\tau = 0$ then v_c will be equal to zero. Depending, however, on whether the receiver PN generator lags (is late) or leads (is early), the control voltage v_c will be of one polarity or the other to increase or decrease the VCO frequency, moving it always in the direction to reduce τ. This fine synchronization scheme, called an *Early–Late gate* will sustain synchronization even in the presence of disturbances, provided that $|\tau|$ remains less than T_H. Otherwise synchronization will be lost, and the whole process beginning with course synchronization will start all over again.

17.9 ACQUISITION (COARSE SYNCHRONIZATION) OF A DS SIGNAL

The *Serial-Search* technique carried out by the arrangement of Fig. 17.9-1 is used to establish acquisition of a DS signal. During the acquisition process data is not superimposed on the chipped carrier so that the received signal is

$$v_i(t) = \sqrt{2P_s}\, g(t) \cos(\omega_0 t + \theta) \tag{17.9-1}$$

Initially, the switch S is in position 1 making connection to a fixed voltage source (corresponding to logic 1) which enables the AND gate. The oscillator, operating at the chipping frequency f_c provides the clock waveform to drive the PN generator. Let us assume that the receiver PN generator and the transmitter generator are not in synchronism, i.e., not in the same state at the same time. Then the product waveform $v_r(t)$ is itself a DS waveform which encompasses a wide spectral range. Because its power is spread over a wide spectrum, its power spectral density is small. Accordingly little power will be available at the output of the bandpass filter (centered at f_0) and the output of the envelope detector will be correspondingly low. The envelope detector output is integrated for a time nT_c,

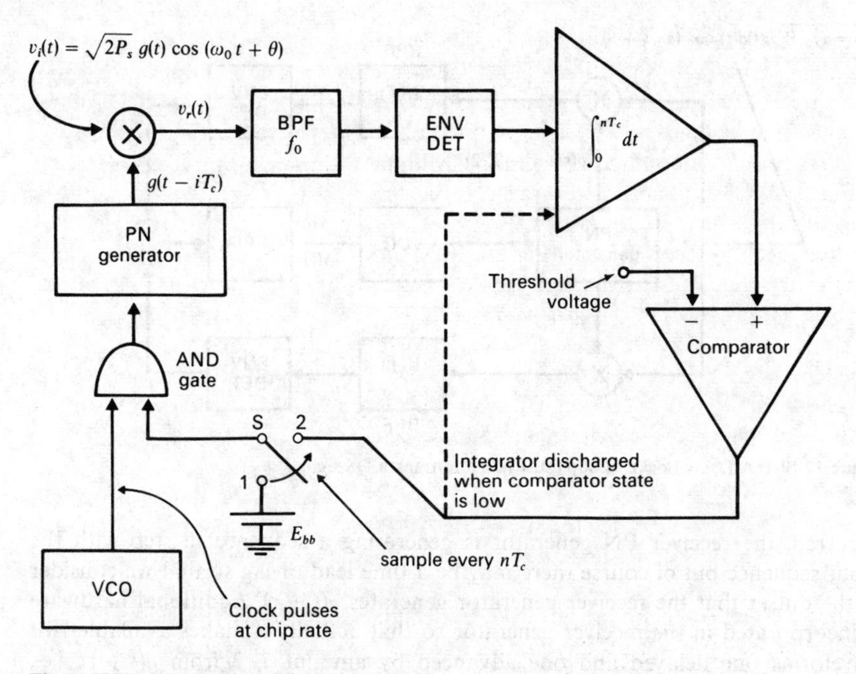

Figure 17.9-1 Direct sequence acquisition circuit.

that is some number n of clock cycles depending on the noise. In the absence of synchronization, the integrator output will not be large enough to exceed the comparator reference voltage. In this case, and by mechanisms not shown, the switch S will be thrown to position 2, disabling the AND gate and stopping the sequencing of the PN generator. In this way the receiver PN generator will then be allowed to step back some cycles with respect to the transmitter generator. Then the accumulated output of the integrator is *dumped* and the switch is thrown back to position 1 to try again. By this repeated trial and error process, acquisition will eventually be established as evidenced at the output of the comparator. For at acquisition, the product $g(t)g(t - iT_c)$ becomes $g^2(t) = 1$ since $i = 0$ and the input v_r to the bandpass filter is simply $\sqrt{2P_s} \cos(\omega_0 t + \theta)$. The envelope detector output is now high, the integrator output is high and the comparator voltage goes to its high state. Acquisition having been established, synchronization continues by tracking (fine synchronization) the signal.

17.10 TRACKING (FINE SYNCHRONIZATION) OF A DS SIGNAL

A circuit which is often used to track a DS signal is the Delay Locked Loop (DLL) shown in Fig. 17.10-1. The input signal in this case involves both the chipping waveform $g(t)$ and the data stream $d(t)$. Since we assume acquisition has

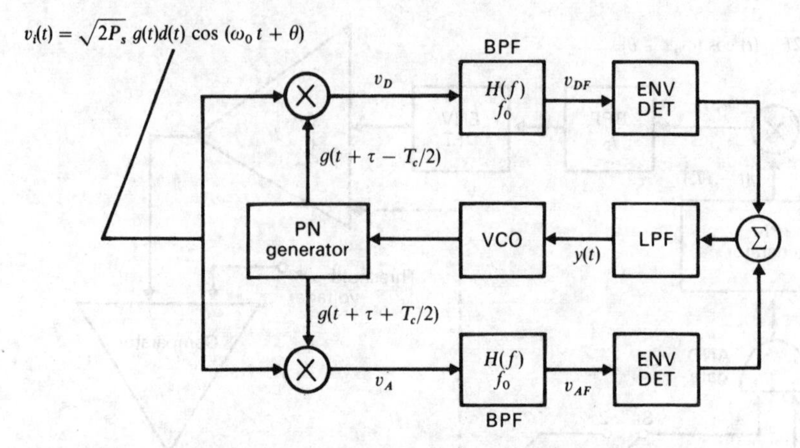

Figure 17.10-1 A Delay Locked Loop (DLL) used to track a DS signal.

occurred, the receiver PN generator is generating a sequence in step with the input sequence but of course there may be a time lead or lag so that we consider at the outset that the receiver generator generates $g(t + \tau)$. Additional hardware is incorporated in the receiver generator so that actually it makes available two waveforms one delayed and one advanced by amount $T_c/2$ from $g(t + \tau)$, i.e., $g(t + \tau + T_c/2)$ and $g(t + \tau - T_c/2)$ as indicated in Fig. 17.10-1.

The waveforms v_D and v_A in Fig. 17.10-1 are then given by

$$v_D(t) = \sqrt{2P_s}\, g(t)g(t + \tau - T_c/2)\, d(t) \cos(\omega_0 t + \theta) \tag{17.10-1}$$

and

$$v_A(t) = \sqrt{2P_s}\, g(t)g(t + \tau + T_c/2)\, d(t) \cos(\omega_0 t + \theta) \tag{17.10-2}$$

These signals are passed through identical bandpass filters each having a bandwidth equal to $B = 2f_b$. Thus the bandwidth of each filter is much less than the bandwidth needed to transmit a PN sequence and only the average value of the product $g(t)g(t + \tau + T_c/2)$ is passed. The filter outputs can then be written as

$$v_{DF}(t) \simeq \sqrt{2P_s}\, \overline{[g(t)g(t + \tau - T_c/2)]}\, d(t) \cos(\omega_0 t + \theta) \tag{17.10-3}$$

and

$$v_{AF}(t) \simeq \sqrt{2P_s}\, \overline{[g(t)g(t + \tau + T_c/2)]}\, d(t) \cos(\omega_0 t + \theta) \tag{17.10-4}$$

We recognize that the average value of the product of the PN sequence and a shifted version of the same sequence is the autocorrelation function (see Sec. 17.4). That is

$$R_g(\tau \pm T_c/2) = \overline{g(t)g(t + \tau \pm T_c/2)} \tag{17.10-5}$$

The envelope detectors extract the envelopes of $v_{DF}(t)$ and $v_{AF}(t)$ thereby removing the data $d(t)$. The result is

$$v_D(t) = |R_g(\tau - T_c/2)| \tag{17.10-6}$$

and

$$v_A(t) = |R_g(\tau + T_c/2)| \qquad (17.10\text{-}7)$$

The average voltages $v_D(t)$ and $v_A(t)$ are then subtracted and low pass filtered to extract as much noise as possible. The input voltage to the VCO, $y(t)$, is

$$y(t) = |R_g(\tau - T_c/2)| - |R_g(\tau + T_c/2)| \qquad (17.10\text{-}8)$$

Equation (17.10-8) is sketched in Fig. 17.10-2 at a function of τ. Note if τ is positive, a positive voltage will appear at the VCO input, increasing its frequency and

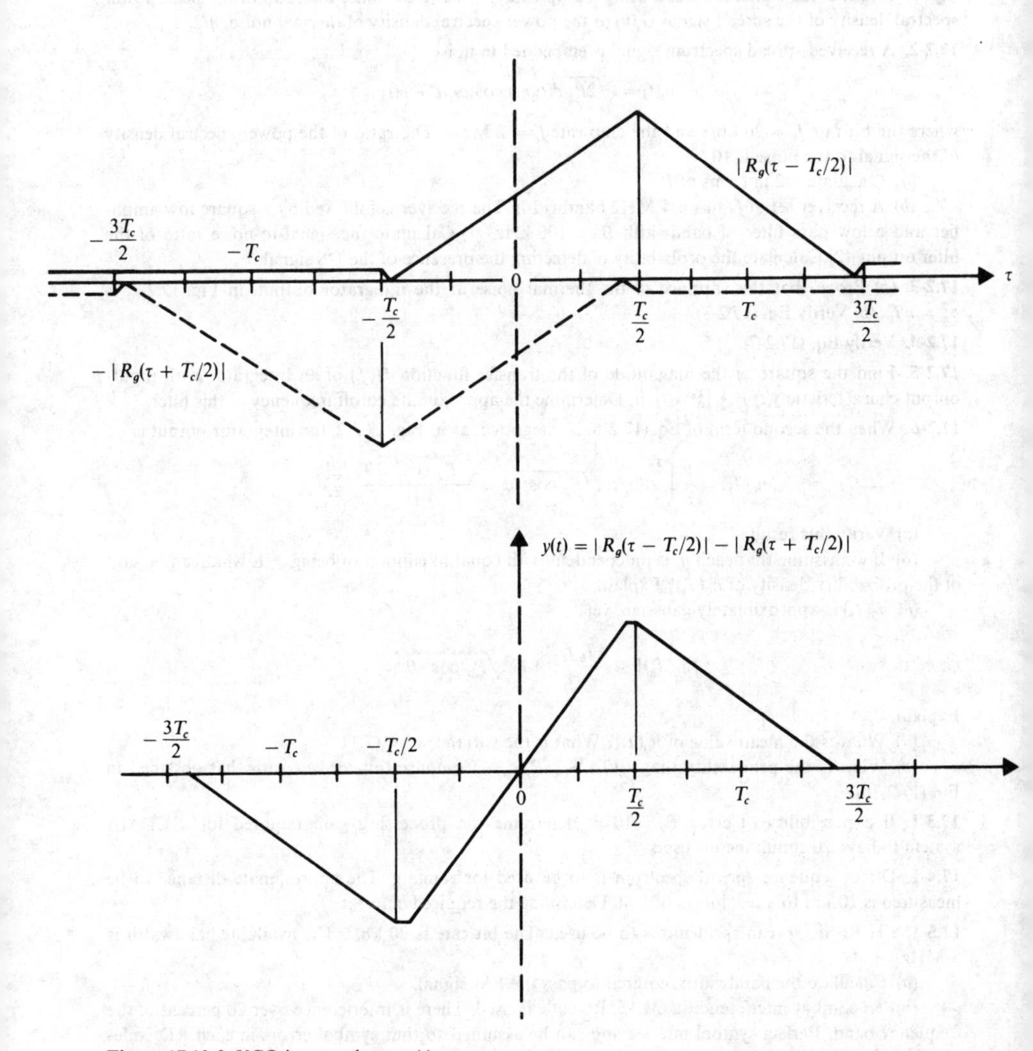

Figure 17.10-2 VCO input voltage, $y(t)$.

thereby decreasing τ. Similarly if τ is negative a negative voltage is generated at the VCO input, decreasing the VCO rate and thereby increasing τ.

Finally, we note by way of overview, that the FH and DS synchronization circuits presented above are only examples of many which are used. Also, it is to be noted that some spread spectrum systems combine both FH and DS.

PROBLEMS

17.2-1. Consider that the signal-to-noise ratio of a received signal not employing spread spectrum is $E_b/\eta = 10$ dB. If the signal is spread using a chip rate $f_c = 10^4 f_b$ calculate the ratio of the peak power spectral density of the spread signal $G_s(\theta)$ to the power spectral density of thermal noise, $\eta/2$.

17.2-2. A received spread spectrum signal is embedded in noise

$$v_I(t) = \sqrt{2P_s}\, d(t)g(t) \cos \omega_0 t + n(t)$$

where the bit rate $f_b = 20$ kb/s and the chip rate $f_c = 2$ Mb/s. The ratio of the power spectral density of the signal to the noise is 10^{-3}.

(*a*) Calculate $\eta/2$ in terms of P_s.

(*b*) A receiver set at f_0 has a 4 MHz bandwidth. The receiver is followed by a square low amplifier and a low pass filter of bandwidth $B_0 = 100$ kHz. (1) Calculate the signal-to-noise ratio of the filter output. (2) Calculate the probability of detecting the presence of the DS signal.

17.2-3. (*a*) Prove that the variance of the thermal noise at the integrator output, in Fig. 17.2-2, is $\sigma_n^2 = \eta T_b$. (*b*) Verify Eq. (17.2-3).

17.2-4. Verify Eq. (17.2-7).

17.2-5. Find the square of the magnitude of the transfer function $H(f)$ of an integrator with input/output characteristic $y(t) = \frac{1}{2}\int_0^{T_b} x(t)\, dt$. Determine the approximate cutoff frequency of this filter.

17.2-6. When the second term of Eq. (17.2-6) is integrated, as in Fig. 17.2-2, the integrator output is

$$n_0(T_b) = \frac{1}{\tau}\int_0^{T_b} dtg(t)\sqrt{P_J \cos^2 \theta} \simeq \frac{\sqrt{P_J \cos^2 \theta}\, T_b}{\tau} \sum_{i=1}^{f_c/f_b} g_i$$

(*a*) Verify this result.

(*b*) If we assume that each g_i is independent with equal likelihood of being ± 1, what can be said of the probability density of $n_0(T_b)$? Explain.

(*c*) $n_0(T_b)$ is approximately gaussian, yet

$$|n_0(T_b)| \leq \frac{AT_b}{\tau}\frac{f_c}{f_b}, \quad A = \sqrt{P_J \cos^2 \theta}$$

Explain.

(*d*) What is the mean value of $n_0(T_b)$? What is the variance of $n_0(T_b)$?

(*e*) What is the probability that $n_0(T_b) > \sqrt{P_s}\, T_b/\tau$. Compare this answer with that obtained in Eq. (17.2-10).

17.3-1. If a possibility of error $P_e = 10^{-3}$, determine the processing gain required for a CDMA system to have 20 simultaneous users.

17.4-1. Direct sequence spread spectrum is to be used for ranging. The approximate distance to be measured is 10 km to a resolution of 1 m. Determine the required chip rate f_c.

17.5-1. FH/BFSK spread spectrum is to be used. The bit rate is 20 kb/s. The available bandwidth is 8 MHz.

(*a*) Calculate the bandwidth required to pass the FM signal.

(*b*) To combat interference, a (31,15) RS code is used. There is interference over 20 percent of the frequency band. Perfect symbol interleaving can be assumed so that symbol errors in each RD codeword can be considered independent.

(1) Calculate the maximum number of hopping frequencies used.
(2) How many bits are there in each hop?
(3) What is the hopping rate?

17.6-1. Draw a PN generator as in Fig. 17.6-1 for $N = 5$. Using Table 17.6-1, show how the parity generator is constructed.

17.6-2. An alternative PN generator is shown in Fig. P17.6-2.

$$b_i = \begin{cases} \text{connected to the feedback loop,} & i = N - 1 - j \\ \text{connected to ground} & i \neq N - 1 - j \end{cases}$$

where j represents each subscript of Q_j shown in Table 17.6-1 for the value of N used.

If $N = 5$, show that connecting b_0 and b_2 to the feedback path, and b_1 and b_3 to ground yield the same sequence as Prob. 17.6-1.

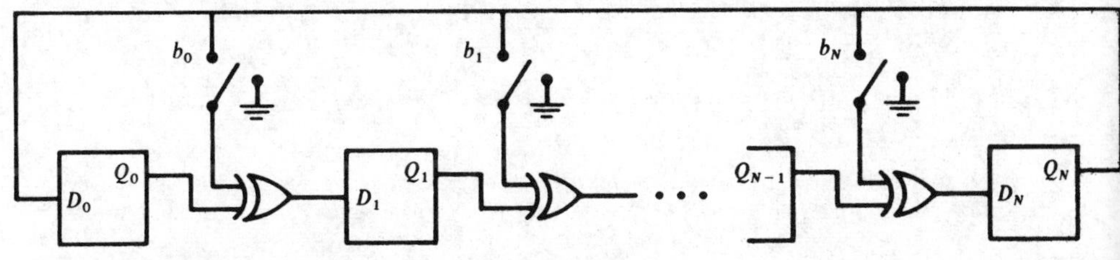

Figure P17.6-2

17.6-3. Verify that Fig. 17.6-2 are mirror images of one another.

17.6-4. For $N = 13$, plot the probability that a cluster of length λ occurs as a function of the length λ. Comment on the density if $N \to$ infinity.

17.6-5. Verify Eq. (17.6-5).

17.6-6. The power spectral density of an unknown DS signal is observed. The spectrum at $f = 0$ is 1 mW, the spectrum goes through zero at 20.47 MHz from the carrier frequency of 1 GHz and the spacing between spectral lines is 10 kHz.

What is the received power, the chip rate, and the number N of shift registers forming the PN sequence?

17.6-7. Verify analytically the result shown in Fig. 17.6-3.

17.6-8. Verify Eq. 17.6-9.

17.7-1. The camp and wait acquisition circuit shown in Fig. 17.7-1 waits at frequency $f_2 + f_0$ for f_2 to arrive. The hopping rate is 1000 hops/s. If the input signal-to-noise ratio $P_S T_H/\eta = 10$ dN calculate the probability of not detecting the signal when f_2 arrives. State assumptions.

17.10-1. Verify Eqs. (17.10-6) and (17.10-7).

17.10-2. Verify Fig. 17.10-2.

(f) Calculate the maximum number of hop patterns required [illegible].
(g) How many bits are there in each hop [illegible]?
(h) [illegible] hand the hopping rate?

17.6-1 Draw a PN generator [illegible] Table 17.6-1 [illegible] [illegible] pattern [illegible].

17.6-2 A [illegible] PN generator is shown in Fig. P17.6-2.

[illegible] of the feedback loop [illegible]
[illegible] connected to ground [illegible]

[illegible] represents each [illegible] shown in Table 17.6-1 for [illegible].
[illegible] show that connecting [illegible] and [illegible] for each path [illegible] ground yield the same sequence as [illegible].

Figure P17.6-2

17.6-3 Verify that [illegible] are [illegible].

17.6-4 [illegible] the probability that a [illegible] function if the [illegible] centered [illegible].

17.6-5 Verify Eq. (17.6-[illegible]).

17.6-6 The power spectral density of an unknown [illegible] signal is observed, [illegible] is [illegible] through [illegible] the carrier frequency [illegible] GHz and [illegible] spacing between [illegible] is 10 Hz.

[illegible] What is the received power, the chip rate and the number of [illegible] forming the PN sequence?

17.7-1 Verify that [illegible] the result shown in Fig. 17.7-[illegible].

17.7-2 Verify Eq. (17.7-[illegible]).

17.7-3 The [illegible] with [illegible] was at frequency [illegible]. The hopping rate is [illegible]. If the [illegible] noise [illegible] and [illegible] determine the probability of [illegible] detecting the signal when [illegible]. State assumptions made.

17.10-1 Verify Eq. (17.10-[illegible]) and (17.10-[illegible]).

17.10-2 Verify Eq. (17.10-[illegible]).

INDEX

INDEX